朱 雨 生
分子生物学文集

YUSHENG ZHU SELECTED WORKS

朱雨生 著

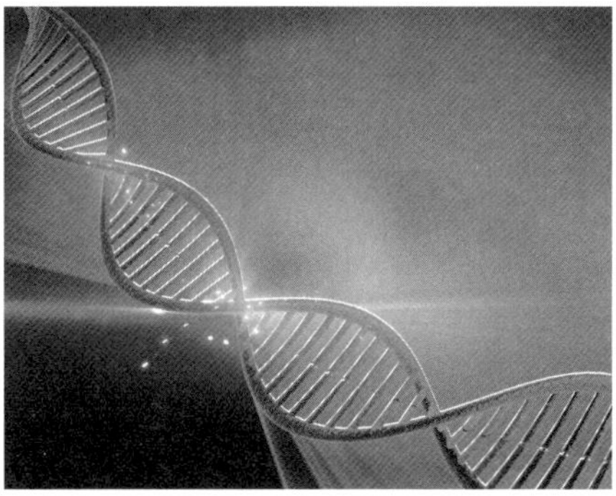

上海交通大学出版社
SHANGHAI JIAO TONG UNIVERSITY PRESS

内容提要

科学兴国造福人类是科学工作者的共同志向。近百年来生命科学有着突飞猛进的发展。作者从事植物微生物生理生化、分子生物学、基因调控、生物工程的研究多年，本书收录了他六十年来国内外著名杂志上发表的论文、总述和普科文章五十余篇，总结了在植物脂肪代谢调控、纤维素酶生产和应用、光合作用的基因调控、艾滋病检测和治疗和医药生物工程等科研工作中的成果、心得和经验，这些研究涉及人类的食品和健康，既有基础研究，又有应用研究，不少研究有原创性和前瞻性。

本书的出版期望能激励年轻学者、大学生和生命科学的爱好者投身于生命科学的研究和科普工作，为富国利民贡献力量。

图书在版编目(CIP)数据

朱雨生分子生物学文集/朱雨生著. —上海:上海交通大学出版社,2019
ISBN 978-7-313-21000-5

Ⅰ.①朱… Ⅱ.①朱… Ⅲ.①分子生物学—文集
Ⅳ.①Q7-53

中国版本图书馆CIP数据核字(2019)第041514号

朱雨生分子生物学文集

著　者:朱雨生
出版发行:上海交通大学出版社　　地　址:上海市番禺路951号
邮政编码:200030　　电　话:021-64071208
印　制:上海万卷印刷股份有限公司　　经　销:全国新华书店
开　本:787mm×1092mm　1/16　　印　张:38.5
字　数:1000千字　　插　页:18
版　次:2019年4月第1版　　印　次:2019年4月第1次印刷
书　号:ISBN 978-7-313-21000-5/Q
定　价:298.00元

此书献给我的父母；
献给我的导师谈家桢、殷宏章和汤佩松院士。

作者简介

朱雨生，1940 年出生于上海，自小热爱科学，在光明中学求学时开展米丘林小组科研活动，获“区三好学生”称号。上海复旦大学生物系植物生理专业毕业，在中国科学院上海植物生理所（现中国科学院上海生命科学研究院植物生理生态研究所）从事植物和微生物生化代谢研究。1980 年到美国马里兰大学攻读博士，三年后获得植物分子生物学博士学位。继后在哈佛大学、伊利诺伊大学和加州伯克利大学分别进行四年博士后研究。毕业后被加州伯克利大学劳伦斯研究所聘为研究员。之后转入工业界，在几个著名的农业和医药生物技术公司，如 Sungene、Chiron 担任高级研究员从事农业生物工程、基因分子诊断、生物制药工作。之后 6 年在罗氏分子诊断公司（RMS）任首席科学家直至退休。退休后兴趣广泛，酷爱旅游摄影艺术，在家中院子继续进行果蔬种植研究，提倡绿色节能、积极健康的田园生活方式。

序 一

很高兴朱雨生先生的论文选集即将出版问世。想趁这个机会，既祝贺他，也写点感想。我俩在中国科学院上海植物生理研究所同事多年。他是复旦大学毕业，我是北京大学毕业，比他迟两年进所工作。他在光合作用室施教耐先生的实验室从事植物代谢，我先在罗士韦先生的植物激素研究室从事植物激素及化学除草剂的应用和植物组织培养，1978 年以后研究室组调整，随后即在细胞生理研究室工作。

“文革”期间，使我国的基础研究受到极大的摧残，植物生理研究亦然。得益于遗传学和分子生物学的发展，植物生理研究在当时也正在发生迅速的变化。记得在“文革”后期罗宗洛老所长在与所内部分中青年科研人员的一次座谈会上说过，希望大家关注遗传学和分子生物学的发展，但并非要求大家都成为一位遗传学家或分子生物学家，而是要了解它们对植物生理研究的影响，善于把遗传学和分子生物学的概念、思路、最新的技术用于植物生理的研究中去。“文革”结束后召开的第一次全国植物生理学大会上，殷宏章先生就预示未来的植物生理在微观方面将向分子、细胞水平发展，在宏观方面将与生态环境更好地结合。1977 年在上海召开的一次全国植物生理学著名教授、科学家制订全国本学科的长远规划会议，我所仅派出我与朱雨生两位代表中青年研究人员参加此次规划讨论，从此我与朱先生才有比较多的接触。当时我注意到北京植物研究所的汤佩松教授对朱先生特别关注，一起商洽起草代谢方面的规划。事后才知道，朱先生进了植生所后继续他在复旦大学研究工作的思路，在油菜脂肪代谢调节研究上有很重要的发现。论文被当时的殷宏章所长发现推荐到《中国科学》杂志上以英文全文发表，引起学界很大的反响。此文也给北京的汤佩松教授注意到了。汤先生是中国植物生理学的元老、院士，也是美国科学院通讯院士。他在 1979 年作为团长率领中国植物学代表团访问美国，在国际植物生理学大会上做一小时主讲“中国植物生理学发展的五十年”中，特别提到上海朱先生等团队的研究是呼吸代谢方面最优秀的。那次全国规划制订会后，朱先生又草拟了一份有关植物生理学今后发展方向的文稿《高等植物酶合成的调节——基因表达》。他认为，今后植物生理学的各个领域，包括光合作用、营养、激素、发育、水分和细胞组织等都应引入分子生物学的最新成果，研究基因的调控才能有所突破。后来殷宏章所长曾推荐此文在全所高级研究人员中传阅。四十年过去了，特别是随植物分子遗传国家重点实验室的建立，植生所面貌已发生极大的改观，研究工作基本上都是用遗传学和分子生物学的观念和技术研究植物重要生命活动的机理、作物生物技术及分子育种。

朱先生从小热爱科学，思路活跃，常有原创和前瞻性想法。“文革”时期基础研究停顿，他马上设法与施教耐先生等商榷改纤维素酶应用研究，将自然界光合作用产生的稻草麦杆甘蔗渣通过微生物发酵变成葡萄糖和酒精。他从拟订规划，打报告给院部，组织全国大协作，到具体实验全程参加，如找寻菌种、建立方法、提高酶活，并应用到农（猪饲料）工（酒精厂、食品厂）医（多酶片）科（细胞脱壁后的原生质体培养和细胞杂交）各方面，取得很大成果，并获得国家科技奖，后来也获得美国纤维素酶权威学者的高度重视，受邀去演讲和撰写综述。

1979 年，根据中科院和英国皇家学会的协议，我去英国作为期两年的访问学者，朱先生也在 20 世纪 80 年代初作为访问学者去美国。惊讶的是他到了美国后，从一个已经做了近 20 年研究的访问学者，改读博士研究生，重新当学生。当年他已经 40 岁，上课备考用英语，还要做研究。可喜的是，三年内他修完全部课程，全部得 A，且完成了五篇高质量论文。这与他一贯的努力学习、积极工作、奋发图强、思路清晰有关，也得益于他熟练的实验技术，善于仔细观察，及时总结的一丝不苟的工作作风。

朱先生在美国学习工作期间，仍不忘祖国的植物生理研究。我们到美国开会参观，他总热情给予帮助。他曾协助马里兰大学到中国科技大学举办暑期分子生物学训练班，他亲自讲课演示实验。他也多次回所做学术报告，把国外新的研究成果经验介绍给大家。

朱先生在马里兰大学获得博士学位后又进入美国最著名的学府如哈佛大学、伊里诺伊大学和加州伯克利大学做博士后，在光合作用基因的光氧调控研究上都有新的重大发现，并用实验指出了一位诺贝尔奖得主在同类研究中由于培养条件的不对导致错误的结论。

虽然朱先生对基础研究十分重视，但是他时刻想把基础研究成果应用到生产实践中为人类造福。所以，他从加州伯克利大学辞去研究员职位后进入工业界，如农业生物工程公司 Sungene 和 Chiron、Roche 等著名医药大公司。他作为高级研究员或首席科学家带领团队从事生物制药工作，研发新药，为民造福。

我到美国出差曾几次拜访过他，知道他住旧金山湾区一个依山面海的“观海山庄”，风景优美。他在自家院子里种植果蔬，观察生长发育结果，还做些研究。最近朱先生告诉我，他退休后又搞起老本行植物生理了。并提倡利用太阳能、节能环保的生活方式。我相信这种自然的原生态的退休生活也是大家向往的。

本书的出版总结了朱先生五十多年来在科学研究上的成果、心得和经验。不少论文具有开创性和前瞻性。他数十年前写的《生物钟》《生物半导体》《植物的电世界》等科普文章，生动活泼，至今还有可读性和预见性。期望本书的出版能激励年轻学者、大学生、生命科学的爱好者，热爱生命科学，投身于生命科学的研究和科普工作，为富国利民贡献力量。

许智宏

中国科学院院士

第三世界科学院院士

历任北京大学校长、北京大学生命科学学院教授

中国科学院上海植物生理生态所研究员等职

现任北京大学现代农学院院长

2018. 2. 20

序　二

去年有一天，朱雨生来访，他兴奋地告诉我，2017年诺贝尔医学奖得主是美国三位科学家关于生物钟的研究。他翻看了他五十多年前为上海《科学画报》写的科普文章《生物时辰节律——生物钟》中就强调和预测生物钟的研究将为农业、医学带来莫大的好处。由此，他想把过去发表过的论文、综述和科普文章收集整理，出版成文集，总结他科学生涯中的点滴经验，也许对读者有用。我很支持他的想法。

朱雨生1962年从复旦大学毕业分配到中国科学院上海科学研究院植物生理生态研究所（植生所），在我们光合作用室物质转化组，跟施教耐先生工作。当时每个进所的大学毕业生都要经过一个外语测试，选考俄语或英语，每人给予两小时的时间。朱雨生在中学、大学主要学俄语，所以他选了俄语。一小时不到，他把俄语试题答完了，问考官再给一份英语试卷，一小时不到又做完了。隔日考试成绩出来了，朱雨生是所有新职员（复旦、北大等）中成绩最好的，而且同时考了两门外语。朱雨生对外语有点天分，他在1957年考入复旦大学时也是经过外语分班预测，他是二千多名入学新生中唯一一个从一年级直接跳级到二年级读俄语快班的学生。他在大学只学了一年英语，短时的日语，就能顺利阅读英语和日语文献。进了植生所后，他两年内上夜校修完了德语和法语并顺利毕业。别人问他为什么要学这么多外语？他回答，要做科研就必须了解最新行情，知道世界各国发展现况，才能制定出自己研究方向，寻找最可靠方法，赶超别人。后来又知道，朱雨生是以第一高分在竞争激烈的1957年考入复旦大学生物系的，以现在的话来讲，可说是过去的“高考状元”吧。

朱雨生在复旦大学受薛应龙教授指导，毕业时就发表了关于植物落花落果生理机制的论文，他进所后又将此思路延用到当时的脂肪代谢课题上，得到施先生的支持。他日夜扑在实验室里，研究终于有了结果，发现有一种天然类黄酮控制油菜种子中糖变油的代谢途径。此发现立即受到当时殷宏章所长的重视，并推荐朱雨生全文以英文发表在《中国科学》上。在讨论作者排名时，朱雨生将施教耐先生放第一，他说因为施先生是导师。此文发表后引起学术界很大反响，尤其受植物生理学界元老、北京植物所所长汤佩松教授关注。后来他俩经常学术上洽商，成了难得的知交。

朱雨生做植物物质代谢研究初露头角，但“文革”时期基础理论停顿，他转研究纤维素酶十余年，也取得很大成功，两次全国纤维素酶学术大会请他主讲。他在光合作用室没有搞光合作用，但是他对光怎么影响植物、怎么促使二氧化碳和水变成葡萄糖的机制一直怀有浓厚的兴趣。他在中学时就被一本苏联作者写的《光与植物》所吸引，大四时写了生物钟文章。他认为生物钟是受自然界的光调节。我们光合室成立较早，后来因为发现光照下能量转化的高能态现象，受到国内外关注，并承担国家重点科研项目，主要进行光合作用能量转换的物理化学变化研究。20 世纪 70 年代末由于分子生物学的发展，朱雨生产生了研究叶绿体光合作用分子遗传的兴趣。这也是他初到美国近十年在这领域的研究工作。

他在马里兰大学与孔宪铎教授和分子生物学家 Lovett 教授工作时进行了烟草叶绿体 DNA 的克隆和分析，并首次将烟草和蓝绿藻叶绿体 RUPB 羧化酶大亚基的基因放到大肠杆菌和芽胞杆菌中表达。又发现光能促使马铃薯块茎中淀粉体变叶绿体。他在哈佛大学跟随著名的植物分子学家 Bogorad 教授，开创性地发现红光能激发许多玉米幼苗中与光合作用有关的质体和细胞核基因，而远红光能逆转这些基因的表达，这说明基因表达与光敏色素有关。这一发现似乎与果蝇研究中红光刺激褪黑素合成、促进睡眠和蓝光抑制褪黑素合成、阻碍睡眠的生物钟现象有些关联。接着朱雨生到伊利诺伊大学师从著名的微生物分子生物学家 Kaplan 教授，从事光合细菌的光合作用分子生物学机理研究。朱雨生在哈佛大学发展了一种灵敏的分离分析 mRNA 的方法，他发现光合器光能转化为能量的捕光蛋白和反应中心是在一个操纵子上。而捕光蛋白的量比反应中心蛋白的量多 10 多倍，这是因为它们分属 DNA 上不同位置转录合成而调控的。这实验结果指出了斯坦福大学诺贝尔奖得主 Stanley Cohen 实验室不久前发表的文章的错误，因为他们培养条件上的失误导致得出 mRNA 稳定性是造成两类蛋白质合成不同的结论。

朱雨生在光合细菌基因调节研究上的结果又受到加州大学伯克利分校 Hearst 教授的注意。他是诺贝尔奖得主卡尔文研究室的继承人，任室主任，他的学生 Cech 博士因发现 RNA 的催化作用而获 1992 年诺贝尔化学奖。朱雨生在这一流的研究室而接触了更多优秀科学家。他继续发现光和氧气对光合细菌一系列能量转化基因、叶绿素和类胡罗卜素合成基因的影响，并发现 DNA 的立体构型能影响基因的表达。他的不少论文由卡尔文教授推荐到美国科学院院报(PNAS)上发表。这些研究结果总结在他与 Hearst 教授合写的总述文章中，发表在《植物生物工程》一书中。有意思的是 Hearst 教授与学生建立了深厚友谊，让他钢琴家的老母亲义务教朱雨生学了三年钢琴。

朱雨生去美国后也一直关心国内植物生理学的发展，接待访美同事，帮助采购试剂，曾回所做学术报告，访问老同事。当他知道我每天还骑车上班，他也跟着学。当我陪他参观我所人工气候室后，他回美国后也在自己后院里建一个人工培育箱，用太阳光板发电供光温控制。他还研究用科学方法制作天然果干带给大家分享。希望本书出版能向同行介绍植物分子生物学研究中的某些成果，也鼓励更多学者在科研中寻找乐趣。

沈允钢

中国科学院院士

历任中国科学院上海生命研究院植物生理生态研究所所长

中国植物生理学会理事长等职

2018.03 于上海

前　言

2017年第一批诺贝尔医学奖获得者是研究生物钟的三位美国科学家，他们发现了控制生物钟的基因，生物钟的研究对医学、农业都有重大的意义。这促使我翻阅了1963年6月我在上海《科学画报》上出版的以“雨生”笔名发表的《生物的时辰节律——生物钟》的文章，介绍了在植物、动物和人体中生物钟的普遍存在，指出了对生物钟的深入研究会给农业、医学和人体健康带来莫大好处。想不到过了50多年，我当时的预言，在今年却实现了。在许多朋友的建议下，我就萌生了把过去50多年来在生命科学研究等方面发表过的文章，归纳为综述、论文、科普三个方面编成文集的念头，也许对同行后人会有点参考价值。我有好几项研究当初仅仅开了个头，可后来却成为了国际上研究的热点，例如如何将光合作用的产物纤维素转化成葡萄糖？如何光能启动光合作用有关的许多基因，将光能转化成化学能，变二氧化碳和水为葡萄糖？如何把PCR技术应用到对病毒，特别是HIV病毒的检测和艾滋病药物的研发等等，我已进入耄耋之年，也想总结一下自己人生在科学研究道路上的坎坷历程和乐趣。

我1940年出生在上海。父亲朱鑫如(朱贻铨)江苏丹阳人，1883年生，是20世纪二三十年代上海针织业界著名爱国企业家，在著名作家余忠的一文《申城创业恣风流——朱鑫如创业上海滩疏记》(镇江朱氏文化研究文集第一辑，2018年团结出版社)中有详细介绍。母亲是诚实善良的家庭主妇。我自小喜欢自然世界，特别是花花草草的植物，以后又对医药有兴趣，与生命科学结上不解之缘。初中的植物课考试得100分，促进了对植物学的喜爱，组织了米丘林小组，开始了对植物研究的兴趣，1956年被评为上海市邑庙区三位“三好”学生之一。以后顺利考上了复旦大学生物系，毕业后进入中国科学院上海植物生理研究所工作了18年。1980年托国家改革开放政策之福，成为几批首次送往美国的访问学者中的一员。我有幸去了美国，40岁的当年从头开始在马里兰大学成了攻读博士的研究生，后来斯坦福大学也接受我，但我还是决定守信，去了巴尔的摩县马里兰大学。初到美国的经历，被“巴尔的摩太阳报”和马里兰大学校刊分别报道(见附录)。三年后发表5篇论文，顺利毕业进入哈佛大学、伊利诺伊大学和加州大学伯克利分校分别做博士后。之后，进入农业生物工程、医学生物工程和医药界，从事农作物转基因和药物开发研究，直到2006年退休。我的人生历程归纳在为上海市光明中学建校120周年所写的《热爱科学　缅怀光明》的一篇自传中。

我一生研究植物和医药，与人类的食物和健康有关。我在高考竞争最激烈的1957年以第一志愿，生物系第一高分，考上了复旦大学生物系；我也是全校唯一一个从一年级俄语班直接跳级到二年级俄语快班的学生；最正规的研究是四年级在复旦大学薛应龙教授的指导下进行的。在研究工作上，新方法的创立改进，往往导致新的发现。这类例子在科学史上屡见不鲜，卡尔文光合碳循环的发现归功于同位素碳-14的发现和应用，农业上许多新品种的育种成功取决于分子杂交技术的应用，我一生的轨迹也大致如此。我发现在复旦大学生物楼门口种植的大叶黄杨，可以用作叶枝离体培养，如此就可方便地研究植物落叶落花过程中离层形成的生理生化及代谢变化。目标是通过研究离层形成的机制来减少落花落果，提高农作物产量。我的研究成果以两篇毕业论文形式分别发表在"复旦大学学报"和"植物生理学报"上。我有幸在复旦的五年聆听著名的遗传学家谈家桢的遗传学课程，与他结下了师生之缘，也有幸成了复旦重点培养对象。毕业后在当时困难时期中国科学院上海植物生理研究所在上海只招一个名额的情况下，我被选中进中科院，开始了我一生的科研生涯。

进了中科院后，在"物质代谢组"施教耐先生门下工作。当时有一个新课题：油料作物的脂肪代谢。我马上就想到把我大学期间的课题研究思路继续用到这项研究上，得到施先生支持。我日夜扑在实验室里，常常工作到深夜12点之后，进不了宿舍，只好爬墙进院子回宿舍睡觉。上苍不负有心人，我在研究上有了重要发现，发现黄酮类在油菜种子脂肪形成时的调控机制，两年里先后在"植物生理学报"上发表5篇论文，有一篇被所长殷宏章院士推荐到"中国科学"杂志上以英文全文发表。这篇文章的发表引起了我的另一贵人——北京植物研究所汤佩松所长的关注。他是中国科学院院士，又是美国科学院通讯院士。后来到上海来参加全国科学规划会议时，与我一见如故，成了挚友和导师。1978年他作为团长率领中国植物学代表团访美，他在美国植物生理年会上作题为"中国植物生理学发展的五十年"大会报告时，特别推荐我们上海这项工作是代谢方面最成功的一项。这也正是汤先生的推荐促成了我到美国做访问学者。

1966年"文革"开始了，气氛变了，我们的理论研究被终止了，要求科技人员下乡下厂，为工农兵服务。我先后下放到崇明岛两年种田养猪，一年到上海郊县五七干校种田改造思想，等到后来要求我们"抓革命，促生产"回到实验室后，我与施教耐先生和另一位实习研究员商讨后，提出"纤维素酶研究"课题，即利用微生物生化工程将农作物纤维废物如稻草麦杆、甘蔗渣等变成葡萄糖和粮食，变废为宝。我与生化所、上海工业微生物所两位研究人员一起到江浙各地调研，回来后植生所领导让我执笔起草总结归纳这项研究的意义、目标方向和具体步骤，上报中国科学院院部。结果这一提议被中科院审核同意，成为全国科学院、农业部、教育部和轻工业部大协作项目。我跑了许多南方山区森林寻找这种微生物，回到实验室与许多研究人员进行分离、生化研究，提高活性，并将试验成果应用到食品、制酒、农业、养猪和医药上。我两次被邀在全国纤维素酶研究会上做大会报告，发表了多篇学术论文(见2篇中文综述和8篇论文)，也获得国家科技奖。后来我的上司施教耐先生，因为脂肪代谢和纤维素酶以及其他由他主导的研究成果被选为中科院院士，这是在我去美国后的事了。我去美国后的第二年参加国际纤维素酶大会，我们的工作受到美国学者的极大兴趣，应邀到新西泽州的Rutguss大学做报告，并应邀以英文综述发表在"酶和微生物技术"杂志上(本书英文的综述一文)。有意思的是，我当时曾提出分离真菌纤维素酶基因，转移到制酒酵母菌中，这样酵母便能利用纤维素废料发酵直接生产酒精。当时我带着这一设想到美国，想进一步试验，但是到美国不能如愿以偿，只能做导师规定的课题。去年读到一条新闻云：一位美国大学华裔教授成功研究了将真菌纤维素酶基因转入了酵母菌。这又是40多年后的事了。

40岁到了美国，进了马里兰大学后成了改读分子生物学的博士研究生，在孔宪铎和

Lovett 教授指导下进行叶绿体 DNA 研究。系主任看了我简历，说我“可免 GRE 考试，免硕士，直接读博”。我在三年中，除了修完全部课程均得 A 外，还发表了 5 篇论文，进行了烟草叶绿体基因的图谱分析，首次将植物和蓝绿藻叶绿体基因放到细菌中表达，而且发现光能促成土豆中的淀粉体变成叶绿体。

三年获博士学位后，哈佛大学有名的 Bogorad 教授接受我做他的博士后研究生。他曾是美国植物生理学会会长，访问中国的美国植物学代表团团长。他给予我很大的影响，我们经常做实验到深夜，还见他在办公室继续写东西。在哈佛这段期间，对我学术成长有很大影响，接触到顶级的科学家和教授，开阔了自己眼界。我在植物光基因研究上也是做了开创性发现。Bogorad 在外出报告时经常会提到我在光基因研究中的开创工作：一个短暂的红光照射能激发玉米幼苗中许多与光合作用有关的质体和细胞核基因的表达，而远红光能抑制这些基因的表达，因此，这些基因是受光敏色素调控的。这些研究的成功归功于我研发出一个极可靠的分离 mRNA 的方法，此方法有助于以后在光合细菌基因调控研究中许多的发现。有趣的是最近对生物钟的研究也发现红光和蓝光对褪黑素的产生有相反的作用从而控制睡眠。这一研究也间接地与生物钟挂上了钩。

1984 年我到中西部的伊利诺伊大学香槟分校微生物系做博士后，师从犹太藉教授 Kaplan。在研究光合细菌的光能转换基因调控时发现了与诺贝尔奖获得者 Stanley Cohen 在同类研究中不同的结果，指出了他们因实验条件的不对导致错误的结论。之后我又进入加州大学伯克利分校卡尔文研究室，师从 Hearst 教授，继续研究光合细菌的基因受光和氧气调控，多篇论文由诺贝尔奖获得者卡尔文推荐到“美国科学院院报”发表（见本书的论文集），毕业后任研究室研究员。

1987 年农业生物工程开始发展，引起了我骨子里喜欢植物的兴趣，加入了 Sungene（太阳基因）公司的团队，带领一个小组将苏云金杆菌中抗虫基因转到棉花中以减少农药使用。我在分离该菌时发现一个新种，能产生专一性杀死农业害虫的毒蛋白，此新种以我的姓命名（Z1），发表在 1998 年 *Appl. Envir. Microbiol* 杂志上（见论文集）。几年后由于工作变动，我的兴趣转向医药，在几家生物制药公司任高级研究员，其中一家公司 Chiron 应聘报名的 500 余名中只有我被选中。任职后我组织一个团队研发药物疫苗，研发了用 PCR 技术检测艾滋病毒极灵敏的方法，并由此发现一种治疗艾滋病药物。之后几年一直在跨国医药公司 Roche Molecular Systems（罗氏公司）任首席科学家，带领一个团队研发疾病分子诊断方法，直至 2006 年退休。大部分工作涉及公司机密不能发表，仅发表 7 篇一般的技术文章，也收集在本书论文集中。

科学工作者的研究离不开科学的普及，我中学期间在“向科学进军”的号召下，爱读“科学画报”“知识就是力量”两种杂志，发表和翻译些俄文文章。进大学后受邀为“科学画报”写科普文章。我写了《生物钟》《植物电世界》《生物半导体》和《试管中生成的植物》等几篇科普文章，试图将生物与化学、物理结合而兴起的生物化学和生物物理新方向传播给读者。虽然是 50 多年前写的，但文字流畅，现在仍有可读性。这 4 篇科普文章收集在本书的科普文章类。

目前生命科学的发展一日千里。如果说 20 世纪是物理学时代，21 世纪则是生命科学的时代。生命科学的发展离不开 DNA 的发现、基因的分析，即分子生物学的发展。在 1977 年制订全国植物生理学科学规划时，全国的专家云集上海，当时中科院植物生理所的中青年中，我和许智宏被选上参加，我和汤佩松先生首次会见，我提出了今后植物生理学的研究一定要引入分子遗传和基因分析才能有所发展和有所作为，汤先生对此非常支持，我与他合作由我执笔写了《高等植物酶合成的调节——基因表达》一文，发表在1979年“植物生生理通讯”上，此文受到当时殷宏章所长的重视，并建议在植物生理所高级研究人员中传阅。此文收录在本书的综述中。后来又为植物生理所在十年动乱后招收的首批研究生的专题讲座中做首次演讲。现在40多年过去了，植物生理所的面貌全变了样，证实了我们的预见：传统的生理研究全改成分子生物学、分子育种。由优秀的、大多数在国外受过严格训练的博士做课题带头人，可以预测，这些变化将为植物生理的基础研究和农作物育种带来新的研究成果。

我自小喜欢自然，热爱科学，在科学研究中寻找乐趣，同时也对艺术、音乐、美术、园艺、旅游、摄影广有兴趣。退休之后，积极提倡节能、环保、简朴、接近自然的田园生活。期望本书的出版能激励年轻学者，大学生和生命科学的爱好者投身于生命科学的研究和科普工作，为富国利民贡献力量！

目　　录

第一篇　综　　述

高等植物酶合成的调节——基因表达 …… 3

Reviews　Induction and regulation of cellulase synthesis in *Trichoderma pseudokoningii* mutants EA_3 - 867 and N_2 - 78 …… 19

Organization and Expression of Genes for Photosynthetic Pigments-Protein Complexes in Photosynthetic Bacteria …… 41

The *Nicotiana* Chloroplast Genome …… 71

国外纤维素酶研究概况 …… 79

纤维素酶研究进展 …… 94

第二篇　植物生理和代谢调节(1961—1965)

棉铃脱落与碳水化合物化谢的关系 …… 109

植物呼吸代谢的生理意义Ⅰ．器官脱落与呼吸途径的联系 …… 119

植物的脂肪合成及其调节Ⅰ．油菜种子形成期间葡萄糖降解的途径 …… 129

植物的脂肪合成及其调节Ⅲ. 油菜籽实形成过程中代谢途径的变化与内源抑制剂的调节作用 …… 140

植物的脂肪合成及其调节Ⅵ. 油菜种子脂肪合成与 HMP 途径关系的同位素证据 …… 155

植物的脂肪合成及其调节Ⅶ. 大豆种子形成过程中物质累积与呼吸代谢的关系 …… 166

植物的脂肪合成及其调节Ⅸ. $NADPH_2$ 的氧化速率对油菜种子 HMP 途径和脂肪形成的调节作用 …… 176

The Changes of Metabolic Pathways and the Regulatory Action of an Endogenous Inhibitor in the Developing Siliques of *Brassica Napus* Linn …… 188

第三篇 微生物发酵工程及纤维素酶在工农医上的应用(1967—1980)

纤维素酶固体曲的制取及其在酒精生产上的试用 …… 207

二株高活力纤维素分解菌 EA_3－867 和 N_2－78 的获得及其特性的比较 …… 225

木霉纤维素酶的诱导形成及其调节Ⅰ. 槐糖对木霉 EA_3－867 纤维素酶形成的诱导作用 …… 241

木霉纤维素酶的诱导形成及其调节Ⅱ. 槐糖对木霉 EA_3－867 洗涤菌丝体纤维素酶形成的诱导作用及降解物阻遏现象 …… 254

木霉纤维素酶的诱导形成及其调节Ⅲ. 葡萄糖母液和槐豆荚提取液对纤维素酶诱导效应的分析 …… 272

木霉纤维素酶的诱导形成及其调节Ⅳ. 高产变异株纤维素酶合成调节的变化——酶活提高原因的初步分析 …… 281

木霉纤维素酶的诱导形成及其调节Ⅴ. 拟康氏木霉(*Trichoderma pseudokoningii* Rafai) N_2－78 洗涤菌丝体纤维素酶诱导形成过程中核酸代谢的变化 …… 293

木霉纤维素酶的诱导形成及其调节Ⅵ. 拟康氏木霉(*Trichoderma pseudokoningii* Rafai) N_2－78 被槐糖诱导形成的纤维素酶组分的分离、纯化及性质 …… 301

第四篇　光合作用和叶绿体基因表达的光氧调控(1980—1986)

Nicotiana Chloroplast Genome I. Chloroplast DNA Diversity ······ 315

Nicotiana Chloroplast Genome Ⅱ. Chloroplast DNA Alteration ······ 326

Nicotiana Chloroplast Genome III. Chloroplast DNA Evolution ······ 336

Nicotiana Chloroplast Genome Ⅴ. Construction, Mapping and Expression of Clone Library of *N. otophora* Chloroplast DNA ······ 348

Nicotiana Chloroplast Genome 7. Express in *E. coli* and *B. subtilis* of tobacco and Chlamydornonas chloroplast DNA sequences coding for the large subunit of RuBP carboxylase ······ 360

Light-Induced Transformation of Amyloplasts into Chloroplasts in Potato Tubers ······ 367

Phytochrome Control of Levels of mRNA Complementary to Plastid and Nuclear Genes of Maize ······ 375

The Organization of the Maize Plastid Chromosome: Properties and Expression of Its Genes ······ 388

Effects of Light, Oxygen, and Substrates on Steady-State Levels of mRNA Coding for Ribulose-1, 5-Bisphosphate Carboxylase and Light-Harvesting and Reaction Center Polypeptides in *Rhodopseudomonas Sphaeroides* ······ 406

Origin of the mRNA Stoichiometry of the *puf* Operon in *Rhodobacter Sphaeroides* ······ 423

Regulation of Expression of Genes for Light-Harvesting Antenna Proteins LH-I and LH-II; Reaction Center Polypeptides RC-L, RC-M, and RC-H; and Enzymes of Bacteriochlorophyll and Carotenoid Biosynthesis in *Rhodobacter Capsulatus* by Light and Oxygen ······ 442

Oxygen-Regulated mRNAs for Light-Harvesting and Reaction Center Complexes and for Bacteriochlorophyll and Carotenoid Biosynthesis in *Rhodobacter Capsulatus* during the Shift from Anaerobic to Aerobic Growth ······ 453

Oxygen and Light Regulation of Expression of Genes for Light Harvesting (LH-I, LH-II), Reaction Center (RC-L, RC-M, RC-H), Pigment Biosynthesis and a Transcriptional Role in the Protective Function of Carotenoids in *Rhodobacter Capsulatus* ······ 469

Transcription of Oxygen-Regulated Photosynthetic Genes Requires DNA Gyrase in *Rhodobacter Capsulatus* ······ 475

第五篇 植物转基因和生物防治(1987—1988)

Separation of Protein Crystals from Spores of *Bacillus thuringiensis* by Ludox Gradient Centrifugation ······ 489

第六篇 医药生物工程药物制造和分子诊断(1989—2006)

More False-Positive Problems ······ 497

The Use of Exonuclease III for Polymerase Chain Reaction Sterilization ······ 500

Pilot Study of Topical Dinitrochlorobenzene (DNCB) in Human Immunodeficiency Virus Infection ··· 502

Quantitative Analysis of HIV-1 RNA in Plasma Preparations ······ 510

A Simplified Method for Quantitation of Human Immunodeficiency Virus Type 1 (HIV1) RNA in Plasma: Clinical Correlates ······ 524

Quantitative Restriction Fragment Length Polymorphism: A Procedure for Quantitation of *Diphtheria* Toxin Gene CRM197 Allele ······ 534

An Easy and Accurate Agarose Gel Assay for Quantitation of Bacterial Plasmid Copy Numbers ······ 543

第七篇 科 普 文 章

植物的电世界 ······ 559

生物半导体 ······ 564

生物的时辰节律——生物钟 567

试管中生成的植物 571

附　录

UMBC receives Chinese grad. student 579

East meets West at UMBC lab 581

热爱科学　缅怀光明 584

后记 594

编者的话 596

第一篇

综　　述

高等植物酶合成的调节——基因表达*

一、序言

植物体许多重要的生理过程如发育、分化，对环境条件的反应等无不与基因的活动相联系。高等植物细胞全能性的发现，证实了植物细胞中含有全套的遗传信息，至于细胞中特异的遗传信息的表达，则是受内部和外部环境的制约，并在时间和空间上加以调控的结果。基因的表达，即基因对性状和生理功能的控制是以代谢活动为基础的。具体地说，基因控制酶蛋白的合成、酶控制代谢反应、多个代谢反应综合的集中的表现，便是性状和生理功能。对于酶、代谢和生理功能的相互关系，汤佩松在10多年前曾提出了“代谢的控制和被控制”观点，从整体及组织水平，论述了作为一个生活着的植物中结构、功能与代谢相互作用的辩证关系[5]。近来又作了进一步阐述[6]。最近，Steward 也持类似的观点。事实上，随着分子遗传学的发展和代谢调节知识的积累，使我们可以在分子水平上，从基因和功能的联系上来认识植物体内许多重要的生理现象。本文中，我们将根据最近十多年来，高等植物中酶合成调节控制研究的进展，从分子水平(基因、酶)到细胞、组织、器官的整体水平，结合发育、激素、光敏色素及环境因素的相互作用，将前文的观点作进一步的补充、改进和发展。由于大部分关于基因对酶蛋白调控的知识是来自微生物，故首先概述微生物中酶合成调节控制的操纵子学说及其近来的发展。

二、Jacod - Monod 操纵子学说及其发展

1961年Jacob和Monod根据对大肠杆菌β-半乳糖苷酶诱导的研究，提出了基因对蛋白合成调节控制的操纵子(operon)模型，根据这一模型，乳糖操纵子由三个结构基因(Structural gene) Z、Y和A，和一个操纵基因O (operator)所组成。Z、Y、A分别编码三种蛋白，即β-半乳糖苷酶(是四聚体，每条肽链由1 172个氨基酸组成，分子量** 135 000)、半乳糖苷渗透酶(是膜结合蛋白，分子量30 000，相当于260个氨基酸残基)和硫代半乳糖苷转乙酰基酶(功能不清，是二聚体，每条链含268个氨基酸残基，分子量是32 000)。各结构基因的表达，即产酶与否，是受操纵基因严格控制的。操纵基因又受调节基因(regulatory gene) i所调节。i产生阻

朱雨生，中国科学院上海植物生理研究所；汤佩松，中国科学院植物研究所。

* 本文承殷宏章、薛应龙、余叔文、唐锡华、沈允钢、邱国雄等同志提出宝贵意见，谨此致谢。

本文发表于1979年“植物生理学通讯”，No (1)：48 - 62。

** 分子量为旧名，现已改为相对分子质量，简称分子质量。——编注

遏物(repressor)与 O 专一性地结合,使结构基因停止转录。当加入诱导剂,它与阻遏物结合,改变后者的构象,使阻遏物不能与 O 结合,转录和蛋白合成便能进行。在阻遏酶情况下,正常产生的阻遏物是不活化的,与终点产物结合而活化,才能阻遏酶的合成。此假说提出后,Gilbert 和 Müller Hill (1966)分离出阻遏物是蛋白质(为四聚体,每个亚基约含 330 个氨基酸残基,分子量是 38 000,系一种变构蛋白);Beyreuther 等(1970)又分析了阻遏蛋白亚基的全部氨基酸顺序;Scife 和 Beckwith 根据基因分析的结果,弄清了启动子 P 在 i 和 O 之间,只有当 c-AMP 和 CAP 蛋白(c-AMP 的受体蛋白,分子量 45 000,由 2 个相同的亚基组成)复合物与之结合,RNA 聚合酶才能在 P 的另一个位置上结合进行转录。降解物阻遏(catabolite repression)(即葡萄糖效应)是由于葡萄糖导致 c-AMP 含量的下降,故启动子不能启动,酶合成停止。最近,由于操纵基因的分离和全部核苷酸顺序测定的成功(见图 1),不仅证实了操纵子学说的正确性,而且开始在分子、亚分子水平上认识基因间的相互作用。目前已有 20 多种操纵基因的核苷酸顺序已经弄清(关谷刚男,1977)。乳糖操纵子是在 proB(合成脯氨酸基因)和 tsx(对唑菌体 T_6)位点之间。LacZ、LacY 和 lacA 三个结构基因约长 5 100 对核苷酸,laci 约长 1 000 个核苷酸对。此时,RNA 聚合酶的核心酶(α,β,β'),便与结合碱基顺序(B)结合,在结合部位,碱基对部分分开,形成前启动复合物(preinitiation-complex),然后,沿着结构基因转录,形成 mRNA,最后,在 rho 因子参与下,在终止信号处转录停止。在整个转录过程中可能还有必需的蛋白因子 π 和 D 参加。以后,以形成的 mRNA 为操纵基因核苷酸顺序的分析,揭示了它一个显著的特征,即含有许多旋转对称的碱基结构。RNA 聚合酶是与启动子上第—6—12 的核苷酸顺序相结全,从第 1 位开始转录。阻遏物是同 RNA 聚合酶转录起始点向转录方向算起第 11 个碱基为中心(・表示)的旋转对称的操纵基因区段结合,阻止转录。C-AMP 和 CAP 蛋白复合物是与操纵基因第—61 和—62 碱基对为中心(・・表示)的旋转对称区结合,使—40 附近的 G—C 对丰富的强杂化区松弛,因而,使 RNA 聚合酶容易与启动子结合。关于 RNA 聚合酶与启动子结合和启动的机制,提出了如下的模型(见图 2)。启动子由识别碱基顺序(R)、结合碱基顺序(B)和转录启动点组成。RNA 聚合酶的 σ 因子先识别启动子上的

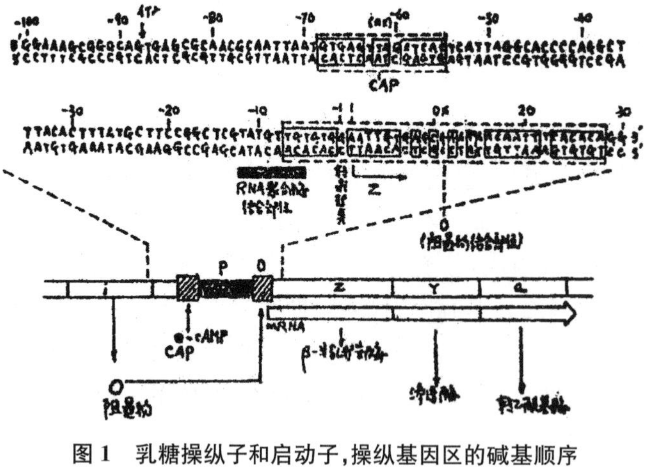

图 1　乳糖操纵子和启动子,操纵基因区的碱基顺序(关谷刚男,1977)

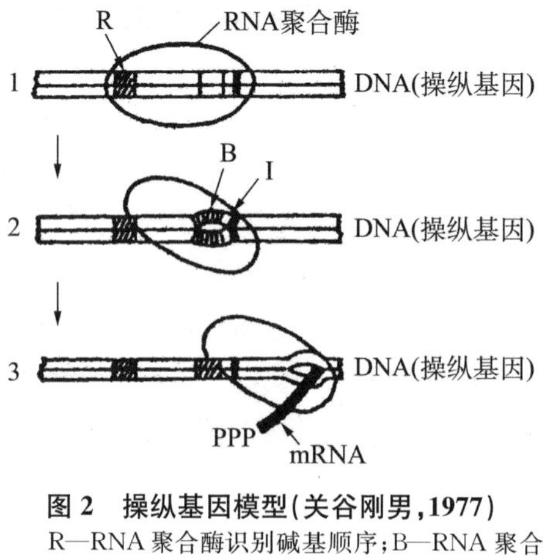

图 2　操纵基因模型(关谷刚男,1977)

R—RNA 聚合酶识别碱基顺序;B—RNA 聚合酶结合碱基顺序;I—转录开始点

识别碱基顺序(R),形成识别复合物(recognition complex-模板,在 rRNA、tRNA、多种酶系和蛋白因子(F_1、F_2、TF_1、TF_2、R_1、R_2)以及能量因子 GTP 参与下,经过启动、肽链延长(转位、易位)和终止,翻译成肽链。合成的肽链进一步形成高级结构和空间构型,或与辅基结合,或做种种修饰(如切去部分肽段等),最后才变成有活性的酶蛋白。

目前,在许多材料中证实了操纵子的存在。如在大肠杆菌中发现的操纵子有 23 种,已证明操纵基因和启动子的有 8 个以上。结构基因可集中,也可分散。调节基因(如大肠杆菌碱性磷酸酯酶操纵子)和启动子(如鼠伤寒色氨酸合成酶的操纵子)也可有多个。控制方式可以是负控制,也可能是正控制(如阿拉伯糖操纵子)。最近证明,克氏肺炎杆菌固氮酶竟由 7 个操纵子以及至少 14 个结构基因组成。我们在真菌木霉纤维素酶诱导形成的研究中,推测可能由三个结构基因分别编码纤维素酶的三个组分——C_1 酶、C_x 酶和 β-葡萄糖苷酶,它们能被槐糖(诱导剂)同步诱导。突变种的调节基因可能发生改变,而导致纤维素酶产量的增加。本霉纤维素酶的形成虽也受诱导和降解物阻遏调节,但细节上可能与原核生物不同[1,2],也有人推测,高等植物中有操纵子存在,最近发现高等真核生物基因活性调节的方式更复杂,不能简单地用操纵子学说解释。根据以上基因对酶合成的调节模型,有人将基因表达的调节点做了归纳(见表 1)。

表 1 基因表达可能的调节点[10]

转录控制 DNA 模板(即基因)用于转录的有效性 转录的启动(模板的识别,RNA 聚合酶与模板的结合) 转录的速度(RNA 聚合酶分子的数量和活性) 转录的终止和 mRNA 的释放
翻译控制 mRNA 的加工、"成熟"及降介(即 mRNA 用于翻译的有效性)mRNA 的运送 t-RNA 和氨酰基- tRNA 合成酶的有效性 起始复合物的形成 起始 肽链形成和易位 蛋白合成的终止和释放 mRNA 的继续使用或降介
翻译后控制 蛋白结构(一级、二级、三级或四级) 蛋白的激活或失活 转换率(合成速度和降解速度的快慢)

三、真核生物基因表达的特点

真核生物由于有细胞核的分化,其基因的表达远较原核生物复杂。

1. 转录和翻译的地点分离

在原核生物(如细菌)中,DNA 转录产生的 mRNA 分子上直接被核糖体附着,翻译成蛋白，转录和翻译是密切联结的。事实上已分离到转录和翻译同时进行的复合体，而且发现RNA聚合酶与核糖体的相互作用。而真核生物中，除了叶绿体和线粒体的个别例子，一般转录是在细胞核中，翻译是在细胞质中进行。细胞核中形成的不均一RNA（HnR－NA－mRNA前体），先与称为信息传递体（informofer）的蛋白结合，穿过核膜，运送到细胞质，加工成mRNA分子，并与蛋白质结合，构成信息体（informosome），再与核糖体结合，翻译成蛋白质。在小麦种子胚的细胞核和细胞质中已证明有信息体的存在，而且细胞质中信息体有游离的和同多聚核糖体结合的两种形式，且随着种子的发育而变化。其次，原核生物中mRNA是短寿命的（1～2分钟或更短），以适应迅速的代谢变化。真核生物mRNA是长寿命的（3～12小时，或更长），而且增加了mRNA加工修饰等转录后调节的可能性。其结构特点是，5′-末端具有7-甲基鸟嘌呤（即 M^7G (5′) ppp (5′) Np……）的帽子，尾部3′-末端具有polyA结构（50～200个腺苷酸）（组蛋白除外）；身体部分由两个非密码区和中间一个密码区组成，真核细胞mRNA的一般构造如图3所示（Proudfoot和Brownlee，1976）。

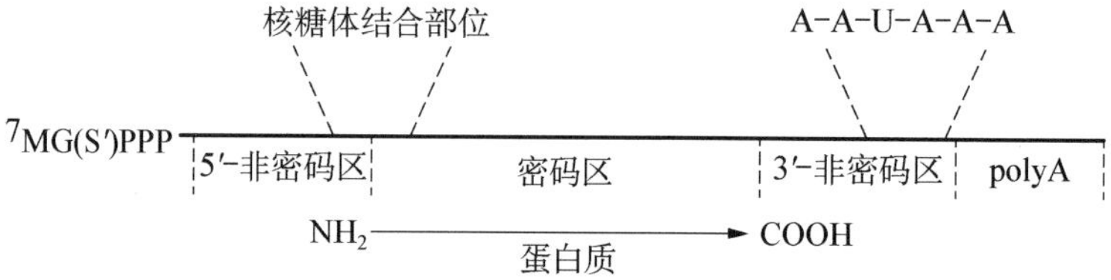

图3 真核细胞mRNA的构造

帽子部分对于mRNA的运送、稳定性或翻译可能有关。尾部polyA促进mRNA从核运送到细胞质中；polyA结构与mRNA分子的稳定性有关，当切去polyA的兔球蛋白mRNA注射到海胆卵中，短时间便分解；而且，polyA可同特殊的蛋白及内质网结合，可能也参与蛋白合成的调节。在3′-非密码区均有一段共同的核苷酸顺序A-A-U-A-A-A，功能不清。兔球蛋白β-链的mRNA一级结构已弄清楚，由438个核苷酸组成密码区，151个核苷酸构成非密码区。图4是人球蛋白α-链的mRNA构造。

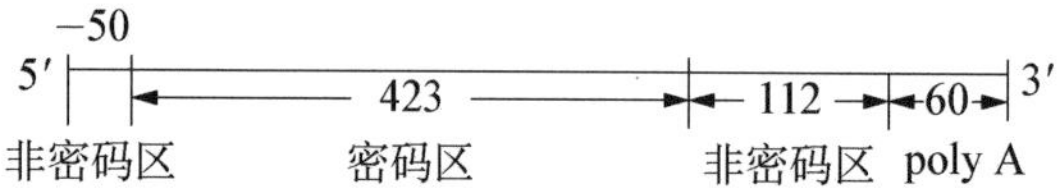

图4 人球蛋白α-链的mRNA构造

目前，高等植物中豆血红蛋白、纤维素酶、淀粉酶和RuDP羧化酶等的mRNA已分离成功。

2. 基因组的复杂性

进化过程中基因增大。哺乳动物的基因组比细菌大700～1 000倍。高等植物的基因组比细菌大很多，可达到 2×10^{12} Da。如果编码一个分子量为 10^6 Da的多肽，则代表着细胞中含 2×10^6 个基因，据估计，其中可能只有 2×10^4 基因是编码蛋白的。一般高等生物分化过程中产生的蛋白种类是大肠杆菌的10～100倍。那么剩余的DNA有什么功能？一个回答是，Britten发现真核细胞基因组中有重复性顺序（占10%～50%，或更高）；其次，是近发现在真核生物的结构基因中间有许多插入性顺序，推测其功能是调节性的。

3. DNA与核蛋白有密切的关系

虽然最近在大肠杆菌中发现有DNA、RNA和特殊蛋白结合的拟核体（nucleoid），但真核细胞中二者的关系要密切得多，DNA与蛋白构成染色质。根据最新的概念（Kornberg，

1977)，染色质的组成单位是核小体。核小体由组蛋白 H_{2a}、H_{2b}、H_3 和 H_4 各 2 个分子组成椭圆形的核心(50×110A)，DNA 双链(140 个碱基对)沿着其短轴，盘旋于此核心上(见图 5)。核小体与核小体之间由含 15～100 个碱基对的 DNA 链连接，此段 DNA 分子具有种族特异性，组蛋白 H_1 和非组蛋白的酸性蛋白即附着在此 DNA 分子上。核小体串珠链受 H_1 的影响，进一步盘旋成稳定的染色质超螺旋结构。当此结构松弛时，基因即可表达。除了组蛋白外，细胞核中非组蛋白的酸性蛋白也有重要的调节作用。DNA 与这些特殊的蛋白质群的相互作用极强，对形态发生可能起主要的调节作用。据此，XecиH 提出真核生物二种转录调节水平的假说：一级水平指染色质中 DNA 排列的紧密度，即染色质 DNA 的松弛原则上可决定某些基因的变化，但某些基因直接的活化要靠第二级水平——特殊的调节蛋白的调节。

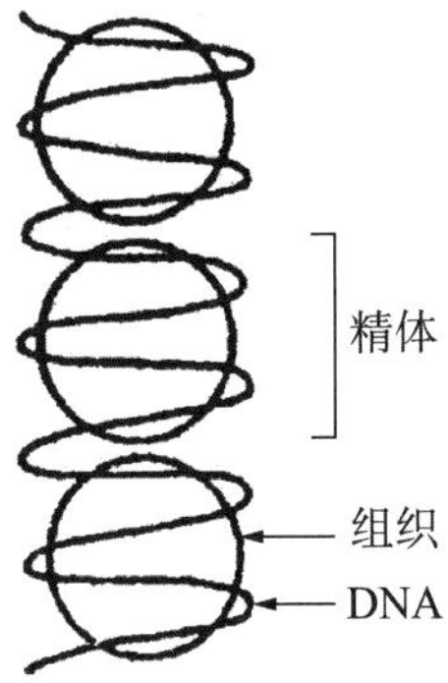

图 5　染色质结构模式图(Kornberg 1977)

4. 基因调节的方式更复杂多样

由于真核生物细胞核的分化，细胞器结构的复杂化，细胞与组织的特异化和分工加强，真核生物无论是发育分化或对环境条件的适应远较原核生物复杂。因而，要求基因调节的方式更复杂多样，包括各种细胞内与细胞间的调节，细胞区域化(compartmentation)的调节，各种化学和物理因素的调节。同时，参加调节的"元件"更多(如激素、神经介质等)，其信息含量更高，空间和时间的层次更纷繁，线路更复杂。而调节的线路虽多条，相互交错，但由于受基因组精密的控制，而互不干扰或发生差错(否则即罹病)。除了基因水平上的缓慢但深刻的调节方式外，还有建立在膜透性变化和酶的变构基础上的快速和瞬间的调节。同时，基因表达的产物、生理功能及外界环境对基因表达又有各种反馈调节。根据真核细胞中基因调节的多线路系统，Britten 和 Davidson 提出了真核细胞基因调节的模型假说[9]。认为诱导剂(包括激素)先作用于"感觉基因"(sensor gene)，激活邻近的"集成基因"(integrator gene)，后者产生激活性 RNA，作用于"受体基因"(receptor gene)，从而启动一整套的"生产基因"(producer gene)。集成基因有多套，生产基因也有多套，相互有专门的线路相通。这种调节方式的复杂性可解释高等生物中调节现象的复杂性。

四、高等植物基因活性的调节

高等植物基因的表达既受细胞内环境(代谢物 pH、离子、细胞区域化等)，又受细胞外环境(光、温、水、矿质、O_2 和 CO_2 等)的调节。激素因情况不同可视作内环境或外环境。现分述如下[11]。

1. 细胞内调节

1) 转录水平的调节

(1) 底物诱导：高等植物中底物诱导最著名的例子是硝酸还原酶、亚硝酸还原酶。其次，还有矮牵牛花青素合成的酶，以及在各种百合中的胸腺嘧啶核苷激酶，它催化胸腺嘧啶核苷和 ATP 形成胸腺嘧啶核苷- 5 -磷酸，后者是合成 DNA 的的原料。只有在百合花粉发育和长花百合离体花芽特定的发育阶段，才有此激酶的诱导现象，这是一种受未知的时相控制的底物诱导现象。总的说来，高等植物由于其自养性，底物诱导现象不如微生物中那么重要。

（2）终点产物阻遏：铵对水稻根的硝酸还原酶，精氨酸对大豆培养物的糖氨酰琥珀酸合成酶、葡萄糖对麦胚 G－6－P 酯酶和 FDP 酯酶均有阻遏作用。已查明，小麦萌发种子中磷酸盐对植质酶的阻遏作用是由于磷酸盐阻止小麦盾片中 mRNA 的合成。甘蔗中葡萄糖对转化酶的阻遏作用则是由于 mRNA 分子的破坏。

（3）组蛋白的阻遏作用：20 世纪 60 年代，Bonner 发现组蛋白能抑制豌豆子叶无细胞系统合成贮藏蛋白。后 Fellenberg 和 Bopp 发现离体豌豆下胚轴新根的形成（被 IAA 诱导），高凉菜的损伤栓内层及树瘿的形成受不同来源的组蛋白抑制。当组蛋白经过化学修饰（乙酰化、氧化、磷酸化或热变性后），其抑制活性便减弱。槲寄生的组蛋白可强烈抑制动植物肿瘤细胞的生长，抑制的部位在转录。此外，在发育过程中组蛋白也发生若干质和量的变化。组蛋白对基因的阻遏作用曾解释为组蛋白与 DNA 结合，阻碍转录。但组蛋白缺乏特异性，最有力的证据是豌豆幼苗和小牛胸腺这两种极不相同的材料，其细胞核组蛋白一个片段氨基酸顺序有惊人的相似之处（更有两个氨基酸不同）。虽然，有人提出组蛋白也许同一种受体 RNA 结合，再专一性地作用于 DNA 特定的部位，总的看来，组蛋白是一种与控制基因转录有关的非特异性调节因子。

（4）酸性蛋白的调节：酸性蛋白不同于组蛋白，在结构功能上有惊人数量的异质性，分子量从 10 000～150 000 Da 以上。功能上分为三类，即结构蛋白、酶类（DNA 和 RNA 聚合酶和蛋白修饰酶系等）和基因调节蛋白。前两类数量多，种族特异性不大，第三类在种间、组织间和分化发育期间存在特异性，因而开始受到人们的重视。其作用机理可能是，某种特异的酸性蛋白与受组蛋白抑制的 DNA 特定位置相结合，通过本身磷酸化，带上负电荷，排斥带负电荷的 DNA，而与带正电荷的组蛋白牢固缔合，结果使 DNA 上组蛋白-酸性蛋白复合物被置换出来，DNA 便能转录。

2）翻译水平的调节

现已证明，微生物和某些高等生物某些诱导酶的形成是在转录外的水平上调节的[11]。如细胞核中大的 RNA 分子前体转化为小的成熟 mRNA（称小瀑布调节，cascade regulation）；或者 mRNA 通过甲基化和“修剪”弯成有活性的 mRNA；mRNA 可能与“荫蔽”蛋白结合（如伞藻）；或 mRNA 分子结构改变，如分子一端打圈，形成短的双股结构，使核糖体不能附着或脱离；或虽然核糖体能附着，由于缺乏蛋白合成的一个或几个控制因子而不能翻译，在诱导条件下被活化。Sussman 则提出，每个真核细胞的 mRNA 分子的 5′末端是一套多余的碱基顺序，当核糖体附着在 mRNA 上，一种核糖体蛋白起着核酸内切酶的作用，切去此多余顺序而活化翻译。高等植物如落花生、棉花、小麦等不少植物中已证明有长寿命的 mRNA 存在。棉花子叶中蛋白酶 mRNA 是在胚形成过程中预先合成的，一直到种子萌发时才活化表达，此外，甘蔗中葡萄糖和果糖对转化酶的阻遏作用证明是发生翻译水平上。

3）翻译后的调节

豌豆中的葡萄糖苷酶原可被蛋白酶作用而激活；在未萌发莴苣种子中，某些磷酸酯酶活性由于用胰蛋白酶或去污剂 Triton X－100 刺激而增加 5～6 倍；许多种子的一些酶由于抑制剂存在而处于抑制状态，种子萌发时，由于抑制剂被破坏而活化。我们在油菜花蕾和幼荚中发现一种葡萄糖－6－磷酸脱氢酶和 6－磷酸葡萄糖酸脱氢酶的天然抑制剂——类黄酮，在籽实成熟过程中，由于抑制剂消失，磷酸戊糖途径中该两种脱氢酶活化，产生的 NADPH 推动脂肪酸的合成[4,19]。

2. 细胞间调节——激素

动物细胞间调节是靠神经系统和内分泌。植物虽无神经系统，但可借助维管束系统，由化学信息——植物激素进行细胞间和器官间的调节。目前，植物体内发现的五大激素的共同特点是：都不是蛋白质（不同于许多重要的动物激素），分子不大，结构较简单（均由疏水基团和亲水基团组成），而作用谱却很广。它们是在特定的发育阶段，或在外界环境信号的作用下，在特定的部位产生，并在靶细胞中发挥作用。生长素、赤霉素和细胞激动素是形态发生不专一的诱导物，脱落酸则是不专一的抑制剂，前面三种“正”激素显然是激活基因的活动，而“负”激素（如脱落酸）必然是抑制基因的活动。关于植物激素与基因活性的关系，曾提出三种可能性[11]：

(1) 基因组总的激活或失活。即正激素在特定的发育阶段激活特定组织中所有可激活的基因，而负激素作用正相反。

(2) 激素只激活或抑制一个基因，然后通过代谢反应再引起基因的次级激活或失活。

(3) 激素先刺激或抑制一个中心代谢反应，再引起基因谱的激活或失活（见图 6）。

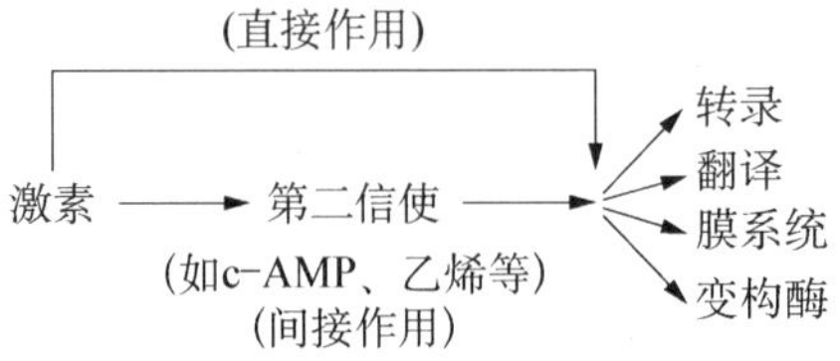

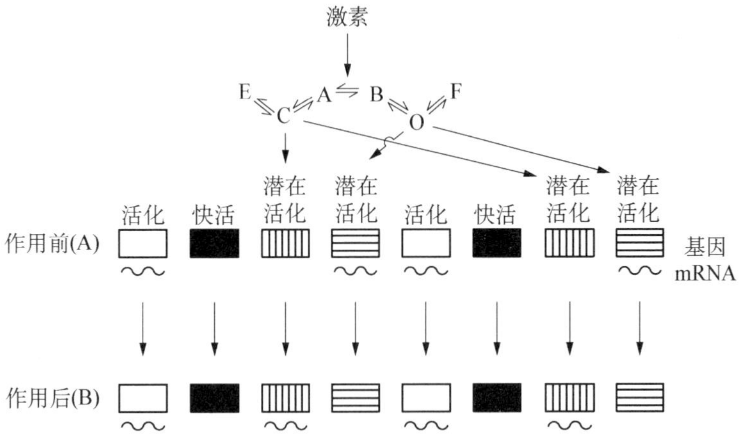

图 6　激素作用模型[11]

根据 Jacob－Monod 操纵子模型，基因的激活或失活标志着阻遏物的失活或激活。激素可能作为效应子（像诱导物或终点产物），先与阻遏物结合，改变其构象，从而使基因激活或失活。当然也可能通过影响组蛋白与 DNA 的结合，而影响基因的活性。在高等植物中激素作为效应子可以作用于多个基因，而单个基因则往往又是受到多因子（好几种激素）控制的。近来发现许多激素是通过如下第二号系统作用的。

在动物中，Sutherland 证明，激素激活膜结合的腺苷酸环化酶活性，使 ATP→c－AMP，c－AMP 可刺激多种酶的活动，如许多激酶，从而使某些蛋白质磷酸化，如组蛋白的磷酸化则导致基因的去阻遏。植物激素（如赤霉素）也可能存在同样的机制。此外，乙烯可能是另一种第二信使，它是植物所特有的，并普遍存在于所有细胞中。IAA 所引起的顶端优势就是通过诱导乙烯而引起的。目前，关于激素、基因和酶合成的关系，以赤霉素诱导α-淀粉酶和生长素诱导纤维素酶的研究较为深入[13,15]。

发芽的大麦、小麦和黑麦种子用赤霉素处理，便能诱导出α-淀粉酶、蛋白酶、核酸酶、磷酸酯酶和过氧化物酶。以后，用胚乳切片和糊粉层的离体培养，均证明赤霉素诱导糊粉层中α-淀粉酶的形成。洗去赤霉素，α-淀粉酶合成便停止。诱导过程中加放线菌素D，则α-淀粉酶合成大大受抑制。同时又发现，在赤霉酸作用下，大麦糊粉层的PolyA-RNA增加了。从赤霉酸处理的大麦糊粉层中成功地分离出mRNA(比对照增加10倍)，并能在麦胚无细胞系统中翻译成α-淀粉酶蛋白。mRNA含量的增加，可能是mRNA分子形成(转录)，或是mRNA加工(翻译)，或是降低mRNA分解。关于mRNA加工，已发现赤霉酸可影响大麦糊粉层和玉米盾片的核酸甲基化，同时也可能调节5′末端的7-甲基鸟嘌呤结构，从而影响翻译活性。桃谷和加藤研究GA_3诱导α-淀粉酶同功酶的变化：推测α-淀粉酶的结构基因至少10个以上，而调节基因也必然有多个。此外，还有学者认为GA_3通过c-AMP或直接影响组蛋白和基因活性。根据赤霉素对糊粉层水解酶诱导的现象，van Oerbeck提出了种子萌发过程中激素对生长发育调节的模式[11]：种子萌发时，水分进入种子和胚，胚被活化，合成各种mRNA，诱发赤霉素形成，赤霉素进入糊粉层，诱导各种水解酶，如淀粉酶则水解淀粉；核酸酶则水解核酸，使细胞激动素游离；蛋白酶则水解蛋白，释放的色氨酸可合成生长素(IAA)。细胞激动素和IAA作用于胚，前者(在某些材料中参与少数氨基酸t-RNA组成)诱导细胞分裂，后者诱导细胞生长，在纤维素酶和果胶酶帮助下，胚冲破种皮，先长出胚根，后长出胚芽鞘。IAA在两器官中依重力而分布，由于二者敏感性不同，根向地生长，芽鞘向上生长。当芽鞘冲出地面，光合器便分化，种子阶段就完成。

IAA促进细胞壁的扩张和诱导纤维素酶的形成。IAA或2,4-D处理菜豆下胚轴，可使纤维素酶活力增加100倍，mRNA增加10倍，从中分离到纤维素酶的mRNA，用免疫电泳证明能在麦胚无细胞系统中翻译成纤维素酶蛋白[24]。IAA调节可发生在转录，也可发生在翻译水平上。纤维素酶分解细胞壁，使壁松弛，利于生长。乙烯起相反的作用，它诱导富羟脯氨酸的结构蛋白，后者与壁多糖交联结合，降低壁的弹性和延伸性[17]。生长素和乙烯提供了对细胞生长的双元调节模式。

生长素处理豌豆茎切段或高浓度2,4-D处理大豆下胚轴，除了促进细胞的膨胀外(处理后一天)，到后期(3～5天)还促进不定根的形成。细胞学的变化是最初皮层薄壁细胞破坏，产生腔隙以后，从维管束中生长出根原基，穿过腔隙，最后冲破皮层，形成不定根。激素的这一系列生理效应的分子基础可能是生长素提高染色质样板的有效性，增加RNA聚合酶Ⅰ和蛋白合成起始因子的活性，诱导mRNA形成，后者与核糖体结合，合成纤维素酶，以破坏皮层薄壁细胞的细胞壁。同时，细胞内含物分解，释放出细胞分裂因子(如细胞激动素)，后者又进一步诱导形成层的细胞分裂，形成根原基和长出不定根[11]。

植物器官脱落过程中，离区的纤维素酶活性大大提高。用放线菌素D和环己烯亚胺处理可阻止纤维素酶的形成，叶柄脱落也被延迟了。因此，纤维素酶在离层形成过程中可能是新合成的。纤维素酶参与植物器官离层形成也受激素控制。在菜豆叶柄离区中分离到二种纤维素酶——酸性PI和碱性PI。前者与组织分化(如木质部分化)有关，受生长素控制；后者与细胞壁分解有关，参与离层形成，受乙烯控制。生长素抑制脱落是因为促进酸性PI和纤维素酶形成，乙烯促进脱落是因为促进碱性PI纤维素酶的形成。

近年来，除了在基因水平上研究激素的调节作用外，还提出细胞膜的假说来解释激素作用的快速反应（反应可快到十分钟，甚至更短）。当生长素直接同原生质膜结合时（植物激素由于均有非极性基团，易渗入膜的拟脂层中）引起膜构象的改变和透性的变化，增加离子流动（ion-flux）（如 K^+ 进入细胞，H^+ 外流），产生生物电位变化，或一系列代谢变化，发生快速反应，如生长素活化原生质膜上的 APP 酶，引起 H^+ 从细胞质流向细胞壁，活化细胞壁水解酶。或直接打断壁中酸不稳定共价键或氢键，使壁弹性增加，利于生长。如果膜上有受体分子，也可能由于膜构象的改变，使受体分子释放一种因子，运送到靶细胞的核仁中，活化基因，增加 RNA 聚合酶Ⅱ的活性，促进 mRNA 合成，导致生长反应，这是通过基因活性的慢反应（见图 7），这种因子已在大豆下胚轴、洋葱茎、豌豆和玉米茎中分离出来。目前已证明许多激素如生长素、赤霉素、激动素，均有专一性的激素受体，后者大多是蛋白[13,21]。许多事实说明，植物激素可能与动物胆固醇激素一样，先与激素的受体蛋白结合，从而影响染色质的模板活性。

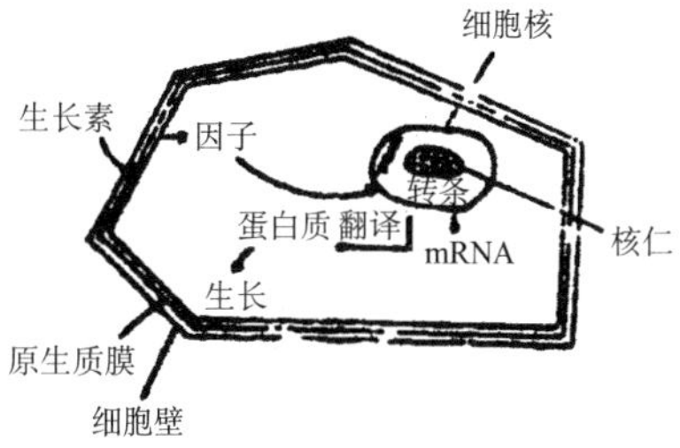

图 7　生长素作用的假设模型

3. 外界环境的调节

植物是个开放系统，它没有高等动物那种恒定的内在调节（如体温）和躲避恶劣外界条件的运动能力。外界环境对植物生命有着广泛而深刻的影响。植物体是发展出一套接受环境信号，经过转换放大变成代谢的语言，从而做出各种反应的机构。

1）光

光是外界环境中最重要的物理信号之一，它能调节植物的生长、发育、空间定向以及许多生理过程。光合作用、光形态发生和向光性组成了植物的光生理学。植物体内有感受不同波长光的色素受体，并通过不同的机制发出瞬间和较长期的反应。叶绿素吸收红光和蓝光进行光合作用，光下形成的 NADH 和 NADPH，在二硫化物还原酶作用下，能使卡尔文循环中许多含-SH 的酶激活；叶黄素和胡萝卜素吸收蓝光后，借助顺式—反式转换，通过同-SH 作用，或同酶的变构中心非共价结合；影响亚基的结构排列等而改变酶的活性；调节蛋白质代谢；磷酸吡哆醛吸收紫外光，影响与赖氨酸 ε-氨基的结合及酶活性；含 FMN 的氨基酸氧化酶吸收蓝光能参与对呼吸的调节。以上例子都是光对酶活性的调节（见图 8）。

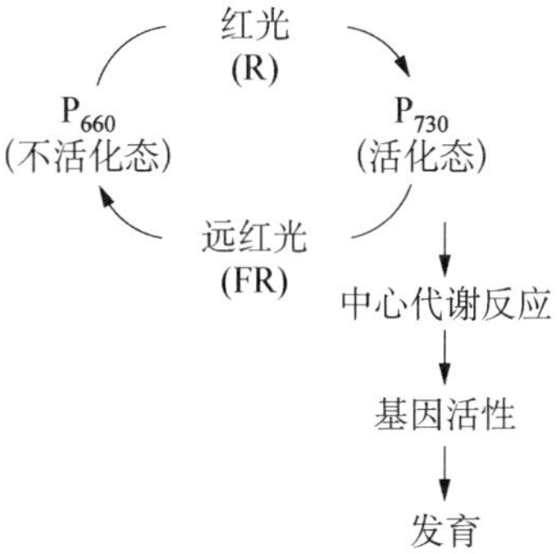

图 8　光敏色素藉红光和远红光控制代源和发育

光敏色素能从周围环境中感知光信号（红光和远红光），通过两种色素状态的可逆性转化，翻译成控制发育，包括形态发生的代谢反应语言而起作用。光敏色素是一种开链四吡咯，它同特殊的蛋白结合，广泛存在于高等植物中，参与各种重要的生理过程，如种子萌发、光周期现象、器官形成、花青素合成、叶绿体的定向、鳞茎的形成、叶子脱落、芽的休眠、性别表现等等，还同生物钟有关，而且调节苯丙氨酸-氨解酶（PAL）、RuDP 羧化酶等数十种酶的活性[18]。比较光敏色素生理学和植物激素生理学，揭示了二者在分子水平上的作用有惊人的共同之处。而且，在发育中有些激素可代替光（或温度）的作用。光敏色素吸收红光，从不活化的 P_{660} 转化成有活性的 P_{730}，P_{730} 可起着图 6 中激素一样的效应因子（effector）的作用，影响一个中心代谢反应，从而对基因组进行一系列的激活和抑制，导致复杂的生理反应。苍耳是短日照植物，在黑夜中插入红光不开花，插入远红光便开花。有人认为是由于红光下，P_{730} 抑制了合成“开花素”的基因。远红光下则变成 P_{660} 不活化态(见图 9)。

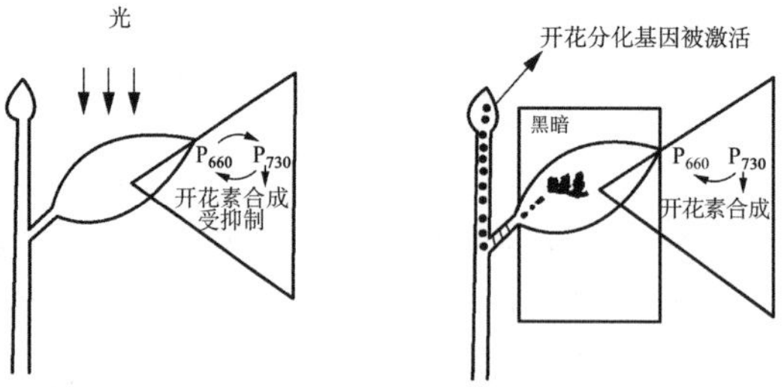

图 9 光下和暗中短日照植物光敏色素的转化与开花的关系图解[11]

Mohr (1972)研究芥菜幼苗花青素形成时就指出，光可能通过光敏色素激活基因。他发现，当照光(远红光)时，花青素便形成。加入放线菌素 D 可阻止花青素的形成，但放线菌素 D 要在照光前或照光时加入才有效，如果照光后 6 小时加入，则只能延迟花青素形成，而不能阻止它。所以，关键性的转录过程是在照光开始时进行的。花青素合成是一个复杂的代谢反应，由多个酶和基因参加。20 世纪 60 年代 Zucker、Mohr、Zenk 发现光能刺激花青素合成中的关键酶——PAL。PAL 的光诱导在马铃薯、芥菜、小黄瓜、豌豆、红甘蓝等材料中做了广泛的研究(Zucker, 1972)。PAL 分子量约 3×10^5，其光诱导过程可分成延迟期(90 min 左右)、酶活直线上升期和酶活迅速下降期。用放线菌素 D 和蛋白合成抑制剂可阻止光诱导，光下含 PolyA 的 RNA 增高，故光可能在转录或翻译水平上，诱发 PAL 的 mRNA 和酶蛋白的新形成，但也有人认为光是阻止 PAL mRNA 的分解。在酶活下降期，加环己烯亚胺，PAL 不下降，可见，酶活下降是由于 PAL 失活系统的合成而引起。此现象在转化酶、硝酸还原酶、中性磷酸酯酶的诱导中也发现。Engelsma (1970)发现，小黄瓜 PAL 下降到对照水平时，在低温(4℃)放 24 小时，再回到 25℃(暗中)，则 PAL 第二次上升，且此酶活上升不受环己烯亚胺抑制。所以，光下形成的 PAL 可能可逆地与蛋白类抑制剂结合，在低温下解离。后来，他在小黄瓜切下的下胚轴扩散物中找到了这种蛋白抑制剂。

PAL 还能被紫外线诱导，一种解释是呋香豆素等被激活，变成 DNA 的插入因子而诱导 PAL 的合成。

PAL 和肉桂酸羟基化酶及对一香豆酸:CoA 合成酶三个酶一起参与酚类合成。在芹菜和荞麦等材料中发现，光同时诱导这三个顺序相关的酶，说明光诱导的酚类合成操纵子至少由三个结构基因构成。

RuDP 羧化酶是另一个受光敏色素调节的酶，它由分子量 55 000 的 8 个大亚基和分子量 12 000 的 8 个小亚基组成。二种亚基的 mRNA 已经分离成功，在麦胚无细胞系统中分别合成两种亚基。大亚基由叶绿体 DNA 控制合成，小亚基由细胞核 DNA 控制合成，该酶的遗传控制已由 Kung[14] 作了精彩的评论。RuDP 羧化酶的形成受光调节。暗处培养的豌豆芽鞘用红光照射 5 分钟，RuDP 羧化酶活力升高 15～90 倍，若再照远红光，活力大大减少。用特殊抗体和同位素已证明，RuDP 羧化酶的光刺激是由于酶蛋白新合成的结果。

色素吸收光信号到光形态反应的表达时间可很短促(甚至几秒钟)。对于这类快速反应，

Hendricks 等(1965)提出了光敏色素作用的膜假设。即光敏色素是膜成分,在光下构象发生变化,从而改变膜的透性,引起离子流动和一系列反应。光敏色素的构象变化从免疫化学和圆二色谱及光旋转色散法得到证明。这种假设与动物中的蛙的光受体模型(光—视紫红质—腺苷酸环化酶—膜透性)是类似的。有人将基因和膜假设结合起来,认为光最初的作用部位是膜,释放出代谢物,然后,直接间接地使特异的基因阻遏或去阻遏[20]。

近来,还有一种“光神经肌肉系统”的假说,认为光敏色素是通过乙酰胆碱而使植物发生快速反应的。大麦黄化幼苗根用红光照射 1 天,可观察到次生根生长的抑制(慢反应)。而在照射数分钟内,可发生根尖吸附在带负电玻板上的所谓 Tanada 效应(快反应)(Tanada, 1968)。照射远红光,则可使根尖从玻板上脱离。Tanada 效应是由于根尖表面生物电位的变化而引起的:根尖表面带正电时(如 H^+)吸附;带负电时解吸。在菜豆幼苗根中发现反应链是:光→光敏色素→乙酰胆碱→膜透性变化→离子流动→生物电→Tanada 效应。高等植物对环境的快速反应(包括含羞草等敏感植物的运动)似与动物神经传导有不少相似之处。

2) 温度[17]

Harris 和 James (1969)发现,蓖麻种子在低温下,不饱和脂肪酸(如十八碳烯酸)形成增加。在冬天,高等植物细胞的类脂也有同样的变化。在芽胞杆菌中观察到,温度是诱导特异的脱饱和酶的重要因素。当温度从 30℃降到 20℃时,诱导出催化棕榈酸变成 5 -十六碳烯酸的 $\triangle^5$ 脱饱和酶,此酶在 30℃测不到,而到 20℃才活化。类似的现象在酵母中也发现。最近,对大肠杆菌膜磷脂脂肪酸形成的研究,揭示了不饱和脂肪酸变化的生理意义,它降低磷脂液晶结构的相交温度,从而保证在低温下维持膜的正常透性和生物合成功能。低温也许是通过激活基因,合成脱饱和酶,影响脂肪酸代谢和膜的结构,产生抗低温的防御反应。

3) 水分

种子吸水后 10 分钟便开始蛋白合成。Marcus (1966 年)发现小麦胚吸水后核糖体活性、多聚核糖体含量和蛋白合成迅速增加。Chen (1968 年)指出,胚吸水最初 24 小时是利用现存的 mRNA 合成蛋白;吸水 24～48 小时是合成原有的 mRNA;48 小时后则合成新的 mRNA;由于吸水,基因才能表达,使 mRNA 得以形成或翻译,导致激素和各种水解酶的合成以及完成种子萌发一系列过程。

在缺水或盐渍条件下,植物体内会发生激素、酶和代谢一系列的变化,以做出防护性的适应反应,如小麦叶切下后 4 小时,由于失水,脱落酸(ABA)含量增加 40 倍,并证明 ABA 是新合成的。ABA 影响保卫细胞对 K^+ 的透性,导致气孔的关闭(Hsia, 1973)。大麦绿叶在缺水时,烯醇式磷酸丙酮酸羧化酶活力显著上升,使 C_4 途径加强,以适应缺水环境。当严重缺水时,叶片中脯氨酸含量明显增高,补充水后,脯氨酸含量则下降(Barneet 等,1966)。脯氨酸是植物抗旱能力增强的重要化学指标,脯氨酸的累积显然是以谷氨酸为前体、新合成的结果(Hsia, 1973)。此外,缺水也引起细胞中淀粉酶、核酸酶等活力的上升和硝酸还原酶,PAL 活力的下降。缺水所引起的一系列生理性防御反应很可能也是通过基因、激素、酶和代谢的调控而实现的。

4) 矿质

正如前面所述,硝酸还原酶和亚硝酸还原酶是高等植物中底物(矿质)诱导基因表达最著名的例子。水稻根浸于硝酸盐溶液中,经过数小时的延迟期,根中便形成硝酸还原酶。除去硝酸盐,或加酪蛋白水解液,或加 NH_4^+,则酶诱导马上停止,酶的诱导还被放线菌素 D 所控制。

可见，该酶的诱导有 mRNA 合成和酶蛋白的新合成。光能促进小麦叶片对硝酸盐的还原，而且激动素和赤霉素可代替光效应[15]。另外，铵盐能诱导水稻幼根的谷氨酸脱氢酶。此酶有两种：一种分布在线粒体上，一种分布在细胞质中，只有后一种能被 NH_4^+ 诱导。

矿质元素激活基因活性的另一个可能的例子是离子泵的形成。细胞对离子的主动吸收（如 K^+、Na^+）的一种方式是通过专一性离子泵（如 K^+ 泵、Na^+ 泵）的诱发形成。离子泵是分布在膜表面的活性蛋白。在许多植物细胞内，ATP 酶起着细胞内 Na^+ 和细胞外 K^+ 交换的离子泵作用，此过程需要 ATP 为能量[20]。组织切片（块茎）放入水中不吸收离子，经一段时间漂洗，才能吸收离子，此现象暗示离子吸收有去阻遏现象。离子吸收受遗传和激素的控制。离子的主动吸收可能与基因激活有关。

无机离子能直接影响染色质的结构，如游离细胞核在低浓度 K^+ 中，显示出分散染色质的特征，合成 RNA 的速度增加。Na^+ 和 K^+ 对染色体上的特异区带有特异的反应。离子浓度还影响核糖体中蛋白质间的结合，从而影响翻译[20]。K^+ 浓度，尤其是 K^+ 梯度，可能在控制分化中起重要的作用。此外，微量元素（如锌）的缺乏，可干扰 RNA 合成，而影响酶形成和生长[18]。

5）损伤

高等植物不同于动物，是固定生长，易受风、雨、机械损伤和虫害、兽害等伤害。植物发展了一套防御机构，邻近损伤部位细胞的呼吸、RNA 和蛋白合成等代谢活性显著增加，形成本质层和栓皮层的保护层，以防止病菌入侵和水分、养料的渗漏，并产生抗真菌的毒素（酚类）。这一系列反应中，PAL 是关键酶，它催化酚类和木质素的形成。田中喜之报道（1978），山芋块根受损伤后 PAL 和酸性转化酶活性迅速增强，二者都是新合成的。PAL 诱导有 3 小时的延迟期，6 小时达到高峰，而 10 小时后则下降。PAL 的下降一方面由于酶合成的降低，另一方面由于 PAL 失活系统的新合成，后者不是蛋白，位于液胞膜上，PAL 在此失活，进入液胞，再被蛋白酶水解破坏。酸性转化酶有类似的变化，而且其活力水平也受合成速度和抑制剂含量的控制。但此抑制剂是分子量为 19 500 的蛋白。细胞中酸性转化酶的水平，通过合成速度的变化来进行粗调节，通过特异的抑制剂的消长，进行细调节。组织受损伤时，IAA 先合成，所以伤害反应的信号（也许是氧分压的改变）可能先通过激素的放大作用，再激活基因和代谢反应（如酚类、木质素代谢），产生一系列的防御反应。类似的机构在动物中也存在。关于氧分压引起代谢类型和种子萌发时形态、功能的变化在水稻材料中已做过报道[7,23]。

4. 发育分化过程中基因的调节[8,10]

细胞分化是特别复杂的生理过程。从酶、细胞器和膜成分的结构和内含物变化，可以看出，分化是在特殊时间和特殊空间，细胞中特异性蛋白的出现和消失。细胞虽然含有整套的基因，但百分之九十左右的基因是处在不活化状态，在分化的各个阶段，某些基因相继活化。而且，不同类型的细胞、器官是以不同类型的酶和结构蛋白为特征的。Bonner (1965)利用豌豆子叶和茎端染色质作为体外合成 RNA 的模板，再以 RNA 为信使，在体外合成蛋白质。他发现子叶染色质转录形成的 RNA 能作为模板合成球蛋白——子叶中主要的贮藏蛋白；而茎端染色质转录形成的 RNA 却不能。这说明不同的细胞中有特异的 DNA 指导特异的蛋白合成，所以细胞分化归根结底是不同的基因表达。

既使存在争论，但广泛相信，真核细胞基因表达也是受结构基因、调节基因和操纵基因的调节。事实上有充分证据说明发育（如细胞分化、种子休眠萌发、开花、受精……）的改变伴随着基因活性（转录）的改变。如菜豆胚柄细胞的巨大染色体在特定的发育阶段，特定的部位膨大(活化区),低温和放线菌素D处理可使它收缩,同时,RNA合成停止。此培养物在激素(萘乙酸和激动素)诱导分化前后的mRNA(已分离)和蛋白图谱有明显的差异。在种子休眠的打破,以及茎外植体用IAA诱导根形成时,染色质中作为合成RNA模板的DNA有效性明显增加了。花芽的形成也与特异的mRNA合成有关。暗处理的矮牵牛子叶中合成的mRNA与不受诱导影响的子叶中的mRNA,在碱基成分上是不同的(Yoshida, 1964),故认为诱导开花激素需特异的蛋白和mRNA的合成。

花粉授粉受精过程涉及细胞的识别和一系列基因活性的变化(Netlancourt, 1977)。授粉过程包括花粉在柱头上的附着、吸水、花粉壁局部破裂,和花粉粒萌发、花粉管伸出、穿透角质层、进入柱头等多个过程。整个过程可在数十分钟内完成。花粉粒与柱头间存在的种属间的识别是由于二者分泌一种由S-基因决定的识别蛋白——S-蛋白。不同种属间的S-蛋白结构不同。当同种花粉与柱头辨认后,就激活一系列与花粉管生长、穿透壁障碍和进入柱头有关的基因和酶系,如几丁质酶、葡聚糖水解酶、酯酶、蛋白酶、淀粉酶、酸性磷酸酯酶、核酸酶等等。van de Donk将受粉柱头的RNA提取出来注入到海胆卵中,可得到一种活性蛋白,由两条分子量为11 000和6 000的多肽组成,它能促进花粉管的伸长。精细胞与卵细胞的结合受精,当然是更复杂的基因激活过程。

植物细胞的发育分化有时可在翻译水平上调节。如单细胞藻类——伞藻的分化(Hämmerling, 1963)。伞藻的巨型细胞从接合子发育,伸长形成长约5 cm的伞柄。从伞柄基部长出假根,细胞核位于一个假根中,几个星期后,形成伞帽,伞帽形成就是分化的过程,它包括许多酶的顺序出现和消失、蛋白和特异多糖的形成。除去细胞核,同样能形成正常的伞帽,且酶的消长、蛋白和多糖合成也是正常的,这是因为与伞帽分化有关的mRNA是长寿命的,在分化过程中活化。筛管细胞在分化过程中可失去细胞核,并可继续合成蛋白达数月之久。在种子萌发过程中,至少某些蛋白mRNA也是预先形成,在转录后的水平上进行调节。一个很有说服力的例子是棉花种子萌发时子叶中两个酶合成的调节,这两个酶是羧肽酶和异柠檬酸裂解酶,分别参与蛋白水解和脂肪转化为糖的反应,以将子叶中贮藏物质分解供胚胎发育利用。Ihle和Dure (1972)证明这两种酶的mRNA早在胚形成期便已合成。棉花种子胚胎发育可分为三个时期:①胚胎发育至85 mg左右(约30天);②胚胎同母体脱离,并继续发育到125 mg (30多天);③子房组织(种皮)干燥脱水死亡,胚胎重时急剧减少。现发现羧肽酶和异柠檬酸裂解酶的mRNA在第①期中期便形成,但不能表达,这是因为到第②期,胚周围的子房组织产生ABA,并可能通过一种接受蛋白或抑制性RNA,阻止其mRNA的翻译。到第③期,子房凋萎死亡,虽ABA破坏,但由于迅速脱水,而使mRNA不能翻译,一直到种子吸水萌发时才开始表达。看来这种转录和翻译在时间上分离可能是发育过程(包括动物受精卵的发育)中较普遍的现象。同时,这个例子生动地说明了激素、基因、酶和发育之间精巧的调节作用。

细胞分化过程中特异蛋白的合成和消失,固然受基因决定,但必须在细胞内和细胞外环境条件(光、温度、激素、代谢物、细胞间接触和位置等)信号作用下才能实现。悬浮培养的单细胞不能分化,但是愈伤组织块便能诱导分化,因此,细胞分化的前提是同其他细胞接触。由合子发育成空间上分布不同的不同类型的胚细胞,说明细胞能识别位置。细胞接受光的量和角度(外部信息)和化学物质(如激素、蔗糖)浓度梯度(内部信息)可能提供这种位置信息。生长素和蔗糖梯度在控制维管束分化时,起着重要的作用。蔗糖也参与质体的发育。激素对器官分化,光对开花的影响更是熟知的事实。

发育是各种代谢反应错综交叉、相互作用(有时间进程)的结果。发育的顺序性可用代谢反应的顺序性来解释。对于春化引起开花发育的阶段性和顺序性有人用图 10 表示[11]。

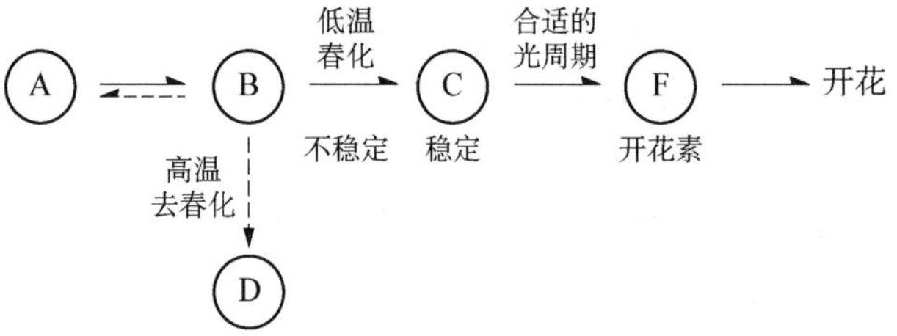

图 10 开花发育的阶段性和顺序性

在发育的春化阶段由 A 变成不稳定的 B,在高温下破坏变成 D(去春化),在低温下完成春化,变成稳定的 C,再在合适的光周期下,形成开花素(F),导致开花。

由于发育的模式是由基因预先决定的,在发育的不同阶段是由不同的基因起作用。正是某些基因的"开"和"关"带来了发育顺序性和连续性。外界环境和细胞内环境可能提供基因"开"和"关"的信号。

五、结语

现代遗传学的发展使遗传学的概念大大地扩大了。它不仅研究亲代与子代间遗传信息的传递,而且也研究当代遗传信息的流动、调控和表达。遗传学和生理学两门学科的相互渗透,使得二者在认识植物生命现象时是缺一不可的。从本评论中可见,分子生物学和分子遗传学的研究成果对植物生理学的渗透,使植物生理学的面貌焕然一新,为植物生理学的发展带来了新的思想、观点和方法,注入了新的生命力,反过来,植物生理学又为分子生物学的研究提供了丰富的材料、五花八门的现象,为分子生物学的发展开辟了广阔的园地。

生命区别于非生命的特征乃是生物能对体内各种反应进行调节控制,负熵增加,并对外界刺激做出反应。植物体内的生理过程正是受到严格调节控制的,往往与基因的活动相联系。近一二十年来,分子遗传学的进展,使植物生理学中两个最重要的研究领域——发育生理和环境生理,在基因表达这一分子基础靠拢起来了。生物膜则是各种生理现象另一个会合点。植物代谢的研究则越出了单纯研究代谢途径的具体反应,而进入到研究各代谢途径在内的联系,和在分子水平上(基因对酶的控制、酶的变构、修饰等)及整体水平上(代谢与生理功能的联系)的调节。代谢反应的联系和调节正是基因到性状表现、生理功能、发育分化之间的桥梁。我们过去关于"代谢的控制和被控制"的论点还应包括"酶的控制(对代谢)和被控制(被基因)"的内容。光通过光敏色素对酶形成和光形态发生(花青素形成、开花等)的调节;赤霉素通过对许多水解酶的诱导,影响碳水化合物、蛋白质和核酸代谢,导致种子萌发;生长素诱导纤维素酶,影响纤维素(分解)代谢而引起的生长反应或离层形成;缺水条件下,通过脱落酸、烯醇式磷酸丙酮酸羧化酶和脯氨酸的变化,以及对 C_4 代谢途径的调节,所作出的防护反应;损伤诱导 PAL,通过酚类和木质素代谢而引起的防御反应;以及低温通过脱饱和酶影响脂肪酸代谢而作出的抗低温反应等等,都是在外界环境影响下,基因控制酶、酶控制代谢和代谢控制生理功能有力的证据。叶绿体的光合作用则是在叶绿体和细胞核遗传器的双重控制下,在外界条件(光等)

诱发下，在各级水平（基因、膜、酶、产物等）上进行调节控制的突出例子。

基因、酶、代谢、生理功能、发育及与环境条件的关系如图 11 所示。

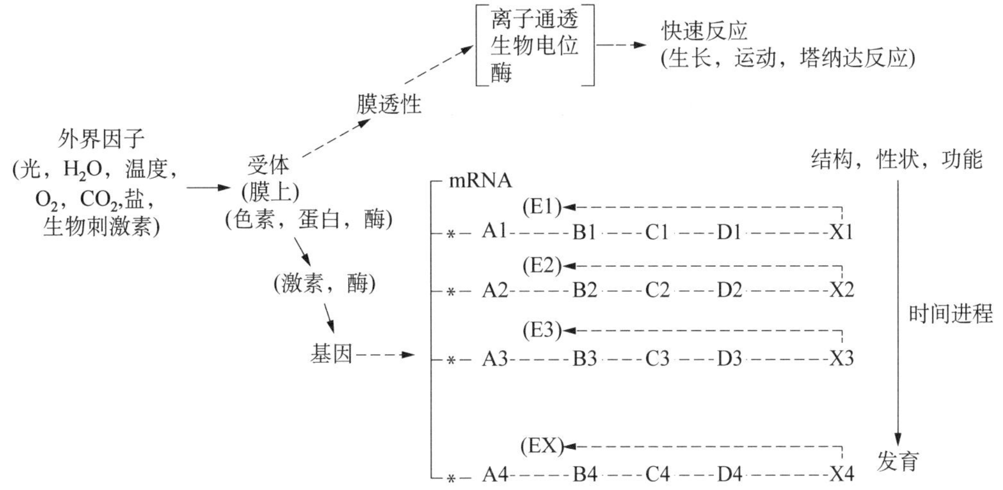

图 11　高等植物中外界环境，基因、酶、代谢、生理功能及发育相互关系调节控制（基因表达）的设想图

这里有许多个代谢反应（从 A—X），每个代谢反应有许多酶参加。有些反应是顺序诱导的。每个反应结果表现为结构性状和生理功能，它们对代谢也可起反馈作用（虚线表示）。X 是代谢产物（包括激素、蛋白等）。一个反应的结果（X）诱发次一个反应的开始（通过激活基因或酶）。所有反应的时间进程便是发育。外界环境和激素信号往往由膜上一个受体分子接收，直接或间接影响基因，经过转录和翻译，控制酶的合成；也可能通过膜透性的改变，影响离子流动、生物电位，或引起酶构象的变化，发生快速反应（如某些光形态调节和激素调节）。外输界环境的信号如光、离子、伤害等也可能先诱发激素进行次级诱发。激素也可诱导别的激素和通过第二信号系统，进行信号的多级放大。植物体的生理过程是极其复杂的，各种代谢反应又是错综交叉的。植物生命过程之所以能有条不紊地进行而不发生差错，一方面固然是靠细胞中基因预先编好的在生物进化中形成的程序指令工作，另一方面，正是靠着植物体多途径、多线路、多层次，各级水平上的多元调节系统（基因水平和酶水平，转录和翻译，酶合成和分解，膜透性和酶的修饰、变构，诱导和反馈，各种粗调节和细调节，细胞核和细胞质、细胞间的调节，细胞区域化调节，细胞、器官、整体间的调节，各种环境因子的信号指示，对各种环境信号的专一性受体以及信号的多级放大）保证了植物正常的生长发育，并迅速对环境条件作出适应性反应，以顺利完成其个体发育和种的进化。

参考文献

[1] 朱雨生、谭常。1978：木霉纤维素酶的诱导形成及其调节。Ⅱ、槐糖对 EA_3 - 867 洗涤菌丝体纤维素酶的诱导作用及降解物阻遏现象，植物生理学报，4，1：1。

[2] 朱雨生。1978：木霉纤维素酶的诱导形成及其调节。Ⅳ、高产变异株纤维素酶合成调节的变化——酶活提高原因的初步分析。植物生理学报 4，2：143

[3] 吴相钰、郑光植、梁峥、汤佩松，1964：小麦叶片中硝酸盐盐和亚硝酸盐的光还原。植物生理学报 1，

1:15。
[4] 施教耐、朱雨生、谭常。1965:植物的脂肪合成及其调节。Ⅲ、油菜籽实形成过程中代谢途径的变化与内源抑制剂的调节作用。植物生理学报,2,2:105
[5] 汤佩松,1965:代谢途径的改变和控制及其与其他生理功能间的相互调节——高等植物呼吸代谢的"多条路线"观点,生物科学动态,第3期第1页。
[6] 汤佩松,1978:植物体线粒体中电子传递途径的改变和调节——再论呼吸代谢多条路线。全国细胞培养与体细胞杂交线粒体呼吸代谢与杂种优势会议论文集。P. 159 甘肃省科技情报所,兰州大学生物系合编。
[7] 汤佩松、王发珠、池访杰,1959:植物呼吸代谢的研究。Ⅲ、不同氧分压对水稻萌发时器官形成、物质和能量转化效率的影响。植物学报。8:188。
[8] Bonner, J., 1963. The molecular biology of development. clarender press, oxford.
[9] Britten, R. J. and Davidson, E. H., 1969. Gene regulation for higher cells: A theory. *Science*, 165 (3891):349.
[10] Bryant, J. A., 1976. Molecular aspects of gene expression in plants. Academic press.
[11] Hess, D., 1975. plant physiology. Spring-verlag.
[12] Ihle, J. N. and Dure, L. S., 1972. The developmental biochemistry of cotton seed embryogenesis and germination. Ⅲ. Regulation of the biosynthesis of enzymes utilized in germination. *J. Biol Chem.*, 247:5048.
[13] *Jacobsen, J. V.*, 1977. *Regulation of ribonucleic acid metabolism by plant hormones.* Ann. Rev. plant physiol., 28:537.
[14] Kung shain-dow, 1977. Expression of chloroplast genomes in hugher plants. *Ann. Rev. plant physiol*, 28:401.
[15] Marcus, A., 1972. Enzyme induction in plants. *Ann. Rev* plant physiol, 22:313.
[16] Marme, D., 1977. phytochrome: membranes as possible sites of primary action. *Ann. Rev.* plant physiol, 28:173.
[17] Milborrow, B. V., 1973. Biosynthesis and Its Control ln plants. p. 127. Academic press.
[18] Price. C. A., 1970. Molecular Approaches to plant physiology. McGraw-Hill Book Company.
[19] Shih, C. N., Y. S. Chu and C. Tan, 1966. The changes of metabolic pathways and the regulatory action of an endogenous inhibitor in developing silliques of Brassica napus Linn. Scientia Sinica 15,3:379 - 393.
[20] Smith. H., 1977. The molecular biobogy of plant cells Blaclwell Scientific publications.
[21] Steward, F. C., 1976. In perspectives in experimental biology. 1:9, Sunderland, ed. pergamon press, N. Y.
[22] Tang. P. S. and H. Y. Wu, 1957. Adaptive formation of nitrate reductase in rice seedlings *Nature*. 179: 1355.
[23] Tang. P. S., Y. L. Tai and C. K. Lee, 1957. Studies on plant respiratory pathways in rice seedling and respiration as an adaptive physiological function of Iiving plant. *Scientia Sinica*. 5:509.
[24] Verma, D. P. S. et al., 1975. Regulation and in vitro translation ofmessenger ribonucleic acid for cellulase from auxintreated pea epicotyls. *J. Biol. Chem.*, 250(3):1019.

Reviews

Induction and regulation of cellulase synthesis in *Trichoderma pseudokoningii* mutants EA_3-867 and N_2-78*

Two mutants, EA_3-867 and N_2-78, with high cellulase yields were obtained from wild strains of Trichoderma pseudokoningii Rifai, 1096 and Mo_3, respectively, by mutagenic treatments with a linear accelerator, ^{60}Co, u. v., nitrosoguanidine (NTG) and diethylsulphate (DTS). The mutants grew slowly to produce small colonies on agar plates with synthetic medium. On agar plates of peptone-yeast extract, the small colonies were as large as those of wild strains. The cellulase activities of these mutants in Koji extracts, shake flask culture filtrates, and enzyme preparations were markedly higher than those of their parents. The mutant N_2-78 reached quite high cellulase activity level when cultured for 60 h in shake flasks in a simple medium containing milled straw, wheat bran, mineral salts plus waste glucose molasses. The cellulase saccharifying activities on CMC, filter paper and cotton, were 255, 8.2 and 13.4 mg glucose/ml enzyme, respectively, or 11, 4.3 and 6 times more than those of its parent Mo_3.

The cellulase synthesis of EA_3-867 and N_2-78 was strongly induced by sophorose, isolated from pods of Sophora japonica L., and was inhibited by glucose, sugar phosphates, glycerol and organic acids. We conclude that cellulase synthesis of the mutants is regulated by catabolite repression as well as by induction. The increase in cellulase production by both mutants results from changes in the regulatory systems for cellulase synthesis, i.e. the mutants showed higher sensitivity to inducer and lower susceptibility to catabolite repression than did the wild types.

A cellulase preparation of Trichoderma pseudokoningii Rifai N_2-78 induced by sophorose was fractionated by DEAE-Sephadex A-50 and Sephadex G-100 column chroma tography, selective inactivation and polyacrylamide gel electrophoresis. The components C_1(exo-β-1,4-glucanase), C_x(endo--(β1,4-glucanase) and β-glucosidase were separated, and their molecular weights were estimated to be 67,000, 62,000 and 42,000 respectively.

Zhu Y S, Wu Y Q, Chen W, Tan C, Gao J H, Fei J X and Shih C N, Shanghai Institute of Plant Physiology, A cademia Sinica, China, Enzyme Microb. Technol. 4:1-12.

* Present address: Department of Biological Sciences, University of Maryland Baltimore County, Catonsville, Maryland 21228, USA.

本文发表于1982年"Enzyme Microb. Technol". 4:1-12.

The homogeneity of C_1 *was verified by polyacrylamide gel electrophoresis, immuno electrophoresis and ultracentrifugal analysis. It is a glyco protein and is rich in glycine, aspartic acid, threonine, serine and glutamic acid. The* C_1 *showed a strong synergistic action with* C_x *in the degradation of cotton, Avicel and Walseth cellulose.*

A PolyA-RNA, induced by sophorose in N_2*-78 mycelium, was isolated by oligo(dT)-cellulose affinity chromatography.*

Keywords: Cellulose; cellulase; Trichoderma pseudokoningii; enzyme induction; catabolite repression; mutation; sophorose; glucose molasses

Introduction

The cellulase research laboratory of the Shanghai Institute of Plant Physiology has been interested in cellulose conversion by enzymes from Trichoderma pseudokoningii, particularly the isolation and characterization of hyper producing cellulase mutants, for several years. Several papers on this work have been published in Chinese.[1-9] This review summarizes these papers in English.

Cellulose, which is a photosynthetic product and a renewable energy resource, occurs in nature in vast quantity as an agricultural, industrial and urban waste. The annual production of rice straw, wheat stem, corn cob and shells of cotton seed in China alone has been estimated to be 4×10^8 tons. Therefore, cellulose is a very important potential food and energy resource. Several approaches have been made to increase cellulase production in our laboratory. The first was to find the hypercellulolytic Trichoderma strains by screening of wild types and selection of further induced mutants. The second was to investigate the induction and regulatory mechanism of cellulase synthesis in Trichoderma. This knowledge is the basis of increasing cellulase production by a change of culture conditions as well as by genetic manipulations. This paper presents results obtained from the study of cellulase induction and regulation using two Trichoderma mutants. The approaches include ① cellulase induction by sophorose, ② catabolite repression, ③ inductive effect of glucose molasses, ④ changes in the regulatory mechanisms of cellulase synthesis in two mutants, ⑤ separation, purification and characterization of cellulase induced by sophorose and ⑥ isolation of a PolyA-RNA induced by sophorose in washed mycelium of Trichoderma.

Materials and methods

Organism

Trichoderma pseudokoningii Rifai mutants EA_3-867 and N_2-78 were obtained from the wild strains 1096 and Mo_3, in a series of mutational steps (see Figure 1).[1] Stock cultures were maintained on yeast extract-peptone agar (0.1% yeast extract, 0.2% peptone, 2%

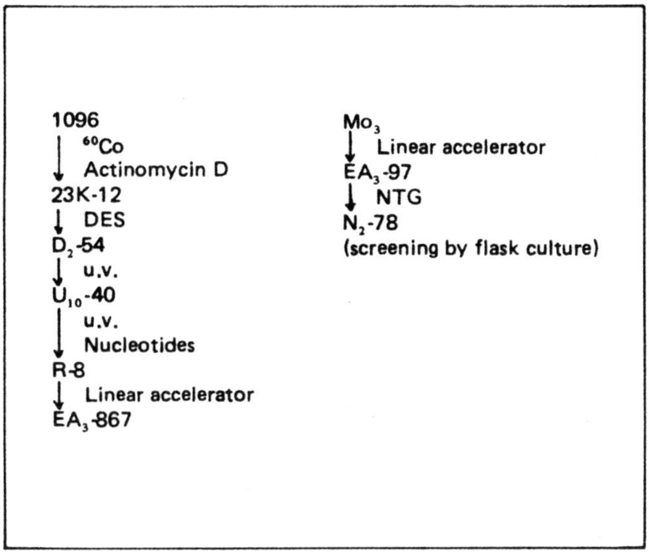

Figure 1 Geneology of Trichoderma pseudokoningii Rifai mutants[1]

cellulose powder, 2% agar) or potato agar (20% potato extract, 2% cellulose powder, 2% agar) and grown at 30℃ for 4—5 days.

Culture

The glycerol medium was used to produce mycelium in the induction study.[10] The medium consists of (%): glycerol, 0.6; $(NH_4)_2SO_4$, 0.14; KH_2PO_4, 0.2; urea, 0.03; $MgSO_4 \cdot 7H_2O$, 0.03; $CaCl_2$, 0.03; peptone, 0.1; micro elements (ppm): $FeSO_4 \cdot 7H_2O$, 7; $MnSO_2 \cdot H_2O$, 1.56; $ZnCl_2$, 1.67; $CoCl_2$, 2; pH = 5.3. In an experiment designed to study cellulase induction, a variety of carbon sources were substituted for glycerol. 1 ml of spore suspension (10^6) of Trichoderma was inoculated on 100 ml glycerol medium, and cultured in a reciprocating shaking bath (108 strokes per min) at 30℃ for 48 h. Then, the mycelium was harvested on a filter paper by suction filtration and washed with 200 ml 0.01 M phosphate buffer (pH=5.0).

Induction

The washed mycelium was resuspended in 50 ml 0.01 M* phosphate buffer (pH=5.0). Then, 2.5 ml of the mycelium suspension (—10 mg dry weight) was transferred to 2.5 ml induction medium (glycerol medium minus glycerol and peptone) and 0.1 ml of sophorose (sophorose was isolated and purified from pods of Sophora japonica L,[5] final concentration: 5×10^{-5}—5×10^{-4} M) in a 25 ml culture flask, and incubated in a reciprocating shaker (108 strokes per min) at 36℃. The culture filtrate (0.1 ml) was harvested at different times of

* M,克分子浓度,为非法定的物质的量浓度。1 M=1 mol/dm³。——编注

incubation for the cellulase assay. Under these conditions mycelium could not grow but could synthesize cellulase rapidly following induction by sophorose. The incubation period usually did not exceed 24 h.

Preparation of enzyme extract

The suction filtrate through filter paper was used as extracellular cellulase. The filtered mycelium was washed and resuspended in 15 ml 0.1 M phosphate buffer (pH = 5.0), sonicated (20 k cycles) for 5 min, and then centrifuged at 4,000 g for 20 min. The supernatant fraction was collected as intracellular cellulase.

Preparation of cellulase for separation and purification

The spore suspension was inoculated on 1 litre glycerol medium in a 5 litre flask, and grown at 28℃ for 48 h. The mycelium was washed with 2 volumes phosphate buffer, transferred to the induction medium, and cultured with sophorose (5×10^{-5} M) for another 24 h. The culture filtrate was precipitated with ammonium sulphate (80% saturation), desalted through Sephadex G-25, and lyophilized. The cellulase powder was then dissolved in acetate- NaOH buffer (0.1 M, pH=5.0) for separation.

Assay

The procedure suggested by Mandels[11] was used to assay CMC (30 min), filter paper (FP) (60 min) and cotton (24 h) cellulase at 50℃, except that in our early study CM cellulase was assayed at 40℃ for 30 min in a system consisting of 2.0 ml 0.63% CMC solution and 0.5 ml diluted enzyme (see Tables 2, 3, 5 and 6; Figure 2—Figure 7 below). The number of units (C_x) equals the inverse of the dilution to produce 0.5 mg glucose. FP cellulase, cotton cellulase and, in some cases, (see Table 4 and Figure 8 below) CM cellulase were expressed as mg glucose produced per ml enzyme in the above defined incubation time (60 min, 24 h and 30 min, respectively). Avicelase activity was assayed on the basis of glucose produced from 0.25 ml 0.5% Avicel as substrate incubated with 0.25 ml enzyme solution at 50℃ for 4 h.[9] Cellulase activity on phosphoric acid-swollen cellulose was determined by incubating 0.5 ml enzyme solution with 0.5 ml 0.1 M acetate buffer (pH = 4.5) and 4 ml 0.5% Walseth cellulose at 50℃ for 2 h.[9] Cellobiase was estimated by glucose produced in the hydrolysis of 2.0 ml 10^{-3} M cellobiose (in 0.02 M acetate buffer, pH = 5.0) with 0.5 ml culture filtrate at 40℃ for 2 h. β-Glucosidase was determined by incubating 0.2 ml diluted enzyme solution with 0.1 ml 0.5% p-nitrophenyi-β-glucoside (PNPG) in 0.1 M acetate buffer (pH=4.5) at 50℃ for 30 min. The β-glucosidase activity was expressed as μg of released p-nitrophenol/ml. Residual sugars were measured by the phenol-sulphuric acid method,[12] glycerol was measured by colorimetry[13] and protein was determined by the procedure of Lowry et al[14] RNA was measured by the orcinol reaction.[15] For the determination of amino acid content, 4.07 mg purified C_1 cellulase was hydrolysed with 5.7 N* HCl at 110℃ for 24 h and assayed with an Hitachi KLA-3B amino acid analyser.

Chromatography

Charcoal chromatography was used to isolate sophorose from glucose molasses or the pods of Sophora japonica L.[6] Paper chromatography was used to identify sophorose and other sugars.[6] DEAE-Sephadex A-50 and Sephadex-100column chromatography were used to separate cellulase components. Oligo(dT)-cellulose chromatography was employed for the isolation of mRNA.[2]

Electrophoresis

Cellulases were separated by polyacrylamide disc gel electrophoresis using 7% polyacrylamide gel.[16,17] The gel was stained with Amido Black and destained in 7% acetic acid. According to the colour of protein bands in stained gel, the unstained control gels were cut into slices which were extracted with distilled water, dialysed against water and lyophilized.[9] To estimate the molecular weights of cellulase components, 5%, 6%, 7% and 8% polyacrylamide gels were used with bovine serum albumin, peroxidase, pepsin, amylase, lactate dehydrogenase and creatine kinase as refer ences. The SDS-gel electrophoresis[2] was also used to determine the molecular weight of C_1. Cellulose acetate film electrophoresis, polyacrylamide gel electrophoresis and immunoelectrophoresis were employed to determine the purity of C_1 cellulase.[9] Antiserum against cellulase was prepared by injecting rabbits with 2 mg of partly purified cellulase in a Freund's adjuvant (1:1 v/v) for the first injection, 0.75 mg and 1 mg of cellulase for the second and third injections, respectively, at 10 day intervals.

Ultracentrifugation

The Schlieren peak of C_1 cellulase (10 mg/ml) was photographed 10 min after reaching 48,000 r/min at 20℃ in an analytical ultracentrifuge.

Results

Comparison of different carbon sources as cellulase inducers

Although Trichoderma mutant EA_3-867 grew at different rates on a variety of carbon sources, only sophorose, cellobiose and cellulose could induce cellulase production (see Figure 2). EA_3-867 could utilize monosaccharides and glycerol very well, but could not produce significant amounts of cellulase. After the mutant was cultured for 48 h, these monosaccharides were usually depleted. The growth of Trichoderma on glycerol medium

* N,当量浓度,非法定的物质的量浓度。1 N=(1 mol/L)÷离子价数。——编注

may roughly be divided into the lag phase (0—16 h), the log phase (16—48 h) and the stationary phase (>48 h). When cellulose or cellobiose (0.5%) were used as sole carbon sources, the mutant grew slowly and began to produce cellulase after 2 or 3 days in culture. The mutant, however, could produce cellulase after 18 h in culture on sophorose (0.5%) as the sole carbon source. On the basis of mycelial growth and cellulase production, all the tested carbon sources are divided into three types.

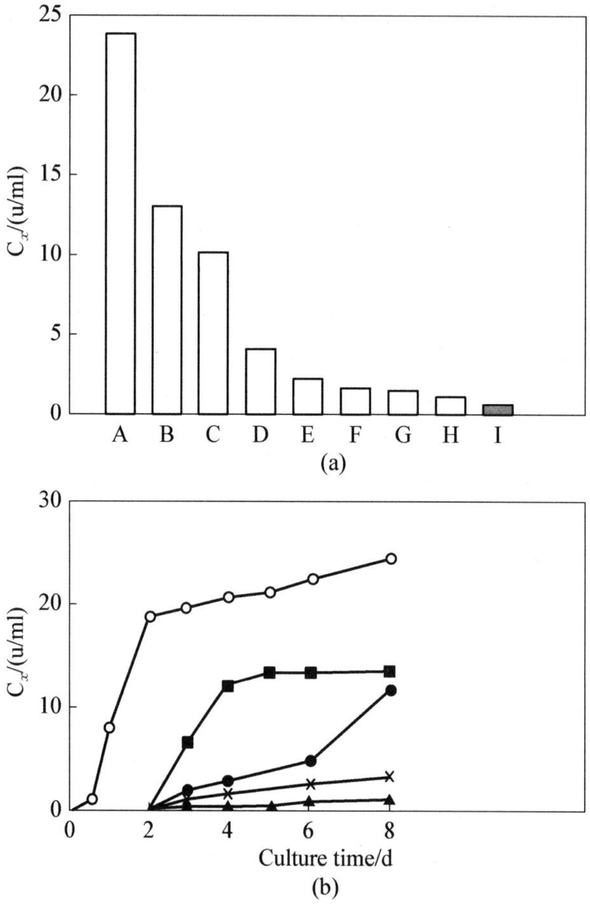

Figure 2 Effect of carban sources on cellulase activity and time coarse

(a) Comparison of cellulase activity of Trichoderma EA_3-867 on the 8th day after being cultured on diffarent carbon sources (0.5%).[5] A, Sophorose (SR); B, cellobiose (CB); C, cellulose powder (CP); D, filter paper (FP); E, CMC; F, cotton (cot); G, mannose (Man); H, Lactose (Lac); I, glucose (Glu), glycerol (Gly), fructose (Fru), galatose (Gal), xylose (Xyl) and sucrose (Suc). (b) Effect of culture time on cellulase production: o, sophorose (SR); ■, cellobiose (CB); ●, cellulose powder (CP); ×, CMC; ▲, glucose, glycerol

(1) Sophorose, as the strongest inducer of cellulase, is characterized by a stimulation of early and rapid cellulase production.

(2) Cellulose (cellulose powder, filter paper, cotton), cellulose derivatives (CMC) and

hydrolysates (cellobiose), which were not good carbon sources for mycelial growth, could induce cellulase production to different degrees. EA_3-867 produced cellulase on cellobiose more rapidly than on cellulose, but more slowly than on sophorose.

(3) Sucrose, maltose, as well as glucose, fructose, xylose, galactose, glycerol, which were good carbon sources for Trichoderma mycelium growth, could not induce cellulase production except for a trace of constitutive cellulase.

Effect of sophorose, cellobiose and cellulose added to glycerol culture upon cellulase induction

An insignificant amount of cellulase was produced by EA_3-867 either on glycerol medium or on the same medium supplemented with cellobiose (5×10^{-4} M) after being cultured for two days. However, a considerable amount of cellulase could be induced by cellulose (0.5%) or sophorose (5×10^{-4} M) added to the glycerol medium on the first day or after two days, respectively.[5] The level of cellulase synthesis induced by sophorose depended on the time at which sophorose was added to the medium, but the time of initiation of cellulase production was not related to the time of adding sophorose. In spite of the time of addition of sophorose to the medium, cellulase was always produced after two days in culture, as glycerol in the medium was depleted. This implies the involvement of catabolite repression in the regulation of cellulase synthesis.

Cellulase components induced by sophorose and cellulose

Cellulose and sophorose supplementation of the glycerol medium induced the extracellular and intracellular cellulase components C_x and cellobiase (see Figure 3). C_1 cellulase induction was also shown in another experiment.

The spore suspension of EA_3-768 was inoculated on glycerol medium with and without cellulose (0.5%, added at 0 h) or sophorose (10^{-4} M, added after two days in culture) supplementation, cultured at 33℃ to reach maximum cellulase production. The extracellular and intracellular cellulases were extracted from the culture filtrates and mycelia, respectively, precipitated by acetone at −20℃, dissolved in acetate buffer (0.1 M, pH= 4.8), and separated by electrophoresis on polyacrylamide disc gel(see Figure 4). All gels showed the very similar pattern of proteins (Zones I—VI). According to the assay of cellulase activity of each gel band, bands III, IV and V were identified as C_1 (cotton ceUulase), C_x (CM cellulase) and β-glucosidase, respectively. Clearly, C_1, C_x and β-glucosidase were strongly induced by cellulose, and especially by sophorose. In the glycerol control, a very small amount of C_1, C_x and β-glucosidase, presumed to be constitutive cellulase, was detected. It is of interest that the level of intracellular cellobiase in the glycerol control was distinctly higher than the extracellular one.

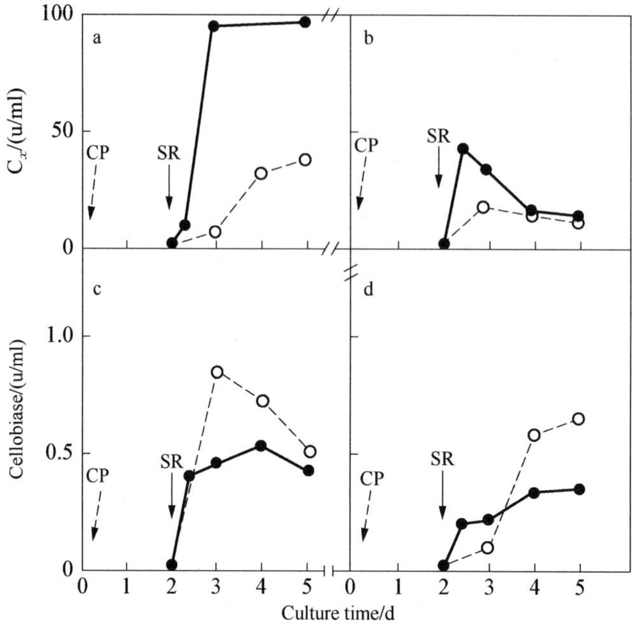

Figure 3 Comparison of distribution of extracellular (a, c) and intracellular (b, d) cellulase C_x, cellobiase of EA_3-867 induced by sophorose and cellulose added to glycerol medium. ●, 10^{-4} Ms sephorose (SP) added after 2 days; o, 0.5% cellulose powder (CP) added at zero time[5]

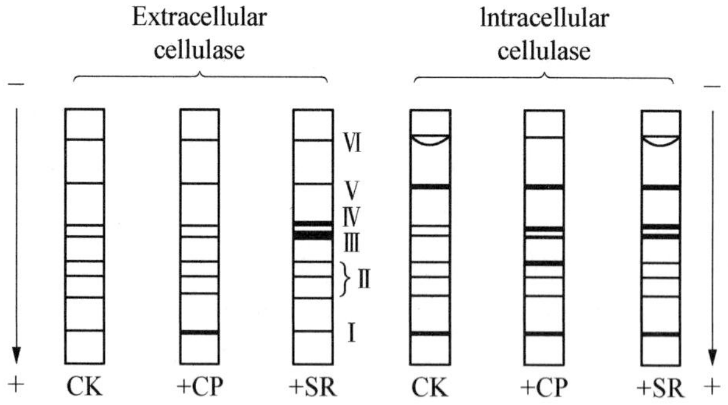

Figure 4 Schematic comparison of polyaceylamide gel electrophoresis patterns of Trichoderma EA_3-867 induced by sophorose (SR). cellulose powder (CP) and no inducer (CK).[5] Electrophoresis conditions: cellulase 400 μg. current 2mA/tube, time 1.5 h. Ⅲ. Cotton cellulase; Ⅳ. CM cellulase; Ⅴ. β-glucosidase

Cellulase induction by sophorose in washed mycelium

The synthesis of cellulase in washed mycelium was markedly induced by sophorose. In mycelium cultured with sophorose on an induction medium for 24 h, cellulase production of Trichoderma approached a maximum with a lag period of 2—3 h (see Figure 5).[6] The washed mycelia of different wild strains (1096, Mo_3, G109, D-92) and mutants (EA_3-867, N_2-78, $W_2$4, 4030, G023, G58) of Trichoderma, and the mutant EA_3-867 ofvarious ages (even as young as the short hyphal filaments 15 h old from inoculation) were all able to produce cellulase in which C_1 and C_x cellulase were predominantly extracellular, whereas cellobiase was mainly intracellular (see Figure 3).[6]

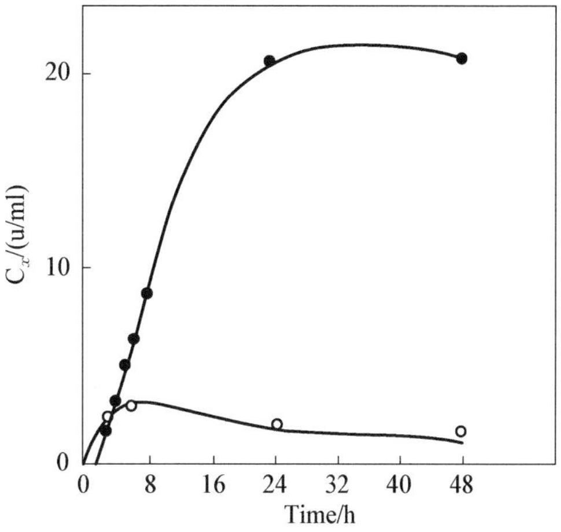

Figure 5 Cellulase synthesis of mycelium of Trichoderma EA_3-867 induced by 5×10^{-4} M sophorose.[6] ●, Extracellular cellulase; ○, intracellular cellulase

The formation of cellulase was affected by the inducer, mineral elements (N, P, Fe^{2+} Mn^{2+}), temperature, pH, aeration and the physiological conditions of the mycelium.[6] The cellulase synthesis stopped soon after the sophorose was removed from the medium. The earlier the sophorose was removed, the less the cellulase was induced. The optimal concentration of sophorose was 10^{-5}—10^{-3} M. No cellulase was induced in washed mycelium by cellulose (cellulose powder, rice straw cellulose, 0.5%) or cellobiose (10^{-4}—10^{-2} M). Table 1 presents the nutrient requirements for mycelial growth and cellulase induction. The optimal temperature of induction was 30℃. The range of pH used for induction was 3—6. The production of cellulase was markedly stimulated by O_2 and strongly depressed either by the absence of O_2 or by inhibitors of respiration such as iodoacetate (IA), sodium fluoride (NaF), malonate, soidum azide (NaN_3) or 2, 4-dinitrophenol (DNP) (see Figure 6). These data indicate the involvement of respiration in cellulase production either by the supply ofintermediates or ATP. The induction of cellulase in washed mycelium was also greatly

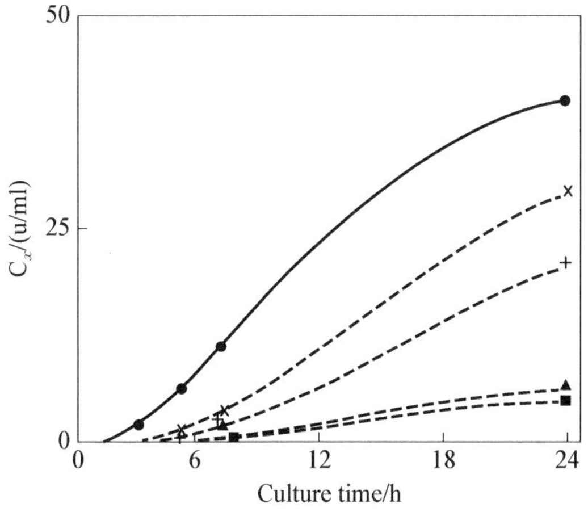

Figure 6 Inhibitory effect of respiratory inhibitors upon cellulase synthesis of washed mycelium of Trichoderma EA_3-867 induced by 5×10^{-4}M sophorose.[6] Mycelium: 2.2 mg/ml, ●, control; ×, 10^{-2}M NaF; +, 5×10^{-5}M DNP; ▲, 10^{-2}M IA; ■, 10^{-1}M malonate, 10^{-2}M NaN_3

inhibited by inhibitors of nucleic acid synthesis, e. g. actidione, 5-fluouracil and 6-azauracil.[6]

Table 1 Requirements of nutrients for Trichoderma mycelium growth and cellulase induction[6]

Nutrients	Growth medium	Induction medium
Carbon source	+	—
N	+	+
P	+	+
Peptone	+	—
Mg^{2+}	+	—
Ca^{2+}	+	—
Fe^{2+}	+	+
Mn^{2+}	—	+
Zn^{2+}	+	—
Co^{2+}	—	—

+, necessary; —, not necessary.

Catabolite repression

Cellulase synthesis in Trichoderma is regulated by catabolite repression as well as by induction. When sophorose was added to the glycerol medium after 48 h culture of EA_3 - 867, the cellulase was produced after 12 h, reached its maximum after 24 h, and then decreased. However, the cellulase increased again by new supplementation of sophorose on the fourth day of culture, an observation explained by the depletion of sophorose in the medium via metabolism, as shown the sophorose determination. Glycerol, however, distinctly repressed the cellulase production induced by sophorose.[5] More Convincing evidence for the regulation of cellulase synthesis by catabobte repression was the finding that cellulase synthesis of washed mycelium was inhibited to a different degree by various, easily utilizable carbon sources, such as glucose, fructose, galactose, ribose, glucose-6-phosphate, 6-phosphogluconate, gluconate, glycerol, as well as organic acids (10^{-3}—10^{-2} M) of the tricarboxylic acid cycle (acetate, pyruvate, citrate, x-ketoglutarate, malate, succinate, fumarate or oxaloacetate)[6] The cellulase induction usually recovered from repression after 24 h of culture. It is of interest that the catabolite repression of glucose could not be overcome by cyclic AMP (see Table 2).

Cellulase induction by glucose molasses and extracts of pods of Sophora japonica L.

Since cellulase synthesis is regulated both by induction and catabolite repression, the principle of increasing cellulase production should be based on the induction and derepression at the optimal conditions for rapid growth of mycelium as well as cellulase synthesis.

Table 2 Catabolite repression of cellulase synthesis in washed mycelium by glucose and effect of c-AMP[6]

	Cx(u/ml)			
	3 h	5 h	7 h	24 h
Control, no addition	1.1	3.8	6.8	21.6
Glucose (10^{-2} M)	0	0	1.9	27.2
Glucose (10^{-2} M) +c-AMP(5×10^{-2} M)	0	0	1.4	25.2
Glucose (10^{-2} M) +c-AMP(5×10^{-2} M)	0	0	1.2	22.8

Mycelium: 45 h old, 2.2 mg/ml, sophorose: 5×10^{-4} M

Although sophorose can strongly induce cellulase production, the commercial application of this rare sugar is not practical. Sophorose is a byproduct in glucose production. It was thus likely that the glucose molasses should contain more sophorose as an impurity. [7] The glucose molasses used for cellulase induction in our experiments was the waste liquor after the first crystallization of glucose in glucose production from corn starch by hydrochloric acid hydrolysis (in a Shanghai glucose refinery). It has been shown that glucose molasses has strong inductive activity for cellulase production in washed mycelium, probably in a similar manner to sophorose plus glucose. Cellulase production on glucose molasses does not commence until glucose is consumed; the remaining sophorose then exerts its inductive effect (see Table 3). However, no cellulase was induced by glucose molasses at higher concentration (1%) as a result of the catabolite repression by glucose itself.

Table 3 Induction of cellulase synthesis by glucose molasses in washed mycelium of Trichoderma EA_3 - 867 and N_2 - 78[7]

Inducer	C_x(u/ml)		
	4 h	6 h	24 h
Strain EA_3-867			
Glucose molasses (0.1%)	0.6	1.2	27.6
Glucose molasses (0.5%)	0	0	38.0
Glucose molasses (1.0%)	0	0	0
Sophorose (10^{-4} M)	1.0	2.4	32.0
Sophorose (10^{-4} M) +glucose (10^{-3} M)	1.0	2.4	35.6
Sophorose (10^{-4} M) +glucose (10^{-2} M)	0	0	34.8
Strain N_2-78			
Glucose molasses (0.1%)	0	1.0	15.2
Glucose molasses (0.5%)	0	0	24.8
Glucose molasses (1.0%)	0	0	0

(continued)

Inducer	C_x (u/ml)		
	4 h	6 h	24 h
Sophorose (10^{-4} M)	0. 8	1. 4	13. 2
Sophorose (10^{-4} M) +glucose (10^{-3} M)	0. 8	1. 6	18. 4
Sophorose (10^{-4} M) +glucose (10^{-2} M)	0	0	26. 4

Mycelium: EA_3-867, 2. 14 mg/ml; N_2-78, 0. 58 mg/ml.

The molasses were separated by charcoal and paper chromatography into four components: glucose (R_G 1. 00), sophorose (R_G 0. 42), gentiobiose (R_G 0. 26), and an unknown triose (R_G 0. 08). Only sophorose was identified to be responsible for cellulase induction. Gentiobiose only had feeble inductive activity.[7] The extracts of pods of Sophora japonica L, after hydrolysis with diluted H_2SO_4, had similar ability to induce the ceUulase production (see Figure 7). It was shown that the extract contained sophorose, glucose, fructose, sucrose and other oligosaccharides.

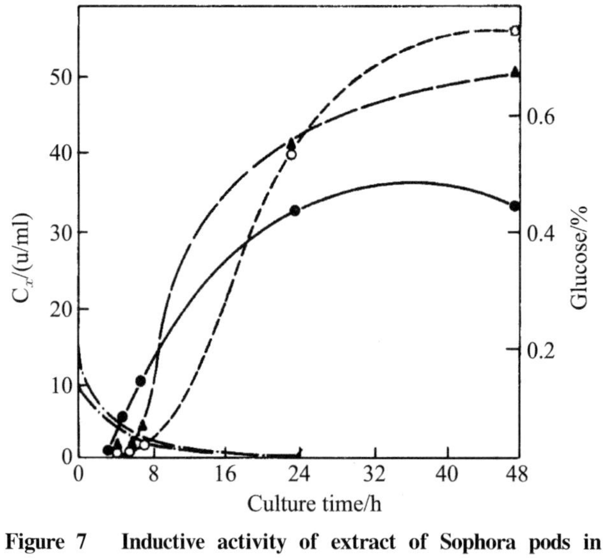

Figure 7 Inductive activity of extract of Sophora pods in cellulase formation of washed mycelium of Trichoderma EA_3-867.[7] Mycelium: 2. 9 mg/ml. ○, 5×10^{-4}M sophorose + 10^{-2}M glucose; ▲, 0. 2 ml extract of SophoN pods (0. 4%); ●, 5× 10^{-4} M sophorose; - • -, reducing sugar

In order to increase cellulase production and shorten culture time, it is suggested that in the early period of culture, or log phase of Trichoderma (first 48 h), optimal conditions for mycelial growth should be provided to promote rapid use of sugar. After 48 h in culture, Trichoderma grows to the stationary phase. Cellulase may then be induced by a strong inducer (such as sophorose) without catabolite repression at the optimal condition for cellulase synthesis. In fact, glucose molasses is an ideal inducer. When glucose molasses was added to a simple medium, the mutant N_2-78 reached a quite high activity level in only 60 h of culture. In contrast, Trichoderma may require one to two weeks to reach the maximum of cellulase production. [18,19]

Changes in the regulatory mechanism of cellulase synthesis in the mutants

Two mutants, EA_3-867 and N_2-78, showed much higher cellulase yields than wild parents in Koji extracts, shake flask culture filtrates, and enzyme preparations. The mutant N_2-78 showed a quite high activity level when cultured in shake flasks on a simple medium containing milled straw, wheat bran, mineral salts plus waste glucose molasses for 60 h, and the highest saccharifying activities of cellulase on CMC, filter paper and cotton reached 255, 8. 2 mg and 13. 4 mg glucose/ml, respectively, or 11, 4. 3and 6 times more than those ofits parent Mo_3 (see Table 4). Crude N_2-78 cellulase preparation of a 1% (w/v) enzyme solution in 0. 01 M citrate buffer, pH=4. 8, in a Warburgshaker (80 strokes/min) at 50℃ was able to degrade filter paper (Whatman no. 1, 1 cm^2) into fine fibres after only 9 min.

Table 4 Comparison of activities of cellulase from mutants and wild strains of Trichoderma in flask cultures[1]

Strains	Cellulase activity on CMC		Cellulase activity on filter paper /(mg glucose/ml)	Cellulase activity on cotton /(mg glucose/ml)
	Saccharifying activity /(mg glucose/ml)	Liquefying activity/(u/ml)		
Wild type				
1096	15. 3	405	1. 02	1. 5
Mo_3	17. 4	230	1. 54	1. 9
Mutant type				
EA_3-867	97. 0	1378	3. 60	9. 9
N_2-78	255. 0	4056	8. 20	13. 4

Seven ml mycelium seed suspensions were inoculated into 100 ml of a rich straw medium in 500 ml flasks, and cultured in a rotary shaker (104 strokes/min) at 30℃ for 60 h. The enzyme filtrates were used to assay cellulase activities. The rice straw medium consists of (%): milled rice straw, 1; wheat bran, 0. 5; $(NH4)_2 SO_4$, 0. 2; $KH_2 PO_4$, 0. 1; $CaCO_3$, 0. 1; glucose molasses, 1. The mycelium seed suspensions were obtained by culturing the spare suspensions of different strains on a seed medium at the same condition for 32 h. The seed medium was the same as ricestraw medium except that rice bran was used in place of wheat bran. CMC saccharifying activity, FP and cotton cellulases were assayed by the procedure suggested by Mandels, [17] and expressed as mg glucose produced per ml in 30 min, 60 min and 24 h incubation, respectively. CM cellulase (liquefying activity) was determined by the decrease in CMC viscosity and expressed as units per ml per min. [1,20]

Morphologically, both mutants produced smaller colonies on agar plates containing synthetic medium or potato glucose medium, and grew more slowly than their parents. But on agar plates containing peptone-yeast extract medium, there were no differences in colony size among the strains. [8]

Besides growth and morphology, the kinetic characteristics and components of cellulase in the mutants and wild strains were also compared. The optimal pH and temperature for cellulase activities (CM cellulase, FP cellulase) were the same.[1] No differences in cellulase components were observed, at least in the patterns of polyacrylamide gel electrophoresis, except that the protein bands of cellulase of the mutants had a darker colour than that of their parents.[8] It is suggested that the changes in the mutants probably do not occur in the structural genes, but only in the regulatory genes which may be involved in higher yield of cellulase.

Since the regulatory systems of cellulase in Trichoderma consist of the induction and catabolite repression, the sensitivity of the mutants and parents to inducer (sophorose) and repressor (glucose) were compared in a washed mycelium test. Cellulase of the mutants could be induced by lower sophorose (10—100 times less than in the parents) (see Table 5), and were repressed by higher glucose (almost 10 times higher than in the parents) (see Table 6). It is evident that the increase in cellulase production of both mutants, which have been shown to be partially derepressed, results from the change in induction as well as in catabolite repression.

Table 5 Comparison of sensitivity of washed mycelia of different strains of Trichoderma to sophorose induaion of cellulase[8]

Strains	Sophorose concentration/M	C_x/(u/ml)	
		4 h	6 h
1096 (wild)	10^{-7}	0	0
	10^{-6}	0	0
	10^{-5}	0	0
	10^{-4}	0.5	3.5
	10^{-3}	2.0	5.9
EA_3-867 (mutant)	10^{-7}	0	0
	10^{-6}	0	0
	10^{-5}	0.5	2.4
	10^{-4}	1.2	3.2
	10^{-3}	1.0	3.0
Mo_3 (wild)	10^{-7}	0	0
	10^{-6}	0	0
	10^{-5}	0	0
	10^{-4}	2.4	5.7
	10^{-3}	3.0	6.2

(continued)

Strains	Sophorose concentration/M	C_x/(u/ml)	
		4 h	6 h
N_2 - 78 (mutant)	10^{-7}	0	0
	10^{-6}	0	3.0
	10^{-5}	1.3	4.6
	10^{-4}	1.5	5.5
	10^{-3}	2.1	6.0

Mycelia: 1096, 2.4 mg/ml; EA_3-867, 1.9 mg/ml; Mo_3, 2.5 mg/ml; N_2-78, 1.7 mg/ml. Sophorose: 10^{-4} M.

Table 6 Comparison of sensitivity of washed mycelia from different strains of Trichoderma to glucose repression[8]

Strain	Glucose concentration/M	4 hours		6 hours	
		C_x/(u/ml)	Inhibition/%	C_x/(u/ml)	Inhibition/%
1096 (wild)	0	0.70	0	1.8	0
	10^{-3}	0.60	14	1.2	33
	10^{-2}	0	100	0	100
EA_3-867 (mutant)	0	1.0	0	2.4	0
	10^{-3}	1.0	0	2.4	0
	10^{-2}	0	100	0	100
Mo_3 (wild)	0	0.6	0	1.0	0
	10^{-3}	0.4	33	0.4	60
	10^{-2}	0	100	0	100
N_2-78 (mutant)	0	0.8	0	1.4	0
	10^{-3}	0.8	0	1.6	0
	10^{-2}	0	100	0	100

Mycelia: 1096, 2.9 mg/ml; EA_3-867, 2.1 mg/ml; Mo_3, 2.5 mg/ml; N_2-78, 1.1 mg/ml. Sophorose: 10^{-4} M.

Separation, purification and properties of cellulase components of N_2-78

Since sophorose coordinately induces cellulase C_1, C_x and β-glucosidase of washed mycelium of Trichoderma to high levels (see Figure 8), this preparation was used as starting material for purification.

After DEAE-Sephadex A-50 chromatography C_1, C_x, and β-glucosidase were separated. The C_1 protein was further purified by preparative gel electrophoresis. The C_x component and β-glucosidase were separated by passing through Sephadex G-100. Selective inactivation was also used.[9]

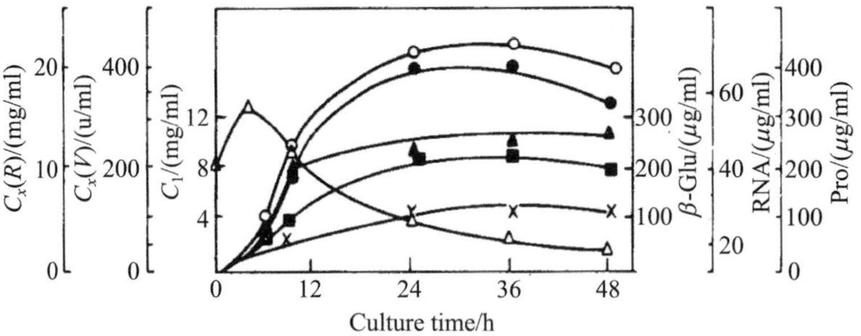

Figure 8 Changes in extracellular cellulase, soluble protein, and RNA of mycelium during induction of washed mycelium of N_2-78 by sophorose. 2 Mycelium: 1.5 mg/ml; sophorose: 10^{-4} M. C_1 is expressed by cotton cellulase, C_x (*R*) and C_x (*V*) represent CM cellulase measured by reducing sugar increase or by viscosity decrease, respectively. ○, C_x (*V*); ●, C_x (*R*); ■, C_1; ▲, β-glucosidase; ×, extracellular soluble protein; △, RNA

The molecular weights of C_1, C_x and β-glucosidase were estimated to be 67,000, 62,000 and 42,000, respectively, on polyacrylamide gel[9] The C_1 component was shown to be homogenous by polyacrylamide gel and cellulose acetate film electrophoresis, immunoelectrophoresis and ultracentrifugal analysis (see Figure 9). Component C_1 was a glycoprotein containing 16 amino acids (mainly glycine, aspartic acid, threonine, serine and glutamic acid). The C_1 component showed a strong synergistic action with C_x component in the degradation of cotton, Avicel and Walseth cellulose. The hydrolysates of Walseth cellulose by C_1 action were mainly cellobiose and a trace of glucose. This supports the concept that C_1 action is mainly as a cellobiohydrolase. The stability of these components to high temperature was lowered when they were separated.

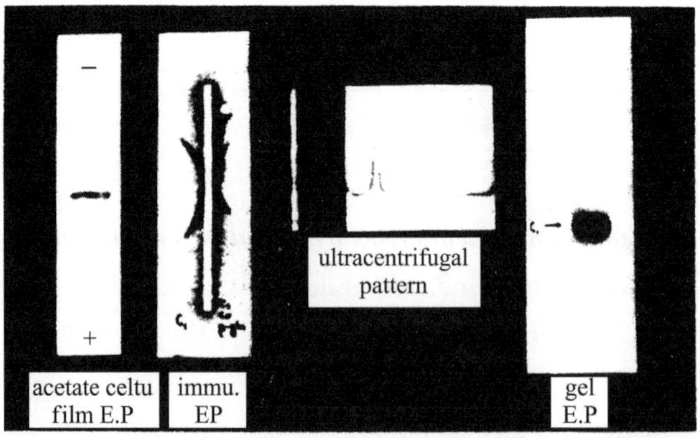

Figure 9 Identification of C_1-cellulase by polyacrylamide gel electrophoresis, cellulase acetate film electrophoresis, immuno electrophoresis and ultracentrifugal analysis[9]

Isolation of PolyA-RNA from mycelium of N_2-78 induced by sophorose

Since mRNA plays an important link in gene expression of cells, the change in mRNA content in mycelium during cellulase induction was determined. The content of RNA reached a maximum after 6 h induction (see Figure 8). The total RNA was extracted with phenol for 6 h at low temperature (<5℃) from 100 ml of washed mycelium which had been incubated with sophorose. RNase was inhibited by diethylcarbonate. Total RNA was then separated into three peaks (Ⅰ, Ⅱ and Ⅲ) by oligo(dT)-cellulose affinity chromato graphy (see Figure 10). The A_{260}/A_{280} ratios of peaks I, II and III were 2.24, 2.14 and 2.16, respectively. Peak I was rRNA, comprising 90% of the total RNA. Peak III, the polyA-RNA, showed the typical UV absorption spectrum of RNA. It was eluted with Tris buffer at low ionic strength or by water. It is important to note that the content of polyA-RNA in mycelium induced by sophorose was higher than the control, as shown by affinity chromatography (see Figure 11) and polyacrylamide gel electrophoresis.[2]

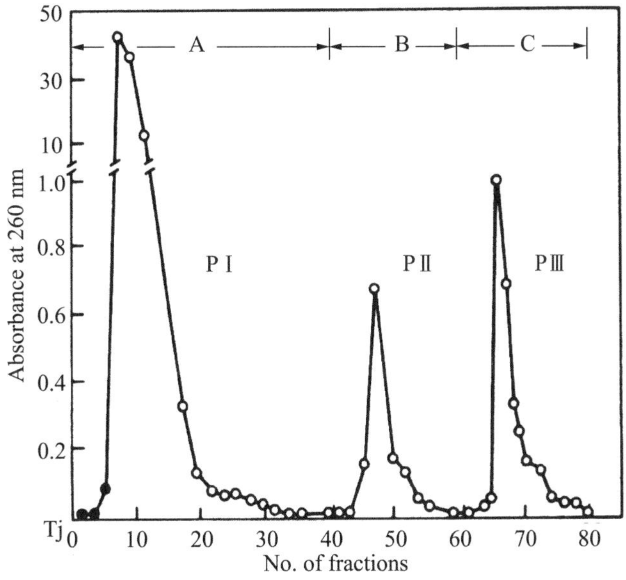

Figure 10 **Chromatography of total RNA of mycelium of Trichoderma N_2-78 induced by sophorose through oligo(dT)-cellulose column[9] A, 0° C, 0.01 MTris, pH=7.5, 0.5 M KCl; B, 54℃, 0.01 M Tris, pH=7.5, 0.5 M KCl; C, 54℃, 0.01 M Tris, pH=7.5**

Discussion

Regulation of enzyme synthesis and gene expression in eukaryotes is one of the most

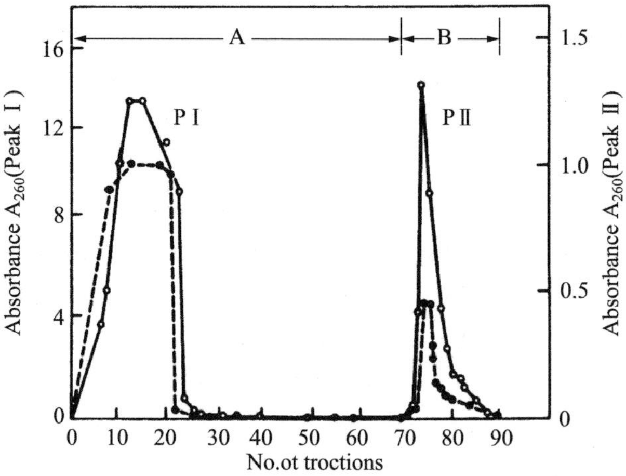

Figure 11 **Chromatography of total RNA of mycelium of Trichoderma N_2-78 induced and non-induced by sophorose through oligo (dT)-Cellulose column.[9] ●, Control; ○, sophorose. A, 0.01 MTris, 0.5 M KCl, pH=7.5; B, H_2O, pH=7.5**

important branches of modern molecular biology. The importance of the study of the synthesis and regulation of cellulase, a typical inducible-extracellular enzyme produced by the simplest eukaryotes (Trichoderma) should be emphasized for its potential significance in solving energy problems as well as for its being a convenient material in the study of gene expression in eukaryotes. Several reviews have dealt with this subject.[3,4,11,19,21-24] However, knowledge of cellulase synthesis and regulation is still limited.

What is the true and natural inducer of cellulase? How does the mycelium of Trichoderma recognize the presence of cellulose substrate in the medium? Mandels and Reese, in their early work,[25] have shown that of various carbon sources tested, only cellulose, cellobiose, lactose and reagent glucose were able to induce cellulase production. Cellobiose, a hydrolysate of cellulose, was presumed to be the natural inducer,[26] The induction activity of cellobiose was low; more cellulase could be induced by cellobiose only at a high concentration of cellobiose (1% or more), by the decrease of metabolism, or the supply of surfactant.[27,28] Therefore, it is questionable whether cellobiose is the native inducer. On the other hand, Mandels and Reese[10,29] have demonstrated that sophorose as an impurity in reagent glucose could strongly induce cellulase production. This discovery is exciting and could contribute profoundly to the cellulase induction research. It prompted us to conduct a series of investrgations of cellulase synthesis and regulation since we had isolated two mutants of Trichoderma, EA_3-867and N_2-78, with high cellulase yield in 1973.

Since cellulase was induced by cellobiose only at higher concentration during growth, no cellulase induction was observed in the washed mycelium test, a standard and reliable test for inducer, it is more likely that some product of cellobiose conversion during mycelial growth, but not cellobiose itself, is the true inducer. Sophorose is more effective in cellulase induction, even at a concentration of 0.3 μg/ml (the ratio of sophorose to mycelium dry weight was 1 : 5,000, with a lag of 2—3 h). Moreover, the induction of cellulase by sophorose and cellulose was quite similar both in the extracellular or intracellular components and in their pattems of polyacrylamide gel electrophoresis. Therefore, sophorose might be a true inducer of cellulase, as proposed recently by others.[30] Sophorose was presumed to be formed by transglucosylation of β-glucoside from cellobiose in vivo,[6,30] a phenomenon similar to β-glucosidase in E. coli,[31] in which the natural inducer allolactose is formed by a similar mechanism. It is not yet known why organisms have developed a mechanism to form the true inducer by modifying substrates (e.g. lactose) or more direct products (e.g. cellobiose from cellulose). It may be needed for a more complicated control of metabolism. On the other hand, on a medium containing a rapidly utilizable carbon source such as glycerol or glucose, which are not inducers, Trichoderma EA_3-867 produced traces of cellulase. Thus, the concept of constitutive cellulase was proposed.[6] The induction of cellulase by insoluble cellulose is presumed to occur as follows: constitutive cellulase, on the cell surface or released into the medium, hydrolyses cellulose to cellobiose which may enter the cell and be converted to sophorose in vivo by transglycosylation. Sophorose enters the nuclei, and activates the transcription process in nuclei whereby the mRNA is synthesized. The latter is transported to the cell membrane. As a consequence, a great amount of inducible cellulase is synthesized there and secreted into the extra-cellular environment.

The cellulase synthesis of Trichodermn is regulated not only by the induction system, but also by catabolite repression, a common phenomenon in other fungi and bacteria.[7,32-35] This is a self-regulation process by which the cells avoid wastage of resources and energy in synthesizing unnecessary cellulase in the presence of other rapidly utilizable carbon sources until the latter have been used. It is of interest that the catabobte repression of glucose could not be overcome by cyclic-AMP, indicating that cellulase regulation in the eukaryote Trichoderma is different from the enzyme regulation in prokaryotes, in which catabolite repression is due to the decrease in the c-AMP level, which acts as a necessary effector in the initiation process of the operon replication.[35]

The coordinate induction of a complete cellulase system (C_1, C_x, and β-glucosidase) induced by sophorose suggests that these three functionally related enzymes may be controlled by the same regulatory gene, although it is not necessary for three structural genes to be linked at the same chromosome. The result presented here is different from other reports,[30] in which no complete cellulase system was found to be induced by sophorose. It raises the interesting question of whether different species of Trichoderma have different genetic control mechanisms in cellulase synthesis.

Cellulase synthesis in washed mycelium was inhibited by the inhibitors of nucleic acid metabolism (5-fluouracil, 6-azauracil) and protein synthesis (actidione). Thus, the metabolism of nucleic acid, especially the regulation of mRNA synthesis, plays an important role in cellulase synthesis. As yet, no information on mRNA in Trichoderma has been reported. In the higher eukaryotes, it has been found that mRNA molecules contain a polyA-RNA sequence at the 5'-end, which is the basis for isolation of mRNA by oligo(dT)-cellulose affrnity chromatography. In a microorganism, however, the short lifespan of mRNAs makes it difficult to isolate them. The data presented in our work indicate that most of the extracellular proteins induced by sophorose were cellulases. It is thus likely that the mRNA of cellulase should be enriched when induced by sophorose. By adding the inhibitor of nuclease, fast extraction at lower temperature, a polyA-RNA has been obtained in washed mycelium induced by sophorose. The ratio A_{260}/A_{280} for polyA-RNA was 1.5 and the yield was 1% of total RNA. Analysis of polyacrylamide gel electrophoresis has shown that its molecular size was in the size range of mRNA.[2] It is important that the content of polyA-RNA in mycelium increased after the induction by sophorose. This increased polyA-RNA may be mRNA of cellulase, and its biological activity should be examined further. Furthermore, cell wall-bound ribosomes were observed in the sophorose-induced mycelium by electronmicroscopy (data not shown here). The biological implication of this phenomenon is not yet clear.

Our knowledge of the regulation of cellulase synthesis has been used to increase cellulase production and genetic analysis of the mutants. Inducer or gratuitous inducers are often used to increase enzyme yields. Examples are xylose for glucose isomerase,[35] sucrose monopalmitate for invertase,[27] and starch, glycogen and maltotriose for amylase.[37] The use of sophorose as an inducer of cellulase has been considered impractical because of its high price.[22] However, the present study suggests that the waste glucose molasses may provide an available and inexpensive resource of sophorose. In our pilot fermenter (1000 litre), waste glucose molasses was effectively used to increase cellulase yields and to shorten the culture time to 60 h.

Two mutants of Trichoderma, EA_3-867 and N_2-78, were compared with their parents on the basis of regulation of cellulase synthesis. The cellulase in the mutants did not change either in kinetics or electrophoretic pattern compared with their parents except that their yields increased considerably, suggesting that only the regulatory gene, but not the structural gene, had changed in the mutants. This suggestion was supported by the discovery that the mutants showed increased sensitivity to inducer and decreased sensitivity to catabolite repression. One possible explanation for the increased sensitivity to inducer in the mutants is the change in the repressor structure which combines with the inducer more tightly as in lac operon mutants. Similarly, the cellulase in the mutants could be induced by much less sophorose than in the parent strains. Although the sensitivity of mutants EA_3-867 and N_2-78 to catabolite repression decreased, cellulase synthesis in the mutants was still repressed by glucose at high concentration (10^{-2} M). Thus, the mutants are considered as partially derepressed strains. Mutations that effect catabolite repression may occur in the promotor, or the genes for CRP protein or adenyl cyclase. Moreover, the mutation of genes in the main metabolic pathway would indirectly affect the degree of catabolite repression.[38] From our results, the latter possibility may hold, as the mutants grew slowly and showed small colonies on synthetic medium, but grew as well as wild type strains on peptone-yeast extract medium. This suggests that the decreased catabolite repression may be due to the retardation of growth and the block in some metabolic reaction which synthesizes necessary growth factor(s) present in peptone and yeast extracts.

The study on cellulase synthesis and regulation in Trichoderma is not only helpful in understanding basic biological process such as gene expression in eukaryotes but it also has practical significance. The cellulases of EA_3-867 and N_2-78 are utilized to isolate protoplasts from plant cells, to increase alcohol production, to make medicines to aid the digestion of plant foods, to improve feed efficiencies in pig feeding, partially to replace alkaline treatment in textile production, and to improve quality in the food industry.[3]

Acknowledgements

The authors wish to express their thanks to Drs Mandels M, Reese E T, Burchard R P, Kung S D, and Schwartz M for their helpful discussions, especially to Drs Mandels M and Kung S D for their encouragement and hospitality during the writing of this paper. We thank Mr Fu X C and Mrs Chen V F for some valuable help in this study, and Mrs Combs A and Miss McCuUough-Green C for their secretarial help.

References

[1] Cellulase Research Laboratory, ShanghaiInstitute of Plant Physiology, Academia Sinica and Shanghai Distillery no. 2. *Acta Microbiologia Sinica* 1978, 18, 27

[2] Wu, Y. Q., Zhu, Y. S., Chen, W., Gao, J. H. and Fei, J. X. *Acta Phytophysiologia Sinica* 1979, 5, 335

[3] Zhu, Y. S. *Appl. Microbiol. (China)* 1975, 2, 7

[4] Zhu, Y. S. *Adv. Biol. Sci. (China)* 1978,6, 29

[5] Zhu, Y. S. and Tan, C. *Acta Microbiologia Sinica* 1978, 18, 320

[6] Zhu, Y. S. and Tan, C. *Acta Phytophysiologia Sinica* 1978,4, 1

[7] Zhu, Y. S. and Tan, C. *Acta Phytophysiologia Sinica* 1978,4, 19

[8] Zhu,Y. S. *Acta Phytophysiologia Sinica* 1978,4, 143

[9] Zhu, Y. S., Wu, Y. Q., Gao, J. H. and Fei, J. X. *Acta Phytophysiologia Sinica* 1980, 6, 1

[10] Mandels, M., Perrish, F. W. and Reese, E. T. *J. Bacteriol.* 1962, 83, 400

[11] Mandels, M. and Weber, S. *Adv. Chem. Ser.* 1969, 95, 391

[12] Dubois, M., Gilles, K. A., Hamilton, J. K., Robert, P. A. and Smith, F. *Anal. Chem.* 1956, 28, 350

[13] Lambert, M. and Neish, A. C. *Can. J. Res.* 1950, 28, 83

[14] Lowry, O. H., Rosebrough, N. J., Farr, A. L. and Randall, R. J. *J. Biol. Chem.* 1951, 193, 265

[15] Chargaff, E., Davison, T. N. in *The nucleic acid: Chemistry and Biology* Academic press, New York, N. Y., 1955, vol. 1, p. 301

[16] Hedrick, J. L. and Smith, A. J. *Arch. Biochem. Biophys.* 1973, 135, 587

[17] Weber, K. and Osbom, M. *J. Biol. Chem.* 1969, 224, 4406

[18] Reese, E. T. *Biotechnol. Bioeng. Symp.* 1975,5, 77

[19] Ryu, Dewey D. Y. and Mandels, M. *Enzyme Microb. Technol.* 1980, 2, 91

[20] OsmundsvŌg, K. and Goksøyr, J. *Eur. J. Biochem.* 1975,57, 405

[21] Eveleigh, P. E. and Montenecourt, B. S. *Adv. Appl. Microbiol.* 1979, 25, 60

[22] Mandels, M. and Sternberg, D. *J. Ferment. Technol.* 1976,54, 267

[23] Mandels, M. , Sternberg, D. and Andreotti, R. E. in *Symp. on Enzymatic Hydrolysis of Cellulose* (M. Bailey T-M,ed), 1975, p. 81

[24] Pathak, A. N. and Ghose, T. K. *Process Biochem.* 1973, 8, 35

[25] Mandels, M. and Reese, E. T. J. *Bacteriol.* 1957, 73, 269

[26] Mandels, M. and Reese, E. T. J. *Bacreriol.* 1960, 79, 816

[27] Reese, E. T. , Lola, J. E. and Parrish, F. W. *J. Bacteriol.* 1969, 100, 1151

[28] Reese, E. T. and Maguire, A. J. *Dev. Ind. Microbiol.* 1971, 12, 212

[29] Mandels, M. and Reese, E. T. *Biochem. Biophys. Res. Commun.* 1959, 1, 338

[30] Sternberg, D. and Mandels, G. R. *J. Bacteriol.* 1979, 139, 761

[31] Jobe, A. and Bourgeois, S. *J. Mol. Biol.* 1972, 69, 397

[32] Horton, J. C. and Keen, N. T. Crzn. *J. Microbiol.* 1966, 12, 209

[33] Stewart, B. and Leatherwood, J. W. *J. Bacteriol.* 1976, 128, 609

[34] Suzuki, H. , Yamane, K. and Nisizawa, K. *Adv. Chem. Ser.* 1969, 95,60

[35] Takasaki, Y. , Kosugi, Y. and Kanbayashi, A. in *Fermentation Advances* (Perlman, D. , ed) Academic Press, New York, N. Y. 1969, p. 561

[36] Perlman, R. L. , Crombrughe, B. D. and Pastan, I. *Nature* (*London*) 1969, 223, 810

[37] Banks, G. T. and Hockenhull, D. J. D. Progr. *Indust. Microbiol.* 1967, 6, 95

[38] Oarke, P. H. in *2nd Int. Symp. Genetics Indust. Microorganisms* (Macdonald, K. D. , ed.) Academic Press, 1976, p. 15

Organization and Expression of Genes for Photosynthetic Pigments-Protein Complexes in Photosynthetic Bacteria

Photosynthesis, the conversion by living organisms of light into chemical energy, is a fundamental biological process on earth. One of the major goals of plant biotechnology is to alter genetically the photosynthetic apparatus of crop plants to increase the efficiency of photosynthesis and ultimately to design an artificial photosynthetic system for harvesting solar energy. This strategy depends on a full understanding of the structure, organization, and function of the photosynthetic apparatus. Over the last few years a great deal of knowledge about the basic physics, chemistry, and molecular biology of photosynthesis has accumulated (Govindjee 1982; Sybesma 1984; Staehelm and Arntzen 1986; Steinback et al. 1985). However, the molecular events involved in biosynthesis, assembly, and regulation of the photosynthetic apparatus are not well understood.

Photosynthetic bacteria, genus Rhodobacter, including Rhodobacter capsulatus and Rhodobacter sphaeroides (formerly Rhodopseudomonas capsulata and *R. sphaeroides*, respectively) afford an unusual opportunity to study the molecular events of photosynthesis. These purple nonsulfur bacteria have a photosynthetic system that is less-complex than that of green plants and algae. Unlike higher photosynthetic organisms, they possess only one photosystem and lack the ability to oxidize water to oxygen. Their photochemical reaction center (RC) is similar to, but simpler than, that of photosystem II, and can thereby provide fundamental information about the light reactions in more complex organisms (Kaplan and Arntzen 1982). For example, a new model of photochemical function of photosystem II components D_1 and D_2 has resulted from a study of DNA sequence homology between genes coding for RC subunits L and M in *R. capsulatus* and genes coding for the 32-kd thylakoid membrane polypeptides in spinach and tobacco (Hearst and Sauer 1984a and 1984b; Hearst 1986). Study of Rhodobacter has been greatly advanced by several well-developed genetic techniques, particularly for *R. capsulatus*, including the use of the gene transfer agent

---

Zhu Yu Sheng, Hearst John E. 本文发表于 1989 年"Plant Biotechnology" (Kung, S. D. & Arntzen C. J. eds), Chapter 11: 257 - 287, Butterworth Publishers.

We are grateful to Dr. S. Kaplan for the use of his unpublished data. We thank D. Cook and G. Armstrong for their comments in preparation of this review. Part of this work was supported by the U. S. Department of Energy and by a grant from the National Institutes of Health.

(Marrs 1974; Solioz et al. 1975), marker rescue (Taylor et al. 1983), transposon mutagenesis (Youvan et al. 1982; Zsebo and Hearst 1984), and interposon mutagenesis (Scholnik and Haselkorn 1984). Photosynthetic mutants are viable because of the remarkable metabolic versatility of purple nonsulfur bacteria. Rhodobacter can grow either chemoheterotrophically or photoheterotrophically, depending on oxygen tension and light intensity in the environment (Kaplan 1978; Drews 1978; Drews and Oelze 1981). This dual lifestyle makes the organism an excellent laboratory tool with which to study the regulation of photosynthetic genes by light and oxygen as well as the formation and development of the photosynthetic membrane. Thus, in comparison with algae and higher plants, both the biophysical chemistry and the genetics of the photosynthetic apparatus are more advanced in the genus Rhodobacter.

The initial experiments that led to the discovery of the photosynthetic gene cluster of *R. capsulatus* were conducted by Marrs et al. They utilized the gene transfer agent, a particle resembling a generalized transducing phage, to map a gene cluster for bacteriochlorophyll (Bchl) and carotenoid (Crt) biosynthesis (Yen and Marrs 1976; Solioz et al. 1975; Yen et al. 1979). Using the R factor mobilization technique, they isolated an R-prime plasmid, pRPS404, bearing 46 kb of chromosomal DNA that contained most of the photosynthetic genes of *R. capsulatus* (Marrs 1981). This key step, achieved by genetic methods, is the basis for studying the molecular biology of photosynthesis in *R. capsulatus*. A subsequent breakthrough in the study of the photosynthetic genes of *R. capsulatus* was achieved by Hearst et al. when they reported the complete DNA sequence of the structural genes encoding the subunuits RC-L, RC-M, and RC-H (designated *puf*L, *puf*M, and *puh*A, respectively) as well as the light-harvesting (LH) I polypeptides LH-I β and LH-I α (*puf*B and *puf*A, respectively) (Youvan et al. 1984a and 1984b). Subsequently, the DNA sequence of two of the three LH-II poly peptides, LH-II β and LH-II α (*puc*B and *puc*A, respectively), was determined by Youvan and Ismail (1985). Additionally, 17 genes for Bchl and Crt biosynthesis have been aligned with a detailed restriction map (Taylor et al. 1983; Zsebo and Hearst 1984; Zsebo et al. 1984). Paralleling this achievement, the *R. sphaeroides* genes for RC-L, RC-M, RC-H (Williams et al. 1983 and 1984; Donohue et al. 1986a), LH-I and LH-II (Kiley et al. 1987; Kiley and Kaplan 1987) have also been cloned and sequenced. Recently, the DNA sequences of the genes for cytochrome c_2 (cycA)(Daldal et al. 1986; Donohue et al. 1986b) and the iron sulfur ubiquinol bc_1, oxidoreductase complex (*fbc*) have been determined in both *R. capsulatus* and *R. sphaeroides* (Gabellini and Sebald 1986; Davidson et al. 1987). Even more recently the structural genes for the Form I ribulose bisphosphate carboxylase, phosphoribulose-kinase, and cytochrome c′ (cydA) have also been identified (Kaplan, personal communication). Furthermore, the sequences of the genes for the B880 holochrome from Rhodospirillum rubrum (Bérard et al. 1986) and the genes for H, L, and M subunits of the RC (Michel et al. 1985 and 1986) from Rhodopseudomonas viridis have been determined. In summary, thus far the structural genes for LH complexes (except the γ subunit of LH-II), RCs, and secondary electron carriers, as well as a few other genes involved in carbon dioxide fixation have been isolated and subjected to detailed structural analyses in both *R. capsulatus* and *R. sphaeroides*.

At the protein level, Deisenhofer et al. (1985a) have recently presented a detailed crystal structure of the RC from R. viridis at 3 Å resolution. A three-dimensional structure of the *R. sphaeroides* RC (Allen and Feher 1984) has also been reported. This approach will lead to significant progress in our understanding of how protein structure facilitates primary photochemistry in the RC.

In addition to gene and protein structure, studies on transcriptional and posttranscriptional regulation of the photosynthetic genes, namely, the synthesis and processing of mRNA, are becoming an important branch in the study of the molecular biological properties of photosynthetic bacteria. Although early studies indicated the importance of transcriptional regulation (Biel and Marrs 1983 and 1985; Clark et al. 1984; Klug et al. 1984), the isolation and characterization by Northern blots and S-1 nuclease analysis of discrete mRNAs for LH-I, RC-L, and RC-M were shown only by Belasco et al. (1985) for *R. capsulatus*, and by Zhu and Kaplan (1985) and Zhu et al. (1986b) for *R. sphaeroides*. Recently, we have described a more-detailed study of the regulation by oxygen and light of the genes coding for the structural proteins and pigment biosynthetic enzymes in *R. capsulatus* (Zhu and Hearst 1986; Zhu et al. 1986a).

In this chapter we have reviewed the recent progress in research aimed at the organization and regulation of photosynthetic genes primarily in *R. capsulatus* and *R. sphaeroides*. For other reviews regarding the biosynthcsis and assembly of the photosynthetic apparatus in purple photosynthetic bacteria, see the following articles: Drew 1978 and 1985; Kaplan 1978; Drews and Oelze 1981; Kaplan and Arntzen 1982; Youvan and Marrs 1984 and 1985; Drews et al. 1986.

Metabolic Versatility

Purple nonsulfur bacteria normally inhabit muddy lake bottoms and sewage lagoons. They have a remarkable ability to adapt metabolically in response to changes in oxygen tension, light intensity, and available nutrients in their environment. Such changes are likely to occur in the mud of ponds and ditches; in the littoral zones of lakes, rivers, and seas; and in all kinds of sewage Iagoons (Imhoff et al. 1984).

The multiple modes of growth seem to parallel the evolution of the earth's atmosphere from an initial reducing environment of hydrogen, ammonia, and methane through the synthesis of organic acids and depletion of ferrous iron in the oceans, to the development of an oxidizing environment through the action of the cyanobacteria. Typical purple nonsulfur bacteria, *R. capsulatus* and *R. sphaeroides*, are capable of six distinct modes of growth; anaerobic photoautotrophy growing on hydrogen and carbon dioxide under light; anaerobic photoheterotrophy in the light using organic compounds and carbon dioxide; anaerobic chemoheterotrophy in the dark, utilizing a variety of organic carbon sources plus a terminal electron acceptor such as DMSO; dark aerobic chemoautotrophy, using hydrogen as the sole source of energy and reducing power; aerobic chemoheterotrophy growing on a diverse group of carbon compounds and generating ATP through oxidative phosphorylation; and anaerobic dark fermentation in the absence of oxygen, utilizing substrate-level phosphorylation. Among the various environmental variables, oxygen tension and light intensity, which change frequently in the vertical ecological zone, are the most important in regulating these metabolic shifts (Drews 1978; Kaplan 1978; Drews and Oelze 1981; Kaplan and Arntzen 1982; Chory et al. 1984; Drews et al. 1986; Youvan and Marrs 1984 and 1985).

When growing chemoheterotrophically, purple nonsulfur bacteria contain a typical Gram-negative outer membrane and a cytoplasmic membrane. Under these conditions cell growth is supported by an aerobic respiratory chain with components structurally and functionally similar to those found in mitochondria (Zannoni and Baccarini-Melandri 1980). Upon the removal of oxygen, the cell develops an extensive intracytoplasmic membrane (ICM) that comprises the photosynthetic apparatus. The amount of ICM per cell and the composition of the ICM are functions of the light intensity (Drews 1978 and 1986; Kaplan 1978). Increase in the light intensity depresses the formation of the photosynthetic apparatus. The quick response of facultative phototrophic bacteria to environmental changes provides an attractive system for studying the regulation of the expression of photosynthetic genes by oxygen and light.

Photosynthetic Apparatus

In most photosynthetic bacteria the photosynthetic apparatus is localized in the ICM in structures known as chromatophores (Schachman et al. 1952). These double-membraned vesicles or tubules are clustered to form grainlike stacks and bundles. For *R. capsulatus* and *R. sphaeroides* the ICM is the result of invaginations of the cytoplasmic membrane in response to a lowering of the oxygen tension below 10 mmHg. The ICM contains three major pigment protein complexes that function in the capture of light energy and electrochemical gradient formation: LH-II (B800 - 850), LH-I (B870), and RC (see Figure 1).

The LH-II antenna complex consists of three polypeptides — α (12 kd), β (10 kd), and γ (14 kd) — three molecules of Bchl, and one molecule of Crt. Two molecules of Bchl, associated with the α subunit, and one molecule of Bchl, associated with the β subunit, are believed to be responsible for the absorption maxima at 850 nm and 800 nm, respectively (Cogdell and Crofts 1978; Sauer and Austin I978). The Crt is probably associated with the β subunit (Webster et al. 1980). Cross-linking experiments support the idea that the LH-II complex in the membrane has an oligomeric (possibly a tetrameric) structure (Drews et al. 1986). An α and a β polypeptide dimerize with α-helical structures crossing through the

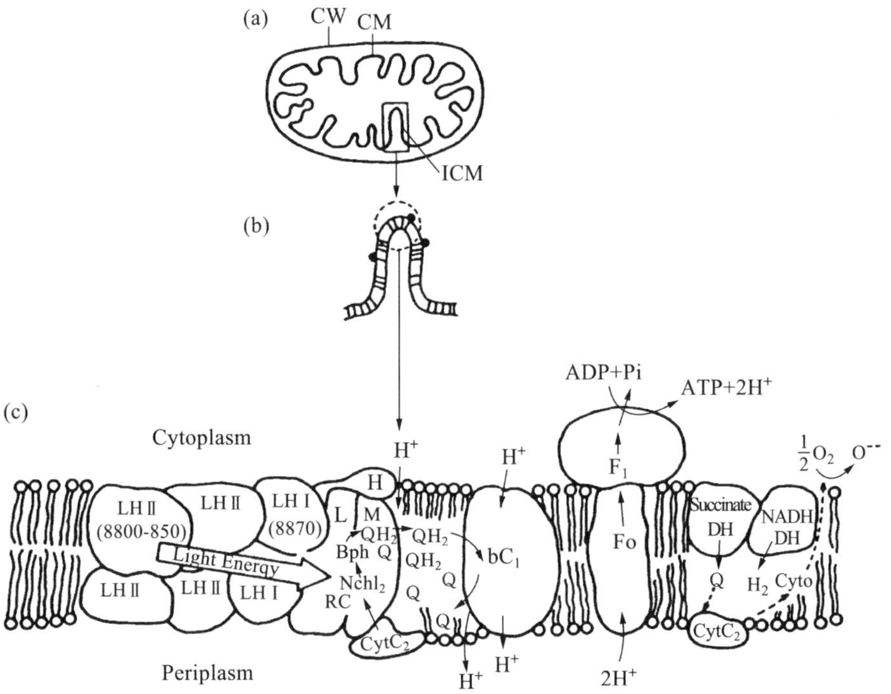

Figure 1 **Model of the topology of the photosynthetic membrane of purple bacteria. (a) A cross-section of a representative cell grown photosynthetically. Cell wall (CW), and intracytoplasmic membrane (ICM) are indicated. (b) A magnified ICM portion. (c) A model of the spatial arrangement of functional units and the photosynthetic electron transport chain in ICM. Light energy absorbed by the antenna LH-II-Bchl is transferred through the LH-I-Bchl complex to the RC, which is composed of H, L, and M subunits. As a result, a Bchl dimer in the RC is excited and gives up an electron to bacteriopheophytin (Bph). Bph in tum reduces a bound quinone molecule (Q) entering a Q cycle within the adjacent membrane-bound cytochrome bc_1 oxidoreductase that pumps protons across the membrane. The electrons finally return to the RC through cytochrome C_2 located on the periplasm of the ICM. The proton gradient across the membrane drives ATP synthesis via an ATPase (F_1-F_0) in the membrane. The respiratory electron transport chain from succinate dehydrogenase (DH) and NADH DH through terminal oxidase Cyto to oxygen is also indicated, although it is not active in the photosynthetically grown cells.**

membrane. This dimer is considered to be the basic structural unit of the LH complexes. The oligomerization of the α-β dimers is proposed to optimize the efficiency of exciton migration by correctly orienting the chromophores. The LH-II γ subunit is less hydrophobic and exposed on the cytoplasmic surface of the membrane. It does not bind Bchl (Drews et al. 1986). Although the function of the LH-II γ subunit is unknown, it may be necessary for the assembly or stability of the complexes in *R. capsulatus*. However, the LH-II γ subunit is absent in *R. sphaeroides* and other photosynthetic bacteria. The LH-II complex from *R. capsulatus* has been crystallized, but its fine structure has not yet been resolved (Welte et al. 1985).

The LH-I antenna complex (B870) from *R. capsulatus* consists of two polypeptides — α (12 kd) and β (7 kd) — complexed with two molecules of Bchl and one molecule of Crt (Drews 1985). They show very low homology to one another (7%—13%), but are very homologous (76%—78%) to the corresponding polypeptides in *R. sphaeroides*. As with the LH-11 complex, a pair of α and β subunits span the membrane via a central, hydrophobic, α-helical domain. Conserved histidine residues in the hydrophobic regions are thought to bind to a pair of Bchl molecules. The polar C-terminal and N-terminal domains of each subunit are distributed toward the periplasm and cytosol, respectively. This basic structural unit is stabilized by forming oligomers, providing the specific binding and arrangement of the Bchl in the polypeptide-lipid environment for efficient energy transfer (Drews et al. 1986; Zuber 1986).

The RC is an integral membrane protein complex where charge separation occurs. It consists of three polypeptides designated RC-H, RC-M, and RC-L; four molecules of Bchl; two bacteriopheophytins (Bphs); two ubioquinones; one non-heme Fe^{2+}; and one molecule of the carotenoid spheroidene (Feher and Okamura 1978; Cogdell et al. 1976). The exact molecular weights of RC-H, RC-M, and RC-L from *R. capsulatus* based on DNA and deduced amino acid sequence data are 28,534, 34,440, and 31,565 daltons, respectively. The RC-L and RC-M subunits have some similar sequences. Each polypeptide chain consists of five hydrophobic regions that are believed to span the membranes as α-helices. RC-L and RC-M bind Bchl and are involved in the primary photochemistry. The RC-H subunit has only one hydrophobic section near the amino terminal (Youvan et al. 1984a and 1984b) and has its carboxyl terminal domain exposed on the cytoplasmic surface of the membrane. At least in *R. sphaeroides* and *Rhodospillum rubrum* the H subunit can be removed without loss of photosynthetic electron transport (Feher and Okamura 1978; Wiemken and Bachofen 1984). The function of the H subunit is unknown. It has been suggested that it may play a role in initiating the assembly of the L-M complex (Chory et al. 1984) or it may coordinate the interaction between RC and LH-I (Takemoto et al. 1982; Peters et al. 1983). It has been also reported that loss of the H subunit changes proper electron transport rates through Q_A and Q_B (Debus et al. 1985).

Recently, a model of the arrangement of the protein backbone and the prosthetic groups of the RC of *Rps. viridis* has been reported (Deisenhofer et al. 1984, 1985a, and 1985b; Michel and Deisenhofer 1986). The L and M polypeptides, 274 and 320 amino acid residues, respectively, span the membrane with five helical segments similar to those hypothesized in *R. capsulatus* and *R. sphaeroides*. The N-terminals of both the L and M subunits are on the cytoplasmic side of the membrane, while the C-terminals of both proteins are on the periplasmic side of the membrane. The prosthetic groups are embedded in a central cylinder formed by the L and M proteins. The folding of L and M are similar, and they are related to one another by an axis of dyad symmetry oriented normal to the membrane and passing through the Bchl special pair and the Fe^{2+}. The transmembrane helices are connected by segments that are partly helical, β-sheet, or irregular in conformation. These segments presumably lie on the membrane surface and mediate the contact to the H subunit. The H subunit (258 amino acid residues) is the cytoplasmic cover of the L-M complex. It is folded in a globular C-terminal part and a loosely folded N-terminal part containing a prominent transmembrane helix. Most of the H subunit is in contact with M and L polypeptides. One difference between the RCs of *R. viridis* and *R. capsulatus* is the presence of a tightly bound c-type cytochrome on the periplasmic surface of the membrane of *R. viridis*. The N-terminal transmembrane helix of the H subunit interacts with a segment of the cytochrome and acts as a transmembrane clamp.

In addition to the RC and LH complexes, the photosynthetic membrane contains a ubiquinol-cytochrome C_2 oxidoreductase complex. The functional oxidoreductase that is part of the cyclic photosynthetic and respiratory electron chain of *R. capsulatus* and *R. sphaeroides* comprises three main subunits—cytochrome *b*, cytochrome C_1, and a high-potential FeS proteincarrying a cluster of four redox centers (Gabellini et al. 1982). Furthermore, ATPase, another membrane component, couples proton flow across the membrane to the synthesis of ATP. This high-molecular-weight, multisubunit complex consists of two primary domains: F_0 and F_1. The F_1 portion, consisting of five subunits, is hydrophilic and sits on the cytoplasmic surface of the membrane. The F_0 part consists of three polypeptides — denoted a, b, and c — and is buried in the membrane. The c subunit from *R. rubrum* has been isolated and its sequence has been determined (Bachofen and Wiemken 1986).

The proper assembly and packaging of photochemical pigments (Bchl, Crt), cofactors (quinone), and polypeptides in the LH and RC complexes are a prerequisite for efficient energy transfer from molecule to molecule. The detailed description of assembly and topology of the photosynthetic membranes of purple bacteria has recently been published (Drews et al. 1983 and 1986; Bachofen and Wiemken 1986; Donohue and Kaplan 1986; Sprague and Varga 1986; Zuber et al. 1986; Loach et al 1986). Studies on the topography of RC polypeptides using proteolysis, site-specific labeling, and x-ray and neutron diffraction have shown that the RC is asymmetrically arranged (Drews 1985). The RC-H subunit is exposed on the cytoplasmic surface of the membrane; the RC-M and RC-L subunits are buried in the membrane (Deisenhofer et al. 1984; Youvan et al. 1984b). All three subunits of the RC interact. The LH-I complexes are localized around the RC and are structurally and functionally interconnected to the RC. The LH-II complexes are arranged peripherally, interconnecting many RC-LH-I units (Monger and Parson 1977).

The pathway of the photochemical reaction is depicted diagrammatically in Figure 1: Light energy absorbed by the antenna LH-II-Bchl is transformed into mobile electronic singlet states, which migrate as excitons by a random walk through the LH-I complex (Drews et al. 1986). These two LH complexes function to increase the cross-sectional area for absorption of light. The excitation energy is finally captured with high efficiency by RC-

Bchl molecules, where the energy is transduced into a charge-separated state. A photon of light normally causes the oxidation of a Bchl dimer (the special pair) in the RC with the resultant transfer of an electron to a tightly bound quinone molecule (Q_A). Q_A is then oxidized by a secondary quinone (Q_B) to form a stable semiquinone (Q_B^-). A second photoreduction results in a second electron transfer to Q_B, resulting in production of the fully reduced quinol, which is released into the membrane. The reduced quinol is subsequently reoxidized by an adjacent membrane-bound cytochhrome c_1 oxidoreductase that pumps protons across the membrane from the cytoplasmic side to the periplasmic side. The electrons finally retum to the RC through cytochrome c_2, which is located on the periplasm of the ICM. The proton gradient across the membrane drives ATP synthesis via an ATPase in the membrane.

Organization of Genes Coding for LH, RC, and Pigment Biosynthetic Enzymes and Cytochromes

The genetic control of photosynthesis and photosynthetic membrane formation has been an important issue in the last few years. Recombinant DNA technology has made possible the isolation of most of the genes associated with the photosynthetic apparatus of the photosynthetic bacteria. The isolated genes have been subjected to a detailed analysis of nucleotide sequence and structure. Therefore, a great deal of information about the structure and function of the photosynthetic apparatus both in *R. capsulatus* and in *R. sphaeroides* has been obtained.

The study of photosynthetic genes in *R. capsulatus* is aided by an indigenous gene-transfer agent (Yen and Marrs 1976) and isolation of a conjugative R-prime plasmid, pRPS404, containing *R. capsulatus* 46 kb DNA, which complements most mutations affecting the differentiation of the photosynthetic apparatus (Marrs 1981). A procedure for the transposon mutagenesis of pRPS404 in *Escherichia coli* that allows for mapping of the positions of the transposon insertions was developed in our laboratory (Youvan et al. 1982). In order to improve the stability of the mutants generated in this way, a new transposon was created that had the transposition characteristics of Tn5 and the antibiotic resistant characteristics of Tn7 (Zsebo et al. 1984). By screening both the earlier mutants generated by Marrs's laboratory and the transposon mutants generated in this laboratory for enhanced near-infrared fluorescence indicating a defect in the utilization of absorbed photons, two restriction fragments from the R-prime plasmid pRPS404 were found to complement all enhanced-fluorescence mutants defective in LH-I and RC genes. The complete nucleotide sequence (8,867 bp) and a deduced polypeptide sequence for these two restriction fragments showed that they encoded 11 proteins from the photosynthetic gene cluster of *R. capsulatus* (Youvan et al. 1984b). Four structural genes for LH-I β (*puf*B), LH-I α (*puf*A), RC-L (*puf*L), RC-M (*puf*M), and two putative genes, C2397 and C2814, were found in the

Bam*HI*-C-*Eco*RI-B restriction fragment, whereas four putative genes-F108, F460, F1025, and F1696-and one structural gene for RC-H (*puh*A) were found in the *Bam*HI-F fragment (see Figure 2). The *puf*B, *puf*A, *puf*L, *puf*M, and C2397 genes are arranged contiguously and comprise a single operon: Anoxygenically induced mRNA hybridizes to this region (Clark et al. 1984; Belasco et al. 1985) and a specific oligonucleotide (17 mer) complementary to the C2397 gene hybridizes to a 2.6 kb long transcript from the *puf* operon (Zhu and Hearst 1986). From the deduced amino acid sequences of the RC subunits it was concluded that both the L and the M subunits are very hydrophobic proteins with 282 and 307 amino acids, respectively. As mentioned earlier, hydropathy plots suggest that the L and M subunits are transmembrane proteins that may cross the membrane five times. On the other hand, the H subunit is a hydrophilic polypeptide (254 amino acids long) with a very hydrophobic amino terminal of 30 amino acids. All five structural and six putative genes possess a ShineDalgarno sequence (A/C) GGAG (A/G) $N_{3\text{-}10}$ ATG that is complementary to the 3′ terminal of *R. capsulatus* 16S rRNA. However, no *E. coli*-like consensus promoters were found over the entire 9 kb of sequenced DNA. Three potential hairpins have been found between the genes for LH-1 α and RC-L, after the gene for C2397, and between the genes for RC-H and F3981.

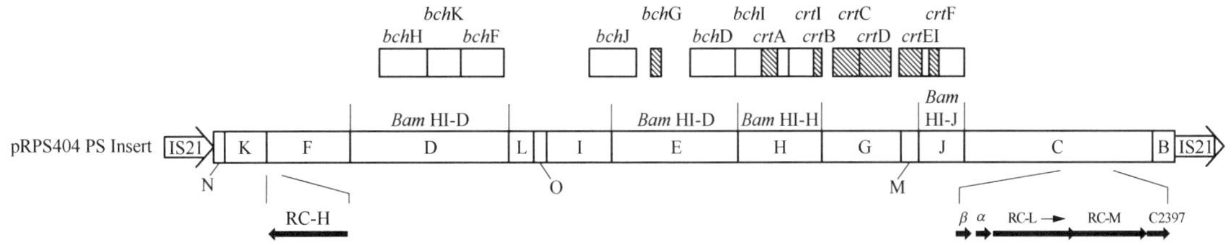

Figure 2 photosynthetic gene cluster map. *Bam*HI restriction sites are shown within a 46 kb photosynthetic gene cluster from *R. capsulatus* carried by the R' plasmid pRPS404. The genes coding for LH-I (β and α) and RC-L and RC-M subunit resides in the *Bam*HI-F fragment. The arrows indicate the direction of transcription. The *Bam*HI fragments between these two genetic loci contain a number of *bch* genes (in *Bam*HI-D and -E) and *crt* genes (in *Bam*HI-H, -G, -M, and -J). The shaded areas indicate the genes determined by cluster point mutations conferring in same phenotype. The blank areas represent genes determined by transposon mutagenesis and complementation. From Y. S. Zhu, D. N. Cook. F. Leach, et al. (1986) *J. Bacteriol.* 168, 1180 - 1188. With permission

The *puf* and the *puh* operons are separated by 35 kb and are transcribed in the opposite direction. A large number of genes (at least 17) associated with Bchl and Crt biosynthetic enzymes are located between the *puf* and the *puh* operons (see Figure 2). Studies of the structure and the regulation of genes for pigment biosynthesis are in progress in our laboratory.

The genes coding for another LH complex and the β (*puc*B) and α(*puc*A) subunits of LH-II were isolated using a deoxyoligonucleotide probe and sequenced (Youvan and Ismail 1985). This completed the sequencing of all seven structural genes coding for the LH and RC polypeptides that bind the pigments and cofactors participating in the primary light reaction of photosynthesis. The *puc*A and the *puc*B genes are at least 11 kb away from *puf*L and *puf*M, or at least 7 kb away from *puh*A. Since the organization of the β and α structural genes for both LH-I and LH-II is identical, and very high homologies exist between the two β-polypeptides and between the two α-polypeptides, it has been suggested that both complexes arose by gene duplication from a single ancestral LH gene. A highly conserved sequence, Ala-x-x-x-His, is believed to bind Bchl (Youvan and Ismail 1985).

Hearst and Sauer (1984a) first reported at the Vlth International Congress on Photosynthesis in 1983 that the RC subunits L and M of *R. capsulatus* share amino acid sequence homology with the Q_B protein of spinach. Particularly, a strong homology has been found between the L subunit and the Q_B protein (residues 196 – 221). It was proposed that this sequence of amino acids is involved in quinone binding and function (You van et al. 1984a and 1984b; Hearst and Sauer 1984a and 1984b).

The *puf*M (Williams et al. 1983), *puf*L (Williams et al. 1984), *puh*A (Donohue et al. 1986a), *puf*B, *puf*A (Kiley et al. 1987a), and *puc*B and *puc*A (Kiley et al. 1987b) in *R. sphaeroides* have also been isolated and sequenced. The same arrangement of genes coding for LH-I (β, α) and RC (L, M) proteins and a striking homology of deduced amino acid sequences of the LH and RC polypeptides were found between *R. sphaeroides* and *R. capsulatus*. The considerable sequence homology between the L and M subunits indicates that their genes arose by duplication of a single gene. The RC-L and RC-M polypeptides of *R. sphaeroides* also share homology with the 32 kd thylakoid membrane protein in spinach (Williams et al. 1983 and 1984) and are proposed to span the membrane five times, like the RC-L and RC-M polypeptides in *R. capsulatus*. One of the histidine residues (His-219) in a hydrophobic segment of the RC-M subunit is a candidate for a Bchl or a quinone binding site (Williams et al. 1983). The genes *puf*B and *puf*A are contiguous with *puf*L and *puf*M. A region about 170 bp upstream from the 5′ end of the structural gene is sufficient for the in vitro expression of LH-I β and α polypeptides (Kiley et al. 1987a). In addition, three regions of dyad symmetry upstream from *puf*B have been found. These inverted repeats may serve as recognition sites for DNA binding proteins in regulating expression of this operon.

Unlike most of the other photosynthetic bacteria, *R. rubrum* contains a single holochrome antenna complex, designated B880, that also consists of β and α subunits. Genes for B880 holochrome have recently been isolated and sequenced (Bérard et al. 1986). The genes for the β and α polypeptides are contiguous and show high homologies to *puf*B and *puf*A in *R. capsulatus*. They are presumed to be transcribed as a single operon. An additional C-terminal tail of 10 and 13 amino acid residues on the α and β subunits, respectively, may serve as a means of orienting these polypeptides in the membrane.

The *puf*L and *puf*M (Michel et al. 1986) and the *puh*A (Michel et al. 1985) genes in *R. viridis* have been cloned using oligonucleotide hybridization and expression vectors, respectively, and sequenced. The arrangement of *puf*L and *puf*M is the same as in *R.*

capsulatus and *R. sphaeroides*. However, due to a deletion of five bases at the end of *puf*L in *R. viridis*, *puf*L and *puf*M do not overlap. Therefore, overlap of these two genes is not necessary for 1 : 1 stoichiometry of the L and M polypeptides (Williams et al. 1984). Another structural feature different from *R. capsulatus* and *R. sphaeroides* is that the gene encoding the RC cytochrome c_2, which is tightly bound to the RC and unique in *R. viridis*, is located at the end of *puf*M and overlaps *puf*M by one base pair. In *R. capsulatus puf*M is followed by an unknown gene C2397 (Youvan et al. 1984b). Furthermore, the M subunit from *R. viridis* has 17 more amino acids than *R. capsulatus* and 18 more amino acids than *R. sphaeroides* at its carboxy-terminal. This extension binds the cytochrome subunit (Deisenhofer et al. 1985a and 1985b).

Cytochrome c_2 is an electron carrier that is common to the photosynthetic and respiratory machinery of *R. capsulatus* (Bartsch 1978). The structural gene for cytochrome c_2(*cyc*A) of *R. capsulatus* has been cloned and sequenced (Daldal et al. 1986). Comparison of the known amino acid sequence of the purified cytochrome c_2 and the nucleotide sequence corresponding to the N-terminal part of the *cyc*A product indicates the presence of a putative 21-amino-acid signal sequence. Cytochrome c_2 may be synthesized as a precursor that is processed during its secretion to the periplasm. Studies of the insertion and insertion-deletion mutations have led Daldal et al. to conclude that cytochrome c_2 is not essential for photosynthetic growth. Donohue et al. (1986b) have analyzed *cyc*A in *R. sphaeroides* and have also found a 21-amino-acid signal sequence. Small (740 nucleotide) and large (920 nucleotide) *cyc*A-specific mRNA species are found in various growth conditions. In contrast to the result of *R. capsulatus* (Daldal et al. 1986), using cartridge insertion coupled with partial deletions of the *cyc*A gene, they have found that cytochrome c_2 obligatory for photosynthetic growth of *R. sphaeroides*.

Recently the genes encoding the three main subunits of the cytochrome bc_1 complex in a strain of *R. sphaeroides* have been cloned (Gabellini et al. 1985). This strain was identified later by Davidson and Daldal (1987) as most likely a strain of *R. capsulatus*. The genes are arranged in the order (5′) *fbc*F, *fbc*B, and *fbc*C, encoding the FeS protein, cytochrome *b*, and cytochrome c_1, respectively. The *fbc* genes are coordinately transcribed in one polycistronic mRNA and constitutively expressed under both photosynthetic and respiratory growth conditions. The nucleotide sequence of 3, 874 bp of cloned chromosomal DNA, including the three structural genes *fbc*F, *fbc*B, and *fbc*C, has been determined (Gabellini and Sebald 1986). The *fbc*F gene codes for 191 amino acids of the FeS protein, starting from a GTG codon that is preceded by an adenine and guanine-rich sequence complementary to the 3′ end of the 16S ribosomal subunit of *R. sphaeroides*. The *fbc*F gene is followed by the *fbc*B gene with a 12 nucleotide space that includes a Shine-Dalgarno-like sequence. The *fbc*B gene codes for 437 residues of the cytochrome *b* subunit. The *fbc*C gene starts 20 bp after the *fbc*B gene and is preceded by a Shine-Dalgarno-like sequence. The *fbc*C gene codes for a 280-residue preapocytochrome c_1 that has a transient leader sequence (21 amino acids).

S-l nuclease protection analysis indicates that the transcription of the *fbc* operon starts approximately 240 bp upstream from the start codon of the *fbc*F gene and terminates 120 bp downstream from the stop codon of the *fbc*C gene. The deduced amino acid sequences, homologies with other similar proteins, and hydropathy plots indicate the following structural features: ① FeS protein, carrying a high-potential 2Fe-2S cluster, has a large hydrophilic domain and is exposed to the aqueous phase; ② cytochrome *b* has nine or ten possible membrane-spanning regions, and four conserved histidine residues are postulated as ligands of the two heme groups; ③ cytochrome c_1 is synthesized as a precursor form including a transient leader sequence of 21 amino acids. The typical sequence that binds the heme covalently is located near the N-terminal corresponding to Cys-55, Cys-88, and His-50.

Similar results of a detailed analysis of the *fbc* operon in *R. capsulatus* were recently reported by Davidson et al. (1987). They designated the *fbc* operon as the photosynthetic electron transport (*pet*) operon. A possible promoter region was found to be located upstream from the first *pet*A structural gene coding for Rieske FeS protein. It is concluded that unlike cytochrome c_2 (Daldal et al. 1986), a functional bc_1 complex is essential for photosynthetic growth.

rRNA from both *R. sphaeroides* (Marrs and Kaplan 1970) and *R. capsulatus* (Yu et al. 1982) consists of 23S, 16S, and 5S species. The 23S rRNA, however, is not stable during isolation and is cleaved into 16S and 14S rRNAs probably by a specific endonuclease (Yu et al. 1982; Klug et al. 1984; Zhu and Kaplan 1985). The rRNA genes of *R. capsulatus* have been cloned (Yu et al. 1982). At least seven rRNA operons are present in the *R. capsulatus* chromosome. The arrangement of 16S, 23S, and 5S rRNAs is the same as that in E. *coli*. However, the spacer (0.91 kb) between the 16S and the 23S rRNA genes is longer than that (0.44 kb) in *E. coli*.

Although ribulose-1, 5-bisphosphate carboxylase (RuBPCase) is not a membrane protein, a great deal of attention has been paid to it because it is the most abundant protein in the cytoplasm and it plays a key role in carbon dioxide fixation in the Calvin cycle in photosynthetic bacteria, algae, and plants. It is of particular interest that two forms of RuBPCase are present in *R. sphaeroides* and *R. capsulatus* (Gibson and Tabita 1977a and 1977b). The form I RuBPCase resembles that found in plants and most other bacteria. It is a large oligomeric protein composed of eight large (52 kDa) and eight small (11 kDa) subunits (Gibson and Tabita 1977b). The form II enzyme is a hexamer of only large subunits (52 kDa) and does not have a small subunit. Recently, the gene for the form II RuBPCase (*rbc*L) was isolated from *R. rubrum* (Somerville and Somerville 1984) and *R. sphaeroides* (Fornani and Kaplan 1983; Quivey and Tabita 1984; Muller et al. 1985). The nucleotide sequence of the *rbc*L gene from *R. rubrum* has been reported (Nargang et al. 1984). No consensus prokaryotic promoter sequence was found in the *rbc*L gene from *R. rubrum*. It is likely that the promoters for highly regulated genes such as those involved in carbon dioxide fixation and the photosynthetic apparatus are quite different from typical bacterial promoters (Quivey and Tabita 1984). A region of dyad symmetry 5′ to the coding region of the *R. rubrum* RuBPCase gene similar to the proposed attenuator regions was found, indicating that additional proteins may be involved in transcription (Nargang et al. 1984).

One of the consequences of the knowledge of structure and organization of nucleotide sequences of the photosynthetic genes in purple photosynthetic bacteria is the ability to mutate the genes that are of interest by site directed mutagenesis and to examine the effect of the resulting mutation on the function of the photosynthetic apparatus (Youvan and Ismail 1985). Another consequence that will be described here in more detail is to facilitate the elucidation of structure and function of the photosynthetic apparatus in plants and algae as well as in purple bacteria. Proteins can be studied based on sequence homologies that may shed some light on functionally important structural features (Youvan et al. 1984a and 1984b; Hearst and Sauer 1984a and 1984b; Hearst 1986). Amino acid sequence comparison between photosynthetic membrane proteins of purple bacteria and those of blue-green algae and higher oxygen evolvers remains an extraordinarily valuable and insightful exercise. A simple generalization is that there is strong sequence homology between proteins of identical function in all bluegreen algae and higher plants. The purple bacteria are very different but have small regions of homology with plant proteins. Such homologous regions provide special insight regarding the essential functional sites in the corresponding proteins.

A weak, but significant, sequence homology exists between the L and M subunits of the RC of *R. capsulatus* and the 32 kDa thylakoid membrane protein (also known as D1, Q_B, herbicide-binding protein) of the RC of photosystem II in higher plants and algae (Zurawski et al. 1982; Youvan et al. 1984b; Hearst and Sauer 1984b; Hearst 1986). Hearst and Sauer (1984b) proposed a model suggesting that a highly conserved sequence of amino acids between the L and M subunits of the RC of *R. capsulatus* and 32 kDa thylakoid membrane protein of spinach and tobacco is involved in quinone binding and function. Figure 3(A) reveals that a sequence PFHMLG - - - -F- - - - - -AMHG-LV-S is common to both the L subunit of *R. capsulatus* and the 32 kDa (D1) proteins of spinach, *Nicotiana*, *Chlamydomonas*, and *Euglena* and starts at the one hundred seventy-second amino acid from the N-terminal of the L subunit (282 amino acids) and at the one hundred ninety-sixth amino acid from the N-terminal of the 32 kDa (D1) protein (353 amino acids). Later another photosystem II RC protein, the D2 protein, was also found to possess sequence homologies to the D1 protein and the L and M subunits (Rochaix et al. 1984; Alt et al. 1984; Holschuhk et al. 1984). A weaker homology between the *R. capsulatus* M subunit (from the one hundred ninety-eighth amino acid of a 306-amino-acid molecule) and the pea and *Chlamydomonas* D2 protein also reveals a conserved sequence PFHM- - - -G-L-A-LLC- - -GATV (see Figure 3(B)). The data strongly indicate that the L and M subunits and the 32 kDa protein arose from a common precursor. The fact that selection pressure preserved such a precise amino acid sequence over three billion years (Olson 1981) suggests that this region

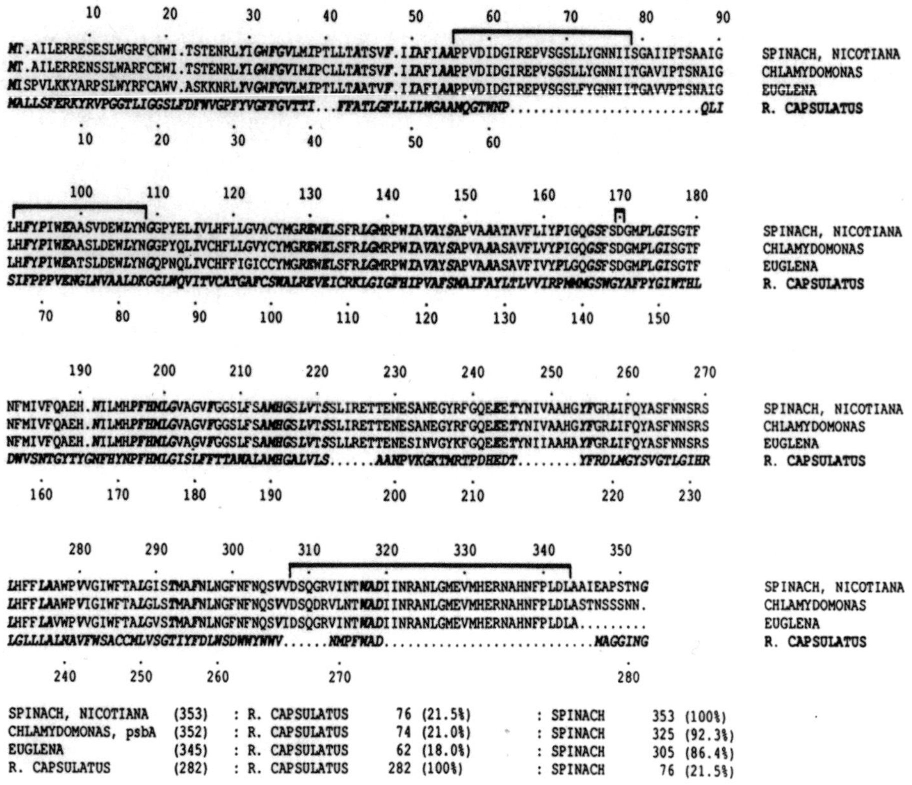

(a)

(b)

Figure 3 (a) An amino acid sequence alignment of the RC-L subunit of *R. capsulatus* and the D1 proteins of spinach, *Nicoliana*, *Chlamydomonas*, and *Euglena* that are encoded by the psbA gene of the chloroplast genome. The italic bold letters that appear in all of the four lines represent the homologous sequences among these organisms. The brackets indicate the potential regions involved in oxygen evolution. There are 76 matches and 21.5% homology between RC-L and the spinach, *Nicotiana* Dl; 74 matches and 21% homology between RC-L and the *Chlamydomonas* Dl; and 62 matches and 18% homology between RC-L and the *Euglena* D1. (b) An amino acid sequence alignment of the RC-M subunit of *R. capsulatus* and the pea and *Chlamydomonas* D2 proteins that are encoded by the *psb*D gene of their respective chloroplast genome. The italic bold letters that appear in all of the three lines illustrate the homologous sequences. There are 64 matches and 20.8% homology between RC-M and the pea D2, and 60 matches and 19.5% homology between RC-M and the *Chlamydomonas* D2.

is functionally very important. Subsequently, the crystal structure of the RC from *R. viridis* implicated the conserved histidine residues in forming the site for the non-heme Fe^{2+} (Deisenhofer et al. 1985a and 1985b). The discovery of sequence homologies between the L and D1 proteins and between the M and D2 proteins suggests that the D1 and D2 proteins are involved in photochemical activity, a hypothesis in conflict with the old concept that two other larger polypeptides of 47—51 kDa and 43—45 kDa are the core of the photosystem II.

This hypothesis is further supported by the finding that both the L subunits and D1 proteins bind the photo affinity-labeled herbicide azido atrazine (Pfister et al. 1981; DeVitry and Diner 1984). Through the use of a tritium-labeled quinone alkylating reagent, Worland et al. (1987) have demonstrated the affinity labeling of a 38 kDa protein that they suggest may be D2 in spinach photosystem II complexes. In addition, the x-ray structure analysis has shown that both L and M subunits are needed to establish the primary electron donor, which is made up of two nearly parallel Bchl moleucles (the special pair) and the electron-accepting quinone-iron com-plex. All the evidence supports the proposal of Hearst (1986) and Michel and Deisenhofer (1986) that the D1 and D2 proteins form the core of the photosystem II RC and the L and M subunits form the RC core of purple photosynthetic bacteria in similar ways. The L subunit is assumed to be the equivalent of D1, and M to be the equivalent of D2.

Figure 3(a) provides a sequence alignment that clearly accentuates the additional sequences in the higher plants that are not found in the corresponding protein in the purple bacteria. The region of particular interest to us at the present time is the carboxy terminal of the D1 or herbicide-binding protein. This region is highly conserved among the oxygen evolvers and might be associated with the oxygen-evolving manganese center. Evidence for such a hypothesis comes from several sources. First, Metz et al. (1987) have identified a D1 mutant that is incapable of oxygen evolution. Second, D1 in higher plants is rapidly turned over because of photolability, while the L subunit in purple bacteria is quite stable (Kyle et al. 1984; Ohad et al. 1984; Bishop 1987). This photolability might well be the consequence of the presence of the oxygen-evolving site. Finally, the amino acid composition of this carboxy terminal region is unusual, containing three acidic residues, four asparagines, two arginines, two methionines, one phenylalanine, and one proline in the twenty-one residues from position 322 to position 342 on the alignment in Figure 3(a), as indicated by the bracket. This suggests a function for this region that must operate on the periplasmic side of the photosynthetic membrane, where oxygen evolution is known to occur, if the analogy with the purple bacteria RC structure of Michel and Deisenhofer (1986) is valid.

Other regions of this protein that are believed to be in the periplasm and that may also be involved with the site of oxygen evolution include amino acids 56 to 76 and 92 to 108, the aspartic acid at position 170, and the amino acids from 308 to 321, as marked with brackets in Figure 3(a).

Regulation of Expression of the Genes Coding for LH, RC, Bchl, and Crt Biosynthesis

The biosynthesis of the photosynthetic apparatus requires a coordinated expression of the genes for LH and RC pigment complexes in response to environmental stimuli, that is, oxygen tension and light intensity. In addition, these components must be correctly assembled into the membrane. In this section we focus on the aspects of internal and extemal regulation of the photosynthetic genes.

Coordination and Interaction of the Genes for LH, RC, Bchl, and Crt Biosynthesis

The expression of the photosynthetic genes for LH and RC proteins and the accumulation of pigments are tightly coupled and interdependent. Mutation within these genes causes a pleiotropic effect. It has been shown that a mutation in any of the genes for LH-I, RC-H, RC-L, Bchl, or Crt biosynthetic enzymes results in different degrees of reduction in the amount of mRNAs from the rest of the genes (O'Brien et al., unpublished observations; Zhu and Hearst 1987). O' Brien et al. have used a series of transposon mutants created in this laboratory (Zsebo and Hearst 1984) that carry their defect within or near the structural genes for the RC to study the expression of these genes during photosynthetic development. Insertions within ORF F1696, which is located upstream from the gene for RC-H, caused severe defects in the photosynthetic apparatus. These defects included reduced levels of mRNA for LH-I, RC-L, RC-M, and RC-H, along with decreased levels of Bchl mRNA and diminished absorption by LH-II. Since we have been unable to detect transcription from the putative F1696 gene (Zhu and Hearst 1986), it seems logical to assume that these insertions blocked the regulatory sequences of the gene for RC-H. Inactivation of the gene for RC-H by an internal mutation resulted in a total loss of LH-II absorption and a dramatic decrease in the amount of the RC-L and RC-M proteins as determined by Western blots. Mutations within the gene for the RC-H protein seemed to have a more dramatic effect than those within the gene for the RC-L protein. Such a difference may have resulted from more stringent requirements for the RC-H protein than for the RC-L and RC-M proteins in photosynthetic development (O'Brien et al., unpublished observations). On the other hand, mutations in bch (bchC) or *crt* (*crt*I) genes substantially reduced levels of mRNA for LH-II, LH-I, RC-L, RC-M, and RC-H. A close link was found to exist between the genes for LH-II proteins and Crt biosynthesis (Zhu and Hearst 1987). We have also performed experiments to determine whether the decreased levels of mRNA for LH and RC proteins in these mutants are due to reduced transcription or to increased degradation of mRNA. A comparison of the decay rate of mRNAs for LH-I, LH-II, RC-L, RC-M, and RC-H after treatment with proflavin as an inhibitor of transcription reveals that the mRNAs in the mutants are at least as stable as those in wild-type, although they are present at very low levels. Therefore, the reduction in mRNA levels in these mutants is due to transcriptional control. We suggest that proper and accurate assembly of the photosynthetic components in the ICM is necessary for transcription. Absence of one of these components (either proteins or pigments) may turn off the transcription of the corresponding genes.

Environmental Stimuli

Oxygen

In purple nonsulfur bacteria a decrease in oxygen tension is the signal for a switch of metabolism from aerobic mode to photosynthetic mode. Upon the removal of oxygen, the cells develop an extensive ICM, mainly consisting of LH-I, LH-II, and RC proteins. This process is accompanied by an immediate biosynthesis of Bchl (Drews and Oelze 1981; Chory et al. 1984). The single chain, however, between variation of oxygen tension in the medium and regulation of gene expression is unknown.

Recent studies (Clark et al. 1984) using Southern hybridization have shown that the transcripts from the genes for LH-I (β, α), RC-L and RC-M in *R. capsulatus* are greatly enhanced (fortyfold) in cells grown at low oxygen tension, whereas the transcripts from the genes coding for RC-H and Bchl biosynthetic enzymes respond to a lesser extent, and transcripts encoding several carotenoid biosynthetic enzymes seem unresponsive to changes in oxygen concentration. Lac fusion experiments support the observation that the rate of Bchl synthesis is regulated transcriptionally to a small degree (Biel and Marrs 1983) and that the *crt* genes are not directly regulated by oxygen tension (Biel and Marrs 1985). The steady state concentration of mRNA specific for LH-II and the level of rRNA also increase after lowering oxygen tension (Klug et al. 1984). Klug et al. (1985) have demonstrated that a maximum in the concentration of mRNA for LH-I and RC proteins occurs 30 minutes after induction under low oxygen concentration, while the maximal expression of mRNA for LH-II proteins appears to lag the LH-I/RC mRNA maximum by about 25 minutes. This differential expression of the genes for LH-I and LH-II agrees with the sequential accumulation of LH-I and LH-II complexes in the membrane of *R. capsulalus* under decreasing oxygen tension (Schumacher and Drews 1978). These observations support the transcriptional control of the genes for LH and RC proteins. None of these studies, however, showed discrete mRNA species until Northern hybridization and S-1 nuclease analysis, using highly defined probes, were applied to *R. capsulatus* (Belasco et al. 1985) and *R. sphaeroides* (Zhu and Kaplan 1985).

Belasco et al. (1985) have observed two anaerobically induced transcripts of 0.5 kb and 2.7 kb from the *puf* operon of *R. capsulatus*. the 2.7 kb transcript coding for LH-I (β, α), RC-L, and RC-M, is less stable than the 0.5 kb transcript coding for LH-I (3, a) only. Similar observations were made in *R. sphaeroides* (Zhu and Kaplan 1985). In addition, we have found that these two mRNAs (0.5 kb and 2.6 kb) respond differently to oxygen concentration and light intensity. The 2.6 kb mRNA is more sensitive to oxygen, whereas the 0.5 kb mRNA is more responsive to light intensity. The response of the 0.5 kb transcript to light intensity can be functionally rationalized, since it acts as a template for LH-I antenna proteins. None of these mRNAs are significantly affected by the redox potential of the organic acids used in the medium. In contrast to the membrane proteins, the cellular level of mRNA specific for the large subunit of RuBPCase (form II), a cytoplasmic protein and a key enzyme in carbon dioxide fixation, is markedly affected by both oxygen concentration and the redox potential of the substrate. We have found that the expression of the *rbc*L gene in *R. sphaeroides* is completely turned off by high oxygen tension under conditions where the mRNAs for LH and RC are still observable.

The regulation by oxygen of the mRNAs for LH-II (β, α), LH-I (β, α), and RC (-L, -M, -H) and for the Bchl and Crt biosynthetic enzymes in *R. capsulatus* has recently been investigated in our laboratory using Northern and dot hybridization (Zhu and Hearst 1986 and 1987a; Zhu et al. 1986b) (see Figure 4). We have demonstrated that while the small transcript (0.5 kb) codes only for LH-I (β, α), the large transcript (2.6 kb) codes for LH-I (β, α), RC-L, and RC-M, as well as an ORF C2397. These five genes thus comprise a single operon. The C2397 gene is contranscribed and coregulated with *puf*L and *puf*M in response to oxygen and light. It is 249 bp long and contains the information for a small hydrophobic protein. The function of the C2397 gene product is as yet unknown. However, it may be functionally associated with the LH-I and RC genes, based on their tight cotranscription and coregulation as one operon. We have found that the gene for RC-H, *puh*A, has two transcripts of 1.2 kb and 1.4 kb, that are initiated within the ORF F1696 and whose ratio is light intensity dependent. The *puh*A gene is coordinately regulated with

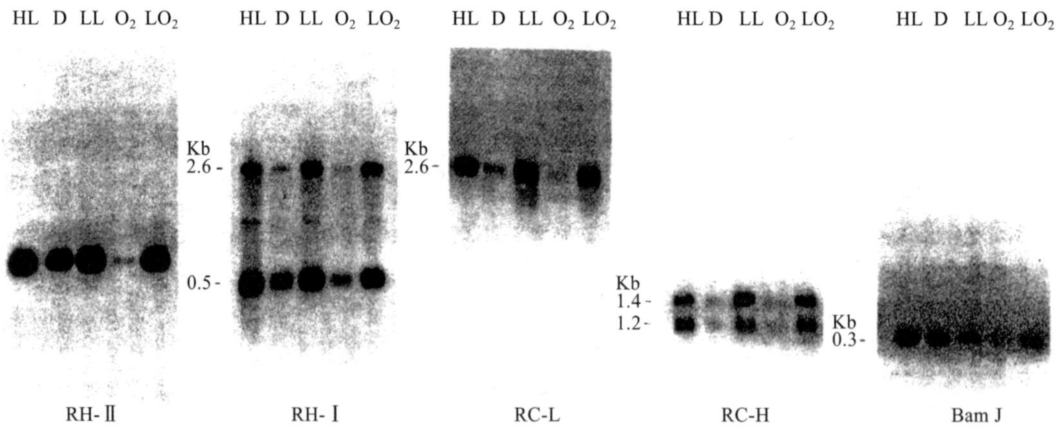

Figure 4 Northern hybridization of *R. capsulatus* RNA with cloned probes for LH-II, LH-I, RC-L, RC-H, and the *Bam*HI-J fragment of pRPS404 containing *crt*F, *crt*E, and an unidentified gene. Cells were grown under different conditions. HL, high light (30 W/m^2) D, a shift of HL-grown cells to the dark for one hour; LL, low light (5 W/m^2); O_2, aerobic conditions; LO_2, semiaerobic conditions. Adapted from Y. S. Zhu and J. E. Hearst (1987). With permission

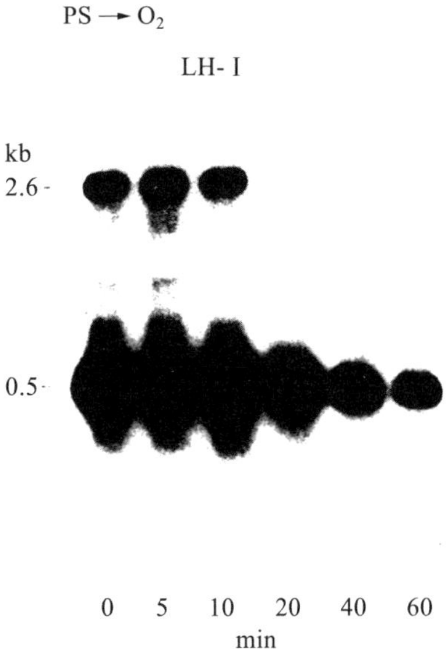

Figure 5 Northen hybridization of *R. capsulatus* RNA and DNA probes specific for the LH-I polypeptides. The mRNAs are degraded when the photosynthetic cells are exposed to oxygen at various times.

*puf*L and *puf*M in response to oxygen (Zhu and Hearst 1986), an observation different from that of Clark et al. (1984). It is important to note that the proteins RC-L, RC-M, and RC-H are present in a virtual stoichiometry of 1 : 1 : 1 in the membrane. Recently, Donohue et al. (1986) made the similar observation in *R. sphaeroides* that the *puh*A gene produced 1.13 kb and 1.4 kb transcripts. The mRNA specifying the LH-II polypeptides is 0.5 kb long and is more abundant and more sensitive to oxygen change, indicating that the genes for the LH-II and LH-I/RC are independently regulated (Zhu and Hearst 1986). Furthermore, we have found that oxygen accelerates the degradation of the mRNA for LH-II (Zhu et al. 1986a) (see Figure 5). In *R. sphaeroides* high oxygen tension (25%) also accelerates the degradation of the mRNAs for LH-I, RC-L, and RC-M (Zhu et al. 1986b). We hypothesize that oxygen represses the transcription of the photosynthetic genes and stimulates the degradation of their mRNAs. Such a combination of transcriptional and posttranscriptional regulations provides the cells with a mechanism to shift to a new metabolic mode quickly in response to oxygen. Recently, a similar result has been reported for the nif genes of *Klebsiella pneumoniae* (Collins et al. 1986).

One discrepancy between our observation and that of Clark et al. is that we have found only a threefold to fourfold or a sixfold increase in the mRNA levels for LH-I and RC, depending on light intensity, in contrast to the fortyfold increase they reported when comparing anaerobically and aerobically grown cells. A similar threefold to fourfold increase in mRNA levels has been observed in *R. sphaeroides* (Zhu and Kaplan 1985). One argument for such discrepancy is the method utilized to normalize the results to the total RNA. If the rRNA undergoes a dramatic change under different conditions, it would indeed affect the results. Klug et al. (1984) reported that the total rRNA per cell of *R. capsulatus* increased sevenfold in the transition period of 140 minutes from high to low oxygen tensions. However, it is difficult to understand how they could have gotten sevenfold difference in rRNA amounts when they applied equal amounts of total RNA of which the majority was rRNA for hybridization. As a matter of fact, an early study by Gray (1967) demonstrated that the total RNA, DNA, and protein per cell of *R. sphaeroides* remain constant during the transtiton from high to low oxygen tesion. Our investigation in *R. capsulatus* confirms Gray's result. The total RNA amounts for the photosynthetic cells and for the aerobiccells are $2.0\times10^{-2}\pm0.3\times10^{-2}$ and $2.1\times10^{-2}\pm0.4\times10^{-2}$ pg per cell, respectively, averaged over four experiments. These values are in reasonable agreement with the measurements by Gray (1967). Since the growth rate is only slightly different under our anaerobic and aerobic conditions. The discrepancy of the rsults among different groups may be due to the experimental differences in methods, including the growth conditions, the accuracy of measurements, and the specificity of the probes used for hybridization. In our experiments we have used specific probes for LH and RC proteins. as well as Northern hybridization to show the discrete mRNA species, which provides direct information about specific mRNA size, quality, etc. Clark et al. used Southem hybridization and a large probe for the entire region, including DNA sequences other than the *puf* operon. We expect that the difference in methodology has caused many discrepancies between the results. We will not attempt to discuss all of them in this review.

The genes for pigment biosynthesis appear to be regulated by oxygen in somewhat different ways. While the levels of *bch* mRNA from the *bch* J, G, D, H, K, and F genes are lower under steady-state aerobic growth conditions than under anaerobic photosynthetic growth conditions, the extent of difference is less than that for the LH and RC mRNAs under those conditions. However, the levels of mRNA from *bch* genes decrease even more under steady-state low oxygen (semiaerobic) conditions than under aerobic conditions. When the *R. capsulatus* cells are shifted from anaerobic to high-oxygen conditions, the mRNAs from *bch* genes decrease rapidly during the first 20 minutes (Zhu et al. 1986a). These results were obtained using several *Bam*HI fragments containing more than one *bch* gene for studies, as were earlier results (Biel and Marrs 1983; Clark et al. 1984). The same results, however, were observed recently when we used a series of probes specific for *bch*C and *bch*A for hybridization (Zhu and Hearst, unpublished observations).

In contrast to the genes for LH, RC, and Bchl biosynthesis, *crt* genes respond to oxygen in an opposite fashion. We have found that the amount of mRNA from *crt* genes, relative to the amounts of mRNA for LH and RC proteins, actually increases during the shift from anaerobic to aerobic conditions (Zhu et al. 1986a) (see Table 1). Although the *Bam*HI-H and the *Bam*HI-J fragments contain several *crt* genes, the *Bam*HI-M fragment is an internal region specific for *crt*E gene (Zhu and Hearst 1987). On the whole the results are consistent. While most photosynthetic genes are turned off by oxygen, the *crt*A gene responsible for the oxidation of spheroidene to spheroidenone, is activated by oxygen, because a transposon mutation of *crt*A gene abolished the twofold increase in mRNA level by oxygen. Furthermore, the activation of *crt*A gene by oxygen and light was also observed when a subfragment of the *Bam*HI-H-fragment specific for *crt*A gene was used as a hybridization probe (Zhu and Hearst, unpublished observations). The reaction catalyzed by an oxygenase encoded by *crt*A may scavenge toxic singlet oxygen, which is formed in the photosynthetic apparatus. These results appear to disagree with Biel and Marrs' argument (1985) that the *crt* genes are not directly regulated by oxygen.

Table 1 Effects of Light and Oxygen on the mRNA Levels of Genes for LH, RC, Bchl, and Crt Biosynthesis in *R. capsulatum*

Genes	*mRNA/%*			
	HL	*D*	*LL*	O_2
*puc*B, A	55	19	100	2
*puf*B, A	59	30	100	17
*puf*L	77	19	100	21
*puf*M	72	20	100	23
*puh*A	57	16	100	17
*bch*J, G, D (*Bam*E)	81	20	100	62
*bch*H, K, F (*Bam*D)	84	30	100	118
*crt*A, I, B, *bch*I (*Bam*H)	125	39	100	131
*crt*C, D (*Bam*G)	132	67	100	72
*crt*E (*Bam*M)	I32	91	100	95

[1] Cells were grown under various conditions (HL, D, and LL were anaerobic conditions) and mRNA levels were measured by dot hybridization using various specific probes. The relative amounts of mRNA are expressed as percentages of the mRNA levels for low-light conditions (set as 100%). HL, high light; D, shift from HL to dark for one hour; LL, low light; O_2, aerobic growth.

Light Intensity

Light intensity is another external signal that regulates the development of the photosynthetic apparatus. The photosynthetic bacteria have developed mechanisms to adapt to different light intensities in order to maintain a more constant rate of photosynthetic activity. Cells grow competently in an environment of dim light, and high intensity light represses the formation of the photosynthetic apparatus. Earlier studies indicated that the cellular Bchl content of *R. sphaeroides* varied in inverse proportion to the incident light intensity (Cohen-Bazire and Kunisawa 1963). The molar ratio of LH-I to RC Bchl is fixed at about 20 to 25 (Agaard and Sistrom 1972). A shift of *R. sphaeroides* cells from high light (30 W/m^2) to low light (3 W/m^2) causes an immediate cessation of cell growth, Bchl accumulation, and protein synthesis. After a short period of adaptation, the cells resume their growth, Bchl synthesis commences at a new and higher rate, and two LH proteins and two additional high-molecular-weight photosynthetic membrane proteins increase threefold to fivefold (Chory and Kaplan 1983). The increased synthesis of LH antenna proteins as a result of reduction in light intensity results in an increase in the number and size of photosynthetic units per ICM and the number of ICMs per cell (Drews and Oelze 1981; Drews 1986). Chory and Kaplan (1982) have developed an in vitro transcription-translation system in which the levels of translatable mRNA are correlated with the increased levels of ICM proteins.

The expression of the genes coding for pigment-binding proteins in *R. sphaeroides* is also inversely correlated to the light intensity (Zhu and Kaplan 1985). When *R. sphaeroides* cells were shifted from the light to the dark, the levels of mRNA for LH-I, RC-L, and RC-M increased in the first five minutes after shifting because of derepression of transcription, although they eventually declined, presumably as a result of shortage of energy generation. Recently we have found that decreased light intensity results in increased levels of mRNAs for LH-I, LH-II, and RC-L, -M, and -H, as well as Bchl biosynthetic enzymes in *R. capsulatus* (Zhu and Hearst 1986). An interesting result is that the *crt* genes (A, I, B, C, D, E) in *R. capsulatus* again respond to light intensity in a way opposite to the other photosynthetic genes: increased light intensity leads to increased mRNA levels from these *crt* genes (Zhu and Hearst 1986) (see Table 1). Since photo-oxidative killing occurs only in high light and oxygen (Drews and Oelze 1981), and since it is well documented that Crt functions both to harvest light as an auxiliary pigment and to protect the photosynthetic apparatus in all photosynthetic organisms (Cogdell 1978; Halliwell 1984), we proposed that the transcriptional response of *crt* genes to high light is essential to the function of protecting cells from photo-oxidative damage. It is interesting to note that a very abundant transcript (0.4 kb) from *Bam*HI-J, whose location does not correspond to the genetically mapped *crt*F or *crt*E genes in this fragment as shown by Armstrong and Hearst (unpublished observations), also increases (threefold) in response to high light (Zhu and Hearst 1986).

In contrast to the pigment-binding protein genes, the *rbc*L gene is positively regulated by light intensity. The increase in light intensity results in an enhancement of both the levels of mRNA and enzymatic activity of RuBPCase in *R. sphaeroides* (Zhu and Kaplan 1985). Obviously, the gene for RuBPCase represents another type of regulation in *R. sphaeroides*. The mechanism by which cells adapt to light intensity is as yet unknown. From the results described here we suggest that the adaptation may take place at the transcriptional level. Another speculation is that variations of light intensity regulate the biosynthesis of the pigment complexes via the redox state of enzymes or regulatory proteins (Drews and Oelze 1981). Recently a sequence region of at least 300—700 bp 5′ of promoter function of the *puf* operon was found to be essential for light regulation (Kaplan, personal communication).

Regulation of the *puf* Operon

The LH-I complexes are found in the photosynthetic membrane in a fixed molar excess (approximately twelvefold) over the RC complex (Chory and Kaplan 1983). However, each of the two polypeptides within an LH-I complex (β and α) or within the RC (L and M) are present in a 1 : 1 stoichiometry. As these four genes plus C2397 are closely linked in a single operon both in *R. sphaeroides* and *R. capsulatus*, it would be of particular interest to elucidate how this stoichiometry and differential expression are achieved. Belasco et al. (1985) have reported that a 2.7 kb mRNA containing information for the LH-I(β, α) and RC (-L, -M) proteins and a small 0.5 kb mRNA containing information for LH-I (β, α) in *R. capsulatus* are derived from the same transcriptional initiation site. They came to the conclusion that the ratio of the small to large transcripts encoding the LH-I and RC proteins in *R. capsulatus* results predominantly from the selective degradation of a specific segment of the initial large transcript at the 3′ end of the molecule, resulting in a stabilized small transcript. The final ratio of small to large transcripts ((10—20) : 1) would then reflect the relative 3′ processing activity.

In *R. sphaeroides* Zhu and Kaplan (1985) have also observed the presence of two transcripts in a similar (10—20) : 1 ratio. Additionally, it has been demonstrated that the half life of the large transcript is 9 minutes and that of the small transcript is 20 minutes (Zhu et al. 1986b). The difference in the stability of the two mRNAs indicates that some type of posttranscriptional control is involved in the differential expression of *puf*B and A and *puf*L and M, but it cannot be the major event responsible for the (10—20) : 1 ratio of small to large transcripts. Further, we have demonstrated that the two transcripts have different 5′ terminals as measured by both S-I nuclease protection analysis and the primer extension technique (Zhu et al. 1986b). The small transcript has a 5′ end 104 bp upstream from the first ATG codon of *puf*B, whereas the large transcript has a 5′ end 75 bp upstream from the same start codon. A model is presented as follows: When the RNA polymerase binds to the sequence 104 bp upstream from the ATG start codon of *puf*B, the transcription terminates between *puf*A and *puf*L; whereas when the RNA polymerase binds to the sequence 75 bp upstream from the start codon of *puf*B, it reads through the first termination site between *puf*A and *puf*L, and then stops at the second termination site at the end of the *puf* operon. The mRNA secondary structure of the intercistronic region between the *puf*B and A and *puf*L and M is compatible with a rho factor-dependent termination signal (Zhu et al. 1986b). Based on the model described it is suggested that a combination of differential transcriptional initiation coupled with transcription termination led to the (10—20) : 1 ratio of small to large transcripts. These results seem to conflict with those of Belasco et al. (1985) who found that both the large and the small messengers initiate from the same initiation site about 35 bp downstream of the *Eco*RI site. This conclusion is based mainly on the S-I mapping in which only one strong band was observed under low-oxygen conditions, although another band corresponding to an initiation site 70 bp downstream of the *Eco*RI site was noticed. Since this minor band accounted for only 1%—2% of the *puf* mRNA 5′ ends, it was neglected (Belasco et al. 1985). By carefully maintaining the growth under strictly anaerobic photosynthetic conditions instead of low-oxygen conditions, which may cause the degradation of the labile large mRNA, we have been able to obtain similar results in *R. capsulatus* and in *R. spharoides* using the S-l mapping technique. Two bands were detected with a correct ratio of (10—20) : 1 and an approximately 35 bp space between their 5′ ends (Zhu and Hearst, unpublished observations). We suggest that the transcripts of the *puf* operon in *R. capsulatus* also have two 5′ ends.

Since the 5′ ends determined by S-1 nuclease analysis may not be the intiation sites, the authors did not exclude the possibility that the 5′ ends observed might themselves be the products of processing of a primary transcript in *R. sphaeroides* (Zhu et al. 1986b). It was also suggested that if this is the route by which these two transcripts are derived, the mechanism of posttranscriptional processing of the primary *puf* transcript of *R. sphaeroides* is far more complex than that envisioned for *R. capsulatus*, because the 5′ end of the small transcript (0.5 kb) mapped upstream of that of the large transcript (2.6 kb) cannot be simply processed from the large transcript (Zhu et al. 1986b). Recently, Beatty et al. (1986) reported that the stable *puf* transcripts might be derived from short-lived precursors. Marrs' group (Bauer et al. 1987) has described an ORF upstream of the *puf* operon, which is designated *puf*Q. Recently they have (Bauer et al. 1988) hypothesized that the *puf*Q gene product is the carrier protein for Bchl synthesis.

The consensus sequence of the promoter of the *puf* operon and the mechanism controlling the termination and antitermination are unknown as yet. It has in general been accepted that the regulatory sequences reside in the 5′ region upstream of the structural genes. Kiley et al. (1987a) have reported that a promoter region for transcription intitiation signal is within the region between 100 bp and 200 bp proximal to the first codon of *puf*B in *R. sphaeroides*. Beatty et al. (1985) have defined a DNA sequence that seems to be necessary for oxygen-regulated transcriptional initiation of the *puf* operon in *R. capsulatus*. Marrs' group has also reported that the *puf* operon may have two promoters: an oxygen-regulated promoter and a constitutive promoter upstream of *puf*Q (Bauer et al. 1988) using lac fusion. Recently, we have found that this Q region includes two ORFs, both controlled by oxygen and light (Zhu et al., unpublished observations). These two ORFs share strong homologies with the corresponding sequences in *R. sphaeroides* (Kiley and Kaplan, personal communication). The fact that the "regulatory gene or genes" are regulated by oxygen, raises a further question: what is the fundamental mechanism for oxygen regulation?

Since most of the photosynthetic genes are coordinately expressed in response to oxygen, there might be a common mechanism for their coordinate expression, although some of these genes also could be regulated individually. It has been suggested that DNA supercoiling may control the prokaryotic gene expression (Yamamoto and Droffner 1985; Kranz and Haselkorn 1986). Recently we have shown that the expression of the photosynthetic genes in *R. capsulatus* in response to oxygen requires the action of the topoisomerase, gyrase (Zhu and Hearst 1988). The supercoiling of the DNA in bacteria is determined by a dynamic balance between gyrase, which supercoils the DNA, and topoisomerase I, which relaxes the DNA. We have found that the synthesis of Bchl, LH-I and LH-II/Bchl complexes, and the expression of most photosynthetic genes coding for LH, RC, and Bchl biosynthesis as well as for form II RuBPCase is immediately inhibited by novobiocin and coumermycin at various concentrations (5—100 μg/ml) (Zhu and Hearst 1988). However, the constitutively expressed *puf* mRNA under aerobic growth conditions, as we mentioned early, is not inhibited by novobiocin. These results imply that the DNA gyrase is necessary for the anaerobic expression of those genes. The oxygen-activated *crt*A gene and other *crt* genes whose transcription is relatively enhanced by oxygen were inhibited by the antibiotics to a far smaller extent. The expression of Q gene also requires gyrase. At the present time we believe these gyrase effects to relate to a need for DNA unwinding during heavy transcription as opposed to the regulation of the initiation of transcription. Although transcriptional regulation plays a major role, we believe that the posttranscriptional and posttranslational regulations, including the assembly of the gene products in the membrane (Zhu et al. 1986a and 1986b; Zhu and Hearst 1987), are also required to elaborate the efficient and complex regulation of the photosynthetic apparatus in purple bacteria in response to environmental signals. A proposed mechanism of the regulation of the photosynthetic genes of purple nonsulfur bacteria at the transcriptional, posttranscriptional, and posttranslational levels is summarized in Figure 6.

Conclusion

In this chapter, we have attempted to provide an overview of the organization and expression of the genes for the photosynthetic apparatus in photosynthetic prokaryotes as revealed in combined studies utilizing genetics, biochemistry, and molecular biology. Over the past few years, several laboratories have made efforts toward isolating and sequencing the genes coding for the membrane components, LH and RC proteins, electron carriers, and cytoplasmic components. It can be predicted that in the future all of the genes involved in photosynthesis, including many genes for Bchl and Crt biosynthesis, will be isolated and sequenced. The function of the photosynthetic apparatus in higher organisms can be determined by comparing their DNA sequences to those of the photosynthetic prokaryotes, as was done for Q_B-binding protein and proteins involved in oxygen evolution. Another important direction in this field is to study the developmental and environmental regulation of photosynthetic genes in response to oxygen and light. It appears that transcriptional control plays a major role in most cases, although posttranscriptional and posttranslational regulation may also be important in some cases. Recently, such studies have been pursued extensively in *R. capsulatus* and *R. sphaeroides*. We have found that different photosynthetic genes can be regulated either negatively or positively by oxygen concentration and light intensity. However, the detailed events and the exact mechanism of oxygen and light regulation, and the interactions and communication among photosvnthetic genes are still obscure. We expect that in the near future more attention will be paid to characterizing the regulatory soquences and franscriptional machinery, to defining the moleadar mechanism of transcriptional and posttranscriptional regalation, and so investigating the interactions of regulatory proteins, DNA and mRNA such in formation will eventually facieitate the development of a biotechnology for increasing the efficiency of photosynthesis and food production.

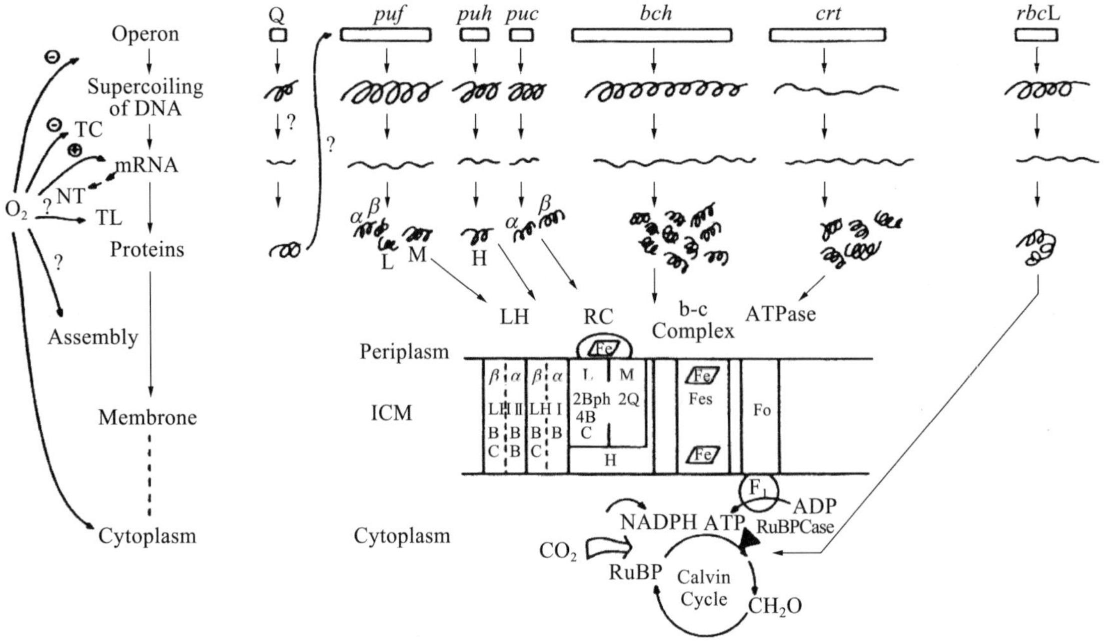

Figure 6 Schematic diagrams illustrating transcriptional (TC), posttranscriptional, and posttranslational (TL) regulation of photosynthetic genes in *R. capsulatus*. In development of the photosynthetic apparatus photosynthetic genes are expressed coordinately and differentially in response to environmental signals such as oxygen and light intensity. Current information indicates three levels of regulation: transcriptional regulation of mRNA synthesis, posttranscriptional processing of mRNA, and posttranslational assembly and insertion of pigment-protein complexes. The photosynthetic apparatus in ICM is comprised of LH complexes (LH-I, LH-II), RC polypeptides (L, M, H), cyctochrome bc_1 complexes, and ATPase. The polypeptides LH-I (β,α); RC subunits L, M, and H, and LH-II (β,α) are encoded by the *puf*, *puh* and *puc* operons, respectively. Bchl (B) binding to LH andRC is the pigment participating in light harvesting (in LH) and photochemical reaction (in RC). A large number of enzymes involved in ①Bchl biosynthesis are encoded by various *bch* genes. Pigments Crt (C) in LH and RC play a role both in light harvesting and in protection of the photosynthetic apparatus from photooxidation. The enzymes catalyzing Crt biosynthesis are encoded by various *crt* genes arranged as a cluster (see Figure 2). *rbc*L is the gene coding for the large subunit of form II RuBPCase, which is a cytoplasmic enzyme catalyzing carbon dioxide fixation in the Calvin cycle driven by reducing power NADPH and ATP derived from the photochemical reaction and coupled with ATPase in the ICM (see Figure 1). Oxygen has a major effect on the transcription of photosynthetic genes. Under anaerobic conditions the photosynthetic genes are activated. The exact mechanism for this activation has not been determined. Some *crt* genes, such as *crt*A, are activated by oxygen (Zhu et al. 1986a; Zhu and Hearst 1987). Oxygen also affects the stabilty of mRNA, accelerating the degradation of some mRNA (i. e., LH-II) to produce nucleotides (NT). It is not clear if oxygen acts on the translational process or assembly and stability of the proteins in the membrane. There is some evidence supporting the theory that oxygen can directly affect the activity of enzymes such as those catalyzing tetrapyrrole synthesis (not shown in this diagram) (Lascelles 1975). The transcriptional, posttranscriptional, and posttranslational regregulatory sequences and transcriptional machinery, to defining the molecular mechanism of transcriptional and posttranscriptional regulation, and to investiguating the interactions of regulatory proteins, DNA, and mRNA. Such information will eventually facilitate the development of a biotechnology for increasing the efficiency of photosynthesis and food production.

---

① To whom correspondence should be addressed

References

[1] Agaard, J., and Sistrom, W. R. (1972) *Photochem. Photobiol.* 15, 209 - 255.

[2] Allen, N. P., and Feher, G. (1984) *Proc. Natl. Acad. Sci. USA* 81, 4795 - 4799.

[3] Alt, J., Morris, J., Westhoff, P., and Herrmann, R. G. (1984) *Curr. Genet.* 8, 597 - 606.

[4] Bachofen, R., and Wiemken, V. (1986) in *Encyclopedia of Plant Physiology: Photosynthesis III*, vol. 19 (Staehelin, L. A., and Arntzen, C. J., eds.), pp. 620 - 631, Springer-Verlag, New York.

[5] Bartsch, R. G. (1978) in *The Photosynthetic Bacteria* (Clayton, R. K., and Sistrom, W. R., eds.), pp. 249 - 279,, Plenum Press, New York.

[6] Bauer, C. E., Eleuterio, M., Young, D. A., and Marrs, B. L. (1987) in *Proceedings of the VII International Congress of Photosynthesis*, vol. 4 (Biggins, J., ed.), pp. 699 - 705, Martinus Nijhoff, Boston.

[7] Bauer, C. E., Young, D. A., and Marrs, B. L. (1988) *J. Biol. Chem.* 263, 4820 - 4827.

[8] Beatty, J. T., Adams, C. W., and Cohen, S. N. (1985) in *V International Symposium on Photosynthetic Prokaryotes*, Grindelwald, Switzerland, September 22 - 28, 1985, Abstract, p. 100.

[9] Beatty, J. T., Adams, C. W., and Cohen, S. N. (1986) in *Microbial Energy Transduction* (Youvan, D. C., and Daldal, F., eds.), pp. 27 - 29, Cold Spring Harbor Laboratory, Cold Spring Harbor, NY.

[10] Belasco, J. G., Beatty, J. T., Adams, C. W., Gabain, A. V., and Cohen S. N. (1985) *Cell* 40, 171 - 181.

[11] Bérard, J., Béanger, G., Corriveau, P., and Gingras, G. (1986) *J. Biol. Chem.* 261, 82 - 87.

[12] Biel, A. J., and Marrs, B. L. (1983) *J. Bacteriol.* 156, 686 - 694.

[13] Biel, A. J., and Marrs, B. L. (1985) *J. Bacteriol.* 162, 1320 - 1321.

[14] Bishop, N. I. (1987) in *Proceedings of the VII International Congress of Photosynthesis*, vol. 3 (Biggins, J., ed.), pp. 795 - 798, Martinus Nijhoff, Boston.

[15] Chory, J., Donohue, T. J., Varga, A. R., Staehelin, L. A., and Kaplan, S. (1984) *J. Bacteriol.* 159, 540 - 554.

[16] Chory, J., and Kaplan, S. (1982) *J. Biol. Chem.* 257, 15110 - 15121.

[17] Chory, J., and Kaplan, S. (1983) *J. Bacteriol.* 153, 465 - 474.

[18] Clark, W. G., Davidson, E., and Marrs, B. L. (1984) *J. Bacteriol.* 157, 945 - 948.

[19] Cogdell, R. J. (1978) *Philos. Trans. R. Soc. Lond.* B 248, 569 - 579.

[20] Cogdell, R., and Crofts, A. (1978) *Biochim. Biophys. Acia* 502, 409 - 416.

[21] Cogdell, R., Parson, W. W., and Kerr, M. A. (1976) *Biochim. Biophys. Acta* 430, 83 - 93.

[22] Cohen-Bazire. G.. and Kunisawa R (1936) *I Cell Biol* 16 401 - 419.

[23] Daldal, F., Cheng, S., Applebaum, J., Davidson, E., and Prince, R. (1986) *Proc. Natl. Acad. Sci. USA* 83, 20I2 - 2016.

[24] Davidson, E., and Daldal, F. (1987) *J. Mol. Biol.* 195, 25 - 29.

[25] Davidson, E., Rook, S., and Daldal, F. (1987) in *Proceedings of the VII International Congress of Photosynthesis*, vol. 3 (Biggins, J., ed.), pp. 713 - 716, Martinus Nijhoff, Boston.

[26] Debus, R. J., Feher, G., and Okamura, M. Y. (1985) *J. Biochem.* 24, 2488 - 2500.

[27] Deisenhofer, J., Epp, O., Miki, K., Huber, R., and Michel, H. (1984) *J. Mol. Biol.* 180, 385 - 398.

[28] Deisenhofer, J., Epp, O., Miki, K., Huber, R., and Michel, H. (1985a) *Nature* 318, 618 - 624.

[29] Deisenhofer, J. R., Michel, H., and Huber, R. (1985b) *TIBS* 3, 243 - 248.

[30] DeVitry, C., and Diner, B. (1984) *FEBS Lett.* 167, 327 - 331.

[31] Donohue, T. J., Hoger, J. M., and Kaplan, S. (1986a) *J. Bacteriol.* 168, 953 - 961.

[32] Donohue, T. J., and Kaplan, S. (1986) in *Encyclopedia of Plant Physiology: Photosynthesis III*,

vol. 19 (Staehelin, L. A., and Arntzen, C. J., eds.), pp. 632 - 639, Springer-Verlag, New York.
[33] Donohue, T. J., McEwan, A. G., and Kaplan, S. (1986b) *J. Bacteriol.* 168,962 - 972.
[34] Drews, G. (1978) *Curr. Top. Bioenerg.* 8,161 - 199.
[35] Drews, G. (1985) *Microbiol. Rev.* 49,59 - 70.
[36] Drews, G. (1986) *TIBS* 11,255 - 257.
[37] Drews, G., Kaufmann, N., and Klug, G. (1986) in *Molecular Biology of the Photosynthetic Apparatus* (Steinback, K. E., Bonitz, S., Arntzen, C. J., and Bogorad, L., eds.), pp. 211 - 222, Cold Spring Harbor Laboratory, Cold Spring Harbor, NY.
[38] Drews, G., and Oelze, J. (1981) in *Advances in Microbial Physiology*, vol. 22, pp. 1 - 93, Academic Press, Orlando, FL.
[39] Drews, G., Peters, J., and Dierstein, R. (1983) *Ann. Microbiol. (Inst. Pasteur)* 134B, 151 - 158.
[40] Feher, G., and Okamura, M. Y. (1978) in *The Photosynthetic Bacteria* (Clayton, R. K., and Sistrom, W. R., eds.), pp. 349 - 382, Plenum Press, New York.
[41] Fornani, C. S., and Kaplan, S. (1983) *Gene* 25,291 - 299.
[42] Gabellini, N., Bowyer, J. R., Hurt, G., Melandri, A., and Hauska, G. (1982) *Eur. J. Biochem.* 126,105 - 111.
[43] Gabellini, N., Harnisch, U., McCarthy, J. E., Hauska, G., and Sebald, W. (1985) *EMBO J.* 2,549 - 553.
[44] Gabellini, N., and Sebald, W. (1986) *Eur. J. Biochem.* 154,569 - 579.
[45] Gibson, J. L., and Tabita, F. R. (1977a) *J. Bacteriol.* 132,818 - 823.
[46] Gibson, J. L., and Tabita, F. R. (1977b) *J. Biol. Chem.* 252,943 - 949.
[47] Govindjee, ed. (1982) *Photosynthesis, vols.* II and III, Academic Press, Orlando, FL.
[48] Gray, E. D. (1967) *Biochem. Biophys. Acta* 138,550 - 563.
[49] Halliwell. B. (1984) *Chloroplast Metabolism. The Structure and Function of Chloroplasls in Green Leaf Cells*, pp. 18 - 206, Oxford University Press.
[50] Hearst, J. E. (1986) in *Encyclopedia of Plant Physiology: Photosynthesis* III, vol. 19 (Staehelin, L. A., and Arntzen, C. J., eds.), pp. 382 - 389, Springer-Verlag, New York.
[51] Hearst, J. E., and Sauer, K. (1984a) in *Advances in Photosynthesis. Proceedings of the Vlth International Congress on Photosynthesis*, vol. III (Sybesma, C., ed.), pp. 355 - 359, Martinus Nijhoff/Dr. W. Junk, Boston.
[52] Hearst, J. E., and Sauer, K. (1984b) *Z. Naturforsch.* 39C, 421 - 424.
[53] Holschuhk, K., Bottomley, W., and Whitfield, P. R. (1984) *Nucleic Acids Res.* 12,8819 - 8834.
[54] Imhoff J. F., Truper, H. G., and Pfennig, N. (1984) *Int. J. Syst. Bacteriol.* 34,340 - 343.
[55] Kaplan, S. (1978) in *The Photosynthetic Bacteria* (Clayton, R. K., and Sistrom, W. R., eds.), pp. 809 - 839, Plenum Press, New York.
[56] Kaplan, S., and Arntzen, C. J. (1982) in *Photosynthesis III* (Govindjee, ed.), pp. 65 - 151, Academic Press, Orlando, FL.
[57] Kiley, P. J., Donohue, T. J., Havelka, W. A., and Kaplan, S. (1987) *J. Bacteriol.* 169,742 - 750.
[58] Kiley, P. J., and Kaplan, S. (1987) *J. Bacteriol.* 169,3268 - 3275.
[59] Klug, G., Kaufmann, N., and Drews, G. (1984) *FEBS Lett.* 177,61 - 65.
[60] Klug, G., Kaufmann, N., and Drews, G. (1985) *Proc. Natl. Acad. Sci. USA* 82,6485 - 6489.
[61] Kranz, R. G., and Haselkorn, R. (1986) *Proc. Natl. Acad. Sci. USA* 83,6805 - 6809.
[62] Kyle, D. J., Ohad, I., and Arntzen, C. J. (1984) *Proc. Natl. Acad. Sci. USA* 81,4070 - 4074.
[63] Lascelles, J. (1975) *Ann. N.Y. Acad. Sci.* 244,334 - 347.
[64] Loach, P. A., Parkes, P. S., Miller, J. F., Hinchigen, S., and Callahan, P. M. (1986) in *Molecular*

Biology of the Photosynthetic Apparatus (Steinback, K. E., Bonitz, S., Arntzen, C. J., and Bogorad, L., eds.), pp. 197 - 209, Cold Spring Harbor Laboratory, Cold Spring Harbor, NY.
[65] Marrs, B. (1974) *Proc. Natl. Acad. Sci. USA* 71, 971 - 973.
[66] Marrs, B. (1981) *J. Bacteriol.* 146, 1003 - 1012.
[67] Marrs, B., and Kaplan, S. (1970) *J. Mol. Biol.* 49, 297 - 317.
[68] Metz, J. G., Pakrasi, H., Arntzen, C. J., and Seibert, M. (1987) in *Proceedings of the VII International Congress of Photosynthesis*, vol. 4 (Biggins, J., ed.), pp. 679 - 682, Martinus Nijhoff, Boston.
[69] Michel, M., and Deisenhofer, J. (1986) in *Encyclopedia of Plant Physiology: Photosynthesis III*, vol. 19 (Staehelin, A. C., and Arntzen, C. J., eds.), pp. 371 - 381, Springer-Verlag, New York.
[70] Michel, H., Weyer, K. A., Gruenberg, H., et al. (1986) *EMBO J.* 5, 1149 - 1158.
[71] Michel, H., Weyer, K. A., and Lottspeich, F. (1985) *EMBO J.* 4, 1667 - 1672.
[72] Monger, T. G., and Parson, W. W. (1977) *Biochim. Biophys.* Acta 460, 393 - 407.
[73] Muller, E. D., Chory, J., and Kaplan, S. (1985) *J. Bacteriol.* 161, 469 - 472.
[74] Nargang, F., McIntosh, L., and Somerville, C. (1984) *Mol. Gen. Genet.* 193, 220 - 224.
[75] Ohad, I., Kyle, D. J., and Arntzen, C. J. (1984) *J. Cell Biol.* 99, 481 - 485.
[76] Olson, J. M. (1981) in *Origins and Evolution of Eukaryotic Intracellular Organelles* (Frederick, J. F., ed.), pp. 8 - 19, New York Academy of Sciences, New York.
[77] Peters, J., Takemoto, J., and Drews, G. (1983) *Biochemistry* 229, 5660 - 5667.
[78] Pfister, K., Steinback, K. E., Gardner, G., and Arntzen, C. J. (1981) *Proc. Natl. Acad. Sci. USA* 78, 791 - 795.
[79] Quivey, R. G. Jr., and Tabita, F. R. (1984) *Gene* 31, 91 - 101.
[80] Rochaix, J. D., Dron, M., Rahire, M., and Malnoe, P. (1984) *Plant Mol. Biol.* 3, 363 - 370.
[81] Sauer, K., and Austin, L. A. (1978) *Biochem.* 17, 2011 - 2019.
[82] Schachman, H. K., Pardee, A. B., and Stanier, R. Y. (1982) *Arch. Biochem. Biophys.* 38, 245 - 260.
[83] Schumacher, A., and Drews, G. (1978) *Biochim. Biophys. Acta* 501, 183 - 194.
[84] Scolnik, P. A., and Haselkorn, R. (1984) *Nature* 307, 289 - 292.
[85] Solioz, M., Yen, H. C., and Marrs, B. L. (1975) *J. Bacteriol.* 123, 651 - 657.
[86] Somerville, C. R., and Somerville, S. C. (1984) *Mol. Gen. Genet.* 193, 214 - 219.
[87] Sprague, S. G., and Varga, A. R. (1986) in *Encyclopedia of Plant Physiology: Photosynthesis III*, vol. 19 (Staehelin, L. A., and Arntzen, C. J., eds.), pp. 603 - 619, Springer-Verlag, New York.
[88] Staehelin, L. A., and Arntzen, C. J., eds. (1986) *Encyclopedia of Plant Physiology: Photosynthesis III*, vol. 19, Springer-Verlag, New York.
[89] Steinback, K. E., Bonitz, S., Arntzen, C. J., and Bogorad, L., eds. (1985) *Molecular Biology of the Photosynthetic Apparatus*, Cold Spring Harbor Laboratory, Cold Spring Harbor, NY.
[90] Sybesma, C., ed. (1984) in *Advances in Photosynthesis. Proceedings of the VIth International Congress on Photosynthesis*, 1983, Brussels, Belgium, Martinus Nijhoff/Dr. W. Junk, Boston.
[91] Takemoto, J. Y., Peters, J., and Drews, G. (1982) *FEBS Lett.* 142, 227 - 230.
[92] Taylor, D. P., Cohen, S. N., Clark, W. G., and Marrs, B. L. (1983) *J. Bacteriol.* 154, 580 - 590.
[93] Webster, G., Cogdell, R., and Lindsay, J. (1980) *FEBS Lett.* 111, 391 - 394.
[94] Welte, W., Wacker, T., Leis, M., et al. (1985) *FEBS Lett.* 182, 260 - 264.
[95] Wiemken, V., and Bachofen, R. (1984) *FEBS Lett.* 166, 155 - 159.
[96] Williams, J. C., Steiner, L. A., Feher, G., and Simon, M. (1984) *Proc. Natl. Acad. Sci. USA* 81, 7303 - 7307.

[97] Williams, J. C., Steiner, L. A., Ogden, R. C., Simon, M. I., and Feher, G. (1983) *Proc. Natl. Acad. Sci. USA* 80, 6506 - 6509.
[98] Worland, S. T., Yamagishi, A., Isaacs, S., Sauer, K., and Hearst, J. E. (1987) *Proc. Natl. Acad. Sci. USA* 84, 1774 - 1778.
[99] Yamamoto, N., and Droffner, M. L. (1985) *Proc. Natl. Acad. Sci. USA* 82, 2077 - 2081.
[100] Yen, H. C., Hu, N. T., and Marrs, B. L. (1979) *J. Mol. Biol.* 131, 157 - 168.
[101] Yen, H. C., and Marrs, B. (1976) *J. Bacteriol.* 126, 619 - 629.
[102] Youvan, D. C., Alberti, M., Begusch, H., Bylina, E. J., and Hearst, J. E. (1984a) *Proc. Natl. Acad. Sci. USA* 81, 189 - 192.
[103] Youvan, D. C., Bylina, E. J., Alberti, M., Begusch, H., and Hearst, J. E. (1984b) *Cell* 37, 949 - 957.
[104] Youvan, D. C., Elder, J. T., Sandlin, D. E., et al. (1982) *J. Mol. Biol.* 162, 17 - 41.
[105] Youvan, D. C., and Ismail, S. (1985) *Proc. Natl. Acad. Sci. USA* 82, 58 - 62.
[106] Youvan, D. C., and Marrs, B. L. (1984) *Cell* 39, 1 - 3.
[107] Youvan, D. C., and Marrs, B. L. (1985) in *Molecular Biology of the Photosynthetic Apparatus* (Steinback, K. E., Bonitz, S., Arntzen, C. J., and Bogorad, L., eds.), pp. 173 - 181, Cold Spring Harbor Laboratory, Cold Spring Harbor, NY.
[108] Yu, P.-L., Hohn, B., Falk, H., and Drew, G. (1982) *Mol. Gen. Genet.* 188, 392 - 398.
[109] Zannoni, D., and Baccarini-Melandri, A. (1980) in *Diversity of Bacterial Respiratory Systems*, vol. 2 (Knowles, D., ed.), pp. 183 - 202. CRC Press, Boca Raton, FL.
[110] Zhu, Y. S., Cook, D. N., Leach, F., et al. (1986a) *J. Bacteriol.* 168, 1180 - 1188.
[111] Zhu, Y. S., and Hearst, J. E. (1986) *Proc. Natl. Acad. Sci. USA* 83, 7613 - 7617.
[112] Zhu, Y. S., and Hearst, J. E. (1987) in *Proceedings of the VII International Congress of Photosynthesis*, vol. 4 (Biggins, J., ed.), pp. 717 - 720, Martinus Nijhoff, Boston.
[113] Zhu, Y. S., and Hearst, J. E. (1988) *Proc. Nail. Acad. Sci. USA* 85, 4209 - 4213.
[114] Zhu, Y. S., and Kaplan, S. (1985) *J. Bacteriol.* 162, 925 - 932.
[115] Zhu, Y. S., Killey, P. J., Donohue, T. J., and Kaplan, S. (1986b) *J. Biol. Chem.* 261, 10366 - 10374.
[116] Zsebo, K. M., and Hearst, J. E. (1984) Cell 37, 937 - 947.
[117] Zsebo, K. M., Wu, F., and Hearst, J. E. (1984) Plasmid 11, 182 - 184.
[118] Zuber, H. (1986) in *Encyclopedia of Plant Physiology: Photosynthesis III*, vol. 19 (Staehelin, L. A., and Arntzen, C. J., eds.), pp. 239 - 251, Springer-Verlag, New York.
[119] Zuber, H., Sidler, W., Füglistaller, P., Brunisholz, R., and Theiler, R. (1986) in *Molecular Biology of the Photosynthelic Apparatus* (Steinback K. E., Bonitz, S., Arntzen, C. J., and Bogorod, L., eds.), pp. 183 - 195, Cold Spring Harbor Laboratory, Cold Spring Harbor, NY.
[120] Zurawski, G., Bohnert, H. J., Whitfield, P. K., and Bottomley, W. (1982) *Proc. Natl. Acad. Sci. USA* 79, 7699 - 7703.

The *Nicotiana* Chloroplast Genome*

The wide range of application of tobacco as the ideal experimental material for various studies has long been recognized. Therefore, it is not surprising that tobacco has often been referred to as the "E. coli" of the plant kingdom. It possesses many advantages as well as unique features. Tobacco belongs to the genus Nicotiana. There are over 60 species in this genus, many of which can be crossed to produce viable hybrid plants. In fact, interspecific hybridization served as the major force of speciation in this genus. An excellent taxonomic treatment of this genus has been presented by Goodspeed[1] showing Nicotiana to be an exceptionally suitable material for genetic and evolutionary studies. This is demonstrated in the study of ribulose 1,5-bisphosphate carboxylase (RuBPCase)[2,3].

The Nicotian a system also offers unique opportunities in the study of the genetics and evolution of the chloroplast genome as compared to the other systems such as Oenothera. In Oenothera, chloroplasts frequently exhibit biparental transmission, whereas Nicotiana exhibits strictly maternal inheritance. By taking advantage of the information available on Nicotiana rapid progress has been made in the study of Nicotiana chloroplast genome. A brief summary of such a study is presented below.

Isolation and Restriction Analysis

A. Isolation and Purification

A newly devised procedure utilizing liquid nitrogen[4] is used for isolation of chloroplasts and purification of ct-DNA. This method minimizes the damage of nuclei, and therefore greatly reduces contamination with nuclear DNA. The yield of ct-DNA from this procedure is also substantially higher.

Tobacco seedlings of about I inch are transplanted into 5-inch soil pots in the greenhouse. The plants are allowed to grow for 4 to 8 weeks (depending on the seasons after transplantation). The leaves are harvested after being placed in the dark for 2 days in order to deplete them of starch.

---

* Y. S. Zhu and S. D. Kung, Department of Biological Sciences, University of Maryland Baltimore County, Catonsville, Maryland 21228 (U. S. A.)

此文发表于 1982 年"Plant Molecular Biology Newsletter" Ⅲ:95 - 100.

After deveining, 100 to 500 g of freshly harvested leaves are homogenized with liquid nitrogen in a metal waring blender. After the liquid is evaporated and the powder is warmed at room temperature to −5℃, three weight volumes of isolation buffer (A) (0.33 M Sorbitol, 5 mM MgCI, 50 mM Tris-HCI (pH 8.0), 0.1% BSA and 0.mM mercaptoethanol) is added to the powder with rapid mixing. The suspension is filtered through miracloth and centrifuged at 1,500 g for 15 min. The crude chloroplast pellets are then gently resuspended in the same buffer and purified through a discontinuous silica sol gradient[5] containing 0.24 M sucrose throughout. A clear band of chloroplasts, which appear to be mostly intact, as observed under microscope, is located in the middle layer, whereas the nuclei and mitochondria are distributed in the bottom and top layers respectively. The chloroplast band is collected, mixed with an equal volume of buffer A, and pelleted at 12,000 g for 30 min. The pellet is resuspended sequentially in 3.6 ml of buffer B (50 mM Tris, 20 mM EDTA, pH 8.0), 0.9 ml of 2% Na-Sarkosyl (in 50 mM Tris, 10 mM EDTA, pH 8.0), and 1.5 ml of 4 M CsCI (in 50 mM Tris, 10 M EDTA, pH=8.0) in order to lyse chloroplasts.

The chloroplast lysate is pelleted at 1,200 g for 30 min, after which the supernatant is removed and supplemented by 2.6 g of CsCI and 0.1 ml of ethidium bromide (10 mg/ml). Another extraction of the pellet with lysing buffers listed above always provides a higher yield of ct-DNA. 7—8 ml of the supernatant is layered on to 3 ml of CsCI (ρ=1.74) in nitrocellulose tubes, then centrifuged at 40,000 rpm for 18 hours at 20℃. After centrifugation the single fluorescent ct-DNA band is visualized by ultraviolet illumination. The ct-DNA band is collected and pooled, and the ethidium bromide is removed by shaking with 0.01 M EDTA-saturated isoamyl alcohol. The DNA is dialysed at 4℃ against three 500 ml changes of buffer C (50 mM Tris, 20 mM NaAc, 2 mM EDTA, 18 mM NaCI) over a 24 hours period. The ct-DNA concentration is usually between 10 and 100 μg/ml after dialysis. The ct-DNA is precipitated with 2 volumes of ethanol and resuspended in buffer D (10 mM Tris, I mM EDTA, pH 8.0) to ct-DNA concentration of 1 μg/μl. This ct-DNA is ready for restriction pattern analysis and molecular cloning.

B. Restriction Enzyme Analysis

The purified ct-DNA is digested with various restriction enzymes, as directed by the supplier. It is known that ct-DNA contains a variable amount of restriction sites for different enzymes ranging from 10—40[6]. These subsets of fragments can be conveniently separated from one another according to size by gel electrophoresis.

Since the large fragments migrate very slowly in the agarose gel, a low percentage gel should be used. It is a general guide that if there are less than 15 fragments generated from a higher plant chloroplast genome by a given restriction enzyme, then a 0.7%—0.8% gel is adequate. If the number of fragments are over 25, then a 1.0%—1.5% gel is suitable. The intensity of some bands are higher than others because of the existence of multiple copies having similar or identical sizes. The exact number of copies in such bands can be estimated from densitometer tracings.

Properties and Organization

Nicotiana ct-DNA is a double-stranded structure having a molecular. weight of (95—10)× 10^6 daltons[7,8]. It has an average buoyant density of 1.698—1.700 g/cm^3, corresponding to a GC content of 38%—41%. In most cases the buoyant density of ct-DNA is indistinguishable from that of the corresponding nuclear DNA. Ct-DNA, however, possesses some unique features. To date, no detectable 5-methyl cytosine has been found in the ct-DNA from Nicotiana or any plants, whereas the corresponding nuclear DNA invariably contains 4%—5%. Another unique property of ct-DNA is the ease with which it can be renatured. This is a reflection of the degree of simplicity and homogeneity of ct-DNA, which is a much less complicated molecule than nuclear DNA both in structural complexity and in genetic content.

Chloroplast DNA is circular in form and has a contour length of 50 μm. There are multiple copies per organelle ranging from a few to over 100, existing in different forms: circles, supercoiled circles and circular dimers[8]. Each circle can be divided into two inverted repeat regions separated by a small and a large single-copy sequence[9]. The inverted repeats are usually 20—25 kilobases (kb) in length and contain the genes for ribosomal and transfer RNAs[8]. The single-copy regions contains the genes for many chloroplast proteins and transfer RNAs. The coding capacity of ct-DNA may well exceed 100 polypeptides of 4×10^4 daltons each. However, only a few polypeptides have been identified as the product of chloroplast genes incuding the large subunit (LS) of RuBPCase[7].

Diversity and Evolution

Based on a number of biochemical and functional considerations ct-DNA of higher plants has long been thought to be highly conserved. This concept originated from observations that ct-DNA exists in multiple copies per chloroplast, contains similar coding information and exhibits uniform physico-chemical properties. This view should be re-examined in light of current evidence. The results obtained from restriction enzyme analysis revealed a wide range of variability of ct-DNA[10,11]. Nicotiana ct-DNA exhibits a high degree of diversity but an overall similarity. For instance the individual restriction pattern is species specific for any given species, while the general configuraton is characteristic of the genus Nicotiana[12].

Among the many restriction enzymes used, EcoRI produces the largest number of fragments and therefore has the highest resolving power to uncover differences, Of the 40 fragments generated by EcoRI enzyme the first 10 are worth noting[6]. Fragments 3 are stable and present in all 40 Nicotiona species examined so far. In contrast, fragments 4 and 5 are extremely variable one or both can be altered; one or both may be absent. It is such a combination of diversity and similarity in fragment pattern that forms the basis of species specificity and overall identity of Nicotiana ct-DNA.

The fragment pattern generated by BamHI revealed a similar degree of diversrty in Nicotiana ct-DNA. For example, N. gossei and N. otophora ct-DNAs differ in 13 of the 27 BanHI bands. This matches the extent of variation in their EcoRI fragments in which 18 of the 40 bands are different. Therefore, both EcoRI ard BamHI can be used to measure accurately the degree of diversity in Nicotiana ct-DNA.

Alteration of DNA is the main force of evolution. Judging from the wide diversity of Nicotiana ct-DNA it is evident that there have been considerable changes through the course of evolution. It has been demonstrated in Nicotiana that the mechanism of ct-DNA alteration involves point mutation, inversion, deletion, duplication and rarely recombination[13]. Point mutations are primarily responsible for the observed gain and elimination of many restriction sites. They occur rather frequently in relation to other mechanisms and are clustered in one region which can be described as a hot spot[6]. A large (11 kb) and a small (0. 5 kb) deletion have also been discovered[14]. Whether these segments contain any structural genes is not known. The example of ct-DNA duplication is the existence of the large inverted and the small scattered repeats.

The limited differences in the restriction patterns generated by Sma I provide excellent markers for phylogenetic studies. It is a better system than the isoelectric focusing patterns of fraction I protein with relation to identifying evolutionary steps in ct-DNA. By using a single restriction enzyme, Sma I, eliminations and sequential gains of its recognition sites during the course of ct-DNA evolution are clearly demonstrated[6].

Cloning and Gene Library

If study of individual chloroplast genes is desired, it is advantageous to construct a chloroplast genome library, from which a variety of interesting genes can be isolated, purified, amplified and stored for further studies.

When the ct-DNA of N. otophora is completely digested with EcoRI, BomHI and Sma I, at least 40,28 and 15 fragments are generated, respectively. BamHI was used to digest ct-DNA for constructon of a chloroplast genome library, as there is a single BamHI site within the tetracycline resistance gene of pBR322.

A "shotgun" procedure was employed for cloning. A mixture of N. otophora ct-DNA and pBR322 was digested with BamHI, ligated, and introduced into E. coli HB101. The transformed E. coli cells were first selected on ampicillin plates and then screened for tetracycline sensitivity by duplicate plating on ampicillin and tetracycline plates. A total of 207 recombinants (Amp^r, Tc^s) were obtained from 2002 colonies (Amp^r). After digestion with BamHI, the chimeric plasmids carrying ct-DNA inserts of various sizes in the recombinants were further analyzed electrophoretically and compared with pBR322 and BamHI-digested N. otophora ct-DNA. Most of the BamHI fragments of ct-DNA (23/28) have been clones in E. coli HB101. pBR322 carries from one to four different BamHI fragments of ct-DNA. The largest inserted ct-DNA fragment (9 kb) is twice the size of pBR322. Fragments larger than this, such as Bam 1,2 and 3, seem to encounter difficulty in being inserted into pBR322. These fragments, from ct-DNA of N. otophora, were separated and recovered from a low-gelling-temperature agarose gel, then ligated to pBR322 and introduced individually into E. coli HB101. Thus, with the exception of Bam fragment 8, a clone library of the entire N. otophora chloroplast genome was constructed[15].

Mapping and Expression

A. Mapping of ct-DNA and location of genes for the LS of RuBPCase and 16s, 23s and 5s rRNA

The SmaI and BamHI restriction sites in N. otophora of ct-DNA were mapped based on the data from the following: ① primary digestion: ct-DNA was digested with SmaI and BamHI, individually or in combination; ② secondary digestion: each SmaI fragment was digested with BamHI or vice versa; ③DNA hybridization of SmaI fragments with BamHI fragments; ④available information on SmaI sites mapped or N. tabacum[16]; ⑤comparison of SmaI and BamHI restriction patterns from different species of Nicotiana. From the results of the mapping procedures we determined that N. otophora is 160 kb in length. The inverted repeats are 22.5 kb in length separated by a small (26 kb) and a large (89 kb) single copy region.

The 16s, 23s and 5s rRNA genes of N. tabacum have been identified to reside on EcoRI fragments 8, 4 and 29, respectively[17,18]. These fragments were used as probes to locate rRNA genes in N. otophora ct-DNA. The rRNA genes were proved to be located in the inverted repeats (Bam 11, 5 and Sma 4, 11, 14), the same as was discovered in other plants[8,19,20]. Similarly, the gene for the LS was identified in the large single copy region (EcoR 4, Sma I and Bam 20,24) by hybridization using a spinach RuBPCase LS gene as the probe. All of these structural features of N. otophora ct-DNA are similar to that of N. tabocum[16,21,22] and other plants, indicating these major chloroplast genes are highly conserved in terms of their location in ct-DNA.

B. Expression of ct-DNA in E. coli and B. subtilis

We are interested in the expression of developmentally and environmentally regulated chloroplast genes in Nicotiana. As a first step in approaching this goal, we took advantage of the recombinant DNA techniques which make possible the expression of eukaryotic genes in bacteria. Efficient expression in bacteria of a non-bacterial gene requires the assistance of a native bacterial promoter and ribosomal binding sites. Since chloroplasts have procaryotic features, ct-DNA is assumed to be easily expressed in bacteria. In order to study the transcription of ct-DNA in bacteria a maxicell (CSR603) technique was employed. All the plasmids from the gene library containing ct-DNA inserts were subcloned in E. coli CSR603. Irradiation with UV destroys chromosomal DNA, whereas plasmids can survive. Therefore, whenever ^{35}S-methionine is applied to irradiated cells, the plasmid-encoded polypeptides can be detected. Though a few chloroplast genes (in Bam 7, 12 and 16) were expressed in the maxicells (as demonstrated by synthesizing 43, 29 and 13.5 kb polypeptides), 90% of the ct-DNA genome was not transcribed and translated[15]. The failure of most chloroplast genes to be expressed may be due to misrecognition of promoters and ribosome binding sites, or expression that is not strong enough to be detected by this method. In addition, some genes may be split by BamHI. One example is the gene for the LS of RUBPCase which is cleaved by BamHi to Bam 20 and Bam 24.

We are especially interested in the RuBPCase as it is the major protein in chloroplasts and a key enzyme in photosynthesis. In Nicotiana, as in other higher plants, it occurs as an oligomer of eight large subunits (LS) (MW 55,000) and eight small subunits (SS) (MW 15,000). The LS, containing the catalytic sites, are coded by the chloroplast genome and are synthesized in the chloroplast. The SS, whose function remains uncertain, are coded by the nuclear genome and synthesized in the cytoplasm as a precursor of higher molecular weight (MW 20,000) that is cleaved and transported into the chloroplast. RuBPCase was first crystallized from N. tabacum, providing an attractive model for studying nucleus-chloroplast cooperation and gene regulation in plants[23].

Although BamHI, as well as pstl, cuts within the LS gene, the unique restriction fragments generated by cleavage of ct-DNA with Smal, Sall, Hindlll or EcoRI contain intact LS genes necessary for expression studies, as demonstrated by hybrid zation using the spinach LS gene as a probe[24]. The fragments Sal 6(14 kb) and Hind 2(11 kb) containing the intact LS gene, were isolated from total N. otophora ct-DNA and cloned into E. coli using pBR325 and pBR322 as the respective vectors. Cloned fragments Sal 6 (in plas mid PRC22) and Hind 2 (in plasmid PRCZI) were shown by hybridization to contain the LS gene.

For detection of the LS gene product a simple in situ immunoassay was employed. The clones CSR603 (PRCZI) and CSR603 (PRC22), contaning the N. otophora LS gene, were inoculated and incubated on a N-Z bottom agarose plate overnight at 37℃. CSR603 cells containing pBR322 and pBR325 were used as controls. After incubation, the colonies were lysed on the plate with chloroform vapor and lysozyme. A channal was made in the center of the plate into which was added antiserum against the LS of RuBPCase. After incubaton for 1 day a sharp immunoprecipitation line was formed between CSR603 (PRCZI) and the central channel, demonstrating that this clone had produced the LS polypeptide. This result was confirmed by immunoprecipitation of extracts of these clones with antiserum against LS in an Ouchterlony double diffusion test. The LS gene from spinach, N. tabacum, and

Chlamydomonas were also expressed in E. coli as shown by this method[25].

We provided the first demonstration of expression of Chlamydomonas LS gene in B. subtilis. This LS gene was previously cloned in pBR322 (pLM 401) by L. Mets (personal communication). Either the anterior or posterior portions of the LS genes were cloned into B. subtilis using pPL608 as the vector. Both portions were expressed in B. subtilis as demonstrated by in situ immunoassay and Ouchterlony assay[21].

Concluding Remarks

The study of Nicotiana ct-DNA reveals that it exhibits a high degree of variability in the restriction fragmet patterns. These patterns are species specific and therefore it can be used as a genetic marker. This offers unique opportunity to study the evolution of organelles. The understanding of the mechanism of molecular evolution of Nicotiana chloroplast genome will also provide a clear clue of the origin of chloroplasts. Our current work on the cloning and sequencing of the genes of the LS of RuBPCase and others from various Nicotiana species has firmly established a strong basis toward reaching this goal.

We thank Mrs. Merkle-Lehman for her assistance in preparing this manuscript. This investigation was supported by NIH grant CM22746 - 01 and U. S. Dept. of Agriculture cooperative agreement 58 - 3204 - 0 - 157 from the tobacco laboratory.

References

[1] Goodspeed, T. H. 1954. The Genus Nicotiana, Walthan, Mass. : Chronica Botanica. pp. 283 - 314.

[2] Vehimiya, H. , K. Chen and S. G. Wildman. 1977. Stadier Symp. (University of Missouri, Columbia) Vol. 9:83 - 100.

[3] Chen, K. and S. G. Wildman. 1981. PI. Syst. Evol. 138:89 - 113.

[4] Rhodes, P. R. and S. D. Kung. 1981. Can. J. Biochem. 59:911 - 915.

[5] Walbot, V. 1977. Plant Physiol. 60:102 - 108.

[6] Kung, S. D. , Y. S. Zhu, G. F. Shen. 1982. Theor. App. Genet. 61:73 - 79.

[7] Kung, S. D. 1977. Ann. Rev. Plant Physiol. 28:401 - 437.

[8] Bedbrook, J. R. and R. Kolodner. 1979. Ann. Rev. Plant Physiol. 30:593 - 620.

[9] Koller, B. and H. Delius. 1980. Mol. Gen. Genet. 178:261 - 269.

[10] Frankel, R. , W. R. Scowcroft and P. R. Whitfeld. 1979. Molec. Gen. Genet. 169:129 - 135.

[11] Vedel, F. , F. Quetier and M. Bayer. 1976. Nature 263:440 - 442.

[12] Rhodes, P. R. , Y. S. Zhu and S. D. Kung. 1981. MoI. Gen. Genet. 182:106 - 111.

[13] Kung, S. D. , Y. S. Zhu, K. Chen, G. F. Shen and V. A. Sisson. 1981. MoI. Gen. Genet. 183:20 - 24.

[14] Shen, G. F. , K. Chen and S. D. Kung. 1982. MoI. Gen. Genet. 187:12 - 18.

[15] Zhu, Y. S. , E. J. Duvall, P. L. Lovett and S. D. Kung. 1982. MoI. Gen. Genet. 187:61 - 66.

[16] Jurgensen, J. E. and D. P. Bourque. 1980. Nucleic Acids Rev. 8:3505 - 3516.

[17] Sugiura, M. and Kusuda, J. 1979. Molec. Gen. Genet. 172:137 - 141.

[18] Takaiwa, F. and Sugiura, M. 1980. Genet. 10:95 - 103.

[19] Bogorad, L. J. R. Bedbrook, D. M. Coen, R. Kolodner and G. Link. 1978. In "Chloroplast and

development", Ed. G. Akoyunoglou, Amsterdam Elsoevier. pp. 541 - 55 1.

[20] Bogorad, L. 1979. In "Genetic engineering principles and methods", Vol. I. , Ed. J. K. Setlow and A. Hollaender. Plenum Press. pp. 181 - 203.

[21] Fluhr, R. and M. Edelman. 1981. Molec. Gen. Genetc. 181:484 - 490.

[22] Seyer, P. , K. V. Kowallik and R. G. Herrmann. 1981. Current Genetics 3:189 - 204.

[23] kung, S. D. 1976. Science 191:429 - 434.

[24] Erion, J. L. , J. Tarnowski, H. Weissbach and N. Brot. I981. Proc. Natl. Acad. Sci. , U. S. A. 78: 3459 - 3463.

[25] Zhu, Y. S. , P. L. Lovett, D. Williams and S. D. Kung. 1982. Science (submitted).

国外纤维素酶研究概况 ①

一、序言

国外纤维素酶发现很早，1906 年 Seilliere 在蜗牛的消化液中发现有纤维素酶，能分解天然纤维素。后来，在第二次世界大战期间，美国曾研究过南方丛林中军用麻袋、帐篷的防腐防霉问题；20 世纪 50 年代，纤维素酶工作重点转向纤维素酶本身的性质、作用方式、培养条件、测定方法等的研究；20 世纪 60 年代，工作进展较快，除了对酶本身的分离、纯化、诱导、成分、作用等继续作理论上的研究外，在酶制剂的生产和应用上也都取得了较大的成绩。美国化学协会在 1962 年和 1968 年分别召开了二次纤维素酶讨论会，并出版了两本专集：《纤维素及有关物质酶水解的进展》和《纤维素酶及其应用》。日本在 1961 年成立了纤维素酶研究会，同时召开了第一次年会，以后每年召开一次年会，到 1968 年共开了 9 次。此外，在加拿大、澳大利亚、英国、印度、苏联、匈牙利等国也都在进行这方面的研究。

国外之所以对纤维素酶工作有如此大的兴趣，其主要原因是因为随着世界人口大幅度的增长，对粮食的要求越来越迫切，希望通过非传统的农业方法来解决人类粮食来源。纤维素酶在充分利用农副产品废料、开辟粮食新来源中将具有巨大的潜力。纤维素是世界上最丰富的原料之一，而且每年更新，用之不尽，取之不完。它占地球上植物总量的三分之一，如果将这部分纤维充分利用起来变为糖类或蛋白，作为动物饲料或人类食品，将为人类的粮食问题作出重大的贡献。此外，世界各国随着工业化程度的进一步提高，污物的处理也成为十分头痛的事情。据说有些国家堆积的废物中有一半是纤维素，在城市固体垃圾中，废纸是其中的主要成份。现在清除废物的方法是焚烧，产生的烟雾会污染城市环境。如果能用纤维素酶分解，既可处理废物，又可得细菌蛋白。但是，目前国外纤维素酶现有的水平还很难解决变纤维废物为粮食的这一问题，近两三年来，工作上也无重大突破，其中最主要的问题是纤维素酶的酶解效率不高，还远不及淀粉酶。有人计算过，纤维素酶分子的转化率，如果以可溶性纤维寡糖或纤维素衍生物为底物，反应速度为 5～29 个键/秒酶分子（无论是内间或外端型的 C_x 酶），对天然固体纤维素分子的反应速度就要更低了，而 α-淀粉酶是 300 个键/秒酶分子，β-淀粉酶是 4 000 个键/秒酶分子，相差十几倍至成百倍，也大大低于一般酶的水平即 1 000 个键/秒酶分子（最高是碳酸酐酶 10^5 个键/秒酶分子，固氮酶较低为 10 个键/秒酶分子）。在我们自己的工作中也碰到这个问题。例如，酒精发酵时用淀粉酶曲 7%作用于淀粉质原料，半小时可分解 90%左右，而分解纤维原料达到同样的程度，就要加纤维素酶曲 100%，作用两天。

上海植物生理研究所纤维素酶组，此文发表于 1975 年“应用微生物”2 期 1－10 页。

① 此文章由朱雨生执笔。

如何来提高酶解效率呢？现在大家都认为提高纤维素酶的酶解作用应该从两方面着手。一方面是提高酶的活力；另一方面是改变纤维底物的结构，使之对酶的作用更为敏感。本文主要想从这两方面看看国外近年来这方面工作的进展情况，最后概要谈谈国外提高纤维素酶解作用的几个途径。

二、纤维素酶的诱导和作用

1. 纤维素酶的分布

纤维素酶在自然界分布是广的，某些原生动物、腔肠动物、蠕形动物、节肢动物、软体动物、棘皮动物、昆虫、爬行动物等均有。这些生物生活在陆地、河泊或海洋中，部分或全部利用陆上或水中的纤维为食料。在反刍动物的瘤胃中有共生的纤维分解细菌和原生动物，在猪的大肠中也有共生的纤维分解细菌。在高等植物中也有纤维素酶，其作用是使细胞壁松弛，与种子发芽、细胞生长有关，也有人认为它与器官脱落时的离层形成有关。

在微生物方面，细菌、真菌、放线菌都有，据说最近(1972)有人在个别酵母中也有发现。真菌中活力较高的是木霉、黑曲、青霉、根霉和漆霉斑等。细菌中有纤维杆菌、球形生孢纤维黏菌、白色瘤胃小球菌等。目前研究得较多的是绿色木霉和纤维杆菌。

2. 胞外酶和胞内酶

纤维素酶分解的底物是纤维素，而纤维素是不溶性的，因此，纤维素首先要在细胞外分解才能为细胞利用。现在知道真菌的纤维素酶大多是胞外酶，也就是分泌到细胞外面来的，而且是诱导酶，即只有外加纤维素，才能合成纤维素酶。例如，绿色木霉在大多数碳源上都能生长，但只有在培养基中加入纤维素或其他含 β(1→4)糖苷键的葡聚糖和少数纤维寡糖时，纤维素酶才会合成和分泌到体外，在细菌中有胞外酶，也有胞内酶或细胞结合酶——分布在细胞里面。King 在反刍动物中证明有表面定位的酶，估计 25%～30%的纤维素酶活性是由细菌表面引起的。西泽俊一等(1969，1971)在萤光假单胞菌中发现有三种纤维素酶产生，其中二种是胞外酶(A 和 B)，一种是胞内酶(C)。这三种酶形成的条件不一样，作用的方式也不一样。加 0.5%纤维素或槐糖，能强烈刺激 A 和 B 的形成(占 90%)。加纤维二糖(0.5%)，则诱导 C 酶的形成(占 90%以上)。如果降低纤维二糖浓度在 0.01%～0.05%，则也可大大促进外酶的形成，使“内型合成”转为“外型合成”。此现象不仅纤维二糖如此，D-葡萄糖和 D-木糖也如此。作者又进一步分析了三种酶在细胞内的分布，发现 A、B 分泌到胞外，而 C 在细胞内分布于细胞壁上，即称表面结合酶。其方法是用溶菌酶和 EDTA(在等渗蔗糖溶液中)处理，释放出来的就是细胞壁上的酶。现在在大肠杆菌中发现十多种酶是表面结合的，如碱性磷酸酯酶、核糖核酸酶 I、UDPG 焦磷酸酯酶。在细胞内还有一种分布在细胞质或细胞质膜上的酶称为细胞内酶，这可用超声波处理而得到。作者发现 β-葡萄糖苷酶就是属于这一类酶。因此，萤光假单胞菌利用纤维素的过程是先合成和分泌外酶 A 和 B 到介质中去，将纤维素分解成纤维二糖和纤维三糖，进入细胞，诱导出 C 酶。C 酶只作用于纤维三糖，而不作用于纤维二糖，这样将纤维三糖分解为纤维二糖和葡萄糖。纤维二糖进入细胞质内，被 β-葡萄糖苷酶作用，形成二个分子的葡萄糖而被利用(见图 1)。

3. 纤维素酶的诱导形成

纤维素是不溶性的，而纤维分解菌在含纤维素的培养基中能诱导纤维素酶的形成，那么要问：固体的不溶性的纤维素既然不能进入细胞，如何给细胞以合成纤维素酶的信息呢？或者更广泛的问：固体的不溶底物如何诱导该底物分解酶的形成？这个问题在20世纪50年代左右就提出来了，作了不少工作，厘清了一些问题，但还有许多不清楚的地方。现在，一般认为这些底物首先被细胞壁上少量的组成酶（结构酶）分解为小分子，小分子进入细胞再诱导大量的酶形成。这种情况下，底物分解产物——小分子常常起着诱导剂的作用。如在点青霉（Penicilluim chrysogenum）中半乳糖醛酸对多聚半乳糖醛酸的诱导（Phaff, 1947）；某些真菌中木糖对多聚戊糖酶的诱导（Simpson, 1954）；黑曲中麦芽糖对淀粉酶的诱导（Tanabe 和 Tennom, 1953）；烟曲霉和漆斑霉中 N-乙酰基葡糖胺对几丁质酶的诱导（Reese）等。

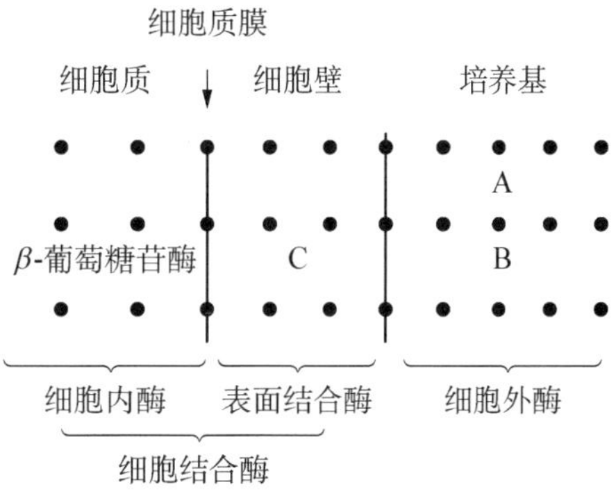

图 1　萤光假单胞菌中纤维素酶在细胞内外的分布

Mandels 和 Reese 在 1960 年左右，用绿色木霉为材料，发表多篇文章。他们认为，纤维素酶形成的天然诱导剂是纤维二糖。但是纤维二糖做诱导剂，远远及不上天然纤维素。例如球磨的棉花或木材纤维（S. F.）诱导的酶单位分别为 30 单位/毫升和 29 单位/毫升，而纤维二糖只诱导 1 单位/毫升（见表 1）。这是为什么？他们认为，这是由于绿色木霉中存在着葡萄糖效应。即葡萄糖或某些能迅速利用的碳源抑制降解酶合成的效应，又称为分解代谢产物阻抑（catabolite repression）。纤维二糖能迅速为细胞所利用，从而抑制纤维素酶的形成。其次，它对纤维素酶还有反馈抑制作用。但是，只要控制纤维二糖的浓度（降低浓度或用纤维二糖辛乙酸酯代替）或减慢代谢速度（不适温度，减少通气，无机盐过量或不足），纤解酶的产量便大大提高。前面谈到的萤光假单胞菌中降低纤维二糖的浓度也证明可大大刺激纤维素外酶的形成。他们认为当真菌在纤维培养基上生长时，先由少量的组成酶分解少量的纤维素，产生纤维二糖，后者进入细胞，诱导纤维素酶的形成。纤维二糖是纤维素分解的最终产物，它既起着诱导剂作用，又起着阻抑和抑制作用，其作用如图 2 所示。

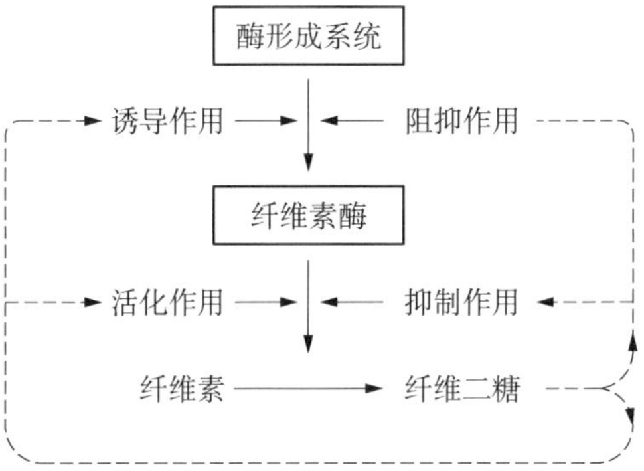

图 2　纤维二糖的作用图解

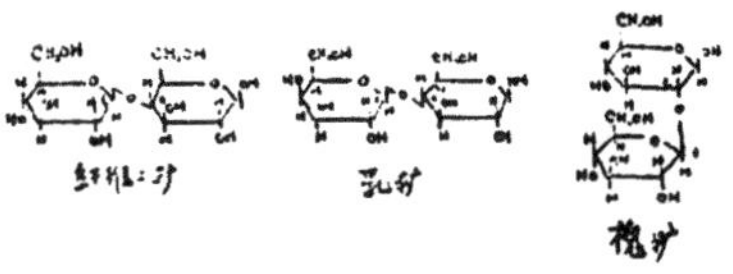
图 3　几种诱导剂的化学结构

此外，Mandels 他们发现结构与纤维二糖类似的乳糖（4-0-β-D-半乳糖苷葡萄糖）和含纤维二糖一半的三聚糖（3^2-0-β-葡糖基纤维二糖和 2^1-0-a-葡糖基纤维二糖，6^2-0-β-葡糖基纤维二糖）也是良好的诱导剂，如图 3 所示。

表1中特别引入注目的是槐糖的作用。它对纤维素酶的刺激作用比纤维二糖大得多，达到2 800单位/毫克，是目前绿色木霉最强有力的诱导荆。不过要说明的是表中的数字是离体的菌体诱导试验，培养时间短(1天)，槐糖的作用是使产酶提前，培养1天便达高峰，而纤维素做底物，产酶慢，所以两者相差就很大，时间延长，两者可接近。

表1 若干碳源对绿色木霉纤维素酶合成的诱导作用

诱导剂	酶单位/毫克诱导剂	
	C_1	(1→4)-B-葡聚糖酶
一、葡萄糖		
棉纱	9.8	5.4
木纤维(Solkafloc)	8.8	10.0
滤纸	7.0	8.5
CMC (DS0.5)	—	0.6
地衣聚糖	—	7.8
燕麦葡聚糖	—	0.8
大麦葡聚糖	—	6.6
二、寡糖		
纤维二糖	2.5	3.8
纤维三糖	—	0.6
纤维六糖	—	0.4
2^1-a-葡糖基纤维二糖	—	2.3
3^2-β-葡糖基纤维二糖	—	2.0
6^2-β-葡糖基纤维二糖	—	18.0
乳糖	1.0	4.8
槐糖	1 200.0	2 800.0
三、非诱导剂		
葡萄糖(试剂级，未纯化)	—	0.3
”(活性炭纯化)	0	0
淀粉	0.03	0.04
甘油	0.02	0.03

槐糖最初是在试剂级的葡萄糖中发现的，试剂级葡萄糖能刺激纤维素酶形成，但用活性炭纯化后就不再有刺激作用，后来发现里面混有少量槐糖杂质(0.006%)，而槐糖的量只要极微就有刺激作用(万分之几的浓度)。

槐糖只对绿色木酶和萤光假单胞菌等少数菌有诱导。

槐糖的衍生物大多数无活力，如四乙基苯亚甲基甲基 a-槐糖、四乙酰基甲基 a-槐糖、甲基-a-槐糖。唯四乙酰基槐糖稍有活力，可能体内有去乙酰酶，使它变成槐糖。

槐糖在结构上类似纤维二糖，都是二个葡萄糖接起来的双糖，所不同的是纤维二糖是β-1,4结构，槐糖是β-1,2结构。当二个葡萄糖以β1→3、β1→6结合，则完全无诱导活力。为什么β1→4，特别是β1→2活力特别高，而β1→3、β1→6无活力？现在还不清楚。

Mandels和Reese一直认为纤维二糖是纤维素酶的天然诱导剂，但是情况远远不是如此简单。原来他们发现在纤维二糖为碳源的绿色木霉培养物中(2～3天)含有一种未知的比纤维二糖强得多的诱导剂，最初分离到一个三聚体，是葡基转移到纤维二糖的6^2位置上(葡萄糖β1→6葡萄糖β1→4葡萄糖)，以后又发现这个三聚物中混有使它活性更强的诱导剂，其含量极微，用层析法测不出，有可能是2^2-β-葡糖基纤维二糖，此三糖中含有槐糖的一半，但至今纯为推测，无法肯定。这说明纤维二糖进一步转化另一个真正的未知诱导剂。

西译俊一等以绿色木霉的洗濯菌体为材料，研究槐糖的诱导作用，他们认为槐糖是真正的诱导剂，因为用双标记同位素法证明，槐糖促进纤维索酶蛋白的含成，但是他们还没有证据证明绿色木霉中确实能从纤维素变成槐糖来诱导。此外，他们发现放线菌素D和嘌呤霉素能强烈抑制纤维素酶的诱导形成，尤其是嘌呤霉素，说明纤维素酶的诱导形成包括了从DNA到mRNA的转录和从mRNA到酶蛋白的翻译过程。葡萄糖能竞争性地抑制槐糖对纤维素酶的诱导，即存在所谓的分解代谢产物阻抑，葡萄糖的作用与放线菌素D和嘌呤霉素抑制进行比较，说明葡萄糖的作用点是在纤维素酶蛋白的翻译水平，它减慢从mRNA到酶蛋白的翻译速率。

总之，在纤维素酶的诱导方面作了一些工作，但还有许多不清楚的地方：固体纤维素是否仅仅通过可溶性水解产物进行诱导？为什么槐糖只对少菌有诱导，对其他菌无诱导？真正的诱导剂是什么？诱导的步骤和过程怎样？绿色木霉纤维素酶的诱导是否服从Jacob-Monod在E. Coli中关于β-半乳糖苷酶的诱导和操纵子模型？等等。

4. 纤维素酶的分泌和表面活性剂的作用

纤维素酶诱导形成后必须分泌到体外，才能分解外界的纤维素。关于这种分泌机制还很不清楚，只知道酶的分泌是一个主动的生理过程，并非自溶释放，因为只有活细胞才能分泌，死细胞不能，用菌丝体自溶得到的酶活性很有限。此外，纤维素酶的分泌与细胞膜的透性显然有着密切的关系。许多表面活性剂能影响膜的透性，结果也影响酶的分泌。表面活性剂的作用如表2所示。

表2 表面活性剂对绿色木霉QM6[a]产酶的影响

表面活性剂	相对作用		
	在1%纤维二糖中产酶	在3%纤维二糖培养基中诱导剂的产酶	菌丝诱导测定中的产酶
对照	1	1	1
油酸钠	100	10	0
亚油酸钠	100	10	0
Tergitol NPX	1(2)	10	0
Digitonin	70	10	0
Tween 80	18	2	2

由表2可知，加油酸钠和亚油酸钠，产酶增加100倍，另外可使诱导剂含量增加10倍，Reese和Maguire认为表面活性剂的作用之一是改变膜的透性，使细胞膜失去选择性的透性作用，或者未尝不可说，让细胞发生轻微的中毒。因为细胞的选择性透性正是生命的标志之一。现在知道许多表面活性剂都有毒性，不同表面活性剂表现毒性的浓度不同。油酸钠在1～2毫克/毫升时稍有毒，digitonin是0.05毫克/毫升稍有毒，正是在这种边缘浓度，菌的生长稍稍减慢，产酶最高。但是毒性不是唯一的因素，如Tween80刺激酶分泌的浓度是无毒的。另一些表面活性剂则不管有否毒性，对产酶均无影响。另外，他们认为表面活性剂的另一个重要作用是增加细胞内诱导剂的含量，这个诱导剂是什么？不知道。它是纤维二糖的转移产物，量极微，只能用菌丝体诱导试验才能测出。作者认为表面活性剂的作用可能是使纤维二糖容易与膜-结合的酶-如转移酶接近，从而形成诱导剂，来刺激酶的合成。在细胞内纤维素酶的合成是受严格的代谢控制的。细胞内合成的纤维素酶浓度高到一临界水平时，由于产物的阻抑作用，使得酶的合成关闭，这对细胞来说是一种经济的调节方式。如果增加酶的分泌、排出、使细胞内纤维素酶的临界水平永远达不到，酶的合成就成为连续的过程，这样，促进酶释放的因素可以大大增加酶的产量。这方面，表面活性剂提供了一个有用的手段。现在至少有10种左右的细胞外酶(如蛋白酶等)的释放能被表面活性剂促进，其中最有效的是Tween80和蔗糖单棕榈酸酯。Reese等试图用表面活性剂结合诱导剂，增加产酶和缩短产酶的时间。现在QM9123二周酶活为220单位/毫升，菌体(20毫升)，加1毫克槐糖，10毫克葡萄糖和40毫克Tween40，培养26小时达到酶活72单位/毫升。作者认为改变培养条件，有希望在26小时达到或超过天然纤维素诱导的水平。

所以酶的诱导形成和分泌机制的研究，将会改革工艺过程，提高产酶，提供理论根据。但迄今对这两方面的知识还是很不完全的，有待进一步研究。

5. 纤维素酶的作用。

1) 提取、分离和纯化

目前生产纤维素酶的国家有美国、日本、西德、荷兰等。日本过去多用固体曲生产，近年来不少改为深层培养。美国是用深层培养。酶液提取后，通过透析硫酸铵沉淀，或有机溶剂沉淀(丙酮、酒精和2-丙醇等)。再通过分子筛Sephardex凝胶，DEAE-Sephadex离子交换树脂和不同型号的纤维素，电泳，可将不同的酶组分分开，或与其他蛋白分开。近十多年来，由于这些新技术的应用，使我们对纤维素酶本身有更进一步的了解。部分酶已获结晶，但目前的结晶还不纯，除含有纤维素酶外，还含有淀粉酶、聚木糖酶等杂质。迄今，还没有人获得无其他酶杂质的纯纤维素酶结晶。

纤维素酶各组分的分离步骤如下：

① 先经Sephadex G-75(分子筛)，去掉低分子量的 C_x 酶；

② 再通过DEAE-Sephadex(阴离子交换树脂)，将 C_1 酶分出来；

③ 最后通过SE-Sephadex(如C-50，是阳离子交换树脂)，将 C_x 和纤维二糖酶分开。

2) 组分和作用

纤维素酶是复合酶，20世纪50年代初Reese提出纤维素酶的作用方式：

$$\text{天然纤维} \xrightarrow{C_1} \text{直接纤维素} \xrightarrow{C_x} \text{纤维二糖} \xrightarrow{\beta\text{-葡萄糖苷酶}} \text{葡萄糖}$$

到20世纪60年代以后，应用分子筛和离子交换树脂等分离技术，分离出不同组分，这些组分如下：

组成	作用底物
1. C_1 2. 内间-(1→4)-β-葡聚糖酶 } C_x 3. 外端-(1→4)-β-葡聚糖酶 } 4. β-葡萄糖苷酶(或纤维二糖酶)	结晶纤维素 β-(1→4) 葡聚糖 纤维二糖

现在认为结晶纤维素经 C_1 酶作用，形成可起反应的纤维素，再经过内间和外端 β-1,4-葡糖酶作用，形成葡萄糖和纤维二糖，纤维二糖经 β-葡萄糖苷酶作用形成葡萄糖(见图 4)。

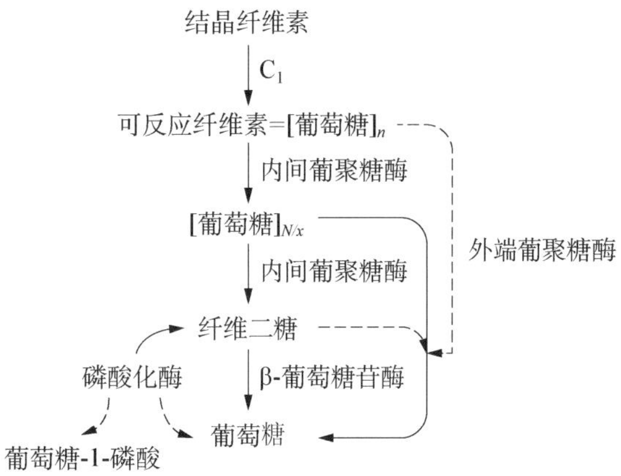

图 4　纤维素酶各组分作用

C_1 酶作用的本质还不清楚，它使结晶纤维素打开和水化，变为非结晶纤维素，使 C_x 进一步作用。在绿色木霉和 Tramates sanguinea 无细胞制剂中含有大量的 C_1 酶，目前对 C_1 酶的测定方法很不统一，有的用浊度法测定，有的是测棉花的分解或张力变化，也有的测其他种种底物的分解，因而种种不同的命名法，如使棉线张力减少的"A"酶，使滤纸崩溃的"滤纸活力"等。Mandels 和 Reese 认为用浊度法测定微晶纤维的浊度变化来代表水合纤维素酶的活力可能最接近 C_1 的性质。

C_1 酶和 C_x 酶之间有协同作用。C_1 酶在柱层析时只有一个峰(含少量 C_x 酶)，但用等电焦点法可分到二个峰，各含少量的 C_x 酶，所以 C_1 酶中总混有 C_x 酶，当用镰刀菌的 C_1 酶单独作用于棉纤维时，只有 7%溶解，C_1+C_x 可提高到 50%，而 C_1+C_x+β-葡萄糖苷酶，则棉纤维溶解可达 59%。

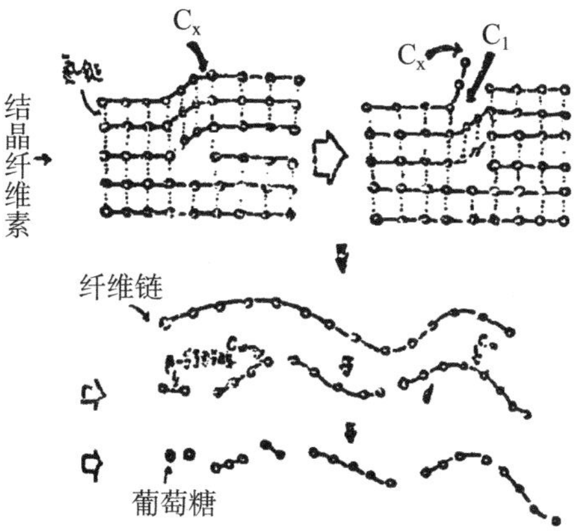

图 5　C_1、C_x、β-葡萄糖苷酶作用于纤维素

对于 C_1 酶和 C_x 酶的协同作用有人是这样解释的：首先由 C_x 酶在微纤维上一些特定的薄弱的位置把纤维分子链打断，C_1 酶接着深入到纤维分子链把氢键打开，这样，就使一段段纤维断片从微纤维上脱落下来。脱落的纤维素断片再经 C_x 酶和 β-葡萄糖苷酶作用，分解为葡萄糖(见图 5)。也有人认为 C_1 是一种亲和因子，它促使水解因子(C_x)结合到固体纤维素上，进行水解。

目前对 C_1 酶的了解仅限于假说阶段，由于纤维素高分子结构十分精细，同时 C_1 酶"固相"作用非常复杂，这无疑给 C_1 酶研究带来很多困难。

内间(1→4)-β-葡聚糖酶首先作用于较长的链，

使黏度迅速下降。内间的键易分解，二端的键较难分解。在底物溶解度的限度内(DP3－6)，随着DP的增加，水解速度也增加，它不能水解纤维二糖。纤维素的化学改变影响对酶的敏感性，一般增加纤维素的溶解度，水解速度也加快；但取代基的存在将有效防止邻近它的键的水解。取代度增加，水解速度也降低，但只要有相邻的二个葡萄糖基不被取代，就能被酶水解。因为内间酶与底物结合至少要有二个没有改变过的葡萄糖单位，至于酶作用的键则是邻近的一个葡萄糖苷键。

外端(1→4)－β－葡聚糖酶能连续不断地将单个葡萄糖从纤维素的非还原端一个一个切下。在绿色木霉中的纤维二糖酶可能属于这一类酶。它们能迅速水解内间酶作用产物4～7个葡萄糖单位的纤维寡糖，在纤维变糖中起着重要的作用。它们作用的专一性很强，对纤维素衍生物或具有混合键及分枝链的葡聚糖作用很有限。

β－葡萄糖苷酶、水解纤维二糖或其他分子较小的纤维寡糖变成葡萄糖。分解二聚糖和三聚糖很快，分子再大，水解速度就迅速下降，β－葡萄苷酶除作用β(1→4)键外，还能作用β(1→1)、β(1→2)、β(1→3)、β(1→6)键，它还有葡萄糖基转移作用，能被葡萄糖内脂强烈抑制。

C_1、C_x和β－葡萄糖苷酶也许可以再分，因为它们分别柱层析，等电焦点法式电泳可进一步分成几个峰。这些酶可能起着同功酶的作用。

3) 分子量和氨基酸组成

不同酶组分的相对分子质量作了测定(见表3)。

表3 几种菌纤维素酶组分的相对分子质量和大小

菌名	C_1	内间-(1→4)-β葡聚糖酶	外端-(1→4)-β葡聚糖酶	β-葡萄糖苷酶	相当于球形直径(Å)	相当于椭圆形长×宽=6 Å)
Irpex lacteus		10 000 50 000				
漆斑霉		5 300 30 000 55 000 49 000			55 70 65 80	30×190 35×220 35×210 40×250
Penicillium chrysogenum Penicillium notatum		32 000 35 000				
Polyporus versicolor		11 000				
康氏木霉		51 000 26 000 50 000			65 35	35×210 20×120
绿色木霉	60 000 61 000	52 000 12 600	76 000	50 000	65	35×120

绿色木霉的C_1酶可能是一种糖蛋白，相对分子质量为60 000。内间-(1→4)－β－葡聚糖酶分子量变动较大，从5 300～55 000。较小的颗粒可能是“正常”酶的亚基。外端－β－葡聚糖

酶和 β-葡萄糖苷酶的相对分子质量分别为 76 000 和 50 000。有人认为葡萄糖苷酶的分子通常大于内间酶。

对一些内间-(1→4)-β-葡聚糖酶的氨基酸组成作了分析。在某些场合，碱性氨基酸占优势，有些则是酸性氨基酸占优势。天门冬氨酸、甘氨酸、丝氨酸、苏氨酸和谷氨较多，含 S 氨基酸(半胱氨酸)通常很少(少于总量的 2%)。另一个值得注意的现象是不同的酶组分中都含有多聚糖。含量从 1%～50%，如 β-葡萄糖苷酶主要含多聚甘露糖，还有微量多聚半乳糖。这些糖与酶的不同结合，也导致了酶组分的多样性。

三、纤维底物结构与酶作用的关系

纤维素酶的作用底物是纤维素，纤维的结构很复杂，它直接影响到酶解的效率，下面先谈谈纤维的结构。

1. 纤维的结构、成分和微纤维结构的几种假设

最典型的天然纤维是棉花和木材纤维，前者是表皮细胞延伸而成。典型的木材纤维是二头尖的纤维细胞，相互间靠果胶质结合在一起，成熟的纤维细胞是死的，中间空腔，四周为细胞壁，细胞壁分初生壁和次生壁，次生壁分三层：S_1、S_2、S_3(分别称外层、中层和内层)。S_1 和 S_3 较薄，S_2 较厚(见图 6)。S_1 和 S_3 呈低螺旋，S_2 呈高螺旋。在棉花中 S_3 不明显，初生壁外有角质层，木纤维初生壁外是细胞间层。

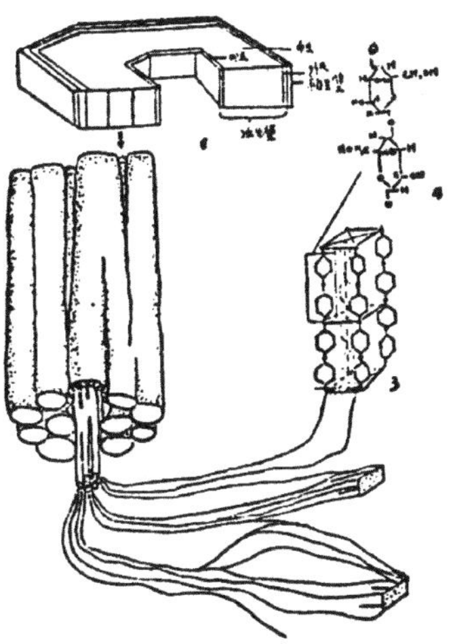

图 6　次生细胞壁亚显微结构 (A. Fahn, 1967)

①带有次生壁的一部分细胞；②表示胶束和纤维素分子排列的微纤维束；③二个纤维素单位晶格；④二个葡萄糖残基

棉花纤维很纯，90%左右是纤维素，其余为果胶质等。木材纤维中纤维素占 40%以上，半纤维素和木质素占一半以上，一般细胞从外到内是纤维素，半纤维素含量增加，相反，果胶质、木质素逐渐减少。

在次生壁中纤维素和其他细胞壁成分聚集成束状排列，称为微纤维，许多微纤维粘合在一起组成微纤维束。每个微纤维中纤维素分子间靠氢链连接，而构成不同的区域，有些区域分子排列整齐，形成结晶区，有些区域分子排列不整齐，就形成无定形区。

图 7 是微纤维三种最新的模型。a 型的微纤维横断面为 50×100 Å，中间长方形的是纤维素整齐排列的结晶蕊，它与横断面方向垂直，周围为排列不规则无定型区(称无定型鞘)所包围，结晶蕊可连续或不连续，不连续处也是无定形区。

b 型认为微纤维是由一些基本微纤维构成的，含 15～40 个纤维素分子，最小横切面为 35 Å。在微纤维的纵向上有几处是排列不整齐的，即为无定形区，其余整齐区为结晶区。

最新的假设是 c 型，认为纤维素分子先形成扁平带状，再摺合成紧密的螺旋状。以上这些模型都是根据 X 光衍射和电子显微镜观察推测的，而且纤维都是处理过的，处理方法不同，聚合体大小也不同(100～1 000 Å)，所以不同的作者得出的结果不一样，目前尚停留在假说阶段。

纤维分解菌要分解纤维首先必须将酶分泌到外界环境中，扩散进入纤维的细微毛细管中，

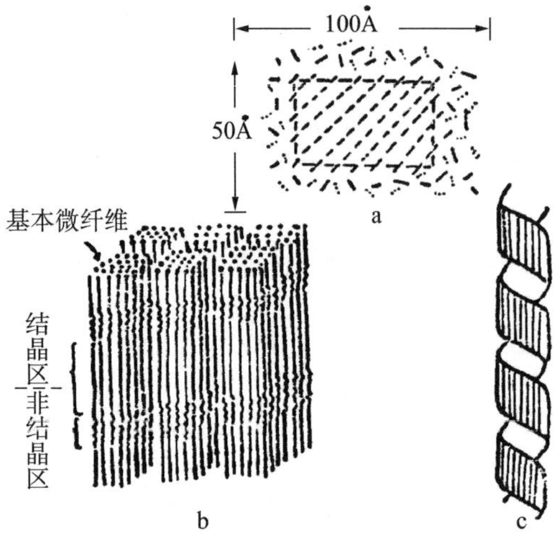

图 7 关于微纤维结构的几种假设

作用于微纤维，使酶和底物形成络合物，从而发生酶解作用，即

酶＋底物 ⇆ 酶、底物络合物 → 酶＋产物

任何影响酶的扩散和酶与底物结合的结构因素，都将影响底物对酶的敏感性，下面谈谈影响酶解作用的结构因素。

2. 纤维结构对酶作用敏感性的影响

1）纤维含水量

纤维材料最有效的防腐方法是干燥，降低含水量。水在纤维分解中很重要，它有三个作用：①使纤维膨胀，打开纤维的细微结构，以便于纤维素酶进入；②它在微生物和纤维间形成一层介质，供酶扩散；③它是纤维素水解对水分子的原料。

2）酶分子大小和扩散性与纤维毛细结构的关系

纤维素酶分子的大小、扩散性，以及纤维毛细结构的大小直接影响酶扩散的酶与底物的结合，所以与酶解作用有很大的关系。木材和棉花纤维中的毛细管腔分为二类。

（1）粗毛细管，如细胞腔、纹孔等，在光学显微镜下可看见，直径为 2 000 Å 到 10μ 或更大。

（2）细胞壁毛细管。如无定形区微纤维间和纤维素分子间的空间，这些空间充填着半纤维素，木质素和各种有机无机成分。当缺水时，大多数细胞壁毛细管是关闭的，吸水后，又重新开放，形成所谓“瞬间毛细管”。在水膨胀的木材、棉花和木浆纤维中细胞壁毛细管的平均值和最大值分别为 10 Å 和 35 Å、5 Å 和 75 Å、25 Å 和 150 Å，当全部为水饱和时，细胞壁毛细管达到最大，有的直径超过 200 Å，一般低于此值。总的来看，细胞壁毛细管的直径从几个 Å 到 100 Å。

在粗毛细管中暴露的总表面积是很大的，接近 1 平方米/克木材或棉花，膨胀后可到 300 平方米/克材料。1 克木材或棉花可容纳 3×10^{15} 个随机排列的酶分子（200×35 Å），相当于 3 毫克酶蛋白/克木材或棉花。

关于酶分子的形状，Whitaker 在漆斑霉中做了许多工作。此纤维素酶是不称的球蛋白，相对分子质量约 63 000，象雪茄烟形状，长 200 Å，最宽处为 33 Å。从相对分子质量、沉降系数、扩散系数等比较，它和血清蛋白（一种球蛋白）很相似。

从表 3 中看到一些纤维素酶的分子，没有一个直径是超过木材或棉纤维的粗毛细管的，它们很容易扩散到粗毛细管中作用于暴露的纤维素分子。此时，吸附作用是限制因子，但进入细胞毛细管较困难，两者直径接近或细胞壁毛细管还略小些，而酶分子在扩散时与其他溶质分子一样，取两种运动方式：转动和移动。这样毛细管的直径要大于酶分子的最大直径（包括运动的最大直径）时，酶才能无阻碍地前进。此外，酶不同于无机催化剂，它对底物有专一性的亲和力，所以当它通过毛细管时，与纤维素分子的亲和力也将减低酶分子的扩散速度。这些结构特点，说明了纤维分解较其他碳水化合物（如淀粉）分解慢的原因之一，以及进行预处理，使纤维毛细管结构膨胀与增强酶解的必要性。

3）结晶度和聚合度

结晶区的结晶度越高，酶解越难，当采用溶剂中再沉淀（如铜氨溶液中得到再生纤维素）、机械破坏（如振动磨）和离子辐射等方法改变或破坏结晶度时，就可增加底物对酶的敏感性。

此外，聚合度与酶解也有关系，纤维素分子长度变化很大，从每个分子含 15 个葡糖基至 8 000～10 000 个葡糖基。聚合度越高，分解越难，从图 8 中可见不同聚合度的失重情况，Poria monticola 引起的红腐与聚合度有很大的关系，从曲线延伸下来，可求得 10%纤维素易分解，此数值刚好与无定形区的含量一致，Polyporus versicolor 引起的白腐与聚合度关系略小。

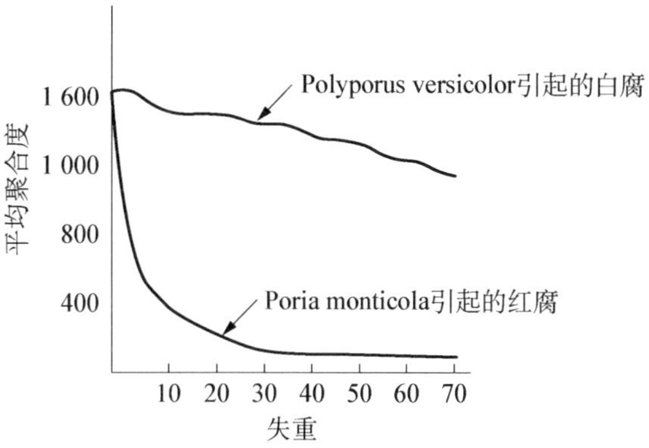

图 8　不同聚合度与失重的关系

4）结晶区单位晶格的大小

纤维素结晶有四种，即Ⅰ、Ⅱ、Ⅲ、Ⅳ。Ⅰ是天然纤维素结晶。Ⅱ是再生纤维素结晶，如黏胶纤、玻璃纸和碱膨胀的棉花，Ⅲ和Ⅳ是用无水乙胺或高温。处理得到的，四种结晶的三度空间大小不一（见表 4）。

表 4　不同纤维素结晶的单位晶格大小

轴	纤维素Ⅰ	Ⅱ	Ⅲ	Ⅳ
a	8.35 Å	8.1 Å	7.7 Å	8.1 Å
b	10.3 Å	10.3 Å	10.3 Å	10.3 Å
c	7.9 Å	9.1 Å	9.9 Å	7.9 Å
a 和 c 轴的交角	84°	62°	58°	90°

纤维素晶格的大小和形状与酶的作用也有关系，最近 Rautila 将绿色木霉分别培养在四种结晶纤维素的培养基中，发现在培养基中含某一种结晶纤维素，则产生的纤维素酶分解这结晶的活化能最低，即最易作用。可见，酶的活力中心结构是同纤维素特殊的晶格结构相配合的。

5）葡萄糖链的构型和立体强度

King 认为结晶区难分不仅因为酶不易接近纤维素分子，而且由于它的葡萄糖链的构型和立体强度。在这些区域中葡萄糖链呈所谓椅式构型，按 Thoma 和 Koshland 在 β-淀粉酶中提出的“诱导适应机制”，酶与底物作用时，不仅有酶的运动，还有底物的配合运动。在溶菌酶中发现酶反应时，底物分子要发生空间构型的变化——底物形变和所谓张力效应，当纤维素结晶度降低，水合度增加时，纤维素链上葡萄糖环的构型和各个方向调整的可能性也增加，从而提高了纤维素对酶解的敏感性。现在这种说法还仅属假设，从酶反应的立体专一性角度来看，是一个很有趣的问题。

6）纤维中其他成分的影响

（1）木质素和半纤维素：它们包围着纤维素或充填在毛细管空腔和无定形区，影响纤维素酶和底物的接触，要分解木材纤维，则需木质素酶和半纤维素酶一起参与作用，或打断它们与纤维素的联系，某些木质素的烷基-苯基、C－C键可被漆酶一类的多酚氧化酶分解，这一裂解在某些自腐菌分解天然木材时起着重要作用。当用碱处理，除去木质素和半纤维素时，也可大大提高酶解效率。

（2）无机成分：纤维中约含1%灰分，它包含纤维分解菌生长所必须的一切成分。故有人设想用螯合剂（chelating agents）除去灰分以防止棉花、木材的腐烂。一般来说，汞、铜、银、铬和锌均具有抑制作用，而镁、钴、钙、锰有刺激作用。

（3）可抽提物质：指溶于中性溶剂如丙酮、乙醚、乙醇、苯和水中的物质，它们的影响有以下几个方面：

（a）生长刺激物质，如维生素（尤其是B_1）和某些供生长的可溶性糖。

（b）有毒物质，特别是酚类，抑制菌的生长和发育。

（c）有些物质堆积在细胞壁的精细毛细管结构中，不利于酶对纤维素分子的接近。

（d）某些纤维素酶的天然抑制剂，直接抑制酶水解的速度和强度。有不少植物能抵御纤维分解菌的作用就是因为含有天然抑制剂，主要是酚类化合物，如丹宁和白色素，在500种被试验植物中，有17%每克干材料中含10个单位以上的抑制剂，分布部位很广，从叶、木质部到花、果实、种子中均有。

日本用缩合丹宁处理绳索防止腐烂，用柿子汁处理纸防腐，我国西南地区制纸伞时涂青柿子汁，恐怕也都是利用柿子中天然抑制剂的作用。

7）取代基团的影响

纤维素分子的羟基氢被甲基、乙基、羟乙基、羧甲基取代，形成纤维素衍生物，这些基团取代上去后，使纤维素结晶度下降，溶解度增加，一般取代度在0.5～0.7时全部溶解，这时酶解最快，取代度再大，就会影响酶和纤维素的结合，降低酶解，直至不能分解。现在国外想用高取代的纤维素来防腐，不过，这会产生料想不到的麻烦，因为污物中的纤维废物将无法分解而造成“纤维”灾害。

总之，纤维素对纤维素酶的敏感性，在很大程度上取决它是否能与微生物的酶相接触。酶和底物的直接接触是水解的先决条件。任何影响或限制纤维素酶接近纤维素的结构特征，都将降低纤维素酶酶解的敏感性。

根据以上纤维素酶本身和底物结构二者对酶解作用的影响。下面谈谈国外提高酶解效率的几个途径。

四、国外提高酶解效率的途径

1. 酶活力方面

1）菌种筛选诱变

（1）T. V. QM_6a 曾是国外活力较高的菌种，最适条件下生长7～10天，每毫升培养液活力为 100Cx 单位。最近他们用高能电子照射,从中选出一株 QM9123(0.05 百万 R),酶活比母株提高 1 倍。

(2) 日本外山信男通过诱变得到黑曲 T_5,既具有纤维素酶活力,又具有细胞分离酶活力,其纤维素酶活力与绿色木霉一样。

2) 改变培养条件

寻找最适培养条件,除了一般最适培养基、温度、pH 外,还寻找诱导剂和添加各种生长促进剂,发酵时加蛋白胨(0.1%~0.2%)、维生素 B_1 或棉子粉(0.25%~1%)常常会增加酶的产量。在固体纤维素为碳源的培养基中,加少量可溶性糖(如葡萄糖)也能促进产酶。Cupla 等固体培养基中加醋酸(0.1%)和抗坏血酸(0.1%),可提高产酶 1 倍左右。在固曲生产纤维素酶时采用麸皮粕、油粕等提高产量,像这类添加附加物以提高产酶的例子很多。另一值得注意的动向是添加表面活性剂,前面谈过的 Reese 等用油酸钠和亚油酸钠使纤维二糖中培养的绿色木霉 CMC 酶活提高 100 倍,产酶时间大大缩短,不过还没有超过在天然纤维素中培养的活力,他们的目的是要使活力达到和超过天然纤维素,而培养时间大大缩短(1 天)。在发酵技术方面,Mandeis 等采用半连续发酵法生产纤维素酶,10 立升培养液培养 6 天后先收集 2 立升酶液,并补加 2 立升新鲜培养液,这样连续补料和取酶。而且他们设计了一个发酵糖化连续系统,从发酵糖液来培养酵母。美路易斯安那大学用甘蔗渣生产缅菌蛋白时采用共生发酵法,即纤维杆菌(生产菌)中再加一只产碱扦菌,后者分泌大量的 β-葡萄糖苷酶、分解纤维二糖,使纤维二糖不会累积,解除对纤维酶形成的阻抑作用,结果细菌生长快 5 倍。

3) 改革酶提取工艺

过去酶的提取和纯化都用硫酸铵或有机溶剂沉淀。日本外山信男改用丹宁沉淀法,可提高酶活,而且这种方法不要用大量的有机溶剂和硫铵,工艺简单,价格便宜。其大致方法是先加丹宁酸($C_{76}H_{52}O_{45}$,相对分子质量 1 701.25),使丹宁酸和酶结合,形成沉淀,再加水溶性聚乙二醇(PEG),或聚乙烯吡咯酮、胶等回收酶。他们将复合物干燥制成酶制剂,使用时悬浮在 PEG 等缓冲液中。后来,他们将粗制的丹宁-酶复合物先用低聚合度(平均相对分子质量 200)的 PEG 水溶液处量,以除去杂质,再用高聚合度(平均相对分子质量 4 000 或 6 000)的 PEG 水溶液处量该复合物,离心得酶液。此方比高聚合度 PEG 水溶液一次提取效果要好,纯度要高。也可将丹宁酸-酶复合物用高聚合度的聚乙二醇和聚乙烯醇等几种配合使用,以促进酶活。绿色木霉麸皮提取液用本法(复合体干燥物加等量的 PEG4000)使酶活提高 1 倍,分解红茶、绿茶的效果较好。如红茶残渣对照为 7.1 克,加酶为 7.0 克;加酶和聚乙烯醇为 3.3 克。绿茶对照为 6.1 克,加酶为 3.2 克,加酶和明胶为 2.1 克。此法现已用于淀粉酶、蛋白酶、脂肪酶、果胶酶、糖化酶等的提取工艺。

2. 纤维底物方面

前面已谈过,纤维原料由于其结晶度,和半纤维素、木质素等的混杂,均影响对酶的敏感性。故改变底物晶体结构,使结构膨胀、松散和去半纤维素、木质素的任何处理终将增加酶介的敏感性。目前国外预处理的方法主要有以下方法。

1) 酸碱预处理

碱预处理是目前较有希望的方法,碱处理可使纤维膨胀,去结晶,和去木质素、半纤维素。美国路易斯安那大学研究用甘蔗渣生产细菌蛋白,未经碱预处理的甘蔗渣分解率一般在 15% 以下,经碱处理后可使分解达 80%~90%。

酸法和酶法结合是提高酶解的另一途径，Padilla 和 Hoskins 先用 25% H_2SO_4，50℃处理木屑 3 小时，中和，再加 1.25%酶液（米曲霉的纤维素酶 Rohm Haas 出品）作用 3 小时，从 100 克木屑得糖 10.8 克（其中葡萄糖 4.1 克，木糖 6.7 克）。

外山报导（1969）日本已研究出去除木质素的新技术，用绿色木霉 TVC15 可使去木质素的纤维（纯白制品）90%以上分解，再结合黑曲纤维素酶（ANU_2）正在试验将纤维废料如树皮、木屑、废材、秸杆、甘蔗渣等变糖和做饲料，但看来成本太高。

美国和加拿大现已在用碱处理的木屑等纤维，如 NaOH、NH_3（或再加尿素）去木质素，直接作为反刍动物（牛、羊）的饲料，利用反刍动物瘤胃中的微生物分解纤维素，开辟饲料新来源，目前研究很热闹，看来这种方法是解决饲料新来源一个很有希望的方法。

2）高温热磨

碱处理后的纤维虽然敏感性增加，但因膨胀，黏度增加，底物浓度一般无法超过 10%，即使全部转化为糖，糖浓也有限（不超过 10%）。Reese 在小规模试验中用高温热磨法，增加底物浓度，将亚硫酸木浆粕先研磨 70 小时，再用别的磨粉机在 220℃研磨 30 分钟，制成 50%悬浮液，用绿色木霉纤维素酶加 β-葡萄糖苷酶，pH4.5，40℃，作用 5 天，糖浓达 24%，14 天达到 32.5%。现在美国有的试验工厂用底物高温热磨法，再用绿色木霉纤维素酶水解，可使葡萄糖浓度达 10%，经过热磨后，纤维颗粒大大减少，酶解接触面增加，而且破坏晶体结构。

五、结束语

纤维素酶在国外已有 30 年左右的研究历史。在理论和应用方面均做了大量的工作，尤其是近十年来，通过离子交换层析、电泳等分离技术的运用，对酶本身的性质、作用和诱导有了较多的了解，通过筛选和诱变以及培养条件的改变得到活力较高的纤维素酶。美国一研究所，20 多年来集中相当一部分的研究人员以纤维糖化为目标，从事纤维素酶的研究，美路易斯安那大学与国家航行、空间管理局，以及固体废料管理局合作，从甘蔗渣试制细菌蛋白做饲料，1970 年开始中间试验，生产出 300 磅蛋白，价格只有 0.12 美元/磅，他们期望在 5 年中能生产供人食用的蛋白质。据 Das 和 Ghose 的计算，采用浓缩酶和分子筛膜分离器等新工艺进行纤维连续糖化，设想每日加工 100 吨废纤维的厂子，可生产 43.6 吨葡萄糖和 10.9 吨纤维二糖，每公斤葡萄糖的成本仅 0.171 美元，厂规模越小，成本越大。据认为，目前设备上还存在问题。日本这十多年进展也较快，工作较全面，在酶制剂生产和应用上领先，但在应用上限于食品加工、医药等，最近外山提到他们也将以纤维变粮食为他们的奋斗目标。看来，国外现在的趋势都想从纤维变糖或蛋白，以解决粮食和污物处理问题。但是，从目前的菌种活力水平来看，要解决这样的大问题，还存在较大的困难，尤其从纤维糖化再培养酵母这条路子来解决粮食问题困难较多，直接利用细菌蛋白也许是较快和较有希望的途径。要解决纤维变粮的问题，今后还必须在菌种、培养条件以及酶的作用机制等方面作进一步的工作。

参考文献

[1] 外山信男世界为におけるセルラーゼの应用，化学と生物（1969），7，10：630
[2] 外山信男最近のセルラーゼ研究の进步[综说]，发酵工业杂志（1969），47，11：714
[3] Cowling，E. B.，Structural features of ce-lulose that influence its Susceptibiiity to enzyma-tic hydrolysis.

in Aduances *in Enzymatic Hydrofysis of Cellulose and Related Materia's*, ed. Reese, E. T., Pergamon Press, London. (1963)
[4] Das, Kand Ghose. T. K., Economic evalation of enzymic utilization of waste cellu losicmatorials, *J. Applicd Chemistry and Biotchnology*, (1973)23,11:829.
[5] Jurasek, L., Ross Coloin J. and Whitaker, D. R., Mocrobio'ogicai Aspects of the Formationand Degradtion of Cellulosic Fibers. *Aduances an Applied Microbiology*, (1967),9:131.
[6] Nisizawa, T., Suzuki, H., Nakayama, M. ind Nisizawa. K., Inductive Formation of Cel-lulase by Sophorose in *Trichoderma viride*, *J*, *Biochcm.*, (1971),70,3:375
[7] Nisizawa, T., Suzuki, H. and isizawa, K. Cataboite repression of Cellulase Formation in *Trichoderma virdie*, *J. Biochcm.*, (1972)71,6:999

纤维素酶研究进展*

纤维素是地球上最丰富、最庞大的可更新资源之一，与人类的生存有着密切的关系。据估计，照射到地球上的太阳能约有0.1%被绿色植物通过光合作用而固定，每年可净产生植物有机物(15～20)×10^{10} 吨，其中纤维素便占了$\frac{1}{3}\sim\frac{1}{2}$。纤维素是植物细胞壁的主要成分，同木质素、半纤维素、果胶质一起构成植物的"骨骼"，起着保护和支撑的作用，但究其化学组成，却是由最简单的葡萄糖分子组成。可惜，至今纤维素的利用只限于造纸、纺织、木材和做化学试剂，部分作为反刍动物的饲料，大部分的纤维素主要作为农业废物(如稻草、稻壳、麦杆、花生壳、玉米蕊、棉子壳、甘蔗渣等)、食品加工废物(果皮、果渣等)、木材废物(木屑、树皮)，及城市废物(40%～60%固体废物是垃圾和废纸)等而被浪费掉了。据统计，全世界每年纤维素产量是10^{11} 吨，这些庞大的，每年更新的光合作用产物——纤维素，虽然无法为人类和大多数动物食用，但如果能用微生物方法，作为发酵原料，转化为葡萄糖和单细胞蛋白等，这将为人类的食品开拓新的取之不尽的来源，从广义上说，这也是太阳能利用的问题。鉴予以上原因，世界各国对纤维素资源作为新的能源十分重视，纤维素酶的国际会议日益频繁。下面罗列这十多年来纤维素酶专业会议的情况。

表1列出了纤维素专业会议，足见20世纪70年代以来，随着世界能源的短缺，对于纤维素作为新兴的可更新能源利用的重视。

表1　纤维素专业会议列表

年份	会 议 名 称	地点
1962	纤维素及其有关物质酶水解的进展	美国[1]
1968	纤维素酶及其应用	美国[2]
1972	第四届国际发酵讨论会	日本[3]
1974	纤维素作为化学和能源讨论会	美国[4]
1915	日本发酵协会年会	日本[5]
1975	纤维素酶水解讨论会	芬兰[6]
1975	木材的微生物生物转化	美国[7]
1976	纤维素酶对纤维素的分解机理	美国[8]
1977	纤维素生物转化为能量、化学试剂和微生物蛋白的国际讨论会	印度[9]

* 朱雨生，中国科学院上海植物生理研究所，此文发表于1978年"应用微生物"1期29－40页。

纤维素酶由于涉及酶对固体纤维素的催化作用，酶蛋白诱导形成等酶化学和分子生物学的基本问题，所以它也是研究现代生物学的基础理论问题一个理想的材料。纤维素酶在自然界的分布十分广泛，从细菌、真菌、某些无脊椎动物，到高等植物都有，故具有普遍的生态意义。在反刍动物中涉及瘤胃和纤维分解细菌及原生动物之间的共生关系；在许多植物罹病过程中（包括木材腐朽）涉及植病发生的问题；值得注意的是，纤维素酶广泛分布在高等物中，与植物许多重要的生理过程，如生长、分化、激素作用等有着十分密切的关系。最近，Verma 等报道，纤维素酶是高等植物中第一个被提取出它的 mRNA，并完成体外翻译的酶[10]。植物的细胞壁及其纤维素的化学组成是区别于动物的重要特征，因此，纤维素及纤维素酶的生物进化也是一个饶有趣味的问题。

我国国内从事纤维素酶的研究只有 10 多年的历史，曾受到各方面的重视，在菌种[11]和培养、应用等方面曾取得一定的进展。纤维素酶已成功地应用于植物细胞脱壁试验应用于酒精发酵提高出酒率[15]、助消化剂试制、糖醛渣生产饲料酵母[16]、化纤降聚和酶法打浆[11]和粗饲料分解等工作中。但由于底物结构的复杂和菌种活力的限制，目前的应用大多停留在试验阶段。而且，基础理论研究的薄弱也限制了应用工作的发展。作者对国外纤维素酶研究概况曾作过一般性的评述[17]。本文进一步就近几年来的资料，尤其是基础理论方面的研究进展，加以扼要的评论。

一、纤维素酶的作用机理

1. 纤维素酶的组分

纤维素酶发现至今已有 60 余年的历史，但是，纤维素酶从化学上加以研究，只是近十多年来，由于应用了葡聚糖凝胶、离子交换树脂、电泳和等电聚焦等现代酶分离技术，得到高度纯化的酶制品后，才有可能。在纤维素酶发现后的四十多年中，大量的工作是关于纤维素酶的生物来源和对底物作用的研究。到 20 世纪 50 年代早期，Reese 等提出了纤维素酶 C_1-C_x 的概念[18]，即天然（结晶）纤维素先经 C_1 酶作用变成直链纤维素，再经 C_x 酶作用变成纤维二糖，最后在 β-葡萄糖苷酶作用下，形成葡萄糖。虽然，这个假设曾受到不少作者的支持，还有人提出氢键酶、膨胀因子等等假设，但由于用于研究的纤维分解菌不分泌 C_1 酶，故 1950—1964 年均集中于 C_x 酶的研究。1964—1965 年开始，由于以绿色水霉和康氏木霉为材料，采用了高分辨力的酶分离技术，真正分离到 C_1 酶，从此，对纤维素酶的研究有了迅猛的发展，使我们对 C_1 酶的木质以与 C_x 酶作用的关系，开始有了一些初步的了解。

据 King 报道（1965），绿色木霉分解纤维素需要四个酶：C_1、内切-葡聚糖酶、外切-葡聚糖酶和乙酶（一种热稳定的葡聚糖酶[19]。

Selby 等（1967）根据 Reese 的假设，分离 C_1 组分，将康氏木霉的纤维素酶分成三种：C_1、C_x 和纤维二糖酶[20]。C_1 酶得到提纯，C_1 和 C_x 间有使棉花溶解的协同效应。当康氏术霉和绳状青霉的 C_1 和 C_x 酶交互重组时，同样获得协同效应。

Wood（1969，1972）在链霉中也将纤维素酶分成三种：C_1、C_x 和 β-葡萄糖苷酶，同时将 Selby 的 C_1 酶进一步用等电聚焦电泳纯化[21]。当它同 C_x 一起时可溶解脱脂棉。它不能显著作用 CMC 或结晶纤维素，但能单独作用磷酸膨胀的纤维素，除了生成少量的葡萄糖外，只生成纤维二糖。这个 C_1 组分是从非还元端的末端逐个切下纤维二糖。Wood 认为 C_1 酶是一种特殊类型的水解酶，称它为β1，4-葡聚糖纤维二糖水解酶。Selby和Wood都主张，木霉纤维素酶组分为二种：内切型和外切型（相当于C_x酶和C_1酶）

与此同时，日本西泽俊一等[22]和瑞典的Pettersson等[23]（1972—1974），以绿色木霉纤维素酶制剂Onozuka为材料，得到高度纯化的微晶纤维素酶（分解微晶纤维素产生纤维二糖）和CMC酶（随机作用CMC，使黏度低），二者也有协同作用，分别相当于Wood的外切-纤维素酶和内切纤维素酶。

中国科学院微生物研究所纤维素酶组（1976）在绿色木霉X_2-85的麸曲提取液中分离得到纯化的C_1酶[24]，用聚丙烯酰胺凝胶电泳和超离心鉴定，都为均一蛋白，它对CMC、β-葡萄糖苷和纤维二糖都不表现活性，但能分解微晶纤维素、磷酸膨胀纤维素和脱酯棉，主要产物是纤维二糖。看来，也是支持Wood的C_1酶是葡聚糖-纤维二糖水解酶的观点，并发现C_1酶似乎有聚集作用，推测聚合作用与巯基有关。

我组在拟康氏木霉的培养滤液中也分离到C_1、C_x和β-葡萄糖苷酶的活性，相互有协同作用（未发表）。

所以，现在一般趋向认为纤维素酶至少由以下成分组成（1975）[4]：

（1）内切-β1，4葡聚糖酶（老的C_x酶）（E. C. 3. 2. 1. 4）：它有作用随机度不同的好几个成分，其中之一也许首先作用于"结晶"纤维素。已经确定比活性>60单位/毫克（底物为CMC，SOT），测定方法是CMC黏度下降。

（2）外切-β1葡聚糖酶，又分为①β1，4葡聚糖葡萄水解酶：是从纤维素酶链的非还元端逐个切下单个葡萄糖分子。鉴定方法是以CMC为底物测定形成的葡萄糖；②β1，4葡萄糖纤维二糖水解酶（CBH）（即老的C_1酶）：它从纤维素酶的非还原端，逐个切下纤维二糖。对纤维素的亲和力很大。以磷酸膨胀纤维素为底物，测定形成的纤维二糖。比活力是毫克。

（3）β-葡萄糖苷酶（纤维二糖酶）（E. C. 3. 2. 1. 21）：作用于纤维二糖的比活力是70单位/毫克。

2. 纤酶素酶的性质及作用方式

关于纤维素酶结构的研究很少。葡萄穗霉的纤维素酶由14种氨基酸组成[25]。Whitaker等在漆斑霉中得到的纤维素酶物理特性如下：扩散系数为$(5.6\pm0.1)\times10^{-7}$；沉降系数为$(3.7\pm0.05)\times10^{-3}$内黏度为$0.087\pm0.001$；相对分子质量49 000[26]。流体力学性质测定，证明酶蛋白是扁平的，轴比5～10，推测酶分子呈雪茄烟状，长约200 Å，宽约33 Å。如果它以同样方向附着在伸长的纤维素分子上，则可盖着约20个纤维二糖分子。C-端是甘氨酸，测不到N-端氨基酸。Eriksson等研究点青霉电泳均一的β-内切葡聚糖酶[27]，相对分子质量为78 000，氨基酸组成已确定，此酶由一条肽链组成，肽链内部靠二个半胱氨酸残基连结，形成一个硫桥，它可被汞离子强烈抑制，加半胱氨酸可恢复活性，说明残基与活性有关。它不被外肽酶分解，表明其结构年固。但可用内肽酶使之失活。Ikeda等从黑曲中得到纯化的酸性纤维素酶，最适pH 2.3～2.5，沉降系数3.27S，相对分子质量46 000，在水中和8 mol/L尿素溶液中a-螺旋含量11%，酶分子含氮14.37%，由378个氨基酸组成，每个蛋白分子含12个葡萄糖胺残基和10个阿拉伯糖残基，N-端的氨基酸残基是封闭的[28]。纤维素酶大多为糖蛋白，二个极端的例子是，绿色木霉C_1酶含糖高达50%，而绳状青霉的C_1酶无糖。糖与蛋白可共价结合，或呈可解离的络合物，并导致了纤维素酶的多样[29]，但是否是同功酶，看法尚不一致。一般C_1酶相对分子质量在50 000左右，C_x酶相对分子质量变动较大，从10 000多至70 000

多，平均的也 50 000 左右。

纤维素酶的作用条件因菌种和作用的底物而不同。真菌纤维素酶的作用 pH 的酸性，一般在 4～5 之间。康氏木霉分解 CMC 最适 pH 为 4.5，分解纤维为 pH 5.0[29]。拟康氏木霉分解 CMC 最适 pH 是 4.4，分解滤纸是 4.8，作用的最适温度分别为 60℃和 50℃。在最短 pH 条件下，酶的耐热性增高[12]。康氏木霉在 70℃以上[29]，拟康木霉在 75℃以上[12]，纤维素酶开始失活。

关于纤维素酶作用的方式，过去流行先 C_1 酶作用，后 C_x 酶用的观点。近年来，C_1 酶高提纯的成功不仅对 C_1 酶和 C_x 酶的概念作了修改，而且对它们作用的方式也提出了不同的看法，即主张先由 C_x 酶随机作用天然（结晶）纤维素链的薄弱点，提供许多可反应的链端，然后由 C_1 酶从纤维素分子的非还元端，逐段切下纤维二糖，后者最终后经 β-葡萄糖苷酶作用形成葡萄糖[29]，反对 Reese C_1 概念，提出先 C_A 酶后 C_1 酶作用的理由如下。

（1）C_1 不能单独分解天然纤维素（或分解很少）只有同 C_x 一起时，才能溶解天然纤维素（如脱脂棉）。

（2）C_1 作用于天然纤维素（如棉花）后，用电镜（包括扫描电镜）和红外光谱分析，均未发现结构有明显变化。用 C_1 处理后的为品纤维素的结晶度反而略有增加。同时，仔细干燥的溴化钾圆片（Disc）也表明 C_1 处理后氢键没有变化。

（3）纤维素酶作用天然纤维素的早期物理变化（如断片化即短纤维形成，张力损失，膨胀因子）都与 C_x 有关，与 C_1 酶无关。

（4）先 C_1 作用，后 C_x 上作用分解棉花的程度远不如先 C_x 后 C_1 作用的为强。

因而认为现在是应该抛弃 C_1 酶是水解功能酶的概念，用新的概念来重新规定纤维素酶作用机制的时候了。

西泽俊一等以 CMC 酶和微晶纤维素酶，来代表 C_1 和 C_x，提出先 CMC 酶，后微晶纤维素作用的类似的观点[21]，如图 1 所示。

3. 纤维素酶作用的化学模拟

Halliwell 曾发现棉纤维经 0.4% H_2O_2 和 10.2 mmol/L $FeSO_4$ 作用，pH 4.2～4.3，在无菌条件下，几天后断裂成很短的纤维，在这种 $FeSO_4$-H_2O_2 系统中，棉花溶解的过程同绿色木霉滤液中得到的结果类似，其他亚铁盐和铁盐同样有效[30]。他指出 H_2O_2-$FeSO_4$ 系统也许代表着自然界纤维素酶水解作用的机制，可能类似于织物的光解。许多微生物具有形成 H_2O_2 的能力（如通过葡萄糖氧化酶的作用），尤其在 Fe^{++} 存在时，能作用结晶纤维素，从而引起类似酶水解的变化（如碱膨胀、碱溶性、聚合度降低），H_2O_2 可能同 C_x 联系起来作用，提出这可能是某些微生物利用纤维素的方法。Koenigs 也指出，某些木材腐朽真菌并不分泌 β-1,4 葡聚糖纤维二糖酶，但分解纤维素能力很强[31]。早

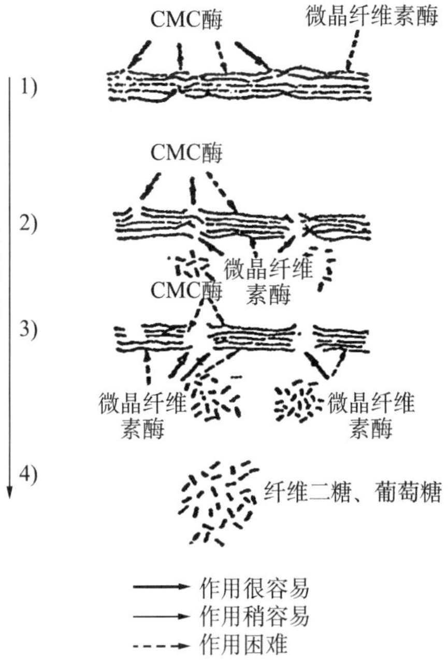

图 1　CMC 酶和微晶纤维素酶作用图谱

期发现这些真菌在葡萄糖培养基中能形成 H_2O_2，可分解多糖，包括降解纤维素。H_2O_2 和 Fe^{++} 能引起棉花的降聚，增加棉纤维对绿色木霉纤维素酶的敏感性。H_2O_2 和 Fe^{++} 系统有三个普遍的特征。

（1）引起纤维素强烈的降聚和碱溶性的迅速增加。

（2）H_2O_2 和 Fe^{++} 对于分子内位置的攻击受某些因素的调节，从而使之带有一定程度的专一性。如 H_2O_2/Fe^{++} 比率高时，纤维素中葡萄糖残基的 C_6 -碳氧化为糖醛酸；低比率时，则 C_2 和 C_3 碳氧化为羧基，最后吡喃环裂开。反应性质也受 pH 调节。

（3）从生态观点出发，此系统也是合理的，因为反应是在温和的温度和 pH 下完成，且产物无毒性。

最近，Reese 在一篇总结中强调：许多纤维分解菌显然不同于木霉，很少或不分泌纤维素酶，使得继续探索纤维素分解的其他方式变得迫切了。H_2O_2 及过氧化物和阴离子自由基是今后可循的二个方面[4]。

这类纤维素酶作用模拟的工作是很有意思的，当然对于阐明纤维素酶作用的机理以及生产应用上的意义，还有待实践的考验。

回顾以上纤维素酶作用模型的种种假设，看来 C_1、C_A 的观点还是为大家普遍接受的，但 C_1、C_x 的木质和作用方式赋予了新的概念。同时。仍有不少问题尚待解决，如：①纤维素酶进一步纯化和结晶，是否由亚基组成？②酶蛋白-级结构和高级结构的确定。③活力中心是什么？④在底物表面和酶分子间是否存在物理性络合物，有多少明确的多肽是处在络合状态？⑤为什么纤维素酶大多是糖蛋白？其中碳水合物是稳定酶的构型，还是起别的什么作用？⑥内切- β1，4 葡聚糖酶（C_x）常有大小、电荷、组成不同的多个成分，相互是什么关系？有什么生理意义？⑦纤维素酶作用的模拟？H_2O_2-Fe^{++} 系统的生理、生化及生态意义？只有回答以上这些问题，才能真正解开纤维素酶结构与功能关系之谜。

4. 纤维素酶的抑制剂

纤维素酶被纤维二糖和 Methocel 完全抑制[29]，某些蛋白试剂如卤素、重金属和去垢剂也可使之失活。某些作者报道作用于游离—SH 基的抑制剂也抑制纤维素酶[27]。产物的抑制随菌种而不同。如乳糖对细小青霉纤维素酶有抑制，对其他菌不抑制。但纤维二糖对许多菌种的纤维素酶都是良好的抑制剂。葡萄糖的抑制作用较弱，30％浓底的葡萄糖对绿色木霉的纤维素酶只抑制 40％[32]。β-葡萄糖苷酶被葡萄糖内酯抑制。纤维素-纤维素酶系统的产物抑制的研究还证明，较大的可溶性寡聚物抑制最大，随着产物分子的减小，抑制迅速减弱[29]. 甲基纤维素也抑制绿色木霉和漆斑霉培养物对 CMC 的分解力[33]。但对其他菌中的纤维素酶不太敏感。

无机离子如 Ag^+、Mn^{2+}、Zn^{2+}、Cu^{2+} 及硼酸钠、硫代硫酸钠对绿色木霉起着抑制剂作用，而 NaF、NH_4OH、HCl、Cu^{++}、Co^{++}、Na^+、Mg^+、Cd^{2+} 起着激活剂作用[32,34]。

绿色木霉的纤维素酶也被某些蛋白水解酶（如胰蛋白酶、无花果蛋白酶）所抑制[29]。对绿色木霉、漆斑霉和 Pestalotiopsis weterdijki 纤维素酶起抑制作用的天然抑制剂，主要属于酚类、丹宁和多聚花青素（花白素）。许多植物的叶、木质部、花、果实和种子中均含有天然抑制剂。在柿子未成熟果实、Metrosideros Phlgmorp 木质部、截叶铁扫帚叶、赤按木质部、葡萄（Vitisrotundifolia）叶、月桂树叶、果实和技条，以及番石榴中含量均较高，可能与植物抗病有关[35]。

5. 纤维素酶作用的动力学

天然纤维的结构非常复杂[17]，除了含有纤维素外，不含有半纤维素、木质素和果胶质。同时，纤维素酶又是复合酶，至少包含 C_1 和 C_x 酶，所以纤维酶作用于天然纤维的动力学模型至少由二种底物(结晶型和无定型纤维素)和至少一种杂质(木质素)及二个酶(C_1、C_x)组成。即使单纯的纤维素种类也很多。表 2 例举了几种纤维素酶作用底物的种类和聚合度[36]。

表 2 几种纤维素酶作用底物的种类和聚合度

种类	例子	聚合度
结晶纤维素	水合纤维素、微晶纤维素	100～200
天然纤维素	棉花，脱脂棉	8 000～100 000
“纯化”纤维素	滤纸，纤维素粉	200～700
无定形(膨胀纤维素)	碱化纤维素(25% NaOH 中膨胀)	4 000
	Walseth 纤维素(85% H_3PO_4 中膨胀)	370～2,140
纤维素衍生物(可溶性)	翔甲基纤维素(CMC)甲基纤维素	
纤维寡糖	纤维二糖至纤维六糖	2～6

纤维素水解程度随着酶和纤维素接触时间的平方根而变化，水解速度随结晶度增加而下降。纤维素水解的速度受吸附速度或酶分子渗透的限制。在破坏的结晶底物表面和可接触表面的平方间存在着函数关系

$$\frac{dA}{dt}=a(S_0)^2 \qquad (1)^{[29]}$$

式中，A 为被破坏底物表面，S_0 为可接触底物表面，t 为接触时间，a 为表面去活化的速度常数。

这表明，酶作用不单单是一个表面现象，但在反应物，由于侵蚀和断裂，总的粒子数显著增加。

根据对纯的和天然纤维素糖化的研究指出，纤维素最初的酶解是假一级反应，所用下式表示

$$t=\frac{1}{K}\ln\frac{S_0}{S_0-S} \qquad (2)$$

式中，t 为反应时间(分)，K 为速度常数(分$^{-1}$)，S 为纤维素浓度(克/毫升)，S_0 为最初纤维素浓度(克/毫升)。

根据稻壳-纤维素酶系统估计，活化能的数量级为 14 870 卡/克分子，是在液相中扩散，或化学反应的数量级范国内[29]。

由于产物的抑制，使得反应动力学复杂化。根据对于磷酸膨胀纤维素所作的无定形纤维素水解动力学的研究，提出了一个酶一种底物作用的反应式[37]：

$$v=\frac{\mathrm{d}(p)}{\mathrm{d}t}=\frac{K_2X_{1m}K_1(E_0)(S)}{1+K_3(P)+K_1(E_0)+K_1X_{1m}(S)} \tag{3}$$

式中,V 是反应速度,S 和 P 分别为底物和产物浓度,X_{1m} 是酶和底物形成络合物的饱和浓度,K_1 代表酶和底物形成络合物的速度常数比$\frac{k_1}{(k_{-1})}$,K_2 是酶和底物络合物形成产物的速度常数,K_3 是酶与产物形成非活性络合物的速度常数比$\frac{k_3}{(k_{-3})}$。可见,纤维素水解速度与底物浓度、酶浓度、底物与酶形成络合物的饱和浓度,以及酶与底物反应形成产物,和产物与酶结合的反应速度常数有关。在同时具有结晶和无定形纤维素的底物中,反应动力学就更为复杂,得到平均反应速度公式如下:

$$\frac{(P)}{t}=\frac{K_2X_{1m}K_1(E_0)}{X_{1m}K_1-K_3}-\frac{1+K_1(E_0)+K_3(s_0)}{X_{1m}K_1-K_3}\left\{\frac{1}{t}\ln\frac{(S_0)}{(S)}\right\}$$

这里 S_0 为起始底物浓度,$S=S_0-P$

纤维素酶作用动力学的研究将有助于纤维素酶作用机理的阐明和糖化工艺的合理设计。这类报道开始在文献中增多。

二、纤维素酶的诱导合成及调节控制

木霉真菌是迄今报道形成和分泌胞外纤维素酶活力最高,成分最全面的菌种[29,33]。但有时细菌,特别是反刍动物中纤维分解细菌的纤维素酶活性很高[36]。Smith 等报道 Ruminococcus album 纤维素酶部分被 O_2 抑制[38]。这也许可解释反刍动物中分离活性纤维素酶的困难。我组分离的中温好氧产黄色纤维单胞杆菌分解纸浆纤维的能力也很强,而且同另一株伴生菌(腐臭假单胞杆菌)间有协同作用[39]。

木霉菌种的纤维素酶活力可以通过诱变而提高[11,12,40]。诱变后高产变异株的特点是菌落小、生长差、孢子形成少[12]。美国 Natick 实验室的 QM_{60}(绿色木霉)经线性加速器处理,得到酶活增加 1 倍的 QM9123[40],又经$_{60}$钴、紫外线和亚硝基胍等处理,得到活力更高的 QM9414。据 Reese 1975 年报道,该菌株是目前国际上液体培养活力最高的菌株,在最佳培养条件下,培养二周,摇瓶中达到的纤维素酶最高活力是:CMC 活力为 100 单位/毫升,滤纸分解活力为 5.0 毫克葡萄糖/毫升酶·小时,棉花活力是 9 毫克葡萄糖/毫升酶·24 小时[6]。我组选育了二株高活力拟康氏木霉变异株 EA_3 - 867 和 N_2 - 78。N_2 - 78 在适宜的培养条件下,只要培养 60 小时,用同样方法测定,达到的纤维素酶活力为 CMC 活力:47 单位/毫升,滤纸分解活力:8.2 毫克葡萄糖/毫升酶·小时,棉花活力:13.4 毫克葡萄糖/毫升酶·24 小时,可溶性蛋白:2.4 毫克/毫升[11]。

真菌纤维素酶形成的早期研究,指出培养条件的重要性[34]。培养条件下不仅影响酶的数量,也影响到质的变化。Ca^+、Mg^+ 和某些微量元素如 Co^+、Fe^+、Mn^+、Zn^+ 对于达到最高的酶量是必须的[36,41], N、P 和 K 都是需要的,最知的 N/纤维素之比为 1/35,有机氮如尿素、氨基酸、蛋白胨和酪蛋白能增加纤维素的利用速度,同时也增加了酶合成的速度[29]。加表面活性剂如 Tween80、蔗糖棕榈酸-酯和油酸钠等显著增加纤维素酶、淀粉酶、嘌呤核苷酶和苯甲酰酶的活性[42]。苯乙醇也有促进作用[43]，可能均与影响膜的透性有关。余永年等曾报道过康氏木霉的生理特性[44]。由于真菌的生长和产酶同时进行，环境因素的影响往往难以分析。我们采用甘油培养基中生长和静止细胞诱导产酶的方法，将生长和产酶分开分析，详细研究了各种条件对生长和产酶的影响[45]。证明了生长需要C源和N、P、Ca、Mg及Zn离子，并将产酶条件归纳为①诱导剂；②菌丝体活跃的代谢；③无机盐如N(尤其是NH_4^+)、P及Fe、Mn离子；④适宜的温度(30℃左右)、pH (2～5)和通气。O_2能刺激产酶，呼吸抑制剂能抑制产酶，因此，纤维素酶形成与呼吸代谢关系甚为密切。木霉菌具有旺盛的糖酵解-三羧循环呼吸途径。同时，产酶与核酸代谢有更直接的关系。纤维素酶的合成包括转录和翻译的过程。

纤维素酶是典型的诱导酶[15,29,33]，即只有外加纤维素底物或其衍生物时，纤维素酶才会形成。但是，对于真正的诱导剂是什么，看法不一。Mandels和Reese发现绿色木霉能在纤维素、纤维二糖、乳糖和葡萄糖中生长产酶，但认为纤维二糖——纤维素的水解产物——是纤维酶真正的诱导剂[33,46]。不过，纤维二糖的诱导活性很低，远远及不上纤维素，在采用较高浓度(1%或更高)及降低代谢速度，或添加表面活性剂时，才能得到较高的酶活性[5,41,47]。曾认为，纤维二糖诱导活性低是由于它是易利用碳源而引起降解物阻遇[41]。又推测，一种三聚糖(纤维二糖的转化产物)可能是纤维素酶的真正诱导剂[41]，但因量太微，难作鉴定。我们用木霉洗涤菌丝体诱导测定的方法，发现纤维二糖不能诱导纤维素酶的形成[45]。另一方面，度剂级葡萄糖中的槐糖杂质(二个葡萄糖分子以β1→2结合形成的双糖)对绿色木霉的纤维素酶有强力的诱导作用[48,49]，萤光假单胞菌同样有效[48]。槐糖对绿色木霉纤维素酶的诱导作用是由于诱使蛋白新合成[56]。我们从槐豆荚中分离得到的结晶槐糖，对拟康氏木霉和绿色木霉等纤维素酶的形成也有强力的诱导作用，在纤维素酶活力，尤其产酶速度上明显超过纤维素[51]。洗涤菌丝体只要同极微量的槐糖(0.34微克/毫升培养基)接触，经过3小时左右的延迟期，纤维素酶各组分(C_1、C_x和β-葡萄糖苷酶)便开始合成。槐糖和纤维素诱导的纤维素酶组成和聚丙烯酰胺凝胶电泳图谱十分类似，推测二者可能能通过共同的机制诱导纤维素酶的形成。又注意到，虽然纤维二糖不能诱导木霉洗涤菌丝体形成纤维素酶，但木霉在0.5%纤维二糖为碳源进行生长培养时，能形成纤维素酶，推测纤维二糖可能在生长过程中转化为槐糖或其类似物。这种转糖苷作用在真菌中十分普遍[46,52,53]。现已证明，大肠杆菌中半乳糖苷酶的真正诱导剂不是乳糖，而是它的转粮苷产物——别乳糖[54]。底物的转化产物作诱导剂的另一例子是木糖对葡萄糖异构酶的诱导作用[55]。我们又根据木霉的纤维素酶有组成酶(微量)和诱导酶二种，将纤维素酶被固体纤维素的诱导过程描述为[51]：木霉菌丝先分泌少量组成性纤维素酶至培养基中，它能“识别”纤维素，将它分解为纤维二糖，后者可能通过某种转化(在胞内或胞外)，形成真正的诱导剂——如槐糖或其类似物。从而诱使大量诱导性纤维素酶的形成并分泌到胞外，去分解纤维素作为营养。

真菌纤维素酶的形成一方面受诱导机制的调节，另一方面又受降解物阻遏机制的控制[48,56,57]。我们在拟康氏木霉中同样发现后一种调节机制的存在[45,51]。槐糖对纤维素酶合成的诱导作用能被甘油、葡萄糖以及各种其他糖类、糖磷酯，许多有机酸及NAD、NADP和ATP等所阻止，这种阻遏作用不能被c-AMP克服。

诱导和降解物阻遏现象是微生物适应环境、经济而有效地合成酶的一种方式。我们认为，提高纤维素酶产量，从酶形成调节控制的原理来设计，那就是保证菌丝体迅速而健壮地生长的基础上，去阻遏和克服降解物阻遏。木霉液体培养中添加葡萄糖母液使产酶明显加速并增加，就是这个原理运用的具体例证[12,58]。葡萄糖母液是酸水解玉米淀粉生产葡萄糖过程中的废液，含有较高槐糖[53]，并提供前期菌丝迅速生长的碳源（如葡萄糖）。葡萄糖母液为利用高效诱导剂——槐糖提供了廉价的来源。

纤维素酶各组分在细胞中的分布及合成部位的研究报道不多。我们发现拟康氏木霉的 C_1 酶、C_x 酶和 β-葡萄糖苷酶主要分布在胞内。Suzuki 等曾对萤光假单胞菌的纤维素分解变种做过较系统的研究[48,59]，它合成三种纤维素酶（A，B，C）。A 和 B（分解纤维素）分泌到胞外，C（分解纤维三糖和纤维二糖）在壁内，β-葡萄糖苷酶在细胞内。各组分在细胞内外的合理分布保证了细胞外的纤维素分子逐步从外到里分解为被细胞利用的葡萄糖。由于 A 和 B 的比例不同，导致胞外纤维素酶成分的多样化，并在生长过程中发生有规律的变化[60]。作者用酶分子的改造、修饰和蛋白酶的简单作用来解释酶成分的多样化。并证明细胞外纤维素酶是在膜结合的多聚核糖体上形成；而细胞内纤维素酶即纤维二糖酶和 β-葡萄糖苷酶既可在膜结合于多聚核糖体上，也可在游离多聚核糖体上合成。

纤维素酶形成的调节制从遗传上加以研究的例子是 Myers 和 Eberhart 的工作。他们发现，粗糙链孢霉中凡 C_x 酶活力高者，纤维二糖酶活力就高，二者是受同一个调节基因 cell-1 调节。cell-1 对 ccll-1^+ 是隐性，可能类似于大肠杆菌中 β-半乳糖苷酶的调节基因 i。酰基-β-葡萄糖苷酶是受另一个 gluc-l 基因调节[61]。β-葡萄糖苷酶有二种：一种是热不稳定的，存在于菌丝和分生孢子中，它不受 gluc-1 基因调节；另一种是热稳定的，只存在于分生孢子中，才是受 gluc-1 基因的调节[62]。我们发现拟康氏木霉野生菌经诱变后得到的高产变异株中，纤维素酶各组分有同步提高的现象，推测可能受同一个调节基因的控制。高产变异株纤维素酶的活力的升高主要是酶蛋白产量的增加，而非酶的结构和性质变化。推测这是调节基因变化而引起，并证明了高产变异株纤维素酶合成的调节机制发生变化：即对诱导剂敏感性增加和对降解物阻遏敏感性减弱导致了酶的高产[63]。

三、纤维素酶的应用

废纤维糖化生产葡萄糖无疑是纤维素酶应用的主要目标。纤维素的酶解和酸水解相比，具有在常温常压下可进行、不需要昂贵的耐酸设备、反应专一、产品纯净等优点。但纤维素糖化曾认为无法同淀粉的糖化相竞争。不过，近十多年来，纤维素酶研究的进步，包括菌种的选育、培养条件的改进、预处理方法（如高温热磨、化学处理等）和糖化工艺的发展，至少使我们向这一目标大大地迈进了。利用木霉纤维素酶化，5%～20%浓度的纤维素（预处理过）底物，6～24 小时，50%可转化为可溶性。纤维素转化率达 70%～90%[47]。在纤维糖化的连续化系统中，用 10%纯纤维素，反应 40 小时，得到 5%浓度的葡萄糖流出液[5]。在另一个糖化系统中，葡萄糖流出速度为 2～4 克/升/小时，已达到商用淀粉水解速度的一半（4～8 克/升/小时）[36]。纤维糖化工艺在技术上是可行的，经济上能否过关尚待证明[5]。为了最终解决应用问题，必须进一步提高菌种活力，提高底物密度，降低预处理成本（现占生产费用的一半），或选预处理过的城市工业废物（如造纸厂度纸浆），改进糖化工艺（包括连续化、酶的吸附和酶的回收）和开展副产品的综合利用（如纤维二糖、木糖、木质素、菌丝体等），菌种显然是突出的问题之一而受到国内外的重视。应该选育高产、耐热和能分解天然纤维的菌种。但是，对于菌种的潜力和选育的方法必须要有客观的估价。常规的菌种诱变方法曾选育出一些优良的菌株，但似乎难以在现有的水平上有更大的突破。应该寻找新的合理的筛选模型。曾有报道想用杂交[64]和基因工程的方法育种。看来,菌种选育工作的成功要吸收新技术、新方法,包括现代遗传工程的技术(如基因转移、增加基因样板量等)。筛选模型的设计、现代遗传工程的应用在很大程度上要依赖于对纤维素酶形成的生理、生化及遗传控制(如诱导、降解物阻遏等的本质)和纤维素复台酶的生化本质的了解。

提高糖化率的途径,除了菌种和预处理外,发挥现有菌种的潜力,提高酶的产量和酶的作用效率、诸如选用高效诱导剂、采用丰富培养基、同时克服降解物阻遏等,也是不可忽视的方面。即就拟康氏木霉 N_2 - 78 而言,在液体培养基中培养 60 小时达到的胞外蛋白为 2.4 毫克/毫升,相当于 1/4 纤维素转化为胞外蛋白。如何提高纤维素酶的作用效率将是摆在我们面前一个很艰巨的任务。

利用纤维分解细菌,分解废纤维生产单细胞蛋白,可能是纤维素酶应用的一个很有希望的目标。因为纤维分解细菌生长繁殖快,活体转化纤维素的效率高,且纤维素直接转化为菌体蛋白。由于对分解甘蔗渣纤维分解菌历时六年的研究[65],已有专利申请[66]。曾计划在美国旧金山结合城市废物(如废报纸)处理,用 1 千万美元建立年产 6 万吨单细胞蛋白的试验工厂[67]。蛋白的质量介于动物蛋白和植物蛋白之间。目前尚存在若干技术问题(如核酸的去除)和成本问题。我组利用纤维分解细菌分解废纸浆,制取细菌蛋白和四种单核苷酸获得了成功,纤维分解率达 95%以上。

纤维素酶在其他方面的应用试验包括食品加工(大豆加工、粮食加工、海藻加工、制速溶茶、提高酵母和小球藻的消化率等)[36,693,69]、医药工业[69]、饲料加工[69]、造纸工业、废水处理[69],以及木霉孢子做土壤抗菌剂[71]。但是,目前国外纤维素酶应用最有实际意义的例子,是医用助消化剂和植物细胞脱壁[36,72]。国内应用 EA3 - 867 纤维素酶制及已成功地从 20 多种植物中分离出大量完整的原生质体,并从烟草的叶肉组织或愈伤组织培养分化为植物株[73]。国内有相当大的人力从事纤维素酶曲的养猪试验,采用 EA_3 - 867 等菌种,在一定的条件下,有增加猪重的效果,但在生产上应用还有待实践考验并要解决制曲发酵、成本等困难,看来某些单位采用纤维素酶曲直接喂猪的方法可能是一种简便可行的方法。

四、结语

纤维素酶在工、农、医等各方面的应用有着无限宽广的前景。但是,绿色植物为了保护和支撑其器官,在长期生物演化过程中发展起来的构成其"骨骼"的纤维素,由于其组成的多样化(同木质素、半纤维素、果胶质等掺合在一起)、物理结构(结晶度高)和化学结构(β-构型)的坚固性,却造成了今日人类破坏其结构,释放其能量,利用其葡萄糖的艰巨性。纤维素利用的最终解决,包括菌种的选育,需要微生物学、遗传学、生物化学、酶学、生化工程和遗传工程等多学科的配合,同时离不了对纤维素酶形成和作用机理基础理论知识的了解。这些问题的研究又涉及现代生物学中最基本的核心问题,诸如核酸与蛋白质的关系、基因的表达、大分子的结构与功能、代谢的调节控制,以及分子遗传等等。这些问题的阐明,反过来将为纤维素酶的应用带来新的面貌。可以设想,要是能将纤维素酶基因转移到现在发酵工业上的重要生产菌上,无疑将会给发酵工业带来新的革命。近一二十年来,纤维素酶研究虽取得了长足的进步,但遗留下来的疑点还不少,如纤维素酶的结构与功能、纤维素酶作用的模拟、纤维素酶合成的调节控模型、合成的部位、信息传递的过程、真正诱导剂的分离，以及纤维素酶分泌机制等。这些问题的解决，无疑将为现代分子生物学和分子遗传学作出巨大的贡献，而纤维素酶的研究具有取材方便、酶稳定、酶分子不大等优点。同时，由于纤维素酶具有广泛的生态、生理意义，它的研究又渗入到动物生理、植物病理和植物生理等多种学科中去。引人注目的是，近年来对高等植物纤维素酶本质和生理功能的研究将有助于阐明植物某些重要的生理过程（如生长、形态发生）和激素作用的机理。

参考文献

[1] Reese, E. T. (ed): "Advances in Enzymic Hydrolysis of Cellulose and Relaed Materials". *Pergamon Press*, (1963).

[2] Guould, R. F. Cellulase and threir Application Adv. Chem. Series, (1969).

[3] Teru. G. (ed): Proc IV IFS: *Fernicenl. Technol. Today.* (1972).

[4] Wilke, G. R (ed): Cellulose as a chemical and Energy Resurce Symosium. *Biofechnol Bioengincering Symp.* N. 5(1975).

[5] Mandels. M. and Sternberg, D: *J. Fermenl. Technol.* 54, 4: 267(1976).

[6] Baileuy, M. et al (ad): Symp. Enzymic Hydrolyisi of Cellulose, *Finland* (1975).

[7] Liese. W (ed): Biological Transformation of wood by Microorganism. *Springer-Verlag-New York.* (1975).

[8] ACS Centennial Califrornia Section Diamodn jubilee Meeting Division of Microbial and Biochemical Tecghnology (MLCR), *San Francisco* (1976).

[9] Ghose T. K: Iinternational Symposium on Bioconversion of Cellulosic Suvbstances into Energy Chemicals and Microbial Protein Biochem, Eng Res. Cent. *Indian instilulc of Technol, Delhi..* (1977).

[10] Verma, D. P. S. et al: *J. Biol. Chem.* 250, 3: 1019. (1975).

[11] 上海化纤六厂、上海酒精二厂、上海植物物生理研究所纤维素酶组：绿色木霉变异株 EA_3 - 867 的获得及其产酶能力的提高，“微生物育种学术会讨论文集”科学出版社，P89，(1975)。

[12] 中国科学院植物生理研究所、上海酒精二厂：二株高活力纤维分解菌—拟康氏木霉 EA_3 - 867 和 N_2 - 78 的获得及菌种特性的比较，微生物学报，18(1)：27，(1978)

[13] 四川省生物研究所纤维素酶组：绿色木霉纤维素酶高产突变型的诱变和筛选。遗传学报，2：157，(1975)。

[14] 上海植物生理研究所细胞生理室：植物原生质体融合的研究，植物学报，17：245，(1975)

[15] 上海酒精二厂、上海植物生理研究所：纤维素酶固体曲的制取及其在酒精生产上的试用，工业微生物，1974 年第 4 期第 2 页。

[16] 北京市日用化学二厂，中国科学院微生物研究所：固体曲纤维素酶水解糠醛渣及生产酵母的中间试验微生物通报，4：23(1977)

[17] 上海植物生理研究所纤维素酶组：国外纤维素酶研究概况，应用微生物。2：1(1957)。

[18] Reese E. T.; *J. Bacteriol* 59: 485(1950)

[19] King, K. W.: *J. Ferment. Technol.* 43: 79, (1965).

[20] Selby, K. and Maitland, C. C.: *Biochem. J.* 104: 716, (1969).

[21] Wood, T. M.: *Ferment. Technol. Today.* (ed. Terui. G.) p. 711. *Soc. Ferment. Technol.* Japan, Osaka, (1972).

[22] Nisizawa, K. et al: Ibid, p. 719, (1972).

[23] Pettersson, L. G., et al: 同 21, p. 727, (1972).

[24] 中国科学院微生物研究所纤维素酶组：绿色木霉纤维素酶系中 C_1 酶的提纯与性质，微生物学报，16：

240, (1976).

[25] Thomas, R. : *Aust. J. Biol.* 9:159, (1956).

[26] Whitaker, D. R. : *Bull. Soc. Chem. Biol.* 42:1701,(1960)

[27] Eriksson E-E, and Pettersson, B. : Biodeterioration of Material, 2:116, *Applied Sciences Publishers*, Ltd. , London, (1967)

[28] Ikeda, R. et al: *Agr. Biol. Chem.* 37, 1153(1973)

[29] Pathak, A. N. and Ghose, T. K. : *Process Biochemistry*, 8,4:35, (1973)

[30] Halliwell, G: *Biochem.* J. 95:35, (1965)

[31] Koenigs, J. W. : 同 4,p. 151, (1975)

[32] Ghose, T. K. and Kostick, J. A. : *Adv. Chem. Series*, 95:415, (1969)

[33] Mandels, M and Reese, E. T. : *J. Bacteriol.*, 73:269, (1957)

[34] Mandels, M. and Weber, J. : *Adv. Chem. Series*, 195:391, (1969)

[35] 外山信男:发酵工杂志:47,11:714, (1969)

[36] Emert, G. H. et al. :"Food Related Enzyme" (ed. Whitaker, T.), p. 77, (1974)

[37] Huang, A. A. : *Biotechnol. Biocngineering Symp.* No. 5:245, (1975)

[38] Smith, W. R. et al. : *J. Bacteriol.* 114:729, (1973)

[39] 中国科学院植物生理研究所纤维素酶组:纤维分解细菌与伴生菌的分离鉴定及其协同作用,微生物学报,18(2):147, (1978)

[40] Mandels, M. et al. : *App. Microbiol.*, 21, 1:152, (1971).

[41] Reese, E. T. and Maguire, A. : *Dev. Ind. Microbiol.* 12:212,(1971).

[42] Reese, E. T. and Maguire, A. : *Appl. Microbiol.* 9:159, (1967).

[43] Stranks, D. W. : *Can. J. Microbiol.* 19, 12:1523(1973).

[44] 余永年等:纤维分解真菌—康氏木霉(Trichoderma Koningi oud.)的生理特性,植物学报 11:191(1963)

[45] 朱雨生,谭常:木霉纤维素酶的诱导形成及其调节,Ⅱ. 槐树对木霉 EA_3 - 867 洗涤菌丝体纤维素酶形成的诱导作用及降解物阻遏现象,植物生理学报,(1978)(即将刊出)

[46] Mandels, M. and Reese, E. T. : *J. Bact.* 79:816, (1960).

[47] Mandels, M. et al. : Symp. on Enzymatic Hydrolysis of Cellulose, Aulanko, Finland, ed. M. Bailey, T-M, Enari, M. Linko, Helsinki, p. 81, (1975).

[48] Suzuki, H. et al. : *Adv. Chem. Series*, 95:60, (1969)

[49] Mandels, M and Reese, E. T. : *Biochem. Biophys. Res. Commun.* 1:338, (1959)

[50] Nisizawa, T. et al. : *J. Biochem.* 70, 3:387, (1971).

[51] 朱雨生,谭常:木霉纤维素酶的诱导形成及其调节,Ⅰ. 槐树对木霉 EA_3 - 867 纤维素酶形成的诱导作用,微生物学报(1978),(即将刊出)

[52] Buston, H. W. and Jabbar, A. : *Biochem. Biophys. Acta*, 15:543, (1954)

[53] Satoko Toda, et al. : *J. Ferment. Technol.* 46:711, (1968)

[54] Jobe, A. and Bourgeois, S. : *J. Mol. Biol.* 69: 397, (1972)

[55] Wiseman, A. : Basic Life Sciences, 6:1, (1975).

[56] Nisizawa, T. et al. : *J. Biochem.* 71, 6:999, (1972).

[57] Stewart, B. and Leatherwood, J. M. : *J. Bact.* 128, 2:609, (1976).

[58] 朱雨生,谭常:木霉纤维素酶的诱导形成及其调节,Ⅲ. 葡萄糖母液和槐豆荚提取液对纤维素酶的诱导作用,植物生理学报,(1978),(即将刊出)

[59] Yamane, K. et al. : *J. Biochem.* 61:19, (1970)

[60] Yoshikawa, T. et al. : *J. Biochem*, 75:531, (1974)

[61] Myers, M. G. and Eberhart, B. : *Biochem. Biophys. Res. Commun.* 24:782, (1966)

[62] Eberhart, B. et al. : *J. Bact.* 87:761, (1966).

[63] 朱雨生:木霉纤维素酶的诱导形成及其调节,Ⅳ.二株高产变异株纤维素酶合成调节机制的变化——酶活提高原因的初步分析,植物生理学报(待发表).
[64] Morozova, E. S.:同 88, p.193, (1975).
[65] Rockwell, P. J.:"Single cell Proteins from Cellulose and Hydrocarbon, Noyes Data Corporation, Park Ridge, New Jersey, U.S.A. (1976)."
[66] USP 3761355, (1973)
[67] Isu and Bechtel: Chemical and Engineering News, 52, 7:20, (1969).
[68] 食品工业(日),第9期,(1966)(纤维素酶专集)
[69] 长谷川忠男等:食品酵素高分子学概论(下)酵素の利用,株式会社,地人书馆,p.151 (1976)
[70] 外山信男:酦酵工业杂志,41(6):346, (1963)
[71] Toyama, N. and Ogawa, K.: *Biotech. and Bioengineering Symp.* 5:225, (1975).
[72] 西泽一俊:蛋白质·核酸·酵素,22(7):960, (1977).
[73] 上海植物生理研究所细胞生理室:生物化学与生物物理学报,3:341,(1976).

第二篇

植物生理和代谢调节（1961—1965）

棉铃脱落与碳水化合物化谢的关系

一、引言

棉花的落蕾落铃极为普遍，随品种和环境的不同，脱落率为 30%～60%[43]；在不良的环境条件下甚至可达 90%以上，严重地影响了棉花的产量。可见，防止和减少棉花的落蕾落铃是棉花增产的关键问题之一。

棉铃的脱落，在形态、解剖上是由于果柄基部与果枝交接处离层细胞衰老、溶解和分离的缘故[12]。

影响棉铃脱落的因素很多：授粉受精的受阻[9,20]，营养生长和生殖生长比例的失调[9,21]；病、虫的侵害[19,26]；矿质营养的缺乏或比例不平衡[3,40]；光线不足或光周期不符[5,27]；水分过多或过少[3,24,28,33]；温度过高[28]以及土壤中缺乏 CO_2[22]等都会引起棉铃的大量脱落。

棉铃脱落的机制主要有两派理论：一派认为养料——特别是碳水化合物和氮——的供应不足是造成蕾铃脱落的主要原因（Mason，1922[41]；King，1932[38]；Wadleigh，1994[46]；Благовещенский 等，1929[17]；Учатвакин 和 Бородуеина，1953[25]；金成忠等，1954，1956[3,4,5]。另一派则认为生长素是控制器官脱落的主要因素：如 Wetmore 等(1953)直接证明离层的形成与叶片中的生长素含量有关[49]；Addicott (1955)认为决定器官脱落的是离层两边生长素浓度差的缘故[27]；Gawadi 等(1950)指出调节器官脱落的原因是叶片中产生的抑制脱落的生长素和促进脱落的乙稀间的比例[34]；Eaton 等(1953—1955)则提出棉铃的脱落是受棉叶中形成的生长素和棉铃中产生的抗生长素间的平衡所左右[29,30,31]；而最近有人认为棉铃脱落是由于存在脱落素的缘故[48]。

此外，有人认为落铃是由于棉叶吸取力大于棉铃吸取力所致(Новиков，1941[23])；有人认为与棉铃细胞液的浓度有关(Hawkins 等，1934[37])；与棉叶汁液的渗透压和电导性有关(Бляовещепский，1929[16])；而 Учатвакин 等(1953)还发现棉叶酶作用的方向与调节棉铃的脱落有关[25]。

金成忠等(1956)[4]和我们[9]都证明有机养料、特别是碳水化合物与棉铃脱落的关系极为密切，当棉铃缺乏糖类时就会引起脱落。因此研究棉铃中碳水化合物的代谢和离层形成的相互关系是解决脱落机制值得重视的问题。

本文的目的即在于探讨棉铃离层形成与碳水化合物代谢的关系，从而为采取保铃措施提供理论根据。

朱雨生，尤天心，薛应龙。此文发表于 1964 年“复旦学报”9 卷 2 期 263－271 页。
本文 1963 年 10 月 20 日收到。

二、材料与方法

本试验是在复旦大学生物系和宝山县五角场人民公融合作的棉花试验田中进行的。品种是岱字15号。从1961年7—9月进行了多次重复试验。为了便于比较,对棉铃施行了人工脱落和不脱落的处理。棉铃不脱落的人工处理系根据金成忠等(1956)[4]曾采用的在开花当天,对果枝进行环剥,并留二叶一花,剪去其余部分的方法;人工脱落的处理系用上述同样的方法,但除去所有叶片而只留一花。试验完全证明这种处理方法的可靠性,凡经脱落处理的幼铃到第4天就全部脱落。

试验在早晨8时左右进行。选择着生位置和生育状态基本相似的当天开放的花,分别进行上述脱落和不脱落的处理,并挂上标记纸牌。以后每天取样,观察分析。脱落处理的取第1、2和3天的样品,因第4天幼铃已全都脱落;不脱落处理的则取第1、2、3、5和7天的样品;此外还分析了开花前一天及开花当天的材料。

所有样品分成两部分进行分析测定:一部分用显微组织化学的方法;另一部分用化学分析的方法。

1. 显微组织化学的方法

将样品沿果柄中上端切去幼铃及切去连接果柄两边的大部分果枝(每边留下果枝0.5 cm),然后将此材料经70%酒精固定后用徒手切片机制成纵切面的新鲜切片,厚度约20 μm左右,并进行下列测定:

(1) 多糖醛酸的测定[11]。在切片上加2%的间苯三酚酒精溶液二滴,然后用几滴35%的盐酸进行媒染,凡有多糖醛酸处被染成樱红色。

(2) 淀粉的测定[11]。切片用碘-碘化钾溶液处理,淀粉即被染成紫黑色的小颗粒。

(3) 纤维素的测定[11]。切片上滴上碘-碘化钾溶液后,再加60%的浓硫酸水解纤维素,对照则不加硫酸处理,切片上显蓝色的部分即标志着有纤维素的存在。

2. 化学分析的测定

将材料按上述方法切下,然后将树皮(包括表皮、皮层和韧皮部)和木质部分开,置120℃烘箱中烘10分钟以杀死组织,再放到80℃恒温烘箱中烘干,24小时后取出,在干燥器中冷却,将木质部称重;树皮则用研钵磨成细粉,并用金属网过筛。测定前再将样品粉末放在80℃恒温箱中24小时以烘至恒重。用分析天平精确地称取样品,测定其中可溶性糖和淀粉的含量。

(1) 可溶性糖的蒽酮法测定[13]。将样品用80%的热酒精重复提取6次,蒸去酒精,将留下的可溶性糖用水溶解并稀释至一定浓度,和蒽酮硫酸溶液混和及煮沸,然后用71型分光光度计在630 mμm处比色。根据标准曲线即可计算出可溶性糖的含量。

(2) 淀粉的测定[13]。可溶性糖提取后的残渣,用30%的过氯酸在室温下重复水解两次,第1次12小时,第2次24小时。水解后仍按蒽酮法测定糖的含量。

三、试验结果

1. 脱落过程中离层细胞的淀粉转化规律

Lee (1911)[39], Livingston (1950)[40], Greisel (1954)[35]等报道器官脱落过程中,离层细

胞常有淀粉堆积的现象。我们在棉花的研究中观察到在果柄维管束周围的淀粉鞘中含有大量的淀粉，而其他组织除木质部外几乎都不含淀粉。淀粉鞘中的淀粉含量在开花前1天至开花当天有明显的增加，这种情况到开花后第2天无多大变化。以后淀粉含量的变化要视幼铃脱落与否而定。图1表示脱落与不脱落棉铃果柄中淀粉变化动态的规律。

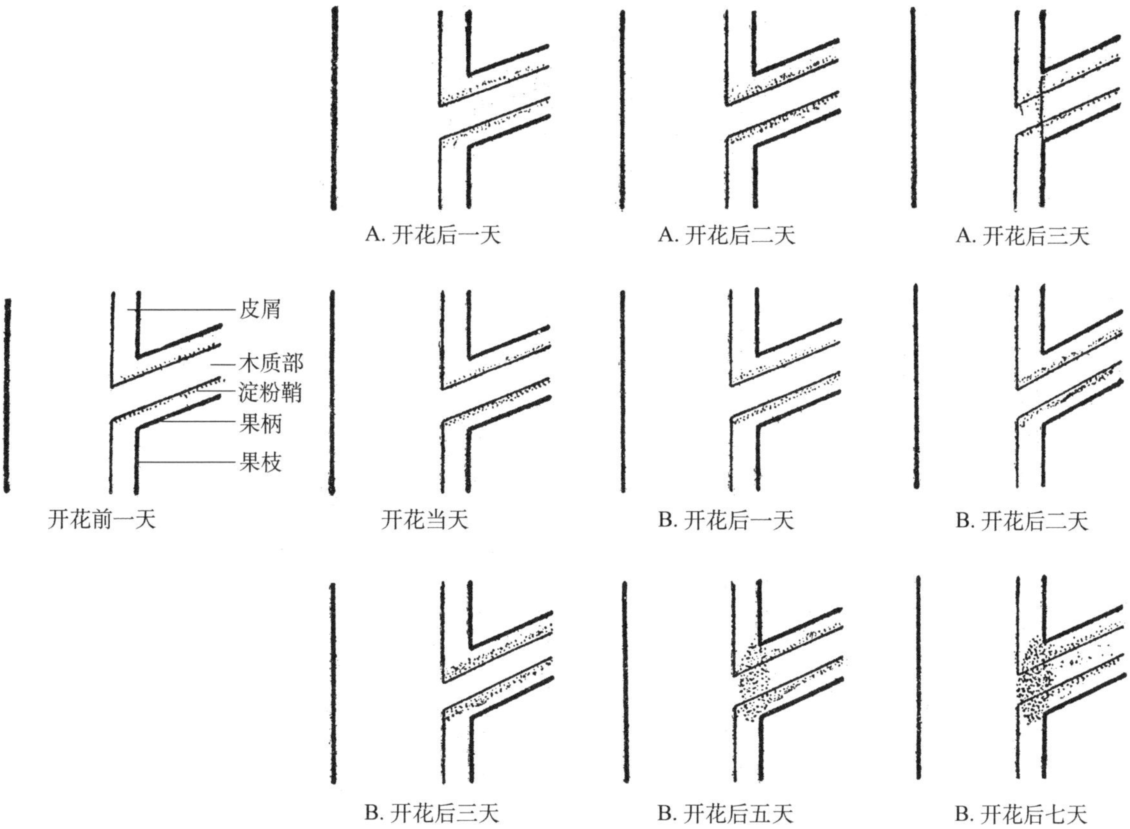

图1　开花前后脱落和不脱落棉铃果柄中淀粉鞘和离层细胞内淀粉含量变化动态

A—代表脱落处理棉铃；B—代表不脱落处理棉铃；黑点表示淀粉粒。

在人工脱落处理的情况下，开花后第3天(脱落的前1天)，淀粉鞘中淀粉显著减少，而此时离层细胞中却突然出现大量淀粉的堆积，随后在离层形成过程中，其中的淀粉又逐渐减少。在不脱落处理的情况下，淀粉鞘中淀粉含量在开花后第3天并无下降的现象，反而有所增加，随后增加更是显著，连髓部也出现大量淀粉贮备。但离层部位，则直到开花后第5天才有淀粉出现，而且和脱落的相比，它的分布范围要宽得多(约30层细胞)，以后也没有消减的现象。

2. 离层形成与碳水化合物代谢的关系

金成忠等(1954,1956)[3,4,5]用环剥、遮光、供糖的试验证明了养料的缺乏是导致棉铃脱落的重要原因。在我们的试验中，用呼吸抑制剂处理阻止叶片中养料向棉铃输送，从而促进了棉铃脱落的事实，也支持了这个观点。

试验时将当天开的花进行环剥，并保留一叶一花。在叶柄上纵向用刀片轻切三条长约1厘米的裂痕，上包纱布条，纱布条末端浸在小试管中。一组试管中加水作为对照，另一组加0.2%的迭氮化钠溶液，以后每天在管中加添蒸发掉的水分，并检查幼铃脱落率。结果表明，经迭氮化钠处理的，显著促进了幼铃的脱落，处理后第 4 天，脱落率达 90%，而对照的脱落率为零，如表 1 所示。

表 1 呼吸抑制剂迭氮化钠处理叶柄对棉铃脱落的影响*

处理	棉铃脱落数(累计)		脱落率/%
	处理后第 3 天	处理后第 4 天	
对照(水)	0	0	0
迭氮化钠(0.2%)	4	9	90

* 每组处理 10 个棉铃。

上述结果是因为迭氮化钠抑制了叶柄的呼吸，从而阻止了叶柄中有机养料(主要是碳水化合物)向幼铃运输的缘故[7]。

为了进一步证明碳水化合物和离层形成的关系，我们还对脱落和不脱落处理的棉铃离层周围树皮的可溶性糖及淀粉作了定量分析比较，结果如图 2 所示。

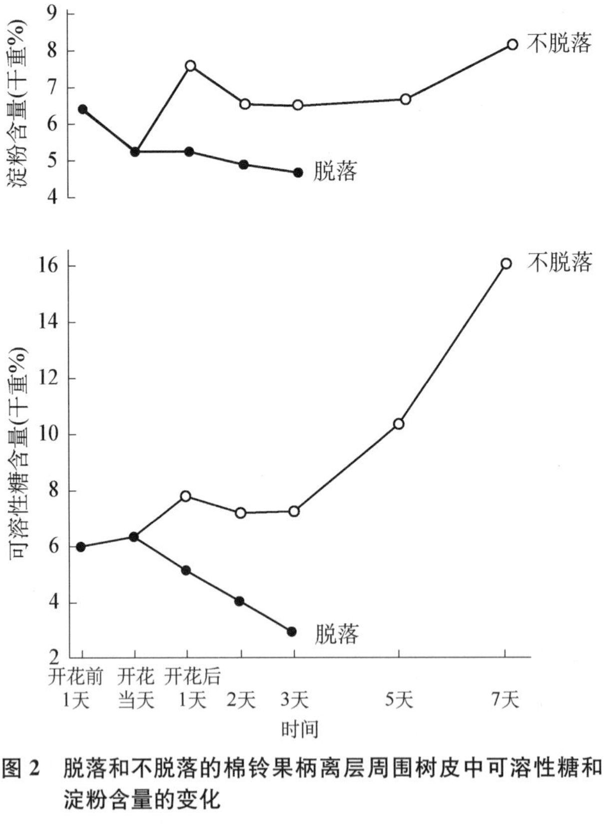

图 2 脱落和不脱落的棉铃果柄离层周围树皮中可溶性糖和淀粉含量的变化

从图中曲线可见：①开花后第 1 天起，不脱落的棉铃，其离层周围树皮中可溶性糖的含量就有增加，第 3 天以后，增加更是迅速；而经脱落处理的棉铃，则开花后第 1 天，可溶性糖便开始下降，与不脱落的相比，约低 50%；以后继续加速下降，至开花后第 3 天(即脱落前 1 天)，则可溶性糖的含量比不脱落的要低 1.5 倍。②开花后第 1 天起，不脱落的棉铃离层周围树皮中的淀粉含量比脱落的要高 40%，第 2、3 天略有下降，以后又逐渐升高；而脱落的棉铃中，淀粉含量一直是逐渐下降的，而且总是低于相应的不脱落的棉铃。

上述这些结果与显微组织化学的测定结果是一致的，即不脱落的棉铃果柄中淀粉含量要较脱落的多。

脱落与不脱落处理的棉铃果柄的木质部重量也有极显著的差别，结果如表 2 所示。

表 2　脱落与不脱落处理的棉铃果柄木质部重量的变化

日期 \ 处理	木质部干重/克	
	脱落	不脱落
开花后 1 天	0.230	0.288
开花后 2 天	0.240	0.324
开花后 3 天	0.245	0.358
开花后 5 天	—	0.475
开花后 7 天	—	0.678
开花后 10 天	—	0.905

从表中可见木质部干重在开花后第 1 天起，不脱落的就比脱落的高，此后，不脱落的棉铃果柄木质部干重增加极为迅速，而脱落处理的，则木质部干重只有略微变化。

此外，从显微组织化学测定中，还观察到脱落处理的棉铃果柄，在离层形成时有多糖醛酸的产生，而不脱落的棉铃中则无多糖醛酸的染色反应（见图 3）。多糖醛酸是细胞中胶层果胶质分解的产物。而且不脱落的棉铃，在开花后第 3 天，果柄的离层细胞就开始纤维化，并日益加深，离层细胞不断老化，以致最后即使剪去棉铃和叶片，切断了养料的供应，果柄也不会产生离层而脱落。

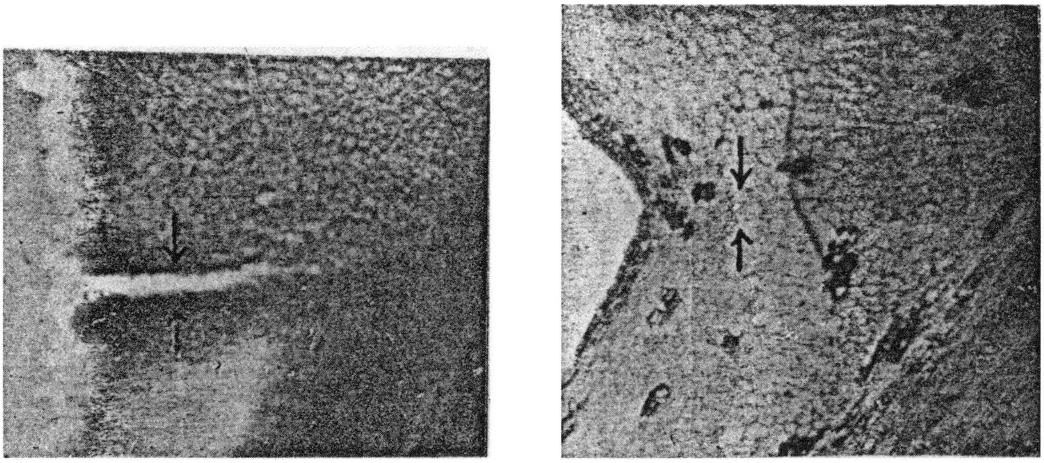

（a）　　　　（b）

图 3　离层形成过程中的果胶物质的转变

（a）脱落的棉铃果柄离层形成时有多糖醛酸的染色反应，箭头暗色部分　（b）不脱落的棉铃果柄离层区域（箭头表示处）无多糖醛酸的染色反应

3. 半乳糖对离层形成的影响

从上述树皮中可溶性糖和淀粉含量变化的结果表明，脱落前不仅棉铃中碳水化合物下降[3]，连离层的周围组织中碳水化合物也是减少的。为了进一步阐明糖类和离层形成的关系，我们采用了切片离体培养的方法，观察半乳糖对离层细胞中胶层果胶物质分解的影响。

离体培养系根据 Robert (1960)报道的离体材料中胶层溶解代谢的研究方法[44]，并加以简化，即培养时不加果胶水解酶，而将当天开的花进行脱落的人工处理，然后在脱落前 1 天(即处理后第 3 天)把材料从植株上剪下，在水中略浸后，取出制成新鲜的纵切片，厚度约为 30～50 μm，并按无菌操作进行组织培养，培养分下列三组：

(1) 水(对照 1)。

(2) 完全培养液加 0.16 mol/L 离子拮抗液(对照 2)。

(3) 完全培养液加 3%(即 0.16 mol/L)的半乳糖溶液。

完全培养液按改进的 White 氏液配制[8]，内含 1.5%的蔗糖，少量氨基酸、维生素及各种矿质离子。离子拮抗液是由 9 份 NaCl∶1 份 $CaCl_2$ 配成。加离子拮抗液是为了抵消处理中加入半乳糖后渗透压增高所可能引起的影响，而离子拮抗液对细胞并无损害作用[15]。

每一处理在三角瓶中分别加入 20 ml 上述各组溶液和 5 片纵切面组织，在(28±1)℃中培养，每隔一定时间在显微镜下观察离层细胞中胶层的溶解情况，以间苯三酚染色法作为鉴定标准，如果有离层形成，则有樱红色反应。结果如表 3 所示。

表 3 半乳糖对棉铃果柄离层轴胞中胶层溶解的影响*

处理	培养时间/h				
	12	40	60	96	108
水(对照 1)	—	—	—	—	—
完全营养液加离子拮抗液(对照 2)	—	—	+	+	+
完全营养液加 3%半乳糖	—	—	—	—	—

注：* +表示有樱红色反应；—表示无染色反应。

(1) 离层形成需要正常的生理环境及营养物质的供应，在水中培养时，由于缺乏这些因素，故离层细胞的中胶层不发生溶解。

(2) 在完全营养液中培养时，由于与正常的生理环境相类似，故发生中胶层的溶解。

(3) 半乳糖有抑制离层细胞中胶层溶解的作用，而蔗糖并无这种效应(对照 2)，可见半乳糖与果胶物质的代谢有着密切的关系。

从表中可见培养在完全营养液中的离体材料经 60 h 后，离层细胞的中胶层便开始溶解，果胶质发生转化；而培养在水中或完全营养液加半乳糖溶液中的材料，则经过 108 h 尚无中胶层的溶解。

四、讨论

器官脱落过程中离层细胞中淀粉含量的变化已有不少研究[35,30,40]。在脱落与不脱落棉铃的比较试验中，也观察到脱落的棉铃在脱落前果柄的离层细胞中有淀粉的累积，但值得注意的是我们发现淀粉鞘中的淀粉却有所减少，因而可以推测离层细胞中的淀粉可能是从淀粉鞘中以可溶性糖的形式运输过来的。以后在脱落过程中，离层细胞中的淀粉又渐渐消减。有意思的是我们看到在不脱落的棉铃离层细胞中也有淀粉的累积，但在时间上略晚些，并且以后不发生淀粉的消失现象。这个事实表明离层的形成需要有能量的供应[1,27]，而淀粉的消耗可能就是通过呼吸提供了能量的来源。试验中所观察到的离层形成过程中可溶性糖的减少，也可能是作为能源而被消耗掉的。

很多研究证明棉铃中碳水化合物的缺乏是导致棉铃脱落的原因[3,4,5.9,17,25,31,41,46]。试验还指出了，在脱落处理的棉铃离层周围树皮中的淀粉和可溶性糖也都比不脱落的低，说明棉铃在脱落过程中离层细胞也处于碳水化合物不足的状态。

除了碳水化合物的变化外，有的报道指出离层细胞中胶层的不溶性果胶钙转化为可溶性的果胶酸是离层产生的原因[32,39]。在我们的研究中，根据显微组织化学的测定，还表明中胶层溶解后有多糖醛酸的产生。这可能是由于果胶酸的进一步分解所致。这种变化，在大叶黄杨叶柄离层形成的研究中，也获得同样结果[1]。这些结果对进一步研究果胶物质及与之有关的物质代谢同离层形成的关系有着重要的意义。我们的试验所证明的半乳糖对离层形成的抑制作用，可能与阻止离层细胞中胶层果胶物质的分解有关。这种现象在其他材料如大叶黄杨叶柄、蚕豆花柄中也都看到[14]。Wood (1960)[50]指出果胶质的基本组成单位是半乳糖醛酸，可溶性果胶进一步水解时有半乳糖醛酸及部分半乳糖和阿拉伯糖产生[36]。而半乳糖、半乳糖醛酸和阿拉伯糖之间则可以通过 UTP(尿嘧啶三磷酸腺核苷酸)进行相互转化[36,42,45]。这些研究成果与在显微组织化学测定中所观察到的离层形成时有多糖醛酸的产生，以及在离体组织培养时半乳糖和半乳糖醛酸能显著抑制离层的形成[14]是相呼应的。

半乳糖在植物体内的代谢与离层形成的关系是值得探讨的课题之一。棉铃脱落过程中碳水化合物酶作用方向的变化已有报道[25]，因而半乳糖很可能与其他糖类的代谢转化有关[36,42]。这些问题的进一步研究可能在阐明器官脱落的机制中提供重要的线索。

结论提要

(1) 本文通过显微组织化学测定，定量化学分析和组织离体培养的方法，研究了棉铃脱落过程中果柄离层细胞中淀粉和可溶性糖的变化动态以及离层形成时果胶物质转化的规律。

(2) 脱落的棉铃在脱落前 1 天，离层细胞中有淀粉的累积，但随着离层的形成，淀粉又趋于消减；不脱落的棉铃，淀粉在离层细胞的累积和分布范围较广，而且以后并无消减的现象。可溶性糖的变化有类似情况，在离层周围的皮层组织中淀粉和可溶性糖也是在棉铃脱落过程中逐渐下降的，而不脱落的棉铃则淀粉和可溶性糖不断增高。

(3) 棉铃的脱落是由于果柄离层细胞中胶层的溶解，中胶层溶解时有多糖醛酸的产生。

(4) 半乳糖具有抑制棉铃果柄离层细胞中胶层溶解的作用。

(5) 本文讨论了离层形成过程中果胶质代谢及碳水化合物转化的意义，并指出半乳糖在棉铃脱落过程中的作用可能与糖类的代谢转化有关。

参考文献

[1] 朱雨生，尤天心，薛应龙，植物器官脱落的生理机制 I. 器官脱落过程中物质与呼吸代谢的动态，中国植物生理学会第一届年会论文(摘要)集 134 页，(1963).

[2] 李曙轩，落叶、落花和落果，科学出版社，(1957).

[3] 金成忠等,棉花的落蕾落铃,植物学报 3: 155～166,(1954).
[4] 金成忠等,有机养料在棉花蕾铃脱落中的作用,植物学报 5: 78～102,(1956).
[5] 金成忠等,光线强度对棉花蕾铃脱落的影响,植物学报 5: 117～136,(1956).
[6] 金成忠等,不受精棉铃的脱落问题,植物学报 5: 69～77,(1956).
[7] 阿法纳西耶娃 M. B.,植物体内营养物质的运输,科学出版社,(1957).
[8] 罗士苇,植物的组织培养,植物生理学通讯 5: 16～25,(1957).
[9] 复旦大学植物生理教研组,受精与棉花落铃的关系,复旦大学自然科学学报 2: 379～388,(1956).
[10] 复旦大学植物生理教研组,植物生理学大学实验讲义(内部资料),(1961).
[11] 麦克林,R. C. 等,植物技术专册,科学出版社,(1958).
[12] 娄成后等,类似生长素药剂对于延迟植物器官的脱落及相关的生理效应,植物学报 3: 167～189,(1954).
[13] 中国科学院植物生理研究所,有关植物生理的研究方法(资料),(1960).
[14] 薛应龙等,植物器官脱落的生理机制Ⅱ.器官脱落与呼吸途径的联系,中国植物生理学会第一届年会论文(摘要)集 135 页,(1963).
[15] 萨比宁 Д. А.,植物营养生理学(上册),科学出版社,(1957).
[16] Бдаговещенский, А. В., Иееледования по физиологии хлопчатника о накоплении веществ раздичнымх сортами хлопчатника, Труды САГУ 8 в, Ботаника, вып, 4～5, Ташкент, (1929).
[17] Бдаговещенский, А. В. 等,Опыт изучения физиологии опадения завязей у хлопчатника, Труды САБУ 8 в, Ботаника, вып, 10, Ташкент, (1929).
[18] Бородулина, А. А., Хлопчатник, 4. физиология и биохимия хлопчатника. Глава 10. Опадение завязей у хлопчатника, Изд. АН. УзССР, (1960).
[19] Демидов, Н. И.. Влияние колюще-соеущих насекомых на опадение плодообразований хлопчатника, Известня УзФАН СССР, No. 6,(1940).
[20] Донов, Г. А. 等,К вопросу об опадении завязей хлопчатника в связи с орошением дождеванием, Докпады ВАСХНИЛ, вып, 9,(1952).
[21] Ивановская, Т. Л., К вопросу об опадении плодовых орянов у американского хлопчатника, Доклады всесоювного совещания по физиологни растений. Вып, 1,(1946).
[22] Меднис, М. П., К вопросу об опадении завязей, Советский хлопок, No. 11～12,(1939)
[23] Новиков, В. А., Причины сбрасываннз бутонов и коробочек у хлопиатника и возможные меры борьбы. ДАН СССР. т. XXXП, No. 2,(1941).
[24] Новиков, В. А., Причины сбрасывания бутонов и коробочек у хлопчатника и возможные меры борьбы. ДАН СССР. т. XXXП, No. 4,(1941).
[25] Учатвакин, Ф. Н. 等,Основные результаты трехлетних исследований по опадению завязей ухдопчатника. Труды института селвского хозяйства. АН. УзССР. вып. 1,(1953).
[26] Черкасова, В. В., Роль сосущих вредитекей вопадений завязей хкопчатника Резукьтаты работы СТАЗРа Союз НИХИ за 1939г., Тамкент, Изд. Союз ИИХИ, (1941).
[27] Addicott, F. T., Physiology of abscission, Ann. Rev. Plant Physiol. 6: 211～238,(1955).
[28] Dunlap, A. A., Light, Drought and heat as factors in cotton boll shedding, Phytopathology, 34: 999. (1944).
[29] Eaton, F. M., Physiology of cotton, Ann. Rev. Plant Physiol. 6: 299～328,(1955).
[30] Eaton, F. M. & Ergle, D. R., The nntritional interpretion of boll shedding in cotton: Seasonal trends in carbohydrate and Nitrogen levels and effects of girdling and spraying with sucrose and urea, Plant Physiol. 28: 503～520,(1953).
[31] Eaton, F. M. & Ergle, D. R., Effectof shade & defoliation on carbohydrates levels & growth, fruiting & fiber properties of cotton plants, Plant Physiol. 29: 39～49,(1954).

[32] Facey, F. , Abscission of leaves in Fraxinus americanna, New Phytologist 49: 103,(1950).
[33] Ferguson, H. , Record cotton fields. Emp. cotton grow rev. XXXII, 3,(1955).
[34] Gawadi, A. G. et. al. , Amer. J. Bot. 37: 172～180.(引自 Addicott, F. T. 见[27]),(1950).
[35] Greisel, W. O. , Phytomorphology 4: 123～132(引自 Addicott, F. T. 见[27]),(1954).
[36] Gibbs, M. , The metabolism of carbohydrates, Ann. Rev. Plant Physiol. 9: 329～378,(1959).
[37] Hawkins, R. S. et al. , Varietal differences in cotton boll shedding as correlated with osmotic pressure of expressed tissue fluids, J. Agr. Res. 48: 149～159,(1934).
[38] King, C. G. & Loomis, H. F. , Agricultural investigation at the United States field station, Sacoton, Arizona, 1925～1930, U. S. D. A. Cir. , 206,(1932).
[39] Lee, E. , The morphology of leaf-fall. Ann. Botany, 25: 51～107,(1911).
[40] Livingston, G. A. (1950) In vitro tests of abscission agents, Plant Physiology 25: 711～721.
[41] Mason, T. G. , Growth and abscission in sea-Island cotton, Ann. Botany 36: 457～484,(1922).
[42] Marrè, E. , Phosphorylation in higher plants, Ann. Rev. Plant Physiology 12: 195～218,(1961).
[43] Nightingale, G. T. & R. B. Farrham (1931) Botan. Gaz. 97: 477～517(引自 Addicott, F. T. 见[27]).
[44] Robert, E. Y. , Possible role of pectic enzymes in abscission, Plant Physiology 35: 157～162,(1960).
[45] Setterfield, G. , et al. , Structure & physiology of cell walls, Ann. Rev. Plant Physiology 11: 299～322,(1961).
[46] Wadleigh, O. H. , Growth status of cotton plant as influenced by the supply of nitrogen, Arkansas, Agr. Exp. Sta. Bull. 466,(1944).
[47] Walhood, V. T. , Abscission in cotton (Doctoral thesis), Univ of Oalif Los Angels, Calif. , (1955).
[48] Wen Chih-lin, Isolation of abscissin, an abscission accelerating substance, Science, 134(3476): 384～385,(1961).
[49] Wetmore, R. H. , et al. (1953) Am. J. Bot. 40: 272～276(引自 Addicott, F. T. 见[27]).
[50] Wood, R. K. , Pectio and cellulotic enzymes in plant disease, Ann, Rev. Plant Physiology 11: 299～322,(1960).

THE RELATION BETWEEN CARBOHYDRATE METABOLISM AND BOLL SHEDDING IN COTTON PLANT

Chu Yu-Shun, You Tien-Shen and Hsueh Ying-Lung

In the present investigation, histochemical, analytical and tissue culture methods were used to study the changes of starch, soluble sugars and pectic substances in the abscission layer zone of fruit petiole during boll shedding. The results of the experiments are summarized in Tables 1—3 and Figs. 1—3, from which the following conclusions may be drawn:

(1) By comparing the shedding with the unshedding bolls, it shows that the accumulation of starch in cells of abscission layer zone is observed in the former one day before dropping, but it is reduced during the formation of abscission layer, while in the latter the starch is still accumulated in much denser state around the abscission layer zone during the boll development. The behavior of soluble sugars is found essentially the same as starch.

(2) As an evidence of our experiment, the dropping of bolls is due to the dissolving of the middle lamella in the abscission layer cells, and consequently, the galacturonic acid is produced. However, the dissolving process is inhibited by the galactose.

The process of the formation of abscission layer is discussed in the light of the above findings, indicating that the action of galactose may be related to the metabolism of carbohydrates and pectic substances during the boll dropping.

植物呼吸代谢的生理意义①

Ⅰ. 器官脱落与呼吸途径的联系

提要

本文通过用大叶黄杨叶柄离体培养的方法证明器官脱落与呼吸途径有着密切联系。利用呼吸途径的中间代谢物,某些与呼吸途径有关的物质以及不同的呼吸途径抑制剂处理结果表明与 HMP 途径有关的一些代谢物如葡萄糖酸、半乳糖醛酸、半乳糖、阿拉伯糖等都有较明显的延迟黄杨叶柄脱落的效果。某些与 HMP 途径有关的物质如 IAA 与抗坏血酸也有类似作用。用各种呼吸途径抑制剂处理结果进一步证明适当浓度的 EMP-TCA 抑制剂延迟了叶柄的脱落,反之,HMP 抑制剂则促进了叶柄脱落。作者等认为器官脱落与否主要决定于离层细胞所进行的呼吸途径。当呼吸途径以 EMP-TCA 为主时促使器官脱落,而当 HMP 途径加强时则延迟或阻止器官的脱落。

一、引言

器官脱落的机制问题,不仅在理论上饶有兴趣,而且在实践中尤为重要,许多植物器官如花、果实、子粒、叶等的脱落常会给生产带来严重的损失(李曙轩,1957)。

许多作者指出器官脱落是由于有机物质的供应不足或失调(金成忠等,1956; Mason, 1922),也有不少作者则认为生长素是调节器官脱落的主要因素(娄成后等,1954; Addicott, 1955; Gardner & Copper, 1943; La Rue, 1936; Shoji 等,1957; Wetmore and Jacobs, 1953)。

关于器官脱落与呼吸代谢的关系,Sampson (1918)最早发现离层组织中氧化酶的活性随年龄而增高;Heimicke (1918)指出离层细胞的过氧化氢酶活性比邻近组织高;Carns (1951)报告离层分离过程中有呼吸升高的现象,表明呼吸与器官脱落的紧密联系。我们的研究结果也表明了器官脱落与呼吸的末端氧化系统有着密切的关系(薛应龙等,1964),至于器官脱落是否与呼吸途径有关,我们尚未看到报导。

本文通过利用与呼吸途径有关的中间代谢物和有机物质以及不同呼吸途径的抑制剂处理

薛应龙,龙天心*,朱雨生**,复旦大学生物学系植物生理教研组。此文发表于 1964 年“植物生理学报”1 卷 1 期 90 - 99 页。

① 1964 年 4 月 13 日收到。

* 现在中国科学院亚热带植物研究所(厦门)。

** 现在中国科学院植物生理研究所(上海)。

本文采用缩写名称如下:HMP,已糖单磷酸酯途径(即磷酸戊糖途径);EMP-TCA,糖酵解和三羧酸循环;G - 6 - P,葡萄糖- 6 -磷酸;6 - PG, 6 -磷酸葡萄糖酸;NADP,烟酰胺腺嘌呤二核苷酸磷酸(辅酶 II);IAA,吲哚乙酸。

的方法探讨大叶黄杨叶柄脱落与呼吸途径的联系。

二、材料与方法

本文试验系用大叶黄杨(Euonymus japonica Thunb L f)带叶柄茎段进行离体培养。离体培养根据 Addicott 等(1948)的方法,并略加改进。每次试验选节位、年龄、大小一致的黄杨枝条,剪取其茎段,每段包括一对对生叶片,切去叶片,保留叶柄,如图 1 所示。

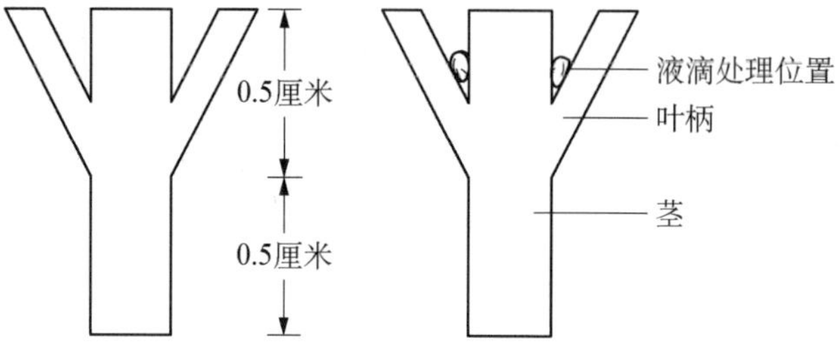

图 1　黄杨离体叶柄培养材料

用直径为 6 厘米的培养皿,内铺 2%约 0.3 厘米厚的一层琼胶,取 10 个上述切好的离体材料分二行插于琼胶内,在 25℃恒温箱中培养,隔一定时间检查叶柄脱落的数目,叶柄经用小镊子轻压而掉下者作为脱落的标准(见图 2)。这样培养的离体材料虽然不用无菌操作,在短期内也能保证无细菌感染。

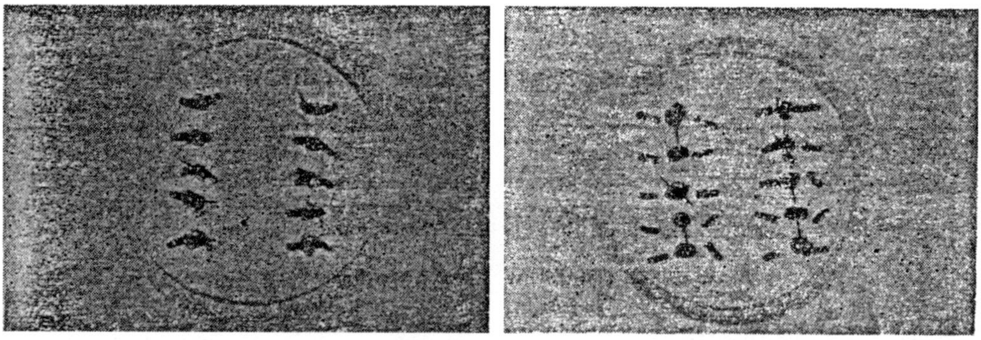

(a)　(b)

图 2　(a) 离体材料培养俯视图,培养初,未脱落　(b) 离体材料培养俯视图,培养 24 小时脱落

试验中用作处理的各种有机物质及呼吸途径抑制剂是以溶液滴在叶腋中,对照则滴蒸馏水(见图 1)。需要注意的是,离体培养的所有试验虽然都是在相同的条件下进行的,但由于材料年龄的不同,因此脱落的绝对数值也是有变化的。

三、试验结果

1. 各种碳水化合物及有关中间代谢物对离体黄杨叶柄脱落的影响

有机养料特别是碳水化合物与器官脱落之间的关系已有不少报道(Barnett 等,1953;

Heimicke，1918）。由试验结果可以看出，用各种碳水化合物及有关中间代谢物处理时也证明有延迟离体黄杨叶柄脱落的效应，但不同的化合物其效应是不尽相同的，如图3所示。

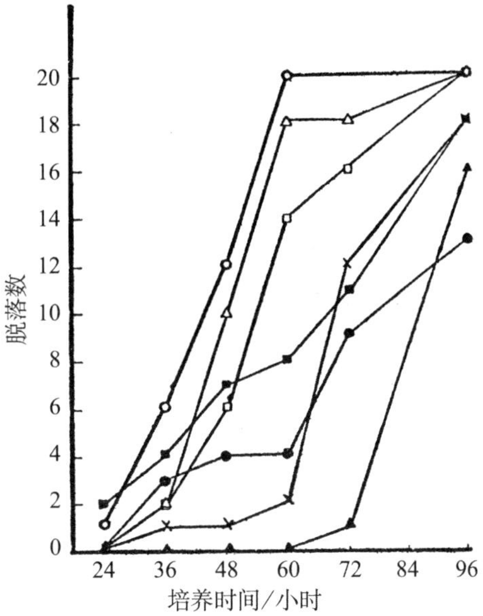

图3　各种碳水化合物对延迟离体黄杨叶柄脱落的效应

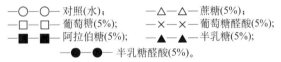

—○—○— 对照(水)；　—△—△— 蔗糖(5%)；
—□—□— 葡萄糖(5%)；　—×—×— 葡萄糖醛酸(5%)；
—■—■— 阿拉伯糖(5%)；　—▲—▲— 半乳糖(5%)；
—●—●— 半乳糖醛酸(5%)。

从图中可见葡萄糖醛酸、阿拉伯糖、半乳糖和半乳糖醛酸都有较明显延迟脱落的效果，对照在培养60小时后叶柄全部脱落，而上列化合物处理的，脱落分别为2、8、0和6，即使时间延迟至96小时仍未全部脱落。呼吸代谢的基本底物葡萄糖及蔗糖作为一般有机营养也略有延迟叶柄脱落的作用，但效果都比上述几种碳水化合物为差。这种影响的规律，在离体培养蚕豆花梗脱落的情况中，也获得类似趋势的结果，如表1所示。

表1　各种碳水化合物对延迟蚕豆花梗脱落的效应

处理 \ 脱落数 \ 时间/小时	48	60	72
对照(水)	16	18	20
蔗糖(5%)	9	16	17
葡萄糖(5%)	9	12	16
葡萄糖醛酸(5%)	13	15	16
阿拉伯糖(5%)	14	14	18
半乳糖(5%)	8	12	17
半乳糖醛酸(5%)	4	5	6

从表 1 中可见效应特别显著的是半乳糖醛酸，对照在培养 72 小时后已全部脱落，而半乳糖醛酸处理的只脱落 6 个，还不到三分之一。离体黄杨叶柄培养 96 小时后也是半乳糖醛酸处理的脱落最少（见图 3）。

2. 吲哚乙酸（IAA）与抗坏血酸对离体黄杨叶柄脱落的影响

IAA 阻止器官脱落的效应已有很多报道（Addicott 等，1951；La Rue，1936；Wetmore 和 Jacobs，1953）。用 IAA 处理也观察到有明显阻止离体黄杨叶柄脱落的效果，而且随着 IAA 浓度增高，阻止脱落的效应越明显，如表 2 所示。

表 2 IAA 抑制离体黄杨叶柄脱落的效应与浓度的关系

时间 \ 脱落数 \ 处理	对照（水）	IAA 浓度（ppm）						
		0.1	0.5	1.0	5.0	10.0	20.0	50.0
24 小时	20	19	16	9	1	0	0	0

进一步用 50 ppm（$1\ \mathrm{ppm}=10^{-6}\ \mathrm{m}$）的 IAA 处理以观察其延迟叶柄脱落的时效，结果对照在培养一天后已基本全部脱落，而 IAA 处理的则经过 10 天只脱落了 65%，如表 3 所示。

表 3 IAA 阻止离体黄杨叶柄脱落的时效

处理 \ 脱落数 \ 时间/天	1	2	3	4	5	6	7	8	9	10
对照（水）	19	20	—	—	—	—	—	—	—	—
IAA (50 ppm)	0	0	2	4	5	8	10	11	12	13

IAA 阻止脱落的效应在有充分糖类供应时，较之单独 IAA 处理的更显著，如图 4 所示。

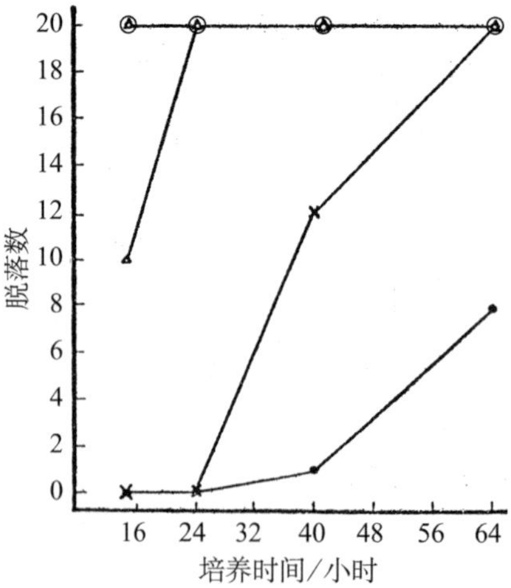

图 4 IAA 和葡萄糖及混合处理对离体黄杨叶柄脱落的影响

注：—○—○— 对照； —△—△— 葡萄糖(2.5%)；
—×—×— IAA(10 ppm)； —●—●— IAA+葡萄糖。

从图中可见单独的葡萄糖效果极微，单独的IAA虽然阻止脱落的效果相当显著，但IAA加葡萄糖时效果尤为显著，前者在培养64小时后，离体黄杨的叶柄已全部脱落，而后者只脱落了40%。这与Aarts所报道的蔗糖能加强萘乙酸阻止羽扇豆(Lupinus polyphyllus)的落花和金成忠等(1956)曾指出的没有糖的供应，生长素不能有效地阻止棉铃的脱落是一致的。

有意思的是，用IAA的前体色氨酸处理黄杨叶柄时，发现也具有类似IAA的作用，对照在48小时已全部脱落，而10^{-3} mol/L的色氨酸到96小时才全部脱落，色氨酸浓度为5×10^{-3} mol/L时只脱落60%，如图5所示。

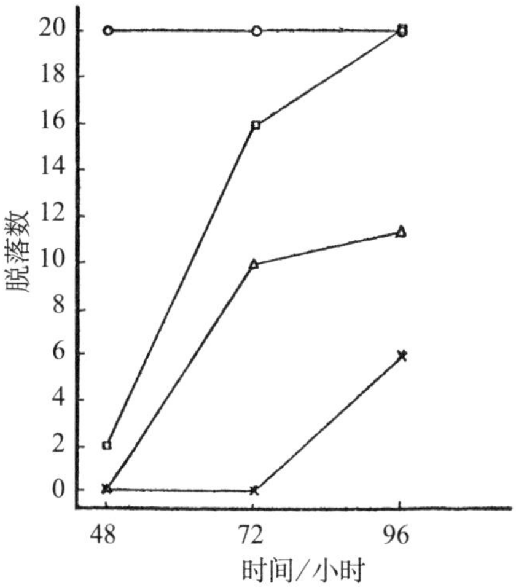

图5 不同浓度的色氨酸对离体黄杨叶柄脱落的影响

注：—○—○— 对照； —□—□— 色氨酸(10^{-3} mol/L)；
—△—△— 色氨酸(5×10^{-3} mol/L)； —×—×— 色氨酸(10×10^{-3} mol/L)。

Marré等(1957)曾指出IAA的作用机制与抗坏血酸的氧化还原系统有密切联系。我们的试验证明用适当浓度的抗坏血酸处理时，也有阻止离体黄杨叶柄脱落的效应，例如浓度较高(1%)时有明显抑制脱落的作用，如表4所示。

表4 不同浓度的抗坏血酸对离体黄榻叶柄脱落的影响

处理 \ 脱落数 \ 时间/小时		36	48
对照(水)		5	20
抗坏血酸	0.01%	7	20
	0.1%	5	19
	0.5%	4	15
	1.0%	0	4

3. 不同呼吸途径抑制剂对离体黄杨叶柄脱落的影响

从上面的试验结果表明离体黄杨叶柄的脱落可能与呼吸途径有密切联系。我们用不同呼吸途径的抑制剂处理来观察对叶柄脱落的影响，获得了极有意义的结果。适当浓度的 EMP-TCA 抑制剂如氟化钠、碘乙酸和丙二酸处理极显著地抑制了离体黄杨叶柄的脱落。例如对照培养 48 小时后叶柄全部脱落，而 0.05 mol/L 氟化钠处理的只脱落了 30%，0.005 mol/L 碘乙酸全部不脱落，0.1 mol/L 丙二酸脱落了 10%；对老的黄杨叶柄这种效应更加明显，如表 5 所示。

表 5　EMP-TCA 抑制剂对离体黄杨叶柄脱落的影响

处理＼结果		48 小时后脱落数
对照(水)20		
氟化钠	0.05 mol/L	6
	0.025 mol/L	20
碘乙酸	0.005 mol/L	0
	0.0025 mol/L	14
	0.001 mol/L	19
丙二酸	0.1 mol/L	2
	0.05 mol/L	18
	0.025 mol/L	20

如果用 HMP 途径抑制剂正磷酸钠（沈善炯等，1959）处理时，则获得了相反的效应，即促进了离体黄杨叶柄的脱落，例如，对照培养 18 小时尚无一个脱落，而 0.1 mol/L 正磷酸钠处理的已 100%脱落，在珊瑚树叶柄和无花果叶柄离体培养中也获得类似结果，如表 6 所示。

表 6　正磷酸钠对几种离体材料叶柄脱落的影响

处理＼脱落数＼时间/小时＼材料	黄杨叶柄		珊瑚树叶柄		无花果叶柄	
	18	24	48	60	48	60
对照(水)	0	20	0	20	0	6
Na_3PO_4(0.1 mol/L)	20	—	20	—	12	20

正磷酸钠的效应并非由于无机磷的缘故，用同样浓度的磷酸氢二钠处理时，丝毫没有促进叶柄脱落的作用，它与对照脱落的情况基本上完全一致，如表 7 所示。

表 7　不同磷酸盐对离体黄杨叶柄脱落的影响

处理＼检查时间/小时	18	24	36
对照(水)	0	0	16
Na_3PO_4(0.1 mol/L)	6	12	20
Na_2HPO_4(0.1 mol/L)	0	0	15

有趣的是正磷酸钠促进叶柄脱落的效果可以被半乳糖醛酸、IAA 和抗坏血酸等所消除，结果如表 8 所示。

表 8　抑制脱落的化合物对 Na_3PO_4 效应的拮抗作用

处理 \ 脱落数 \ 培养时间/小时	24
对照(水)	0
Na_3PO_4(0.1 mol/L)	10
Na_3PO_4(0.1 mol/L)+半乳糖醛酸(2.5%)	2
Na_3PO_4(0.1 mol/L)+抗坏血酸(0.5%)	0
Na_3PO_4(0.1 mol/L)+IAA (50 ppm)	0

从前面的试验中已证明半乳糖醛酸、IAA 和抗坏血酸都有显著阻止黄杨叶柄脱落的效果。

四、讨论

很早就有人指出器官脱落是由于离层细胞中胶层溶解的结果(Lloyd, 1916)；娄成后等(1954)也观察到棉铃脱落和白菜脱帮过程中离层细胞中胶层的溶解。已有报道指出中胶层的溶解是由于不溶性的果胶钙转化为可溶性的果胶酸所致(Facey, 1950; Lee, 1911)。在棉铃脱落过程中不仅看到离层形成时中胶层的溶解，而且还观察到中胶层溶解后有多糖醛酸的产生(朱雨生等，1964)。Wood (1960)的评述中指出果胶物质的基本组成单位是半乳糖醛酸，还指出当可溶性果胶进一步水解时有半乳糖醛酸及部分半乳糖和阿拉伯糖产生。由此可见，加入某些形成果胶物质的中间代谢物可能阻止果胶物质的溶解，从而阻止离层的形成。我们的试验证实了这一点。例如葡萄糖醛酸、半乳糖醛酸、半乳糖、阿拉伯糖等都有延迟离体黄杨叶柄离层形成的效果，而半乳糖醛酸的效果最为显著也是完全可以理解的。在棉花研究中也发现半乳糖有延迟棉铃离层形成的作用(朱雨生等，1964)。果胶物质的代谢已证明与葡萄糖降解的代谢途径有紧密关系(Gibbs, 1959)。从葡萄糖醛酸具有一定延迟离层形成的作用也可证明这种关系。有意思的是蔗糖和葡萄糖并无阻止离体叶柄脱落的效果，可见有机养料的供应并非是影响器官脱落的主要原因，重要的是葡萄糖在降解过程中的中间代谢途径。根据实验结果可以推断直接氧化途径(Hexo monophos phate, HMP)有利于延迟离层的形成。

根据 IAA 和抗坏血酸处理延迟离体叶柄脱落的结果可以进一步证实上述推论。IAA 可以阻止器官脱落早有报道(La Rue, 1936; Shoji 等，1951; Addicott 等，1951; Wetmore 和 Jacobs, 1953)。关于它的作用，有的认为在于加强了有机养料的供应(张德颐，1963; Booth 等，1962)；有的认为在于促进了果胶甲基酯酶(Pectin methylesterase, PME)的活性，因而阻止了果胶物质的溶解，也阻止了离层的形成(Bryan 和 Newcomb, 1954; Yager, 1960)；近年来有人指出 IAA 有增强 HMP 的作用(Shaw 等，1958)。从试验结果看来，IAA 加葡萄糖对延迟叶柄脱落的效果较之单独的 IAA 和葡萄糖都要明显，这与 IAA 加强葡萄糖的 HMP 中间代谢因而阻止果胶物质的溶解是相呼应的。另一方面，已有文献指出 HMP 的中间代谢产物莽草酸所形成的色氨酸是合成 IAA 的前身(Sprinson，1960；Gibson 等，1962)，我们的试验也证明色氨酸延迟叶柄脱落的效果与 IAA 很相似，由此可见 IAA 与 HMP 之间的关系是非常密切的。

关于抗坏血酸，金成忠等(1956)证明棉花在开花受精后抗坏血酸含量显著比不受精的要高，这一事实可能与阻止棉铃脱落有关。试验证明抗坏血酸有延迟黄杨叶柄脱落的效应。已有报道指出抗坏血酸与谷胱甘肽的氧化还原有密切关系(Marré 等，1957)；而谷胱甘肽的氧化还原与 NADP 系统及 HMP 途径的 G－6－P 脱氢酶和 6－PG 脱氢酶都有联系(Axelrod 等，1953；Barnett 等，1953)；而且抗坏血酸的氧化还原也直接与 NADP 系统有关(Beevers，1961；Marré，1961)。由此可见，抗坏血酸延迟离层形成的作用与 IAA 一样，也可能是加强了 HMP 途径的缘故。

从利用不同呼吸途径抑制剂的试验中，更加具体证实了上述的推论。试验证明用 EMP 抑制剂氟化钠和碘乙酸(James，1953；Hackett，1960)及 TCA 抑制剂丙二酸处理有明显地延迟离体叶柄脱落的效果，而 HMP 抑制剂正磷酸钠(Shen San-Chun 等，1959)则促进了叶柄脱落。可见抑制了 EMP-TCA 途径也就阻止了离层形成；而抑制 HMP 途径时则促进离层的形成。特别有意思的是，半乳糖醛酸、IAA 和抗坏血酸都有逆转正磷酸钠促进脱落的效果。这种相互作用也有利于阐明上述推论。

根据上面的讨论，无论是从各种碳水化合物及其中间代谢物，以及与呼吸途径有关的物质或者从呼吸途径抑制剂的试验结果看来，都使我们相信器官脱落与否，主要决定于离层细胞所进行的呼吸途径。离层的形成是由于 EMP-TCA 途径活跃的结果，而 HMP 途径的加强则有利于延迟或阻止离层的形成。

参考文献

[1] 朱雨生、尤天心、薛应龙，1964：棉铃脱落与碳水化合物代谢的关系. 复旦学报(即将刊出).

[2] 李曙轩，1957：落叶落花和落果. 科学出版社.

[3] 金成忠等，1956：有机养料在棉花蕾铃脱落中的作用. 植物学报，5：78—107.

[4] 金成忠等，1956：光线强度对棉花蕾铃脱落的影响。植物学报，5：117—136.

[5] 娄成后等，1954：类似生长素药剂对于延迟植物器官脱落及相关的生理效应. 植物学报，3：167—189.

[6] 张德颐，1963：植物体内光合产物的运输和分配以及调节运输、分配的生理基础. 研究生论文. 中国科学院植物生理研究所.

[7] 薛应龙等，1964：植物呼吸代谢的生理意义. II. 器官脱落过程中呼吸代谢的变化(未发表).

[8] Addicott，F. T. et al.，1949：A method for the study of foliar abscission in vitro. Plant Physiol. 24：537－539.

[9] Addicott，F. T. et al.，1951：Acceleration and retardation of abscission by IAA. Science 114：688－689.

[10] Addicott，F. T.，1955：Physiology of abscission. Ann. Rev. of Plant Physiol. 6：211－238.

[11] Axelrod，B. and R. S. Bandurski，1952：Oxidative metabolism of hexose phosphate by higher plants. Fed. Proc. 11：182.

[12] Axelrod，B. et al.，1953：The metabolism of hexose and pentose phosphate in higher plants. J. Biochem. Chem. 202：619－634.

[13] Barnett，R. C. et al.，1953：Phosphogluconic dehydrogenase in higher plants. Plant Physiol. 28：115－

122.
[14] Beevers, H. , 1961: Respiratory metabolism in plant. Row Peterson & Company.
[15] Booth, A. J. et al. , 1962: Effect of IAA on the movement of nutrients within plants. Nature 194: 204 - 205.
[16] Bryan, W. H. and E. H. Newcomb, 1954: Stimulation of PME actvity of cultured tobacco pith by IAA. Physiol. Plantarum 7: 290 - 297.
[17] Carns, H. R. , 1951: O_2 respiration and other critical factors in abscission. Doctoral thesis Univ. of Calif. 引自 Addicott, F. T. , 1955: Physiology of abscission.
[18] Facey, V. , 1950: Abscission of leaves in Fraxinus americana, L. New Phytol. 49: 103 - 116.
[19] Gardner, F. E. and W. C. Copper, 1943: Effectiveness of growrh substances in delaying abscission of Coleus petioles. Bot. Gaz. 105: 80 - 87.
[20] Gibbs, M. , 1959: The metabolism of carbohydrates. Ann. Rev. of Plant Physiol. 9: 329 - 378.
[21] Gibson, M. I. et al. , 1962: The branch point in the biosynthesis of the aromatic amino acids. Nature 195(4847): 1173.
[22] Hackett, D. P. , 1960: Respiratory inhibitors. Handbuch der Pflanzen-Physiologie XII/2: 23 - 41.
[23] Heimicke, A. J. , 1918: Proc. Amer. Soc. Hort. Sci. 16: 76 - 83. 引自 Addicott, F. T. , 1955: Physiology of abscission.
[24] James, W. 0. , 1953: The use of respiratory inhibitors. Ann. Rev. of Plant Physiol. 4: 59 - 90.
[25] La Rue, C. D. , 1936: The efect of auxin on the abscission of petioles. Proc. Natl. Acad. Sci. 22: 254 - 259.
[26] Lee, E. , 1911: The morphology of leaf fall. Ann. Bot. 25: 51 - 107.
[27] Lloyd, F. E. , 1916: Abscission off lower buds and fruitin Gossypium and itrelation to environmental changes. Trans. Roy. Soc. Can. 10: 55 - 62.
[28] Marré, E. and O. Arrigoni, 1957: Metabolic relation to auxin. I. The effect of auxin on glutathione and the effect of glutathione on growth of isolated plant parts. Physiol Plantarum 10(2): 289.
[29] Marré, E. , 1961: Phosphorylation in higher plants. Ann. Rev. Plant Physiol. 12: 195 - 218.
[30] Mason, T. C. , 1922: Growth and abscisson in sea island cotton. Ann. Bot. 36: 457 - 484.
[31] Sampson, H. C. , 1918: Chemical changes accompanying abscission in Colcus Blumei. Bot. Gaz. 66: 32 - 53.
[32] Shaw, M. et al. , 1958: Some effect of IAA and MH on the respiration and flowering of wheat. Can. J. Bot. 36(2): 233 - 237.
[33] Shen San-chun (沈善炯) and Chen Jaun-pio (陈俊标), 1959: Pentose metabolism and the influence of orthophosphate on the paths of sugar degradation of streptomyces aureofaciens. Scientia Sinica 8(7): 733 - 745.
[34] Shoji, K. et al. , 1951: Auxin in relation to leaf blade abscission. Plant Physiol. 26: 189 - 191.
[35] Sprinson, D. B. , 1960: The biosynthesis of aromatic compound from D-glucose. Adv. carbohydrate Chem. 15: 235. Academic Press.
[36] Wetmore, R. H. and W. P. Jacobs, 1953: Studies on abscission: The inhibitory effect of auxin. Amer. J. Bot. 40: 272 - 276.
[37] Wood, R. K. S. , 1960: Pectic and Cellulolytic enzymes in plant disease. Ann. Rev. Plant Physiol. 11: 299 322.
[38] Yager, R. E. , 1960: Possible role of pectic cenzymes in abscission. Plant Physiol. 35(2): 157 - 162.

THE PHYSIOLOGICAL SIGNIFICANCE OF RESPIRATORY METABOLISM IN PLANTS

I. THE RESPIRATORY PATHWAYS IN RELATION TO ABSCISSION

For studying the effects of various substances on abscission of the leaf petiole of Euonymus japonica Thunb. L. f. the petiole explant test described by Addicott et al. was used with some modifications made by us. The position of treatment of various substances is showed in Figure 1. The notable fact should be mentioned that explant test performed under similar conditions showed a considerable degree of variation due to the differences in age of material and in time.

The results of present investigations justify the following conclusions:

1. Application of the various metabolites related to hexose monophosphate pathway, such as glucuronic acid, galacturonic acid, galactose and arabinose, gave rise to the depression of abscission of the petiole explant obviously.

2. The longer abscission time is observed by application of IAA, tryptophan and ascorbic acid. The effect is somewhat proportional to the concentration applied. It suggests that the action of these substances may be due to the enhancement of the HMP pathway.

3. Treatment with sodium fluoride, iodoacetic acid and malonic acid, the well known EMP-TCA inhibitors, prevent abscission strongly.

4. Treatment with sodium phosphate, presumably as a HMP inhibitor, accelerates abscission demonstrably. The interesting evidence shows that the galacturonic acid, IAA and ascorbic acid can overcome this acceleration of activity.

5. Together with these observations, wich constitute the available evidence, we have come to the conclusion that the abscission is closely related to the respiratory pathways. The predominance of EMP-TCA route accelerates the abscission process, while the prevalence of HMP pathway delays it.

Hsueh Y. L., T. S. You AND Y. S. Zhu(Laboratory of Plant Physiology, Fu Tan University)

植物的脂肪合成及其调节①

Ⅰ. 油菜种子形成期间葡萄糖降解的途径

摘要

成熟期间油菜荚中存在整个糖酵解途径的酶系：己糖激酶、磷酸己糖异构酶、磷酸己糖激酶、醛缩酶、3-磷酸甘油醛脱氢酶、磷酸甘油酸激酶、磷酸甘油酸变位酶、烯醇化酶和丙酮酸激酶；也存在磷酸戊糖支路的全部酶系：葡萄糖-6-磷酸脱氢酶、6-磷酸葡萄糖酸脱氢酶、磷酸戊糖异构酶、磷酸戊糖表异构酶、辅酮酶和辅醛酶。借用化学分析和酶活力测定证明核糖-5-磷酸能转化为景天庚糖、丙糖、果糖-1,6-二磷酸、果糖-6-磷酸、葡萄糖-6-磷酸、3-磷酸甘油酸和丙酮酸，因而证明有完全的戊糖循环在周转。文中还指出，糖酵解途径和磷酸戊糖支路可以通过磷酸丙糖和磷酸果糖相联系。

一、引言

油菜、大豆等是我国重要的油料作物。提高种子含油量、增加食油产量对于国民经济有着极为重要的意义。我们认为，从物质代谢及其调节着手是达到这一目的较为根本和有效的手段。可惜，目前关于油菜、大豆种子脂肪代谢方面的研究工作报道还很少；而涉及脂肪合成调节的探讨尚未见诸报道。为此，近年来我们以油菜、大豆为材料，对油料作物种子形成过程中脂肪代谢与糖代谢的关系及其调节机制作了一些探索。本文先探讨油菜种子接近成熟时葡萄糖降解途径。

二、材料与方法

1. 材料

开花后30～40天的油菜荚取自本所试验农场，品种为胜利油菜，制成丙酮粉，备作测定。

(1) 试剂。所有试剂均系商品。G-6-P、6-PG、F-6-P、FDP、R-5-P、ATP和3-

施教耐，朱雨生，谭常，中国科学院植物生理研究所。
此文发表于1965年“植物生理学报”2卷1期33-43页。

① 1964年9月14日收到。

本文采用简称如下：HMP途径：磷酸戊糖支路。EMP途径：糖酵解途径。TCA循环：三羧酸循环。NADP：烟酰胺腺嘌呤二核苷酸磷酸(即辅酶II)。NAD：烟酰胺腺嘌呤二核苷酸(即辅酶I)。ATP：腺嘌呤核苷三磷酸。ADP：腺嘌呤核苷二磷酸。G-6-P：葡萄糖-6-磷酸。6-PG：6-磷酸葡萄糖酸。R-5-P：核糖-5-磷酸。F-6-P：果糖-6-磷酸。FDP：果糖-1,6-二磷酸。3-PGA：3-磷酸甘油酸。2,3-DPGA：2,3-二磷酸甘油酸。Ru-5-P：核酮糖-5-磷酸。X_u-5-P：木酮糖-5-磷酸。EDTA：二乙胺四乙酸。Tris：羟甲基氨基甲烷。PCMB：对氯汞苯甲酸。

PGA、2,3－DPGA 都是钡盐。NAD 纯度为 100%，NADP 纯度为 75%或 78%。所有钡盐使用前均变成钾盐，即用稀盐酸溶解，加稍过量饱和 K_2SO_4，沉淀下来的 $BaSO_4$ 离心除去，清液用稀 KOH 调 pH 至 7.5，再稀释到所需浓度。

(2) 酶结晶醛缩酶。按 Taylor (1948)的方法，从兔肌中制备，有时由中国科学院生物化学研究所赠送。G－6－P 脱氢酶由本所王业芹等同志供给，从酵母中按 Kornberg 等方法(1955)制备和提纯。

2. 方法

(1) 丙酮粉的制备。按照 Hageman 和 Arnon 的方法(1955)，10 克鲜荚剪碎后加 15 ml EDTA (0.03 mol/L)和磷酸钾(0.1 mol/L)缓冲液(pH 8.2)研磨成细末。缓缓加入 100 ml 用干冰冷却到－15℃至－20℃的冷丙酮，用布氏漏斗迅速抽气过滤，再用 40 ml 冷丙酮洗三次，抽干。丙酮粉在室内放置 1 小时左右，至无丙酮味，然后转移到盛 P_2O_5 的真空干燥器中干燥。全部操作在 0—5℃进行。干燥后的丙酮粉置于小广口瓶中在－15℃下保存。一般保存时间不超过一个月。

(2) 酶液的提取。按照 10 克鲜组织加 25 ml EDTA (0.001 5 mol/L)和磷酸钾(0.01 mol/L)缓冲液(pH 7.2)的比例，向丙酮粉中加缓冲液研磨提取。用四层纱布过滤，弃去残渣。滤液用 10 000×g 冷冻离心 5～10 分钟，测定 G－6－P 脱氢酶和 6－PG 脱氢酶。其余酶测定一般取用 5 000 转/分，离心 20 分钟左右的上清液。离心液用滤纸或玻璃毛过滤，几小时内进行测定(酶液未经透析，含有一定量的内源底物)。

(3) 酶活力测定。葡萄糖-6－磷酸脱氢酶和 6－磷酸葡萄糖酸脱氢酶根据还原型 NADP 在 340 μm 波长处的吸收峰，用 Beckman DU 型或日立分光光度计测定。转化 R－5－P 的酶系的测定是以酶液和 R－5－P 保温，一定时间后，测定 R－5－P 的减少。己糖激酶测定是以葡萄糖和 ATP 与酶液保温，利用内源 G－6－P 脱氢酶，用日立分光光度计测定 340 μm 处的 NADP 的还原作用来测定形成的 G－6－P 量。磷酸己糖异构酶按 Slein (1955)方法，以 G－6－P 为底物测定形成的 F－6－P，F－6－P 按 Roe (1934)法测定。磷酸己糖激酶活力是以 F－6－P 和 ATP 作用后生成的 FDP 表示。醛缩酶按 Sibley (1949)法略作修改，以巴比妥酸为缓冲液，亚硫酸氢钠为固定剂，测定 FDP 分解产物(磷酸丙糖)。3－磷酸甘油醛脱氢酶按 Black 等(1962)方法，根据 NADH 在 340 μm 的光吸收用日立分光光度计测定。3－磷酸甘油酸激酶采用 Axelrod (1953)的方法，以 3－PGA 为底物，酶反应后产生的 1,3－DPGA 变成羟肟酸后测定。从 3－PGA 至丙酮酸形成的酶系是以 3－PGA 为底物，加上各种辅因子，按 Kachmar (1953)法测定形成的丙酮酸。

(4) 糖磷酯纸谱分离(Bandurski 1951)。采用杭州 1 号滤纸(20 cm×30 cm)，先用酸相(甲醇∶甲酸∶水＝80∶15∶5)洗一次。样品于碱相中(甲醇∶NH_3∶水＝60∶10∶30)下行 36 小时(2℃)。显色剂改用钼酸铵(4%钼酸铵∶60%过氯酸∶IN HC1∶水＝25∶5∶10∶60)和 H_2S 显色。凡有糖磷酯处变成蓝色。

(5) 产物分析。己糖用蒽酮法或硫酸-半胱氨酸法(Dishe 等，1949)测定。果糖用 Roe 法测定。戊糖和景天庚糖用甲基间苯二酚法测定(Horecker 等，1953)，加热时间为 40 分钟。戊酮糖按 Dishe 等(1951)的咔唑反应测定。丙糖用二硝基苯肼法测定(Sibley，1949)。G－6－P 的测定是反应液中加入 NADP 和纯化的酵母 G－6－P 脱氢酶，测定 340 μm 处的光吸收。FDP 是借结晶醛缩酶的作用，测定形成的磷酸丙糖量(反应条件同醛缩酶一样)。

三、结果与讨论

1. EMP途径的酶系活力测定

以葡萄糖为底物，加上ATP、Mg^{2+}和NADP，与酶液保温后不久，在340 μm波长处有光吸收现象(见图1)，说明有己糖激酶的存在，所形成的G-6-P在G-6-P脱氢酶作用下，致使NADP还原。

酶液作用于G-6-P，用间苯二酚测得有果糖的形成，表明有磷酸己糖异构酶的活力。以FDP为底物，与酶液保温30分钟后，有磷酸丙糖的产生，标志着有醛缩酶的作用。若以F-6-P代替FDP为底物，在同样的条件下，再加上ATP和Mg^{2+}，在反应液中也能测出磷酸丙糖的增加，说明还有己糖激酶的存在。

以FDP为基质，加上NAD和酶液等共同保温，用分光光度计测定，在340 μm波长处有光密度的增加，表示有还原辅酶I的形成(见图2)。

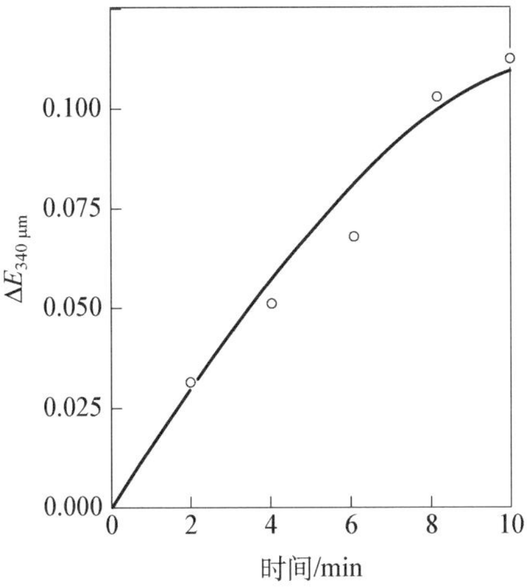

图1　己糖激酶的活力测定

反应液含：甘氨酰甘氨酸125 μmoles，pH 7.5，$MgCl_2$ 20 μmoles，葡萄糖5 μmoles，ATP 2 μmoles，NADP 0.27μmole，酶液0.5 ml，总体积3 ml。空白缺NADP，用日立分光光度计于25℃测定340 μm的光密度变化。

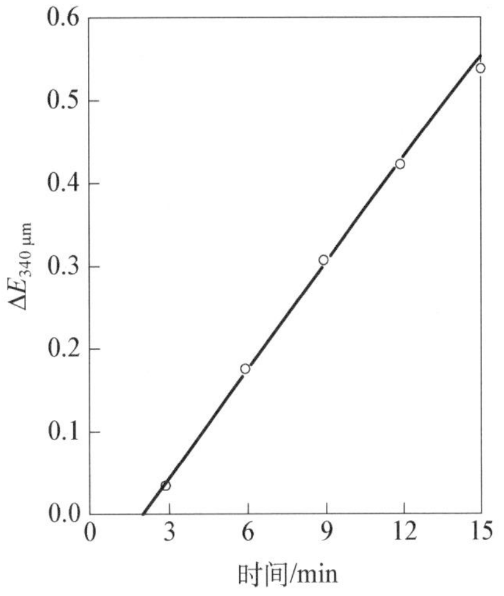

图2　3-磷酸甘油醛脱氢酶的活力测定

反应系统：Tris 150 μmoles，pH 8.5，FDP 10 μmoles，砷酸钠17 μmoles，NaF 10 μmoles，谷胱甘肽15 μmoles，NAD 1 μmole，酶液0.5 ml，总体积3 ml。空白者缺NAD。于室温下用日立分光光度计测定340 μm的光吸收。

这是由于FDP在醛缩酶作用下分解形成磷酸丙糖，再经3-磷酸甘油醛脱氢酶的作用形成还原型的NAD。缺FDP时活力减弱，缺NAF、NAD、谷胱甘肽或砷酸盐时反应不能进行；并且反应能为碘乙酰胺和PCMB所抑制(见表1)，说明油菜荚中3-磷酸甘油醛脱氢酶的性质与微生物(沈善炯等，1957，余微明等，1964)和动物(Cori等，1948)中的相似，都以巯基为它的活力中心。但该酶很不稳定，提取时间稍长或高速离心，活力就大大降低，原因尚未弄清。

表 1 辅因子和 PCMB、碘乙酰胺对 3-磷酸甘油醛脱氢酶活力的影响

反应系统	$\triangle E_{340\ \mu m}$	活力/%
完全*	0.063	100
- FDP	0.017	27.0
- NAD	0	0
-谷胱甘肽	0	0
-砷酸盐	(- Arsenate)	0
- NaF	0.037	58.7
完全**	0.239	100
+PCMB (10^{-3} mol/L)	0.144	60.8
+碘乙酰胺(5×10^{-3} mol/L) (+Iodoacetamide 5×10^{-3} mol/L)	0	0

注：* 反应 9 分钟的计数。
** 反应 10 分钟的计数。
反应系统见图 2 说明。

酶液与 3 - PGA、ATP 和 Mg^{2+} 等保温，有羟肟酸的形成，说明有 1，3 - DPGA 的产生和 3 - PGA 激酶的活力存在(见图 3)，缺 ATP 反应不能进行，缺 Mg^{2+} 时活力大大降低(见表 2)，这也与一般激酶的性质相似。

以 3 - PGA 为底物，加上酶液和辅因子 ADP、Mg^{2+}、2，3 - DPGA 等，反应液中有丙酮酸的产生(见图 4)，说明有 3 - PGA 变位酶、烯醇化酶和丙酮酸激酶的一系列酶活力存在。缺 ADP 时活力下降 83.3%，NaF 能百分之一百的阻止丙酮酸的形成(见表 3)，前者是丙酮酸激酶的辅因子，后者是烯醇化酶的强烈抑制剂。

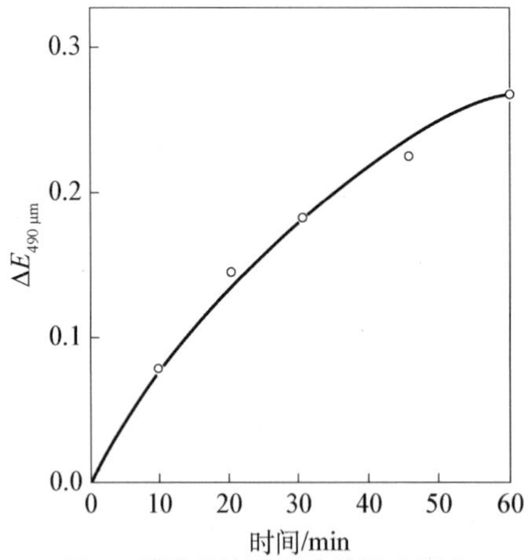

图 3 磷酸甘油酸激酶的活力测定

反应系统：Tris 100 μmoles，PH 7.4，3 - PGA 20 μmoles，ATP 8 μmoles，$MgCl_2$ 6 μmoles，NH_2OH 1 000 μmoles，酶液 0.1 ml，总体积 1.3 ml，38℃，反应 30 分钟。用三氯乙酸终止反应。

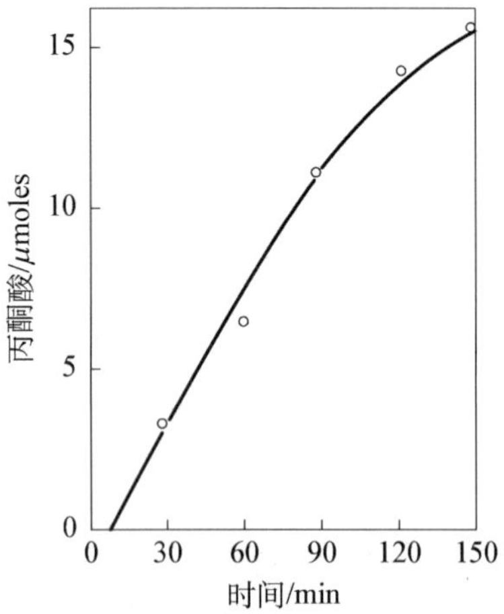

图 4 从磷酸甘油酸至丙酮酸形成的酶系活力的测定

反应系统：Tris 150 μmoles，pH 7.4，3 - PGA 20 μmoles，ADP 10 μmoles，$MgCl_2$ 10 μmoles，2，3 - DPGA 30μg，KCl 25 μmoles，酶液 0.2 ml，总体积 2.0 ml。于 30℃保温 30 分钟，杀死酶液。以二硝基苯肼法测定形成的丙酮酸。

表 2　辅因子对磷酸甘油酸激酶活力的影响

反应系统	$\Delta E_{490\ \mu m}$	活力/%
完全*	0.235	100
- ATP	0.009	3.8
- Mg^{2+}	0.083	35.3

注：* 反应系统见图 3 说明。

表 3　辅因子和 NaF 对从 3 - PGA 至丙酮酸形成的酶系活力的影响

反应系统	形成的丙酮酸/μmoles	活力/%
完全*	1.8	100
- ADP	0.3	16.7
+NaF	0	0

注：* 反应系统见图 4 说明。

至此，在成熟油菜荚中证明有从葡萄糖直至丙酮酸形成的整个 EMP 途径的酶系存在。

2. HMP 途径的酶系活力测定

以 G - 6 - P 或 6 - PG 为底物，加入 NADP 和酶液，在 340 μm 处用分光光度计测得有光密度的增加(见图 5)，说明有葡萄糖- 6 -磷酸脱氢酶和 6 -磷酸葡萄糖酸脱氢酶的活力。该二酶对 NADP 是专一的，NAD 不能代替(见图 6)。

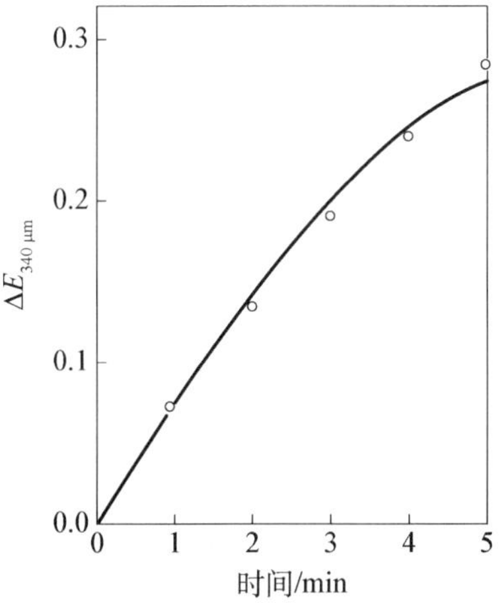

图 5　葡萄糖- 6 -磷酸脱氢酶的活力测定

反应系统：甘氨酰甘氨酸 125 μmoles，pH 7.5，$MgCl_2$ 10 μmoles，G - 6 - P 5 μmoles，NADP 0.27 μmoles(纯度 78%)，酶液 0.5 ml。总体积 3 ml。空白者缺 NADP。反应混合物预先在 38℃水浴中预温 5 分钟，酶液的加入作为反应的开始。用 Beckman DU 型分光光度计测定 340 mμ 的光吸收。

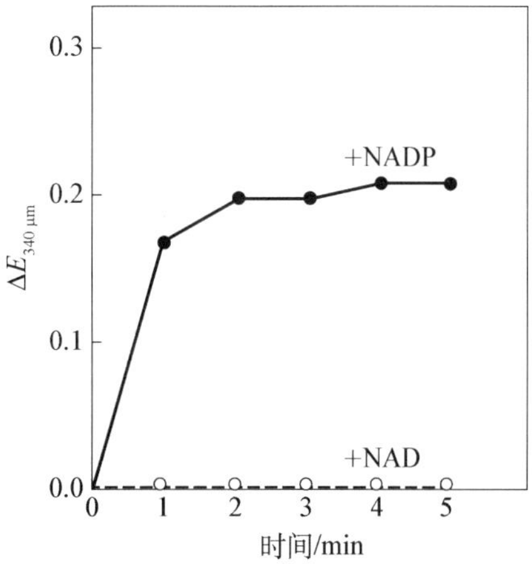

图 6　油菜荚葡萄糖- 6 -磷酸脱氢酶的辅酶专一性

反应系统(见图 5)说明：NADP 0.27 μmole，NAD 1 μmole。

为了检查酶液中是否有 NADP - NAD 转氢酶存在，在葡萄糖- 6 -磷酸脱氢酶反应系统中，当 NADP 还原达到饱和时，加入 NAD，并不能使 340 μm 处的吸收峰继续增加，说明不存在 NADPH - NAD 转氢酶的活力。葡萄糖- 6 -磷酸脱氢酶和 6 -磷酸葡萄糖酸脱氢酶活力能为 PCMB 所抑制(见表 4)，与 Glock 和 Mclean (1953)在鼠肝中描述的性质一样，证明有巯基的存在。

表 4 PCMB 对葡萄糖- 6 -磷酸脱氢酶和 6 -磷酸葡萄糖酸脱氢酶的抑制作用

反应系统	G - 6 - P 脱氢酶*		6 - PG 脱氢酶**	
	$\Delta E_{340\,\mu m}(5')$	抑制/%	$\Delta E_{340\,\mu m}(5')$	抑制/%
对照	0.202	0	0.198	0
+PCMB (10^{-3} mol/L)	0.093	53	0.127	35

注：* 反应系统见图 5 说明。
** 反应系统同 G - 6 - P 脱氢酶，以 6 - PG (5 μmoles)代替 G - 6 - P。

R - 5 - P 与酶液保温后，根据对于甲基间苯二酚的显色反应，证明有 R - 5 - P 的减少(见图 7)。这说明有利用和转化 R - 5 - P 的酶活力存在。缺 NADP 和 Mg^{2+}，或半胱氨酸时活力有所降低(见表 5)，因为转酮反应需 Mg^{2+} 和半胱氨酸。

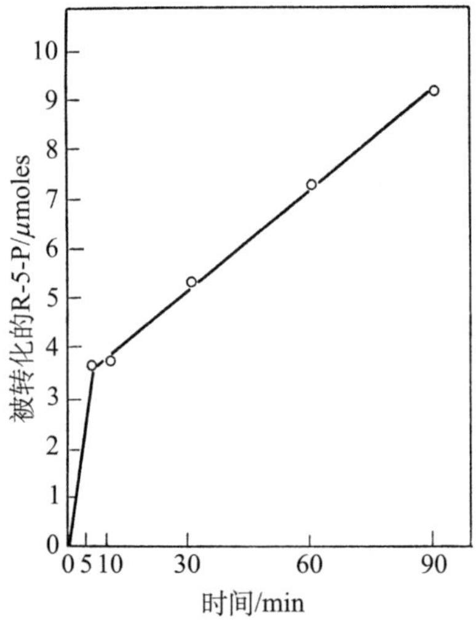

图 7 R - 5 - P“转化”酶系的活力测定

反应系统：R - 5 - P 10 μmoles，甘氨酰甘氨酸 88 μmoles，pH 7.5，$MgCl_2$ 10 μmoles，NADP 0.5 μmole，核黄素 0.05 μmole，半胱氨酸 1.5 μmole，酶液 1.0 ml，总体积 2.5 ml。于 30℃反应半小时，加等体积 20%三氯乙酸终止反应，离心弃去沉淀，清液用甲基间二酚法，根据 670 mμm 和 580 mμm 吸收强度的变化，再按实验中求得的计算公式计算核糖和景天庚糖的含量。光密度变化用 BeckmanDU 型分光光度计或国产 71 型分光光度计测定。

表 5 辅因子对 R - 5 - P“转化”酶系活力的影响

反应系统	转化 R - 5 - P(μmoles)/半小时	活力/%
完全*	0.211	100
- NADP 和 Mg^{2+}	0.079	38.0
-半胱氨酸	0.158	75.4

注：* 反应条件同图 7 相同。

由于酶液中有内源果糖，在利用咔唑反应测定戊酮糖时有干扰，故先将酶液对 1 000 ml 蒸馏水于 0℃透析 28 小时(中途换 3 次)或用 70%饱和度硫酸铵沉淀酶蛋白，再对 1 500 ml 蒸馏水透析 3 小时，然后用甲基间苯二酚和咔唑反应，用分光光度计测定 R－5－P 保温产物的吸收光谱的变化。证明在反应 10 分钟后，670 mμm 的吸收峰下降，540 mμm 的吸收峰升高(见图 8)，前者是核糖的吸收峰，后者是戊酮糖的吸收峰，因而有磷酸戊糖异构酶的活力。

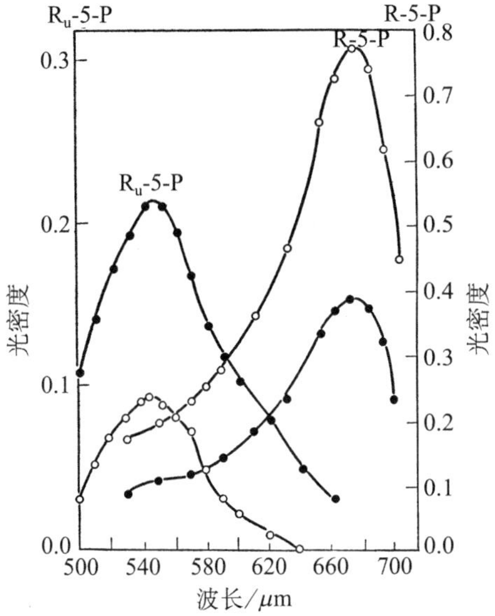

图 8　R－5－P 保温产物的分析(吸收光谱)

○—○ 0 分钟

●—● 保温 10 分钟

反应系统：甘氨酰甘氮酸 88 μmoles，R－5－P 10 μmoles，$MgCl_2$ 10 μmoles，NADP 0.5 μmole，半胱氨酸 1.5 μmoles，核黄素 0.05 μmole，酶液 1 ml(用 70%饱和度硫酸铵提纯过)。总体积 2.6 ml。38℃反应 10 分钟后，加等体积无水酒精杀死。分别用甲基间二酚和咔唑反应测定吸收光谱。

酶液与 R－5－P 共同保温，测得反应液中景天庚糖和丙糖(见表 6)，证明有磷酸戊糖异构酶、磷酸戊糖表异构酶和转酮酶的存在。反应液中也有果糖的产生，这可能由于景天庚糖－7－

表 6　R－5－P 保温产物的分析

	保温时间	
	0 分钟	30 分钟
R－5－P	52.8	26.8
S－7－P	15.6	28.4
果糖	17.3	23.3
FDP	0	3.0
丙糖	0	2.86

注:反应系统：双甘氨肽 725 μmoles (pH 7.5)，$MgCl_2$ 100 μmoles，核黄素 0.25 μmole，半胱氨酸 7.5 μmoles，NADP 2.7 μmoles，R－5－P 52.8 μmoles，酶液 5 ml(含 5.3 mg 蛋白)、总体积 12.5 ml，38℃，保温 30 分钟，加等体积无水酒精杀死测定。

磷酸和磷酸丙糖在转醛酶作用下形成磷酸果糖，或磷酸丙糖通过磷酸丙糖异构酶和醛缩酶的作用再缩合成二磷酸果糖。此外，反应液中还有丙酮酸的产生。可能是 R－5－P 通过转酮转醛作用产生的磷酸丙糖进一步沿共同的酵解途径降解的结果。由此可见，HMP 途径和 EMP 途径可以通过磷酸果糖和磷酸丙糖相互联系起来。

R－5－P 保温不同时间的产物动态变化如图 9 所示，从 0 分钟到 10 分钟，R－5－P 已显著下降，此时景天庚糖、丙糖、己糖和丙酮酸都有上升，说明 R－5－P 循戊糖循环而降解，部分磷酸丙糖进行酵解；从 10 分钟到 30 分钟，R－5－P 继续下降，景天庚糖、己糖和丙酮酸逐渐增加，而丙糖已经开始下降，这表明除了 R－5－P 继续降解外，形成的丙糖进一步沿酵解途径转化为丙酮酸，也可能通过醛缩酶的作用缩合形成己糖；从 30 分钟到 60 分钟，己糖也开始下降；从 60 分钟至 90 分钟则所有产物均有不同程度的下降。推测，这些产物进一步沿酵解和三羧酸循环而氧化，提供合成脂肪酸的醋酸来源和 ATP 的来源。刘承宪等(1964)已证明油菜种子中存在三羧酸循环。

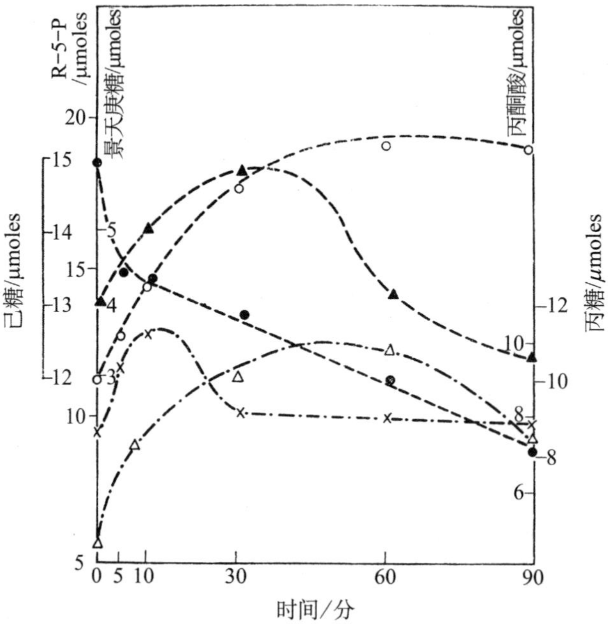

图 9 R－5－P 转化产物的时间变化

反应条件见图 7 说明，保温时间：0 分、5 分、10 分、30 分、60 分和 90 分。核糖、景天庚糖、己糖、丙糖、丙酮酸测定见正文说明。

己糖；●R－5－P；○景天庚糖；×丙糖；△丙酮酸。

将 R－5－P 反应产物减压浓缩，用纸谱进行分离，证明保温后 FDP 的斑点加深，并出现 F－6－P 和 3－PGA 的斑点。用醛缩酶作用于反应液，有磷酸丙糖的产生，也说明有 FDP 的产生。反应液中加入 G－6－P 脱氢酶和 NADP，在 340 μm 的光吸收增加，证明有 G－6－P 的产生；若反应液中加 NAD 的谷胱甘肽、砷酸盐、NaF，则也有 340 μm 光吸收增加的现象，说明有磷丙糖的形成(见图 10)。

若再以 G－6－P 或 F－6－P 为底物，加上辅因子 NADP、Mg^{2+}，与透析过的，或用 70%饱和度的硫酸铵提纯过的酶液(以除去对戊糖测定有干扰的物质)共同保温(30℃)10 分钟,用甲基间苯二酚反应借助分光光度计测定产物的吸收光谱,证明有标志着核糖产生的 670 μm 吸收峰的出现(见图 11),因而在接近成熟的油菜荚中具有完全的 HMP 途径的酶系。

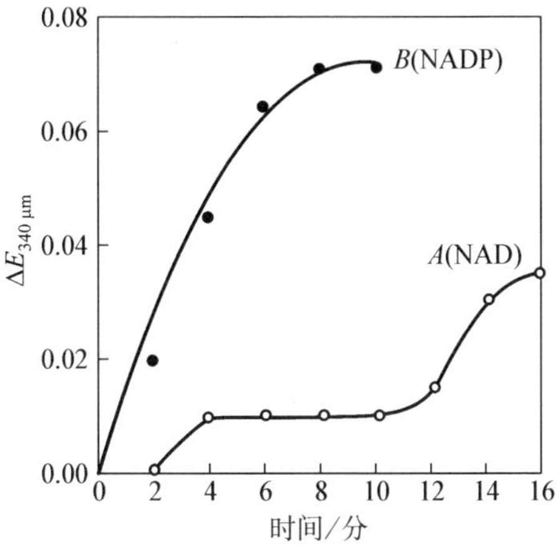

图 10　R-5-P 保温产物对 NADP 和 NAD 的反应

注:反应系统 A. Tris 105 μmoles, PH 7.4, R-5-P 10 μmoles, 核黄素 0.05 μmole, 半胱氨酸 1.5 μmoles, $MgCl_2$ 10 μmoles, 砷酸钠 17 μmoles, NaF 10 μmoles, 谷胱甘肽 15 μmoles, NAD 1 μmoles, 酶液 0.5 ml, 空白缺 R-5-P, 总体积 3 ml; B. 甘氨酰甘氨酸 88 μmoles, PH 7.5, R-5-P 10 μmoles, 核黄素 0.05 μmole, 半胱氨酸 1.5 μmoles, $MgCl_2$ 10 μmoles, NADP 0.5 μmole, 酶液 0.5 ml, 总体积 3 ml, 空白缺 R-5-P。反应温度: 38℃,用日立分光光度计测定光密度的变化。

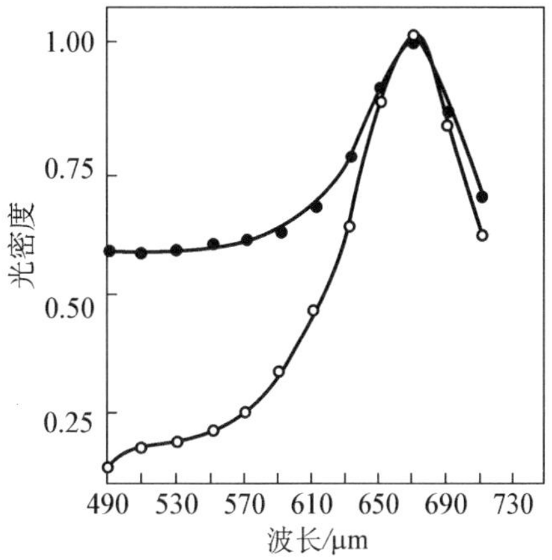

图 11　6-PG 和 F-6-P 保温产物的吸收光谱(甲基间二酚反应)

注:●—● F-6-P 保温产物;○—○ 6-PG 保温产物

反应系统: 甘氨酰甘氨酸 125 μmoles, pH 7.5, $MgCl_2$ 10 μmoles, NADP 0.5 μmole, 6-PG 10 μmoles(或 F-6-P 5 μmoles)。酶液 0.5 ml, 总体积 3 ml。于 38℃反应 10 分钟后加 3 ml 无水酒精杀死。空白预先用无水酒精杀死。反应液用甲基间二酚反应后,用日立分光光度计测吸收光谱。

参考文献

[1] 沈善炯、陈俊标、洪孟民,1957: 金霉菌的糖类代谢 I. Embdem-Meyerhof-Parnas 系统与磷酸已糖的分路代谢. 生理学报,21(3): 302—310.

[2] 余微明、杨常仁、周光宇,1964: 谷氨酸发酵菌—Brevibacterium ketoglutarium nov. sp. 北京 2990-6 的代谢: 葡萄糖的降解途径. 生物化学与生物物理学报,4(1): 46—53.

[3] Axelrod, B. and R. S. Bandurski, 1953: Phosphoglyceryl kinase in higher plants. J. Biol. Chem., 204: 939-948.

[4] Bandurski, R. S. and B. Axelrod, 1951: The chromatographic identification of some biologically important phosprate esters. J. Biol. Chem., 193: 405-410.

[5] Black, C. C. Jr. and T. E. Humphreys, 1962: Effect of 2,4-dichlorophenoxyacetic acid on enzymes of glucolysis and pentose phosphate cycle. plant Physiol., 37: 66-67.

[6] Cori, G. T., M. W. Slein and C. F. Cori, 1948: Crystalline D-glyceraldehyde-3-phosphate dehydrogenase from rabbit muscle. J. Biol. Chem., 173: 605-618.

[7] Dische, z. and E. Borenfreund, 1951: A new spectrophotometric method for the determination of keto sugars and trioses. J. Biol. Chem., 192: 583-587

[8] Dische, z., L. B. shettles and M. Osnos, 1949: New specific color-reacion of hexose and spectrophotometric micromethods for their determination. Arch. Biochem., 22: 169 - 184.
[9] Fairbain, N. J., 1953: A modified anthrone reagent. Chem. and Ind., 4: 86
[10] Gibbs, M., J. M. Earl and J. L. Rithe, 1955: Respiration of the pea plant. Metabolism of hexose phosphate and triose phosphate by cell-free extracts of pea roots. plant physiol., 30: 463 - 467.
[11] Hageman, R. H. and D. I. Arnon, 1955: Changes in glyceraldehyde phosphate dehydrogenase during the life cycle of a green plant. Arch. Biochem. Biophys., 57: 421 - 436.
[12] Horecker, B. L., P. Z. Smyrniotis and Hans Klenow, 1953: The formation of sedoheptulose phosphate from pentose phosphate. J. Biol. Chem., 205: 661 - 682.
[13] Kachmar, J. F. and P. D. Boyer, 1953: Kinetic analysis of enzyme reactions II. The potassium activation and calcium inhibition of pyruvic phosphoferase. J. Biol. chem., 200: 669 - 682.
[14] Kornberg, A. and B. L. Horecker, 1955: Glucose - 6 - phosphate dehydrogenase, in Colowick, s. p. and N. O. Kaplan eds., Method in enzymology, I: 323 - 327, AP., Springer-Verlag, Berlin.
[15] Roe, J. H., 1934: A colorimetric method for the determination of fructose in blood and urine. J. Biol. Chem., 107: 15 - 22.
[16] Sibley, J. A. and A. L. Lehninger, 1949: Determination of aldolase in animal tissues. *J. Biol. Chem.*, 177: 859 - 872.
[17] Slein, M W., 1955: Phosphohexoisomerases from muscle, in Colowick, S. P. and N. O. Kaplan eds., Method in enzymology, I: 300 - 301, AP. Springer-Verlag, Berlin.
[18] Taylor, J. F., A. A. Green and G. T. Cori, 1948: Crystalline aldolase. J. Biol. Chem., 173: 591 - 604.

FAT SYNTHESIS AND ITS REGULATION IN PLANTS

I. ENZYME SYSTEMS OF GLUCOSE DEGRADATION IN RAPE SEEDS DURING RIPENING

C. N. SHIH, Y. S. ZHU AND C. TAN

(*Institute of plant physiology, Academia sinica*)

A series of investigations have been conducted on the fat metabolism and its regulation in plants. Special attention was paid to fat synthesis in relation to carbohydrate breakdown. In this paper, the enzyme systems of glucose degradation in the pods of rape during its development are described.

Evidences were presented showing that the acetone powder preparations of pods, 30th day after flowering, prossessed all the enzymes of EMP pathway; i. e. , hexokinase, phosphohexoisomerase, phosphohexokinase, aldolase, phosphoglyceraldehyde dehydrogenase, Phosphoglycerokinase, phosphoglyceric mutase, enolase and pyruvic kinase. At the same time, the enzymes of the HMP pathway were also detected, to wit: glucose - 6 - phosphate dehydrogenase, 6 - phosphogluconic dehydrogenase, phosphoriboisomerase, phosphoriboepimerase, transketolase and transaldolase. It was inferred that both catabolic pathways, EMP and HMP, were operative in the pods of rape.

植物的脂肪合成及其调节

Ⅲ．油菜籽实形成过程中代谢途径的变化与内源抑制剂的调节作用[①]

提要

油菜籽实形成过程中，EMP和HMP途径的酶活力发生有规律的变化。从花蕾到花，两途径的酶活力皆略有增加，但EMP途径的酶(醛缩酶)活力增加高于HMP途径酶(G－6－P和6－PG脱氢酶)活力的增加。从花到幼荚或幼荚到老荚，EMP途径的酶活力(F－6－P激酶、醛缩酶、3－PGA激酶和从3－PGA至丙酮酸形成的酵解酶系)迅速下降；而HMP途径的酶活力(G－6－P和6－PG脱氢酶，转化R－5－P的酶系)则急剧上升，到油菜荚将近成熟时，维持在一定水平或略有下降。

呼吸抑制剂试验表明，在油菜种子形成过程中，被碘乙酰胺和NaF抑制的呼吸所占的比例逐渐下降，到种子成熟时，又稍有回升。

两途径变化的转折点在花后10至20多天。此时正是油菜种子中可溶性糖含量急剧下降，而脂肪含量迅速上升的时期。

在花蕾、花和幼荚中发现有强烈抑制G－6－P脱氢酶和6－PG脱氢酶的天然抑制剂存在。它除了对油菜结实器官脱氢酶有抑制外，还能抑制从棉花苗根系，大豆种子和酵母中制备的G－6－P脱氢酶，且不能为半胱氨酸或谷胱甘肽所恢复。

天然抑制剂水溶性强，不能透析，也不为饱和硫酸铵所沉淀，对热稳定，表明它是相对分子质量较大的非蛋白质的化合物。纸上电泳证明，该抑制剂带正电。光谱测定发现在260 μm有一吸收峰。紫外线照射后，抑制活力丧失，260 μm的峰也随之消失。红外光谱分析证明抑制剂含羰基和苯环。纸谱斑点在紫外线下显示蓝色萤光，NH_3和$AlCl_3$处理能改变萤光颜色。根据以上性质，推测它可能属于类黄酮或其衍生物。

油菜天然抑制剂主要分布在花蕾、花和幼荚中，且随着年龄增长，荚中含量逐渐减少，乃至20天以上的老荚中几乎全部消失。再联系G－6－P和6－PG脱氢酶以及代谢途径

施教耐，朱雨生，谭常，中国科学院植物生理所。此文发表于1965年“植物生理学报”2卷2期105－118页。

① 1964年12月5日收到。

＊本工作在殷宏章教授指导下完成，特此致谢。

本文采用简称如下：HMP途径，已糖单磷酸途径(即磷酸戊糖支路)；EMP途径，糖酵解途径；NADP，烟酰胺腺嘌呤二核苷酸磷酸；NAD，烟酰胺腺嘌呤二核苷酸；G－6－P，葡萄糖－6－磷酸；6－PG，6－磷酸葡萄糖酸；R－5－P，核糖－5－磷酸；F－6－P，果糖－6－磷酸；FDP，果糖－1，6－二磷酸；3－PG，3－磷酸甘油醛；3－PGA，3－磷酸甘油酸；2，3－DPGA，2，3－二磷酸甘油酸；Ru－5－P，核酮糖－5－磷酸；Xu－5－P，木酮糖－5－磷酸；EDTA，乙二胺四乙酸；G－6－PDH，G－6－P脱氢酶；6－PGDH，6－PG脱氢酶；R－5－P“TE”，转化R－5－P的酶系；3－PGA DH，3－磷酸甘油醛脱氢酶；ATP，腺嘌呤核苷三磷酸。

的转折，我们认为油菜天然抑制剂对于糖代谢途径以EMP为主转入以HMP为主，可能起着内部调节的作用。

一、引言

在油料作物籽实形成过程中，开始是糖类的累积，接着是脂肪大量形成。当种子从一种物质的累积转入另一种性质不同物质的累积时，在中间代谢方面一定会发生改变。植物体内同脂肪合成有关的糖代谢途径，在脂肪形成期间，发生什么变化？这种变化是怎样调节与控制的？目前，至少就植物方面资料，还没有回答。本工作目的即在于探讨油菜籽实形成过程中糖代谢途径的变化，以及其中的调节机制。

近年来，根据 Langdon (1957)、Wakil (1961)和 Abraham 等(1957)在动物、Lynen (1955)在酵母、Stumpf (1959)在植物方面的工作，一般认为 $NADPH_2$ 是脂肪酸生物合成的专一性供氢体。加入 $NADPH_2$ 能大大刺激标记乙酸渗入肝脏脂肪酸(Langdon 1952)和油菜种子脂溶物(朱雨生等，未发表)。由于合成一个 C_{16} 脂肪酸分子需耗费 14 个 $NADPH_2$ 分子(Martin 等，1962)，因而 $NADPH_2$ 是脂肪酸生物合成中重要的速度限制因子之一(Stumpf, 1961)。因为植物体内 $NADPH_2$ 主要来自 HMP 途径(Beevers, 1960)，因此，脂肪酸合成可能与 HMP 途径有联系。由此推想，油料种子从幼小阶段的糖类累积转入成长阶段的脂肪累积的同时，可能伴随这个代谢途径的变化。

前文(施教耐等，1965)中已报道，在油菜籽实中存在两途径的酶活力，本实验将继续测定两途径酶活力在籽实形成过程中的变化，以及呼吸抑制剂(碘乙酰胺和氟化钠)对不同年龄油菜种子呼吸的影响。从获得的资料而言，仍支持了上述见解。同时还发现油菜花蕾和幼荚中存在一种天然抑制剂，似乎对 EMP 途径转入 HMP 途径起着调节作用。

二、试验材料和方法

1. 材料

试验系用大田栽培的“胜利”油菜，在盛花期，每隔一定天数，标记开花样品。经过若干时间后采集花蕾、花和不同年龄的油菜荚，制成丙酮粉，以测定酶活力。

试剂：所有试剂均系商品，G-6-P、3-PGA 系 Light 公司出品；FDP、R-5-P、2,3-DPGA 系 B. D. H. 制品；6-PG 购自上海试剂厂；NADP 系东风厂制备，纯度 75%左右；NAD 是 Floka 产品，纯度 100%。

2. 方法

(1) 丙酮粉制备、酶液提取和两途径酶系活力测定均按施教耐等 1965 年报道的方法。

(2) 呼吸的测定。种子分别放在盛有碘乙酰胺(0.05 mol/L)或氟化钠(0.1 mol/L)称量瓶中，减压渗入，然后用 Warburg 呼吸器测定种子吸氧量。

(3) 抑制剂的分离和初步提纯。将油菜花蕾丙酮粉提取液于水浴中煮沸 5 分钟，离心弃蛋白沉淀。上清液用纸层析和电泳进行分离和提纯。有的试验中，在纸层析和电泳前，花蕾提取液还经氯仿和苯多次萃取，以除去脂溶杂质。

(4) 纸层析。用 $20\times17.5\ cm^2$ Whatman 1 号滤纸，每张点样 5 ml。采用的溶剂为丙酮：甲酸：水(20：10：70)。在 25℃下，下行 24 小时左右。晾干后，从底线开始至溶剂前沿剪成等距离 4 段，依次标以 1、2、3、4，加上点样区共 5 段。各用 5 ml 重蒸馏水浸提。测定浸提液对 G-6-P 脱氢酶或 6-PG 脱氢酶活力的抑制程度。

(5) 电泳。队式电泳槽，电压 400 V。滤纸带 $2\times36\ cm^2$。点样 0.5 ml。室温下，在硼酸缓冲液(pH 9)中，从正极向负极电泳 4 小时。完毕，待滤纸干后，按色泽不同，剪成数段，各用 4 ml 重蒸馏水浸提后，测定对脱氢酶的抑制作用。

三、试验结果

1. 油菜籽实形成过程中 HMP 和 EMP 途径酶系活力的变化

1) HMP 途径的酶系活力

如图 1(a)所示，在籽实形成过程中，G-6-P 脱氢酶和 6-PG 脱氢酶活力上升，尤以花后 10～20 天或 20 多天间上升较快，如 G-6-P 脱氢酶活力增加了 4 倍多。转化 R-5-P 的酶活力从花蕾直到花后 30 天也是逐渐上升的。

2) EMP 途径的酶系活力

(1) G-6-P 异构酶。从花后 8～15 天，活力也是下降，而 15 天以后，一直到 30 天，却有所上升(见表 1 和图 2)，或者在有的试验中维持不变。

表 1　不同年龄油菜荚 G-6-P 异构酶活力的变化

年龄(开花后天数)	活力单位*/克丙酮粉
8	117
15	65
23	129
30	155
40	148

* 活力单位＝每 15 分钟转化 1 μmole G-6-P。

(2) F-6-P 激酶。随籽实的成长，活力迅速下降，至 30 天已降得很低[见图 1(b)]。

(3) 醛缩酶。从花蕾到花或幼荚活力略有上升，以后随油菜荚的长大，活力迅速下降，如以花后 30 天与 10 天比较活力下降达 3 倍[见图 1(b)]。

(4) 3-磷酸甘油醛脱氢酶。该酶活力表现的规律类似于 G-6-P 异构酶，从花后 8 天至 15 天活力略有下降，以后随年龄增长，活力逐渐升高(见图 2)。

(5) 磷酸甘油酸激酶和从"3-PGA 至丙酮酸"的酶系。如图 2 所示，这些酶的活力均随着油菜荚年龄的增长而迅速下降。如 40 天与 8 天相比，3-PGA 激酶和从"3-PGA 至丙酮酸"酶系活力分别下降了 6 倍和 1.7 倍。

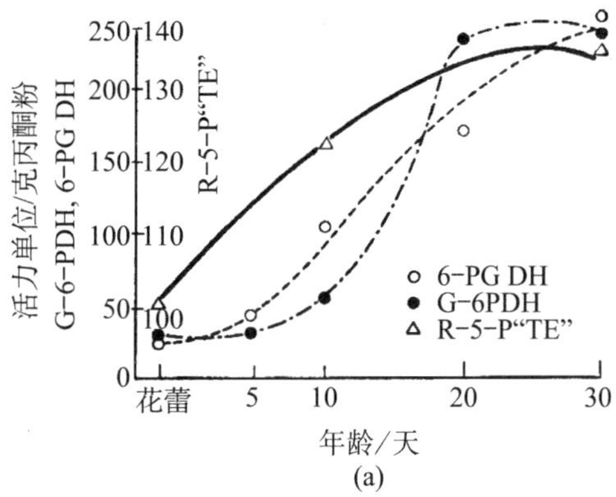

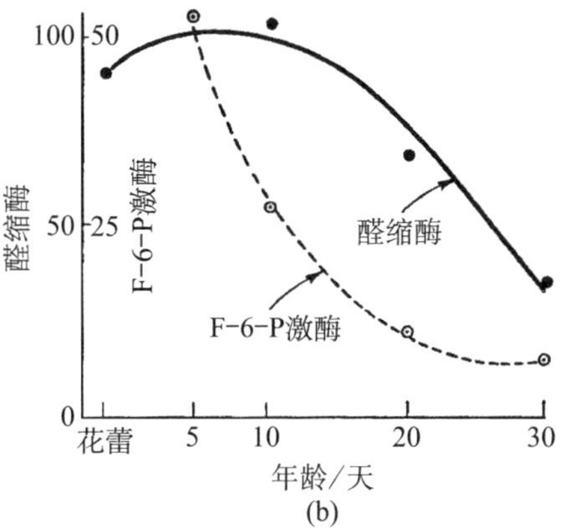

图 1　油菜籽实形成过程中 HMP 和 EMP 途径几种酶的活力变化

反应系统如下：

(a) G-6-P 脱氢酶和 6-PG 脱氢酶：双甘氨肽 125 μmoles (pH 7.5)，$MgCl_2$ 10 μmoles，G-6-P(或 6—PG) 5 μmoles，NADP 0.27 μmole(纯度 78%)，酶液 0.5 ml，总体积 3 ml，空白缺 NADP。反应混合物预先在 38℃水浴中预温 5 分钟，酶液的加入作为反应的开始。任意规定最初 5 分钟在 340 mμm 处光密度增加 0.01 时(用 Beckman DU 型分光光度计，比色杯 1 cm)的酶活力为 1 个单位。

R-5-P"转化"酶：双甘氨肽 88 μmoles (pH 7.5)，R-5-P 10 μmoles，核黄素 1.5 μmoles，半胱氨酸 1.5 μmoles，$MgCl_2$ 10 μmoles，NADP 0.5 μmole，总体积 3 ml。空白缺 R-5-P，酶液 0.5 ml，温度 38℃，并以转化 R-5-P 的微克分子数表示活力单位。

(b) F-6-P 激酶：Tris 缓冲液(pH 8.6)100 μmoles，亚硫酸氢钠(pH 8.6)250 μmoles，F-6-P 4 μmoles，ATP 6 μmoles，$MgCl_2$ 10 μmoles，酶液 0.5 ml，总体积 3 ml。于 38℃保温 15 分钟，测定形成的磷酸丙糖量，并规定使 0.1 μmole F-6-P 磷酸化的酶活力为 1 个单位。

醛缩酶：巴比妥钠酸缓冲液(pH 8.6)100 μmoles，亚硫酸氢钠(pH 8.6)250 μmoles，FDP 12.5 μmoles，酶液 0.5 ml，总体积 3.0 ml。在 38℃反应半小时。加等体积 20%三氯乙酸终止反应。用碱性介质中二硝基苯肼显色法测定形成的磷酸丙糖量，并规定使 1 个微克分子 FDP 分解的酶量为 1 个活力单位。

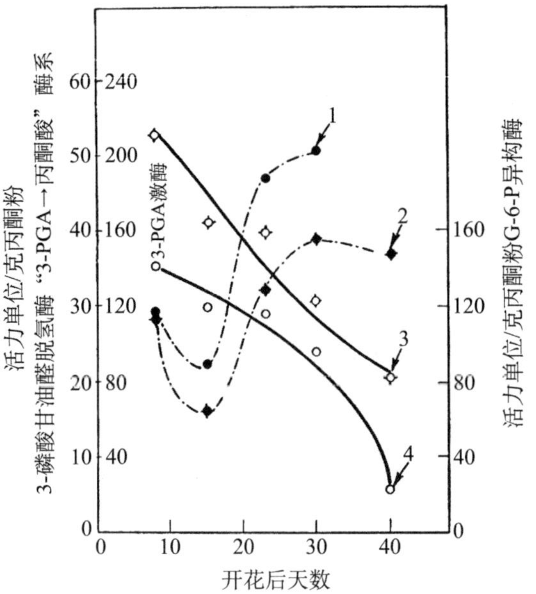

图 2 油菜籽实形成过程中 3-磷酸甘油醛脱氢酶、G-6-P 异构酶、3-PGA 激酶和"3-PGA-丙酮酸"酶系的活力变化

1—3-磷酸甘油醛脱氢酶；2—G-6-P 异构酶；3—"3-PGA-丙酮酸"酶系；4—3-PGA 激酶

反应系统如下：

3-磷酸甘油醛脱氢酶：Tris (pH 8.5) 150 μmoles，FDP 10 μmoles，NaF 10 μmoles，砷酸钠 17 μmoles，谷胱甘肽 15 μmoles，NAD 1 μmole，酶液 0.5 ml，总体积 3 ml，空白者缺 NAD。于室温下用日立分光光度计测定 340 μm 处的光吸收。

3-PGA 激酶：Tris (pH 7.4) 100 μmoles，3-PGA 20 μmoles，ATP 8 μmoles，$MgCl_2$ 6 μmoles，羟氨 1 000 μmoles，酶液 0.1 ml。总体积 1.3 ml，38℃，反应 30 分钟。加 3 ml $FeCl_3$-HCl-三氯乙酸* 终止反应，测定形成的 1,3-二磷酸甘油酸。

($FeCl_3$-HCl-三氯乙酸溶液配法：$FeCl_3 \cdot 6H_2O$ 8.3 g，20 g 三氯乙酸溶于 42 ml 浓 HCl 中，加蒸馏水至 500 ml)

从3-PGA 至丙酮酸形成的酶系：Tris (pH 7.4) 150 μmoles，3-PGA 20 μmoles，ADP 10 μmoles，$MgCl_2$ 10 μmoles，2,3-DPGA 30 μg，KCl 25 μmoles，酶液 0.2 ml，总体积 2.0 ml，于 30℃保温 30 分钟。杀死酶液，用二硝基苯肼测定形成的丙酮酸。

3-磷酸甘油醛脱氢酶和 3-PG 至丙酮酸形成酶每小时使 1 微克分子底物转化的活力规定为 1 个活力单位。

2. 油菜籽实形成过程中呼吸途径的变化

油菜种子呼吸能为碘乙酰胺和氟化钠所抑制，但随着年龄增长，被抑制的呼吸部分逐渐减少(见表 2)。如花后 8 天种子被碘乙酰胺和 NaF 抑制的呼吸部分分别占 62.5%和 61.6%，而 30 天种子分别降为 29.6%和 22.0%。有意思的是，到种子成熟时，抑制的呼吸部分又略有回升的趋势。

表 2 油菜种子形成过程中呼吸途径的变化

种子年龄(开花后天数)	对照* 吸 O_2 μl/小时·克鲜重 O_2	碘乙酰胺(0.05 mol/L)		氟化钠(0.1 mol/L)	
		吸 O_2 μl/小时·克鲜重 O_2	抑制/%	吸 O_2 μl/小时·克鲜重 O_2	抑制/%
8	508.6	190.9	62.5	195.3	61.6
11	535.2	213.8	60.1	324.5	39.4
14	434.6	199.4	54.1	252.7	41.6

（续表）

种子年龄（开花后天数）	对照* 吸 O_2 μl/小时·克鲜重 O_2	碘乙酰胺（0.05 mol/L）		氟化钠（0.1 mol/L）	
		吸 O_2 μl/小时·克鲜重 O_2	抑制/%	吸 O_2 μl/小时·克鲜重 O_2	抑制/%
20	406.0	212.2	47.7	273.8	32.6
23	302.9	211.8	30.1	213.7	29.4
30	282.9	198.4	29.6	219.7	22.0
40	288.0	166.2	42.3	218.2	24.9

* 对照仅用磷酸盐缓冲剂（pH 5.0）处理。

3. 油菜内源抑制剂对代谢途径的调节作用

1）抑制剂的发现

在测定油菜花蕾 G-6-P 脱氢酶和 6-PG 脱氢酶时，发现活力都很低。这到底是酶蛋白本身缺少，还是酶蛋白受到某种因素的限制，以致不能表现活力。试验证实了后一种设想。

当花蕾酶液稀释后，酶活力即突然上升。如从 1 ml 酶液降为 0.5 ml 时，G-6-P 脱氢酶活力增加 5 倍，6-PG 脱氢酶活力增加 4 倍。因而可以设想，在油菜花蕾提取液中存在天然抑制剂，使 G-6-P 脱氢酶和 6-PG 脱氢酶处于不活跃状态。当酶液稀释后，抑制剂浓度冲淡，抑制作用解除，因而酶活力得以表现。

为了进一步证实抑制剂的存在，将花蕾丙酮粉提取液煮沸 6 分钟，蛋白沉淀弃去，上清液加到花蕾酶液中，结果大大地抑制了 G-6-P 和 6-PG 脱氢酶的活力。对于油菜老荚的酶活力也表现出同样强烈的抑制效应（80%～90%），而且随着抑制剂浓度的增加，抑制程度也加大（见图 3）。

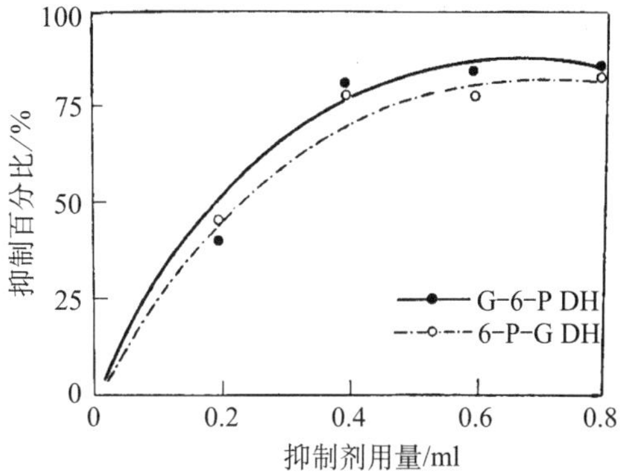

图 3　不同浓度抑制剂对油菜荚 G-6-P 脱氢酶和 6-PG 脱氢酶的抑制作用

表 3　花蕾抑制剂对油菜荚、棉花幼苗根、大豆种子和酵母中 G-6-P 脱氢酶活力的抑制作用

材料	处理	G-6-P 脱氢酶	
		活力 Activity（$\Delta E_{340\,m\mu m}$，5′）	抑制/%
油菜荚	对照＋抑制剂	0.260 0.030	0 88
棉花幼苗根	对照＋抑制剂	0.145 0.020	0 90
酵母	对照＋抑制剂	0.127 0.025	0 80
大豆种子	对照＋抑制剂	0.355 0.095	0 73

油菜花蕾抑制剂作用极为广泛。对于棉花苗根、大豆种子和酵母中纯化的 G－6－P 脱氢酶都有强烈的抑制作用。

2）抑制剂的作用性质

（1）与巯基的关系。花蕾抑制剂的抑制作用不能为半胱氨酸和谷胱甘肽所恢复（见表 4），所以其作用点不是巯基。

表 4 加入谷胱甘肽和半胱氨酸对“花蕾抑制剂”抑制活力的影响

处理	ΔE_{340}（4 分钟）	
	G－6－P 脱氢酶	6－PG 脱氢酶
对照（0.5 ml 酶）	0.155	0.175
＋花蕾抑制剂（0.5 ml）	0.030	0.040
＋谷胱甘肽（0.1 mol/L）	0.025	0.010
＋半胱氨酸（0.05 mol/L）	0.028	0.040

（2）溶解度。将花蕾丙酮粉用不同溶剂提取，提取液减压浓缩，赶去溶剂。剩下残渣溶解于水，测定对脱氢酶的抑制效力。发现抑制剂极易溶于水，也能溶于热酒精、正丁醇和乙醚，但不溶于氯仿和苯（见表 5）。

表 5 “花蕾抑制剂”的溶解度*

G－6－P 脱氢酶来源	溶剂	活力（%）
油菜花蕾	对照	100
	水	36.6
	正丁醇	40.0
	乙醚	52.5
	苯	91.7
	氯仿	75.0
酵母	对照	100
	热酒精	0.4

* 见正文说明。

（3）稳定性。抑制剂对热稳定。酶液煮沸 5 分钟后，清液仍有抑制活力（见表 6），但抑制剂对紫外线敏感，能为紫外线所破坏。

（4）分子大小。抑制剂对 5 000 毫升磷酸钾－EDTA 缓冲液透析 6 小时后，测定其活力，如表 6 所示，保存了大部分活力。说明抑制剂可能主要是一些分子较大的化合物，但它不是蛋白质，因为煮沸去蛋白后，清液仍具有抑制作用。为了进一步证实抑制剂不是蛋白质，用 100%饱和度$(NH_4)_2SO_4$ 将酶蛋白全部沉淀下来，溶解后加到油菜荚酶液中，结果未表现出抑制活力（表 6）。

表 6　煮沸、硫酸铵沉淀和透析对“花蕾抑制剂”活力的影响

试验	处理	G-6-P脱氢酶		6-PG脱氢酶	
		$\Delta E_{340}(5')$	抑制/%	$\Delta E_{340}(5')$	抑制/%
105	对照(0.5 ml 酶)	0.225	0	0.220	0
	+抑制剂(0.5 ml)	0.104	54	—	—
	+煮沸抑制剂(0.5 ml)	0.075	66	—	—
	+蛋白质部分(0.8 ml)	0.185	17	0.170	22
107	对照(0.5 ml 酶)	0.155	0	0.175	0
	+抑制剂(0.5 ml)	0.030	80	0.040	77
	+透析抑制剂(0.5 ml)	0.060	61	0.050	71

(5) 带电性质。抑制剂进行纸上电泳时，从阳极游向负极，故带阳电。移动速度约为4 cm/h。

(6) 萤光现象。抑制剂经纸谱分离后，纸谱上斑点在紫外光下发出蓝色萤光。用 NH_3 或 $AlCl_3$ 处理，能加深斑点黄色和改变萤光颜色。

3) 抑制剂的初步提纯和鉴定

花蕾酶液经煮沸去蛋白后，得到黄色上清液，用丙酮：甲酸：水(20：10：70)进行纸谱分离。单相下行 24 小时后，抑制剂分布在离底线最远的一端，愈靠近底线抑制剂越少(见表 7)。电泳分离后，在纸带上显现出不同的色带，有浅黄、黄、棕黄几种。

表 7　抑制剂在层析纸上的分布*(溶剂相——丙酮：甲酸：水)

纸段		油菜老荚 G-6-P脱氢酶 $\Delta E_{340}(4')$	抑制/%
编号	颜色		
对照	…	0.155	…
0(点样区)	淡黄	0.150	3
1	无色	0.120	23
2	略黄	0.112	27
3	淡黄	0.065	58
4	黄色	0.030	81

* 见正文说明。

不论纸谱或电泳纸上，凡分布有抑制剂处总带黄色。若将经过丙酮：甲酸：水系统层析后，纸上分布有抑制剂的最末一段按黄色和无色两部分剪开，用水溶出，结果黄色部分的抑制活力比无色部分高将近 4 倍。可见抑制剂主要是那些黄色的化合物。

经过丙酮：甲酸：水系统层析分离的抑制剂，测定其吸收光谱，发现在 260 mμm 处有一吸收峰。经紫外光照射 15 分钟后(15 cm，500 W，220 V)，260 mμm 的吸收峰消失，随即其抑制活力也丧失(见图 4)。

花蕾干粉用丙酮多次抽洗，残余物用重蒸馏水提取，煮沸，去蛋白。上清液经氯仿、苯萃取

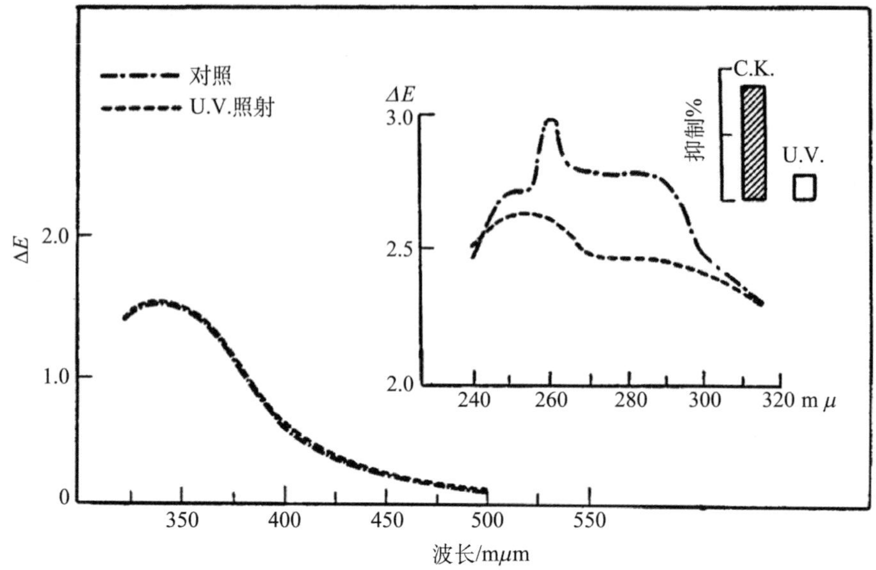

图 4　“花蕾抑制剂”的吸收光谱

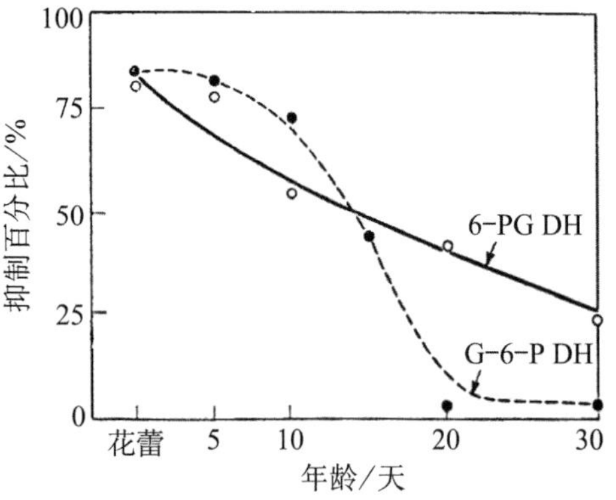

图 5　抑制剂在花蕾和不同年龄油菜荚中的分布

二次，以除去脂溶杂质。水溶液浓缩，用丙酮：甲酸：水纸谱分离，重复二次。纯化后的抑制剂加热浓缩至黄棕色黏状物。用石蜡油为溶剂，用双射式自动记录的红外光谱仪测定红外光谱①。由于样品在石蜡油中不溶解，粒子磨不匀，吸收峰很少，仅在 1 600 cm^{-1} 和 1 110 cm^{-1} 处有二吸收峰，分别相当于羰基和苯环、类黄酮化合物重要的功能团。

4）不同年龄结实器官内抑制剂含量的变化

如图 5 所示，抑制剂主要分布在花蕾、花和幼荚中，随着荚子长大，抑制剂逐渐减少，而老荚中几乎消失。

四、讨论

根据现在一般认识的脂肪酸生物合成的机制，先由乙酰 CoA 羧化形成丙二酰 CoA，再行脱羧缩合而形成长链脂肪酸(Stumpf, 1961)。在缩合还原过程中需要消耗大量的 $NADPH_2$ 作为电子供体。$NADPH_2$ 的来源有二处：一是 HMP 途径最初的二个脱氢反应(G-6-P 脱氢酶和 6-PG 脱氢酶)；二是上清液中与 NADP 连结的异柠檬酸脱氢酶反应(与 NAD 连结的异柠檬酸脱氢酶位于线粒体上并与 TCA 循环及氧化磷酸化相联系)(岛薗顺雄等，1960)。但无疑，HMP 途径在 $NADPH_2$ 的供应上起着更大的作用，因为每氧化一分子 G-6-P 可形成 12 个 $NADPH_2$ 分子。因而推想在脂肪形成与 HMP 途径之间可能存在着某种联系。以前在

① 中国科学院有机化学研究所代为测定。

动物方面，脂肪酸合成与HMP途径的关系有过一些证据(McLean，1958)。本试验目的即在于探讨植物体内脂肪形成与HMP途径的关系。

在油菜籽实形成过程中HMP途径的酶活力(G-6-P脱氢酶，6-PG脱氢酶，转化R-5-P的酶系)均显著上升，而EMP途径的酶活力(F-6-P激酶，醛缩酶，3-PGA激酶和从3-PGA至丙酮酸形成的酶系)却迅速下降，两者变化的转折期在花后10至20多天，此时正是种子中可溶性糖急剧下降，脂肪迅速上升的时期(倪晋山等，1960)。根据酶活力可以为代谢途径强弱的指标之一的说法(Glock和McLean，1954；Gibbs和Earl，1959)，可以推测脂肪合成的加强与HMP途径的增强及EMP途径的削弱相平行。这一结论从对碘乙酰胺和NaF敏感的呼吸部分所占比例在该时期迅速下降得到佐证，因为碘乙酰胺和NaF分别是油菜籽实EMP途径二个重要的酶：3-磷酸甘油醛脱氢酶和烯醇化酶强烈的抑制剂(施教耐等，1965)。此外，油菜籽实形成过程中，与EMP途径密切联系的TCA循环逐渐减弱(刘承宪等，1965)也和上述推论相衔接。

在EMP途径的所有酶活力中，只有G-6-P异构酶和3-磷酸甘油醛脱氢酶较为特殊。二者活力在花开后8～15天虽有下降，但15天之后又逐渐回升。G-6-P异构酶本身是HMP途径的组成成分之一，而从R-5-P形成G-6-P这一完全的HMP循环在前文中已得到证实(施教耐等，1965)。所以G-6-P异构酶在后期的回升或不下降较易理解。至于3-磷酸甘油醛脱氢酶在后期的上升究属何因尚难定论。

目前，HMP途径的存在普遍受到人们的关注。自从30年前Warburg等(1933，1935)首次在红血球溶血和酵母自溶液中发现G-6-P脱氢酶以来，对于大部分动植物和微生物中非酵解途径(即HMP途径)的存在已无可非议。迄今，对于该途径的循环机制，中间产物和酶反应特性都有了比较肯定的了解(Horecker和Mehler，1955；Gunsales等，1955；Racker，1957)。最近十年来，人们的兴趣渐渐转向该途径的生理意义及其调节机制。如Gibbs等(1959)证明组织年龄与HMP途径有关；Shaw和Samborski (1957)、Daly等(1957)发现HMP途径在植物组织免疫中的作用。国内对于金霉素形成(沈善炯等，1957)、器官脱落(薛应龙等，1964)以及花芽分化(屠锦炎等，1963)与HMP途径的关系也有过若干报道。本文就植物脂肪形成与HMP途径的密切关系所提出的证据进一步证明HMP途径在植物体代谢和生理活动中的意义。

HMP途径的调节机制是一个饶有趣味的问题。有大量证据指出HMP途径受NADP的供应所限制，即$NADPH_2$的氧化速度、$NADP/NADPH_2$的比例能影响葡萄糖进入HMP途径的速度。人工电子受体或其代谢反应需要$NADPH_2$的底物加入肝切片(Cahill等，1958；Hers，1957)、红血球(Brin等，1958)、角膜上皮细胞(Kinoshita，1957)、腹水癌细胞(Wenner，1958)、乳腺(McLean，1960)、胡萝卜切片(Rees和Beevers，1960)都能增加[1-C^{14}]葡萄糖放出$C^{14}O_2$的速度。此外磷酸盐(沈善炯等，1958；Kravitz等，1958)、葡萄糖浓度(Racker，1956)、Cu^{2+}对G-6-P竞争的两途径酶反应之米氏常数(McLean，1958a)、6-PG (Parr，1956)等对HMP途径和HMP途径的调节作用也有过若干推测。可惜，目前这些资料多出于动物和微生物，植物方面的工作很少。

油菜籽实形成过程中代谢途径的转折是受什么因素控制呢？在本试验中，根据油菜花蕾丙酮粉提取液稀释一倍后，G-6-P脱氢酶和6-PG脱氢酶活力大大升高，以及花蕾提取液对一系列不同来源，器官中制备的G-6-P脱氢酶都有强烈的抑制作用，证明花蕾中有对脱氢酶起抑制作用的内源抑制剂存在。并根据吸收光谱、萤光特性等一系列性质初步推测它可能属于类黄酮或其衍生物。

近年来，随着植物体内一系列天然抑制剂的发现，对于天然抑制剂的生理调节意义也日益受到重视。Sacher (1962)认为菜豆豆荚外果皮中 IAA 氧化酶抑制剂的存在，具有维持活力和防止衰老的作用。Jacqueline (1962)发现南欧海松(*Pinus pinaster* Sol.)芽和枝条中的抑制剂能阻止冬季枝条的生长，Mayer 等(1962)认为许多果实在未成熟时种子处于休眠是由于有抑制剂的存在。傅家瑞等(1963)也证明水浮莲(大蕈)种子中有几种抑制剂使种子处于休眠状态。Nagao 和 Ohwaki (1955)还推测天然抑制剂起着氧化磷酸化解联剂的作用。油菜花蕾中天然抑制剂具有什么生理意义呢?

正如试验所指出，抑制剂主要分布在花蕾和花中，幼荚中较少，而老荚中几乎消失。再结合不同年龄油菜籽实中 G-6-P 脱氢酶和 6-PG 脱氢酶的活力变化，不难看出，抑制剂含量左右着油菜籽实中 G-6-P 脱氢酶和 6-PG 脱氢酶的活力表现。在籽实形成早期，抑制剂含量高，故脱氢酶处于抑制状态；后期抑制剂消失，故脱氢酶活力得以表现，HMP 途径"活门"打开，因而推想，油菜天然抑制剂控制着籽实形成过程中适应于脂肪形成所引起的代谢变化，对代谢方向从以 EMP 途径为主转入以 HMP 途径为主，起着内部调节的作用。当然本文并不因此排除油菜籽实形成过程中代谢途径的转变还受其他因素调节的可能性。

参考文献

[1] 沈善炯、陈俊标、洪孟民，1957：金霉菌的糖类代谢 I. Embden-Meyerhof-Parnas 系统与磷酸己糖的分路代谢. 生理学报，21(3)：302—310.

[2] 沈善炯，陈俊标，1958：金霉菌的糖类代谢 II. 戊糖代谢和磷酸盐对代谢途径的影响，生化学报，1(1)：69—77.

[3] 施教耐，朱雨生，谭常. 1965：植物的脂肪合成及其调节 I. 油菜种子形成期间葡萄糖降解的途径. 植物生理学报，2(1)：33—43.

[4] 倪晋山，等，1960：油菜的若干生理问题 . 32 页，科学出版社.

[5] 屠锦炎、潘维雄、欧阳光察、薛应龙，1963：开花生理的研究 1. 胜利油菜(*Brassica napus* Linn)生殖器官发育过程中核酸动态与呼吸途径的联系. 中国植物生理学会第一届年会论文(摘要)集，135 页.

[6] 傅家瑞、范培昌，1963：水浮莲(大薸)种子的休眠及抑制物质. 中国植物生理学会第一届年会论文(摘要)集，248 页.

[7] 岛薗顺雄、石川晋次、神谷知弥，1960：生化学讲座 4，中间代谢の化学 I.(岛薗顺雄编集)p. 50.

[8] Abraham, S., K. J. Matthes & I. L. Chaikoff, 1961: Factors involved in synthesis of fatty acids from acetate by a soluble fraction obtained from lactating rat mammary gland. *Biochim. Biophys. Acta*, **49**: 268-285.

[9] Brin, M. & R. H. Yonemoto, 1958: Stimulation of the glucose oxidative pathway in human erythrocytes by methylene blue. J. *Biol. Chem.*, **230**: 307-317.

[10] Cahill, G. F. Jr., A. B. Hastings, J. Ashmore & S. Zottu, 1958: Studies on carbohydrate metabolism in rat Liver slices. X. Factors in the regulation of pathways of glucose. *J. Biol. Chem.*, **230**: 125-135.

[11] Daly, J. M., R. M. Sayre & J. H. Pazur, 1957: The hexose monophosphate shunt as the major respiratory pathway during sporulation of rust of sufflower. *Plant Physiol.*, **32**: 44-48.

[12] Glock, G. E. & P. McLean, 1954: Levels of enzymes of the direct oxidative pathway of carbohydrate metabolism in mammalian tissues and tumours. *Biochem. J.*, **56**: 171-175.

[13] Gibbs, M., 1959: Mechanism of carbon compounds. *Ann. Rev. Plant Physiol.*, **10**: 329 - 378.
[14] Gibbs, M. & J. M. Earl, 1959: Effect of age of tissue on hexose metabolism. I. An enzyme study with pea root. *Plant Physiol.*, **34**: 529 - 534.
[15] Gunsalas, I. C., B. L. Horecker & W. A. Wood, 1955: Pathways of carbohydrate metabolism in microorganisms. *Bact. Rev.*, **19**: 79 - 128.
[16] Hers, H. G., 1957: Le métabolism du fructose, Editions Arscia, Brussels, Belgium.
[17] Horecker, B. L. & A. H. Mehler, 1955: Carbohydrate metabolism. *Ann. Rev. Biochem.*, **24**: 207 - 274.
[18] Jacqlleline, C., 1962: Présence d'une substance de croissance et d'um inhibiteur dans les bourgeons et les pousses de premiére année chez *Pinus pinaster* Sol. Evolution de ces substances au cours de L'année. *C.R. Acad. Sci.*, **254**: 2643 - 2645.
[19] Kinoshita, J. H., 1957: The stimulation of the pentose phosphogluconate oxidation pathway by pyruvate in bovine corneal epithelium. *J. Biol. Chem.*, **228**: 247 - 253.
[20] Kravitz, E. A. & A. J. Guarino, 1958: On the effect of inorganic phosphate on hexose phosphate meta bolism. *Science*, **128**: 1139 - 1140.
[21] Langdon, R. G., 1957: The biosyntheais of fatty acids in rat liver. *J. Biol. Chem.*, **226**: 615 - 629.
[22] Lynen, F., 1953: Functional group of coenzyme A and its metabolic relations especially in the fatty acid cycle. *Feder. Proc.*, **12**: 683 - 691.
[23] Martin, D. B. & P. R. Vagelos, 1962: The mechanism of tricarboxylic acid cycle regulation of fatty acid synthesis. *J. Biol. Chem.*, **237**: 1787 - 1992.
[24] McLean, P., 1958: Carbohydrate metabolism of mammary tissue. I. Pathway of glucose catabolism in the mammary gland. *Biochim. Biophy. Acta*, **30**: 303 - 315.
[25] McLean, P., 1958b: Carbohydrate metabolism of mammary tissue. II. Levels of oxidized and reduced DPN and TPN nucleotide in the rat mammary gland. *Biochim. Biophys. Acta*, **30**: 316 - 324.
[26] McLean, P., 1960: Carbohydrate metabolism of mammary tissue. IV. Facotors in the regulation of pathways of glucose catabolism in the mammary gland of the rat. *Biochim. Biophys. Acta*, **37**: 290 - 309.
[27] Nagao, M. & J. Ohwaki, 1955: The action of transcinnamic acid and 2,3,5 - triiodobenzoic acid in the rice seedling. *Sci. Repts. Tohoku Univ.*, **21**: 96.
[28] Parr, C. W., 1956: Inhibition of phosphoglucose isomerase. *Nature*, **178**: 1401.
[29] Racker, E., 1956: Carbohydrate metabolism in ascites tumor cells. *Ann. N.Y. Acad. Sci.*, **63**: 1017 - 1021.
[30] Racker, E., 1957: Harvey Lect., 51: 143 (Ref. Glock, G. E. & McLean P., 1958: Pathway of glucose utilization in mammary tissue. *Proc. Roy. Soc.* (London) B, **149**: 354 - 362).
[31] Rees, T. Ap & H. Beevers, 1960: Pathways of glucose dissimilation in carrot slices. *Plant Physiol.*, **35**: 830 - 838.
[32] Sacker, J. A., An IAA oxidase-inhibitor system in bean pods. I. Physiological significance and source of the inhibitor. *Amer. J. Bot.*, **48**: 820 - 828.
[33] Shaw, M., D. J. Samborski & A. Oaks, 1958: Some effects of indolacetic acid and maleic hydrazide in the respiration and flowering of wheat. *Can J. Bot.*, **36**: 233 - 237.
[34] Stumpf, P. K., 1961: Biosynthesis of lipids in higher plants. Proc. 5th Internatl. Congr. of Biochem. VII, 74.
[35] Stumpf, P. K. & C. Bradbeer, 1959: Fat metabolism in higher plants. *Ann. Rev. Plant Physiol.*, **10**: 197 - 222.
[36] Wenner, C. E., 1958: The hexose monophosphate shunt in glucose catabolism in Ascites tumor cells.

Cancer Res., **18**: 1105 - 1114.

[37] Wakil, S. J., 1961: Mechanism of fatty acids synthesis. *J. Lipids Res.* 2: 1.

[38] Warburg, O. und W. Christian, 1933: Über das gelbe Ferment und seine Wirkung. *Biochem. Z.*, **266**: 377 - 411.

[39] Warburg, O., Christain, W. und A. Griese, 1935: Wassertoffübertragendes Coferment, Seine Zusam mensetzung und Wirkungsweise. *Biochem. Z.*, **282**: 157 - 205.

FAT SYNTHESIS AND ITS REGULATION IN PLANTS

Ⅲ. THE CHANGES OF METABOLIC PATHWAYS AND THE REGULATORY ACTION OF AN ENDOGENOUS INHIBITOR IN THE DFVELOPING SILIQUES OF *BRASSICA NAPUS* LINN.

C. N. Shih, Y. S. Zhu AND C. Tan

(*Institute of Plant Physiology, Academia Sinica*)

In the previous article, we showed that in the siliques of B. *napus* glucose can be oxidized via both EMP and HMP pathways. A series of work was then initiated to investigate the relative activities of these two pathways in relation to fat accumulation during silique development and maturation. The following results are presented in this paper:

(1) Activities of the isolated enzymes of EMP and HMP pathways vary with the stages of development of the siliques. From budding to flowering, the activities of both pathways increase, but the activities of enzymes of EMP pathway increase relatively more. From flowering to fruiting and to silique maturation, the activities of EMP enzymes decline rapidly, whereas the activities of HMP enzymes increase sharply, and maintain at a higher level or only show a slight decrease till the approach of full maturation.

(2) Experiments with inhibitors showed that the fraction of respiration sensitive to iodoacetamide and NaF, decreases with the age of the seeds, but shows a small increase again when maturation approaches.

(3) The metabolic transition from the predominance of EMP pathway to the predominance of HMP pathway in the siliques takes place between 10 to 20 or more days after flowering. From data of other authors, this period corresponds to the time of rapid fat accumulation.

(4) A natural inhibitor of G-6-P dehydrogenase and 6-PG dehydrogenase was found in the reproductive organs. It is present mainly in the buds, flowers and young siliques, and is nearly absent in the old siliques. There appears to be a correlation between the presence of the inhibitor, the activities of glucose degradation pathways and fat formation. When the inhibitor eventually disappears, enzymes of the HMP pathway become activated and fat is actively synthesized. It seems that the inhibitor plays an important role in the regulation of carbohydrate breakdown and fat synthesis.

(5) The exact nature of this inhibitor has not yet been ascertained. It is soluble in water, n-butanol, ethyl ether and hot alcohol; and is insoluble in benzene and chloroform. It is non-dialyzable, not precipitated by saturated $(NH_4)_2SO_4$ solution and stable to heat.

These properties indicate that it is a substance with higher molecular weight and is non-proteinous in nature.

(6) On paper chromatogram, the inhibitor is yellow in color, and shows a blue fluorescence under ultra-violet light. When exposed to the fumes of concentrated ammonia or treated with $AlCl_3$ solution, its color become more intense and the flurescence changes.

(7) Infra-red spectrophotometric analysis indicates that it contains a carbonyl and a phenyl group. Spectral determinations show that it has an absorption maximum in the region around 260 mμm. Exposure to ultra-violet light abolishes the absorption peak and the inhibitory activity.

(8) From the evidences stated above, the inhibitor is presumed to be a flavanoid or its derivative.

植物的脂肪合成及其调节

Ⅵ. 油菜种子脂肪合成与HMP途径关系的同位素证据①

提要

本试验以 $G-1-C^{14}$、$G-6-C^{14}$ 和 $G\text{-}u\text{-}C^{14}$ 为呼吸基质，测定油菜种子在形成过程中所放出的 $C^{14}O_2$ 和$\frac{C_1}{C_6}$及$\frac{C_1}{C_u}$比的变化，从而推测其糖代谢途径的变化，并对 C-1、C-6 标记葡萄糖的转化产物作了分离和比较。试验结果表明，在油菜种子形成过程中，$\frac{C_1}{C_6}$比逐渐升高，接近成熟的种子$\frac{C_1}{C_6}$比接近2或大于2，当种子完全成熟时，$\frac{C_1}{C_6}$比又下降。$\frac{C_1}{C_u}$比的变化也表现类似的规律。由于30天种子中测不出葡萄糖脱氢酶的活力，所以脂肪累积盛期，$\frac{C_1}{C_6}$比的升高无疑是HMP途径加强的结果。此外，接近成熟的种子保温4小时后，标记葡萄糖转化为醇溶物（糖、有机酸、氨基酸）、脂溶物、蛋白质、淀粉和粗纤维，其中 $G-6-C^{14}$ 参入有机酸、脂溶物、蛋白质、淀粉和粗纤维的脉冲数均高于 $G-1-C^{14}$，这与 $C^{14}O_2$ 自 C-1 优先放出的结果相符，可以作为接近成熟种子中存在HMP途径的另一旁证。对于HMP途径的定量计算表明，接近成熟的种子，HMP途径约占47%。

一、引言

在前文中（施教耐等，1965a），我们曾经报道在油菜种子中存在着葡萄糖降解的两条途径：糖酵解（即EMP）和磷酸戊糖支路（即HMP）。在油菜籽实形成过程中EMP途径的酶活力显著下降，而HMP途径的酶活力急剧上升；对碘乙酰胺和NaF敏感的呼吸部分随着种子形成而减少（施教耐等，1965b）。因而推测，油菜种子脂肪形成与HMP途径有着密切的联系。本试验以定位标记和全标记葡萄糖为底物，测定油菜种子在形成过程中$\frac{C_1}{C_6}$及$\frac{C_1}{C_u}$比的变化并对标

朱雨生，刘承宪，施教耐，中国科学院植物生理研究所。此文发表于1965年"植物生理学报"2卷3期185-194页。

① 1964年12月5日收到。

＊本工作在殷宏章教授指导下完成，谨以致谢。谭常同志协助本工作。

本文采用简称如下：$G-1-C^{14}$、$G-6-C^{14}$ 和 $G\text{-}u\text{-}C^{14}$ 分别为C-1、C-6标记和全标记的葡萄糖；EMP-TCA，糖酵解和三羧酸循环；HMP，磷酸戊糖支路；G-6-P，葡萄糖-6-磷酸；6-PG，6-磷酸葡萄糖酸；Tris，羟甲基氨基甲烷；NAD，烟酰胺腺嘌呤二核苷酸（即辅酶Ⅰ）；NADP，烟酰胺腺嘌呤二核苷酸磷酸；NADPH，还原型烟酰胺腺嘌呤二核苷酸磷酸。

记葡萄糖的转化产物作了分离和测定,试验结果进一步证实了上述的论点。

二、材料与方法

1. 材料

油菜种子主要取自大田,品种为胜利油菜。花后10～20天种子称为年幼种子,30天左右种子称为接近成熟的种子,40天种子称为成熟种子。

试剂:G-1-C^{14} 比放射性为3.55 μc/μmole,G-6-C^{14} 比放射性为30.5 μc/μmole。G-u-C^{14} 比活性是3.5 μc/mg。经纸谱检定,上述标记葡萄糖均为层析纯。乙酸-2-C^{14} 比放射性是0.67 μc/μmole。NAD纯度为98%。NADPH由中国科学院生物化学研究所伍钦荣先生供给。

2. 方法

1) 标记葡萄糖的饲喂和 $C^{14}O_2$ 的计数

油菜荚剥壳后,小心取出种子(注意不破损种皮)。迅速称取大小颜色均匀的种子2克,置于容量为50 ml的康惠皿(Conway dish)中。皿内加0.01 mol/L、pH 6.5的磷酸缓冲液4.6 ml,中央井内放入玻璃小碟1只,内加20% KOH 0.3 ml,皿缘涂上凡士林。最后向皿内加标记葡萄糖0.4 ml(相当2 μc左右)。立即盖上玻璃盖,摇匀,用橡皮圈扎紧,以免漏气。在恒温箱中保温,中途摇几次。待保温时间结束后,用扁头镊子取出康惠皿中的玻璃小碟,于红外灯下烘干,用盖革氏计数管计数。预备试验表明上述的反应条件是令人满意的。皿中不必另加非标记葡萄糖,否则会冲淡标记葡萄糖的浓度,降低标记 CO_2 的放出。至于不同标记葡萄糖含有非标记葡萄糖的量虽然不等,但比活性均很高,非标记葡萄糖的量与种子葡萄糖含量相比可以忽略。虽在各次试验中,喂入不同标记的葡萄糖的放射性强度尽可能相等,但有时脉冲数可能有些上下,所以均换算成喂入脉冲数相等时的 $C^{14}O_2$ 产率,以便比较。所有试验均为二次结果的平均值。

2) 种子中标记葡萄糖转化产物的分离

醇溶物——上述喂过标记葡萄糖的种子洗去种皮表面的放射性葡萄糖,用80%热酒精杀死。磨碎种子,用80%热酒精提取6～7次,每次酒精6～7 ml,最后于红外灯下浓缩干,用石油醚洗三次,待干后加水4 ml溶解,吸0.1 ml铺样记数,剩余的3.9 ml作为糖、有机酸和氨基酸分析之用。

糖、有机酸和氨基酸——基本上按照Canvin等(1961)的方法,将剩下3.9 ml醇溶物通过Zerolite 225(H^+型,200/400 mesh,0.7×7 cm^2)阳离子交换树脂,用50 ml 80%酒精冲洗,流下的糖和有机酸再通过Dowex-1×8(甲酸型,200/400 mesh,0.7×7 cm),仍用80%酒精冲洗,洗下者为糖。后用50 ml 2N NH_4OH 洗下阳离子交换树脂上的氨基酸和用4N甲酸洗下阴离子交换树脂上的有机酸。最后糖、有机酸和氨基酸分别浓缩至干,用80%酒精溶解,铺样记数。

脂溶物——剩下的种子残渣用丙酮提取6～7次,与上述石油醚洗液合并,得到脂溶物。

蛋白质——上述残渣用0.4% NaOH抽提4次,清液加三氯乙酸,使浓度成5%,得到蛋白质沉淀。加一定量的NaOH,吸出一部分铝碟上烘干,测定放射性强度。

淀粉——基本上按Pucher等(1948)的方法。剩下的残渣加4 ml水煮沸20分钟,使淀粉糊化，再加 60%过氯酸(4 ml)，使总浓度成 30%。摇匀，立即离心。清液加 7.5 ml NaCl (20%)，再加 2.5 ml 碘-碘化钾溶液。将沉淀下来的淀粉-碘复合物离心，用 NaCl (2%)-酒精 (70%)洗一次，再用 80%酒精洗 2 次。最后，用少量酒精洗出，于铝碟中烘干记数。

粗纤维——残渣加 2 ml 浓硫酸，使之糊化，取 1.0 ml 铺在玻璃碟中测定放射性强度。

3) 葡萄糖脱氢酶

按照 McLean (1958)的方法，用日立分光光度计测定在 340 mμm，波长处 NAD 的还原。酶液制备是相当于 1 克鲜重的丙酮粉(丙酮粉制备法见施教耐等 1965a)加 9 ml 0.15 mol/L 含 1.6×10^{-4} mol/L $KHCO_3$ 的 KCl 研磨，离心，取上清液。反应液中包含双甘氨肽(pH 7.5) 125 μmoles，葡萄糖 100 μmoles，NAD 0.3 μmoles，酶液 0.5 ml，空白者缺 NAD，总体积 3 ml。于室温下测定。

4) 葡萄糖醛酸还原酶

参照 Isherwood 等(1961)的方法。反应液含 Tris，pH 7.5，120 μmoles，KCl 300 μmoles，葡萄糖醛酸 10 μmoles，烟碱酰胺 40 μmoles，NADPH 0.5 μmoles，酶液 0.5 ml(制备方法同上)。空白缺葡萄糖醛酸，总体积 3 ml。用日立分光光度计测定 NADPH 在 340 mμm 吸收峰的减少。

三、试验结果

1. 种子形成过程中$\frac{C_1}{C_6}$和$\frac{C_1}{C_u}$比的变化

油菜种子具有吸收标记葡萄糖和通过呼吸放出标记 CO_2 的能力。测定不同年龄油菜种子自 G-1-C^{14} 和 G-6-C^{14} 释放 $C^{14}O_2$ 的能力和$\frac{C_1}{C_6}$比表明，在油菜种子形成过程中 C-1 和 C-6 的放出都在增加，但 C-1 放出之增加远大于 C-6，故$\frac{C_1}{C_6}$比逐渐上升，以 30 天种子与 10 天种子比较，$\frac{C_1}{C_6}$比增加 2.5～5 倍，种子快要成熟时$\frac{C_1}{C_6}$比又下降(见表 1)。$\frac{C_1}{C_u}$比在种子形成期间的变化表现类似的规律(见表 2)。

表 1 油菜种子形成过程中$\frac{C_1}{C_6}$比的变化

试验	种子年龄/天	释放 $C^{14}O_2$/cpm		$\frac{C_1}{C_6}$比
		G-1-C^{14}	G-6-C^{14}	
No. 1	10	106	217	0.49
	20	212	257	0.82
	30	502	251	2.00
	40	623	447	1.39

（续表）

试验	种子年龄(天)	释放 $C^{14}O_2$/cpm		$\frac{C_1}{C_6}$比
		G-1-C^{14}	G-6-C^{14}	
No. 2	10	69	169	0.41
	35	446	423	1.05
No. 3	10	87	219	0.40
	30	661	348	1.90

注：反应条件：No. 1：30℃，5 小时；No. 2：30℃，4 小时；No. 3：30℃，5.5 小时。

表 2　油菜种子形成过程中$\frac{C_1}{C_u}$比的变化

试验	种子年龄/天	释放 $C^{14}O_2$/cpm		$\frac{C_1}{C_u}$比
		G-1-C^{14}	G-u-C^{14}	
No. 4	10	140	196	0.71
	20	181	140	1.29
	25	150	93	1.61
No. 5	10	243	289	0.84
	15	162	170	0.95
	25	364	300	1.21
	35	806	983	0.82

注：反应条件：No. 4：35℃，4 小时；No. 5：37℃，4 小时。

测定不同时间种子呼吸放出 $C^{14}O_2$ 的数量，并计算$\frac{C_1}{C_6}$和$\frac{C_1}{C_u}$比的变化，发现不论种子年龄如何，$\frac{C_1}{C_6}$比均随呼吸时间（三小时内）的延长而增高（见表 3）；而$\frac{C_1}{C_u}$比相反，发生有规律地递降（见表 4）。很可能是由于随着时间的延长，氧气减少，无氧呼吸（EMP——能释放 C-3，C-4）加强，因为 G-u-C^{14} 会有 C-3、C-4 位置上标记的碳原子，在无氧呼吸中变成 $C^{14}O_2$ 而释放。

表 3　油菜种子呼吸时$\frac{C_1}{C_6}$比的变化

种子年龄/天	$\frac{C_1}{C_6}$比			
	第一小时	第二小时	第三小时	第四小时半
10	0.28	0.45	0.53	0.63
30	1.17	2.08	3.07	2.65

注：反应条件：30℃。

表 4 油菜种子呼吸时$\frac{C_1}{C_u}$比的变化

种子年龄/天	$\frac{C_1}{C_u}$比			
	第一小时	第二小时	第三小时	第四小时
10	1.06	—	0.94	0.91
30	1.87	—	0.87	0.74

注:反应条件:35℃。

表 1 和表 3 中一个值得注意的现象是年幼种子中 C-6 的放出远大于 C-1,因此$\frac{C_1}{C_6}$比小于 1,如 10 天种子$\frac{C_1}{C_6}$为 0.4～0.5。看来 C-6 的放出不是通过 EMP 途径形成乙酸,再循 TCA 循环而氧化,如表 5 所示,种子利用与葡萄糖之 C-6 标记相当的甲基标记的乙酸释放 $C^{14}O_2$ 的能力极有限。

表 5 油菜种子利用乙酸-2-C^{14} 释放 $C^{14}O_2$ 的能力

种子年龄/天	绎放的 $C^{14}O_2$/cpm
10	35
30	45

注:反应条件:30℃,5.5 小时。

为了进一步确定 C-6 优先放出的现象是否由于年幼种子中存在糖醛酸循环,测定了非标记葡萄糖醛酸对 G-1-C^{14} 和 G-6-C^{14} 释放 $C^{14}O_2$ 的影响。结果葡萄糖醛酸对 C-1、C-6 释放的影响是不同的,即 C-6 放出减少较小,而 C-1 放出减少了 46%(见表 6)。同时在 10 天种子提取液中也测不出葡萄糖醛酸还原酶。

表 6 葡萄糖醛酸对 10 天油菜种子从葡萄糖 C-6、C-1 放出 $C^{14}O_2$ 的影响

处理	释放的 $C^{14}O_2$/cpm		$\frac{C_1}{C_6}$比
	G-1-C^{14}	G-6-C^{14}	
对照	97	238	0.41
葡萄糖醛酸	59	226	0.26

注:反应条件:30℃,5.5 小时;葡萄糖醛酸:100 mg。

在 30 天种子匀浆中未发现有葡萄糖脱氢酶的活力存在。

2. 标记葡萄糖转化产物分离试验和 HMP 途径的定量计算

喂饲了 C-1-C^{14} 和 G-6-C^{14} 的油菜接近成熟种子的各转化产物经分离后发现,醇溶物(糖、有机酸和氨基酸)、脂溶物、蛋白质、淀粉和粗纤维都带有标记(见表 7)。

表 7 葡萄糖标记碳在油菜种子各组分中的分布

	$G-1-C^{14}$ /cpm	$G-6-C^{14}$ /cpm	$\frac{C_1}{C_6}$比
CO_2	1 405	500	2.81
醇溶物	58 120	78 181	0.74
糖	23 392	26 232	0.89
有机酸	6 840	7 192	0.95
氨基酸	4 896	6 096	0.80
脂溶物	1 140	2 160	0.53
蛋白质	400	792	0.51
淀粉	38	51	0.75
粗纤维	72	112	0.64

注:反应条件:30℃,4 小时。30 天油菜种子 2 克,保温前对半切开,饲喂 $G-1-C^{14}$ 或 $G-6-C^{14}$ 各 5 μc。比放射性相同。中央玻璃碟中 KOH (20%)0.2 ml。

从表中可见,标记 C^{14} 主要分布在醇溶物(糖、有机酸、氢基酸)、脂溶物和 CO_2 中。蛋白质、淀粉和粗纤维等高分子化合物中出现较少。此外,$G-6-C^{14}$ 参入有机酸、氨基酸、脂肪、蛋白质、淀粉和粗纤维的量均高于 $G-1-C^{14}$。若都算成$\frac{C_1}{C_6}$比,则除 CO_2 外,所有产物的$\frac{C_1}{C_6}$之比均小于 1,这是与 30 天种子中 C-1 放出大于 C-6 放出相呼应的。

根据表 7 中所示的 $G-1-C^{14}$ 和 $G-6-C^{14}$ 参入脂肪的强度,可以近似计算种子中 HMP 途径在葡萄糖氧化降解途径中所占的比例。设以 $G-1-C^{14}$ 参入脂肪的量代表 EMP 途径,$G-6-C^{14}$ 参入脂肪的量代表 EMP+HMP 途径,则可算出接近成熟种子中 EMP 途径约占$\frac{1\,140}{2\,160}=53\%$,HMP 途径约占(100-53)%=47%。

四、讨论

自从 Bloom 和 Stettin (1953)利用定位标记葡萄糖和$\frac{C_1}{C_6}$比研究组织的代谢途径以来,这一方法在动物(Brin 和 Yonemoto, 1958、McLean, 1958、Wenner, 1958)、微生物(Inda,等 1961; Wang 等,1958)和高等植物中(Axelrod 和 Beevers, 1956; Gibbs, 1959)得到了广泛的应用。在玉米芽鞘(Beevers, 1956)、根尖(Butt 和 Beevers, 1961),豌豆根、叶、节间和向日葵茎(Beevers 和 Gibbs, 1954),豌豆种子切片(Wager, 1963),番茄、黄瓜、柑桔和橙子的果实(Barbour 等,1958),胡萝卜和马铃薯块茎(Ap Rees 和 Beevers, 1960a, 1960b; Romberger 和 Norton, 1959,1961)等组织中均发现葡萄糖 C-1 的放出大于 C-6;田波和汤佩松(1961)借用$\frac{C_1}{C_u}$比研究受病毒感染的烟草叶组织的代谢,也发现 C-1 放出大于 C-u。从而各位作者都证明了上述组织中有非 EMP-TCA 循环,即葡萄糖醛基首先氧化的 HMP 途径的存在。不少作者(Katz 和 Wood, 1956; Korkes, 1956)对于放射性呼吸测量法的优点和局限性作了讨论。但目前,同位素方法仍不失为研究活体代谢途径简单而有效的方法。

本试验采用 C-1、C-6 和 C-u 标记的葡萄糖,研究油菜种子形成期间代谢途径的变化。试验证明,在种子形成过程中伴随着脂肪的累积,$\frac{C_1}{C_6}$比逐渐增高,在脂肪形成高峰期,$\frac{C_1}{C_6}$比最高,但种子完全成熟,趋于衰老时,$\frac{C_1}{C_6}$比又下降,标志着脂肪累积速度与 HMP 途径的加强相平行。$\frac{C_1}{C_u}$比的变化也表现类似的规律。这一结果与 McLean (1958)在大鼠乳腺中观察到的现象非常相像。由表 1 还可看出,种子形成过程中$\frac{C_1}{C_6}$比的升高主要由于 C-1 放出的增加,如开花后 30 天种子,与 10 天种子相比,C-1 的放大最大的增加近 5 倍。因为另一优先氧化 C-1 的葡萄糖脱氢酶(McLean, 1958)在 30 天油菜种子中测不出,再结合以前观察到的 30 天种子中 G-6-P 脱氢酶和 6-PG 脱氢酶活力较 10 天种子高 4~5 倍(施教耐等,1965b)。我们相信,在脂肪积累盛期,种子 C-1 释放的增加主要是由于 HMP 途径酶活力升高的结果。

值得注意的是,油菜年幼种子(开花后 10 天左右)$\frac{C_1}{C_6}$比远小于 1,因为 G-6-C^{14} 标记碳位置相当于通过 EMP 途径形成乙酸的甲基位置,而种子利用乙酸-C^{14}(甲基-C^{14})释放 $C^{14}O_2$ 的能力却有限,说明 C-6 的优先放出主要不是借助 TCA 循环。而且根据理论上推测,若葡萄糖分子完全循 EMP-TCA 循环氧化,则 C-1 和 C-6 释放的速率相等,$\frac{C_1}{C_6}$比的理论值为 1;若组织中只存在 HMP 途径,且 HMP 不与 EMP-TCA 相联系,以及无重循环时,由于葡萄糖氧化只发生在 C-1,则$\frac{C_1}{C_6}$比的理论值将趋于无穷大;两条途径同时存在时,$\frac{C_1}{C_6}$比大于 1,而不会小于 1。所以$\frac{C_1}{C_6}$比小于 1 暗示着有另一优先氧化葡萄糖 C-6 的反应存在。根据现有的知识,葡萄糖通过糖醛酸循环氧化时,是首先释放 C-6,生成 CO_2 的(Gibbs, 1959)。在植物材料中也有人相继发现该循环的存在(Frinkle 等,1960; Isherwood 等,1954; Loewus 等,1958; Hassan 和 Lehninger, 1956)。但本试验中未发现该循环的重要中间产物——葡萄糖醛酸对 G-6-C^{14} 释放 $C^{14}O_2$ 有显著的稀释效应,加之几乎测不出葡萄糖醛酸还原酶,说明油菜年幼种子中糖醛酸循环或者不存在,或者作用甚微。本试验中 G-6-C^{14} 的比放射性较 G-1-C^{14} 高,吸收率也要高些,但经测定和若根据吸收率加以校正,则$\frac{C_1}{C_6}$比仍小于 1,故 C-6 优先放出的原因究属如何,目前尚无法定论。

分离油菜种子标记葡萄糖的转化产物发现,在保温 4 小时后,标记葡萄糖转化为醇溶物(有机酸和氨基酸)、脂溶物、蛋白质、淀粉和粗纤维。说明接近成熟的油菜种子生理活动很旺盛,涉及的代谢面很广。其中分子较简单的醇溶物、脂溶物和 CO_2 中同位素参入较多,高分子的蛋白质,尤其是淀粉和粗纤维中同位素参入较少。此与 Ap Rees 和 Beevers (1960a)在马铃薯中的观察相类似。此外,有意思的是 G-6-C^{14} 参入有机酸、脂肪、蛋白质、淀粉和粗纤维的脉冲数均高于 G-1-C^{14} 参入的量,这是与 $C^{14}O_2$ 自 G-1-C^{14} 释放强于 G-6-C^{14} 释放相呼应的，并且也与上述作者(Ap Rees 和 Beevers，1960a)观察到的 G-6-C^{14} 参入有机酸、氨基酸、蛋白质、已烷抽提物高于 G-1-C^{14} 基本上相一致，可以作为接近成熟的油菜种子中 HMP 途径存在的另一佐证。

目前，在研究 HMP 途径的生理意义时，对于 HMP 途径的定量研究，对于 HMP 途径在整个代谢途径中所占比例的了解，不仅显得重要，而且很有必要。最初采用$\frac{C_6}{C_1}$比和$\frac{C_1-C_6}{C_1}$比表示 EMP 和 HMP 途径比例的方法，由于需要作一系列的假定，且这些假定在某些场合下又常常不能成立，而未能作为 HMP 途径定量估价的有效手段(Holl-mann，1961；Ap Rees & Beevers，1960a)。近几年来，有些作者对于 HMP 途径的定量计算提出了新的方案，虽然仍带有一定的局限性。Wang 等(1962)在研究果实代谢时，以 C-1、C-6 和 C-3,4 标记的葡萄糖为呼吸基质，根据三种标记葡萄糖中 $C^{14}O_2$ 的放出强度，算出蕃茄果实中有 73%葡萄糖通过 EMP-TCA 氧化，有 27%葡萄糖沿 HMP 途径氧化。在动物组织的研究中，在测定 C-1、C-6 标记葡萄糖释放 $C^{14}O_2$ 的同时，还分析代谢产物如乳酸、乙酰乙酸(Woltenholme 和 Conner，1959)和脂肪酸(Abraham 等，1954)中的放射性强度。Abraham 的方法具有较大的可靠性，他假定 C-1 标记葡萄糖中 C^{14} 参入脂肪酸是葡萄糖通过 EMP 途径转化为乙酸底物的结果，C-6 标记葡萄糖中 C^{14} 参入脂肪酸是葡萄糖通过 EMP 和 HMP 两条途径转化为乙酸底物的结果。这样算得大鼠乳腺中 HMP 途径约占 60%的比例。我们根据类似的原理，分别测定 G-1-C^{14} 和 G-6-C^{14} 参入脂肪的强度，所算得的接近成熟的油菜种子中 HMP 途径的比例为 47%，约占整个代谢途径的一半。虽然这一数值仅具有近似的意义，然而，仍能给予我们 HMP 途径在代谢中所起贡献的数量概念。

参考文献

[1] 田波、汤佩松，1961：感染烟草花叶病毒的叶组织中葡萄糖的氧化代谢，植物学报，**11**：1—8.

[2] 施教耐、朱雨生、谭常，1965a：植物的脂肪合成及其调节 I. 油菜种子形成期间葡萄糖的降解途径. 植物生理学报，**2**：33—43.

[3] 施教耐、朱雨生、谭常，1965b：植物的脂肪合成及其诵节 III. 油菜籽实形成过程中代谢途径的变化与内源抑制剂的调节作用. 植物生理学报，**2**：105—118.

[4] Abraham，S. and P. F. Hirsh，1954：The quantitative significance of glycolysis and non-glycolysis in glucose utilization by rat mammary gland. *J. Biol. Chem.*，211：31-38.

[5] Ap Rees，T. and H. Beevers，1960a：Pathways of glucose dissimilation in carrot slices. *Plant Physiol.*，**35**：830-848.

[6] Ap Rees，T. and H. Beevers，1960b：Pentose phosphate pathway as a major component of induced respiration of carrot and potato slices. *Plant Physiol.*，35：839-848.

[7] Axelrod，B. and H. Beevers，1956：Mechanism of carbohydrate breakdown in plants. Ann. Rev. *Plant Physiol.*，**7**：267-298.

[8] Barbour，R. D.，D. R. Bukler and C. H. Wang，1958：Identification and estimation of catabolic pathways of glucose in fruits. *Plant Physiol.*，**33**：394-400.

[9] Beever，H.，1956：Intermediates of the pentose phosphate pathway as respiratory substances. *Plant Physiol.*，**31**：339-347.

[10] Beever，H. and M. Gibbs，1954：The direct oxidation pathway in plant respiration. Plant Physiol.，*29*：322-324.

[11] Bloom，B. and M. R. Stettin，1953：Pathway of glucose catabolism. *J. Amer. Chem. Soc.*，**75**：5446.

[12] Brin, M. and R. H. Yonemoto, 1958: Stimulation of the glucase oxidative pathway by methylene blue. *J. Biol. Chem.*, **230**: 307 - 317.

[13] Butt, V. S. and H. Beevers, 1961: The regulation of pathways of glucose catabolism in maize roots. *Btochem. J.*, **80**: 21 - 27.

[14] Canvin, D. T. and H. Beever, 1961: Sucrose syntheisis from acetate in the germinating castor bean: Kinetics and pathway. *J. Biol. Chem.*, **236**: 988 - 995.

[15] Frinkle, B. J., S. Kelly and F. A. Loewus, 1960: Metabolism of D-(1 - C^{14})- and D-(6 - C^{14})-glucuronolactone by the ripening strawsberry. *Biochirn. Biophys. Acta*, **38**: 332 - 339.

[16] Gibbs, M., 1959: Metabolism of carbon compounds. *Ann. Rev. Plant Physiol.*, **10**: 329 - 378.

[17] Hasson M. ul and A. L. Lehninger, 1956: Enzymatic formation of ascorbic acid in rat liver extracts. *J. Biol. Chem.*, **223**: 123 - 138.

[18] Hollmann, S., 1964: Non-glycolytic pathways of metabolism of glucose. p. 46, AP. New York-London.

[19] Inda, M., De Issaly, A. Issaly and A. O. M. Stoppant, 1961: Role of the pentose phosphate in Pasteurella mullocida. *Nature*, **191**: 727 - 728.

[20] Isherwood, J. A., J. T. Chen and L. W. Mapsen, 1954: Synthesis of L-ascorbic acid in plants and animals. *Biochem.* J., **56**: 1 - 15.

[21] Katz, J. and H. G. Wood, 1956: The use of glucose-C^{14} for the evaluation of the pathways of glucose metabolism. *J. Biol. Chem.*, **235**: 2165 - 2177.

[22] Korkes, S., 1956: Carbohydrate metabolism. *Ann. Rev. Biochem.*, **25**: 685 - 734.

[23] Loewus, F. A., B. J. Frinkle and R. Jang, 1958: L-ascorbic acid: A possible intermediate in carbohydrate metabolism in plants. *Biochim. Biophys. Acta*, **30**: 629 - 635.

[24] McLean, P., 1958: Carbohydrate metabolism in the mammary tissue. I. Pathway of glucose catabolism in the mammary gland. *Biochim. Biophys. Acta*, **30**: 303 - 315.

[25] Pucker, G. W., C. S. Leavenworth and H. B. Vickery, 1948: Determination of starch in plant tissue. *Anal. Chem.*, **20**: 850 - 853.

[26] Romberger, J. A. and G. Norton, 1959a: Respiratory pathways in freshly cut potato. *Plant Physiol.*, **34**: Suppl. : xii.

[27] Romberger, J. A. and G. Norton, 1959b: Changing respiratory pathways in potato tuber slices. *Plant Phy stol.*, **36**: 20 - 30.

[28] Silva, G. W., W. P. Doyle, and C. H. Wang, 1958: Glucose catabolism in the American coakroach. *Nature*, **182**: 102 - 104.

[29] Wager, H. G., 1963: Pathways of utilization of (1 - C^{14})- glucose and (6 - C^{14})- glucose in slices of peas. *J. Expt. Bot.*, 14: 63 - 81.

[30] Wang, C. H., W. P. Doyle and J. C. Romsey, 1962: Role of hexose monophosphate pathway in tomato catabolism. *Plant Physiol.*, **37**: 1 - 7.

[31] Wang, C. H., I. Stern, C. M. Gilmork, S. Klungsoyr, D. J. Reed, J. J. Baily, B. E. Christensen and V. H. Cheidelim, 1958: Comparative study of glucose catabolism by the radiorespirometric method. *J. Bact.*, **76**: 207 - 216.

[32] Wenner, C. E., 1958: The hexose monophosphate shunt in glucose catabolism in ascites tumor cells. *Cancer ReS.*, **18**: 1105 - 1114.

[33] Woltenholme, G. E., C. M. O. Conner, 1959: Ciba foundation symposium on the regulation of cell metabolism.. J. & A. Churchill Ltd., London.

FAT SYNTHESIS AND ITS REGULATION IN PLANTS

Ⅵ. ISOTOPE EVIDENCE FOR THE RELATION OF FAT SYNTHESIS BY RAPE SEEDS TO THE HMP PATHWAY

Y. S. Zhu, C. H. Liu and C. N. Shih

(*Institute of Plantt Physiology, Academia Sinica*)

Since the fat synthesis in rape seeds during development is accompanied by the increase in the activity of the HMP pathway and the decrease in the activity of the EMP pathway, as previously indicated in studies of enzyme and respiratory inhibitors, it was suggested that the fat synthesis by rape seeds closely depends upon the HMP pathway. In this study, the oxidation of glucose - 1 - C^{14} and glucose - 6 - C^{14} or glucose-u-C^{14} by rape seeds and the radio activity in C_1/C_6 or C_1/C_u at different stages of the seed development were estimated. In addition, the products of conversion of labelled glucose were separated and determined. The isotope evidence further supports above suggestion.

The seeds of different age were incubated in the Conway dish with glucose - 1 - C^{14}, glucose - 6 - C^{14} or glucose-u-C^{14} in phosphate buffer (0.01 mol/L, pH 6.5) at 30℃ (or 35℃, 38℃) for the determination of respiration. After incubation the seeds were immediately killed by adding hot alcohol. Then the lipid-soluble fraction, alcohol-soluble fraction, sugars, organic acids, amino acids, starch, crude cellulose and proteins were separated for the radioactivity assay. The results of experiments indicate that the ratio in C_1/C_6 rises as the seeds are developing, reachs the maximum level in seeds approaching to ripeness, but sharply decreases in full-ripe seeds, parallel to the course of the fat accumulation. The ratio in C_1/C_u shows similar changes during seed development. These changes are very striking and conform in pattern with those obtained in a previous study on the enzymes of the HMP pathway. It seems, that in period of fat accumulation the rise of C_1/C_6 ratio in seeds, being attributed to the relatively greater increase in the contribution of C - 1 to CO_2, is due to the increase in the activity of the HMP pathway, because no activity of glucose dehydrogenase, another enzyme responsible to the release of CO_2 from C - 1 of glucose, was observed in 30 days-old seeds.

The conversion of labelled glucose into lipid-soluble fraction, alcohol-soluble fraction (e. g. sugars, organic acids, amino acids), proteins, starch and crude cellulose was also demonstrated. It is especially interested that the incorporation of glucose - 6 - C^{14} into organic acids, lipid-soluble substances, proteins, starch and crude cellulose is greater than that of glucose - 1 - C^{14}, consistent with the preferential release of $C^{14}O_2$ from glucose - 1 - C^{14} oxidized via the HMP pathway. It may be a by-evidence for the operation of the HMP pathway in maturing rape seeds. The contribution of the HMP pathway to the total glucose oxidation in rape seeds approaching to ripeness, as appreciated on the basis of amounts of incorporation of G - 1 - C^{14} and G - 6 - C^{14} into fat, is about 47%.

植物的脂肪合成及其调节

Ⅶ. 大豆种子形成过程中物质累积与呼吸代谢的关系*

提要

大豆种子形成过程中可溶性物质(磷、氮、氨基酸、糖)逐渐增加,并伴随着有机物质合成过程的加强和贮藏物质(蛋白质、脂肪、淀粉、植质等)的大量累积。不同时期的产物分析表明,在种子成熟期间,开始脂肪和蛋白质同时累积,而尤以脂肪累积为主(花后20～30天),接着转入以蛋白质合成为主的时期(花后30～40天)。

与有机物质的合成与累积相平行的是单粒种子呼吸强度的上升,其中花后10～15天呼吸上升较快,30天左右形成呼吸高峰。不同时期大豆种子对碘乙酰胺的敏感程度不同。花后11～32天,被碘乙酰胺抑制的呼吸(EMP)逐渐上升,32天以后有所下降;相应的剩余呼吸(HMP)表现出类似的变化规律,但从升降速度来看,花后15～32天剩余呼吸增加更为迅速,32天以后下降更为剧烈。大豆种子两途径酶活力的变化趋势与呼吸规律相一致。从开花后10至30天G－6－P脱氢酶、6－PG脱氢酶、G－6－P异构酶、醛缩酶活力均显著增加,30天以后都有下降趋势,但30天以后HMP途径酶活力的下降远较EMP途径剧烈,以至此时EMP途径的相对比重升高。

甲烯蓝和DNP对不同时期大豆种子的呼吸表现出不同的影响。甲烯蓝对各个时期种子呼吸均有程度不等的刺激,但以后期为主。DNP只对前期和后期种子呼吸有促进的表现,尤其是前期。在呼吸高峰期,甲烯蓝和DNP的效应较小。

根据上述结果,表明大豆种子呼吸途径的变化与有机物的合成与转化密切相关,进一步证实了脂肪形成与HMP途径的关系,并指出蛋白合成可能与EMP-TCA途径有关。

一、前言

大豆种子是积累蛋白质和脂肪的器官。成熟的大豆种子中含有蛋白质35％～44％,脂肪15％～25％(EHKeH, 1959)。在大豆种子形成期间,种子粗蛋白、粗脂、蔗糖、还原糖、糊精、

朱雨生,谭常,施教耐,中国科学院植物生理研究所。
本文发表于1965年“植物生理学报”2卷3期195－203页。
1964年12月5日收到。

* 本工作在殷宏章教授指导下完成,特此致谢。吴敏贤同志参加部分分析工作。

本文采用简称如下:EMP,糖酵解途径;TCA,三羧酸循环;HMP,磷酸戊糖支路;G－6－P,葡萄糖－6－磷酸;6－PG,6－磷酸葡萄糖酸;FDP,果糖－1,6－二磷酸;NADP,烟酰胺腺嘌呤二核苷酸磷酸;NAD,烟酰胺腺嘌呤二核苷酸;ATP,腺嘌呤核苷三磷酸;ADP,腺嘌呤核苷二磷酸;DNP,2,4－二硝基苯酚;EDTA,二乙胺四乙酸。

纤维素、半纤维素和灰分等均有规律的上升(Wolf 等,1962)。随着大豆种子脂肪的形成,种子呼吸商升高(Howell, 1958)。可惜,大豆种子形成时期的代谢变化研究很少(Howell, 1958; Howell 等,1958; Bils, 1960)。至于涉及种子呼吸代谢与物质合成关系的知识,则更为贫乏。

在前文中(施教耐、朱雨生、谭常,1965a、b;朱雨生、谭常、施教耐,1965)我们报道了油菜种子脂肪形成与 HMP 途径的关系,推测 HMP 途径供应脂肪合成所需的 NADPH。本试验以大豆为材料,应用呼吸抑制剂和酶活力测定方法,以便进一步证实脂肪形成与 HMP 途径的关系。此外,还探讨了蛋白质合成与 EMP 途径的关系以及呼吸速度的限制因子。

二、材料和方法

1. 材料

采用大田栽培的早熟种大豆,品种为 5 月早。6 月中旬开花,花后隔不同时期采集豆荚样品,并根据标准样品估计荚子年龄。豆荚剥壳后,选择大小均匀的豆粒作测定和分析用:一部分于 105℃烘箱杀死,在 70℃左右烘干保存,待作各项分析之用;另一部分测定呼吸;最后一部分制成丙酮粉,测定酶活力。各项测定结果均以粒子为单位表示。

2. 方法

(1) 化学分析。可溶性糖采用 Fairbain (1953)的方法。用 80%酒精于 70℃提取样品 3~4 次,用蒽酮比色法定糖。残渣用 30%过氯酸在室温下水解提取二次,水解提取液用蒽酮法定糖,即得淀粉含量(McCready 等,1950)。总脂含量按 Huerga 等(1953)法,即用石油醚提取样品 4~5 次(50~60℃水浴中),用二氧六环-硫酸比浊法定脂。蛋白质和可溶氮的测定是用 5%三氯乙酸提取样品,提取液为可溶性氮,残渣则含蛋白氮。分别加浓硫酸和接触剂加热消化,用 Nesseler-Folin 比色法定氮。总氮含量×6.25=蛋白质含量。氨基酸按 Moore 和 Stein (1948)法,用 5%三氯乙酸的样品提取液用茚三酮比色法测定。磷化合物分部方法如下:用 5%三氯乙酸提取样品,提取液直接测定无机磷。提取液用三酸消化后测得的磷减去无机磷为酸溶性有机磷。残渣消化后定磷为酸不溶性有机磷。三种磷的总和为总磷。植质磷测定按 Pons 等(1953)方法稍作修改:用三氯乙酸提取样品。定磷沿用 Bertramson (1943)的钼青比色法。

(2) 呼吸测定。用 Warburg 呼吸器测定种子氧吸收。种子去皮后,迅速称取粒数和重量都一样的样品(每组粒数为 3~10 粒,重量在 0.4~1.0 g,每组样品 1 小时总耗氧量为 150~250 μl)。分别用各处理溶液减压渗入:抽气半小时,一般总共渗入 2 小时。取出种子,用滤纸吸干,迅速置于反应瓶主室中(内盛同样的处理溶液 0.5 ml),中央井加 KOH (20%)0.2 ml。于 30℃测定反应瓶中气压变化,每 15 分钟读数一次,共持续 1 小时。

(3) 丙酮粉的制备和酶活力测定。丙酮粉制备方法见以前介绍(施教耐等,1965a),放在 15℃冰箱中保存,用 pH 7.2 EDTA (0.001 5 mol/L)和磷酸钾缓冲液(0.01 mol/L)研磨提取,在 10 000×g 下高速离心 10 分钟。上清液测定 G-6-P 脱氢酶、6-PG 脱氢酶和醛缩酶;另外取一部分丙酮粉用 pH 9.0 硼酸缓冲液提取,测定 G-6-P 异构酶的活力。G-6-P 脱氢酶和 6-PG 脱氢酶的测定是根据 NADPH 的形成,用 Beckman DU 型分光光度计测定 340 mμm 处的光密度变化,温度 38℃,时间为 1 分钟。醛缩酶是根据 FDP 分解,用二硝基苯肼显色法测定形成的磷酸丙糖量。G-6-P 异构酶活力的测定是用间苯二酚测定酶反应后由 G-6-P 所形成的果糖量。所有酶活力均换算成单位粒子每小时消耗的 G-6-P 微克分子数。具体测定方法见以前报告(施教耐等,1965a)。

三、试验结果

1. 大豆种子形成过程中糖、总脂、蛋白、可溶氮和磷化合物的变化

大豆种子形成过程中可溶性糖、淀粉、脂肪和蛋白含量均升高,但可溶性糖、淀粉和蛋白含量在 30 天以后略有下降(见图 1)。

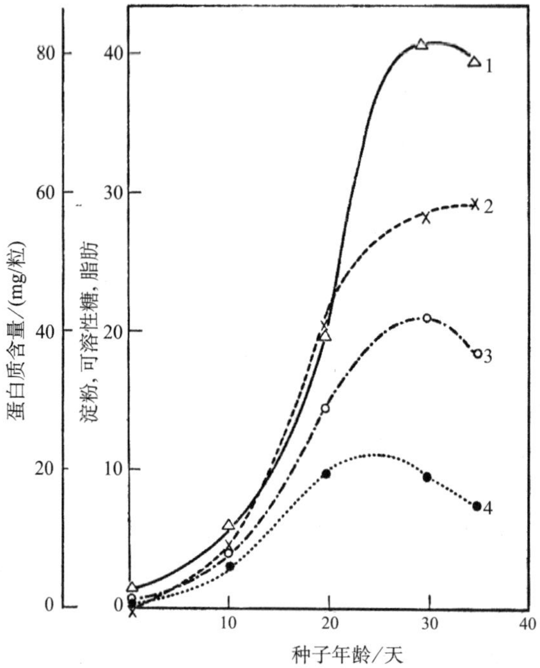

图 1 大豆种子形成过程中可溶性糖、淀粉、脂肪、蛋白含量的变化

1—蛋白质;2—脂肪;3—可溶性糖;4—淀粉

以单位粒子计算,分析不同时期物质累积量的变化可知,脂肪增长量在花后 20～30 天最高,这一时期所形成的脂肪量占成熟种子脂肪含量的一半以上;蛋白质增长量则在 30～40 天最高,占成熟种子蛋白质含量的一半以上(见表 1)。

若以干物质为单位,可看出花后 10～30 天,脂肪含量增加最快,30～40 天蛋白含量增加最快,此外在大豆种子形成期间,由于蛋白质和脂肪积累最快,以至就相对含量而言,淀粉含量反而下降,可溶性糖则变化不大(见表 2)。成熟种子(45 天)中蛋白质含量最高(占干重 40%),脂肪次之(15%),可溶性糖占 10%,淀粉约占 4%。

可溶性氮、氨基酸也随着种子年龄的增长而增加。在花后 10～30 天增加最快,40 天后可溶性氮增加较少,氨基酸迅速下降(见表 3),标志着后期氨基酸向蛋白质的转化过程大大加强。

如表 4 所示,大豆开花后 10～40 天,无机磷、酸溶性有机磷和酸不溶性有机磷都有不同程度的增加,40 天以后无机磷下降,酸溶性和酸不溶性有机磷仍有增加。植质磷在酸溶性有机磷中占相当大的比例,但初期含量极少,成熟种子中含量最高。非植质的酸溶性有机磷在花后

表 1 不同时期种子脂肪蛋白积累量的比较

种子年龄/天	脂肪		蛋白质	
	增长量/mg	占成熟种子(45 天)/%	增长量/mg	占成熟种子(45 天)/%
10～20	3.89	13.4	9.54	12.1
20～30	15.95	54.9	27.04	34.2
30～40	7.57	26.1	41.38	52.4
40～45	1.01	3.5	−1.50	−1.9

表 2 大豆种子形成过程中可溶性糖、淀粉、脂肪和蛋白含量的变化(占干重%)

种子年龄/天	可溶性糖*	淀粉*	总脂	蛋白质
10	8.16	6.12	5.98	24.38
20	8.38	7.02	9.53	25.51
30	10.28	6.98	14.62	27.94
40	9.91	4.61	13.42	38.54
45	9.55	3.75	15.01	40.82

注：* 可溶性糖和淀粉以葡萄糖表示。

表 3 大豆种子形成过程中种子可溶性氮、氨基酸的变化

种子年龄/天	可溶性氮/(mg/粒)	氨基酸/(μmoles/粒)
10	0.24	3.80
20	0.84	13.40
30	1.30	18.30
40	1.59	19.00
45	1.76	14.80

表 4 大豆种子形成过程中磷化合物的动态/(mg/粒)

种子年龄/天	无机磷	酸不溶性有机磷	酸溶性有机磷			总磷	有机磷/无机磷
			总量	植质磷	非植质磷		
10	0.027	0.024	0.025	0.001	0.024	0.076	1.8
20	0.085	0.080	0.170	0.038	0.132	0.335	2.9
30	0.126	0.182	0.587	0.336	0.237	0.895	6.1
40	0.146	0.251	1.106	0.835	0.271	1.503	9.1
45	0.097	0.271	1.41	0.793	0.348	1.509	14.3

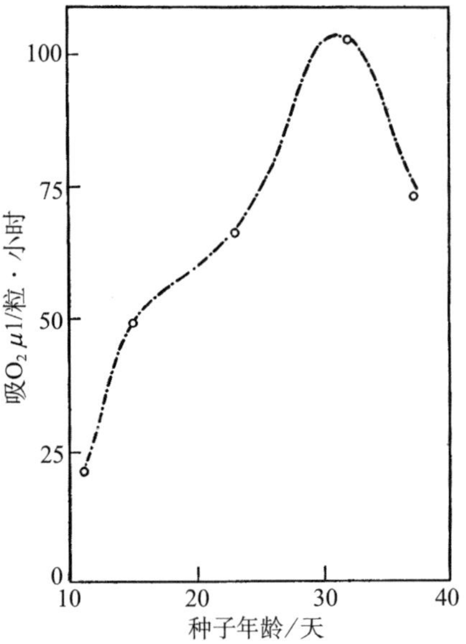

图 2　大豆种子形成过程中呼吸强度的变化

20 天之后大量累积。此外，从表中还可看到，有机磷与无机磷之比随种子年龄的增长而增大，说明无机磷合成有机磷的过程随种子形成而加强。

2. 呼吸测定

大豆种子形成过程中呼吸强度渐渐升高，在花后 10～15 天和 30 天左右，单粒种子呼吸强度增加最快，32 天后呼吸急剧下降，因而造成 30 天左右以粒子为单位的呼吸高峰（见图 2）。

在种子形成过程中，呼吸途径也发生了变化。如表 5 所示，花后 11～32 天，被碘乙酰胺抑制的呼吸部分（EMP 呼吸）逐渐增加，32 天后有所下降。对碘乙酰胺不敏感的呼吸（剩余呼吸）以类似的规律而变化，但涨落的速度更为显著，如对碘乙酰胺敏感和不敏感的呼吸从花后 15～32 天分别增加 51%和 385%，从 32 天后分别下降 25%和 36%，说明若以 30 天左右为分界，此之前 HMP 呼吸相对比例升高，此之后 EMP 呼吸相对比例升高。

表 5　大豆种子形成期间呼吸强度的变化和对碘乙酰胺敏感度的比较

种子年龄/天	对照/（吸 O_2 μl/粒·小时）	渗入碘乙酰胺（10^{-2} mol/L）/（吸 O_2 μl/粒·小时）	剩余呼吸/（吸 O_2 μl/粒·小时）
11	20.78	18.25	2.53
15	48.29	40.07	8.22
32	100.50	60.51	39.99
37	70.86	45.30	25.56

甲烯蓝和 DNP 对不同年龄大豆种子呼吸的影响也不同。如图 3 所示，向不同年龄大豆种子渗入甲烯蓝后，呼吸强度均有不同程度的促进（从 11.7%～49.4%），而且随着年龄的增长，促进作用增强。但在 20～30 天增长较慢，37 天促进最为显著。DNP 只对 11 天和 37 天种子呼吸有促进，对中间其余年龄的种子呼吸非但没有促进，还有较小程度的抑制作用。

3. 种子形成过程中两途径酶活力的变化

在大豆种子形成过程中，两途径的酶活力均迅速上升，30 天左右达到高峰，此与呼吸高峰恰相重合，30 天以后活力均下降（见表 6）。

对照不同时期两途径酶活力的变化可知，从 10～30 天 HMP 途径与 EMP 途径的酶活力同时增长，以 30 天同 10 天活力比较，各酶活力增长情况是：G－6－P 脱氢酶约 20 倍，6－PG 脱氢酶约 10 倍，G－6－P 异构

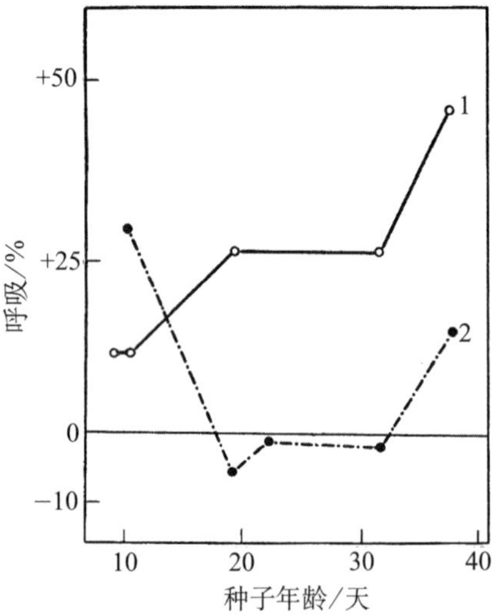

图 3　甲烯蓝和 DNP 对不同年龄大豆种子呼吸的影响

＋代表促进%；－代表抑制%

表 6 大豆种子形成期间两途径酶活力的变化(利用 G－6－P μmoles/粒・小时)

种子年龄/天	G－6－P 脱氢酶	6－PG 脱氢酶	G－6－P 异构酶	醛缩酶
10	0.58	0.58	3.97	1.09
20	6.82	2.71	11.88	10.67
30	11.79	6.54	12.16	26.66
40	0.81	0.58	9.70	21.32

酶 2 倍多,醛缩酶 25 倍多。从 30 天到 40 天,两途径的酶活力均下降,但代表 HMP 途径的酶活力下降远快于 EMP 的酶,如 G－6－P 和 6－PG 脱氢酶分别下降 14 倍和 10 倍多,而 G－6－P 异构酶和醛缩酶只下降 1/4 和 1/5 左右,因而后期 EMP 途径的酶占据优势。

四、讨论

根据不同时期大豆种子成分分析的结果,大致可将种子形成期分为三个时期:前期(花后 20 天以前)、中期(花后 20～30 天)和后期(30～40 天)。前期是胚的发育、种子体积的增大和结构建成期,可溶物物质增长最快。后两个时期是子叶中有机物质的积累与贮藏期:脂肪积累以中期为主,蛋白质积累以后期为主。鉴于不同时期物质合成的强度、方向和内容都不同,与此相适应的是各个时期的呼吸强度和呼吸途径也发生相应的变化。归根到底,呼吸代谢和能量代谢是与物质代谢相一致的,从而在体内构成一个精巧而完善的代谢网。

在大豆种子形成过程中,单位粒子的呼吸强度不断增加。花后 10～15 天和 30 天左右增长最快,32 天以后,可能由于种子含水量的迅速下降而导致呼吸急剧降落,以至形成 30 天左右明显的呼吸高峰(以单位种子计算)。不少作者在果实成熟期观察到呼吸高峰的出现(梁厚果,1961a;吕忠恕,1964; Hillerd 等,1953; Hulme 和 Neal, 1957),并认为这是与果实内蔗糖合成有关(梁厚果,1961),Roman 等(1958)指出单位果实的呼吸高峰是与蛋白质合成相适应的。根据本试验结果,30 天左右正是脂肪和蛋白质合成很强烈的时期,花后 20～40 天有机物质的累积量占整个种子形成期的 80%以上,所以 30 天左右单位种子呼吸量的高峰可能与子叶中脂肪、蛋白质等合成相适应,这一点从花后 20 天以后酸溶性非植质有机磷大量累积得到引证,因为这部分有机磷化合物大部分为高能磷化合物。至于花后 10～15 天呼吸增长较快可能与胚呼吸相关,因为大豆花受精后 12 天正是子叶、胚根和胚轴完全分化期(何孟元,1963), Norman (1963)也提到花后 15 天之前,为胚胎形成期。而胚胎发育与器官建成需要呼吸能源(Beevers, 1960),可惜本试验中并未分别测定胚和子叶的呼吸。

甲烯蓝是呼吸链上还原型辅酶(NADPH, NADH)的人工电子受体(Cantarow 和 Schepartz, 1962),DNP 是氧化磷酸化的解联剂(James, 1953),根据本试验结果,甲烯蓝对物质大量累积期种子呼吸的促进远小于对成熟末期(37 天)种子呼吸的促进,以及 DNP 对物质大量累积期种子呼吸非但没有促进,还略有些抑制,使我们猜想,在物质大量累积期,大豆种子呼吸基本上不受电子传递和氧化磷酸化限制,此原因即在于此时有机物质大量合成,还原型辅酶和 ATP 迅速用于合成过程,氧化型辅酶和 ADP 的再生迅速,从而保证了呼吸链上电子的畅流。至于花后 10 天左右和 37 天左右,从甲烯蓝和 DNP 的效应判断,可以推测呼吸受电子传递和氧化磷酸化的限制：前一时期以受氧化磷酸化牵制为主；后一时期以电子传递的限制为主。

在大豆种子形成过程中，呼吸途径发生了有意思的变化。对碘乙酰胺敏感和不敏感的呼吸在花后 11～32 天均有上升，但后者增长更迅速，32 天后二者均有下降，但后者下降更快。借用一般认为碘乙酰胺所抑制的呼吸部分略可代表 EMP 呼吸，剩余呼吸代表 HMP 呼吸，可以认为 32 天前，HMP 途径呼吸增长较快，32 天后 HMP 途径呼吸下降较快。再结合脂肪和蛋白质积累的不同时期分布，说明有机物质合成与呼吸途径之间有着不可分割的关系，即花后 20～30 天，HMP 途径增长较快，这时脂肪积累最多；花后 30～40 天，EMP 途径的相对比例上升，此时蛋白质合成最强烈，这就再次证实了我们以前在油菜的一系列工作中所提出的脂肪形成与 HMP 途径相关的论点，而且还证明蛋白质合成可能与 EMP-TCA 途径相联系。初步推测，这可能由于葡萄糖通过 EMP 途径和接着的 TCA 循环降解，提供合成蛋白质所需的氨基酸的骨架来源。

参考文献

[1] 朱雨生、刘承宪、施教耐，1965：植物的脂肪合成及其词节 VI. 油菜种子脂肪合成与 HMP 途径关系的同位素证据. 植物生理学报，**2**：185.

[2] 吕忠恕，1964：果实的呼吸作用. 植物生理学通讯，**1**：41.

[3] 何孟元，1963：大豆的胚胎学研究. 植物学报，**11**：318.

[4] 施教耐、朱雨生、谭常，1965：植物的脂肪合成及其调节Ⅲ. 油菜籽实形成中代谢途径的变化及内源抑制剂的调节作用. 植物生理学报，**2**：105.

[5] 梁厚果，1961a：白兰瓜呼吸作用的研究Ⅰ. 果实呼吸的气体交换和氧化酶更替的规律性. 植物学报，**9**：219.

[6] 梁厚果，1961b：白兰瓜呼吸作用的研究Ⅱ. 果实呼吸作用与糖分及含氮物质代谢的关系. 兰州大学学报，**3**：97.

[7] Beevers, H., 1960: Respiratory metabolism in plants. pp. 185 – 197. Row, Peterson & Company, New York. Bertramson, B. K., 1943. Studies on the cereleomolybdate determination of phosphorus. *Soil Sci.*, **53**: 135.

[8] Bils, R. F., 1960: Biochemical and cytological changes in the developing soybean. Ph. D. Thesis, Univ. of Illi-nois (Chmura & Howell, 1962).

[9] Cantarow, A. and B. Schepartz, 1962: Biochemistry. Third ed. Philadelphia, London.

[10] Chmura, T. and R. W. Howell, 1962: Respiration of developing and germinating soybean seeds. *Physiol. Plantarum*, **19**: 341.

[11] Fairbain, N. J., 1953: A modified anthrone reagent. *Chem. and Ind.*, **4**: 86.

[12] Hillerd, A., J. Bonner and J. B. Biale, 1953: The climacteric rise in fruit respiration as controlled by phos-phorylative coupling. *Plant Physiol.*, **28**: 521.

[13] Howell, R. W., 1958: Respiration of unmature soybean seed as related to synthetic activities. *Agron. Abstr.* 54.

[14] Howell, R. W., F. I. Collins and V. E. Sedgwick, 1959: Respiration of soybean seeds as related to weathering losses during ripening. *Agron. J.*, **51**: 667.

[15] Howell, R. W., and J. L. Cartter, 1958: Physiological factors affecting composition of soybean. Ⅱ. Response of oil and other constituents of soybean to temperature under controlled conditions. *Agron. J.*, **50**: 604.

[16] Huerga, J. De La, M. D. Charlotte, B. S. Yesinick and M. D. Han Popper, 1953: Estimation of total

serum lipids by a turbidimetric method. *Am. J. Clin. Path.*, **23**: 1163.
[17] Hulme, A. C. and G. E. Neal, 1957: A new factor in the respiration climacteric of apple fruits. *Nature*, **179**: 1192.
[18] James, W. O., 1953: Plant respiration. Clarendon Press, Oxford.
[19] Kock, F. C. and T. L. McMeekin, 1942: A new direct nesslerization micro-kjeldahl method and a modification of the Nessler-Folin reagent of ammonia. *J. Amer. Chem. Soc.*, **46**: 2066.
[20] McCready, R. M., Guggolz, V. Silviera and H. S. Owens, 1950: Determination of starch and amylose in vege-tables. *Anal. Chem.*, **22**: 1156.
[21] Moore, S. and W. H. Stein, 1948: Photometric ninhydrin method for use in the chromatography of amino acids. *J. Biol. Chem.*, 176: 367.
[22] Norman, A. G., 1963: The soybean. p. 104, A. P. New York-London.
[23] Pons, W. A. Jr., M. F. Stansbury and C. L. Hoffpauir, 1953: An analytical system for determinating phos-phorus compounds in plant materials. J. A. O. A. C., *36*: 492.
[24] Roman, K. S., H. K. Pratt and R. N. Robertson, 1958: The relationship of high-energy phosphate content, pro-tein synthesis, and the climacteric rise in the respiration of ripening avocado and tomato fruits. *Aust. J. Biol. Sci.*, 11: 329.
[25] Walf, A. C., J. B. Park and R. C. Burrell, 1942: A study of the chemical composition of soybeans during maturation. *Plant Physiol.*, 17: 289.
[26] Енкен, В. В., 1959: Соя, ст. 331. Г. Изд. Сельск. лит. Москва.

Fat Synthesis and Its Regulation in Plants

Ⅶ. The Accumulation of Organic Compounds in Soybean Seeds During Seed Development in Relation to The Respiratory Pathways①

To elucidate the relationship between the accumulation of organic compounds and the respiratory pathways is of great significance. This work was conducted to investigate the relation of fat and protein synthesis to the HMP and EMP-TCA pathways respectively, using the seeds of soybean of various developmental ages as experimental material on account of its ability to accumulate both fat and protein extensively. In order to appreciate the different respiratory pathways the experiment with respiratory inhibitor and the enzyme study on the HMP and EMP pathways were performed. The results are summarized in Table 6 and Figure 3, from which the following conclusions may be drawn:

(1) During the development of soybean seeds the amount of soluble substances including inorganic phosphorus, acidsoluble organic phosphorus, soluble nitrogen, amino acids and sugars progressively increases, followed by the intensification in the activity of the synthesis of organic compounds and extensive accumulation of storage compounds (protein, oil, starch and phytin). Analyzing the products of seeds at different stages of development showed that the fat synthesis predominates within 20—30 days after flowering, more than half of fat content of ripe seeds being accumulated within this period, while the protein synthesis proceeds mainly within 30—40 days after flowering, 52.4% of total protein content of ripe seeds being accumulated within these ten days.

(2) In parallel with the course of the synthesis and accumulation of organic compounds in seeds the respiratory intensity per seed rises rapidly, especially from 10 to 15 days after flowering, with a respiratory peak near 30 days. It is believed that the respiration of seeds provides the necessary materials and energy for the synthesis of organic compounds. The sensitivity of respiration to iodoacetamide of soybean seeds at different developmental ages is different from each other. The respiratory fraction (EMP) inhibited by iodoacetamide (0.01 mol/L) rises progressively from 11 to 32 days after flowering, but declines after 32 days. The residual respiration (HMP) varies in the same way, but more sharply. The tendency of the activities of enzymes of two pathways is in good agreement with that of respiratory

① Y. S. Zhu, C. Tan and C. N. Shih (Institute of Plant Physiology, Academia Sinica)

experiment.

(3) The effects of methylene blue and dinitrophenol (DNP) on the respiration of seeds at different stages of the development were compared. Methylene blue shows different degrees of stimulatory effect on respiration of seeds at various stages, but mainly in the later period. DNP stimulates respiration of seeds only in early and later period, particularly in the latter case. In period of respiratory peak, the effect of methylene blue is relatively small and no effect of DNP has been observed.

On the basis of the results obtained, it seems most reasonable to assume that the synthesis and conversion of organic compounds by seeds closely coincide with the changes in the respiratory pathways. It is interesting to show that not only the conclusion on the relation of fat synthesis to the HMP pathway presented in our previous papers has further been justified and confirmed, but also the dependence of protein synthesis on the EMP-TCA pathway has also been demonstrated. The respiratory peak and regulatory factors in respiration were also discussed in this paper.

植物的脂肪合成及其调节

Ⅸ. $NADPH_2$ 的氧化速率对油菜种子 HMP 途径和脂肪形成的调节作用*

提要

本工作对 PMS、MB、TPTZ 等人工电子受体对 C－1、C－6 标记葡萄糖释放 $C^{14}O_2$ 的影响作了研究，并观察促进脂肪形成对葡萄糖 C－1、C－6 氧化的影响。实验结果表明所采用人工电子受体都表现出一定的生化活性。首先三种电子受体均有接受 $NADPH_2$ 的电子并加以传递的能力，其次三种电子受体均能促进种子呼吸，其性质虽有所不同，但均表明主要是促进 HMP 部分。亚砷酸可促进脂肪形成，并同时促进葡萄糖 C－1 的释放达相应的倍数。这些结果再次说明了脂肪形成与 HMP 途径的密切联系。

一、前言

在前文中，已有一系列证据证明油菜种子中，葡萄糖降解除沿经典 EMP-TCA 途径外，还循 HMP 途径而氧化(施教耐等，1965a；朱雨生等，1965)，HMP 途径随着种子形成而加强，产生大量 $NADPH_2$，从而有利于脂肪形成(施教耐等，1965b；朱雨生等，1965)，为什么在种子形成过程中 HMP 途径随着脂肪形成而加强？HMP 途径受什么因素调节和控制？无疑，这是代谢调节中的一个基本理论问题，有待探讨和阐明。

近年来，积累了丰富的资料，表明 HMP 途径受 NADP 的控制。加快 $NADPH_2$ 的氧化速率(增加 NADP 的浓度)提高了 NADP 的浓度，从而提高了$\dfrac{NADP}{NADPH_2}$的比例，有利于 HMP 途径的进行。本文试对几种人工电子受体对 C－1、C－6 标记葡萄糖释放 $C^{14}O_2$ 的影响作了研究，并观察促进脂肪形成对葡萄糖 C－1、C－6 氧化的影响。实验结果表明，油菜种子 HMP 途径受 NADP 的限制，加速 $NADPH_2$ 的氧化，能促进 HMP 途径的进行，并指出脂肪形成，

朱雨生，马玉琛，施教耐，中国科学院植物生理研究所。
此文发表于 1965 年“植物生理学报”2 卷 4 期 362－373 页。
1965 年 10 月 18 日收到。

* 本工作在殷宏章教授指导下完成，特此致谢。

本文采用简称如下：HMP 途径，己糖单磷酸途径(即磷酸戊糖支路)；EMP 途径，糖酵解途径；TCA 循环，三羧酸循环；NADP，烟酰胺腺嘌呤二核苷酸磷酸(即辅酶 II)；$NADPH_2$ 途径，还原型烟酰胺腺嘌呤二核苷酸磷酸；C－1－C^{14}，[C－1]标记的葡萄糖；G－6－C^{14}，[C－6]标记的葡萄糖；PMS，二氮蒽甲硫酸；MB，甲烯蓝；TPTZ，三苯基四唑氯化物；DNP，2，4－二硝基苯酚；G－6－P，葡萄糖－6－磷酸；6－PG，6－磷酸葡萄糖酸；R－5－P，核糖－5－磷酸；G－6－DH，G－6－P 脱氢酶，6－PG DH，6－PG－脱氢酶。

$NADPH_2$ 氧化和 HMP 途径运转相互之间的调节作用。

二、材料与方法

1. 材料

(1) 油菜种子。取自大田，部分来自温室。品种为胜利油菜，年龄分开花后 10～15 天(年幼种子)和 30 天(成熟种子)两种。

(2) 试剂。$G-1-C^{14}$ 和 $G-6-C^{14}$ 为商品，比放射性分别为 19.7 μc/mg 和 170 μc/mg，PMS 由光合作用室沈巩懋等制备供给，MB 和 TPTZ 为商品，分别配成溶液后，pH 都调至 6.0。G-6-P、6-PG 和 R-5-P 是钡盐，使用前用 1N HCl 溶解，加稍过量 10% K_2SO_4，离心弃去硫酸钡沉淀，清液用 1N KOH 中和，再稀释到所需要的浓度。NADP 纯度为 75% 左右。

2. 方法

(1) 标记葡萄糖的饲喂和 $C^{14}O_2$ 的测定。油菜荚取下后迅速剥壳，小心取出种子(0.5～1.0 g)，放入 50 ml 容量的康惠皿中，皿中加 pH 6.5 的磷酸缓冲液和其他试剂，总体积 5 ml。中央井放玻璃小碟一只，内加 KOH (20%)0.2 ml，在皿缘涂上凡士林，最后向皿中加标记葡萄糖(0.5～2.0 ml)，迅速盖上玻片，用橡皮圈扎紧，于 30℃恒温箱中保温 4 小时。待保温结束后，取出玻璃小碟子红外灯下烘干，用盖革氏计数管计数，测定 $C^{14}O_2$ 的强度。另外，吸干皿中溶液，用丙酮杀死和固定种子，备作脂溶物测定。脂溶物按下法测定：用研钵研碎种子，用丙酮提取 4～5 次，合并后在红外灯下蒸去丙酮，再用石油醚溶出脂溶物，铺样记数。标记葡萄糖的吸收率是根据保温前后溶液放射性强度的差别算出。根据喂入的标记葡萄糖在 0.2 ml 20% KOH 中(烘干)的放射性强度，种子对标记葡萄糖的吸收率和放出的 $C^{14}O_2$ 放射性强度算出 $C^{14}O_2$ 的回收率；根据标记葡萄糖的喂入量，吸收率和脂溶物的放射性强度，算出脂溶物的回收率。

(2) 标记醋酸的饲喂和脂溶物的测定。30 天种子 0.25 g 对半切开，置于离心管中，加磷酸缓冲液(pH 5.5)30 μmoles 和其他试剂成所需浓度(以水为对照)，总体积为 0.75 ml，于 30℃左右保温 2 小时。按 Bligh 和 Dyer (1959)的方法分离脂溶物：加 0.05 ml 2N HCl 杀死种子，用匀浆器磨碎，加 3 ml 氯仿-甲醇(1∶2)摇匀，再加 1 ml 氯仿摇匀分层，最后加 1 ml 水摇匀离心，取出下层氯仿层 0.2 ml 铺样记数，得到脂溶物的放射性强度。

(3) 呼吸的测定。采用标准的瓦氏呼吸器技术。迅速称取 0.1 g 种子置于瓦氏瓶主室中，内加磷酸缓冲液(pH 5.5)100 μmoles，电子受体浓度同上述同位素试验一样，以水为对照，总体积 1.5 ml。中央井加 KOH (20%)或水 0.2 ml，于 30℃，测定一小时内种子耗 O_2 量或 CO_2 释放量。

(4) 丙酮粉的制备和酶液的提取。见前报道(施教耐等，1965)，10g 种子加 15 ml EDTA (0.3 mol/L)和磷酸钾(0.1 mol/L)缓冲液(pH 8.2)研碎，缓缓加入 100 ml 用干冰冷却的冷丙酮(－15℃～20℃)，用布氏漏斗迅速抽气过滤，再用 40 ml 冷丙酮洗三次，抽干，在盛 P_2O_5 的真空干燥器中干燥，于低温(－20℃)下保存。制备酶液时，按 10 g 鲜组织加 25 ml EDTA (0.0015 mol/L)和磷酸钾(0.01 mol/L)的比例，研磨抽提丙酮粉，用纱布过滤，滤液用 4 000 转/分速度于低温离心 20 分钟，取上清液测定酶活力。

(5) 酶的测定。葡萄糖-6-磷酸脱氢酶和6-磷酸葡萄糖脱氢酶是根据NADP还原后在340 mμm处出现吸收峰,用日立EUP-2A型分光光度计测定。

(6) 糖测定。1 g鲜种子磨碎后,用80%热酒精提取4～5次,蒸去酒精,用水溶出残物,用蒽酮比色法(Fairbairn, 1953)测定可溶性糖含量,用Roe的间苯二酚比色法测定酮糖。以可溶性糖一酮糖=醛糖(施教耐等,1965a)。

三、实验结果

1. 葡萄糖C-1氧化限制步骤的分析

葡萄糖通过HMP途径氧化时,C-1首先氧化成CO_2释放。有若干证据证明最初二步脱氢反应起着限制HMP途径的作用(Glock和McLean, 1954; Butt和Beevers, 1961),以G-1-C^{14}为底物,根据种子对其吸收率,释放$C^{14}O_2$的强度可以计算出呼吸放出$C^{14}O_2$的回收率,再根据种子中游离糖的含量(醛糖),可以计算出葡萄糖C-1(也是醛糖)的氧化速度。如表1所示,15天种子C-1氧化速度为每小时每克鲜重氧化0.56微克分子G-6-P, 30天种子C-1氧化速度几乎增加1倍,达0.97微克分子。但与G-6-P脱氢酶的活力比较,不论年幼种子或成熟种子,葡萄糖C-1氧化速度均远小于G-6-P脱氢酶,说明种子中葡萄糖C-1氧化潜力没有发挥,未能以全速进行,一定是酶因素之外的其他因素的限制(见表1)。本文将以此为基点,沿此线索,追究其限制因素。

表1 葡萄糖C-1氧化速度与G-6-P脱氢酶活力的比较

种子年龄/天	C-1-C^{14}吸收/%	$C^{14}O_2$释放/%	醛糖含量 μmoles/g fr. wt. *	C-1氧化速度 G-6-P μmoles/g fr. wt. **	G-6-P脱氢酶 G-6-P μmoles/g fr. wt.
15	27.3	1.75	127	0.56	8.7
30	25.8	4.10	95	0.97	18.8

* ft. wt. =鲜重; ** g h. =克小时。

2. 人工电子受体试验

在HMP途径最初二步脱氢反应中,辅酶NADP被还原,因而除了酶活力外,$NADPH_2$的氧化速度,NADP的重新产生显然是保证脱氢反应和C-1氧化继续进行的必要条件。以下试述$NADPH_2$的几种人工电子受体对种子呼吸和C-1氧化的影响。

(1) 人工电子受体的电子传递能力。如图1(a)所示,在人工电子受体PMS (10^{-3} mol/L)、MB (10^{-3} mol/L)或TPTZ (10^{-2} mol/L)存在的情况下,酶液与G-6-P(或6-PG)、NADP、Mg^{2+}保温,不会有$NADPH_2$的产生,其原因就是G-6-P脱氢酶(或6-PG脱氢酶)作用所产生的$NADPH_2$均迅速为PMS、MB和TPTZ所氧化。由图1(b)中可见,待G-6-P脱氢酶反应一段时间,中途加入人工电子受体,则所形成的$NADPH_2$的[2H]可以全部传递给电子受体,从而在340 mμm处的吸收峰消失。上述结果都表明PMS、MB和TPTZ可作为$NADPH_2$的人工电子受体。

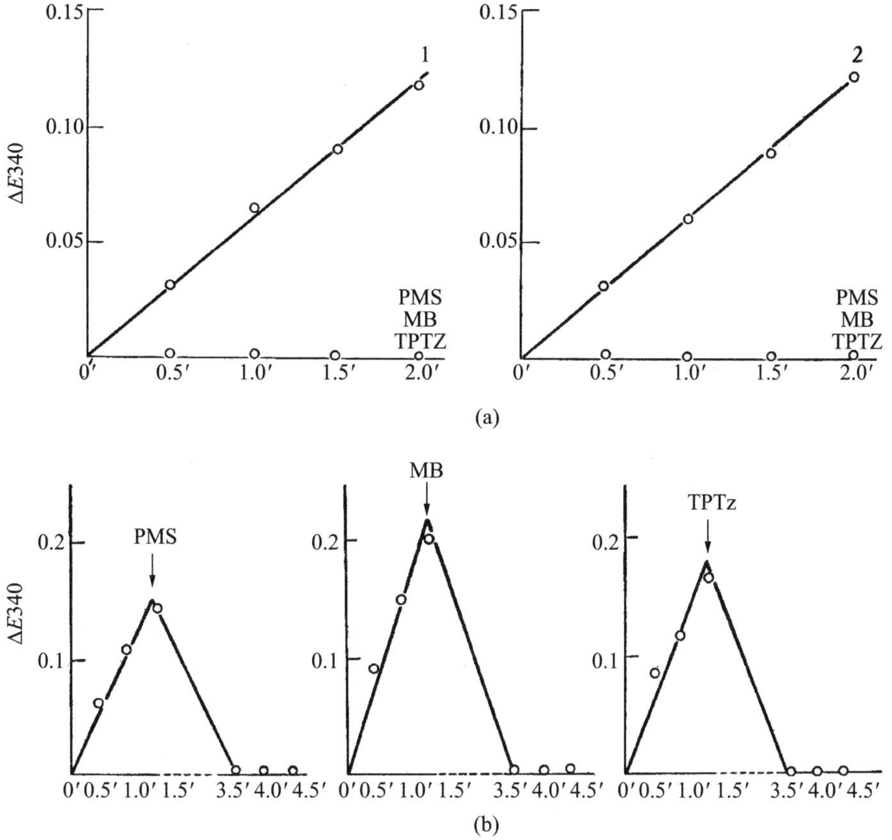

图 1 PMS、MB 和 TPTZ 的电子传递能力

反应条件：双甘氨肽 125 μmoles，$MgCl_2$ 10 μmoles，G-6-P(或 6-PG) 5 μmoles，NADP 0.5 μmole，酶液 0.5 ml，PMS 10^{-3} mol/L，MB 10^{-3} mol/L，TPTZ 10^{-2} mol/L，总体积 3 ml，空白者缺 NADP，用日立分光光度计测定 340 mμm 光吸收

1——对照 G-6-P 脱氢酶；2——对照 6-PG 脱氢酶.

(2) 人工电子受体对呼吸的影响。PMS、MB 和 TPTZ 处理，均促进种子消耗 O_2 和释放 CO_2（见表 2～表 4）。其中 PMS 的促进作用最明显，其次是 TPTZ，MB 的作用最小。TPTZ 对 $C^{14}O_2$ 释放的促进作用远大于对消耗 O_2 的促进，故呼吸商显著升高，如二次试验中升高到 1.55 和 1.89，而且 TPTZ 处理的种子中，由于 TPTZ 的还原，形成一种红色的沉淀。PMS 和 MB 对呼吸商影响不大，或略有上升。

表 2 PMS 对油菜种子(10 天左右)呼吸的影响

试验	处理	O_2 吸收/ μl O_2/g hr.	CO_2 释放/ μl CO_2/g hr.	R. Q.
A	对照 +PMS/(10^{-3} mol/L)	530.6 697.4	727.1 991.0	1.37 1.42
B	对照 +PMS/(10^{-3} mol/L)	305.2 853.2	255.0 1154.0	0.84 1.35

反应条件：磷酸缓冲液 100 μmoles，pH 5.5，30℃。

表 3 MB 对油菜种子(10 天左右)呼吸的影响

试验	处理	O_2 吸收/ μl O_2/g hr.	CO_2 释放/ μl CO_2/g hr.	R. Q.
A	对照/ +MB (10^{-3} mol/L)	603.6 626.4	734.4 873.3	1.22 1.39
B	对照/ +MB (10^{-3} mol/L)	466.2 498.4	531.8 593.0	1.11 1.19

反应条件同表 2。

表 4 TPTZ 对油菜种子(10 天左右)呼吸的影响

试验	处理	O_2 吸收/ μl O_2/g hr.	CO_2 释放/ μlCO_2/ghr.	R. Q.
A	对照/ +TPTZ (10^{-2} mol/L)	412.5 439.0	548.7 829.4	1.33 1.89
B	对照/ +TPTZ (10^{-2} mol/L)	522.8 574.1	699.4 891.0	1.34 1.55

反应条件同表 2。

(3) 人工电子受体对 G-1-C^{14} 和 G-6-C^{14} 释放 $C^{14}O_2$ 的影响。

① 对标记葡萄糖吸收的影响：PMS 和 MB 处理对种子(不论种子的老幼)吸收的标记葡萄糖没有显著影响，结果与 McLean (1960)在大鼠乳腺中的结果一致，但与 Butt (1961)注意到的 PMS 抑制玉米根尖对葡萄糖的吸收不同。TPTZ 显著地抑制种子对标记葡萄糖的吸收(见表 5 和表 6)，此现象则与 Butt 所观察的相仿。

表 5 PMS、MB 和 TPTZ 对幼小种子从 G-1-C^{14} 和 G-6-C^{14} 释放 $C^{14}O_2$ 及转化为脂溶物的影响

试验	处理	葡萄糖的吸收/%		由葡萄糖释放的 $C^{14}O_2$/%		$\frac{C_1}{C_6}$	由葡萄糖转化的脂溶物/%	
		G-1-C^{14}	G-6-C^{14}	G-1-C^{14}	G-6-C^{14}		G-1-C^{14}	G-6-C^{14}
A	对照	22.5	34.1	4.0	3.0	1.3	1.03	0.82
	PMS/(10^{-3} mol/L)	20.3	34.3	10.5	2.4	4.4	0.73	0.70
	MB/(10^{-3} mol/L)	21.4	32.7	6.4	3.2	2.0	0.51	0.73
	TPTZ/(10^{-2} mol/L)	14.5	28.6	8.4	3.5	2.4	0.56	0.88
B	对照	27.3	26.4	1.8	1.8	1.0	0.84	1.28
	PMS (10^{-3} mol/L)	25.4	29.6	7.2	1.7	4.5	0.46	0.84
C	对照	16.4	30.1	3.9	2.1	1.9	1.19	0.50
	MB/(10^{-3} mol/L)	19.7	30.1	5.4	2.0	2.7	0.48	0.38
	TPTZ/(10^{-2} mol/L)	12.4	25.9	7.0	1.7	4.1	0.63	0.67

反应条件：A. 反应液含磷酸缓冲液(pH 6.5)860 μmoles，种子(10 天以下)1g，G-C^{14} 各 2μc，总体积 5 ml，30℃，4 小时。
B. 磷酸缓冲液(pH 6.5)900 μmoles，种子(15 天)1 g，其余同 A。
C. 种子(10 天)1 g，其余同 A。

表 6 PMS、MB 和 TPTZ 对成熟种子从 $G-1-C^{14}$ 和 $G-6-C^{14}$ 释放 $C^{14}O_2$ 及转化为脂溶物的影响

试验	处理	葡萄糖的吸收/%		由葡萄糖释放的 $C^{14}O_2$/%		$\frac{C_1}{C_6}$	由葡萄糖转化的脂溶物/%	
		$G-1-C^{14}$	$G-6-C^{14}$	$G-1-C^{14}$	$G-6-C^{14}$		$G-1-C^{14}$	$G-6-C^{14}$
A	对照	25.8	28.2	4.1	2.0	2.1	0.98	0.95
	PMS (10^{-3} mol/L)	26.3	28.7	11.6	1.7	6.8	0.60	0.83
B	对照	18.0	36.5	3.4	1.1	3.4	2.94	1.71
	PMS (10^{-3} mol/L)+丙二酸(5×10^{-2} mol/L) Malonate	18.0	31.1	13.5	0.8	13.5	1.19	1.67
	MB (10^{-3} mol/L)	17.8	32.9	3.8	1.0	4.8	1.90	1.58
	TPTZ (10^{-2} mol/L)	13.8	14.4	6.9	1.5	4.3	2.07	2.16

反应条件：A. 种子(30 天)1g，B. 种子(30—35 天)1 g，其余同表 2。

② 对 $C^{14}O_2$ 释放的影响：从表 5 和表 6 可见三种电子受体均促进年幼或成熟种子的 $G-1-C^{14}$ 释放 $C^{14}O_2$，而对 $G-6-C^{14}$ 释放 $C^{14}O_2$ 影响不大，因而 $\frac{C^1}{C_6}$ 比升高。从促进效果来看，PMS 的效应最大，年幼种子经 PMS 处理后，C-1 释放促进 1.5～3 倍，但对 C-6 释放略有抑制，$\frac{C^1}{C_6}$ 比从 1 左右上升到 4 以上，从表中还可以看出 PMS 所刺激的呼吸不能为丙二酸所抑制；TPTZ 作用次之，使年幼种子 C-1 氧化加快 1 倍左右，甲烯蓝影响较小，年幼种子经处理后，C-1 氧化促进 38%～60%。若比较三种电子受体对不同年龄种子 C-1 释放的影响，则可看到三种电子受体对年幼种子 C-1 释放的促进大致比成熟种子的促进略大，PMS、TPTZ 和 MB 使成熟种子 C-1 释放促进的倍数约分别为 1.8、1.0 和 0.1(见表 6)。这也许表明，成熟种子中 C-1 氧化的限制比年幼种子小。

③ 对标记葡萄糖参入脂溶物的影响：三种电子受体不论处理年幼种子或成熟种子，均使 $G\text{-}l\text{-}C^{14}$ 参入脂溶物减少，对 $G-6-C^{14}$ 参入脂溶物影响不大。

(4) DNP 对标记葡萄糖释放 C-1、C-6 的影响。为了证实上述三种电子受体对 C-l 释放成 $C^{14}O_2$ 的促进归于本身的电子传递能力，而非氧化磷酸化解联作用，将种子和 DNP 一起保温，观察对种子氧化 $G\text{-}l\text{-}C^{14}$ 和 $G-6-C^{14}$ 释放 $C^{14}O_2$ 的影响，如表 7 所示，DNP 的作用与电子受体不同，对葡萄糖 C-1 氧化不促进，对 C-6 释放影响也不大，$\frac{C_1}{C_6}$ 比略下降，$G\text{-}l\text{-}C^{14}$ 参入脂溶物增加，而 $G-6-C^{14}$ 参入脂溶物降低，因而排除了电子受体因起氧化磷酸化解联作用，增加磷酸或磷酸接受体的利用力，而促进 C-1 氧化的可能性。

(5) 人工电子受体对乙酸-C^{14} 参入脂溶物的影响。成熟种子对半切开后，与乙酸-C^{14} 保温，分离脂溶物，测定几种电子受体对乙酸-C^{14} 参入脂溶物的影响，结果如表 8 所示。经 PMS、MB 和 TPTZ 处理后，乙酸-C^{14} 参入脂溶物都受到程度不等的抑制，其中加 PMS 或 TPTZ 的抑制，大于 MB。三种电子受体都是 $NADPH_2$ 的氧化剂。竞争脂肪酸合成所必须的

表 7 DNP 对油菜种子从 G-1-C^{14} 和 G-6-C^{14} 释放 $C^{14}O_2$ 及转化为脂溶物的影响

试验	处理	葡萄糖的吸收/%		由葡萄糖释放的 $C^{14}O_2$/%		$\frac{C_1}{C_6}$	由葡萄糖转化的脂溶物/%	
		G-1-C^{14}	G-6-C^{14}	G-1-C^{14}	G-6-C^{14}		G-1-C^{14}	G-6-C^{14}
A	对照	28.6	33.5	18.0	4.5	4.0	6.54	7.85
	DNP (2×10^{-5} mol/L)	25.6	34.0	13.1	4.9	2.7	9.35	6.96
B	对照	23.0	31.4	39.1	9.1	4.3	13.80	15.60
	DNP (2×10^{-5} mol/L)	20.1	29.3	40.0	10.0	4.0	17.00	12.50

反应条件：A 与 B 相同，种子(30 天)0.5g，对半切开，G-1-C^{14} 和 G-6-C^{14} 各 1 μc，其余同表 2。

表 8 PMS、MB 和 TPTZ 对乙酸-2-C^{14} 参入脂溶物的影响

处理	试验 A	试验 B
	脂溶物/cpm	脂溶物/cpm
对照	520	1 000
PMS (10^{-3} mol/L)	470	860
MB (10^{-3} mol/L)	510	940
TPTZ (10^{-2} mol/L)	400	820

反应条件：见正文说明。

$NADPH_2$，它们的存在致使 $NADPH_2$ 的无效氧化，因而减少乙酸-C^{14} 参入脂溶物。

3. 天然电子受体试验

油菜种子中含有大量脂肪，在乙酰 CoA 缩形成长链脂肪酸时，需要消耗大量的还原型 NADP，因而乙酰 CoA 在合成脂肪酸时可作为 $NADPH_2$ 的天然电子受体，加速脂肪酸的合成必然会引起代谢发生如加入人工电子受体一样的变化。如表 9 所示，10^{-3} mol/L 亚砷酸钠大大刺激乙酸-C^{14} 参入脂溶物(达 2～3 倍以上)，与此同时观察对 G-1-C^{14} 释放 $C^{14}O_2$ 的影响，从表 10 可知，葡萄糖 C-1 氧化也相应增加 2～3 倍，对 G-6-C^{14} 氧化无甚影响，$\frac{C_1}{C_6}$ 比上升至 12 左右，对标记葡萄糖的吸收表现出一定程度的抑制。看来亚砷酸钠对 C-1 释放的促

表 9 亚砷酸钠对乙酸-2-C^{14} 参入脂溶物的影响

试验	处理	脂溶物/cpm
A	对照	520
	亚砷酸钠/10^{-3} mol/L	1 790
B	对照	1 000
	亚砷酸钠/10^{-3} mol/L	2 180

反应条件：见正文说明。

表 10 亚砷酸钠对种子从 G-1-C^{14} 和 G-6-C^{14} 释放 $C^{14}O_2$ 的影响

试验	处理	葡萄糖的吸收/%		由葡萄糖释放的 $C^{14}O_2$/%		$\frac{C_1}{C_6}$
		G-1-C^{14}	G-6-C^{14}	G-1-C^{14}	G-6-C^{14}	
A	对照	26.3	32.0	27.0	6.0	4.5
	亚砷酸钠/(10^{-3} mol/L)	14.8	23.4	74.7	6.2	12.1
B	对照	23.8	41.7	25.2	3.7	6.8
	亚砷酸钠/(10^{-3} mol/L)	11.9	30.8	55.9	4.5	12.4

反应条件：A. 种子(30 天)1 g,对半切开,磷酸缓冲液(pH 6.5)840 μmoles, G-C^{14} 各 2μc, 30℃,4 小时。
B. 种子(35 天)1 g,对半切开,其余同 A。

进不是由于对 HMP 途径的直接影响,10^{-3} mol/L 亚砷酸钠对 G-6-P 脱氢酶和 6-PG 脱氢酶活力无显著影响。

四、讨论

油菜种子与其他油料作物种子一样,在种子形成过程中,先是糖分的积累,后来大量转化为脂肪。为什么油料作物种子在形成过程中糖分会转化为脂肪? 这种转化是受什么因素调节与控制的? 这些问题已引起部分研究者的注意(Stumpf, 1963),在以前若干篇报道中发现在油菜种子形成过程中,与脂肪累积相平行的是 HMP 途径的加强,后者所产生的 $NADPH_2$ 是合成脂肪酸所必不可少的条件。进一步值得深思的问题是为什么在脂肪形成过程中 HMP 途径会加强,它又是受什么因素调节与控制的? 一般认为 HMP 途径最初的两个脱氢反应是其关键性的步骤。脱氢反应中包含二个主要的因素: ①脱氢酶(G-6-P 脱氢酶和 6-PG 脱氢酶)的活力;②NADP 的供应。增加脱氢酶的活力或提高 NADP 的含量均能加强 HMP 途径。影响酶活力的因素可能有两个: ①酶蛋白本身的浓度;②酶的激活或抑制。在前文中(施教耐等,1965a)主要探讨酶因素对 HMP 途径的调节,而且发现在油菜花蕾中存在一种 G-6-P 脱氢酶和 6-PG 脱氢酶的天然抑制剂在调节着酶的活力。本文就脱氢反应中的第二个因素即 NADP 的供应,探讨其对 HMP 途径的调节作用。

本试验所采用的三种人工电子受体都表现出一定的生化和代谢活性,首先,三种电子受体均有接受 $NADPH_2$ 电子的能力,因而起着电子传递体的作用。三者对种子呼吸都有促进,但性质上有所不同。PMS 氧化 $NADPH_2$ 是直接的和非酶促的反应,MB 和 TPTZ 要通过心肌黄酶的作用接受电子。此外,PMS 和 MB 还原后能自动氧化,而 TPTZ 还原后形成的 Formazan(甲臜)沉淀不能在空气中重新氧化(Butt & Beevers, 1961),因而 PMS 和 MB 起交替的电子载体作用,对 CO_2 释放和 O_2 吸收同时促进,而 TPTZ 促进 CO_2 释放,对吸收 O_2 促进很少,致使呼吸商显著升高。同位素试验进一步表明电子受体对呼吸的促进主要是活化呼吸的 HMP 部分,因对葡萄糖 C-1 的氧化显著促进,且此促进不能为丙二酸(TCA 循环的抑制剂所抑制),对葡萄糖 C-6 氧化则不影响,$\frac{C_1}{C_6}$比升高。此现象在肝脏(Cahill 等,1958)、腹水癌细胞(Wenner, 1958)、乳腺(McLean, 1960)、马铃薯切片(Ap Rees 等,1960)、玉米根尖(Butt & Beevers, 1961)也有同样的报道。

PMS、MB 和 TPTZ 对葡萄糖的吸收与 $C^{14}O_2$ 的释放有着性质和程度上的若干差异,说明三种电子受体的作用不完全一样，不能全归为有利于 $NADPH_2$ 的氧化(Smith, 1952; Novikoff, 1959; Gaur 和 Beevers, 1959)，但根据三种电子受体均促进葡萄糖 C-1 氧化，仍然有理由认为三者有着共同的作用点，并暗示出油菜种子中 HMP 途径的参与受还原型 NADP 氧化速率的限制。

不少研究者注意到加入使 $NADPH_2$ 氧化的代谢底物——天然电子受体，也能大大促进葡萄糖 C-1 的氧化，Holzer 等(1960)观察到酵母细胞中加入铵盐，能使 G-6-P 进入 HMP 途径的氧化速度加速，因该条件下，α-酮戊二酸转化为谷氨酸是一个氧化 $NADPH_2$ 的还原氨化过程。Butt 和 Beevers (1961)发现硝酸盐处理玉米根尖能大大刺激葡萄糖 C-1 的氧化，因为硝酸盐的还原也需要 $NADPH_2$，此结果为王业芹等(1964)在小麦的工作中所证实。Kinoshita (1957)在牛角膜上皮细胞中加 $NADPH_2$ 的受体丙酮酸(在乳酸脱氢酶作用下形成乳酸)时，观察到 C-1-C^{14} 放出 $C^{14}O_2$ 的量增加 8 倍。Wenner (1958)在腹水癌细胞中也观察到同样的现象。McLean (1960)应用胰岛素刺激大鼠乳腺脂肪形成，也发现对葡萄糖 C-1 氧化的促进。亚砷酸钠能显著地刺激油菜种子的脂肪形成(刘承宪等)。在本试验中亚砷酸钠刺激脂肪形成的同时，葡萄糖 C-1 的氧化以相应的倍数而促进，对 C-6 氧化则没有影响，表现出与上述人工电子受体类似的作用，说明在合成脂肪时，脂酰 CoA 起着 $NADPH_2$ 的天然电子受体的作用，由此得出结论，脂肪合成的加强，有利于 $NADPH_2$ 的氧化，有利于 NADP 的再生，有利于 HMP 途径的加强，因而在脂肪合成、$NADPH_2$ 氧化和 HMP 途径加强间存在着相互影响、相互促进的偶联关系。

本工作结合以前天然抑制剂的结果，对油菜种子形成过程中 HMP 途径与脂肪形成间的调节可作如下的推测(见图 2)：在花蕾中，由于天然抑制剂的存在，G-6-P 脱氢酶和 6-PG

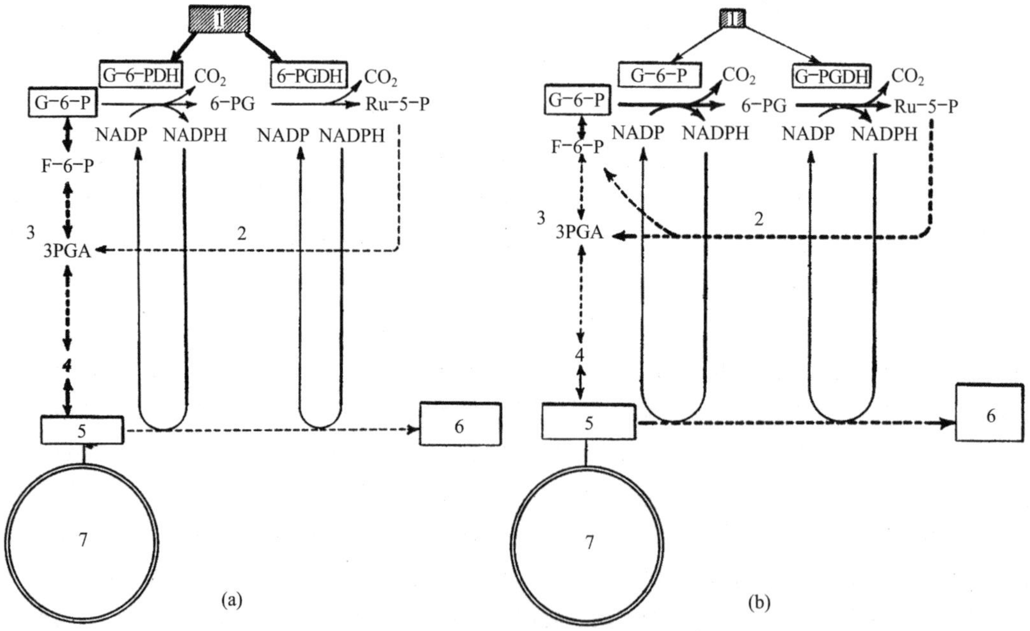

图 2 油菜种子发育过程中脂肪合成调节

(a) 幼小种子由于存在着抑制 G-6-P 脱氢酶和 6-PG 脱氢酶的天然抑制剂，EMP 途径占优势，而 HMP 途径活力低，因此限制了脂肪酸的合成 (b) 成熟的种子由于天然抑制含量降低，HMP 途径活力显著增加，促进了脂肪酸的合成，并使 $NADPH_2$ 的重新氧化加强，因而形成了正反馈现象

1—天然抑制剂 Natural inhibitor; 2—戊糖支路 HMP pathway; 3—酵解途径 EMP pathway; 4—丙酮酸 Pyruvic acid; 5—乙酰 CoA Acetyl-CoA; 6—长链脂肪酸 Long chain fatty acids; 7—三羧酸循环 TCA cycle

脱氢酶处于抑制状态，HMP途径受阻，$NADPH_2$ 供应不足或由于其他原因，脂肪不形成。随着子房受精、种子形成，抑制剂减少乃至消失，脱氢酶活化，HMP途径作用，$NADPH_2$ 产生，加上其他因素(如三羧酸循环的减弱等)，脂肪开始形成，$NADPH_2$ 的利用加快，NADP的再生迅速，再促进HMP途径。反过来，又促进 $NADPH_2$ 的产生和脂肪的合成，从而在脂肪合成与HMP途径间建立起"正反馈"的调节机制，而 $NADPH_2$ 的氧化还原成为两者间的衔接点。当然上述推测并不排除其他因素在脂肪代谢中起调节作用的可能性，实际上，生物体的调节，必然是多因素的，而且是相互联系、相从依存、相互影响的。

参考文献

[1] 王业芹、张德颐、汤玉玮，1964：高等植物的氮素代谢 VII. 硝酸盐和铵盐对小麦根中糖降解途径的调节. 实验生物学报，9：255—260.

[2] 朱雨生、刘承宪、施教耐，1965：植物的脂肪合成及其调节 VI. 油菜种子脂肪合成与HMP途径关系的同位素证据. 植物生理学报，2：185—194.

[3] 施教耐、朱雨生、谭常，1965a：植物的脂肪合成及其调节 I. 油菜种子形成期间葡萄糖降解的途径. 植物生理学报. 2：33—43.

[4] 施教耐、朱雨生、谭常，1965b：植物的脂肪合成及其调节 III. 油菜籽实形成过程中代谢途径的变化与内源抑制剂的调节作用. 植物生理学报，2：105—118.

[5] Ap Rees, T. and H. Beevers, 1960: Pathways of glucose dissimilation in carrot slices. *Plant Physiol.*, 35: 830 - 838.

[6] Bligh, G. G. and W. J. Dyer, 1959: A rapid method of total lipid extraction and purification. *Can. J. Biochem., Physiol.*, 37: 911 - 918.

[7] Butt, V. S. and H. Beevers, 1961: The regulation of pathways of glucose catabolism in maize roots. *Biochem. J.*, 80: 21 - 27.

[8] Cahill, G. F., Hastings, A. B., Ashmore, J. and S. Zottu, 1958: Studies on carbohydrate metabolism in rat liver slices. X. Factors in the regulation of pathways of glucose metabolism. *J. Biol. Chem.*, 230: 125 - 135.

[9] Dickens, F. and H. McIlwain, 1938: Phenazine compounds as carriers in the hexosemonophosphate system. *Biochem. J.*, 32: 1615 - 1625.

[10] Fairbain, N. J., 1953: A modified anthrone reagent. *Chem. and Ind.*, 4: 86.

[11] Gaur, B. K. and H. Beevers, 1959: Respiratory and associated response of carrot discs to substituted phcnols. *Plant Physiol.*, 34: 427 - 432.

[12] Glock, G. E. and P. McLean, 1954: Levels of enzymes of the dircct oxidative pathway of carbohydrate metabolism in mammalian tissues and tumours. *Biochem. J.*, 56: 171 - 175.

[13] Holzer, H. and I. Witt, 1960: Beschleunigung des oxydatiaen pentosephosphatcyclus in Hefezellcn durch ammoniumsalze. *Biochim. Biophys. Acta*, 38: 163 - 164.

[14] Horecker, B. L., Smyniotis, P. Z., Hiatt, H. H. and H. Klenow, 1953: The formation of sedoheptulose phosphate from pentose phosphate. *J. Biol. Chem.*, 205: 661 - 682.

[15] Kinoshita, J. H., 1957: The stimulation of the phosphogluconate oxidation by pyruvate in bovine corneal epithelium. *J. Biol. Chem.*, 228: 247 - 253.

[16] McLean, P., 1960: Carbohydrate metabolism of mammary tissue. III. Factors in the regulation of pathways of glucose catabolism in the matmmary gland of the rat. *Biochim. Biophys. Acta*, 37: 296 - 309.

[17] Novikoff, A. A., 1959: In Subcellular particlcs, p. 14, ed. by Hayashi, T., New York, Ronald Press Co.

[18] Roe, J. H., 1934: A colorimetric method for the determination of fructose in blood and urine. *J. Biol. Chem.*, 107: 15 - 22.

[19] Smith, G. F., 1952: The mechanism of the tetrazolium reaction in corn embryos. *Plant Physiol.*, 27: 445 - 456.

[20] Stumpf, P. K., 1963: Some problems in the control of lipid synthesis in higher plants. in G. Popiak and J. K. Grant ed.: The control of lipid metabolism, Biochemical society sym., no. 24, London and New York.

[21] Wenner, C. E., 1958: The hexosemonophosphate shunt in glucose catabolism in ascites tumors cells. *Cancer Res.*, 18: 1105 - 1114.

Fat Synthesis and Its Regulation in Plants

Ⅸ. The Regulatory Action of the Oxidation Rate of $NADPH_2$ in the HMP Pathway and Fat Synthesis in Rape Seeds*

Studies were made on the action of artificial electron acceptors (PMS, MB, TPTZ) on the production of $C^{14}O_2$ by developing rape seed incubated with glucose labelled at carbon 1 and carbon 6. Observations were made on the rate of oxidation of glucose C-1 and C-6 during stimulated fat synthesis.

Results showed that all the three artificial electron acceptors have biological activity. They all can accept electron from $NADPH_2$ and enhance transfer. They all stimulate respiration, by acting in the HMP pathway, though differ in certain aspects. Arsenite can also stimulate fat synthesis and concurrently $C^{14}O_2$ production from C-1 labelled glucose.

The results supply additional evidence to the interrelationship between fat formation and the HMP pathway.

* Y. S. Zhu, Y. S. Ma and C. N. Shih (Institute of Plant Physiology, Academia Sinica)

The Changes of Metabolic Pathways and the Regulatory Action of an Endogenous Inhibitor in the Developing Siliques of *Brassica Napus* Linn*

Abstract

Activities of the isolated enzymes of the EMP and HMP pathways vary with the stages of development of siliques. From budding to flowering, the activities of both pathways increase, but the activities of enzymes of the EMP pathway increase relatively more than those of the HMP pathway. From flowering to fruiting and to silique maturation, the activities of the EMP enzymes decline rapidly, whereas the activities of the HMP enzymes increase sharply, and remain at that level or only show a slight decrease till the approach of full maturation.

Collateral evidence has been obtained in respiration studies with inhibitors.

The shift from the metabolic predominance of the EMP pathway to the HMP pathway in the siliques takes place between 10 to 20 or more days after flowering, coincident with the time of rapid fat accumulation.

A natural inhibitor of G－6－P dehydrogenase and 6－PG dehydrogenase was found in the reproductive organs. It is present mainly in the buds, flowers, and young siliques, and is nearly absent in the old siliques (>20 days). There seems to be a correlation between the

---

C N Shih(施教耐), Y S Zhu(朱雨生), and C Tan(谭常), Institute of Plant Physiology, Academia Sinica.

此文发表于1966年"Scientia Sinica"15卷3期379－393页.

∗ First published in Chinese in Acta Phytophysiol. in., Vol. 2, No. 1, pp. 33－43, 1965.

Abbreviations used in this paper are: EMP, Embden-Meyerhof-Parnas; HMP, hexose monophosphate; TCA, tricarboxylic acid cycle; NADP, oxidized form of nicotinamide adenine dinucleotide phosphate; $NADPH_2$, its reduced form; NAD, oxidized form of nicotinamide adenine dinucleotide; $NADH_2$, its reduced form; G－6－P, glucose－6－phosphate; 6－PG, 6－phosphogluconate; R－5－P, ribose－5－phosphate; Xu－5－P, xylulose－5－phosphate; F－6－P, fructose－6－phosphate; FDP, fructose, 1,6－diphosphate; 3－PG, 3－phosphoglyceraldehyde; 3－PGA, 3-phosphoglycerate; 2,3－DPGA, 2,3－diphosphoglycerate; EDTA, ethylene diamine tetraacetate; G－6－P DH, G－6－P dehydrogenase; 6－PG DH, 6－PG dehydrogenase; R－5－P′TE′, enzymes utilizing R－5－P; 3PGA DH, 3－PG dehydrogenase; ATP, adenosine triphosphate.

activity of this inhibitor on the glucose degradative pathways and seed fat formation. Preliminary investigations indicate that the inhibitor is probably a flavonoid or its derivative.

Introduction

It is well known that during the seed development of oil-bearing plants fat formation proceeds after sugar accumulation. In the seed, when the process of accumulation of one substance has shifted to that of another, there must be some changes in the intermediary metabolism. During rapid fat formation, what changes occur in the degradative pathways of glucose, and how these changes are regulated and controlled, are questions still open at present.

In recent years, following the works of Langdon[1], Wakil[2] and Abraham et al. [3] in animals, Lynen[4] in yeast, Stumpf et al. [5] in plants, it is generally considered that in the biosynthesis of fatty acids, $NADPH_2$ is a specific electron donor, and the addition of $NADPH_2$ can noticeably stimulate the incorporation of labelled acetate into fatty acids in livers[1] and lipids in rape seeds (Zhu et al., unpublished). Since the synthesis of one molecule of $C_{[16]}$ - fatty acid requires 14 molecules of $NADPH_2$, it appears likely that $NADPH_2$ is one of the important rate-controlling factors in the fat synthesis in vivo. The HMP pathway is the primary source of $NADPH_2$ in plant, therefore, the synthesis of fatty acids is probably related to this pathway and the shift of sugar accumulation to fat accumulation of oil seed is likely accompanied by relative increase of the activity of the HMP pathway.

In a previous article[6] we had already shown that in the siliques of Brassica napus glucose can be degraded either via the EMP or the HMP pathways. In the present study, the patterns of the enzyme activities of these two pathways during the development of the siliques and the effect of respiratory inhibitors (iodoacetamide and sodium fluoride) on the respiration of seeds of different degrees of maturation have been investigated. The data obtained confirm the above assumption. In the mean time, a natural inhibitor of G - 6 - P dehydrogenase and 6 - PG dehydrogenase has been found in the buds and young siliques. Since its presence varies with the age of siliques, it seems to play a role in the regulation of transition of the respiratory pathway and fat synthesis.

Materials and Methods

Brassica napus Linn. var. "*Victory*" was used as experimental material. During the blooming period, flowers were marked every five days. After definite time intervals, buds, flowers, and siliques of different ages were collected and made into acetone powders for subsequent estimation of the enzyme activities.

Preparation of acetone powder. The procedure for preparing acetone powder essentially followed the method of Hageman and Arnon[7]. Fresh siliques (10 g) were cut to pieces with a stainless steel scissors in a glass motar. After the addition of 15 ml of a mixture of EDTA (0.03 mol/L) and phosphate (1.1 mol/L) buffer, pH 8.2, the tissue was ground until homogenized, and transferred to a 500 ml beaker. Cold acetone (−15° to −20℃, 100 ml) was then added slowly with stirring, and the slurry was filtered immediately by suction in a Büchner funnel. The precipitate was washed with 40 ml cold acetone for three times, left in the air about 1 hour until free from acetone odour, and dried in vacuo for 12 hour over P_2O_5. The dry povrder was then stored in a bottleat −15℃.

Enzyme extract. Acetone powder extracts were prepared by grinding the powder in a glass motar with a mixture of EDTA (0.001,5 mol/L) and phosphate (0.01 mol/L) buffer, pH 7.2(25 ml buffer solution for 10g of fresh tissues). The slurry was filtered through four layets of cheese-cloth, and the resulting filtrate was centrifuged in the cold at 10,000×g for 5 to 10 min. The filtrate is designated as enzyme extract.

Assay of enzymes. G-6-P DH and 6-PG DH were determined spectrophotometrically by measuring the reduction of NADP at 340 mμm with a Beckman DU spectrophotometer.

R-5-P conversion enzymes were determined by following the disappearance of R-5-P with time of incubation at 30℃.

Phosphoglucoisomerase was measured by the method of Slein[8], F-6-P formed from G-6-P was determined by the resorcinol method of Roe[9].

Phosphofructokinase was detected by determining the quantity of phosphotriose formed with the endogenous aldolase from F-6-P and ATP.

Aldolase activity was estimated by the procedure of Sibley and Lehninger[10], the phosphotriose formed was determined by dinitrophenyl hydrazine reaction in alkaline medium.

Phosphoglyceraldehyde dehydrogenase activity was determined by the method of Black and Humphreys[11], in which the $NADH_2$ formed was measured spectrophotometrically at 340 mμm with a Hitachi EPU-2A spectrophotometer.

Phosphoglyceric kinase activity was determined by the method of Axelrod and Bandurski[12], the 1,3-diphosphoglycerate formed is trapped by hydroxylamine.

Enzymes involved in the conversion of 3-PGA into pyruvic acid were detected by determining the pyruvic acid formed from 3-PGA with the method modified by Kachmar and Boyer[13].

Measurement of respiration. Seeds of different developmental stages were placed in weighing bottles containing iodoacetamide (0.05 mol/L) or sodium fluoride (0.1 mol/L) and infiltrated under reduced pressure for about 2hr in a desiccator. Oxygen uptake was measured with the conventional Warburg apparatus at 30℃, at pH 5.0, without exogenous substrate.

Separation and purification of the inhibitor. Extract of acetone powder of rape buds was boiled for 5 min in a water bath. After removal of the precipitated proteins by centrifugation, the supernatant was further fractionated by paper chromatography or paper electrophoresis. In some experiments the supernatant was washed several times with chloroform or benzene to remove lipid impurity before being subjected to chromatography or electrophoresis.

Paper chromatography. Samples (5 ml) were applied to a 20 cm × 17.5 cm sheet of Whatman No. 1 filter paper and developed by the descending technique with acetone: formic acid : water (20: 10: 65) for 24 hour at 25℃. After drying, the paper was cut equally into four sections from the starting line to the solvent front. Including the original region there were all together five sections, designated as 0, 1, 2, 3, 4 respectively. Each section was extracted with 5 ml redistilled water and the inhibitory effect on G-6-P DH or 6-PG DH was estimated.

Paper electrophoresis. A sample (0.5 ml) was applied to a 2 cm × 36 cm strip of filter paper held horizontally with ends dipped into borate buffer (pH 9). A potential of 400 V was applied for four hours at room temperature. At the end of electrophoresis, the paper was allowed to dry in air and cut into several sections according to colour. After extraction by 4 ml of redistilled water, the inhibitory effect of each section was estimated as above.

Experimental Results

1. *Changes of the Enzymic Activities of the HMP and the EMP Pathways during the Development of Rape Siliques*

(1) Activities of the enzymes of the HMP pathway. As shown in Figure 1(a), the activities of G-6-P DH and 6-PG DH increase during the development of siliques, particularly between the age of 10 to 20 or more days after flowering. At this period, the activity of G-6-P DH increases more than four times. The activities of enzymes utilizing R-5-P also increase gradually from the stage of budding to 30 days after flowering.

(2) Activities of the enzymes of the EMP pathway.

① Phosphoglucoisomerase. The activity of phospfioglucoisomerase decreases from 8 days to 15 days after flowering, but increases thereafter till the age of 30 days (see Table 1 and Figure 2). In some experiments it remains more or less at a constant level.

② Phosphofructokinase. As the silique grows the activity of phosphofructokinase decreases rapidly and approaches a very low level at the age of 30 days after flowering [see Figure 1(b)].

③ Aldolase. The activity of aldolase shows a slight increase from the stage of budding to flowering or to young siliques. Thereafter as the siliques grow older, the activity declines rapidly, so that the activity in 30 day siliques is about one third of that of 10 days old (Figure 1(b)).

④ 3-phosphoglyceraldehyde dehydrogenase. The changes of the activity of this enzyme appears to be similar to that of phosphoglucoisomerase: a slight decline takes place from 8 days to 15 days old after flowering, then a rapid increase up to 30 days (see Figure 2).

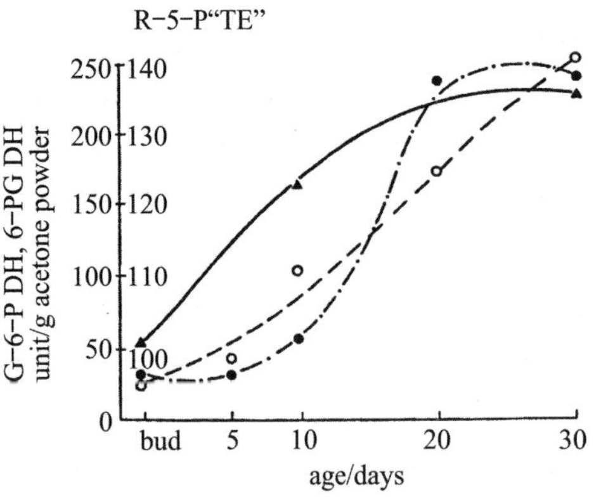

(a)

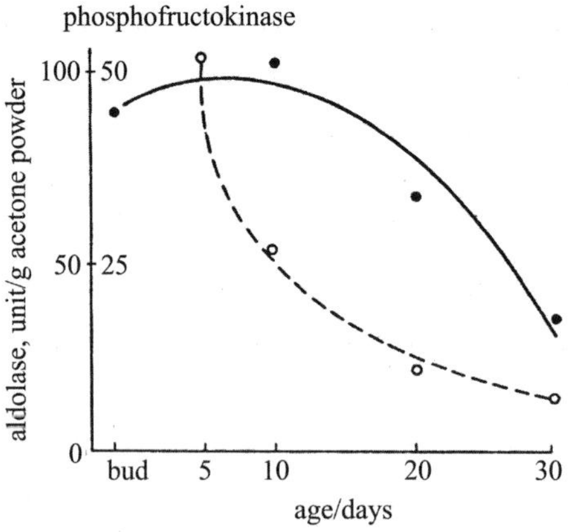

(b)

Figure 1　Changes of activities of some enzymes of the HMP and EMP pathways during development of rape siliques

Reaction systems:

(a) G-6-P dehydrogenase and 6-PG dehydrogenase: Glycylglycine 125 μmoles (pH 7. 5), $MgCl_2$ 10 μmoles, G-6-P (or 6-GP) 5) μmoles, NADP 0. 27 μmole (purity 78%), enzyme 0. 5 ml. Total volume 3 ml. Blank without NADP. Reaction mixture was previously incubated for 5 min in 38℃ water bath and the reaction was started on addition of enzyme. One unit of enzyme activity is arbitrarily defined as the optical density increase by 0. 01 at 340 mμm in the first 5 min. (Beckman DU type spectrophotometer and 1 cm light path cuvette.) Enzymes utilizing R-5-P: Glycylglycine 88 μmoles (pH 7. 5), R-5-P 10 μmoles, riboflavin 1. 5 μmoles, cystein 1. 5 μmoles, $MgCl_2$ 10 μmoles, NADP 0. 5 μmole, total volume 3 ml. Blank omitted R-5-P. Enzyme 0. 5 ml, temp. 38℃. Unit of enzyme activity is defined as amount of converted R-5-P in μmoles.

(b) Phosphofructokinase: Tris buffer (pH 8. 6)100 μmoles, sodium bisulphite (pH 8. 6)250 μmoles, F-6-P 4 μmoles, ATP 6 μmoles, $MgCl_2$ 10 μmoles, enzyme 0. 5 ml. Total volume 3 ml. The reaction mixture was incubated at 38℃ for 15 min and the quantity of phosphotriose produced was determined. One unit of enzyme activity is defined as amount of enzyme phosphorylating 0. 1 μmoles F-6-P.

Aldolase: Barbital sodium buffer (pH 8. 6)100 μmoles, sodium bisulphite (pH 8. 6)250 μmoles, FDP 12. 5 μmoles, enzyme 0. 5 ml. Total volume 3 ml. Reacted at 38℃ for 30 min. Reaction was stopped by adding equal volume of 20% trichloroacetic acid. Phosphotriose produced was determined by dinitrophenyl hydrazine reaction in alkaline medium. One unit of enzyme activity is defined as amount of enzyme degradating 1 μmole FDP.

Table 1　Change of Activity of Phosphoglucoisomerase with the Age of Rape Silique

Age (days after flowering)	unit/g acetone powder
8	117
15	65
23	129
30	155
40	148

Unit=converted 1 μmole G-6-P/15 min.

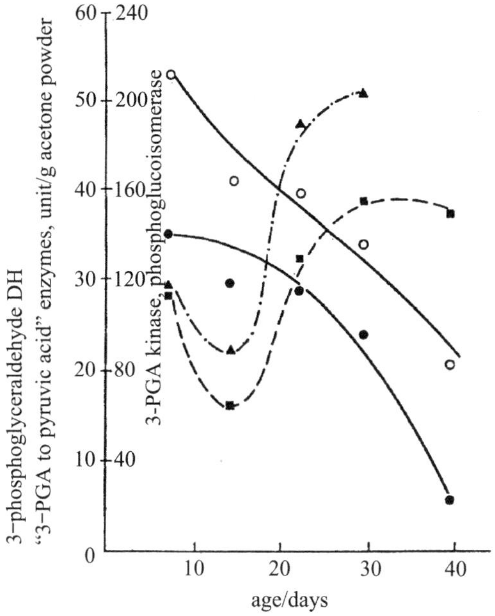

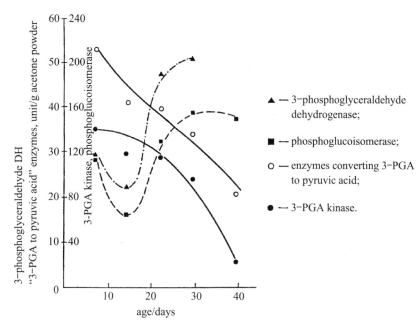

Figure 2 **Changes of the activities of phosphoglucoisomerase, phosphoglyceraldehyde dehydrogenase, 3 - PGA kinase, and enzymes involved in converting 3 - PGA to pyruvic acid during the development of rape siliques**

Reaction systems:

Phosphoglyceraldehyde dehydrogenase: Tris (pH 8. 5)150 μmoles, FDP 10 μmoles, NaF 10 μmoles, sodium arsenate 17 μmoles, glutathione 15 μmoles, NAD 1 μmole, enzyme 0. 5 ml. Total volume 3 ml. Blank omitted NAD. Light absorption was measured spectrophotometrically at 340 mμm with Hitachi EPU - 2A spectrophotometer at room temperature.

3 - PGA kinase: Tris (pH 7. 4)100 μmoles, 3 - PGA 20 μmoles, ATP 8 μmoles, $MgCl_2$ 6 μmoles, hydroxylamine 1,000 μmoles, enzyme 0. 1 ml. Total volume 1. 3 ml. Reacted at 38℃ for 30 min.

Reaction was stopped by adding 3 ml $FeCl_3$ - HCl-trichloroacetic acid. The 1,3 - diphosphoglyceric acid produced was determined ($FeCl_3$ - HCl-trichloroacetic acid: $FeCl_3$ - $6H_20$ 8. 3 g, trichloroacetic acid 20 g were dissolved in 42 ml concentrated HCl and distilled water was added to 500 ml).

Enzymes involved in conversion of 3 - PGA to pyruvic acid: Tris (pH 7. 4)150 μmoles, 3 - PGA 20 μmoles, ADP 10 μmoles, $MgCl_2$ 10 μmoles, 2,3 - DPGA 30 μg, KCl 25 μmoles, enzyme 2 ml. Total volume 2 ml. Incubation was proceeded at 30℃ for 30 min when enzymes were killed. Pytuvic acid produced was determined with dinitrophenylhydrazine. One unit of the enzymes is defined as amount of enzyme converting 1 μmole of substrate per hour.

⑤ 3 - PGA kinase and enzymes involved in the conversion of 3 - PGA into pyruvic acid. As shown in Figure 2, the activities of these enzymes decline steadily as the siliques grow. In comparison with the 8 - day siliques, the activities of these enzymes in the 40 - day siliques drop down 6 times and 1. 7 times respectively.

2. ***Changes of Respiratory Pathways during the Deveiopment of Rape Seeds***

The respiration of rape seeds is inhibited by iodoacetamide and sodium fluoride, but the inhibition gradually decreases with the age of seeds (see Table 2).

In the 8 - day old seeds, the fractions of respiration inhibited by iodoacetamide and sodium fluoride were 62. 5% and 61. 6% respectively, while in the 30 - day old seeds, they dropped to 29. 6% and 22. 0%. However, it is interesting to note that during the ripening period the inhibited fraction tends to rise again.

Table 2 Changes of the Respiratory Pathways of Rape Seeds during Its Formation

Age of Seed (days after flowering)	Control*	Iodoacetamide (0.05 mol/L)		Sodium Fluoride (0.1 mol/L)	
	O_2 Absorbed/hour/g (fr. wt.)	O_2 Absorbed/hour/g (fr. wt.)	Inhibited/%	O_2 Absorbed/hour/g (fr. wt.)	Inhibited/%
8	508.6	190.9	62.5	195.3	61.6
11	535.2	213.8	60.1	324.5	39.4
14	434.6	199.4	54.1	252.7	41.6
20	406.0	212.2	47.7	273.8	32.6
23	302.9	211.8	30.1	213.7	29.4
30	282.9	198.4	29.6	219.7	22.0
40	288.0	166.2	42.3	218.2	24.9

* The control was treated with phosphate buffer (pH 5.0) only.

3. *The Regulation of Metabolic Pathways by an Endogenous Inhibitor*

(1) Discovery of inhibitor. In assay of G-6-P DH and 6-PG DH of rape bud by the usual procedure, the activities were found to be very low. But they increased markedly when the diluted extract was used. If the amount of extract used in the reaction mixture was reduced from 1 ml to 0.5 ml, the activity of G-6-P dehydrogenase was enhanced by 5 fold, and that of 6-PG dehydrogenase by 4 fold. These observations led us to the tentative conclusion that an endogenous inhibitor is present in the extract of rape buds and it somehow renders the G-6-P DH and 6-PG DH inactive. When the extract is diluted, the concentration of inhibitor is reduced, and the inhibitory effect disappears, resulting in a much higher enzyme activity.

To get a direct proof of the presence of the inhibitor, another experiment was carriedout by boiling the extract of bud acetone powder in a water bath for 5 min. After centrifuging off the denatured protein, the supernatant was added to the enzyme extract of rape bud. A profound inhibition of the activities of G-6-P DH and 6-PG DH was then noticed. A similar inhibition effect was observed by adding this supernatant to the enzyme extracts of older siliques, and the degree of inhibition increases with the amount of the added supernatant (see Figure 3). The same results were obtained with the purified G-6-PDH of yeast, cotton roots, or soybean seeds (see Table 3).

(2) The nature of the inhibitor.

① Relation to sulphydryl group. The inhibitory effect of the natural inhibitor on G-6-P DH and 6-PG DH could not be relieved by cysteine or glutathione (see Table 4), showing that the site of action is not of a sulphydryl nature.

② Solubility. By extracting the bud acetone powder with various solvents, evaporating off the solvents under reduced pressure, and resuspending the residues in water, the solubility properties of the inhibitor could be ascertained. It was found that the inhibitor is

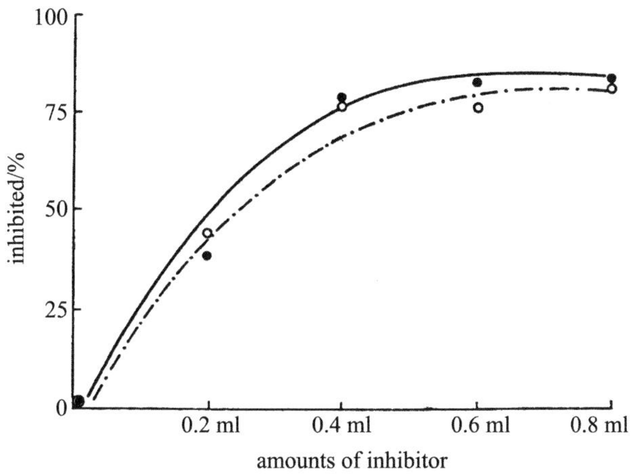

Figure 3 Inhibition of G－6－P dehydrogenase and 6－PG dehydrogenase at different concentrations of the rape silique inhibitor

●—G－6－P DH；○—6－PG DH

Table 3 Inhibition of G－6－P Dehydrogenase of Different Origin by the Natural Inhibitor of Rape Silique

Materials	Treatment	G－6－P Dehydrogease	
		Activity($\Delta E_{340\ m\mu m,\ 5'}$)	Inhibited/%
Rape silique	control	0.260	0
	+inhibitor	0.030	88
Cotton root	control	0.145	0
	+inhibitor	0.020	90
Yeast	control	0.127	0
	+inhibitor	0.025	80
Soybean seed	control	0.355	0
	+inhibitor	0.095	73

Table 4 Effect of Glutathione and Cysteine on the Activity of Natural Inhibitor

Treatment	$\Delta E_{340\ m\mu m}$, (4′)	
	G－6－P Dehydrogenase	6－PG Dehydrogenase
Control(0.5 ml enzyme)	0.155	0.175
+Inhibitor (0.5 ml)	0.030	0.040
+Glutathione (0.1 mol/L)	0.025	0.010
+Cysteine (0.05 mol/L)	0.028	0.040

soluble in water, n-butanol, ethyl ether, and hot alcohol, but slightly soluble in chloroform, and insoluble in benzene (see Table 5).

Table 5 Solubility of Natural Inhibitor*

Sources of G-6-P Dehydrogenase	Solvents	Enzyme Activity/%
Buds of rape	control	100
	water	36.6
	n-butanol	40.0
	ethyl ether	52.5
	benzene	91.7
	chloroform	75.0
Yeast	control	100
	hot alcohol	0.4

* For details, see explanation in text.

③ Stability. The inhibitor is heat stable: it is not affected by heating in boiling water for 5 min (see Table 6). On the other hand, it is quite sensitive to ultra-violet light; its action would be abolished on a 15-min exposure (see Figure 4).

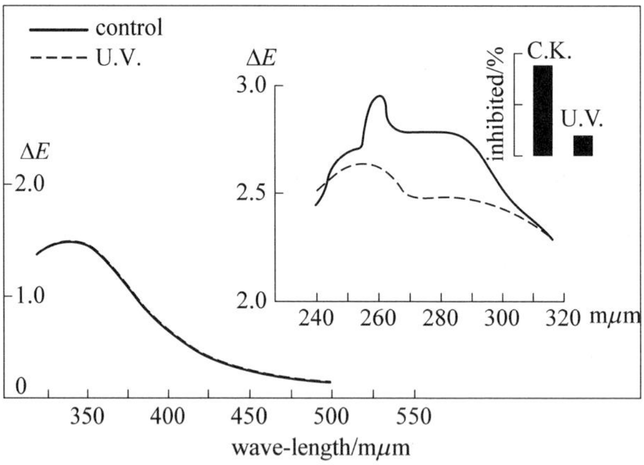

Figure 4 Absorption spectra of the natural inhibitor

④ Molecular size. When the inhibitor solution was dialysed against phosphate-EDTA buffer for 6 hours, no significant loss of activity was found (see Table 6). Thereforethe inhibitor is likely to be a compound of large molecular weight, but nonproteinous innature, as evidenced by its resistance to boiling and precipitating with 100% saturated a mmonium sulphate (see Table 6).

⑤ Electric charge. Electrophoresis showed that the inhibitor migrates from anode to cathode, bearing therefore a positive charge.

Table 6 Effect of Boiling, Precipitation by Ammonium Sulphate, and Dialysis on the Activity of Natural Inhibitor

Experiment	Treatment	G-6-P Dehydrogenase		6-PG Dehydrogenase	
		$\Delta E_{340\,m\mu m}$, (5′)	Inhibited /%	$\Delta E_{340\,m\mu m}$, (5′)	Inhibited /%
105	control (0.5 ml enzyme)	0.225	0	0.220	0
	+inhibitor (0.5 ml)	0.104	54	—	—
	+boiled inhibitor (0.5 ml)	0.075	66	—	—
	+protein portion (0.8 ml)	0.185	17	0.170	22
107	control (0.5 ml)	0.155	0	0.175	0
	+inhibitor (0.5 ml)	0.030	80	0.040	77
	+dialysis inhibitor (0.5 ml)	0.60	61	0.050	71

⑥ Fluorescence. On paper chromatogram the inhibitor appeared yellow in colour, and showed a blue fluorescence under ultra-violet light. When it was exposed to the fumes of concentrated ammonia or treated with $AICl_3$ solution, the colour became more intense or the fluorescence changed from blue to green.

(3) Purification and identification of the inhibitor. A yellow supernatant was obtained after removing the protein by boiling the enzyme extract of rape buds. The supernatant was subjected to paper chromatography using the acetone: formic acid: water (20 : 10 : 65) solvent system. At the end of 24 hour, the inhibitor was located mainly near the solvent front (see Table 7).

Table 7 Locations of Inhibitor on Paper Chromatogram*

Sections of Paper		G-6-P Dehydrogenase of Older Rape silique $\Delta E_{340\,m\mu m}$ (4′)	Inhibited /%
No.	Colour		
Control	—	0.155	—
0 (Original region)	pale yellow	0.150	3
1	colourless	0.120	23
2	slightly yellow	0.112	27
3	pale yellow	0.065	58
4	yellow	0.030	81

* For details, see explanation in text.

Both on paper chromatogram and on electrophoretogram, the areas where the inhibitor was located which were always yellow in colour. If the last section (designated as No. 4 in Table 7) of the paper chromatogram was divided into yellow and colourless regions, the inhibitory activity of the coloured region was four times higher than the colourless. These experiments showed that the inhibitor is a yellow compound.

Spectral examination of the partly purified inhibitor showed that it has an absorption maximum in the region round 260 mμm. Exposure to ultra-violet light for 15 min abolishes the absorption peak and the inhibitory activity (see Figure 4).

After being washed several times with acetone, the dry powder of rape buds was then extracted with redistilled water, boiled, and centrifuged to remove protemous impurities. The supernatant was further washed with chloroform or benzene twice to remove lipidimpurities. The concentrated extract was then purified twice by paper chromatography. A yellow-brown viscous residue was obtained after the eluate was concentrated by heating. This residue was then ground briefly with a drop of liquid paraffin, the resulting paste was transferred to an infrared window, and pressed into a thin layer with a second window. An infrared absorption spectrum was. obtained by an automatically recording double beam infrared spectrophotometer. Because of the difficulty met in preparing a uniformly thin layer sample, only two absorption peaks. 1,600 cm^{-1} and 1,100 cm^{-1}, corresponding to carbonyl and phenyl groups respectively, were observed. These are important functional groups of flavonoid compounds.

(4) Content of the inhibitor in the reproductive organs of rape at various developing stages. As shown in Figure 5, the inhibitor mainly occurs in the buds, flowers, and young siliques; it gradually diminishes as the siliques grow older and almost disappears in the old siliques.

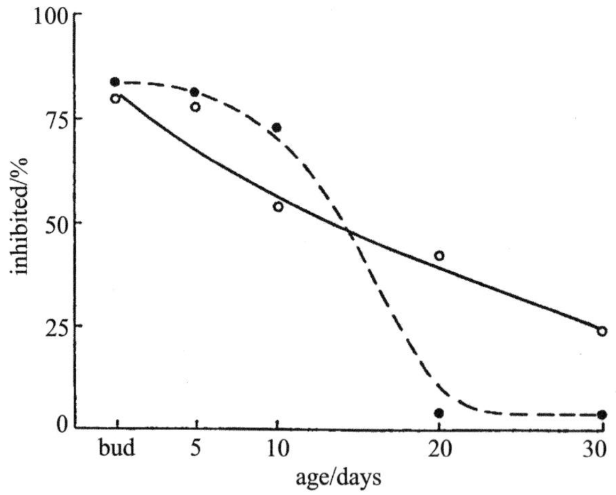

Figure 5 Distribution of the natural inhibitor in rape sdiques of difeerent ages

○—6-PG DH;●—G-6-P DH

Discussion

It is an interesting fact that in rape seeds, as in other oil-bearing seeds, the carbohydrates store is abruptly and rapidly converted into fats usually at the age of 10—20 days or more days after flowering. This observation reported in a variety of seeds suggests that there are some factors controlling the "turn about" of metabolism and synthesis of fats in the maturation process[14]. Of these factors one may be the activity of the TCA cycle[15] and another the generation of $NADPH_2$, as discussed in this paper.

In the light of the present knowledge it is generally contended that fatty acids were synthesized in the following way: acetyl-CoA is first carboxylated in the presence of CO_2, ATP, Mn^{2+}, and biotin, forming malonyl-CoA, then decarboxylated and condensed to form higher fatty acids[16]. It is well known that in the process of fatty acid synthesis a large amount of $NADPH_2$ is needed as a specific electron donor for the reduction of the carbonyl to methylene groups. The chief sources of $NADPH_2$ are two dehydrogenation reactions: first, the two primary reactions of the HMP pathway (G-6-P DH and 6-PG DH); second, the reaction of the extramitochondrial NADP-linked isocitrate dehydrogenase (the NAD-linked isocitrate dehydrogenase is located in the mitochodria and is coupled with the TCA cycle and oxidative phosphorylation)[17]. Obviously, the HMP pathway plays a more important part in the supply of $NADPH_2$, since 12 molecules of $NADPH_2$ can be generated from a molecule of G-6-P. Evidences on the relationship between fat formation and the HMP pathway have been reported in animal tissues, but this approach has not yet been made with tissues of higher plants.

There are considerable experimental evidences for the existence of a pathway of glucose catabolism other than the EMP or glycolytic route in developing rape seeds. In a preliminary account[6] we have shown that besides all the enzymes of the EMP pathway the rape siliques contain a very active G-6-P DH and 6-PG DH as well as enzymes concerned with the breakdown of R-5-P. The identification of several intermediates of the HMP pathway and the ability of the cell-free extract of the siliques to resynthesize hexosemonophosphate from R-5-P indicate the existence of the complete cycle in this tissue[6]. In later experiments in which the tape seeds were incubated with glucosel-C^{14} or glucose-6-C^{14}, we found that the aldehyde carbon of the glucose is converted more rapidly into $C^{14}O_2$ than the carbon atom 6. It was further calculated from the relative incorporation of C^{14} from specifically labelled glucose molecules into lipid that in rape seeds approaching maturation approximately 40% of glucose utilization proceeded via the HMP pathway[18].

It is shown in the present experiments that the activities of enzymes of the HMP pathway including G-6-P DH, 6-PG DH, and enzymes utilizing R-5-P in the rape siliques rise noticeably with the development of rape seeds whereas the activities of enzymes of the HMP pathway, namely phosphofructokinase, aldolase, 3-PGA kinase, and enzymes involved in the conversion of 3-PGA into pyruvic acid decline rapidly. The metabolic shift from the predominance of the EMP pathway to that of the HMP pathway in siliques takes place between 10 to 20 days or more days after flowering, and corresponds to the period when soluble sugars in the seeds decrease sharply and fats increase markedly[19]. Taking the activities of the isolated enzymes as a criterion of the intensity of a metabolic pathway in vivo, as suggested by Glock and McLean[20] and Gibbs and Earl[21], we may conclude that the advance of fat synthesis in rape seeds is correlated with the increase in the activity of the HMP pathway and the decrease in the activity of the EMP pathway. This conclusion is supported by the fact that the fraction of respiration sensitive to iodoacetamide and NaF, powerful inhibitors of two important enzymes of the EMP pathway, 3 - phosphoglyceride dehydrogenase and enolase[6], declines rapidly with the age of the siliques. Further confirmation comes from feeding experiment with specifically labelled glucose, in which it was demonstrated that the ratios C_1/C_6 and C_1/C_u of the liberated CO_2 rise with the development of seeds and reach maxima on approaching maturation parallel with the course of fat accumulation[18]. The finding that the activity of the TCA cycle which is closely linked to the EMP pathway declines noticeably during the development is also in line with the above conclusion[15].

Among the enzymes of the EMP pathway only phosphoglucoisomerase and 3 - phosphoglyceraldehyde dehydrogenase behave somewhat differently from others. Their activities decrease from 8 to 15 days after flowering, but increase gradually again after 15 days. As phosphoglucoisomerase is involved in the HMP pathway, its enhancement in the later period is to be expected. The similar behaviour of 3 - phosphoglyceraldehyde dehydrogenase is, however, difficult to understand.

Since the discovery of G - 6 - P DH in erythrocyte haemolysate and yeast autolysate by Warburg et al. [22,23] nearly 30 years ago, a great amount of evidence has been accumulated on the contribution of the HMP pathway in the breakdown of glucose in many animal and plant tissues. The cycle mechanism, the metabolic intermediates, and the properties of enzyme of the pathway have been studied rather tliorouglily[24-26]. In the last 10 years, the interest of investigators has turned to the physiological role and the regulatory mechanism of the HMP pathway. It has been demonstrated, e. g. , that the age of tissues is related to the HMP pathway, thus, Gibbs et al. [21] observed that while immature tissues respire exclusively or to a large extent via the EMP glycolytic pathway, but as the tissue ages and diffetentiates, the HMP pathway becomes increasingly important. Shaw and Samborski[27], and Daly et al. [28] stressed the significance of the HMP pathway in the immunity of plant tissues. Other workers reported that aureomycin formation in *Streptomyces aurefaciens*[29], the abscission of organs of plants[30], and the differentiation of flowering buds[31], may also be correlated to the HMP pathway. The evidence presented in this investigation as well as in our other report[18] for the intimate association of lipogenesis and the HMP pathway further emphasized the significance of the HMP pathway in the metabolism and physiological activities of the higher plants.

The regulation and control of the HMP pathway of carbohydrate metabolism in organisms are of great interest. There is growing evidence showing that the supply of NADP is one of the factors controlling the activity of the HMP pathway in many tissues. The rate of

$NADPH_2$ reoxidation and the ratio of $NADP/NADPH_2$ can affect the rate of channelling of glucose into the HMP pathway. As has been shown in our laboratory, the presence of artificial electron acceptors such as phenazine methosulpbate, methylene blue, and triphenyltetrazolium, or the supply of substrates which require $NADPH_2$ for their metabolism (e. g. , the synthesis of fatty acids), can increase the rate of the relative formation of $C^{14}O_2$ from glucose - 1 - C^{14} by rape seeds[32]. Similar effects have been observed in other tissues such as liver slices[33,34], erythrocytes[35], corneal epithelium[36], aspites tumour cells[37], mammary gland[38], and carrot slices[39]. In addition to NADP, the regulatory effects of phosphate[40,41], glucose concentration[42], Cu^{2+} content, the Michaelis constants of enzymes of two pathways competiting for G - 6 - P[43], and 6 - PG[44] on the HMP pathway and EMP pathway have also been frequently discussed.

The most interesting result of the present investigation is the finding of a natural inhibitor of G - 6 - P DH and 6 - PG DH in the reproductive organs of rape seed. Its presence was evidenced by the facts that activities of both G - 6 - P DH and 6 - PG DH rose markedly when the extract of acetone powder of the buds was diluted, and that the extract of the buds could strongly inhibit the G - 6 - P DH derived from different sources. On the basis of its solubility and stability properties, absorption spectra, fluorescence under ultra-violet, etc. , the inhibitor was presumed to be a flavonoid or its derivative.

Experiments showed that the natural inhibitor is present in buds, flowers, and young siliques, but nearly absent in the old siliques. These results, together with the changes of the activities of G - 6 - P DH and 6 - PG DH in rape reproductive organs during seed development, led us to believe that the inhibitor may take part in controlling the activities of G - 6 - P DH and 6 - PG DH and therefore regulate the rate of fat formation. In the early periods of seed development, the content of inhibitor is high enough to block the dehydrogenase of the HMP pathway and no fat is synthesized. Later, when the inhibitor decreases and disappears and the enzymes of the HMP pathway become active, the "valva" of the HMP pathway becomes to be opened, and fat is actively synthesized, as a result of the abundant supply of $NADPH_2$ generated. Thus, it seems that the inhibitor may play an important role in the regulation of carbohydrate breakdown and fat synthesis. Such mechanism of regulation is quite different from that discovered in animal tissues by Tepperman and Tepperman[45], who suggested that the hepatic G - 6 - P DH increase seen on refeeding and after fructose feeding probably represents denovo protein synthesis. Here, according to our findings, it seems more likely that, as the seeds develop, the natural inhibitor decreases and eventually disappears, thus releasing the G - 6 - P DH and 6 - PG DH from inhibition, and accelerating the operation of the HMP pathway. This results in an increase of generation of $NADPH_2$, which together with other unknown factors, e. g. , the formation or activation of the enzyme complex required for fatty acid synthesis, as suggested by Stumpf[14], would stimulate the synthesis of fatty acids. The increased speed of fatty acid synthesis would in turn accelerate the oxidation of $NADPH_2$ to NADP, which further quickens the entrance of glucose into the HMP pathway, resulting thus in a positive "feedback". Such mechanism, schematically postulated in Figure 6, could explain why the carbohydrates are suddenly and increasingly converted into fats during the development of oil-bearing seeds. It must be emphasized that the mechanism of regulation of fat synthesis in oil-bearing seeds is very complicated. The simple interpretation given here does not claim to be adequate or even correct. It only serves as a starting clue to approach the complex and interlocking problems of seed metabolism and lipogenesis, the theoretical and practical importance of which is obvious.

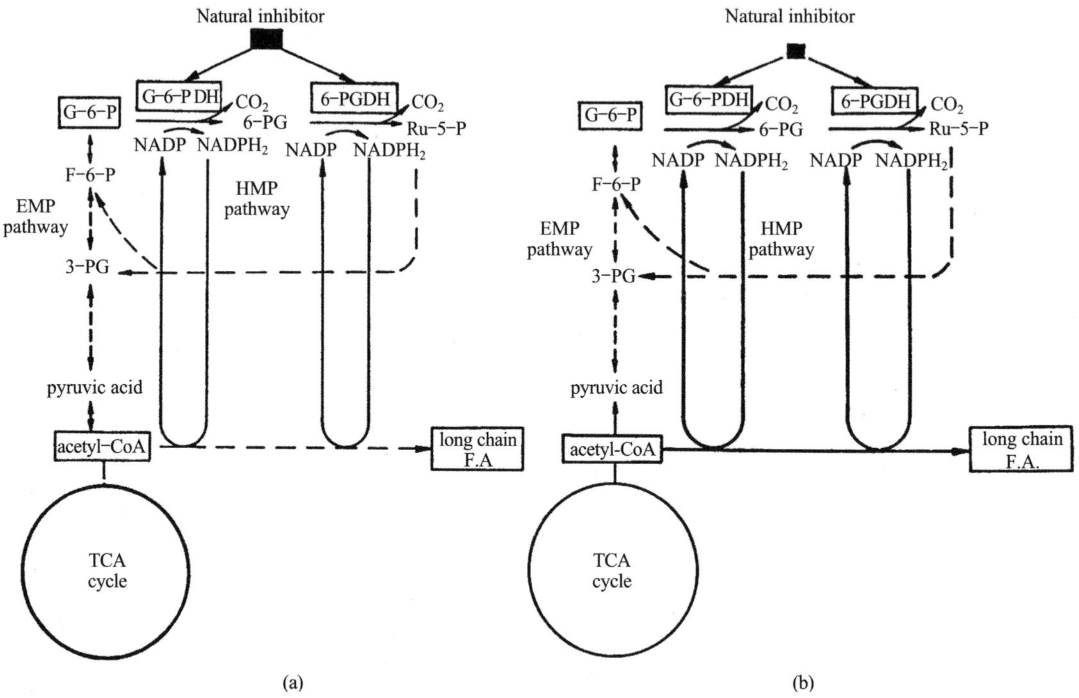

Figure 6 A postulated scheme of the regulation of fat synthesis in rape seeds during development

(a) Youth seeds in which the EMP-TCA pathway predominated and the activity in the HMP pathway is low owing to the presence of the natural inhibitor, which blocks the G-6-P DH and 6-PG DH with the result of very limited fatty acid synthesis; (b) Matured seeds in which the activity in the HMP pathway greatly increases with the diminution of the natural inhibitor, enhancing the fatty acid synthesis. In the same time, the $NADPH_2$ reoxidation is promoted, thus a positive feedback is established between the HMP pathway and fatty acid synthesis.

The authors wish to express their gratitude to Prof. Ying Hung-chang for his advice and encouragement throughout the course of this work, and to Comrade C. F. Chen of the Institute of Organic Chemistry for the measurement of infrared absorption of the natural inhibitor.

References

[1] Langdon, R. G. 1957 J. *Biol. Chem.* 226,615.
[2] Wakil, S. J. 1961 J. *Lipid Res.*, 2,1.

[3] Abraham, S., Mathes, K. J., & Chaikoff, I. L. 1961 *Biochim. Biophys. Acta*, 49,268.
[4] Lynen, F. 1953 *Feder. Proc.*, 12,683.
[5] Stumpf, P. K. & Bradber, C. 1959 *Ann. Rev. Plant Physiot.*, 10,197.
[6] Shih, C. N., Chu, Y. S., & Tan, C. 1965 *Acta Phytophysiol. Sin.*, 2,33.
[7] Hageman, R. H. & Arnoti, D. I. 1955 *Arch. Biochem. Biophys.*, 57,421.
[8] Slein, M. W. 1955 in Colowick, S. P. & Kaplan, N. 0. (Eds.), *Method in Enzymol.*, 1,300, Berlin: A. P., Spring-Verlag.
[9] Roe, J. H. 1934 J. *Biol. Chem.*, 107,15.
[10] Sibley, J. A. & Lehninger, A. L. 1949 J. *Biol. Chem.*, 177,859.
[11] Black, C. C., Jr. & Humphreys, T. E. 1962 *Plant Physiol.*, 37,66.
[12] Axelrod, B. & Bandurski, R. S. 1953 J. *Biol. Chem.*, 204,939.
[13] Kachmar, J. F. & Boyer, P. D. 1953 J. *Biol. Chem.*, 200,669.
[14] Stumpf, P. K. 1963 in Grant, J. K. (Ed.) *The Control of Lipid Metabolism*, pp. 29 – 36, London and New York: A. P.
[15] Liu, C. H., Wu, M. H., & Shih, C. N. 1965 *Acta Phytophysiol.* Sin., 2,127.
[16] Stumpf, P. K. 1961 *Proc. 5th Internatl. Congr. of Biochem.*, 7,74.
[17] Shimazono, H., Ishikawa, S., & Kamiya, T. 1960 *Lecture on Biochem.*, 4,50. Kyoritsu Shippan Kabushiki Kaisha.
[18] Chu, Y. S., Liu, C. H., & Shih C. N. 1965 Acta *Phytophysiol. Sin.*, 2,185.
[19] Ni, Chin-san et al. 1960 *Some Physiotogical Problems in Brczssica napus*, p. 32, Peking: Science Press.
[20] Glock, G. E. & McLean, P. 1954 *Biochem.* J., 56,171.
[21] Gibbs, M. & Earl, J. M. 1959 *Pl. Physiol.*, 34,529.
[22] Warburg, O., Christain, W. 1933 *Biochem.* Z., 266,377.
[23] Warburg. O., Christain, W., & Griese, A. 1935 *Biochem.* Z., 282,157.
[24] Horecker, B. L. & Mehler, A. H. 1955 *A. Rev. Biochem.*, 24,207.
[25] Gunsalas, I. C., Horecker, B. L., & Wood, W. A. 1955 *Bact. Rev.*, 19,79.
[26] Racker, E. 1957 Harvey Lect., 51,143 (Ref. Glock, G. E. & McLean, P. 1958 *Proc. R. Soc.*, B, 149,354).
[27] Shaw, M. & Samborski, D. J. 1957 *Can. J. Bot.*, 35,389.
[28] Daly, J. M., Sayre, R. M., & Pazur, J. H. 1957 *Pl. Physiol.*, 32,44.
[29] Shen, San-chium, Chen, Jaun-pao, & Hong, Mang-ming 1957 *Acta Physiol. Sin.*, 21,302.
[30] Hsueh, Y. L., You, T. S., & Chu, Y. S. 1964 *Acta Phytophysiol. Sin.*, 1,90.
[31] Tu, C. Y., Pen, W. H., Ouyoung, K. C., & Hsueh, Y. L. 1963 *Proc. 1st Ann. Meet. of Chin. Ass. of Pl. Physiol.*, p. 135.
[32] Chu, Y. S., Ma, Y. S., & Shih, C. N. 1965 *Acta Phytophysiol. Sin.*, 2,362.
[33] Cahill, G. F., Jr., Hastings, A. B., Ashmore, J., & Zottu, S. 1958 *J. Biol. Chem.*, 230,125.
[34] Hers, H. G. 1957 *Le metabolism du fructose*, Editions Arscia, Belgium: Brussels.
[35] Brin, M. & Yonemoto, R. H. 1958 *J. Biol. Chem.*, 230,307.
[36] Kinoshita, J. H. 1957 *J. Biol. Chem.*, 228,247.
[37] Wenner, C. E. 1958 *Cancer Res.*, 18,1105.
[38] McLean, P. 1960 *Biochim. Biophys. Acta*, 37,296.
[39] Ap Rees, T. & Beevers, H. 1960 *Pl. Physiol.*, 35,830.
[40] Shen, San-chium & Chen, Jaun-Pao 1958 *Acta Biochim. Sin.*, 1,69.
[41] Kravitz, E. A. & Guarino, A. J. 1958 *Science*, 128,1139.

[42] Racker, E. 1956 Ann. N. Y. *Acad. Sci.*, 63,1017.
[43] McLean, P. 1958 *Biochim. Biophys. Acta*, 30,303.
[44] Parr, C. W. 1956 *Nature*, 178,1401.
[45] Tepperman, H. M. & Tepperman, J. 1963 in Weber, G. (Ed.) *Advances in Enzyme Regulation*, Vol. 1, p. 121. Oxford, London, New York, Paris: Pergamon.

第三篇

微生物发酵工程及纤维素酶在工农医上的应用（1967—1980）

纤维素酶固体曲的制取及其在酒精生产上的试用*

一、前言

我厂以山芋干等粮食为原料生产酒精。这些原料中，除含60%～70%的淀粉外，尚含有3%～5%的纤维素，近年来，为落实毛主席的“广积粮”指示，我厂又使用了含淀粉30%～35%，纤维素20%～25%的金刚刺，代替粮食原料生产酒精。历年来，原料中的这些纤维素成分，一直未被利用，白白地浪费了。

在毛主席革命科研路线指引下，1969年开始，我厂和上海植生所协作，进行了纤维素酶应用于酒精发酵提高出酒率的试验，至1970年，初步达到：加1%～2%的固体纤曲。提高酒精得率1.5%左右的效果[1,2]。

当时，因用曲量大，投产尚有困难。从1970年5月份开始，由上海植生所、上海化纤六厂和我厂，组织三结合协作组，进行高酶活菌种的选育工作，经过一些诱变处理后，得到二株新菌种 EA_3-867 和 N_2-78，分解纤维素的酶活提高一倍以上，性能较稳定[3,4]。

在上级公司和本厂党支部的直接领导下，努力贯物执行毛主席关于学习理论、反修防修、安定团结和把国民经济搞上去等一系列重要指示。我们发挥了以工人为主体的三结合攻关小组的作用，发扬苦干加巧干的精神，用较短的时间，摸索了新菌种的固体培养条件，提高了曲的酶活。

接着，又将新菌种的纤曲用于发酵对比试验。从小型、中型、大型发酵对比试验证明：使用新菌种用曲量降到0.3%，提高出酒率1.5%左右，降低黏度2～4倍，已部分投入了大生产使用。

现把试验过程和结果初步整理如下。由于我们的工作做得不多、不细，错误和不完善的地方一定不少，因此，请同志们提出宝贵意见，以便进一步完善、提高，使这一技术能够早日推广应用，为落实毛主席关于“广积粮”的指示做出应有的贡献。

二、纤曲的制造

1. 工艺流程

绿色木霉菌种(砂土管)→斜面→三角瓶→种曲→厚层通风曲。

---

* 上海酒精二厂，上海植物生理研究所。此文发表于1976年“工业微生物”4期1－30页。

2. 菌种的培养

1）菌种的选育系谱

（1）1096 $\xrightarrow[\text{放线菌素 K}]{\text{钴 60}}$ 23K－12 $\xrightarrow{\text{硫酸二乙脂}}$ D_2－54 $\xrightarrow{\text{紫外线}}$ U_{10}－40 $\xrightarrow[\text{核苷酸}]{\text{紫外线}}$ R－8 $\xrightarrow{\text{高能电子流}}$ EA_3－867[3]。

（2）木-3 $\xrightarrow{\text{高能电子流}}$ EA_6－97 $\xrightarrow{\text{亚硝基胍}}$ N_2－78[4]。

注：1096 和木-3 系绿色木霉野生种。

2）菌种的分离

（1）分离用培养基：

① 0.1%酵母膏，0.2%蛋白胨，2%纤维素粉*，2%琼脂，0.15%胆酸钠。

② 10%～20%土豆汁，2%纤维素粉*，2%琼脂，0.15%胆酸钠。

③ 5°～7°Bix 麦芽汁，2%纤维素粉*，2%琼脂，0.15%胆酸钠。

④ 0.2% KH_2PO_4，0.14%（NH_4）SO_4，0.1% KNO_3，0.05% $MgSO_4$，$7H_2O$，0.03% $CaCl_2$ · 0.000 5% $FeSO_4$，$7H_2O$，0.000 2% $CoCl_2$，0.000 17% $ZnCl_2$，0.000 16% $MnSO_4$，0.000 05% $CuSO_4$，2%纤维素粉*，2%琼脂，0.15%胆酸钠。

*纤维素粉用磨碎的新华一号滤纸代替亦可。

（2）分离操作：

用玻珠制成单孢悬浮液，适当稀释后分别浇于上述任一种培养基的培养皿中，30℃培养 40～60 小时，挑取较正常的单个菌落入斜面培养。

3）菌种的培养

（1）斜面（或茄子瓶）的培养：培养基采用分离用培养基，减去胆酸钠。从培养皿或斜面接入孢子，置 30℃，培养 5～7 天，待孢子转绿，即可取出，检查，正常者置冰箱备用。

（2）三角瓶（或铝盒）培养：取稻草粉 900 克，麸皮 100 克。$(NH_4)_2SO_4$ 25 克，水 2 200～2 500 毫升，充分拌匀，分别装入三角瓶或铝盒中（若 500 毫升三角瓶，每瓶装干料 15～20 克；若 1 000 毫升三角瓶，每瓶装干料 30～35 克；若铝盒，装料厚度以 5 厘米左右为宜）以双面绒布（或八层纱布）扎口，并复以牛皮纸以防潮湿，1.0 公斤/厘米^2 杀菌 1 小时，冷却后，将斜面或茄子瓶中的孢子接种（或孢子悬浮液）摇匀，置 30℃培养。第二日扣瓶一次（彻底摇匀），并开始保湿。第四天再扣瓶一次，（扣成小块）并停止保湿，只保干温。第五或第六日成熟，孢子长满，并转绿，即可出房，检查，无杂菌供备用。

（3）种曲培养（若用量不多，可用三角瓶或铝盒代替）。

3. 厚层通风曲的培养

1）设备

（1）斗式提升机和绞龙：和一般酒精厂磨粉车间的原料提升机相同。

（2）转锅：蒸稻草屑用，上海大明铁工厂出品，圆柱形，使用时可转动，直径 1.2 米左右，高 1.2 米，容积约 1 米^3。（内有夹层）功率 1.1～1.7 kW－4 P，转速 0.5～1.0 转/分。

（3）杨夫机：和糖化曲车间的相同。

（4）曲箱：水泥池。长×宽×高＝3.0×1.9×0.4（米^3），底面斜度：4 度。

（5）鼓风机，离心鼓风机，型号：8－18－19#5。

2）工艺与操作

(1) 称取造纸厂下脚稻草屑 200 千克，麸皮 10～20 千克。

(2) 将硫酸铵 5 千克、硫酸 800 毫升，溶于 200 升水中。(估计杀菌时的蒸汽冷凝水约 120 升，所以实际共加水约 320 升。料∶水≈1∶1.5)。

(3) 由斗式提升机将(1)项麸皮，稻草屑提入绞龙，并加入(2)项的水，拌匀，落入转锅，直至料、水全部入转锅。

(4) 转动转锅，一边转运，一边进蒸汽，开始杀菌，待压力到 1.5 公斤/厘米^2，保压 1 小时(中间使转锅经常转动)。

(5) 放汽、降压、冷却到 38～40℃，将 0.5%～0.8%的种曲倒入转锅，再转动，使匀。然后，经杨夫机打松，进箱，并立即通风，使品温在 30℃左右，开始培养。

(6) 进箱后，每隔四小时左右，通风一次，每次约 10 分钟。约 10 小时后，品温开始上升，以后每一小时左右通风一次，通风时间长短视品温、生长时期和原料湿度等而定，一般在生长中期和后期，生长迅速，品温升高快；或原料潮湿，通风较差时，通风时间宜长些。整个培养过程中控制依据以品温为主，前期在 30～33℃，中期在 31～35℃，后期在 32～37℃。曲房室温控制在 30℃左右，湿度保持在 88%～96%之间，前期湿度略高，后期湿度略低些。

(7) 进箱后 60～66 小时，活力趋于平稳，即可出曲。

另：若无高压杀菌条件，而用常压，则以 2.0～2.5 小时为宜。若用帘子做曲，则与种曲同，但不用翻帘。

3）与纤酶活力关系较大的几个因素

(1) 制曲原料和酶活的关系：纤维素酶，是一种诱导酶。所以原料中需含有纤维素。而且，不同的纤维素，对纤酶活力的诱导效果各异。各种原料对比试验的结果表明，用稻草屑制纤曲，酶活最高。但即使是稻草粉，不同来源、不同品种，或不同粗细都会影响酶活。本厂做曲使用的是造纸厂的稻草下脚。对稻草下脚要求新鲜、清洁、不腐败变质。为促进生长，可加入 5%～10%的麸皮。同时，为了防止细菌的感染，加适量的硫酸，调节 pH 到 4～5。

(2) 原料加水和酶活的关系：无论孢子发芽或菌丝生长、酶的形成，都需要适量的水分，所以水分和酶活关系颇大。但水分太多，通气不良，升温太快，易引起污染酸败。水分太少，菌丝生长，发育不良，也得不到较高的酶活。一般以出锅水分 62%～68%为宜(根据原料、品种、粗细及季节等不同情况而定)，其次曲房要保持一定的湿度，过高过低均不宜。

(3) 制曲品温和酶活的关系：品温和酶活关系较大，一般应分阶段有所不同的控制。前期品温较低，在 30～33℃，利于孢子发芽，菌体生长。后期品温略高，在 32～37℃，利于酶的积累。

(4) 通气与酶活的关系：绿色木霉产生和分泌纤维素酶是耗能的过程，需要旺盛的呼吸。厚层通风制曲时，为了保证得到较高的酶活，需要进行适当的通风，通风的目的，一方面是提供呼吸所需的氧气赶走二氧化碳。另一方面是有助于驱散生长过程中产生的热量，控制品温的上升。生长不同的时期，要求的通风量也不同，前期，生长较缓，呼吸较弱，品温上升较慢。可间歇通风，通风时间也可短些，中期和后期(40 小时以后)，生长旺盛，呼吸强烈，品温上升快，通风时间宜长，甚至连续通风。

(5) 生长速度和酶活的关系：诱变菌种 EA_3－867 和 N_2－78 一个重要的特征是菌落小，生长速度慢，纤维素酶活力高(与野生种相比)。如果培养基营养太好，生长太快，则纤维素酶活力反而低。故只要基本的营养得到保证就行，营养不必太好。温度、湿度也要适当控制，生长不宜太快。

（6）杂菌污染和酶活的关系：为了保证制曲的成功，纤曲不受杂菌污染极为重要，纤曲染菌后，往往活力大大下降，甚至木霉全然不能生长。稻草原料中杂菌较多，尤其是枯草杆菌一类，耐高温。若原料杀菌不透，极易生长，品温猛升，原料臭败，木霉长不出。故原料杀菌定要彻底。调节原料 pH 至酸性，也即为了防止细菌生长。曲房用福尔马林或硫磺熏蒸消毒，操作工具用蒸汽消毒，并始终保持清洁。在制曲过程中易染菌的种类主要有水毛霉（如原料曲房温度太大时易见）、黄曲和黑曲等等，染菌对酶活的影响与染菌的时间有关。纤曲生长前期染菌影响较大，后期影响较小。故前期一定要保证不让杂菌生长，从而使木霉生长先占优势。根据我们的经验，把好种子关，保证斜面，种曲不受染菌极为重要，因杂菌生长快，从种子接到大曲中，杂菌很快占优势，导致制曲失败。

4. 纤曲的质量检验

1）微生物的检查

（1）外观：本菌系属绿色木霉的变异株，故斜面或三角瓶的外观正常者，菌丝较稀而短，色泽草绿而均匀（有时出现白色再生菌丝）。

（2）镜检：气生菌丝呈树枝状，有横隔，椭圆形的孢子成簇地长在小梗上。

（3）孢子数：取鲜种曲一克，加水 10 毫升，捣匀后取样，用血球板计数，正常者在 2.0 亿孢子数/克鲜曲以上。

2）纤酶活力的测定

（1）试剂的准备：

① 0.5%CMC 缓冲溶液：取 1%CMC 原液 100 毫升，加水 80 毫升，pH 4.6 的 0.1 mol/L 的醋酸缓冲液 20 毫升，充分摇匀，即可。

② NS 液，称取 3,5-二硝基水杨酸 6.3 克，氢氧化钠 21 克，酒石酸钾钠 182 克，重蒸酚 5 克，无水亚硫酸钠 5 克，搅拌溶于 1 000 毫升蒸馏水中，用棕色试剂瓶保存，稳定 3～5 天，用葡萄糖标定后，即可使用。

③ 1×3 厘米^2 滤纸条：取新华一号滤纸，截成 1 厘米宽，3 厘米长的小纸条备用。

（2）酶液的制取：称取相当于一克干曲的鲜曲，加水 35 毫升，30℃浸泡 1 小时，然后，3 000 转/分离心 5 分钟，上清液即为酶液。

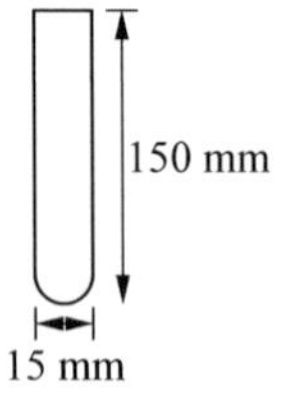

（3）测定：

① 滤纸破碎酶活：（代表 C_1 酶，表示分解天然纤维素的能力）。取上述酶液 4 毫升，加 pH 4.6 醋酸缓冲液 1 毫升于（15×150） mm① 的试管中，摇匀，投入 1×3 厘米^2 的滤纸条一张，放在 40℃水浴锅中作用 2 小时，然后取出试管，顺倒摇二次，观察滤纸的破碎程度，以一点不破为“0”象“□”，以稍破为“+”象“[illegible]”或“[illegible]”。以破碎成小块为“++”，象“[illegible]”或“[illegible]”，以全破为“+++”象“[illegible]”。介于二者之间的尚有“⊥”“卄”“卌”等等。

三角瓶，种子，要求 卌 。

通风曲，要求 卄－卌 。

① 试管（15×150）mm 表示为底面直径 15 mm，百度为 150 mm 的试管。

② CMC 酶活(代表 C_x 酶,表示分解可溶性纤维素的能力)。

取上述酶液,再稀释 20～40 倍(三角瓶种曲,稀释 40 倍,通风曲稀释 20 倍)。然后,吸 0.1 毫升,加 0.5%的 CMC 缓冲溶液 2.5 毫升。40℃作用半小时,取出,加 DNS 液 2.5 毫升,沸水浴中加热 5 分钟,立即冷却。绿光(540 mm)下比色。0・D 要求在 1.0 左右(相当于葡萄糖 400 微克左右)。

③ 有时,还测定滤纸糖和酶液糖等供参考。

滤纸糖:将上述①项的测定液离心,吸取上清液 0.5 毫升,加蒸馏水 2.0 毫升,DNS 液 2.5 毫升,摇匀,沸水浴上加热 5 分钟,立即冷却,绿光下比色。

酶液糖:用酶液代替滤纸糖中上清液,余同上。

三、发酵试验

N_2－78 绿色木霉的纤维素酶应用于酒精发酵试验,共进行了 24 批 500 毫升三角瓶小型发酵试验,5 批 3 000 立升发酵罐的中型发酵试验和 5 批 12 000 立升发酵罐的大型发酵试验。通过对测定方法的改进和试验数据的误差分析,证明 0.3%的 N_2－78 绿色木霉的纤维素酶曲应用于酒精发酵生产,能稳定地提高山芋干(90%)和金刚刺(10%)混合原料的出酒率 1.5%左右,降低发酵醪的黏度 2～4 倍。

1. 测定方法

酒度:准确称取发酵醪 100 克,注入 500 毫升或 1 000 毫升圆底烧瓶内,加入 100 毫升蒸馏水,以小火加热蒸馏,蒸馏速度控制在一小时到一个半小时。蒸出 100 毫升馏出液,以 1/10 刻度的容量百分率酒精计和 1/10 刻度温度计同时测定酒精和温度,查表换算成 20℃时的酒度。

黏度①:发酵醪流完 50 毫升胖肚吸管所需的时间。

纤维素含量:准确称取发酵醪 100 克,烘干至恒重,研磨成粉作样品。精确称取 1～3 克上述样品,加入 75 毫升 80%醋酸,5 毫升硝酸(比重 1.4)与 2 克三氯醋酸配成的混合液。于水浴锅中煮沸 15～20 分钟,然后通过预先准备好的铺有石棉的古氏坩埚过滤。过滤毕,用热的混合试剂小心冲洗烧瓶,将剩余的纤维素全部收集于坩埚中。用乙醇和乙醚冲洗坩埚中的残渣,干燥后称至恒重,灼烧半小时,冷却后称重,与灼烧前重量之差,即为纤维素含量。

固形物:准确称取 100 克醪液,80℃烘干至恒重,称取干物质重量,以百分数表示。

外观糖度、还原糖度、总糖、总残糖、残糖、可溶性物的测定按 1974 年全国酒精协作会议制订的酒精生产半成品检查方法。

2. 加纤曲的发酵试验

1) 500 毫升三角瓶小发酵试验

取 500 毫升三角瓶 12 只,分成 A、B、C、D 四组,每组各三只。在各瓶中,分别称入冷却到 30℃的车间糖化醪 300 克,成熟酒母 30 毫升,然后,按下表 1 加入死、活纤曲,(浸泡过)小心摇匀,装上发酵器,置 30℃,发酵 68 小时,取出测定酒度,得表 2。

---

① 黏度用“秒”来计量酒精工业界测黏度的方法和单位,即发酵醪流完 50 毫升胖肚吸管所需的时间。

表 1 死活纤曲用量表

纤曲用量* / 组号 / 纤曲种类**	A	B	C	D
死纤曲	1.0%	0.7%	0.5%	0%
活纤曲	0%	0.3%	0.5%	1.0%

* 纤曲用量$=\frac{\text{纤曲干重}}{\text{原料干重}}\times 100\%$。

** 将待用的纤曲拌匀，分成二份，一份为活纤曲，一份经高压蒸汽 1.0 公斤/厘米^2 1 小时，破坏酶活，作为死纤曲。

表 2 不同纤曲用量与酒度的关系

批号 / 酒度 / 纤曲用量	1	2	3	4	5	6	7	8	9	10
0%(对照)	8.5 8.5 8.5	8.5 8.6 8.55	9.0 9.2 9.1	8.9 8.9 8.9	8.5 8.5 8.5	8.6 8.6	8.5 8.4 8.45	8.5 8.4 8.45	6.2 6.2 6.2	6.3 6.3 6.3
0.3%	8.7 8.7 8.7	8.6 8.6 8.6	9.2 9.2 9.2	9.1 9.1 9.1	8.7 8.8 8.75	8.9 8.9 8.9	8.6 8.7 8.65	8.7 8 8	6.4 6.3 6.35	6.5 6.5 6.5
0.5%	8.7 8.7 8.7	8.7 8.6 8.65	9.2 9.2 9.2	9.1 9.0 9.05	8.8 8.9 8.85	8.9 9.0 8.95	8.7 8.6 8.65	8.7 8.7 8.7	6.4 6.4 6.4	6.5 6.5 6.5
1.0%	8.7 8.7 8.7	8.6 8.8 8.7	9.2 9.2 9.25	9.1 9.3 9.2	8.8 8.8 8.8	8.9 9.0 8.95	8.7 8.7 8.7	8.8 8.7 8.75	6.4 6.4 6.4	6.6 6.5 6.55

小发酵中，黏度未测定，但从和黏度成平行关系的沉淀情况看，加纤曲后，沉淀情况明显良好，且纤曲用量越多，沉淀越好。

将表 2 数据整理，即得到表 3。

表 3 不同纤曲用量与酒精增产效果的关系

批号 / 效果% / 纤曲用量	1	2	3	4	5	6	7	8	9	10	平均
0%(对照)	0	0	0	0	0	0	0	0	0	0	0
0.3%	2.36	0.59	1.10	2.25	2.95	3.49	2.37	2.37	2.42	3.18	2.31
0.5%	2.36	1.17	1.10	1.69	4.12	4.07	2.37	2.96	3.22	3.18	2.62
1.0%	2.36	1.75	1.65	3.35	3.53	4.07	2.93	3.56	3.22	3.96	3.04

* 效果$=\frac{\text{加纤曲后的酒度}-\text{对照的酒度}}{\text{对照的酒度}}\times 100\%$。

根据几十次小型试验的结果，根据我厂目前的人力、场地等条件，决定用曲量 0.3%，并进行中型、大型试验。

2）3 000 升发酵罐的中型发酵醪试验

在 3 000 升发酵罐中加入 2 500 升左右的糖化醪，加入 10%（V/V）的酒母，以 2 千克/厘米^2 压力的压缩空气强制通风 5 分钟，使醪液均匀，迅速压入另一只 3 000 升发酵罐，待二只罐发酵醪压平后，一罐加入浸泡一小时的新鲜纤曲 1 500 克（折合成干纤曲，相当于原料的 0.3%），另一罐加入相当于 1 500 克鲜曲的经过高压消毒的稻草粉和水作对照。通风 5 分钟，每罐迅速取 3 只样品，进行测定。发酵 68 小时后，两罐同时通风搅拌 5 分钟，同时迅速取 3 只样品，测定外观糖度、黏度、酒度、固形物、纤维素含量以及残糖、可溶性物、总残糖。五批结果如表 4 所示。

表 4　0.3%纤曲应用于五批 3 000 升发酵罐发酵试验效果

<table>
<tr><th colspan="2">批号
项目</th><th colspan="2">1</th><th colspan="2">2</th><th colspan="2">3</th><th colspan="2">4</th><th colspan="2">5</th><th colspan="2">平均</th></tr>
<tr><td rowspan="2">黏度（秒）</td><td>对照</td><td>21.1</td><td rowspan="2">降低 3.4 倍</td><td>17.0</td><td rowspan="2">降低 2.7 倍</td><td>40.7</td><td rowspan="2">降低 4.7 倍</td><td>30.8</td><td rowspan="2">降低 3.0 倍</td><td>29.4</td><td rowspan="2">降低 2.6 倍</td><td></td><td rowspan="2">降低 3.3 倍</td></tr>
<tr><td>加纤</td><td>6.2</td><td>6.3</td><td>10.0</td><td>10.3</td><td>11.2</td><td></td></tr>
<tr><td rowspan="2">酒度/%</td><td>对照</td><td>8.92</td><td rowspan="2">+1.71%</td><td>8.02</td><td rowspan="2">+2.87%</td><td>8.01</td><td rowspan="2">+3.25%</td><td>9.33</td><td rowspan="2">+1.93%</td><td>8.85</td><td rowspan="2">+1.02%</td><td></td><td rowspan="2">+2.16%</td></tr>
<tr><td>加纤</td><td>9.08</td><td>8.25</td><td>8.27</td><td>9.51</td><td>8.94</td><td></td></tr>
<tr><td rowspan="2">纤维素/%</td><td>对照</td><td></td><td rowspan="2"></td><td>1.53</td><td rowspan="2">−15%</td><td></td><td rowspan="2"></td><td>1.81</td><td rowspan="2">−19%</td><td>1.30</td><td rowspan="2">−13%</td><td></td><td rowspan="2">−16%</td></tr>
<tr><td>加纤</td><td></td><td>1.30</td><td></td><td>1.46</td><td>1.13</td><td></td></tr>
<tr><td rowspan="2">纤曲质量</td><td>CMC OD</td><td colspan="2">1.10</td><td colspan="2">1.29</td><td colspan="2">1.46</td><td colspan="2">1.38</td><td colspan="2">0.90</td><td colspan="2"></td></tr>
<tr><td>纸崩种类</td><td colspan="2">卌
500 毫升
三角瓶</td><td colspan="2">卌
500 毫升
三角瓶</td><td colspan="2">卌
500 毫升
三角瓶</td><td colspan="2">卌
帘曲</td><td colspan="2">卌
帘曲</td><td colspan="2"></td></tr>
</table>

3）12 000 升发酵罐的大型发酵试验

糖化醪从底部同时压入二只 12 000 升发酵罐，容量各为 9 000 升，分别接入 10%（V/V）酒母，在一罐内加入事先加水浸泡的鲜曲 20 千克（干纤曲，相当于原料的 0.3%），另一罐加入相当于 20 千克鲜曲的经高压消毒的稻草粉和水。分别搅拌 5 分钟，各取 5 只样品测定。发酵 65 小时后，以同样方法各取 5 只样品测定外观浓度、黏度、酒度、固形物、纤维素、总糖、残糖、可溶性物，结果如表 5 所示。

中型和大型发酵均证明纤曲能提高出酒率，降低发酵醪黏度，而且，发酵旺盛，酵母生长健壮。

3. 发酵结果的误差分析

添加 0.3%N_2 − 78 绿色木霉的纤维素酶曲后，山芋干和金刚刺粉混合原料的固形物明显减少，纤维素减少 13%～27%，出酒率提高 1.5%左右，酒度增加的百分比虽小，但数十余次试验中，只要每批的试验组与对照组条件控制得好，各批酒度增加的幅度虽有变动，但都是酒度增加的正结果，而且运用生物统计的方法，对试验数据（共十批）进行了误差分析，也证明加纤曲后引起酒度增加的差异是显著的。

表 5 0.3%纤曲应用于五批 12 000 升发酵罐的发酵试验效果

批号 项目		1		2		3		4		5		平均
黏度 (秒)	对照 加纤	21.0 68	降低 3.1 倍	19.8 5.11	降低 3.9 倍	13.2 6.4	降低 2.1 倍	7.5 4.4	降低 1.7 倍	8.6 4.2	降低 2.1 倍	2.6 倍
酒度 /(V/V%)	对照 加纤	7.72 7.87	+1.94%	9.02 9.15	+1.44%	9.40 9.46	+0.68%	7.56 7.65	+1.16%	8.15 8.24	+1.10%	1.26%
纤维素 /%	对照 加纤	1.41 1.03	减少 27%	1.08 0.84	减少 24%	1.29 1.02	减少 21%	0.70 0.50	减少 29%	1.28 1.08	减少 16%	−23%
纤曲 质量	CMC (OD) 纸崩 种类	0.85 卌 厚层通风曲		1.12 卌 厚层通风曲		0.84 卌 厚层通风曲		0.96 卌 厚层通风曲		0.97 卌 厚层通风曲		

以第二批 12 000 升发酵罐的大型发酵试验中酒度的测定数据为例(见表 6)。

表 6 第二批 12 000 升发酵罐发酵终了酒度测定的数据

样品号 酒度 处理	1	2	3	4	5	Σ	组内平均$\frac{1}{5}\Sigma$
对　照	8.99	9.06	9.04	8.94	9.08	45.11	9.02
加纤曲	9.16	9.26	9.12	9.18	9.05	45.77	9.15

* 全体总和 90.88，总平均 9.09。

由于试验误差引起的平方差和

对照为：$(8.99-9.02)^2+(9.06-9.02)^2+(9.04-9.02)^2+(9.08-9.02)^2+(8.94-9.02)^2=0.0129$

加纤曲为：$(9.16-9.15)^2+(9.26-9.15)^2+(9.12-9.15)^2+(9.18-9.15)^2+(9.05-9.15)^2=0.0240$

两项相加，用 S_2 表示，反映了试验误差的大小。

$S_2=0.0129+0.0240=0.0369$　自由度为 8

$S_2=0.0369\div8=0.0046$

添加纤曲后引起酒度改变，可用对照和加纤曲的平均酒度与总平均酒度差的平方和表示，由于取样 5 次，将此平方和 5 倍，以 S_1 表示。

$$S_1=5[(9.02-9.09)^2+(9.15-9.09)^2]=0.0425$$

自由度为 1 $\overline{S}_1=S_1$ 比较$\overline{S}_1$与$\overline{S}_2$的大小，可以看出加纤曲引起酒度提高的显著程度。

$$F=\frac{\overline{S}_1}{\overline{S}_2}=\frac{0.0425}{0.0046}=9.24$$

查 F 表：=0.10 时(置信度 99%)，F(1,8)=11.26

[F(1,4)=21.2

F(1,2)=98.50]

=0.05 时(置信度 95%),F(1,8)=5.32

[F(1,4)=7.71

F(1,2)=18.51]

=0.01 时(置信度 90%),F(1,8)=33.46

[F(1,4)=4.54

F(1,2)=8.53]

5.32<9.24<11.26

结论:添加 0.3%N_2-78 绿色木霉纤维素酶曲提高酒度的影响是显著的,如表 7 所示。

表 7 十批中、大型发酵试验酒度测定方差分析表

	1	2	3	4	5	6	7	8	9	10
$\overline{S}_1$	384	793.5	101.4	486	121.5	562.5	422.5	90	202.5	81
$\overline{S}_2$	56	10	95	6	13	49	46	18.5	47	10
自由度 f_1	1	1	1	1	1	1	1	1	1	1
自由度 f_2	4	4	4	4	4	8	8	8	8	2
F	6.86	79.35	10.67	81	9.35	11.48	9.19	4.87	4.31	8.1
显著性	*	***	**	***	**	***	**	*	*	—

*代表差异显著;**代表差异很显著;***代表差异特别显著。

四、结果讨论

纤维素酶应用于酒精、白酒发酵,以提高出酒率的试验报告,各地报道很多。过去,我们曾报道过野生种纤维素酶曲(1%~2%用曲量)能提高酒精出酒率,而且发酵快,发酵透,碘反映良好[1]。2 万立升大型发酵试验证明,2%纤维素酶曲(野生菌种 1096)能提高出酒率 1%~2%,诱变种 867 提高出酒率的效果优于野生菌种 1096[2]。诱变种 N_2-78 酶活又有了进一步的提高[4]。经过 40 多批小、中、大型发酵试验,并对试验数据进行误差分析,证明 0.3%的纤曲用量,即能稳定地提高山芋干(90%)和金刚刺(10%)混合原料的出酒率 1.5%左右,而且表现为加纤维素酶曲后,纤维素含量减少,发酵醪黏度下降的一致规律。

纤维素酶分解纤维素后的主要产物是葡萄糖,还有少量的纤维二糖[5],均能为酵母所利用。山芋干(90%)和金刚刺(10%)的混合原料约含纤维素 5%以上。若纤维素分解率平均以 20%计,则从理论上计算,可提高出酒率 1.7%左右,其次,纤维素酶可分解细胞壁,利于淀粉的释放和利用,有助于出酒率的提高。在原料淀粉利用率低,发酵不完全的情况下,此作用更为明显,这一点可从加酶后,发酵醪碘反应好,淀粉利用完全加以证明。看来,纤维素酶应用于纤维素含量高的野生植物,提高出酒率的效果可能更为显著。

经过多次小、中、大型试验比较,纤维素酶提高原料出酒有一定的变化幅度(1%~

2%，甚至以上），此可能与纤维素酶活力的高低、原料纤维素的含量多少，以及原料淀粉利用的好坏有关。诱变种 N_2－78 的酶系分析表明，该菌株除了能生产活力较高的纤维素酶以外，还能产生活力较高的果胶酶和半纤维素酶（内部资料），从而导致发酵醪黏度的下降。发酵醪黏度降低对酒精生产是很有利的，便于醪液输送，蒸馏时利于热交换，可以节省煤、电、水。

纤维素酶应用于酒精生产，提高出酒率，虽然目前还有一些问题，需要进一步解决，但我们认为，结果是肯定的，技术工艺是可行的，有可能在某些工厂较快的推广。无疑，这一技术工艺的推广应用对贯彻执行毛主席关于“广积粮”的指示具有深远的意义。按照目前提高出酒率 1.5%的水平，以我厂年产酒精 2.3 万吨计算，如在我厂全面推广，每年可节约粮食近 200 万斤。假如全国酒精生产单位都能推广应用，以年产 30 万吨酒精计算，每年可节约粮食 2 520 万斤，这是一个非常可观的数字。这对于贯彻执行毛主席的“广积粮”指示，具有深远的意义。

五、小结

（1）N_2－78 诱变菌种的优点。N_2－78 诱变种是一株纤维素酶活力较高的菌株，其纤维素酶活力较野生菌提高 1 倍以上，厚层通风制曲 66 小时，滤纸酶活（1∶35）可达到 卄 ～ 卅；CMC 酶活（1∶700）可达到 4 毫克葡萄糖/毫升酶。

（2）提高出酒率，节约粮食。经过 40 多批小、中、大型发酵试验，并对试验数据进行误差分析，证明 0.3%的用曲量，即能提高山芋干（90%）和全刚刺（10%）混合原料的出酒率 1.5%左右。提高出酒的原因主要是原料纤维素分解成糖变酒，其次是增加淀粉的释放和利用。

（3）降低黏度，有利于节煤、节电、节水。加纤维素酶曲后，原料纤维素含量减少，发酵旺盛，醪液黏度显著下降（2～4 倍）。便于醪液输送，蒸馏时利于热交换，可以节省煤、电、水。

（4）纤维素酶在酒精生产上的推广应用，变废为宝，节粮节煤，这对于贯彻执行毛主席的“广积粮”指示，具有深远的意义。

六、存在问题与今年打算

（1）继续开展菌种的分离复状培养等试验工作，不同批次发酵罐试验结果对比如表 8～表 17 所示。

（2）探讨适合于不同稻草下脚原料的制曲工艺。

（3）继续提高纤维素酶的活力和出酒率。

（4）开展纤维素酶液体曲的生产工艺的研究。

纤维素酶在酒精生产上应用的前途是无限的，道路是广阔的。让我们在深入学习无产阶级专政理论，普及大寨的热潮中，共同努力，团结战斗，一面推广现有的研究成果，一面开展探索新工艺、新途径的研究。使纤维素酶的应用迅速在更大范围内开花结果，为进一步的提高，打开新局面，闯出新路子，为“广积粮”做出更大的贡献。

表 8　第一批:3 000 升发酵罐发酵试验结果

项目 \ 处理 \ 时间	发酵初始								发酵终了							
	对照				加 0.3%纤曲				对照				加 0.3%纤曲			
	1	2	3	平均	1	2	3	平均	1	2	3	平均	1	2	3	平均
外观浓度(Bx)	17.5	17.5	17.5	17.5	17.5	17.5	17.5	17.5	0.70	0.70	0.73	0.71	0.67	0.67	0.67	0.67
黏度(秒)									21.8	20.4	21.2	21.1	6.0	6.4	6.2	6.2
酒度/%									9.00	8.94	8.82	8.92	9.10	9.12	9.02	9.08
固形物/%				19.70				20.00								

表 9　第　二　批

项目 \ 处理 \ 时间	发酵初始								发酵终了							
	对照				加 0.3%纤曲				对照				加 0.3%纤曲			
	1	2	3	平均	1	2	3	平均	1	2	3	平均	1	2	3	平均
外观浓度(Bx)				15.47				15.47	0.72	0.74	0.73	0.73	0.73	0.73	0.80	0.75
黏度(秒)									17.0	16.9	17.1	6.4	6.3	6.4	6.3	6.3
酒度/%									8.05	7.98	8.03	8.02	8.28	8.24	8.23	8.25
固形物/%				17.15				17.35				7.60				6.80
纤维素/%				1.37				1.44				1.53				1.30

表 10 第 三 批

项目 \ 处理 \ 时间	发酵初始								发酵终了							
	对照				加 0.3%纤曲				对照				加 0.3%纤曲			
	1	2	3	平均	1	2	3	平均	1	2	3	平均	1	2	3	平均
外观浓度(Bx)				17.00				17.00				1.07				1.14
黏度(秒)									40.2	40.9	41.2	40.7	10.2	9.8	9.9	10.0
酒度/%									7.94	8.09	8.00	8.01	8.20	8.20	8.40	8.27
残糖/%				4.2				4.2				0.50				0.56
总残糖/%												2.23				2.33
可溶物/%												1.42				1.42

表 11 第 四 批

项目 \ 处理 \ 时间	发酵初始								发酵终了							
	对照				加 0.3%纤曲				对照				加 0.3%纤曲			
	1	2	3	平均	1	2	3	平均	1	2	3	平均	1	2	3	平均
外观浓度(Bx)				18.00				18.10	0.90	0.85	0.80	0.85	1.00	0.90	0.90	0.93
黏度(秒)									30.4	30.6	31.7	30.9	10.1	10.3	10.3	10.3
酒度/%									9.32	9.36	9.30	9.33	9.49	9.51	9.52	9.51
固形物/%				18.40				18.30				6.06				6.00
纤维素/%				1.07				0.97				1.81				1.46
直接还原糖/%												0.50				0.50
总残糖/%												1.83				1.87
可溶物/%												1.20				1.25

表 12 第 五 批

时间 / 处理 / 项目	发酵初始								发酵终了							
	对照				加 0.3%纤曲				对照				加 0.3%纤曲			
	1	2	3	平均	1	2	3	平均	1	2	3	平均	1	2	3	平均
外观浓度(Bx)				16.75				16.75				—				—
黏度(秒)									28.9	29.2	30.3	29.4	11.0	11.2	11.3	11.2
酒度/%									8.86	8.86	8.84	8.85	8.96	8.97	8.88	8.94
固形物/%				18.48				18.49				6.60				6.30
纤维素/%				1.17				1.40				1.30				1.13
直接还原糖/%												0.47				0.47
总残糖/%												1.63				1.69
可溶物/%												1.49				1.49

表 13 第六批:12 000 立升发酵罐发酵试验结果

项目 \ 处理 \ 时间	发酵初始											
	对照						加 0.3%纤曲					
	1	2	3	4	5	平均	1	2	3	4	5	平均
外观浓度(Bx)						15.20						15.41
黏度(秒)						47.0						47.0
酒度/%												
固形物/%	15.80		15.92		15.88	15.87	16.06		16.08		15.70	15.94
纤维素/%	1.25		1.05		1.13	1.16	1.19		1.17			1.18
直接还原糖/%												
总残糖/%						12.66						12.66
可溶物/%												

项目 \ 处理 \ 时间	发酵终了											
	对照						加 0.3%纤曲					
	1	2	3	4	5	平均	1	2	3	4	5	平均
外观浓度(Bx)	0.77	0.72	0.73	0.73	0.73	0.73	0.67	0.87	0.83	0.77	0.87	0.80
黏度(秒)	21.1	20.6	20.8	21.1	21.4	21.0	6.8	6.6	6.6	6.9	6.9	6.8
酒度/%	7.76	7.66	7.62	7.82	7.72	7.72	7.94	7.80	7.83	7.86	7.92	7.87
固形物/%	6.00		5.88		6.04	5.97	5.96		5.82		5.90	5.89
纤维素/%	1.44		1.37		1.41	1.41	1.03		1.05		1.01	1.03
直接还原糖/%					0.29							0.34
总残糖/%					1.37							1.40
可溶物/%					0.42							0.48

表 14 第 七 批

项目＼处理＼时间	发酵初始							
	对照				加 0.3%纤曲			
	1	3	5	平均	1	3	5	平均
外观浓度(Bx)				17.48				17.37
黏度(秒)				34.5				15.4
酒度/%								
固形物/%	17.88	17.96	18.00	17.95	17.80	18.0	18.04	17.95
纤维素/%	0.99	1.15	1.01	1.05	0.89	1.08	1.02	1.00
直还原糖/%								
总残糖/%								
可溶物/%								

项目＼处理＼时间	发酵终了											
	对照						加 0.3%纤曲					
	1	2	3	4	5	平均	1	2	3	4	5	平均
外观浓度(Bx)	0.74	0.64	0.64	0.64	0.64	0.66	0.63	0.74	0.74	0.74	0.74	0.72
黏度(秒)	19.4	19.8	19.2	20.9	19.6	19.8	5.2	5.1	5.3	5.1	5.0	5.1
酒度/%	8.99	9.06	9.04	8.94	9.08	9.02	9.16	9.26	9.12	9.18	9.05	9.15
固形物/%	5.72		5.62		5.56	5.63	5.52		5.44		5.56	5.51
纤维素/%	1.07		1.08		1.09	1.08	0.86		0.82		0.83	0.84
直还原糖/%						0.28						0.34
总残糖/%						1.31						1.36
可溶物/%						—						—

表 15 第 八 批

时间／处理／项目	发酵初始								发酵终了											
	对照				加 0.3%纤曲				对照						加 0.3%纤曲					
	1	3	5	平均	1	3	5	平均	1	2	3	4	5	平均	1	2	3	4	5	平均
外观浓度(Bx)				17.97				18.15	0.41	0.41	0.41	0.41	0.41	0.41	0.51	0.51	0.51	0.41	0.55	0.50
黏度(秒)				33.1				28.5	13.9	12.3	12.4	13.1	14.0	13.1	6.3	6.4	6.4	6.3	6.4	6.4
酒度/%									9.38	9.45	9.42	9.38	9.36	9.40	9.45	9.49	9.39	9.46	9.52	9.46
固形物/%				18.54				18.47	6.20		6.28		6.24	6.24	6.00		6.06		6.00	6.02
纤维素/%				1.10				1.10	1.29		1.28		1.29	1.29	1.04		1.02		1.00	1.02
直还原糖/%														0.32						0.36
总残糖/%				14.76				13.69						1.36						1.49
可溶物/%														0.71						0.95

表 16 第 九 批

时间	发酵初始												发酵终了											
处理	对照						加 0.3%纤曲						对照						加 0.3%纤曲					
项目	1		3		5	平均	1		3		5	平均	1	2	3	4	5	平均	1	2	3	4	5	平均
外观浓度(Bx)						15.16						15.12						0.55						0.61
黏度(秒)													7.5	7.5	7.4	7.5	7.6	7.5	4.6	4.3	4.4	4.3	4.4	4.4
酒度/%													7.54	7.54	7.47	7.60	7.67	7.56	7.60	7.62	7.60	7.74	7.68	7.65
固形物/%	14.84		14.90		14.86	14.86	14.90		14.90		15.12	14.98			456		4.56	4.56	4.52		4.50		4.46	4.50
纤维素/%						0.63						0.60						0.70						0.50
直还原糖/%																		0.24						0.32
总残糖/%																		1.14						1.20
可溶物/%																		0.45						0.45

表 17 第 十 批

时间	发酵初始												发酵终了											
处理	对照						加 0.3%纤曲						对照						加 0.3%纤曲					
项目	1		2		3	平均	1		2		3	平均	1		2		3	平均	1		2		3	平均
外观浓度(Bx)	16.43		16.42		16.42	16.42	16.42		16.42		16.42	16.42	0.50		0.50		0.50	0.50	0.60		0.50		0.50	0.53
黏度(秒)	16.8		16.0		16.8	16.5	11.0		11.0		10.8	10.9	8.9		8.2		8.7	8.6	4.1		4.2		4.3	4.2
酒度/%													8.18		8.12			8.15	8.25		8.23			8.24
固形物/%	17.44		17.40		17.40	17.41	17.30		17.30		17.20	17.21	6.10		6.10		6.10	6.10	—		5.84		5.80	5.82
纤维素/%	1.19		1.29		1.26	1.25	1.19		1.14		1.34	1.22			1.26		1.30	1.28			1.10		1.06	1.08
直还原糖/%																		0.29						0.34
总残糖/%						12.74						12.74						1.31						1.36
可溶物/%																		0.69						0.73

参考文献

[1] 纤维素酶应用于酒精发酵的初步试验(1969年全国纤维素酶交流材料之四)
纤维素酶无锡协作组 上海东方红酒精厂 1969年11月
[2] 纤维素酶应用于酒精发酵的试验汇报(1973年全国酒精会议交流资料)
上海酒精二厂 1973年2月
[3] 绿色木霉变异株 EA_3 - 867 的获得及其产酶能力的提高
上海化纤六厂 上海酒精二厂 上海植生所纤维素酶组
引自“微生物育种学术讨论会文集(研究工作报告)”微生物育种学术讨论会编,科学出版社出版 (1975年)
[4] 1973年诱变育种工作小结(内部资料)
上海化纤六厂 上海酒精二厂 上海植生所纤维素酶协作组 1974年2月
[5] 绿色木霉变异株 R_8,EA_3 867 纤维素酶对几种纤维材料的糖化效果(1973年轻工部召开纤维素酶研究交流会交流材料)
上海植生所纤维素酶组 1973年

二株高活力纤维素分解菌 EA_3-867 和 N_2-78 的获得及其特性的比较

拟康氏木霉(*Trichoderma pseudokoningii Rifai*)野生型菌株 AS3.3002 和木-3,分别经多种物理化学因素(高能电子、^{60}钴、紫外线、亚硝基胍、硫酸二乙酯等)诱变处理后,得到二株变异菌株 EA_3-867 和 N_2-78。它们的固体曲、液体曲和酶制剂的各项纤维素酶活力测定结果,均比野生型菌株明显提高。N_2-78 菌株经摇瓶培养 60 小时,其 CMC 糖化力为 255 毫克葡萄糖/毫升酶液,滤纸糖化力为 8.2 毫克葡萄糖/毫升酶液,棉花糖化力为 13.4 毫克葡萄糖/毫升酶液,分别比野生型菌株提高 14.6 倍、5.3 倍和 7 倍。由 EA_3-867 和 N_2-78 的固体曲制取的纤维素酶粗制剂,其各项纤维素酶活力测定结果,均较 Onozuka R-10 纤维素酶制剂高。上述野生型菌株及变异菌株均具有果胶酶、半纤维素酶、淀粉酶和少量蛋白酶的活力。变异菌株中果胶酶的活力较野生型菌株高,淀粉酶和蛋白酶活力则较低。同时,变异菌株的形态也与原始菌株有明显差异。上述四株菌的孢子致死温度相同,CMC 糖化、滤纸崩溃和滤纸糖化的最适 pH 相近,最适温度相同,分别为 pH 4.4、60℃,pH 4.8、60℃和 pH 4.8、60℃。

在选育高活力纤维素分解菌时,有些作者采用诱变的方法选育纤维素酶活力高的木霉菌株[1-6]。我们曾报道过用各种物理化学因素处理木霉属菌株 AS 3.3002,得到一株纤维素分解活力较高的菌株 EA_3-867[7]。该菌株在破碎植物细胞壁等方面已被国内有关单位广泛使用[8,9]。后来,我们又以木霉属菌株木-3 为出发菌株,经高能电子和亚硝基胍处理,获得另一株纤维素分解活力更高的诱变菌株 N_2-78。

本文对上述几株木霉属菌的固体培养物、液体培养物(摇瓶),和由它们制成的酶制剂的纤维素酶等酶活力作了分析和比较;并比较了它们的菌落、菌丝和孢子形态,以及生长速度、纤维素酶性质等。试验表明,EA_3-867 和 N_2-78 是二株纤维素酶活力较高的优良菌株。

一、材料和方法

1. 菌种

出发菌株:野生型木霉 AS 3.3002 原编号为 1096 和木-3,由纤维素酶菌种筛选小分队供给,经鉴定,均属于拟康氏木霉(*Trichoderma pseudokoningii Rifai*)。

中国科学院上海植物生理研究所纤维素酶组,上海酒精二厂(朱雨生执笔)。此文发表于 1978 年“微生物学报”18 卷 1 期 27-38 页。

本文于 1977 年 6 月 30 日收到。

* 参加菌种诱变工作的还有上海化纤六厂。广东省微生物研究所邓庄等同志协助进行菌种鉴定和形态观察。

2. 培养基

1）斜面培养基

（1）马铃薯汁培养基（%）：马铃薯 20①，纤维素粉 2，琼脂 2。

（2）酵母膏蛋白冻培养基（%）：酵母膏 0.1，蛋白胨 0.2，纤维素粉 2，琼脂 2。

（3）麦芽汁培养基（%）：5°～7° Brix 麦芽汁，纤维素粉 2，琼脂 2。

（4）合成培养基（克）：每升溶液中含 $(NH_4)_2SO_4$ 1.4，KH_2PO_4 2.0，尿素 0.3（或 KNO_3 1.0），$MgSO_4 \cdot 7H_2O$ 0.3，$CaCl_2$ 0.3，$FeSO_4$ 0.005，$MnSO_4 \cdot H_2O$ 0.001 6，$ZnCl_2$ 0.001 7，$CoCl_2$ 0.002，纤维素粉 20，琼脂 20。

2）平板分离培养基

成分同上，另加 0.12%～0.15%胆酸钠，以限制菌落的扩散生长，便于菌落的分离。但经多次诱变后，菌落直径显著缩小。因此，用变异株为出发菌株时，可不加胆酸钠。

3）固体曲培养基

500 毫升三角瓶装 15～20 克稻草粉（或 9 份稻草粉加 1 份麸皮），加入 2.2～2.5 倍原料重量的水，水中含对原料计的 2.5%硫酸铵，于 30℃培养 3 天。

4）液体培养基

（1）种子培养基（%）：米糠 1，麸皮 0.5，$(NH_4)_2SO_4$ 0.2，KH_2PO_4 0.1，$CaCO_3$ 0.1，葡萄糖一次结晶母液（上海葡萄糖厂供给）。500 毫升三角瓶中装培养基 100 毫升，30℃于旋转式摇床（104 转/分）培养 32～35 小时，作二级培养的种子用。

（2）二级培养基：上述种子培养基中以 1%稻草粉代替米糠，其余相同。接种菌丝体，接种量 6%（V/V），30℃旋转式摇床培养 60 小时。

3. 诱变方法和菌种选育系谱

1）诱变程序

出发菌株⟶单孢纯化⟶斜面培养⟶分生孢子或萌发分生孢子悬浮液的制备⟶诱变处理⟶平板分离培养⟶活菌计数，并移接至斜面培养⟶三角瓶固体培养（初筛）⟶三角瓶固体培养（复筛）⟶斜面传代⟶三角瓶培养（第二次复筛）⟶斜面传代⟶优良菌株保藏。

2）菌种选育系谱

（1）EA_3 - 867：AS 3.3002 $\xrightarrow[\text{放线菌素 D}]{^{60}\text{钴}}$ 23K - 12 $\xrightarrow{\text{硫酸二乙酯}}$ D_2 - 54 $\xrightarrow{\text{紫外线}}$ U_{10} — 40 $\xrightarrow[\text{核苷酸}]{\text{紫外线}}$ R - 8 $\xrightarrow{\text{高能电子②}}$ EA_3 - 867。方法如前所述[7]。

（2）N_2 - 78：木- 3 $\xrightarrow[\text{(0.2 兆拉德)}]{\text{高能电子①}}$ EA_6 - 97 $\xrightarrow[\text{(400r/ml)}]{\text{亚硝基胍}}$ N_2 - 78 $\xrightarrow[\text{(液体培养筛选)}]{\text{单孢分离}}$ N_2 - 78。

30～32℃，处理 80 分钟。

① 每 200 克去皮马铃薯加蒸馏水煮沸半小时后，经纱布过滤，取滤液稀释至 1 000 毫升即成，然后，加纤维素粉和琼脂各 2 克配成。

② 由上海原子核研究所协助，用静电加速器产生的高能电子照射孢子悬浮液。

4. 酶制剂制备

培养 3 天的固体曲，加 10 倍重量的水浸泡 1 小时(30℃)，用尼龙布或合成纤维棉过滤，滤波低温离心 20 分钟(3 000 转/分)，上清液加 3 倍体积冷丙酮(0℃以下)。低温离心 30 分钟(3 000 转/分)，收集沉淀，真空干燥成粗酶制剂(丙酮粉)。

5. 酶活力测定

1) 各种酶液的制备

(1) 固体曲：称取培养 3 天的相当于 1 克干曲的鲜曲，加 35 毫升水，于 30℃左右浸泡 1 小时，离心 5 分钟(3 000 转/分)，上清液即为 1∶35 的固体曲酶抽提液。本文除特别说明外，固体曲酶抽提液，均指此种液体。

(2) 液体曲：培养 60 小时的液体曲，离心 5 分钟(3 000 转/分)，上清液即为液体曲酶液。

(3) 酶制剂：将酶制剂配成 1%、0.5%或 0.1%的水溶液。

2) 酶活力测定方法

(1) CMC 酶解活力。

① CMC 糖化力(以还原糖表示)。

A 法：0.5 毫升适当稀释的酶液，加 2 毫升 0.625%羧甲基纤维素钠盐溶液①(简称 CMC 溶液，原品由上海长虹塑料厂出品，黏度(PM_6)300～600 厘泊。)，50℃水浴中反应 30 分钟，加 2.5 毫升 DNS 试剂终止反应，煮沸 5 分钟，于 540 毫微米波长下比色，由比色读数换算成每毫升酶液作用半小时后产生的葡萄糖毫克数。

B 法：基本上按 Mandels 的方法[10]，0.5 毫升适当稀释的酶液，0.5 毫升 1% CMC 溶液(溶于 0.05 mol/L 柠檬酸-柠檬酸钠缓冲液，pH 5.0)，50℃反应 30 分钟，加 1 毫升 DNS 试剂煮沸 5 分钟，于 540 毫微米波长下比色，由比色读数换算成每毫升酶液作用半小时后产生的葡萄糖毫克数。规定在上述条件下，每分钟产生 1 微克分子葡萄糖为 1 个酶活力单位。

② CMC 液化力[11](以黏度表示)。

2.5 毫升适当稀释的酶液，加 2.5 毫升 0.2 mol/L pH 5.0 的醋酸缓冲液和 5 毫升 1% CMC 溶液，30℃反应 10 分钟，用黏度计测定并求出反应前后反应液的比黏度，按下式求出 CMC 酶活力：

$$\text{CMC 酶活力}=\left(\frac{1}{\text{反应后比黏度}}-\frac{1}{\text{反应前比黏度}}\right)\times\frac{1\,000}{\text{反应时间(分)}}$$

(2) 滤纸酶解活力。

① 滤纸崩溃法(静止法)。

试管(15 厘米×1.5 厘米)内加 4 毫升酶液(固体曲酶提取液或液体曲原液)及 1 毫升柠檬酸(0.05 mol/L)-磷酸氢二钠(0.1 mol/L)缓冲液(pH 4.8)，再投入一张 1 厘米×3 厘米新华 1 号(快速)滤纸，50℃静止保温 1 小时后，观察滤纸崩溃程度。

② 滤纸糖化力。

A 法(静止法)：吸取前法所得反应液 0.5 毫升，用 DNS 试剂定糖，扣去空白后，换算成每毫升反应液在 1 小时产生的葡萄糖毫克数(简称滤纸糖)。

---

① 500 毫升 1% CMC 溶液，加 100 毫升 0.05 mol/L 柠檬酸-0.1 mol/L 磷酸氢二钠缓冲液和 200 毫升水即成，pH 4.4。

B法(静止法)：基本上根据 Mandels 的方法[10]，试管(15 厘米×1.5 厘米)中加 0.5 毫升酶液及 1 毫升柠檬酸缓冲液(0.05 mol/L，pH 4.8)，试管底部预先竖放一张卷曲的 1 厘米×6 厘米的 Whatman 1 号滤纸(约重 50 毫克)，50℃，静止保温 1 小时，用 DNS 试剂法测定形成的还原糖，扣去空白后，换算成每毫升酶液产生的葡萄糖毫克数(简称滤纸糖)。

C法(震荡法)：反应液同前述 B法，但反应在震荡器上震荡(80～90 次/分，振幅 5 厘米)进行。

③ 滤纸崩溃法(震荡法)。

25 毫升容量瓶中注入 4 毫升酶液(酶制剂为 1%浓度)及 1 毫升柠檬酸(0.05 mol/L)-磷酸氢二钠(0.1 mol/L)缓冲液(pH 4.8)，并放入 2 张 1 厘米×1 厘米 Whatman 1 号滤纸，于震荡器上震荡(80 次左右/分，振幅 5 厘米)，观察二张滤纸全部崩溃成糊状所需的时间。再按外山法[12]计算活力单位。

(3) 棉花糖化活力[10]。

带塞试管(11 厘米×1.5 厘米)中，加 1.0 毫升酶液(不稀释)和 1 毫升柠檬酸缓冲液(0.05 mol/L，pH 4.8)，及 50 毫克脱脂棉，于 50℃水浴中，静止保温 24 小时，用 DNS 法测定形成的还原糖，扣去空白后，换算成每毫升酶液作用(24 小时)产生的葡萄糖毫克数(简称棉花糖)。

(4) 果胶酶活力。

① 黏度法：0.5 毫升 1%浓度的酶制剂溶液，加 4.5 毫升水及 5 毫升 1%的果胶溶液(苹果果胶，B. D. H 产品，溶于 0.2 mol/L，pH 5.0 的醋酸缓冲液中)，30℃恒温水浴中反应 10 分钟，测定反应前后黏度的变化，求出相对黏度和比黏度，换算成每毫升酶液的活力单位数。本法不适用液依曲酶液。

② 还原糖法：2 毫升适当稀释的酶液，加 0.5 毫升 0.5%的多聚半乳糖醛酸溶液(Light 产品，溶于 0.1 mol/L，pH 4.5 醋酸缓冲液中)，30℃，反应 30 分钟，用 DNS 法测定形成的半乳糖醛酸，扣去酶液空白，换算成每毫升酶液(半小时)作用形成的半乳糖醛酸毫克数。

(5) 半纤维素酶[13]。

以玉米芯中提取的半纤维素为底物，用甲基间苯二酚法测定酶解形成的木糖，换算成每毫升酶液作用(半小时)产生的木糖毫克数。

(6) 淀粉酶活力。

0.5 毫升适当稀释的酶液，加 2 毫升 1%的可溶性淀粉(溶于 0.02 mol/L，pH 4.8 醋酸缓冲液中)，50℃反应 30 分钟，用 DNS 法测定形成的还原糖，扣去空白，换算成每毫升酶液作用(半小时)形成的葡萄糖毫克数。

(7) 蛋白酶[13]。

以酪蛋白为底物，用福林试剂测定形成的酪氨酸，规定每分钟形成 1 微克酪氨酸为 1 个活力单位(u)。

6. 可溶性蛋白质[14]

以牛血清蛋白为标准，用福林试剂测定。

二、结果

1. 诱变菌株和野生菌株的形态比较

在麦芽汁培养基上，野生型菌株 AS 3.3002 和木-3 的菌丝伸展迅速，菌苔平整，绵毛状。

初浅绿,后变为深橄榄色,背面蜜黄色。匍匐菌丝粗,直径达 9～10 微米,充满颗粒状细胞质。直立气生菌丝较细,约 3 微米,有些扭曲成螺旋状。厚垣孢子稀少,多生于菌丝间,偶有顶生者,近球形,无色,直径 6～10 微米。分生孢子梗长,主枝直径 4.5～5 微米,顶端着生小梗或短侧枝中下部有侧枝,侧枝长短不一,排列不规则,对生,互生,间隔距离不等,瓶梗生于侧枝及主枝上端,多呈直角,排列不规则,单生、互生或双对生,形似酒瓶,但基部明显内缩,而呈纺锤状或梭形,有时向主轴方向弯曲,长 4.5～6 微米,直径 2.5～3.0 微米,有的可长达 10～12 微米。分生孢子大多呈椭圆形或长椭圆形,稍有棱角。电子显微镜观察,孢子纵向有明显削面,孢子顶端圆,基部往往平截,大多(3.5～4.5)微米×(1.8～2.5)微米,最大的可达(5.0～5.8)微米×(2.5～3.0)微米,但长度不超过 6.0 微米。孢子大小和形态变化较大(尤其是木- 3)。这两株菌应属于 Rifai 分类系统中的拟康氏木霉(*Trichoderma pseudokoningii*, *Rifai*)。

变异株 EA_3 - 867 在麦芽汁培养基上较原始菌株 AS 3.3002 生长慢,菌苔较薄,产生孢子少,但分生孢子梗的分枝较多而密,直接着生于主枝的小梗少。分生孢子梗的主梗及侧枝往往弯曲,侧枝稍长。N_2 - 78 与木- 3 比较,生长速度明显降低,菌落变小,菌苔薄,孢子产生慢且少,分生孢子梗的分枝稍多而密,长椭圆形及近圆柱形的分生孢子比例增多,袍子长度增加,可达(6.5～7.4)微米×(1.8～2.5)微米。AS 3.3002 和 EA_3 - 867 菌落反面均为浅绿黄色,木- 3 和 N_2 - 78 则均为蜜黄色。

四株菌在马铃薯-葡萄糖培养基上培养三天后的形态比较如表 1 所示。可见野生型菌株和变异株在菌落形态、生长速度、孢子形成等方面有明显的差异。变异菌株的菌落明显缩小,生长变慢,孢子形成迟而稀少(见图 1)。其中 N_2 - 78 菌落最小,生长最差,孢子最少。EA_3 - 867 的菌落颜色比 AS 3.3002 略淡,木- 3 为草绿色。N_2 - 78 为黄绿色,在马铃薯-纤维素粉或合成培养基-纤维素粉琼脂平板上,EA_3 - 867 和 N_2 - 78 菌落颜色都较野生型菌株浅。诱变菌株在以上几种培养基平板上经常分泌较多的黄色色素,色素的形成与氮源种类无关(硝态氮、氨态氮或有机氮),在液体培养滤液中也常有黄色色素加深的现象,此结果与 Norkrans[15] 报告相似。尚不清楚黄色色素与酶活力的关系。还可看到菌丝、小梗和孢子形态上的一些变化(见表 2): N_2 - 78 菌株较木- 3 菌株菌丝加粗,小梗伸长,孢子变长和变大。同在麦芽汁培养基上一样,经常见到 N_2 - 78 有较长的柱形袍子,EA_3 - 867 有时也有。它们的电子显微镜照片如图 2 所示,这种柱形的孢子是否与纤维素酶活力的提高有关值得进一步研究。AS3.3002 经诱变后,菌丝也加粗,但小梗缩短,光学显徽镜和电子显微镜观察都表明四株菌株的分生孢子壁是光滑的,没有小疣状突起,在显微镜下单个孢子看不出有颜色,成堆时呈绿色。根据以上孢子特征,AS3.3002 和木- 3 不应归入绿色木霉。

四株菌株的分生孢子悬浮液经不同温度处理后,在马铃薯—葡萄糖琼脂平板上,于 30℃ 培养 3 天,观察孢子存活情况,结果表明经 80℃以上处理后,不能萌发生长。

2. 变异菌株和野生型菌株的纤维素酶活力比较

1) 固体曲

在稻草粉固体培养基上,与野生菌株比较,EA_3 - 867,尤其是 N_2 - 78 的菌丝生长差,孢子形成慢而少。比较不同菌株在稻草粉麸皮固体培养基上生长 3 天的酶活力表明(见表 2),变异株比野生型菌株,无论是 CMC 酶解活力,滤纸酶解活力、棉花糖化活力均有不同程度的提高,其中以 CMC 酶解活力增加最多(2～3 倍),其次是滤纸酶解活力(震荡法,1 倍多),静止法测定的滤纸糖和棉花糖增加量较少;可溶性蛋白质略有增加。从两株野生型菌株来看,木- 3

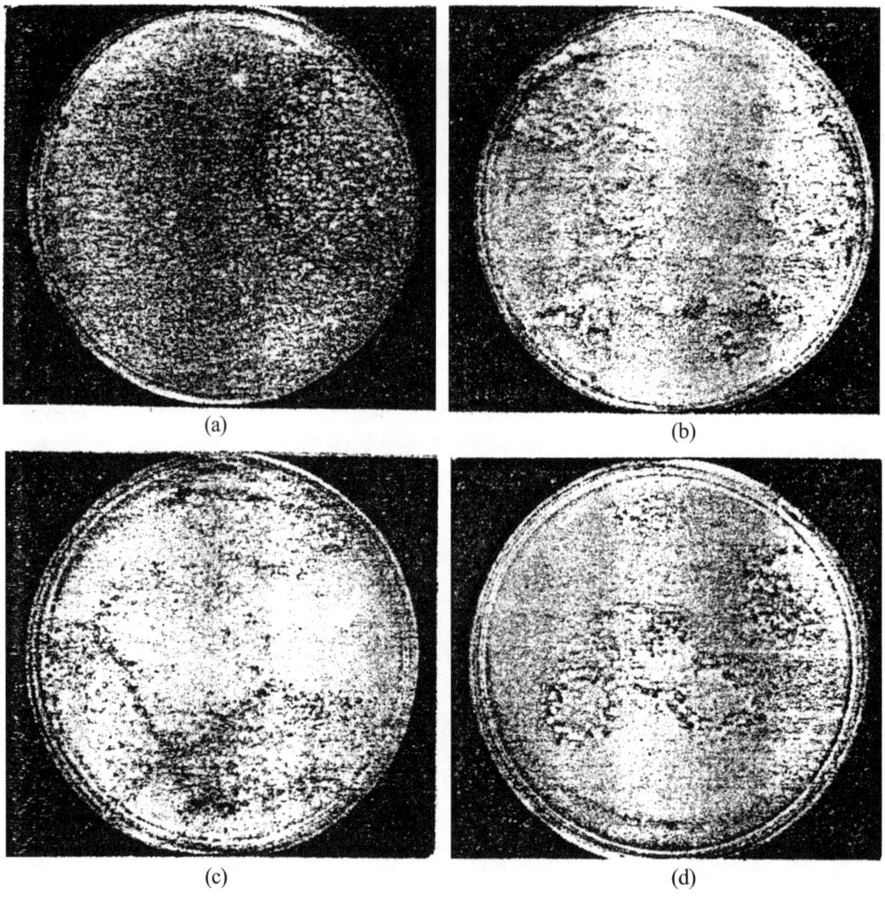

图 1 菌落形态(在合成琼脂平板上 30℃ 培养 4 天)

(a) AS3.3002；(b) EA_3 - 867；(c) 木-3；(d) N_2 - 78

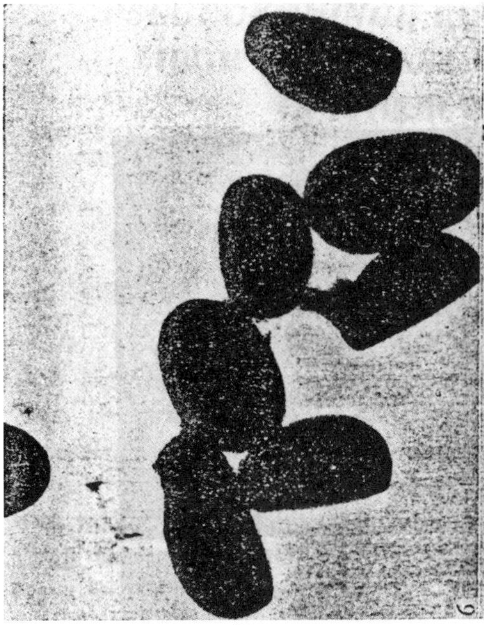

(a)

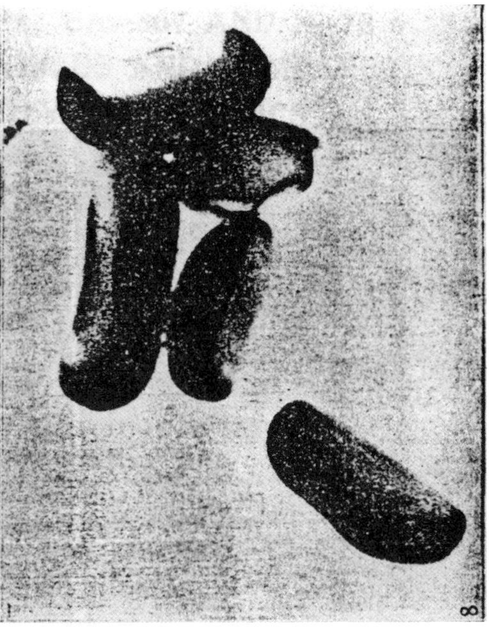

(b)

(c)

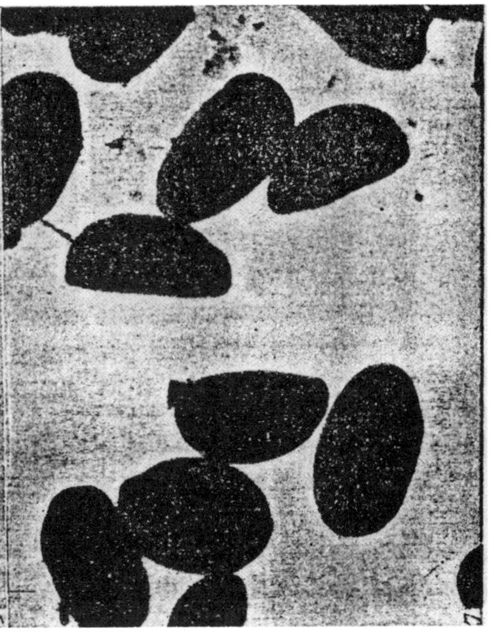

(d)

图 2　孢子及菌丝的电子显微镜照片

(a) AS3. 3002 的孢子及菌丝形态(4 800×)；(b) EA_3－867 的孢子形态(4 800×)；(c) 木-3 的孢子形态(4 800×)；(d) N_2－78 的孢子形态(4 800×)

表 1 AS 3.3002、EA_3－867、木－3 和 N_2－78 的形态比较*

形态描述 / 菌种	菌落				菌丝					分生孢子小梗				分生孢子				颜色	
	直径/cm	生长孢子形成	菌落颜色	色素分泌	形态	基质菌丝分枝角度	气生菌丝分枝角度	直径/μm	分枝方式	形状	分枝方式	长/μm	宽/μm	形状	长/μm	宽/μm	孢壁	单个	成堆
AS 3.3002	4.8	快	深绿	无色	较挺直	直角	锐角	1.5～3.5	互生	瓶形，基部稍窄，中部宽，顶部窄，稍弯曲	互生	8.6～17.0	1.5～2.5	卵形、倒卵形	3.5～5.0	2.0～3.5	光滑	无色	绿色
EA_3－867	0.9	慢	绿	淡黄	较挺直	直角	锐角	3.5～5.0 (7.0)	互生	瓶形，基部稍窄，中部宽，顶部窄，稍弯曲	互生	5.0～12.0	1.5～3.5	卵形、倒卵形、少量柱形	3.5～5.0	2.0～3.5	光滑	无色	绿色
木－3	3.5	快	草绿	淡黄	扭曲	锐角	锐角	1.5～3.5 (5.0)	互生	瓶形，基部稍窄，中部宽，顶部窄，稍弯曲	互生	5.0～10.0	1.5～2.5	卵形、倒卵形	3.5～5.0	2.0～3.5	光滑	无色	绿色
N_2－78	0.8	慢	黄绿	深黄	较挺直	直角	锐角	3.5～5.0	互生	瓶形，基部稍窄，中部宽，顶部窄，稍弯曲	互生	10.0～17.0	1.5～3.5	较多、柱形少量卵形倒卵形	3.5～7.0 (8.5)	2.0～3.5	光滑	无色	绿色

* 培养在马铃薯、葡萄糖琼脂平板培养基上。

表 2 四株菌的固体曲纤维素酶活力比较*

测定项目 / 酶活力 / 酶的来源	CMC 酶解活力					滤纸酶解活力				棉花糖化力/(毫克糖/毫升酶)	可溶性蛋白质/(毫克/毫升酶)
	CMC 糖化力/(毫克糖/克曲)		CMC 糖化力/(毫克糖/毫升酶)		CMC 液化力/(单位/毫升酶)	滤纸崩溃法（静止法）	滤纸糖化力(毫克糖/毫升酶)				
	A 法	B 法	A 法	B 法			A 法（静止法）	B 法（静止法）	C 法（振荡法）		
1096	1 332	1 610	43	44	565	⊥**	0.33	1.92	2.40	7.09	1.25
EA_3－867	3 960	4 900	113	140	1 086	＋＋	0.57	2.63	5.59	12.60	1.45
木－3	1 198	1 417	33	41	386	⊥	0.39	1.54	3.26	9.35	1.27
N_2－78	4 943	5 435	143	155	1 302	＋＋＋	0.72	3.08	6.95	13.70	1.68

* 培养基：见正文。

** 滤纸膨胀软化为＋；滤纸破碎断裂，但仍成形为＋＋；滤纸完全崩溃，呈糊状为＋＋＋。⊥代表滤纸崩溃程度比“＋”差。

的 CMC 活力低于 AS 3.3002，但滤纸活力和棉花糖均高于 AS 3.3002，可能暗示木-3 的 C_1 酶活力高于 AS 3.3002，两株变异菌株中，N_2-78 的纤维素酶对各种底物的酶解活力和可溶性蛋白质含量都比 EA_3-867 的要高。

2) 液体曲(摇瓶)

四株菌株震荡培养 60 小时，培养滤液在外观上有明显的差异，AS 3.3002 呈淡黄色，木-3 呈灰绿色，EA_3-867 呈黄色，N_2-78 呈清澈透明的亮黄色。在培养过程中，菌丝均结成小球，但结球形状、大小、结构都有不同。AS 3.3002、木-3 和 EA_3-867 的菌丝成球较大而松散，有时聚成条状或块状，但 N_2-78 菌丝成球，小而紧密，成鱼籽状。

四株菌株的纤维素酶活力有明显差异。AS3.3002 和木-3 经诱变后，纤维素酶对各种底物的酶解活力都大幅度增长(见表 3)。其中以 N_2-78 的活力最高。液体曲与固体曲相比，野生型菌株各项纤维素酶活力都低于固体曲酶液的活力，而变异株均明显高于固体曲酶液的活力，故变异株在液体培养中活力提高倍数均比固体培养物中高得多。摇瓶中 N_2-78 生长 60 小时，与同样培养的木-3 相比，CMC 活力，滤纸糖(震荡法)增加 10 多倍；滤纸糖(静止法)和棉花糖也分别增加了 7 倍多和 6 倍。本液体培养条件可能更适合这二株变异株(尤其是 N_2-78)产生纤维素酶。此外，在培养滤液中还测定得到较高的果胶酶、半纤维素酶，一定的淀粉酶活力，以及少量的蛋白酶活力。果胶酶活力在变异株中也有所提高，其余几种酶都略有减少。

3) 酶制剂

生长 3 天的固体曲制成丙酮粉。各种酶活力测定的结果列于表 4。变异菌株中纤维素酶活力增加的情况基本上同固体曲类似。CMC 糖化力、CMC 液化力，滤纸糖(震荡法)增加较多，滤纸糖(静止法)和棉花糖增加较少。由四株菌制成的酶制剂中，N_2-78 各项纤维素酶活力仍为最高。

诱变后，除纤维素酶活力显著提高外，可溶性蛋白、果胶酶和半纤维素酶也有一定增加，但淀粉酶和蛋白酶，同液体曲一样，由变异株制成的酶制剂中，活力都减低。

由变异株制成的固体曲、液体曲和酶制剂中，各种酶活力与原始菌株比较如表 5 所示。

由 EA_3-867 和 N_2-78 菌株制成的固体曲酶制剂与 Onozuka R-10 纤维素酶制剂，在相同条件下进行各种活力比较，结果如表 6 所示。除果胶酶活力外，其他纤维素酶活力，由 EA_3-867 和 N_2-78 菌株制成的固体曲酶制剂均高于 Onozuka R-10。

3. 变异菌株和野生型菌株纤维素酶特性的比较

分别测定了由四株菌制成的固体曲的酶液中纤维素酶反应最适条件。图 3 和图 4 是 AS3.3002 菌株的 CMC 糖化力和滤纸糖化力的 pH 曲线和温度曲线。其余三株菌的酶反应 pH 曲线的温度曲线与此完全一样，唯在活力上，总是变异菌株的显著高于野生菌株。

1) CMC 糖化力

几株菌的纤维素酶反应最适 pH 都是 4.4(50℃反应)，但最适 pH 与反应温度有关，如 N_2-78 的纤维素酶在 30℃时最适 pH 为 5.0，40℃时为 4.8，50℃时为 4.4，60～75℃时为 4.0～4.4。用 CMC 糖化力测定的纤维素酶耐热性较高，从 30℃开始，随着温度升高，糖化力迅速增加，最适温度均为 60℃，70℃以上酶活力显著下降。这种酶的耐热性又与反应 pH 有关，即在不适合的 pH 条件下，酶易因温度升高而失活，如 N_2-78 纤维素酶的 CMC 糖化力在 pH 2 或 7 时，如温度在 40℃以上，便开始降低；pH 3 或 6 时，温度 50℃以上，活力开始降低，而在较适宜的 pH 4 与 5 之间时，温度 60℃以上，酶活力才开始降低。

表 3　四株菌的液体曲(摇瓶)酶活力比较

测定项目 / 酶活力 / 酶的来源	CMC 酶解活力			滤纸酶解活力				棉花糖/(毫克糖/毫升酶)	果胶酶活力/(毫克糖/毫升酶)	半纤维素酶活力/(毫克糖/毫升酶)	淀粉酶活力/(毫克糖/毫升酶)	蛋白酶活力/(毫升酶)	可溶性蛋白含量/(毫克糖/毫升酶)
	CMC 糖化力/(毫克糖/克酶)		CMC 液化力/(毫升酶)	滤纸崩溃(静止)	滤纸糖/(毫克糖/毫升酶)								
	A 法	B 法			A 法 静止法	B 法 静止法	C 法 震荡法						
AS3. 3002	18. 4	15. 3	405	⊥$^{-}$**	0. 08	1. 02	0. 76	1. 5	0. 48	7. 3	5. 5	3. 6	0. 96
EA_3 - 867	130. 0	97. 0	1378	++	0. 50	3. 60	3. 62	9. 9	0. 53	6. 0	4. 8	2. 1	1. 65
木- 3	23. 8	17. 4	230	⊥	0. 10	1. 54	0. 88	1. 9	0. 48	5. 8	5. 8	2. 7	1. 00
N_2 - 78	288. 0	255. 0	4056	+++$^{+}$	0. 92	8. 20	10. 60	13. 4	0. 56	4. 5	4. 5	1. 8	2. 40

* 培养条件：见正文。
** 符号说明见表 2。右上角的“+”或“－”号是表示同一等级中较强或较弱的。

表 4　四株菌的纤维素酶粗酶制剂酶活力比较

测定项目 / 酶活力 / 酶的来源	CMC 酶解活力			滤纸酶解活力				棉花糖化力/(毫克糖/毫升酶)	果胶酶		半纤维素酶活力/(毫克糖/毫升酶)	淀粉酶活力/(毫克糖/毫升酶)	蛋白酶活力/(毫升酶)	可溶性蛋白含量/(毫克/毫升酶)
	CMC 糖化力/(毫克糖/克酶)		CMC 液化力(单位/毫升酶)	滤纸崩溃(静止)	滤纸糖化力/(毫克糖/毫升酶)				黏度法/(毫升酶)	还原糖法/(毫克糖/毫升酶)				
	A 法	B 法			A 法(静止)	B 法(静止)	C 法(震荡法)							
AS3. 3002	21 500	15 250	1 435	+**	0. 52	3. 46	4. 44	32. 5	162	1. 10	4. 68	11. 8	5. 6	2. 6
EA_3 - 867	80 000	38 000	3 188	+++$^{+}$	0. 64	5. 40	11. 85	46. 0	322	1. 24	4. 94	7. 2	4. 6	3. 7
木- 3	22 000	16 000	1 088	++	0. 82	3. 76	5. 82	41. 0	93	1. 17	4. 94	10. 0	5. 0	2. 5
N_2 - 78	108 000	69 000	3 617	+++$^{+}$	0. 96	7. 00	18. 00	52. 5	118	1. 15	8. 32	5. 2	2. 6	4. 5

* 粗酶制剂指丙酮粉。CMC 液化力。滤纸活力、棉花糖和可溶性蛋白测定用 0. 5%浓度酶制剂；果胶酶、半纤维素酶、蛋白酶用 1%浓度酶制剂测定。丙酮粉得率(对干曲)为 5%左右。
** 说明见表 2。

表 5　变异株 EA_3－867 和 N_2－78 与野生型菌株 AS3.3002 和木－3 的各种酶制剂活力比较

比较内容 \ 酶的来源 \ 测定结果 \ 测定项目		CMC糖化力 A法	CMC糖化力 B法	CMC液化力	滤纸糖化力 A法静止法	滤纸糖化力 B法静止法	滤纸糖化力 C法震荡法	棉花糖化力	果胶酶 黏度法	果胶酶 还原糖法	半纤维素酶	淀粉酶	蛋白酶	可溶性蛋白质
EA_3－867 比 AS3.3002 增加的倍数	固体曲	2.0	2.0	0.9	0.7	0.4	1.3	0.8						0.2
	酶制剂	2.7	1.5	1.2	0.3	0.6	1.7	0.4	1.0	0.1	0.1	－0.4*	－0.2	0.4
	液体曲	6.0	5.3	2.4	2.6	2.5	3.6	5.6		0.1	－0.2	－0.2	－0.4	0.7
N_2－78 比木－3 增加的倍数	固体曲	3.1	2.8	2.4	0.9	1.0	1.1	0.5						0.3
	酶制剂	3.9	3.3	2.3	0.2	0.9	2.1	0.3	0.3	0	0.7	－0.5	－0.5	0.8
	液体曲	11.1	13.6	16.6	7.3	4.3	11.0	6.0		0.2	－0.3	－0.2	－0.3	1.4

*“－”代表活力降低的倍数。

表 6　EA_3－867 和 N_2－78 纤维素酶制剂与酶制剂 Onozuka R－10 的活力比较

酶来源 \ 测定结果 \ 测定项目	CMC糖化力/(毫克糖/毫升酶)	CMC液化力/(毫升酶)	滤纸崩溃活力(震荡法) 崩溃时间/分钟	滤纸崩溃活力(震荡法) /毫升酶	滤纸糖化力/(毫克糖/毫升酶)	棉花糖化力/(毫克糖/毫升酶)	果胶酶(还原糖法)/(毫克酶/毫升酶)	可溶性蛋白质/(毫克/毫升酶)
Onozuka R－10	210	2 810	26	1 440	2.08	1.80	2.6	3.8
EA_3－876	520	4 500	17	2 200	2.40	2.80	2.6	4.9
N_2－78	700	6 391	14	2 680	2.88	3.12	2.6	5.4

注：以上测定除滤纸糖化力和棉花糖化力是用 0.1%浓度酶制剂外，其余都是用 1%浓度。CMC 糖化力用 B 法，滤纸糖化力用 B 法(静止法)。

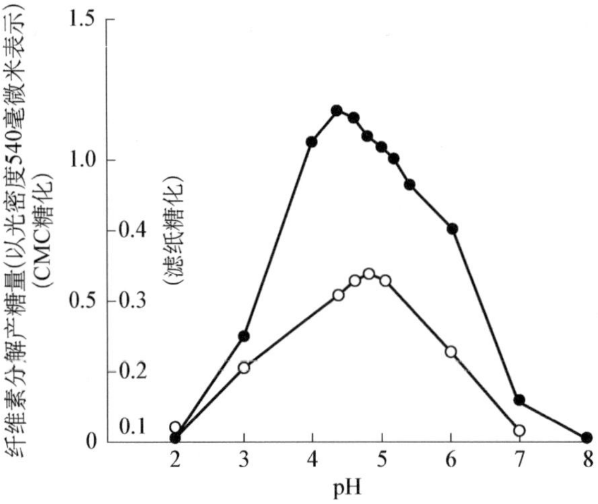

图 3　AS 3. 3002 菌株纤维素酶 CMC 和滤纸糖化活力的 pH 曲线

注：○——○ 滤纸糖化活力；●——● CMC 糖化活力

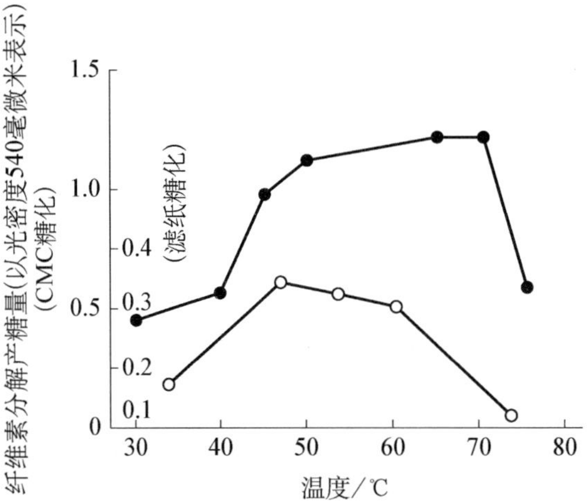

图 4　AS 3. 300 2 菌株纤维素酶 CMC 和滤纸糖化活力的温度曲线

注：○——○ 滤纸糖化活力；●——● CMC 糖化活力

2）滤纸糖化力(A 法)

四株菌株的纤维素酶的最适 pH 均为 4. 8(50℃反应)，最适温度均为 50℃。

3）滤纸崩溃法(静止)

四株菌株的纤维素酶的最适 pH 均为 4. 8(50℃反应)，最适温度均为 60℃。

三、讨论

当前，纤维素酶在应用上的主要问题之一，是要获得活力高的纤维素酶菌种[16]。木霉属是迄今被认为纤维素酶系成分最全面、分解天然纤维素活力最高的一类菌[10,17,18]。但是，对于木霉属的分类过去曾经相当混乱[19]。开始，有人把木霉全部归入绿色木霉(*Trichoderma viride Pers. ex Fries*)一种；但较多的研究者倾向于分成几种，如康氏木霉(*Trichoderma koningi Oud*)和绿色木霉二种，或白木霉(*Trichoderma album Preuss*)，黄绿木霉(*Trichoderma glaucum Abbott*)，康氏木霉和木素木霉(*Trichoderma lignorumTode ex Harz*)。最近，Rifai 根据大量的工作，将木霉属分成 9 个种群[20]。AS 3. 3002 和木- 3 菌株类似 Rifai 分类系统中的长梗木霉或拟康氏木霉，但从分生孢子梗排列较紧密，小梗基部的明显收缩，以及孢子长度不超过 6 微米等几个特征来看，应属于拟康氏木霉。

Mandels 等将野生型菌株 $QM6_a$ 经高能电子照射处理后，分离得到一株纤维素酶活力比原始菌株高二倍的 QM9123[1]。以后，通过进一步诱变处理，得到活力更高的 QM9414[2,17]。国内也有人用紫外线和亚硝酸[5]、^{60}钴[6]等照射木霉，得到纤维素酶活力提高的菌株。得到的 EA_3 - 867 和 N_2 - 78 菌株，纤维素酶活力均有较大幅度的提高，且在使用 4 年多的过程中活力稳定。看来，亚硝基胍、高能电子以及 γ-射线、紫外线可能是木霉属菌种较有效的诱变因子。同时，多次物理化学因素的复合处理，往往可以累积正向变异，使活力逐步提高。出发菌株的选择也有一定意义，如我们曾进一步用物理化学方法诱变 EA_3 - 867，但纤维素酶活力增加不多，要进一步增加酶活力较困难。后来改用纤维素酶活力与 AS 3. 3002 差不多的木- 3 菌株(仅 C_1 酶活力较高)进行诱变，得到 N_2 - 78，其纤维素酶活力提高较多。我们还看到菌落形态、长势与纤维素酶活力间有一定的相关性。就现有资料看，纤维素活力高的菌株都是生长差、菌落小的。其他研究者也观察到类似的现象[1,5]，故挑取小菌落的办法看来可以作为挑选高产变异株的一种参考。

关于纤维素酶活力的测定方法，至今没有统一，美国 Mandels 等主要测定其 CMC 糖化力、滤纸糖化力和棉花糖化力[10,18]。日本外山等主要测定其滤纸崩溃活力[12]，Yamane 等是测定其 CMC 糖化力和液化力[22]。国内许多单位分别采用了上述各种方法，并作了各种修改，因而使方法更加多样化，这造成了无法相互比较活力的情况。方法不统一的原因，首先是因为纤维素酶是一个复合酶，各个酶组分并没有完全搞清楚，对 C_1 和 C_x 概念、看法和作用尚有争论[16,23]；其次是测定方法上的困难。因为纤维素酶作用的底物纤维素是固体的，酶与底物接触以及反应产物扩散所受的限制，常常影响酶活力的表现。在试验中，变异菌株的酶液，用滤纸和棉花为底物，用静止保温的方法测定时，酶活力不能充分表现，而改用震荡法或可溶性 CMC 为底物时，活力得到充分表现，就是这个原因。不过可溶性 CMC 仍是部分可溶的，且纤维素酶对它的分解活力受其中羧甲基取代度的影响。在测定酶活时，要求底物浓度过量，但由于酶对固体底物的作用受到限制，实际上底物总不会有效的过量，因而，所有测定结果都大大受酶液稀释度的影响。只有调整酶液稀释度，使每个测定中反应产物量(即还原糖)相近进行酶活力比较，才是比较合理的[10,18]。本工作中，纤维素酶活力的测定采用了比较多的方法，是为了对变异菌株的纤维素酶活力水平作多方面的考查。CMC 糖化力主要代表外切 β1，4 -葡聚糖酶的活力，CMC 液化力主要代表内切 β1，4 -葡聚糖酶的活力，滤纸崩溃法和棉花糖化力则相对代表 C_1 酶活力，滤纸糖化力代表“纤维素糖化”酶活力[10,12,17,18]。看来，各种方法之间有很好的相关性。在实际使用时，可挑选有代表性的几种方法，如 CMC 糖化力(B 法)、滤纸糖化力(B 法)、滤纸崩溃法(静止)或棉花糖化力。

据报道[16]，QM9414 是迄今国际上液体培养时纤维素酶活力最高的菌株。我们采用他们所用的测定方法，表明 N_2 - 78 菌株纤维素酶活力，在 CMC 糖化力(47 单位/毫升酶)、滤纸糖化力(8. 20 毫克/毫升酶)、棉花糖化力(13. 4 毫克糖/毫升酶)和可溶性蛋白质含量(2. 4 毫克/

毫升酶)等方面都超过 QM9414 的一般水平。与其最高水平比较,虽然 CMC 糖化力略低,但滤纸糖化力和棉花糖化力仍高于 QM9414,而且培养时间大大缩短了。由于没有 QM9414 菌株,无法直接进行比较。从酶制剂活力比较测定结果(见表 6)及过去的工作[9]也可看到,EA_3 - 867 和 N_2 - 78 两菌株的纤维素酶制剂超过日本的 Onozuka R - 10。

本试验中,诱变处理后,变异株中纤维素酶各组分活力的同步提高,可能表明木霉属纤维素酶的各个组分的合成受同一遗传控制系统调节。变异株纤维素酶活力的增高是数量上增加,还是结构上变化,以及诱变后纤维素酶活力提高的原因,有待进一步探讨。

参考文献

[1] Manclels, M., J. Webcr and R. Parizek: *Appl. Microbiol.*, 21: 152,1971.
[2] Mandels, M. et al.: Paper Presented at Army Science Conference, West-point, New York, 1972.
[3] Morozova, E. S.: Symp. on Enzymatie Hydrolysis of Cellulose, p. 193, Aulanko, Finland, 1975.
[4] 西沢一俊: 食品工业 9: 61,1906。
[5] 四川省生物研究所纤维素酶组: 遗传学报,2: 157,1975。
[6] 西北水土保持生物土壤研究所微生物室箱纤组: 微生物育种学术讨论会文集,98 页,北京,科学出版社,1975。
[7] 上海化学纤维六厂,上海酒精二厂,上海植物生理研究所纤维素酶组: 同上,89 页,1975。
[8] 中国科学院遗传研究所五室二组: 遗传学报,1: 57,1974。
[9] 上海植物生理研究所细胞生理室: 植物学报,17: 245,1975。
[10] Mandels, M. and J. Weber: *Adv. Chem. Ser*, 95: 391,1969.
[11] Osmundsvåg, K. and J. Goksøyr: European. *J. Biochem.* 57: 405,1975.
[12] Toyama, N.: Adv. Chem. Ser, 95: 3a9,1969.
[13] 北京大学制药厂编: 微生物学和酶学基本知识,北京,科学出版社,1971。
[14] Tsu-Lee Huang et al.: *J. Ferment. Technol.* 49: 574,1971.
[15] Norkrans, B.: *Adv. Appl. Microbiol.*, 9: 91,1967.
[16] Reese, E. T.: Biotechnol. and Bioengineering Symp., 5: 77,1975.
[17] Pathak, A. N. and T. K. Ghose: *Process Biochemistry*, 8: 35,1973.
[18] Mandels, M.: Biotechnol. and Bioengineering Symp., 5: 81,1975.
[19] 中国科学院微生物研究所: 常见与常用真菌,206 页,北京,科学出版社,1973.
[20] Rifai, M. A.: *Mycol. Papers.* No. 116,1969.
[21] Mandels, M.: Symp. on Enzymatic Hydrolysis of Cellulase, p. 81, Aulanko, Finland, 1975.
[22] Yamaue, K. et al.: *J. Biochem.* 67: 9,1970.
[23] Wood, T. M.: Biotechnol. and Bioengineering Symp., 5: 111,1975.

Isolation of Two Mutant Strains of Trichoderma Pseudokoningii Rifai EA_3 - 867 and N_2 - 78 with High Cellulase Yields and Comparison of Their Characteristics

(1) Two mutants EA_3 - 867 and N_2 - 78 with high cellulase yields were obtained from the wild strains of *Trichoderma pseudokoningii* Rifai AS 3. 3002 and Mo_3 respectively by mutagenic treatments with various physical and chemical agents (high energy electrons, ^{60}Co, ultravioletray, N - methyl - N′- nitro - N - nitrosoguanid-ine, diethyl sulfate, etc.). The cellulase activities of their koji, shake flask cultures, and enzyme preparations were distinctly higher than those of their parents. The mutant N_2 - 78 reached quite high activity level when cultured in shake flasks on a simple medium containing milled straw, wheat bran, nutrient salts plus waste glucose molasses for 60 hours, and showed the highest cellulase saccharifying activities on CMC, filter paper and cotton, namely 255, 8. 2 and 13. 4mg glucose/ml enzyme respectively, or 10. 4 and 6 times more than those of its parent Mo_3. Sueh cellulase activity levels are higher in comparation to the data reported in the literatures.

(2) The enzyme activities of cellulase preparations prepared from koji of $EA_3$867 and N_2 - 78 were compared with the Japanese commercial cellulase preparation "Onozuka R - 10". The cellulase activities of EA_3 - 867 and N_2 - 78 on CMC (saccharifying and liquefying activities), filter paper degradation, and filter paper and cotton saccharification were all higher than those of the "Onozuka R - 10", while pectinase activity was about the same.

(3) Beside cellulases, the strains AS 3. 3002, MO_3, EA_3 - 867 and N_2 - 78 could also produce and secrete active pectinase. Hemicellulase, and some amylase and small amount of proteinase. In mutant strains not only the cellulase aetivities were enhanced, but pectinase was also inereased.

(4) As compared with the wild strains AS 3. 3002 and Mo_3, the mutants EA_3 - 867 and N_2 - 78 showed differences in the size, color and growth rate of the colonies, and in the production and morphology of conidia. They formed smaller colonies, grew more slowly,

---

Cellulase Research Laboratory, Shanghai Institute of Plant Physiology, Academia Sinica and the Shanghai Distillery No. 2

and produced less and longer conidia. But the lethal temperature for conidia of all the four strains was the same.

(5) The enzymatic characteristics of the mutants and wild strains were similar. The optimal pH and optimal temperature for the saccharification of CMC and of filter paper, and for filter paper degradation were 4.5 and 60℃, 4.8 and 50℃, 4.8 and 60℃ respectively for all four strains.

(6) The method of mutation induction, assay and the levels of cellulase activities are discussed.

木霉纤维素酶的诱导形成及其调节*

Ⅰ. 槐糖对木霉 EA_3－867 纤维素酶形成的诱导作用

在槐豆荚提取液中分离到白色针状的槐糖结晶，该糖对拟康氏木霉（*Trichoderma pseudo-konigngii Rafai*）EA_3－867 的纤维素酶（C_1 和 C_x）有强力的诱导作用。在纤维素酶活力，尤其是产酶速度上明显超过纤维素的诱导作用。槐糖的诱导作用与添加槐糖的时间和菌种有关，并受甘油强烈阻遏。在以纤维二糖（0.5%）为碳源培养时，木霉 EA_3－867 也能较迅速地形成纤维素酶，但在 EA_3－867 的甘油培养物中加入纤维二糖（5×10^{-4} mol/L）并不能诱导 C_x 酶。槐糖和纤维素对纤维素酶的诱导作用，无论在诱导胞外和胞内纤维素酶的成分上或从凝胶电泳图上，都十分相似。作者认为木霉 EA_3－867 的纤维素酶形成同时受诱导-阻遏机制调节，并对组成型和诱导型的纤维素酶的作用，以及固体纤维素对纤维素酶可能的诱导机制作了推测。

酶的诱导形成是生物学上的重要现象。研究原核生物诱导酶形成的调节及遗传控制，导致了 Jacob-Monod 操纵子学说的创立和分子遗传学的飞跃发展[1]。变废纤维为葡萄糖的纤维素酶即典型的诱导酶[2—4]，其中真菌木霉属是迄今所知形成和分泌胞外纤维素酶系成分最全面、活力最高的一个属[2, 3]。对纤维素酶的诱导，若干研究者已有过一些报道[2, 3]，但对于固体纤维素如何诱导纤维素酶形成的机制的解释，仍缺乏系统和完整的概念。我组从野生型拟康氏木霉 1096 和木-3 分别诱变得到二株纤维素酶活力较高的菌株 EA_3－867 和 N_2－78[5, 6]。从 1973 年开始，我们对该二株菌纤维素酶的诱导形成及其调节机制做了一些初步的研究。我们认为，阐明纤维素酶诱导形成的机制和调节控制的原理，不仅有着理论上的意义，而且将为改变培养条件提高纤维素酶产量，设计筛选模型选育高产菌种，提供新的线索和手段。

一、材料和方法

1. 菌种

EA_3－867。

2. 试剂

纤维素粉（No. 123）：层析用，Carl Schleicher & Schüll, Dassel 公司产品。滤纸纤维素：WhatmanNo. 1 滤纸，烘干，磨成粉。脱脂棉：国产医用，烘干，磨成粉。CMC（羧甲基纤维素

朱雨生，谭常，中国科学院植物生理研究所纤维素酶组，上海。此文发表于 1978 年“微生物学报”18 卷 4 期 320－331 页。
本文于 1977 年 8 月 2 日收到。
* 本工作得到沈善炯同志的帮助和指导，表示感谢。

钠盐)：黏度 300～600 厘粕，上海长虹塑料厂产品。纤维二糖：B. D. H 公司产品。多聚半乳糖醛酸：Light 公司产品。葡萄糖：上海葡萄糖厂生产的口服葡萄糖，经活性炭吸附，水洗脱，再结晶得到的纯葡萄糖。槐糖：槐豆荚中提取。

3. 槐糖的提取、纯化、结晶和鉴定

参考 Clancy 的方法[7]，并作了修改。称取 750 克洗净晾干的槐豆荚(*Sophora japonica* L.)，加 2 升热蒸馏水浸泡，用组织捣碎机打碎，于沸水浴中搅拌提取 20 分钟，纱布过滤，二次滤液合并，4 000 转/分离心 20 分钟，取清液薄膜浓缩(16 mmHg，70℃，以下同)至 400 毫升淡黄色糖浆。滴加 4 升无水酒精，过夜，以沉淀蛋白和聚多糖。再加 500 毫升乙醚进一步去多糖，过滤或离心，取清液，薄膜浓缩至 250 毫升糖浆。加 8 升 H_2SO_4，溶液(22.5 mmol/L)，于 100℃沸水浴中煮沸 1 小时，以水解槐糖苷。趁热加 40 毫升 $Ba(OH)_2$ 饱和溶液，以中和至 pH 6.0 左右。过滤，滤液再薄膜浓缩至 400 毫升左右，即可作活性炭分离纯化。活性炭型号 769，预先于 150℃左右加热活化 4 小时，悬浮于蒸馏水中装柱，装柱量 540 克(40×6 cm)。糖浆入柱后，先用 35 升蒸馏水洗脱，以除去单糖，以酚-硫酸法测定，不再有糖，洗下为止。接着，用 7%酒精约 12 升洗脱，流速 500 毫升/小时，基本上可将槐糖全部洗下。酒精洗脱液薄膜浓缩至干，加 80%甲醇结晶，再用冰冷的无水甲醇洗涤结晶多次，抽干，60℃烘干，即得白色槐糖结晶制品，于干燥器中保藏。

槐豆荚提取液经活性炭吸附和洗脱后，洗脱液中糖份的分布如图 1 所示。水洗脱得到峰Ⅰ，以 5%、7%、10%、15%和 50%酒精洗脱分别得到峰Ⅱ(分峰尖Ⅱ-Ⅰ和峰肩Ⅱ-2)、Ⅲ、Ⅳ、Ⅴ和Ⅵ。其中峰Ⅰ、Ⅱ-1、Ⅱ-2、Ⅲ、Ⅳ和Ⅴ的得糖量分别为 105、2.46、0.91、0.42、0.36 和 0.27 克。峰Ⅵ中混有大量被洗下的色素。根据预备试验表明，769 活性炭上用水洗脱下来的为单糖，5%～10%酒精洗脱部分主要为双糖，10%以上酒精洗脱下的为叁糖和寡糖。将峰Ⅰ-Ⅵ分别浓缩，进行纸层析分离，用苯胺-二苯胺试剂显色，并与标准糖进行比较(见图 2)。

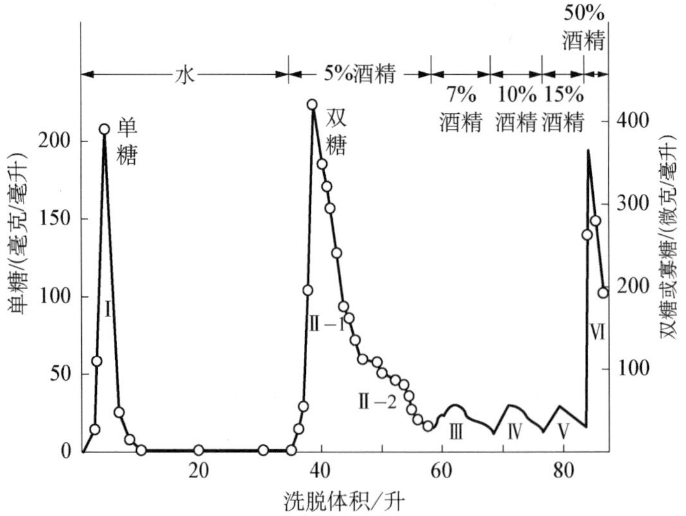

图 1　槐豆荚提取液活性炭

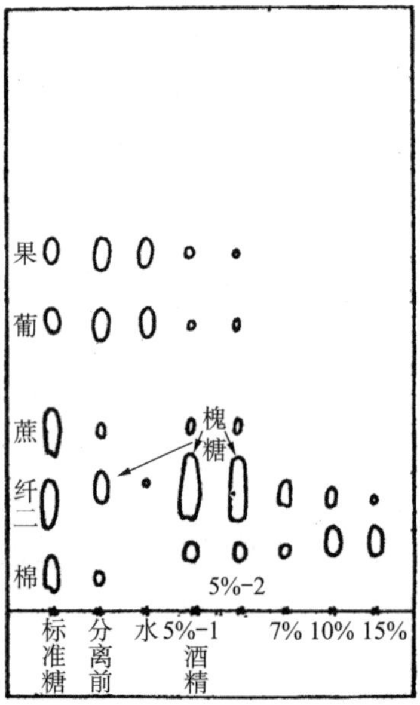

图 2　槐豆荚提取液经活性炭层析分离后各洗脱部分的纸层析图

可看到，水洗脱液中主要含单糖，5%和7%酒精洗脱液中含有大量的未知双糖（位置在蔗糖和纤维二糖之间，R_G 约 0.33）。随着酒精洗脱浓度的增加，未知双糖的含量逐渐减少，而未知寡糖的含量逐渐增多。未知双糖用苯胺-二苯胺试剂显色时，在纸谱上显出特殊的棕黄色，用 $AgNO_3$ 试剂显色时呈咖啡色，而用三苯基四唑化氯试剂显色时，颜色几乎看不出；而且峰Ⅱ和峰Ⅲ的糖只能用酚-硫酸法检出，与 DNS 试剂却不起反应，这说明，此未知双糖是非还原糖。将峰Ⅱ和峰Ⅲ洗脱液合并，浓缩，在80%甲醇中能析出白色针状结晶，得率为0.24%（对槐豆荚鲜重），用显微熔点测定仪测得晶体熔点为192～195℃，而且，对木霉洗涤菌丝体的纤维素酶形成有强烈的诱导作用（待发表）。总之，从以上未知双糖在活性炭上的吸附洗脱行为，纸层析位置和显色情况、结晶的颜色、形状和熔点，以及诱导活性等等，与文献上记载的槐糖特性相符，说明从槐豆荚中提取分离到的双糖结晶就是 2－O－β－D 葡萄吡喃糖基－D－葡萄糖，即槐糖。

4. 液体培养方法

种子斜面培养基(%)：土豆汁 20，纤维素粉 2，琼脂 2；或蛋白胨 0.2，酵母膏 0.1，琼脂 2。

甘油无机盐培养基成分(%)[2]：甘油 0.6，$(NH_4)_2SO_4$ 0.14，KH_2PO_4 0.2，尿素 0.03，$MgSO_4 \cdot 7H_2O$ 0.03，$CaCl_2$ 0.03，蛋白胨 0.1。每升含微量元素 $FeSO_4 \cdot 7H_2O$ 5 毫克，$MnSO_4 \cdot H_2O$ 1.56 毫克，$ZnCl_2$ 1.67 毫克，$CoCl_2$ 2.0 毫克，pH 5.3。在不同碳源试验中，用各种碳源(0.5%)代替上述甘油，其他培养基成分相同。培养中途加槐糖或甘油等的试验中，槐糖和甘油等均先分别灭菌，无菌操作加入。500 毫升三角瓶中装入 100 毫升培养基，孢子悬浮液接种，每 100 毫升培养基中约接入 10^6 个左右的孢子，在往复式摇床(108 次/分)上 30℃震荡培养，至所需时间，取培养滤液测定酶活力。

5. 酶活力测定

1）酶液制备

(1) 胞外酶：用 3 号砂蕊漏斗抽滤培养液，抽滤液为胞外酶。

(2) 胞内酶：未滤去的菌丝体用 100 毫升磷酸盐缓冲液(0.01 mol/L，pH 5)洗涤，抽干，再悬浮在 15 毫升同样的缓冲液中，用超声波(M.S. E)破碎，20 KC，破碎 5 分钟，得到胞内酶。

2）酶活力测定方法

(1) CMC 酶活(C_x)：0.5 毫升适当稀释的酶液，加 2.0 毫升 pH 5 的 0.625% CMC 溶液(1000 毫升 1% CMC，加 200 毫升 0.05 mol/L 柠檬酸-磷酸氢二钠缓冲液和 400 毫升水配成)，于 40℃水浴中保温半小时，加 2.5 毫升 DNS 试剂[8]终止反应，于沸水浴中煮沸 5 分钟，在 540 毫微米波长比色，要求光密度读数在 0.5 毫克葡萄糖左右。按 Mandels 早期定单位的方法[2]，规定在上述反应条件下，每产生 0.5 毫克还原糖，为 1 个 C_x 活力单位*。

(2) 滤纸酶活(F. D)：50 毫升容量瓶内投入二张 Whatman No. 1 滤纸(1×1 厘米^2)，加 4 毫升酶液和 1 毫升醋酸缓冲液(0.1 mol/L，pH 4.6)，于瓦氏呼吸器振荡机上 40℃恒温振荡(100 次/分，振幅 5 厘米)观察滤纸全部崩溃成粉状所需的时间(分钟)，依下面公式计算活力单位(u)：

---

* 关于纤维素酶 C_1、C_x 的概念，近年来有不少争论和变化[9-11]。本文中的 C_x、C_1 仍沿用 Rcese 的概念，以分解 CMC 的酶代表 C_x，分解脱脂棉的酶代表含 C_1。

活力单位/毫升＝250/崩溃时间(分钟)

(3) 纤维二糖酶：0.5 毫升酶液，加 2 毫升纤维二糖溶液(10^{-3} mol/L，溶于 0.02 mol/L 醋酸缓冲液中，pH 5.0)，于 40℃水浴中保温 2 小时，用 DNS 法测定还原糖的增加。单位定法同 C_x。

(4) 淀粉酶：0.5 毫升酶液，加 2 毫升 1%淀粉溶液(溶于 0.02 mol/L，pH 4.8 醋酸缓冲液中)，于 30℃保温半小时，用 DNS 法测定形成的还原糖。单位定法同 C_x。

(5) 果胶酶：2 毫升适当稀释的酶液，加 0.5 毫升多聚半乳糖醛酸溶液(0.5%，溶于 0.1 mol/L，pH 4.5 醋酸缓冲液中)，30℃反应半小时，用 DNS 法测定形成的半乳糖醛酸，每产生 0.5 毫克半乳糖醛酸为 1 个单位。

(6) 半纤维素酶[12]：以玉米芯提取的半纤维素为底物，用甲基间苯二酚法测定酶解形成的木糖。在规定条件下形成 0.5 毫克木糖为 1 个单位。

(7) 蛋白酶[12]：以酪蛋白为底物，测定酶解产生的酪氨酸，规定形成 1 个微克的酪氨酸为 1 个单位。

6. 糖、甘油和生长的测定

用酚-硫酸法[13]测定还原糖或非还原糖。用过碘酸钠氧化甘油，亚砷酸钠终止反应，变色酸显色，在 570 毫微米比色测定甘油[14]。生长测定是将液体培养的菌丝体用 3 号砂芯漏斗抽滤，80℃烘干至恒重，称重。

7. 糖的纸上层析

Whatman No.1 滤纸，溶剂为正丁醇∶醋酸∶水(4∶1∶5)。30℃左右下行约 7 天。用苯胺-二苯胺试剂，硝酸银试剂或三苯基四唑化氯试剂[13]显色。

8. 酶组分的凝胶电泳分离法

(1) 酶制剂制备：胞外或胞内酶液冷冻，缓缓加入 3 倍体积冷却至 0℃以下的丙酮，以沉淀酶蛋白，冷冻离心(－5℃以下，3 000 转/分，20 分钟)，收集沉淀，用无水丙酮脱水，抽滤，真空干燥，即得酶粉。

(2) 聚丙烯酰胺凝胶电泳：基本上按照 Davis[16]的方法。在 10×0.5 厘米玻璃管中制备聚丙烯酰胺凝胶，每管中注入 1%酶制剂(配在 400%蔗糖溶液中)0.04 毫升(相当于蛋白质 200 微克左右)，在 Tris-甘氨酸(pH 8.3)缓冲液中电泳 1 小时半，每管电流 2 毫安，酶蛋白用氨基黑染色，并在 7%醋酸中退去底色。

三、结果

1. 不同碳源中生长和纤维素酶形成的比较

木霉 EA_3-867 在各种碳源中生长时，只有在纤维素或纤维素衍生物中才能产酶，因而是诱导酶(见图 3)。在各种碳源中生长速度不同：一般在单糖中培养 48 小时，单糖基本上耗完，菌丝体重达到 200～400 毫克/100 毫升，在甘油中达到 550 毫克/100 毫升；对双糖的利用较差，培养 48 小时，只有部分双糖被利用，菌丝体一般在 30 毫克/100 毫升以下；在纤维素或纤维二糖(0.5%)中培养时，生长较慢，在培养 2～3 天之后出现纤维素酶活力，以后随着时间而逐渐上升，在纤维素粉中甚至到第 8 天还在继续上升。以槐糖为碳源进行培养时，生长不到 1

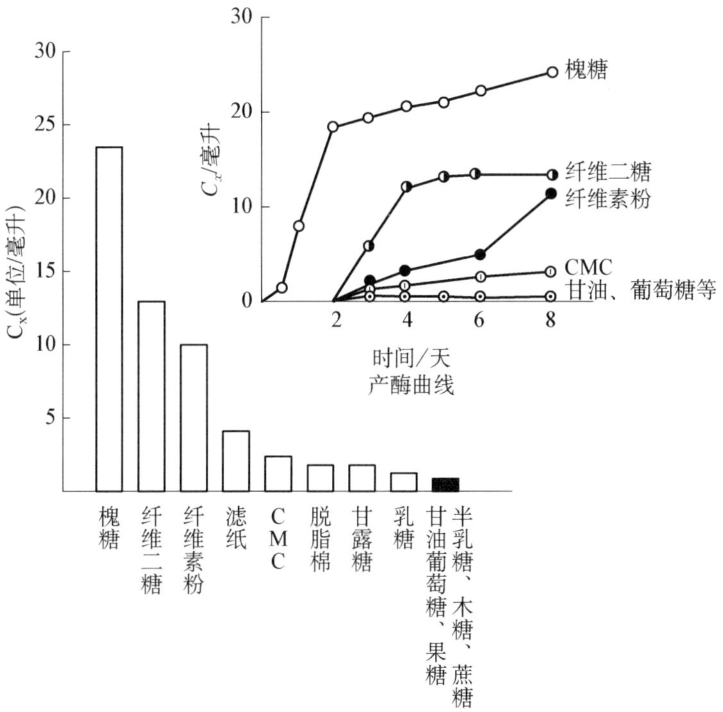

图 3　木霉 EA_3 - 867 在不同碳源中培养 8 天的纤维素酶活力比较

注：500 毫升三角瓶中含 100 毫升各种碳源（0.5%）培养基（配方见方法部分），30℃振荡培养，不同时间取样测定。

天（18 小时）便明显出现活力，第二天即达到相当高的水平，以后活力上升不多，因为 2 天左右槐糖已被消耗光。

EA_3 - 867 在上述不同碳源中培养时的生长产酶情况大致可分三类。第一类槐糖，它是强力诱导剂，培养特点是产酶早，活力高，生长也较快；第二类是纤维素、纤维素衍生物或纤维素分解产物，如纤维素粉、滤纸、脱脂棉、CMC 和纤维二糖。在这些碳源中培养，虽能不同程度的产生纤维素酶，但大部分生长慢，产酶也较槐糖中培养的慢，其中纤维二糖诱导产酶较纤维素粉诱导为快；第三类主要是与纤维素结构无直接关系（葡萄糖除外）的碳源，如甘油、果糖、木糖、半乳糖、蔗糖、麦芽糖等。在这些碳源中（除双糖外）培养，一般生长很快，但纤维素酶活力极低，而且不随培养时间而变化。有些作者报道过甘露糖和乳糖能诱导纤维素酶[17, 18]。

2. 槐糖、纤维二糖和纤维素粉对木霉 EA_3 - 867 甘油培养物纤维素酶形成的影响

1）在甘油培养基中生长、甘油利用和 pH 的变化

EA_3 - 867 孢子悬浮液接种在甘油无机盐培养基中液体振荡培养 7～8 小时，在显微镜下可见到孢子开始膨胀萌发，12 小时有少量菌丝长出，24 小时明显长出肉眼可见的菌丝，36 小时开始生孢子，48 小时后培养液开始转淡绿，生长基本上停止，以后菌丝体自溶减少，至第 7 天，残存的菌丝体已极少（见图 4）。上述 EA_3 - 867 的生长曲线可分为延迟期（12 小时前）、对数期（12～48 小时）、稳定和自溶期（48 小时以后）。

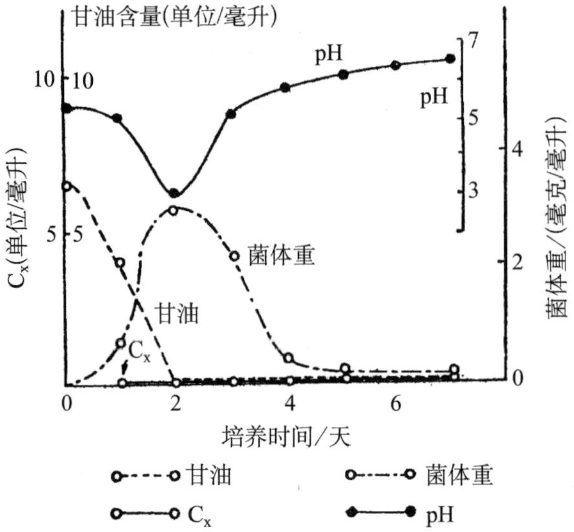

图 4　木霉 EA_3 - 867 在甘油无机盐培养基中液体培养不同时间甘油、菌体重、C_x 和培养液 pH 的变化培养基 50 毫升，30～32℃ 振荡培养

2）甘油培养物中加槐糖、纤维二糖和纤维素对纤维素酶形成的影响

EA_3 - 867 在甘油无机盐培养基中培养，纤维素酶活力极微弱。添加纤维二糖（5×10^{-4} mol/L）不能诱导 C_x 酶产生。但如添加 0.5%纤维素粉或低浓度槐糖（5×10^{-4} mol/L），则纤维素酶能大量产生（见图 5）。但以纤维素为诱导剂时，产酶至少要有三天以上的延迟期；而槐糖可使产酶延迟期缩短为数小时，不论在产酶速度或活力方面均远远超过纤维素。

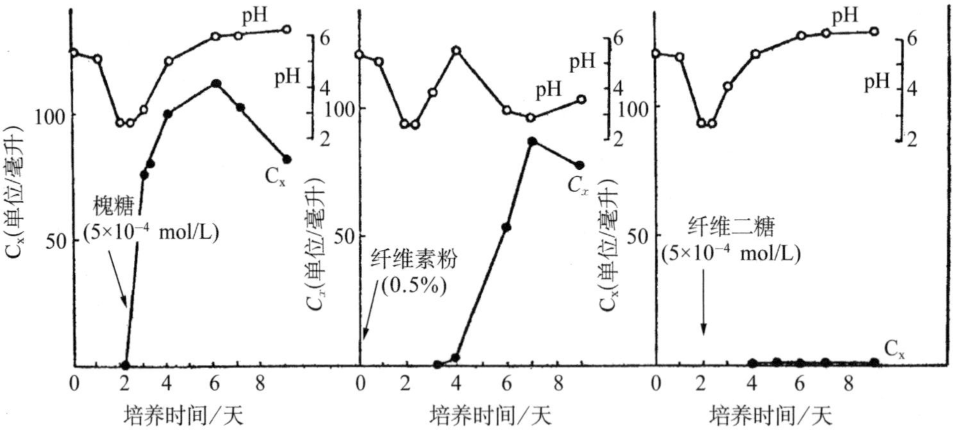

图 5　木霉 EA_3 - 867 在甘油无机盐培养基中加槐糖、纤维素粉和纤维二糖后，纤维素酶诱导形成和 pH 的变化

注：50 毫升培养基，孢子悬浮液接种后，于 29℃ 振荡培养，纤维素粉在 0 小时加入，槐糖和纤维二糖在 48 小时加入，测定不同培养时间培养滤液 C_x 和 pH 的变化。

槐糖诱导纤维素酶合成的能力，与添加槐糖的时间有关；但是产酶的时间，与添加槐糖的时间无关(见图 6)。培养 48 小时加槐糖,诱导产酶最高,24 小时,尤其在开始培养时(0 小时)加入槐糖,产酶会大大减弱;不论槐糖加入的时间多早,纤维素酶均在培养 48 小时后才产生。这些现象可能归因于生长稳定期前甘油对酶形成的阻遏

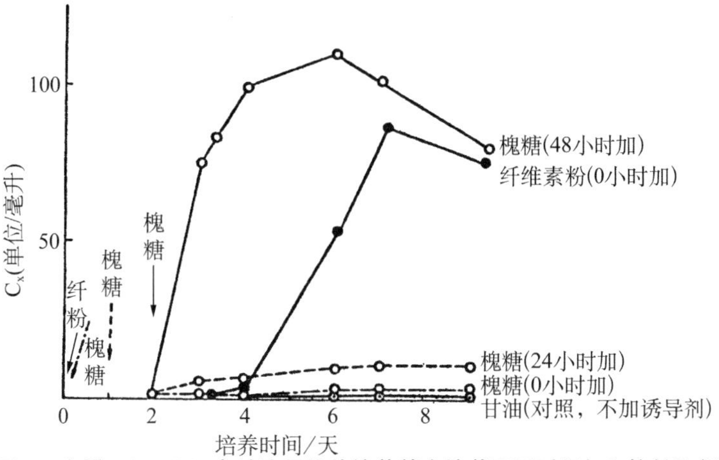

图 6 木霉 EA_3－867 在甘油无机盐培养基中培养不同时间加入槐糖和纤维素粉对纤维素酶形成的诱导作用

注:培养基 5 毫升,槐糖浓度 5×10^{-4} mol/L,纤维素粉 0.5%,29℃,振荡培养。

作用和槐糖本身在培养过程中被消耗,此可以从洗涤菌丝体的诱导试验(待发表)和图 7 的试验中得到证明。从图 7 中可以看到,EA_3－867 甘油培养物在 48 小时加入槐糖,隔 12 小时便明显出现酶活力,24 小时后达到高峰,48 小时后(即培养第 4 天)C_x 活力开始下降,此时,用酚-硫酸法测定,培养液中槐糖已用光。若此时再次补加槐糖,则纤维素酶又恢复合成。但在 48 小时菌龄的甘油培养物中若同时加入槐糖和甘油,则酶的产生便明显受阻,而且到培养第 4 天,即使再补加槐糖,纤维素酶也不再继续合成。

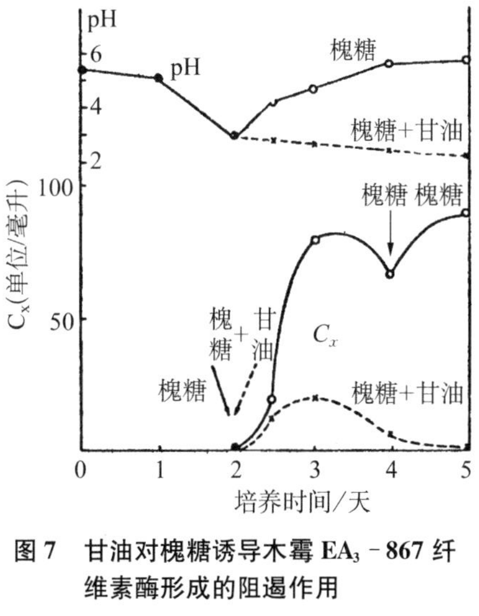

图 7 甘油对槐糖诱导木霉 EA_3－867 纤维素酶形成的阻遏作用

注:培养基 5 毫升,29℃振荡培养 48 小时时加入槐糖(5×10^{-4} mol/L),或槐糖(5×10^{-4} mol/L)和甘油(0.6%)同时加。在培养第 4 天,再次补加槐糖(5×10^{-4} mol/L)。

3. 槐糖和纤维素粉对不同菌种甘油培养物形成纤维素酶的影响

图 8 比较了槐糖和纤维素粉对不同菌种纤维素酶的诱导效应。1096、EA_3－861 和 657 是拟康氏木霉,1096 是野生型,EA_3－867 和 657 是诱变种,其中 657 是白色孢子型。G109 和 D－92 是取自湖北省微生物所的绿色木霉野生型菌株,T17.5－24 是能产纤维素酶的曲霉。不同菌种的孢子悬浮液接种在甘油无机盐培养基中,于摇床 29.5℃振荡培养。纤维素粉在 0 小时加入,槐糖在培养 48 小时加入。可以看出,拟康氏木霉的三个菌株的 C_x 酶均能被槐糖所诱导,白色变异株 657,尤其是高产变异

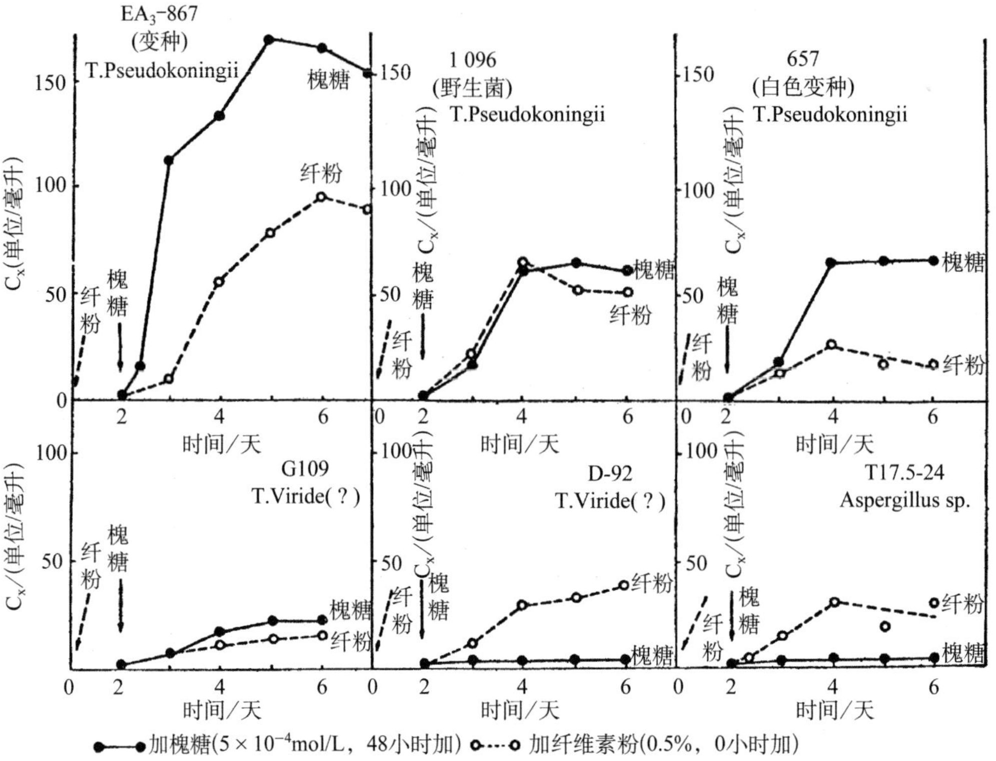

图 8　槐糖和纤维素粉对不同菌种纤维素酶诱导形成的影响

注：甘油无机盐培养基 5 毫升，29.5℃振荡培养。

株 EA_3－867 的产酶速度和活力远远超过以纤维素粉诱导的结果。G109 被槐糖诱导的酶活很低，D－92 和 T17.5－24 几乎不被槐糖所诱导。滤纸崩溃活力的测定也有同样的趋势。这些结果说明不同菌种对槐糖的反应是不同的，同一菌种经诱变后，对槐糖的敏感性也可能不同。

4. 槐糖和纤维素粉诱导的纤维素酶组分的比较

槐糖对 EA_3－867 不同的酶组分具有同样的诱导能力。图 9 是 EA_3－867 甘油培养物中加入槐糖和纤维素粉后，C_x 和滤纸崩溃活力(F. D)的变化。在加入槐糖后，C_x 和 F. D 活力立即同时被诱导出来，并远远超过纤维素粉的诱导结果。在另一批试验中证明，槐糖诱导的纤维素酶能分解脱脂棉，产生葡萄糖，其活力也超过纤维素的诱导结果，表明槐糖同纤维素一样，也能诱导 C_1 酶的形成。此外，两者对 C_x 酶和纤维二糖酶的胞内胞外部分都有诱导作用(见图 10)。而且胞外的 C_x 酶明显高于胞内，可能意味着进入细胞的槐糖，先诱导胞内的 C_x 酶合成，之后立即分泌到胞外。纤维二糖酶在胞内胞外的分布较均匀。Suzuki 等认为，在萤光假单胞菌中槐糖诱导的 C_x 是胞外酶，纤维二糖酶主要是胞内酶[19]。

EA_3－867 孢子接种在甘油无机盐培养基中，分甘油对照(即不加任何诱导剂)、加纤维素粉(0.5%，0 小时加入)和加槐糖(10^{-4} mol/L，48 小时加入)三组样品，于 32～33℃振荡培养，在 C_x 活力达高峰(槐糖组是 6 天，纤维素粉组是 12 天，甘油对照是 6 天)时，分别取出，制备胞外、胞内酶制剂，进行聚丙烯酰胺凝胶电泳(见图 11)。

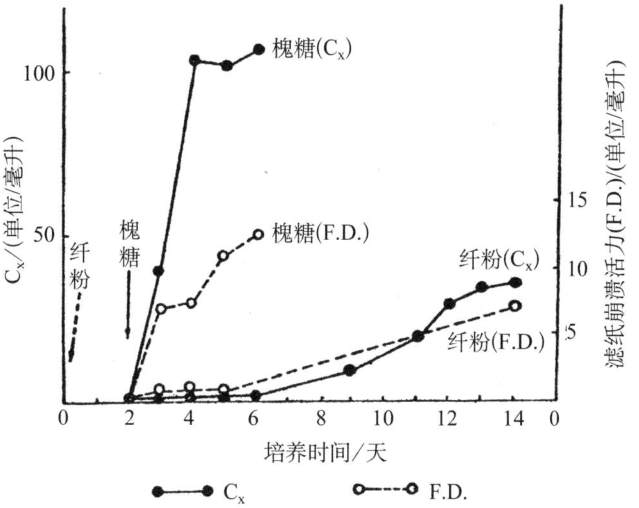

图 9　木霉 EA_3－867 在甘油无机盐培养基中加纤维素粉或槐糖诱导的 C_x 酶和滤纸崩溃活力(F. D.)的变化

注:100 毫升培养基,30℃振荡培养,纤维素粉(0.5%)在 0 小时加入,槐糖在培养 48 小时加入(10^{-4} mol/L)。

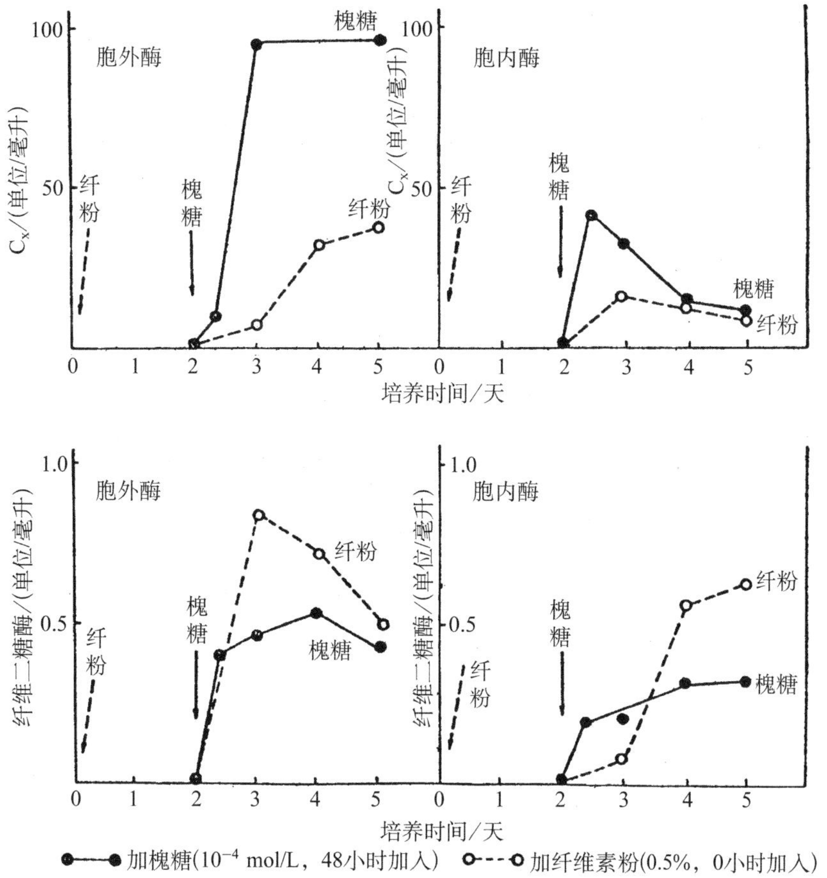

图 10　木霉 EA_3－867 在甘油无机盐培养基中加槐糖或纤维素粉所诱导的 C_x 酶和纤维二糖酶在胞内胞外分布的比较 50 毫升培养基,在 30～32℃培养

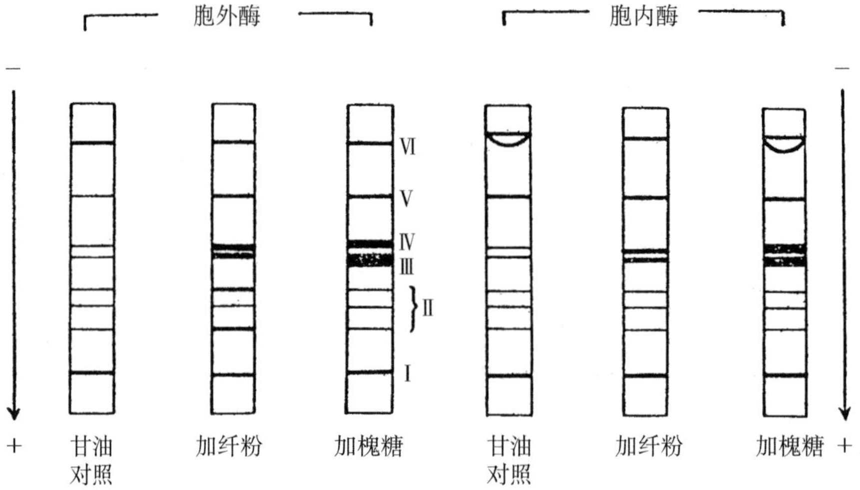

图 11　纤维素粉和槐糖诱导的木霉 EA_3－867 纤维素酶(胞内、胞外)的凝胶电泳图

注:50 毫升甘油无机盐培养基,孢子接种后,于 32～33℃振荡培养,分甘油对照(不加诱导剂)、0 小时加纤维素粉(0.5%)和 48 小时加槐糖(10^{-4} mol/L)三组,在 C_x 活力达高峰时,分别取出,制备胞内、胞外酶制剂,配成 1%浓度(溶于 40%蔗糖溶液中),进行凝胶电泳,每管加酶 400 微克,电流 24 毫安,电泳时间 1 小时半,用氨基黑染色。

三组凝胶电泳图大致都可分为主要的 6 条蛋白区带(Ⅰ～Ⅵ),在加入纤维素粉或槐糖后,蛋白带Ⅲ、Ⅳ和Ⅴ明显加深,尤其是加槐糖后,Ⅲ和Ⅳ最深。据凝胶分段切割测定,蛋白带Ⅲ主要含 C_1 酶(测棉花糖),Ⅳ代表 C_x 酶,Ⅴ代表 β－葡萄糖苷酶。在甘油对照中同样有这些蛋白带,但量极微。

表 1 所列是另一次试验中纤维素粉和槐糖诱导的酶制剂中各种酶活力测定的结果。纤维素粉和槐糖只诱导纤维素酶。果胶酶和淀粉酶——可能是组成酶,量极微,不能诱导半纤维素酶和蛋白酶。凝胶电泳的蛋白带也相似。

表 1　纤维素粉和槐糖诱导的酶制剂活力比较*

诱导剂	C_x(单位/毫升)	滤纸活力		纤维二糖酶/(单位/毫升)	果胶酶/(单位/毫升)	淀粉酶/(单位/毫升)	半纤维素酶/(单位/毫升)	蛋白酶/(单位/毫升)	凝胶电泳峰
		崩溃时间(分钟)	(单位/毫升)						
纤维素粉(1%)	86	68	7.4	1.0	0.9	0.9	0	0	Ⅰ～Ⅵ
槐糖(5×10^{-4} mol/L)	700	15	33.4	0.7	1.0	1.1	0	0	Ⅰ～Ⅵ

注:* 100 毫升甘油无机盐培养基,孢子悬浮液(EA_3－867)接种,30℃振荡培养,纤维素粉在 0 小时加入(1%),槐糖的培养 48 小时加入(10^{-4} mol/L)。在纤维素粉中培养 14 天(产酶高峰)和槐糖中培养 5 天(产酶高峰)的培养滤液制成丙酮粉,均配成 1%浓度,进行各种酶活力的测定比较。凝胶电泳的条件同图 11。

四、讨论

研究真核细胞的基因表达，酶蛋白诱导合成的调节控制，是当今分子遗传学的研究方向之一[20]。纤维分解真菌——木霉能产生诱导性纤维素酶，又是真核生物中最简单的一种，加上取材容易、培养方便、产酶迅速、酶活稳定、测定简便、有实用价值，因而是研究酶诱导形成的良好材料，也是研究真核细胞基因表达的理想课题。同时，随着纤维素酶研究的逐步深入，直接关系到酶产量的纤维素酶的诱导现象正受到人们的重视。早期，Mandels 等发现在不同碳源中只有纤维素、纤维二糖、乳糖和试剂级葡萄糖对纤维素酶有诱导作用[18]，后认为纤维素分解产物纤维二糖可能是真正的天然诱导剂[21]。但是，纤维二糖的诱导活力很低，在采用高浓度纤维二糖（1%或更高）及降低代谢速度，或添加表面活性剂，才能得到较高的纤维素酶活力[22-26]。因此，纤维二糖是否是真正的诱导剂，大家的看法是混乱的[24, 25]，另一方面，迄今所知，槐糖是诱导木霉和个别纤维分解细菌纤维素酶活力最强的诱导剂[19, 24, 25]。大多研究者认为纤维素酶的形成受特殊的诱导剂诱导。但 Hulme 等在研究漆斑霉纤维素酶形成时，认为该菌的纤维素酶不是诱导酶，各种不同碳源，只要控制利用速度，解除降解物阻遏，都能产生纤维素酶[27]。

我们从槐豆荚提取分离到的槐糖结晶，对拟康氏木霉的胞内胞外纤维素酶以及纤维素酶的不同组分都有强力的诱导作用。槐糖诱导的纤维素酶与纤维素诱导的相比，只表现出量（速度和活力高低）的差异，在质上没有什么差别，即两者诱导的酶类、组分或凝胶电泳图谱都十分相似，是否可认为二类诱导剂诱导纤维素酶形成的途径和机制相同。两者间的桥梁的纤维素的分解产物是纤维二糖。后者（β，1－4 连结）同槐糖（β，1－2 连结）是结构颇为相似的双糖。但纤维二糖加入木霉的甘油培养物中不能诱导纤维素酶的形成，洗涤菌丝体的试验也证明了这点（待发表），除非以较高浓度的纤维二糖（0.5%）作为碳源进行生长培养才能迅速产酶，速度恰好介于槐糖和纤维素之间。这似乎只能解释为纤维二糖本身对于纤维素酶（如 C_x）并不是真正的诱导剂，而是它在生长培养过程中的转化产物才是真正的诱导剂。这种代谢转化，包括转化成槐糖或其类似物在其他微生物中已得到证实[21, 28, 29]。据此，我们推测，纤维素对木霉纤维素酶形成的诱导作用可能是通过其分解产物纤维二糖转化为槐糖或其类似物，这种转化需要一个过程，因此纤维素的诱导比槐糖慢（纤维二糖介于两者之间），而两者诱导的方式却类似。

木霉 EA_3－867 在非纤维素类的碳源如甘油、单糖和双糖中生长，能分泌微量的不随时间变化的纤维素酶到介质中去，其凝胶电泳图也表明存在微量的 C_1 和 C_x 酶，说明木霉纤维素酶除了诱导酶（占主要）外，还有一类不受诱导剂影响的组成酶。它们可能是纤维分解菌所固有的遗传特性，分布在细胞表面或分泌到介质中去，它们在识别环境中的纤维素，从而将信息传递给细胞以合成纤维素酶时，可能起着“信号受体”的重要角色。因为纤维素废物是不溶于水的大分子，无法进入细胞。细胞表面或介质中的纤维素酶，一旦遇到纤维素时，便将后者分解为纤维二糖，进入细胞，进一步通过某种机制，形成槐糖或其类似的强力诱导剂，从而去阻遏，诱导纤维素酶形成并分泌到胞外分解纤维素，供生长繁殖用。

本试验指出，木霉 EA_3－867 纤维素酶的形成同时受诱导-阻遏系统的调节，此现象在其他真菌、细菌中也见到[4, 19, 31, 32]。纤维素酶的诱导形成是微生物适应环境的重要方式。另一方面，当培养基中同时存在难分解的纤维素和易利用的碳源(如葡萄糖、甘油等)时，木霉首先利用后者，并借降解物阻遏机制，关闭指导合成纤维素酶的基因，以避免不必要的浪费，直至易利用的碳源用光，降解物阻遏解除，纤维素酶才诱导形成，再分解纤维素为葡萄糖供生长利用。所以，降解物阻遏的调节机制同诱导机制一样，同样是微生物细胞经济而有效地合成酶的一种调节方式。

参考文献

[1] Jacob, F., and Monod, J.: *J. Miol. Biol.*, **3**: 318,1961.

[2] Mandels, M., and Weber, J.: *Adv. Chem. Ser.*, **95**: 391,1969.

[3] Pathak. A. N., and Ghose, T. K.: *Process biochemistry*, 8(4): 35,1973.

[4] Horton, J. C., and Keen, N. T.: *Canad. J. Microbiol.*, **12**(2): 209,1966.

[5] 上海化纤六厂、上海酒精二厂、上海植物生理研究所纤维素酶组: 微生物育种学术讨论会文集，科学出版社; p. 89,1975。

[6] 上海植物生理研究所、上海酒精二厂: 微生物学报,**18**(1): 27,1978。

[7] Clancy, M. J.: *J. Chem. Society*, **11**: 4213,1960.

[8] Summer, J. B.: Laboratory Experiments in Biological Chemistry, Acad. Press, New York, p. 38,1949.

[9] 中国科学院微生物所纤维素酶组: 微生物学报,**16**(3): 240,1976。

[10] Wood, T. M.: Proc. Ⅳ. IFS. Ferment. Technol. Today, p. 711,1972.

[11] Reese, E. T.: Biotechnol. and Bioen-gineering Symp. No. 5, p. 77,1975.

[12] 北京大学制药厂编: 微生物学和酶学基本知识，科学出版社,1971。

[13] Michel Dubois, et al: *Anal. Chem.*, 28: 350, 1956.

[14] Lambert, M., and Neish, A. C.: *Canad. J. Res.*, 28: 83, 1950.

[15] Trevelyan, W. E., et al.: *Nature*, **166**: 444,1950.

[16] Davis, B. J.: *Annals N.Y. Acad. Sci.*, **121**: 404,1964.

[17] Yamane. K., et al.: *J. Biochem.*, 67(1): 9,1970.

[18] Mandels, M., and Reese, E. T.: *J. Bact.*, 73: 269,1957.

[19] Suzuki, H., Yamane, K., and Nisizawa, K.: *Adv. Chem. Ser.*, **95**: 60,1969.

[20] 堀内忠郎尾辻望: 化学增刊,**42**: 61,1970。

[21] Mandels, M., and Reese, E. T.: *J. Bact.*, **79**: 816,1960.

[22] Reese, E. T., and Maguire, A.: *Develop-ments in industrial Microbiol.*, **12**: 212,1971.

[23] Reese, E. T., and Mandels, M.: "Cellulose and Cellulose Derivaties" ed. by Norbert M. Bikales, Part Ⅴ. p. 1079,1971.

[24] Mandels, M., and Sternberg, D.: *J. Ferment. Technol.*, **54**(4): 267, 1976.

[25] Mandels, M., Sternberg, D., and Andreotti, R. E.: Symp. On Enzymatic Hydrolysis of Cellulose. Aulanko, Finland, 12-14, March, 1975. ed. by M. Bailey, T-M. Enari, M. Linko, Helsinki, p. 81, 1975.

[26] Ghose, T. K., Pathak, A. N., and Bisaria, V. S.: Ibid. p. 111,1975.

[27] Hulme, M. A., and Stranks, D. W.: *Nature*, **226**(5244): 469,1970.

[28] Buston, H. W., and Jabbar, A.: *Biochem. Biophys.* Acta, 15: 543,1954.

[29] Satoko teda, et al.: *J. Ferment. Technol.*, 46: 711,1968.

[30] Mandels, M., and Reese, E. T.: *Biochem. Biophys. Res. Commun.*, 1: 338,1959.

[31] Stewart, B., and Leatherwood, J. M.: *J. Bact.*, 128(2): 609,1976.

[32] Nisizawa, T., et al.: *J. Biochem.*, 71(6): 999,1972.

Induction and Regulation of Cellulase Formation in Trichoderma

Ⅰ. The Induction of Cellulase Formation in Trichoderma $EA_3 - 867$ by Sophorose*

A series of investigations on cellulase formation and its regulation in *Trichoderma* have been conducted. In this paper we report the strong induction of cellulase by sophorose in $EA_3 - 867$, a strain of *Trichoderma pseudokonigii* Rifai. Sophorose, a white needle-shaped crystal, was obtained from the extract of the pods of *Sophora japonica* L. The cellulase activities (C_1, C_x and cellobiase) induced by sophorose, especially its initial rate of increase, were considerably higher than those induced by cellulose. The induction of cellulase by sophorose depended upon the time of addition and the organisms used, and was repressed strongly by glycerol. $EA_3 - 867$ produced cellulase on a medium with cellobiose (0.5%) faster than with cellulose, but slower than with sophorose, in the last case cellulase was produced in the medium being cultured only for 18 hours. Cellobiose was, presumably, not a true inducer itself. The induction of cellulase by sophorose and cellulose was quite similar both in the extracellular or intracellular components and in the patterns of polyacrylamide gel electrophoresis of cellulase preparations induced by them. On a medium containing rapidly utilizable carbon source such as glycerol or glucose, which were not inducers, *Trichoderma* $EA_3 - 867$ produced only a trace of cellulase, which ought to be a constitutive cellulase. On glycerol-mineral nutrients supplemented with cellulose or sophorose, $EA_3 - 867$ also. produced a small amount of pectinase and amylase, which seem to be constitutive enzymes also. The constitutive cellulase, present on cell surface or released into the medium, hydrolyzes cellulose to cellobiose. The latter might be converted into sophorose or its analogues through some mechanisms unknown in the cell. The sophorose or its analogues thus produced is a powerful inducer for cellulase. The constitutive cellulase may thus play a role in the convertion of the insoluble cellulose into a soluble and hence recognizable "signal" molecule. There have been considerable evidences which indicates that the cellulase formation in *Trichoderma* $EA_3 - 867$ was regulated by the induction mechanism and by the catabolite repression mechanism as well.

---

* Zhu Yu-sheng and Tan Chang (Cellulase Laboratory, Shanghai Institute of Plant Physiology, Shanghai)

木霉纤维素酶的诱导形成及其调节

Ⅱ. 槐糖对木霉 EA_3 - 867 洗涤菌丝体纤维素酶形成的诱导作用及降解物阻遏现象*

提要

槐糖对拟康氏木酶(*Trichoderma pseudokoningii* Rifai) EA_3 - 867 洗涤菌丝体的纤维素酶形成有强力的诱导作用,经过 3 小时左右的延迟期,24 小时后,纤维素酶便达到高峰。各种木霉菌种和 EA_3 - 867 不同菌令的菌丝体都能被槐糖诱导形成纤维素酶,C_1、C_x 酶主要分布在胞外,纤维二糖酶分布在胞内。菌丝体诱导形成纤维素酶需要诱导剂;无机盐(氮、磷、铁、锰盐),适宜的温度、pH 和通气;以及菌丝体细胞活跃的代谢。槐糖的最适浓度是 10^{-5} ~ 10^{-3} mol/L,诱导的最适温度为 30℃,适宜的 pH 范围为 3~6,最适 pH 为 5。O_2 能刺激纤维素酶的形成,而呼吸抑制剂碘乙酸、氟化钠、丙二酸、迭氮化钠和 3,5 - 二硝基苯酚能强烈抑制菌丝产酶。槐糖对菌丝体纤维素酶的诱导形成能不同程度地被核酸代谢抑制剂如放线菌酮、5 - 氟尿嘧啶、6 - 氮杂尿嘧啶和放线菌素 D 抑制。葡萄糖、甘油、各种糖类、糖磷酸酯、有机酸、辅酶 NAD、NADP 及 ATP** 均能阻遏槐糖对菌丝体纤维素酶的诱导作用。葡萄糖对纤维素酶形成的阻遏作用不能被 c - AMP 介除。

一、引言

拟康氏木霉 EA_3 - 867 的纤维素酶形成能强烈地被槐糖诱导,被甘油阻遏,推测该菌株的纤维素酶形成,与一般的分解代谢酶的形成相同,受到底物诱导和降解物阻遏机制所调节(朱雨生等,1978)。但是,不清楚槐糖的作用是否诱导纤维素酶蛋白的新合成,以及木霉纤维素酶诱导形成的调节控制方式是否同 Jacob 和 Monod 在大肠杆菌中得到的乳糖操纵子模型中的类似(Jacob 和 Monod, 1961)。由于过去研究纤维素酶的形成,大多采用以纤维素为碳源,进行长时间生长培养的方法,生长和产酶同时进行,对取得的结果难以作深入的分析。故本工作采用改进的洗涤菌丝体静止细胞诱导培养的方法,详细研究槐糖诱导纤维素酶形成的条件、过程,与呼吸代谢和核酸代谢的关系,以及降解物阻遏现象。

朱雨生,谭常,中国科学院上海植物生理研究所。此文发表于 1978 年"微生物学报"4 卷 1 期 1 - 18 页。

1977 年 12 月 22 日收到。

* 沈善炯同志对本工作提出了宝贵意见,特致谢意。

** 本文缩写:ATP,腺苷三磷酸;NAD,烟酰胺腺嘌呤二核苷酸;NADP,烟酰胺腺嘌呤二核苷酸磷酸;c - AMP,环腺苷酸;G - 6 - P,葡萄糖- 6 -磷酸;6 - PG,6 -磷酸葡萄糖酸;FDP,果糖- 1,6 -二磷酸;DNP,2,4 -二硝基苯酚;EDTA,乙二胺四乙酸。

二、材料和方法

1. 菌种

拟康氏木霉 EA_3 - 867：同前文（朱雨生等，1978），是自野生型菌株 1096（或编号 AS 3.3002）经过多种物理化学诱变而得到的纤维素酶高产变异株。

2. 试剂

纤维素粉、稻草纤维素、纤维二糖、葡萄糖、甘油等见前文所述（朱雨生等，1978）。槐糖是自槐豆荚中提取、分离、纯化得到的结晶制品（朱雨生等，1978）。放线菌素 D 和放线菌酮由中国科学院上海药物研究所抗菌素室供给。利福霉素是上海制药五厂产品。ATP、NAD、NADP、c - AMP、5 -氟尿嘧啶均为中国科学院生物化学研究所东风试剂厂产品。6 -氮杂尿嘧啶系 Cal. Biochem. 公司产品。

3. 方法

1）静止细胞诱导培养方法

（1）菌丝体生长培养条件。

种子斜面：见前文（朱雨生等，1978）。

生长培养基（%）：甘油 0.6，$(NH_4)_2SO_4$ 0.14，KH_2PO_4 0.2，尿素 0.03，$MgSO_4 \cdot 7H_2O$ 0.03，$CaCl_2$ 0.03，蛋白质 0.1。含微量元素（毫克/升）：$FeSO_4 \cdot 7H_2O$ 5.0，$MnSO_4 \cdot H_2O$ 1.56，$ZnCl_2$ 1.67，$CoCl_2$ 2.0，pH 5.3。500 毫升三角瓶中装入 100 毫升培养基，孢子悬浮液接种（接种量影响生长和产酶量，一般每毫升孢子数约 10^6 个），在往复式摇床（108 次/分）上于 30℃左右（超过 34℃，菌丝体诱导活力下降）震荡培养二天左右，将菌丝体取出，洗去培养基，换用无碳源的培养基，作诱导测定。

（2）菌丝体洗涤方法。

于 3 号细菌漏斗上铺一张滤纸，将上述三角瓶中的培养物全部倒入抽滤，滤去培养基，留在滤纸上的菌丝体用 200 毫升磷酸盐缓冲液（0.01 mol/L，pH 5）洗涤，抽干。菌丝体饼取出，悬浮在 50 毫升上述缓冲液中，取一部分立即作诱导测定（冰箱中长时间放置会导致失活）。另取一部分于 80℃烘干，称干重，并换算成下面每个测定瓶中加入的菌丝体重量（以毫克/瓶表示）。

（3）诱导测定方法。

诱导培养基同生长培养基一样，但不含甘油和蛋白胨，使菌丝不能生长。25 毫升三角瓶或容量瓶中，加入诱导培养基 2.45 毫升（浓度高 1 倍），菌丝体悬浮液 2.45 毫升，槐糖 0.1 毫升（最终浓度 10^{-4} 或 5×10^{-4} mol/L），塞上棉花塞或瓶塞，于往复式摇床或恒温震荡器（100 次左右/分）30℃左右震荡培养。一般菌丝体与槐糖接触，经过 3 小时左右的延迟期，即出现纤维素酶活力，24 小时后可达高峰（见图 1）。由于培养基中无碳源，偏酸性，加入的菌丝量较多，测定时间又短，故不易染菌，整个操作过程不必

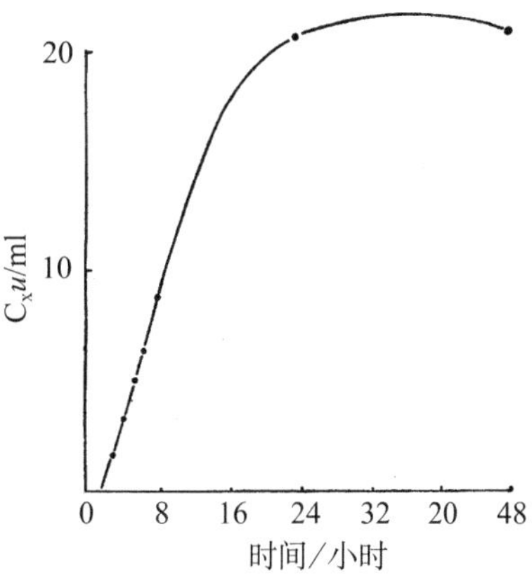

图 1　槐糖对木霉 EA_3 - 867 菌丝体纤维素酶合成的诱导作用

菌令 48 小时；菌丝体重量 8,4 mg/瓶；槐糖浓度 5×10^{-4} mol/L；诱导温度 30℃。

无菌操作。本法也是定性地鉴定纤维素酶诱导剂的有效方法，在上述标准测定中，凡是具有同槐糖一样，能诱导菌丝体形成纤维素酶的物质的，即是诱导剂。

2）纤维素酶活力测试

（1）酶液制备：胞外酶和胞内酶制备方法详见前文（朱雨生等，1978），一般酶活力测定取上述诱导培养滤液。

（2）纤维素酶活力测定方法：以分解羧甲基纤维素的 CMC 酶活力（C_x）作代表。C_x 和纤维二糖酶测定方法见前文（朱雨生等，1978）。在规定条件下每产生 0.5 毫克葡萄糖定为 1 个活力单位（u），每个测定均换算成每毫升滤液的酶活力单位数（u/ml）。

三、试验结果

1. 洗涤菌丝体纤维素酶的诱导形成

1）菌丝体生长条件的影响

（1）碳源。EA_3－867 能在各种碳源中生长（见表 1），但在甘油中生长最好，纤维二糖、麦芽糖、蔗糖和乳糖等双糖中生长较差。在所有碳源中生长的菌丝体都能被槐糖诱导形成纤维素酶，但菌丝体生长量的不同直接影响诱导产酶能力。生长量多的，一般产酶也高。同一碳源中培养的菌丝体，加入到诱导测定中的量越多，产酶量也越高。有意思的是，在纤维二糖中菌丝生长虽差，但诱导活力却较高，此与前文中我们推测纤维二糖在生长过程中可能转化为真正的诱导剂是一致的。

表 1　木霉 EA_3－867 在不同碳源中培养的菌丝体被槐糖诱导形成纤维素酶活力的比较

碳源	菌丝体重/（毫克/50 毫升）	C_x/（u/ml）（7 小时）
甘油	13.7	17.2
木糖	5.7	19.2
半乳糖	8.1	18.2
果糖	9.4	17.6
葡萄糖	4.6.	16.0
甘露糖	4.0	11.8
纤维二糖	1.9	12.2
麦芽糖	0.8	4.4
蔗糖	0.3	2.8
乳糖	0.3	2.6

注：培养基 50 毫升，各种碳源浓度均为 0.5%，唯甘油为 0.6%，孢子悬浮液接种，30℃摇床培养 72 小时，洗涤菌丝体用槐糖诱导，槐糖浓度为 5×10^{-4} mol/L，诱导温度为 30℃，7 小时后测定 CMC 活力（C_x）。

（2）营养盐。氮、磷、镁和钙等元素的无机盐及蛋白胨对于 EA_3－867 菌丝体的生长是必须的。并直接影响到纤维素酶的形成（见表 2）。缺镁时，不能生长；既缺钙又缺微量元素时，生长所受影响不大，但孢子形成加快，菌丝产酶能力却剧减；微量元素中铁和锌对生长是需要的,钴看来对生长有抑制。但是对菌丝产酶而言,生长过程中只需锌离子,其他三种微量元素影响不大,甚至不利,此与其他作者报道一致(Mandels 等,1957)。

表 2　生长培养基成分对木霉 EA_3 - 867 菌丝体生长及其被槐糖诱导产生纤维素酶活力的影响

生长培养基成分	培养基最终/pH	菌丝体重/(毫克/瓶)	C_x/(u/ml)	
			4 小时	7 小时
完全	3	12.1	4.4	14.8
- N	5	9.1	0.9	12.4
- P	5	7.4	0.9	9.4
-蛋白胨	3	4.3	1.1	4.8
完全	3	12.7	3.5	8.4
- Mg	5	0	—	—
- Ca	3	9.1	2.0	6.1
- Fe, Mn, Zn, Co	5	11.1	3.2	7.6
- Ca, Fe, Mn, Zn, Co	5	13.5	1.1	3.6
- Fe	3	11.4	5.0	11.0
- Mn	3	12.7	4.6	9.3
- Co	3	17.8	4.7	9.8
- Zn	3	11.3	2.3	7.2

注:菌令 45 小时;生长温度 29.5℃;诱导剂槐糖浓度 5×10^{-4} mol/L;诱导温度 29～30℃。

2) 诱导条件的影响

(1) 营养盐。槐糖诱导 EA_3 - 867 纤维素酶的形成需要氮、磷、铁、锰离子,其中氮和磷的影响最大,氮素中尤以硫酸铵为重要(见表 3)。曾比较了不同氮源的影响,发现尿素和硝态氮都不如氨态氮。在缺氮培养基中震荡 3 小时后,补加氮素,则最终酶活力可追上对照,而缺氮的影响主要表现在诱导初期,如诱导 6 小时,酶活力只及对照的一半,到 24 小时则相当于对照的 90%,缺磷也是同样的趋势。此可能与诱导过程中动用菌体自身分解产生的氮和磷有关,因为在培养 7 小时后,菌丝体约已减重 1/4 左右。

表 3　诱导培养基成分对木霉 EA_3 - 867 菌丝体被槐糖诱导形成纤维素酶活力的影响

诱导培养基成分	C_x/(u/ml)(7 小时)
(1) 完全	5.5
- N, P Mg, Ca	1.2
- N(硫酸胺,尿素)	2.7
- P	3.0
- Mg	5.6

（续表）

诱导培养基成分	C_x/(u/ml)(7 小时)
- Ca	5.6
- Fe, Mn, Zn, Co	4.2
(2) 完全	14.8
- N(硫酸胺,尿素)	7.0
-硫酸胺	10.2
-尿素	13.2
(3) 完全	7.4
- Fe, Mn, Zn, Co	6.7
- Fe	6.9
- Mn	7.1
- Zn	7.4
- Co	7.4

注：
(1) 菌令 48 小时,菌丝体重,13.1 毫克/瓶。
(2) 菌令 45 小时,菌丝体重,11.2 毫克/瓶。
(3) 菌令 45 小时,菌丝体重,10.2 毫克/瓶。
完全培养基成分见方法部分说明。

生长培养基和诱导培养基中营养盐对槐糖诱导 EA_3 - 867 洗涤菌丝体纤维素酶形成的影响归纳在表 4 中。

表 4 生长培养基和诱导培养基成分对槐糖诱导木霉 EA_3 - 867 菌丝体纤维素酶形成的影响(总结)

成分	生长培养基	诱导培养基
C 源	+	—
N	+	+
P	+	+
蛋白胨	+	—
Mg	+	—
Ca	+	—
Fe	+	+
Mn	—	+
Zn	+	—
Co	—	—

注:“+”代表需要,“—”代表不需要。

(2) 槐糖(诱导剂)浓度。极微量的槐糖(10^{-6} mol/L,相当于 0.34 ppm)便对纤维素酶有明显的诱导作用。浓度升高,诱导活力也增加。7 小时测定时,10^{-5}～10^{-3} mol/L 槐糖作用差不多;24 小时测定时,10^{-3} mol/L 槐糖的诱导活力为最高;48 小时测定时,则以 10^{-2} mol/L 槐糖的诱导活力为最高(见图 2)。这说明诱导时间延长,槐糖被消耗,造成浓度不足。

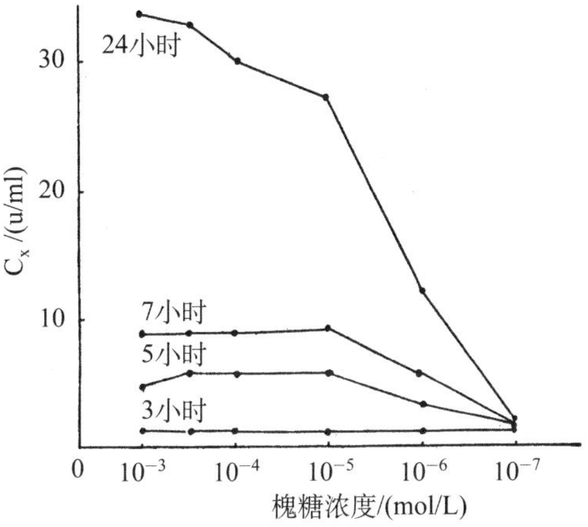

图 2　不同槐糖浓度对木霉 EA_3－867 菌丝体纤维素酶诱导形成的影响

菌令 45 小时;菌丝体重量 14.6 mg/瓶;诱导温度 30℃。

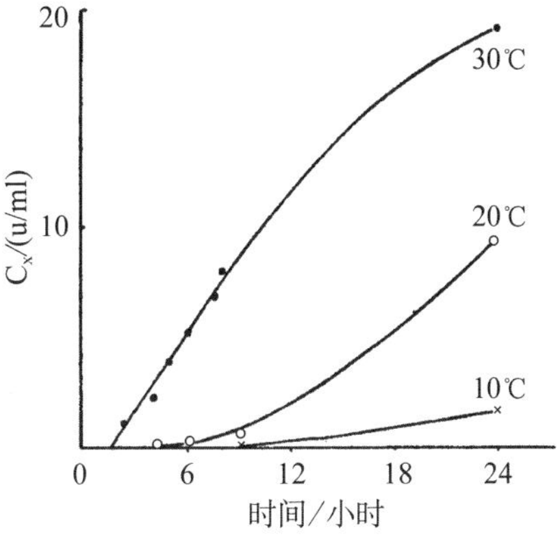

图 3　不同温度对槐糖诱导木霉 EA_3－867 菌丝体纤维素酶形成的影响

菌令 48 小时;菌丝体重 8.4 mg/瓶;槐糖浓度 5×10^{-4} mol/L。

.—. 30℃;○—○ 20℃;×—× 10℃。

(3) 诱导温度。最适宜温度在 30℃左右。随着温度的降低,延迟期便延长,产酶量也降低,$Q_{10}>2$(见图 3)。

(4) 诱导 pH。菌丝产酶的 pH 范围较广,偏酸性,从 pH 3～6 均适宜,最适 pH 在 5 左右(见图 4)。碱性条件显然不利于产酶。

(5) 菌令。孢子萌发不久刚长出肉眼可见的短菌丝(15 小时)时,洗去甘油,进行槐糖诱导测定,便发现此幼令的菌丝已具有合成纤维素酶的能力,表明控制纤维素酶合成的遗传装置和蛋白质合成系统,在生长早期就在菌丝体中形成。随着菌令的增长,菌丝的生长,总的产酶量也随之增高。72 小时后,菌丝体开始自溶,活力便下降(见图 5)。

(6) 培养方式和通气条件。震荡条件下菌丝体诱导形成纤维素酶的产量远远超过静止培养,说明产酶与通气有关。在诱导过程中通入氧气,可以增

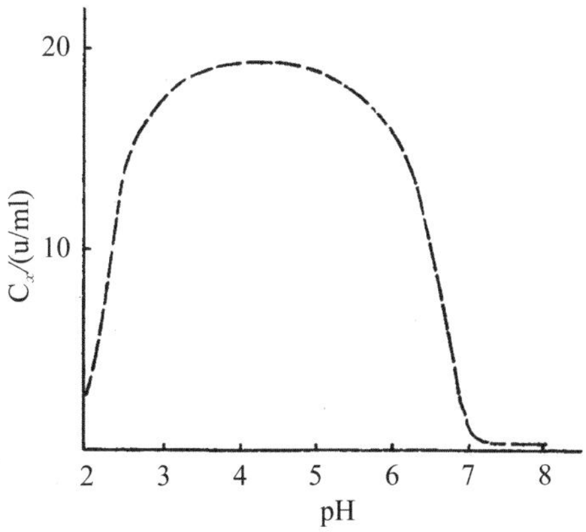

图 4　pH 对槐糖诱导木霉 EA_3－867 菌丝体纤维素酶形成的影响

菌丝体菌令 48 小时;菌丝体重量 11.9. mg/瓶;槐糖浓度 5×10^{-4} mol/L;诱导温度 29℃。菌丝体用不同 pH 的磷酸缓冲液(0.2 mol/L)洗涤,并悬浮在同样的各种 pH 的缓冲液中,诱导 8 小时后,测定 CMC 酶(C_x)活力

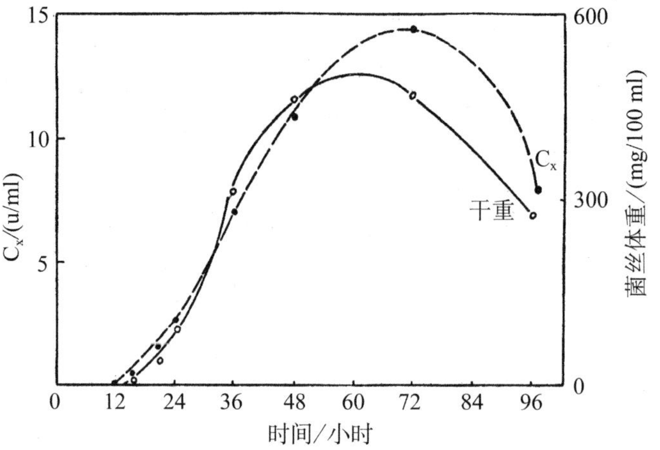

图 5　木霉 EA_3 - 867 不同菌令菌丝体被槐糖诱导形成纤维素酶活力的变化

.----. C_x；○——○ 干重。

加产酶；抽去空气，灌入氩气，则酶不能诱导产生，即使在诱导 3.5 小时后恢复空气，纤维素酶也不再形成(见表 5)。

表 5　O_2 对槐糖诱导木霉 EA_3 - 867 菌丝体形成纤维素酶活力的影响

气相	通气时间/小时	C_x/(u/ml)(7 小时)
空气(对照)	0～7	4.1
O_2	0～7	5.7
Ar	0～7	0
	0～3.5 小时，后 3.5～7 小时恢复空气	0

注：菌令 45 小时；菌丝体重 7.6 毫克/瓶；槐糖浓度 5×10^{-4} mol/L，诱导温度 30℃。

诱导试验在桑氏管中进行。槐糖和培养基加在侧管中，菌丝体悬浮液加在主管中，密封后，抽气 3 分钟，再分别灌入 O_2 或 Ar，以空气作对照。然后，将侧管溶液倾入主管，于瓦氏呼吸器振荡机上保温振荡培养。通 Ar 组有一组在诱导 3 小时半后恢复空气。所有处理在培养 7 小时测定纤维素酶活力。

(7) 诱导中途洗去槐糖的影响。菌丝体与槐糖的持续接触，对于纤维素酶的不断形成是必须的。若在菌丝体与槐糖一起震荡培养的不同时间，用细菌漏斗抽滤，除去槐糖，并用磷酸盐缓冲液洗涤菌丝体，后者悬浮在缺槐糖的诱导培养基中作诱导测定，可以看到，纤维素酶的合成不久便停止(见图 6)。洗去槐糖的时间越早，影响越大。一般在洗去槐糖后的 90 分钟至 120 分钟，纤维素酶的产生便低于对照。

表 6 是诱导 4 小时后，改换培养基或菌丝体对纤维素酶形成的影响。试验 1 中，改换新的培养基(包括槐糖和营养盐)后，产酶明显增加。但换用新的菌丝体，对酶合成影响不大，说明诱导中途(4 小时)进一步影响产酶的主要限制因子是培养基的消耗，而不是菌丝体的衰老。从试验 2 中，则进一步可知，培养基中限制产酶的主要成分是槐糖，而不是无机盐。到诱导后期(24 小时)，菌丝体的衰老和解体将成为限制产酶的重要因素。

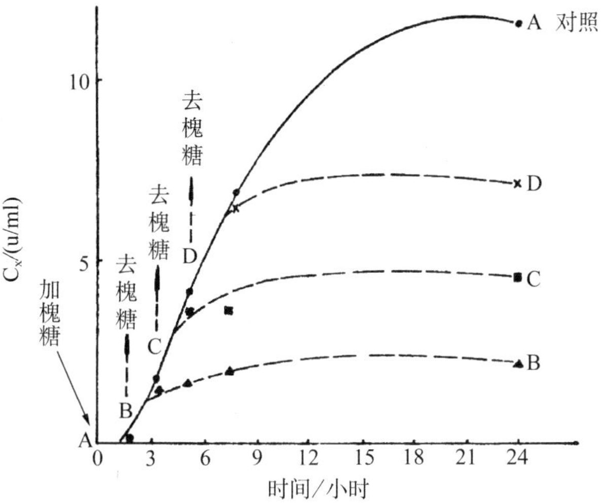

图6 木霉 EA_3－867 菌丝体槐糖诱导纤维素酶形成的过程中不同时间洗去槐糖对酶合成的影响

菌令 48 小时；菌丝体重 12.1mg/瓶；槐糖浓度 10^{-4} mol/L；诱导温度 30℃。

●——● A(对照)；■----■ C 诱导 3 小时洗去槐糖；▲----▲ B 诱导 1.5 小时洗去槐糖；×----× D 诱导 5 小时洗去槐糖。

表6 木霉 EA_3－867 菌丝体被槐糖诱导中途改换培养基对纤维素酶形成的影响

试验	处理	C_x/(u/ml)		
		4 小时	7 小时	24 小时
1	对照	4.4	8.8	22.8
	抽滤后悬浮在原培养基	4.4	8.5	18.4
	抽滤后悬浮在新鲜培养基	4.4	11.1	31.6
	抽滤后滤液中加新鲜菌丝	4.4	3.8	20.0
2	对照	3.4	8.0	25.6
	补加无机盐(4 小时加)	3.4	7.6	25.2
	补加槐糖(10^{-4} mol/L)	3.4	8.5	29.2

注：试验 1，菌令 45 小时；菌丝体重 14.1 毫克/瓶；试验 2，菌令 48 小时；菌丝体重 12.4 毫克/瓶。

槐糖浓度均为 10^{-4} mol/L，诱导温度 30℃，改换培养基(含无机盐和槐糖)是在诱导 4 小时后，将培养物倒入 3 号细菌漏斗中抽滤，再用 5 毫升磷酸缓冲液(0.01 mol/L，pH 5)洗涤二次，将抽干的菌丝体悬浮在新鲜的或原来的培养基中。补加无机盐或槐糖是指诱导 4 小时后，分别补加到培养物中，补加的浓度均相当于原来培养基中的浓度，以补加水作对照。

(8) 不同诱导剂的比较。曾研究纤维素粉、稻草纤维素、纤维二糖及槐糖对 EA_3－867 洗涤菌丝体诱导形成纤维素酶的能力作了比较。在槐糖(10^{-4} mol/L)诱导下，24 小时纤维素酶(C_x)活力达到 22 u/ml，值得注意的是，纤维素粉、稻草纤维素(0.5%)和不同浓度的纤维二糖(10^{-4}～10^{-2} mol/L)都不能诱导形成纤维素酶。

(9) 菌种。各种木霉菌种的菌丝体均能不同程度地被槐糖诱导形成纤维素酶(见表7)，且诱变菌株合成纤维素酶的能力高于野生型菌株。但是，槐糖对具有纤维素分解能力的曲霉T17.5－24及产黄色纤维单胞杆菌(D菌)的诱导能力极低。

表7 槐糖对不同菌种菌丝体诱导纤维素酶形成的比较

菌　种	菌丝体重/(毫克/瓶)	槐糖浓度/(mol/L)	C_x/(u/ml)(7小时)
EA_3－867(拟康氏木霉，从1096诱变而来)	8.4	5×10^{-4}	12.1
657(拟康氏木霉，从1096诱变而来)	10.1		9.2
1096(拟康氏木霉，野生菌)	10.5		9.0
G109(木霉)	10.1		2.1
D－92(木霉)	9.2		2.5
T17.5－24(曲霉)	0.25		0.7
D(细菌)	0.30		0.6
木－3(拟康氏木霉，野生菌)	11.7	10^{-4}	7.0
N_2－78(拟康氏木霉，从木－3诱变而来)	2.8		15.0
G58(长梗木霉，诱变种)	4.3	10^{-4}	12.0
4030(木霉，诱变种)	4.6		3.0
G023(木霉，诱变种)	7.8		13.4
$白_{24}$(拟康氏木霉，从EA_3－867诱变而来)	6.0		9.0

注：菌令42～47小时；诱导温度29.5℃。

本试验中G109和D－92是湖北省微生物所供给的木霉野生菌，G58由广东省微生物所供给，4030和G023由西北水土保持所供给，$白_{24}$由中山大学生物系供给，D菌是本组筛选的产黄色纤维单胞杆菌。细菌培养基为每100毫升中含：$(NH_4)_2SO_4$ 0.2克，Na_2HPO_4 0.14克，KH_2PO_4 0.03克，$MgSO_4\cdot7H_2O$ 0.02克，甘油0.6克，维生素B_1 500微克，pH 7.0。菌体培养后，离心，用40毫升磷酸缓冲液洗涤数次，细菌菌体悬浮在缺甘油的上述培养基中，加入槐糖，进行诱导测定。

3) C_x酶和纤维二糖酶的胞内胞外分布

EA_3－867洗涤菌丝体被槐糖诱导形成的纤维素酶的不同组分在细胞内外的分布有着一定的规律。纤维二糖酶分布在胞内，C_x酶主要分布在胞外(见图7)。但是，在诱导初期(4小时)，胞内的C_x酶略高于胞外，以后胞内C_x酶变化不大，而胞外的C_x酶直线上升。C_x酶在胞内胞外的这种消长变化与生长培养时的规律一致(朱雨生等，1978)，这意味着C_x酶先在细胞内合成，接着迅速地向细胞外分泌。C_1酶(棉花糖表示)也是分泌到胞外。

4) 纤维素酶诱导形成与呼吸代谢的关系

EA_3－867洗涤菌丝体同槐糖保温的同时，加入不同的呼吸抑制剂，如碘乙酸、氟化钠，丙二酸、迭氮化钠和DNP，测定对纤维素酶诱导形成的影响(见图8)，证明各种呼吸抑制剂对菌丝体产酶都有明显的抑制作用。

5) 纤维素酶诱导形成与核酸代谢的关系

EA_3－867洗涤菌丝体与槐糖及各种核酸代谢的抑制剂同时保温震荡，测定对纤维素酶形成的影响。结果如表8、表9和图9所示。从表中可知，放线菌素D、放线菌酮、6－氮杂尿嘧啶

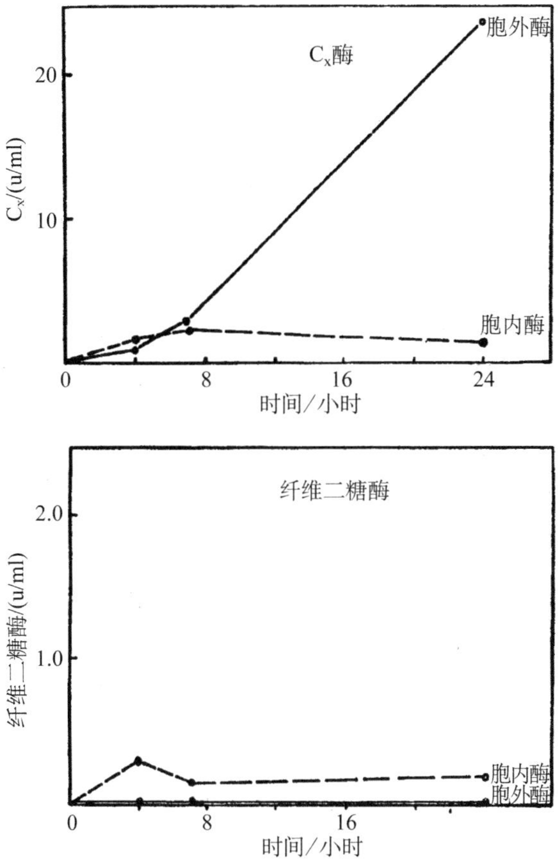

图 7　木霉 EA_3－867 菌丝体被槐糖诱导不同时间 C_x，纤维二糖酶的胞内胞外变化

菌令 45 小时；菌丝体重量 10.8 mg/瓶；槐糖浓度 10^{-4} mol/L；诱导温度 30℃。

○—○ 胞外酶；●⋯● 胞内酶。

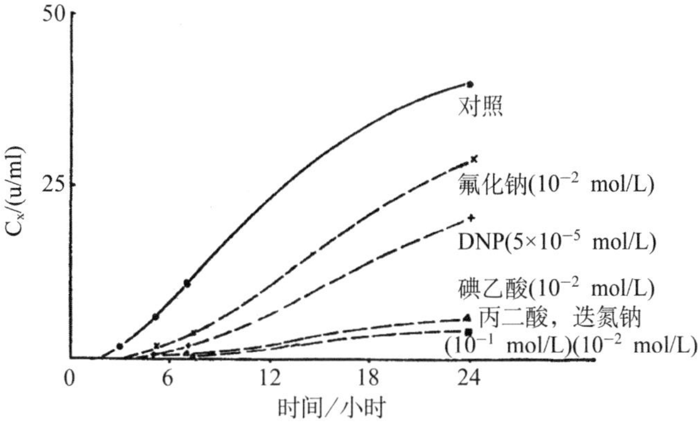

图 8　呼吸抑制剂对槐糖诱导木霉 EA_3－867 菌丝体纤维素酶形成的抑制作用

菌令 45 小时；菌丝体重量 10.9mg/瓶；槐糖浓度 5×10^{-4} mol/L；温度 30℃。

●—●对照；▲⋯▲ 加碘乙酸；×⋯× 加氟化钠；■⋯■ 加丙二酸或迭氮钠；+⋯+ 加 DNP。

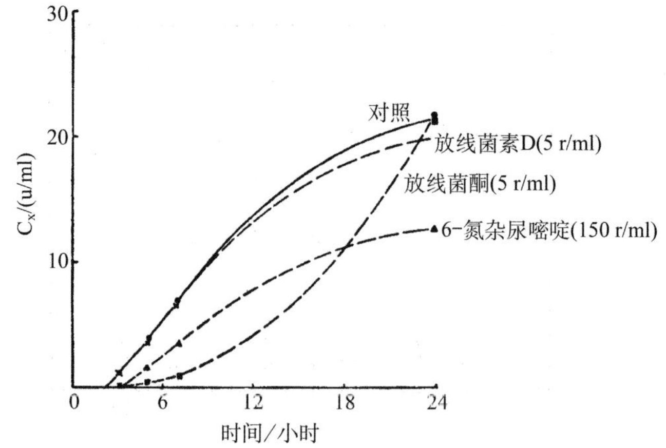

图 9　核酸或蛋白合成抑制剂对槐糖诱导木霉 EA_3 - 867 菌丝体纤维素酶形成的抑制作用

菌令 45 小时；菌丝体重 11.7 mg/瓶；槐糖浓度 5×10^{-4} mol/L；诱导温度 30.5℃。

●——● 对照　▲- - -▲ 加6-氮杂尿嘧啶

×- - -× 加放线菌素D　■- - -■ 加放线菌酮

表 8　不同浓度放线菌素 D 和放线菌酮对槐糖诱导木霉 EA_3 - 867 菌丝体合成纤维素酶的影响

抑制剂	浓度/(微克/毫升)	C_x/(u/ml)		
		3 小时	5 小时	7 小时
对照	—	2.1	4.1	7.6
放线菌素 D	1	2.0	3.6	7.4
"	10	1.8	3.2	7.4
"	100	1.6	2.5	5.5
放线菌酮	1	1.5	2.1	5.6
"	10	0	0	1.5
"	100	0	0	1.5

注：菌令 45 小时；菌丝体重 10.2 毫克/瓶；槐糖浓度 10^{-4} mol/L；诱导温度 30℃。

表 9　核酸和蛋白质合成抑制剂对槐糖诱导木霉 EA_3 - 867 菌丝体纤维素酶合成的影响

抑制剂	浓度/(微克/毫升)	C_x/(u/ml)		
		5 小时	8 小时	24 小时
对照	—	4.8	8.2	23.6
利福霉素	10	4.6	7.3	22.4
"	100	4.2	7.5	20.0
6-氮杂尿嘧啶	150	3.0	5.6	16.4
"	1 500	1.9	3.2	8.4
5-氟尿嘧啶	20	2.8	5.0	15.2
"	200	2.7	4.5	13.2

注：菌令 45 小时；菌丝体重 9.2 毫克/瓶；槐糖浓度 10^{-4} mol/L；诱导温度 29.5℃。

和 5-氟尿嘧啶对纤维素酶的合成均有不同程度的抑制作用。其中放线菌酮的抑制作用最强，在 1～10 微克/毫升的浓度范围，对纤维素酶的合成便有明显的抑制作用。但到 24 小时后，纤维素酶的合成往往可以恢复。EA_3-867 纤维素酶的合成对放线菌素 D 不太敏感，当浓度高到 100 微克/毫升时，才有部分的抑制作用。曾试图用 EDTA（10^{-3} mol/L）处理菌丝体（Perlman 等，1968），以及超声波破碎细胞的方法，以排除透性的影响，但均不能增加 EA_3-867 对放线菌素 D 的敏感性。由表 9 可知，利福霉素对 EA_3-867 纤维素酶的形成无明显的抑制作用。

若 EA_3-867 菌丝体与槐糖预保温 4 小时，让 mRNA 充分形成，然后，加入放线菌素 D，放线菌酮和 6-氮杂尿嘧啶，震荡培养不同时间，测定对纤维素酶合成的影响（见图 10）。结果发现 6-氮杂尿嘧啶，尤其是放线菌酮，对纤维素酶合成有明显的抑制作用，而放线菌素 D 抑制不明显，直至培养 20 小时，才表现若干抑制作用。

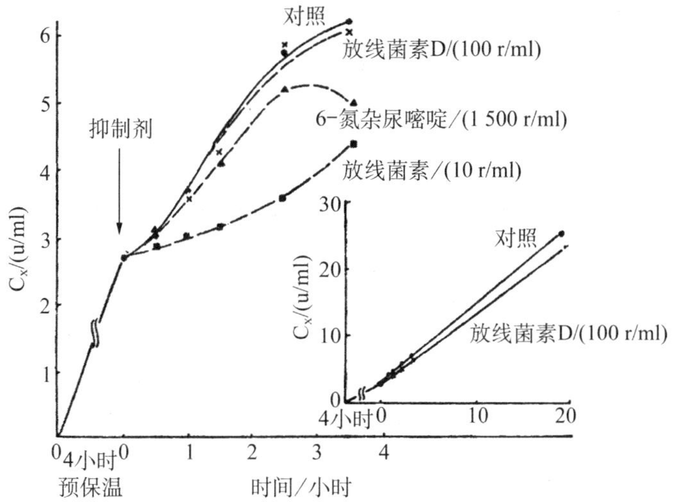

图 10　核酸或蛋白合成抑制剂对槐糖诱导木霉 EA_3-867 菌丝体纤维素酶形成的抑制作用

菌令 45 小时；菌丝体重量 9.1mg/瓶；槐糖浓度 10^{-4} mol/L；诱导温度 29.5℃。菌丝体先同槐糖预保温（震荡）4 小时后，分别加各种抑制剂。
●—● 对照；▲⋯▲ 6-氮杂尿嘧啶；×⋯× 放线菌素 D；■⋯■ 放线菌酮。

2. 降解物阻遏

菌丝体与槐糖保温的同时，加入葡萄糖（10^{-2} mol/l），则对纤维素酶的形成产生明显的阻遏作用（见表 10），表现为延迟期的延长和酶活力高峰的推迟。但是，由于葡萄糖作为生长的碳源，增加了菌丝体的数量，最终的产酶量要高于对照。增加槐糖的浓度不能克服葡萄糖的阻遏作用，说明两者的作用不是竞争性的。但在同样的葡萄糖浓度水平下，槐糖浓度越高，形成的纤维素酶也越多。表 10 中还可看出，加入葡萄糖后，培养基的 pH 迅速地从 5.3 降为 2.5，以后随着酶活力的升高而回升。若在诱导过程中，调节培养基中的 pH，使其稳定在 pH 5 左右，葡萄糖对纤维素酶的产生仍呈现阻遏作用。葡萄糖对纤维素酶形成的阻遏作用不能被加入的 c-AMP（5×10^{-3}～5×10^{-2} mol/L）所解除（见表 11）。

表 10 葡萄糖对木霉 EA_3 - 867 菌丝体在不同浓度槐糖诱导下纤维素酶形成的阻遏作用

诱导剂及阻遏剂	4 小时		7 小时		18 小时		24 小时	
	pH	C_x/(u/ml)	pH	C_x/(u/ml)	pH	C_x/(u/ml)	pH	C_x/(u/ml)
槐糖(10^{-4} mol/L)	5.3	4.3	4.6	11.0	5.1	28.0	5.1	30.0
槐糖(10^{-4} mol/L)+葡萄糖(10^{-2} mol/L)	2.5	0	2.5	4.2	4.4	31.2	5.1	34.4
槐糖(10^{-3} mol/L)+葡萄糖(10^{-2} mol/L)	2.5	0	2.5	4.4	4.0	35.6	5.1	38.8
槐糖(10^{-2} mol/L)+葡萄糖(10^{-2} mol/L)	2.5	0	2.5	4.2	3.5	44.0	3.5	57.0

注:菌令 42 小时;菌丝体重 7.0 毫克/瓶;诱导温度 29.5℃。

表 11 葡萄糖对槐糖诱导木霉 EA_3 - 867 菌丝体纤维素酶的阻遏作用和 c - AMP 的作用

处　理	C_x/(u/ml)			
	3 小时	5 小时	7 小时	24 小时
对照	1.1	3.8	6.8	21.6
+葡萄糖(10^{-2} mol/L)	0	0	1.9	27.2
+葡萄糖(10^{-2} mol/L)　c - AMP(5×10^{-3} mol/L)	0	0	1.4	25.2
+葡萄糖(10^{-2} mol/L)　c - AMP(10^{-2} mol/L)	0	0	1.2	22.8

注:菌令 45 小时;菌丝体重 11.7 毫克/瓶;槐糖浓度 5×10^{-4} M;诱导温度 30.5℃。

除了葡萄糖外,其他各种糖、糖磷酸酯等,如果糖、半乳糖、木糖、核糖、乳糖、麦芽糖、G-6-p、6-PG,以及葡萄糖酸、甘油等对木霉 EA_3 - 867 洗涤菌丝体纤维素酶的诱导形成都有不同程度的阻遏作用(见表 12)。但蔗糖利用很慢,未见明显的阻遏作用,FDP 也是例外。一般在加入阻遏剂后 4 小时左右,便表现明显的阻遏现象,到 24 小时后,阻遏剂基本上耗尽,各种阻遏大部分解除,并由于它们作为碳源,促进了生长,故最终纤维素酶的产量赶上或超过对照。

表 12 糖和糖磷酯等对槐糖诱导木霉 EA_3 - 867 菌丝体形成纤维素酶的阻遏作用

阻遏剂	C_x/(u/ml)	
	5 小时	24 小时
对照	6.1	31.6
葡萄糖	1.6	35.6
葡萄糖酸	1.1	32.0
果糖	5.3	34.8
半乳糖	3.2	26.8
木糖	4.3	33.2
核糖	4.9	39.6
乳糖	3.5	33.6
蔗糖	6.0	31.2

（续表）

阻遏剂	C_x/(u/ml)	
	5 小时	24 小时
麦芽糖	4.0	35.6
甘油	4.5	45.2
G-6-P	5.1	34.4
6-PG	5.5	32.8
FDP	7.0	32.4

注：菌令 45 小时；菌丝体重 12.8 毫克/瓶；槐糖浓度 10^{-4} mol/L；诱导温度 30℃；阻遏剂浓度 10^{-2} mol/L。

EA_3-867 洗涤菌丝体纤维素酶的诱导形成还被三羧酸循环中的各种有机酸（10^{-2}～5×10^{-2} mol/L），如醋酸、丙酮酸、柠檬酸、α-酮戊二酸、琥珀酸、苹果酸、延胡索酸和草酰乙酸所阻遏。辅酶 NADP、NAD 和 ATP 也有阻遏作用（见表 13）。

表 13　三羧酸循环中的有机酸及 ATP、NADP、NAD 对槐糖诱导木霉 EA_3-867 菌丝体形成纤维素酶的阻遏作用

阻遏剂	浓度/(mol/L)	C_x/(u/ml)		
		4 小时	6 小时	24 小时
对照	—	6.5	8.0	29.2
醋酸	10^{-2} 5×10^{-2}	2.9 1.5	5.6 2.0	23.6 7.2
丙酮酸	10^{-2} 5×10^{-2}	5.1 4.9	7.7 7.5	26.4 19.2
柠檬酸	10^{-2} 5×10^{-2}	5.5 5.4	8.0 8.1	28.4 27.6
α-酮戊二酸	10^{-2} 5×10^{-2}	4.7 6.3	7.1 8.6	28.4 29.2
琥珀酸	10^{-2} 5×10^{-2}	4.7 3.8	7.0 5.9	23.6 12.0
苹果酸	10^{-2} 5×10^{-2}	4.6 4.3	7.3 8.6	28.8 29.6
延胡索酸	10^{-2} 5×10^{-2}	4.2 4.1	7.8 7.1	22.4 18.8
草酰乙酸	10^{-2} 5×10^{-2}	5.0 3.6	7.7 6.7	30.0 27.2
ATP	10^{-2} 5×10^{-2}	5.2 5.2	8.0 9.1	27.6 32.8
NADP	10^{-2} 5×10^{-2}	6.0 6.0	9.0 9.3	28.8 29.6
NAD	10^{-2} 5×10^{-2}	5.1 5.4	7.7 8.6	29.2 30.4

注：菌令 45 小时；菌丝体重 11.1 毫克/瓶；槐糖浓度 10^{-4} mol/L；阻遏剂在 0 小时加入。

EA_3 - 867 洗涤菌丝体先同槐糖预保温 4 小时，再加入各种阻遏剂（10^{-2} mol/L），一起震荡培养，90 分钟后，测定纤维素酶的活力，则葡萄糖、各种有机酸和辅酶等对纤维素酶形成仍有阻遏作用（见表 14），但 24 小时后，纤维素酶活力大多又恢复。

表 14 槐糖诱导木霉 EA_3 - 867 菌丝体形成纤维素酶的中途加入三羧酸循环中有机酸和 ATP、NADP、NAD 对产酶的阻遏作用

阻遏剂	C_x/(u/ml)	
	90 分钟	20 小时
对照	5.8	34.0
葡萄糖	5.3	34.4
醋酸	4.2	23.6
丙酮酸	5.3	35.6
柠檬酸	5.5	35.2
α-酮戊二酸	5.3	35.6
琥珀酸	5.3	30.4
苹果酸	5.4	36.0
延胡索酸	5.4	30.8
草酰乙酸	5.5	34.4
ATP	5.4	34.4
NADP	5.2	33.6
NAD	5.3	32.8

注：菌令 45 小时；菌丝体重 13.8 毫克/瓶；槐糖浓度 10^{-4} mol/L；诱导温度 30℃。

菌丝体先同槐糖保温震荡 4 小时（此时酶活 C_x 为 3.9 u/ml），再加入各种阻遏剂，浓度均为 10^{-2} mol/L，并在加入后 90 分钟和 20 小时测定纤维素酶活力。

四、讨论

当下关于酶诱导形成的一系列试验，证明诱导酶的形成是新酶蛋白分子的合成，而不是前酶的激活（Jacob 等，1961；Monod，1958）。Nisizawa 等用双标记同位素法，令人信服地证明绿色木霉纤维素酶的诱导形成是酶蛋白新合成的过程（Nisizawa 等，1971a）。静止细胞的诱导培养法是研究酶的诱导形成的有效方法（Pardee 等，1961；Spiegelman，1948）。若干研究者报道了槐糖诱导绿色木霉洗涤菌丝体纤维素酶形成的方法（Mandels 等，1962；Nisizawa 等，1971；Reese 等，1971），但对于产酶的条件和规律，未见系统的报道。本文的试验结果表明，拟康氏木霉纤维素酶的诱导形成需要：①诱导剂；②营养盐；③适宜的温度、pH 和通气；④菌丝体细胞活跃的代谢。

槐糖是各种木霉菌纤维素酶的强力诱导剂。但槐糖对其他菌的纤维素酶的诱导能力极低，这与其他研究人员报道一致（Mandels 等，1962），萤光假单胞菌是例外（Yamane 等，1970）。Mandels 等曾认为纤维素的水解产物纤维二糖是真正的诱导剂（Mandels 等，1969），但在我们的试验中，纤维素、纤维二糖对洗涤菌丝体静止细胞的纤维素酶没有诱导作用。然而，在以纤维素或纤维二糖为碳源的培养基中进行生长培养时，则 EA_3 - 867 能形成纤维素酶（朱雨生等，1978），所以推测，纤维二糖可能不是纤维素酶（如 C_x 酶）真正的诱导剂，其必须在生长过程中通过代谢转化才能诱导纤维素酶形成。底物乳糖通过转糖苷作用转化为别乳糖才能诱导大肠杆菌的半乳糖苷酶形成，是诱导剂代谢转化形成的另一例证（Jobe 等，1972）。

EA_3 - 867 纤维素酶的形成需要有氮、磷、铁和锰等元素的无机盐参加。其中，氮的影响最大，缺氮后补加铵盐可使纤维素酶的合成达到对照水平，这说明 EA_3 - 867 纤维素酶的诱导形成可能也是酶蛋白新合成的结果。缺氮时仍能合成部分纤维素酶可以归因于菌体蛋白的解体以提供酶合成的原料。Spiegelman 和 Dunn 在研究啤酒酵母对半乳糖的适应时也发现，外加氮源时，新酶的形成首先利用外加的氮；而缺氮时，新酶的形成则靠其他酶蛋白或贮存蛋白的解体（Spiegelman，1947；Spiegelman 等，1948）。

正常生长的菌丝体和菌丝体正常活跃的代谢是纤维素酶诱导形成的另一必须条件。在营养不良和条件不利（如高温）情况下生长的菌丝体产酶量低，正常生长的菌丝体，若用 10^{-3} mol/L 的 EDTA 处理 10 分钟，产酶能力便降低，若用 3×10^{-2} mol/L 的 EDTA 处理，或冰冻处理，则活力全部丧失，可见纤维素酶的形成可能与细胞膜的结构有关。正常生长的菌丝体，如果在不适宜的温度、pH 条件下，酶的形成也受阻。从温度对产酶的影响，求得 Q_{10} 大于 2，说明诱导过程是个生化反应。诱导系统中抽去氧气，酶形成被抑制，增加氧气，则促进产酶，看来，纤维素酶的形成与呼吸代谢（特别是有氧呼吸）有着密切的关系。呼吸抑制剂试验则进一步说明，EA_3 - 867 纤维素酶的诱导形成，必须要由菌丝体旺盛的呼吸，即糖酵解和三羧酸循环参加，以及需要 ATP 的能量供应，末端氧化是由金属氧化酶参加。

EA_3 - 867 洗涤菌丝体纤维素酶的诱导形成不同程度地被放线菌酮、5 -氟尿嘧啶、6 -氮杂尿嘧啶和放线菌素 D 所抑制。其中放线菌素 D 是抑制 DNA 的转录（Zahner，1972），放线菌酮是抑制 mRNA 的翻译（Wisiman 1975），5 -氟尿嘧啶和 6 -氮杂尿嘧啶干扰 RNA 的形成（Pardee 等，1961）。说明 EA_3 - 867 纤维素酶的诱导形成也应包括转录和翻译的过程，需要 mRNA 的参加。此 mRNA 是不稳定的，因为诱导中途洗去槐糖，数小时后，酶的合成便停止。EA_3 - 867 纤维素酶的形成虽然能被放线菌素 D 抑制，但抑制较弱，此与其他作者在绿色木霉中的观察不同（Nisizawa 等，1971a；Nisizawa 等，1972）。放线菌酮的抑制作用较强，但 24 小时后，抑制常常可以解除。EA_3 - 867 对利福霉素不敏感，说明纤维素酶合成不在线粒体上。

拟康氏木霉 EA_3 - 867 纤维素酶的形成不仅受诱导机制的调节，而且受降解物阻遏机制的调节。本试验表明，不仅葡萄糖和甘油，其他各种糖类、糖磷酸酯、三羧酸循环中的许多有机酸，以及辅酶 NAD、NADP 及 ATP 都能阻遏槐糖对洗涤菌丝体纤维素酶的诱导形成。由此推测各种阻遏剂可能通过共同的代谢产物或共同的机制来阻遏酶的形成。其中 FDP 是例外，Moses 等在大肠杆菌中报道过类似的现象（Moses 等，1970），并认为 FDP 可能起着作用子的作用。本试验中值得注意的是，c - AMP 不能解除葡萄糖对纤维素酶形成的阻遏作用。而许多相关报道，大肠杆菌中降解物阻遏是由于降低了细胞中 c - AMP 的含量（Makman 等，1965；Crombrugghe 等，1969），外加 c - AMP 可克服上述阻遏（Perlman 等，1968），并认为 c - AMP 是结合在启动因子上发动基因转录的必要因子（Perlman 等，1968，1969；Pastan 等，1970）。由此表明拟康氏木霉 EA_3 - 867 降解物阻遏机制同原核生物是不同的。

参考文献

[1] 朱雨生、谭常，1978，木霉纤维素酶的诱导形成及其调节、I. 槐糖对木霉 EA_3 - 867 纤维素酶形成的诱导作用，微生物学报，18：（即将发表）

[2] Crombrugghe, B. D. , R. L. Perlman, H. E. Varmus and I. Pastan, 1969. Regulation of inducible enzyme synthesis in *Eschrichia coli* by cyclic adenosin 3′, 5′ - monophosphate. *J. Biot.* Chem. 244: 5828.

[3] Jacob, F. and J. Monod, 1961. Genetic regulatory mechanisms in the synthesis of proteins. *J. Mol. Biol.* 3: 318.

[4] Jobe, A. and S. Bourgeois, 1972. Lac repressor-operator interaction. Ⅵ. The natural inducer of the lac operon. *J. Mol. Biol.* 69: 397.

[5] Makman, R. S. and E. W. Sutherland, 1965. Adenosine 3′, 5′ phosphate in *E. coli*. *J. Biochem. Chem.* 240: 1309.

[6] Mandels, M, and E. T. Reese, 1957. Induction of cellulase in Trichoderma viride as influencel by carbon sources and metals. J. Bact. 73: 269.

[7] Mandels, M. , F. W. Parrish and E. T. Reese. 1962. Sophorose as an inducer of cellulase in *Trichoderma viride*. *J. Bact.* 83: 400.

[8] Mandels, M. and Weber. J. 1969. The production of cellulases. *Adv. Chem. Ser.* 95: 391.

[9] Monod, J. 1958. An outline of enzyme induction. *Rec. Trav. Chim.* Pays-Bas, 77: 569.

[10] Moses, V. and B S. Pamela, 1970. Adenosine 3′, 5′- cyclic monophospbate and catabulite repression in *E. coli*. *J. Biochem.* 118: 481.

[11] Nisizawa, T. , H. Suzuki and K. Nisizawa, 1971 a. "De novo" synthesis of cellulase induced by sophorose in *Trichoderma viride* cells. J. Biocchem. 70: 387.

[12] Nisizawa, T. , H. Suzuki, M. Nakayama and K. , Nisizawa, 1971 b. Inductive formation of cellulase by sophorose in Trichoderma viride. *J. Biochem.* 70: 375.

[13] Nisizawa, T. , H. Suzuki, and Ki Nisrzawa, 1972. Catabolite repression of cellulase formation in *Trichoderma viride*. *J. Biochem* 71: 999.

[14] Pardee, A. B. and Prestidge, L. S. 1961. The initial kinetics of enzyme induction. *Biochem. Biophys. Acta*, 49: 77.

[15] Pastan, I. and I. Perlman, 1970. Cyclic adenosine monophosphate in bacteria. *Science*, 162: 339.

[16] Perlman, R. L. , B. D. Crombrugghe and I. Pastan, 1969. Cyclic AMP regulates catabolite and transient repression in E. coli. Nature, (Lond.) 223: 810.

[17] Perlman, R. L. and I. Pastan, 1968. Cyclic 3′, 5 - AMP: Stimulation of β - galactosidase and tryptophanase induction in *E. coli*. *Biochem. Biophys. Res. Commun.* 30: 656.

[18] Reese, E. T. and Maguire, A. 1971. Increase in cellulase yields by addition of surfactants to cellobiose cultures of *Trichoderma viride*. *Devel. Indust. Microbiol.* 12: 212.

[19] Spiegelman, S. and Dunn, R. 1948. Interactions between enzyme forming synthesis during adaptation. J. *Gen. Physiol.* 31: 153.

[20] Spiegelman, S. 1947. Nuclear and cytoplasmic factors controlling enzymatic constitution. Cold Spring Harbor Symp. Quant. Biol. 11: 256.

[21] Wisiman, A. 1975. Enzyme induction. Basic Life Sciences, Vol. 6, ed. D. V. Parke, Plenum Press, London and New York, p. I.

[22] Yamane, K. , H. Suzuki, M. Hiratani, H. Ozawa and K. Nisizawa, 1970. Effect of nature and supply of carbon sources on cellulase formation in *Pseudomonas fluorescens Var. cellulosa*. *I. Biochem.* 67: 9.

[23] Zahner, H. and W. K. Maas, 1972. Biology of Antibiotics. Springer-Verlag, New York, Heidelberg, Berlin, p. 86.

Induction and Regulation of Cellulase Formation in *Trichoderma*

II. Induction of Cellulase by Sophorose in Washed Mycelia of *T. Pseudokoningii* EA_3 - 867 and its Catabolite Repression*

The formation of cellulase in washed mycelia of T. pseudokoningii EA_3 - 867 was markedly induced by sophorose. The cellulase activity, after shaking the mycelial suspension with sophorose, approached maximum within 24 hour culture with a lag period of about 3 hours. The washed mycelia of various species of *Trichoderma* and that of T. pseudokoningii EA_3 - 867 with various ages were all able to produce cellulase in which C_1, C_x-cellulase was Predominantly extracellular, whereas cellobiase wes intracellular. The formation of cellulase was affected by the inducer, minerai elements (nitrogen, phosphorus, iron and manganese) temperature, pH, aeration, and the physiological conditions of the mycelial cells. The optimal concentration of sophorose was 10^3—10^5 mol/L, optimal temperature was 30℃. the range of pH used for induction was 3—6, and optimal pH was 5. The production of cellulase was stimulated by O_2, but strongly repressed by inhibitors of respiration, such as iodoacetate fluoride, malonate, azide and 2,4 - dinitrophenol. The indution of cellulase in washed mycelia was inhibited also by actidione, 5 - fluouracil and 6 - azauracil greatly, but inhibited by actinomycin D only slightly. Furthermore, various sugars, sugar phosphates, glycerol, organic acids, NAD, NADP and ATP all repressed the inductive cellulase formation. The catabolite repression of glucose could not be overcome by c - AMP added to the medium. From these results it is obvious that the induction of cellulase formation is closely related to the respiration and nucleic acid metabolism of fungus. The mode of cellulase synthesis in this microorganism was discussed.

* Y. S. Zhu and C. Tan, Shanghai Institute of Plant Physiology, Academia Sinica

木霉纤维素酶的诱导形成及其调节

Ⅲ. 葡萄糖母液和槐豆荚提取液对纤维素酶诱导效应的分析

提要

葡萄糖母液对拟康氏木霉 1096、木-3、EA_3-867 和 N_2-78 的洗涤菌丝体的纤维素酶形成,有强力的诱导效应。用活性炭层析分离和纸谱分离证实,葡萄糖母液对纤维素酶的诱导效应主要是由于其中含有槐糖杂质的缘故。龙胆二糖是葡萄糖母液中另一含量较高的杂质,但它对纤维素酶的诱导能力较槐糖低得多。槐豆荚提取液同样具有诱导木霉纤维素酶形成的作用。本文讨论了如何利用木霉纤维素酶形成的调节控制原理来加速和增加纤维素酶的产量。

一、引言

前文中我们指出拟康氏木霉纤维素酶的诱导形成受诱导和降解物阻遏机制的调节(朱雨生等,1978a);纤维素酶的合成与菌丝体呼吸代谢和核酸代谢有着密切的关系,菌丝体的呼吸提供酶合成的物质和能量基础(朱雨生等,1978b)。因此,提高木霉纤维素酶的产量,从酶形成的调节控制原理着手,就是在满足菌丝体正常而迅速生长的基础上,在最适宜的酶合成条件下,进行诱导(去阻遏)和克服降解物阻遏。

槐糖对木霉纤维素酶的形成有强力的诱导作用(朱雨生等,1978a, 1978b; Mandels 等,1960; Nisizawa 等,1971)。但是,这种稀有的糖在生产上如何应用,是值得探讨的问题。我组曾报道,在拟康氏木霉 EA_3-867 的液体培养中,葡萄糖母液有提高产酶的作用(中国科学院植物生理研究所纤维素酶组等,1978)。本文指出,葡萄糖母液——葡萄糖生产过程中的下脚料——对木霉菌丝体的纤维素酶有强力的诱导效应。根据对葡萄糖母液作用原因的分析,证明其诱导效应是由于其中含有槐糖杂质的缘故。本文对槐豆荚提取液的诱导作用也作了研究。

二、材料与方法

1. 菌种

1096 和木-3 系野生型拟康氏木霉,EA_3-867 和 N_2-78 分别为变异株(中国科学院植物生理研究所纤维素酶组等,1978)。

朱雨生,谭常,中国科学院上海植物生理研究所。此文发表于 1978 年"植物生理学报"4 卷 1 期 19-26 页。1977 年 12 月 22 日收到。

2. 试剂

葡萄糖母液系上海葡萄糖厂用盐酸水解精制玉米淀粉生产葡萄糖过程中，经过第一次结晶以后的废糖蜜，俗称葡萄糖头道母液（本文均简称为葡萄葡母液），内含30%～40%葡萄糖，2%左右的NaCl，少量的双糖、叁糖等杂质，以及微量的Ca、Fe、Cu和Si、Mg等离子。“769”活性炭是上海活性炭厂生产的颗粒状活性炭。龙胆二糖是Fluka AG公司产品。槐糖是从槐豆荚中制备得到的结晶（朱雨生等，1978a）。

3. 方法

1）葡萄糖母液的分离

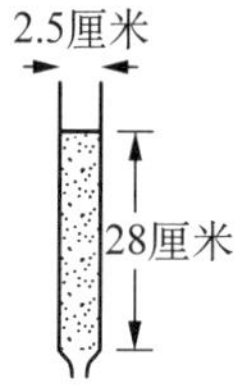

（1）活性炭层析分离。颗粒状活性炭“769”20克，活化后装柱，柱2.5×28厘米①。葡萄糖母液上样量20毫升。用6 000毫升蒸馏水洗脱，使柱中单糖全部洗脱下来。再用1 000毫升7%酒精洗脱，流速5毫升/分钟左右。水洗脱液中含总糖约9克，酒精洗脱液中含总糖约185毫克，非还原糖80毫克。水洗脱液和酒精洗脱液分别浓缩，配成适当浓度，作纸谱分离和诱导活性测定。

（2）纸谱分离和检定。用Whatman 1号滤纸（25×58厘米）点样，溶剂为正丁醇-醋酸-水（4∶1∶5），于室温（20℃左右）下行7天，晾干后，用苯胺一二苯胺②试剂于100℃左右显色检糖。

2）槐豆荚提取液制备方法

20克洗净晾干的槐豆荚（*Sophora japonica L.*），用200毫升热水浸泡，在组织捣碎机中捣碎，于沸水浴中提取20分钟，过滤。在200毫升滤液中，加0.125毫升浓硫酸，沸水浴中水解1小时，加0.7克左右的$Ba(OH)_2$中和至pH 6.0左右。过滤，得到橙黄色滤液，即为槐豆荚提取液。

3）糖含量测定

用酚-硫酸法测定总糖（Michel等，1956），用DNS法测定还原糖（Sumner，1949），非还原糖为总糖与还原糖之差。

4）诱导活性测定

2.45毫升洗涤菌丝体，加2.45毫升无机盐培养基及0.1毫升配成适当浓度的葡萄糖母液，或槐豆荚提取液，或槐糖及葡萄糖，于30℃左右，振荡培养，间隔不同时间，测定滤液中的纤维素酶活力，方法详见前文报道（朱雨生等，1978b）。

5）纤维素酶活力测定

以羧甲基纤维素钠盐（黏度300～600厘粕，上海长虹塑料厂产品）为底物，测定CMC酶活力（即C_x酶），作为纤维素酶活力的代表（朱雨生等，1978）。

三、试验结果

1. 不同浓度葡萄糖母液的诱导作用比较

拟康氏木霉EA_3－867和N_2－78的洗涤菌丝体，在无机盐培养基中，加入不同浓度的葡

① 柱2.5×28厘米，表示底面直径2.5厘米，高度28厘米。

② 苯胺一二苯胺试剂配法：苯胺1毫升，加二苯胺1克，再加11.8毫升85%磷酸和87毫升丙酮，滴入水，直至无沉淀产生为止。试剂要当天使用当天配制。

萄糖母液，恒温震荡，测定纤维素酶的活力。研究发现，葡萄糖母液对纤维素酶的诱导作用与其加入的浓度有密切的关系(见表1)。高浓度(1%)的葡萄糖母液不能诱导菌丝体产生纤维素酶，但浓度降为0.5%、0.1%时，便能诱导产酶。其中0.1%浓度的葡萄糖母液，同0.5%的相比，诱导产酶早，但最终产酶量低。根据对不同浓度葡萄糖母液在诱导过程中含糖量变化的分析结果表明，1%浓度的葡萄糖母液所含还原糖(主要为葡萄糖)。

约为 3×10^{-2} mol/L，诱导培养24小时后，含糖仍在 10^{-2} mol/L，此浓度范围均可明显阻遏纤维素酶的形成(朱雨生等，1978b)；0.5%浓度的葡萄糖母液在保温6小时后，含糖约为 10^{-2} mol/L，酶也不产生，但保温24小时后，所含还原糖降为 10^{-3} mol/L 左右，此浓度对本菌株的纤维素酶已无明显的阻遏作用；0.1%浓度的葡萄糖母液所含的还原糖甚微，在4小时，纤维素酶便被诱导产生，但因葡萄糖母液浓度较低，最终的酶活力低于0.5%者。由此可见，葡萄糖母液对纤维素酶形成的影响包括着未知诱导剂的诱导作用和内含葡萄糖的降解物阻遏作用(即葡萄糖效应)这二个因素的综合作用。前者诱导酶的产生，后者在高浓度下阻止酶的形成。从表1中根据产酶量推算，0.1%葡萄糖母液的作用同 10^{-4} mol/L 槐糖相当，而0.5%葡萄糖母液的作用同 10^{-4} mol/L 槐糖加 10^{-2} mol/L 葡萄糖的复合作用相当。

表1 不同浓度葡萄糖母液对木霉 EA_3－867 和 N_2－78 洗涤菌丝体纤维素酶的诱导作用

菌种	诱导剂	浓度	C_x/(u/ml)		
			4小时	6小时	24小时
EA_3－867	葡萄糖母液	0.1%	0.6	1.2	27.6
	"	0.5%	0	0	38.0
	"	1.0%	0	0	0
	槐糖	10^{-4} mol/L	1.0	2.4	32.0
	槐糖＋葡萄糖	10^{-4} mol/L＋10^{-3} mol/L	1.0	2.4	35.6
	槐糖＋葡萄糖	10^{-4} mol/L＋10^{-2} mol/L	0	0	34.8
N_2－78	葡萄糖母液	0.1%	0	1.0	15.2
	"	0.5%	0	0	24.8
	"	1.0%	0	0	0
	槐糖	10^{-4} mol/L	0.8	1.4	13.2
	槐糖＋葡萄糖	10^{-4} mol/L＋10^{-3} mol/L	0.8	1.6	18.4
	槐糖＋葡萄糖	10^{-4} mol/L＋10^{-2} mol/L	0	0	26.4

注：菌令48小时；菌丝体重 EA_3－867 10.7毫克/瓶，N_2－78 2.8毫克/瓶。

2. 葡萄糖母液中诱导剂的分离和鉴定

葡萄糖母液为何能诱导纤维素酶形成？其所含的未知诱导剂是什么？为此，对葡萄糖母液进行活性炭层析和纸谱分离，并作诱导剂成分的鉴定和活性测定。

1）活性炭层析分离和活性测定

葡萄糖母液经活性炭层析分离后，得到水洗脱液和酒精(7%)洗脱液二部分，分别浓缩至干，并都配成相当于0.1%葡萄糖母液浓度的水溶液，测定葡萄糖母液、水洗脱部分和酒精洗脱部分对拟康氏木霉不同菌株菌丝体纤维素酶的诱导能力(见表 2)。结果发现,主要的活性部分是在 7%酒精洗脱液中,而水洗脱液中活性甚少。根据过去的试验,7%酒精洗脱液中主要含双糖(朱雨生等,1978)。

表 2　葡萄糖母液及其分离成分对拟康氏木霉不同菌株洗涤菌丝体纤维素酶诱导活力的比较

菌种 诱导剂	1096 C_x/(u/ml)	木-3 C_x/(u/ml)	EA_3-867 C_x/(u/ml)	N_2-78 C_x/(u/ml)
槐糖(10^{-4} mol/L)	13.0	11.8	22.8	37.6
葡萄糖母液(0.1%)	9.2	7.6	29.6	43.2
水洗脱液	2.0	2.0	4.0	2.4
7%酒精洗脱液	11.4	14.8	25.6	32.0

注:菌令 48 小时;菌丝体重 1096 11.0 毫克/瓶;木-37.4 毫克/瓶;EA_3-867 7.0 毫克/瓶;N_2-78 4.3 毫克/瓶;诱导温度 30℃;诱导时间 24 小时。

2) 纸谱分离和活性测定

将酒精洗脱部分和水洗脱部分进一步作纸谱分离,结果如图 1 和表 3 所示。

表 3　葡萄糖母液及其经活性炭层析分离后的水洗脱液和酒精洗脱液中四种被检糖的含量(纸谱分离)

RG	糖	含糖量*		
		葡萄糖母液	水洗脱液	酒精洗脱液
1.00	葡萄糖	+++++	+++++	+
0.42	槐糖	微量	微量	++
0.26	龙胆二糖	+++	+	+++
0.08	未知寡糖	+	○	+

注:*用+～+++++表示含糖的多少(定性)。

根据用苯胺—二苯胺试剂的显色斑点判断,葡萄糖母液主要含有四个斑点。RG 分别为(A) 1.00、(B) 0.42、(C) 0.26、(D) 0.08。RG 1.00 和 0.26 的斑点分别相当于葡萄糖和龙胆二糖;RG 0.08 的斑点可能是叁糖(位置同棉子糖相当);RG 0.42 的斑点,根据位置和特殊的显色(棕黄色)断定,是槐糖。四种糖的含量是葡萄糖≫龙胆二糖≫槐糖及未知叁糖。水洗脱液中主要含葡萄糖、少量的龙胆二糖和微量的槐糖;酒精洗脱液中葡萄糖很少,而以槐糖和龙胆二糖两种双糖为主。这四种糖中,只有槐糖是已知的拟康氏木霉纤维素酶的强力诱导剂,酒精洗脱液的诱导活性强,正是由于其中所含槐糖合量高的缘故。

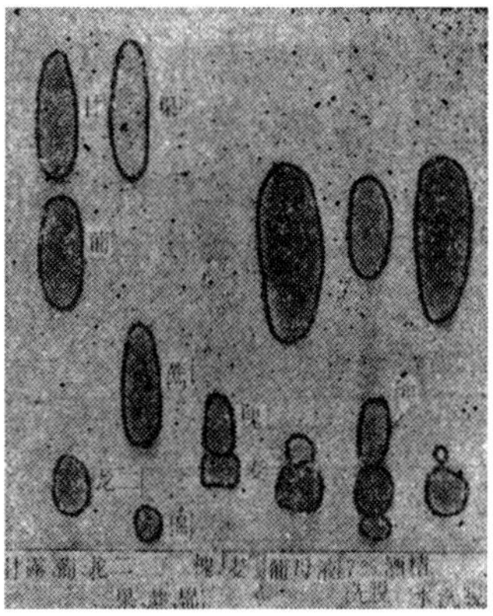

图 1　葡萄糖母液及其分离成分的纸层析图

从图中参照标准糖的斑点位置可知葡萄糖母液中主要含有葡萄糖,龙胆二糖槐糖和微量未知寡糖。

为了进一步确定这四个斑点中糖的作用,将与四个斑点位置相当的邻近未显色的同一张纸谱,依次剪成四段,即 A、B、C、D 分别剪碎,用 80%酒精于 70℃下浸提,离心,上清液于 70℃

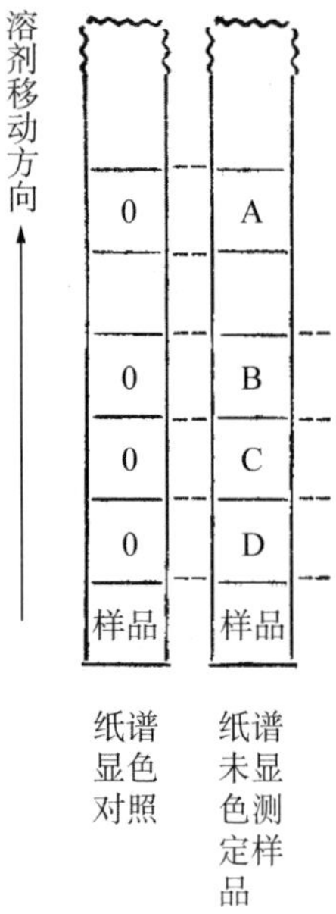

烘干，配成相同的浓度，测定它们的诱导活性(见表 4)。结果证明，B 段的诱导活性最高，是槐糖。A 段(葡萄糖)没有诱导活性。C 段(龙胆二糖)和 D 段(未知叁糖)也有部分诱导活性，很可能是由于糖的纸谱分离不完全而混入了槐糖杂质之故。为此，测定了标准的龙胆二糖(10^{-4} mol/L)对不同木霉菌株洗涤菌丝体纤维素酶的诱导能力(见表 4、表 5)，结果证明，龙胆二糖的诱导活性很低。由此可见，葡萄糖母液对木霉纤维素酶的诱导作用主要由于其中含有槐糖的缘故。

最后，根据葡萄糖母液中的槐糖含量分析，寻求其含量与标准槐糖诱导活性的数量关系。槐糖是非还原糖，葡萄糖母液中非还原糖(主要是槐糖)约 0.4%，则在 0.5%葡萄糖母液中所含的槐糖量应在 10^{-4} mol/L 数量级。同时，根据葡萄糖母液中含葡萄糖 30%～40%计算，0.5%葡萄糖母液中还应含有 10^{-2} mol/L 数量级的葡萄糖。此结果，恰好同表 1 试验中，0.5%葡萄糖母液同 10^{-4} mol/L 槐糖加 10^{-2} mol/L 葡萄糖的诱导活性相仿的推断一致。这种数量关系的吻合，更进一步证实，葡萄糖母液的诱导作用归因于其中的槐糖。

表 4　葡萄糖母液的酒精洗脱液经纸谱分离后四个组分对木霉 EA_3 - 867 和 N_2 - 78 洗涤菌丝体纤维素酶诱导活力的比较

分离样品(组分)	RG	糖	C_x/(u/ml)(24 小时)	
			EA_3 - 867	N_2 - 78
A	1.00	葡萄糖	0	0
B	0.42	槐糖	28.8	14.0
C	0.26	龙胆二糖	14.4	6.0
D	0.08	未知寡糖	3.6	3.0
标准糖	0.42	槐糖(10^{-4} mol/L)	35.2	14.8
	0.26	龙胆二糖(10^{-4} mol/L)	2.0	3.0

注：菌令 48 小时；菌丝体重 EA_3 - 867 6.8 毫克/瓶；N_2 - 78 3.1 毫克/瓶；诱导温度 30℃。

表 5　槐糖和龙胆二糖对拟康氏木霉不同菌株洗涤菌丝体纤维素酶诱导活力的比较

糖	C_x/(u/ml)			
	1 096	木-3	EA_3 - 867	N_2 - 78
槐糖(10^{-4} mol/L)	13.0	11.8	24.8	37.6
龙胆二糖(10^{-4} mol/L)	1.6	2.2	2.0	1.6

试验条件同表 2。

3. 槐豆荚提取液的诱导作用

槐豆荚提取液含有葡萄糖、果糖、蔗糖和槐糖等(朱雨生等，1978a)。在纤维素粉-甘油-无机盐培养基中，于不同时间加入槐豆荚提取液(0.5%)，可使 EA_3 - 867 纤维素酶的产生大大加速，产酶量也明显增加(见表 6)。其作用同槐糖，或槐糖加葡萄糖十分类似。如对照，到第 7

天才产酶,加槐豆荚提取液,或槐糖及葡萄糖后,在培养后2天,便产生纤维素酶。

表6 木霉 EA_3－867 在纤维素粉-甘油-无机盐培养基中培养不同时间加槐糖或槐豆荚提取液对纤维素酶产生的影响

诱导剂	浓度	加诱导剂时间	C_x/(u/ml)				
			1天	2天	3天	4天	7天
对照	—	—	0	0	0	0	8.0
槐糖	10 ppm	0小时	0	7.6	7.8	8.4	20.8
槐糖	10 ppm	24小时	0	8.1	7.5	7.6	19.0
槐糖	10 ppm	48小时	0	1.6	21.6	17.2	21.6
槐糖＋葡萄糖	10 ppm＋0.1%	48小时	0	1.6	11.6	10.0	22.0
槐豆荚提取液	0.5%	0小时	0	3.6	3.5	4.3	18.2
槐豆荚提取液	0.5%	24小时	0	4.6	4.4	5.0	23.0
槐豆荚提取液	0.5%	48小时	0	0	16.0	14.4	22.2

注:试验方法:300毫升三角瓶中装入50毫升甘油-无机盐培养基和1%纤维素粉,在30℃,于往复式摇床振荡培养。槐豆荚提取液是指用稀硫酸水解并经中和的粗提取液。提取比例是荚:水为1:10,每个三角瓶中加2.5毫升提取液,相当于干荚浓度0.5%。

槐豆荚提取液对木霉 EA_3－867 洗涤菌丝体的纤维素酶也有明显的诱导效应。其产酶曲线同样与槐糖加葡萄糖十分类似(见图2)。槐豆荚提取液中含还原糖约0.1%,24小时后基本消耗完,当还原糖降为0.05%以下时,由于不再存在降解物阻遏,纤维素酶便迅速产生。

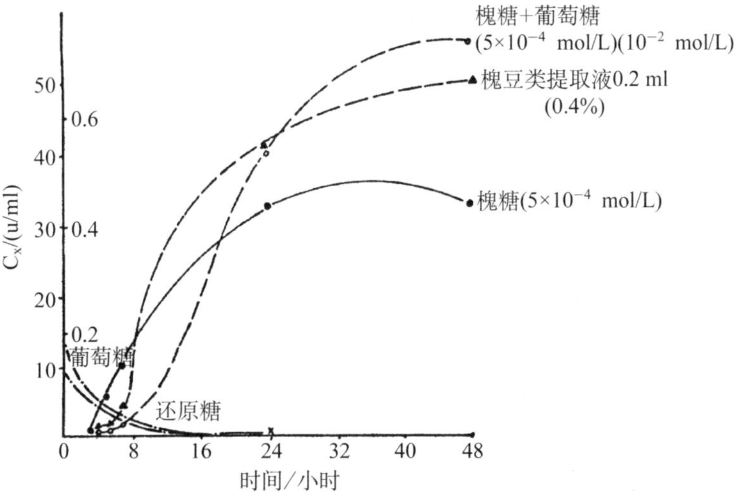

图2 槐豆荚提取液对木霉 EA_3－867 洗涤菌丝体纤维素酶的诱导作用

菌令45小时;菌丝体重量14.6 mg/瓶;诱导温度30℃。

●———● 加槐糖; ×—·—·—× 葡萄糖;
○－－－○ 加槐糖和葡萄糖; ●—·—·—● 槐豆荚中还源糖;
▲－－－▲ 加槐豆荚提取液。

四、讨论

目前，纤维素酶生产上一个较大的问题是酶活低，培养时间长，液体培养竟长达二周(Toyama 等，1975；Reese，1975)。这样，缩短培养时间即是纤维素酶液体曲生产中的关键问题。以前的工作证明，木霉纤维素酶的诱导形成需要四个条件：①诱导剂；②无机盐；③适当的温度、pH 和通气；④正常的代谢活跃的菌丝体(朱雨生等，1978b)。其中①至③是外因，④是内因。木霉纤维素酶的诱导形成——其内因——受底物诱导和降解物阻遏两个控制系统的调节。同时，产酶与生长是一对矛盾过程：生长是产酶的基础，但生长太好，往往由于降解物阻遏现象而阻止酶的产生。如能通过外界条件的改变(即外因)，控制微生物代谢的方向(即内因)；同时处理好产酶与生长的矛盾，有可能多而快地合成和分泌纤维素酶。我们认为缩短培养时间、提高产酶量的合理的纤维素酶液体培养方法，是在木霉生长的前期，应创造有利于生长的一切条件(易利用的碳源、营养盐、温度、pH 及菌丝接种等)，使菌丝体尽可能快和好地生长，并使碳源迅速用完而达到生长稳定期。此过程应在 48 小时前完成，此后不再有降解物阻遏，然后，在酶合成最有利的条件(营养盐、pH、温度、通气等)，让外加高效诱导剂(如槐糖)发挥最大的诱导作用，使纤维素酶大量而迅速地产生。这样，木霉液体曲培养时间缩短到三天左右，在理论上是完全可能的。事实上，木霉 N_2－78 在稻草粉、麸皮和无机盐培养基中，添加葡萄糖母液，只要培养 60 小时，纤维素酶产量便可接近高峰(中国科学院植物生理研究所纤维素酶组等，1978；上海酒精二厂等)。根据以上原理，也可设计将来可能会很有实用前途的、在其他酶制剂生产上(如 α-甘露糖苷酶)已采用的、先生长后诱导的两步液体深层培养法(Faith 等，1971)。

在纤维素酶产生菌的培养过程中，如何使用高效诱导剂，又如何选用廉价的碳源，是酶诱导原理用于实践的关键。利用木糖诱导葡萄糖异构酶(Takasaki 等，1969)，蔗糖棕榈酸一酯诱导转化酶(Reese 等，1969)，淀粉、肝糖和麦芽三糖诱导淀粉酶(Banks 等，1967)等是利用诱导剂提高产酶量的良好例证。Mandels 和 Reese 最初发现葡萄糖能促进纤维素酶的形成(Mandels 等，1957)，以后又证明，葡萄糖的刺激作用是由于其中含有的槐糖杂质(Mandels 等，1962)。但他们最近强调，槐糖虽然是高效诱导剂，然而这种稀有的双糖应用于生产纤维素酶是不实际的(Mandels 等，1976)。曾有研究人员应用含微量槐糖杂质的商用葡萄糖来生产纤维素酶，但酶产量只有纤维素为底物时的 20%左右(Brown 等，1975)。此外，葡萄糖本身虽能促进菌体生长，有利于产酶(朱雨生等，1978b)，但培养基中添加昂贵的葡萄糖似乎也是不切实际的。

葡萄糖母液提供了槐糖和葡萄糖的廉价来源。因为，槐糖既然是玉米酸水解生产葡萄糖过程中由于葡萄糖在酸性环境中缩合而形成的杂质，那么，在葡萄糖结晶后的下脚料——葡萄糖母液中含量应更高。本试验证明葡萄糖母液确实对木霉洗涤菌丝体纤维素酶的形成有强力的诱导作用，而活性炭层析分离、纸谱分离及活性测定，证实了葡萄糖母液的诱导作用，主要由于其中混有槐糖杂质的缘故。同时，葡萄糖母液中含有 30%～40%的葡萄糖，能促进前期菌丝的生长，是廉价的碳源来源。

槐糖用于提高纤维素酶产量的另一可能途径，是槐豆荚提取液。它既含有槐糖(含量可达 0.24%，朱雨生等，1978a)，又含有葡萄糖、果糖等易利用碳源。槐豆荚是提取贵重的植物胶——龙胶的原料。龙胶提取后，经过简单的水解、中和，即可用于木霉液体曲培养，产生纤维素酶。本试验表明，槐豆荚提取液的诱导作用也同槐糖加葡萄糖的作用相仿。

参考文献

[1] 上海酒精二厂，复旦大学微生物生化专业，拟康氏木霉 N_2－78 纤维素酶摇瓶发酵条件的初步研究，复旦学报(待发表)

[2] 中国科学院上海植物生理研究所纤维素酶组、上海酒精二厂，1978。二株高活力纤维分解菌 EA_3－867 和 N_2－78 的获得及其特性的比较，微生物学报。18(1)：27—38.

[3] 朱雨生、谭常，1978a，木霉纤维素酶的诱导形成及其调节 I，槐糖对木霉 EA_3－867 纤维素酶形成的诱导作用，微生物学报．18：

[4] 朱雨生：谭常，1978b，同上，I，槐糖对木霉 EA_3－867 洗涤菌丝体纤维素酶形成的诱导作用及降解物阻遏现象，植物生理学报，4(1)：1—18

[5] Banks, G. T. et al. 1967. Progress Induct. Microbiol. 6: 95.

[6] Brown, D. E., D. J. Halsted and P. Howard, 1975. Studies on the biosynthesis of cellulase by *Trichoderma viride* QM 9123. Proc. Symp. Enzym. Hydrol. Cellulase, Aulanko, Finland, 12－14 March, 1975, ed. Bailey, M., et al. Helsinki, p. 137.

[7] Faith, W. T., C. E. Neubeck and E. T. Reese, 1971. Production and application of enzymes, In Adv. Biochem. Engineering, I. Spinger-Verlag Berlin, Heidelberg, New York (ed. Ghose, T. K., and A. Fiechter) p. 77.

[8] Mandels, M. and E. T. Reese, 1960. Induction of cellulase in fungi by cellobiose. *J. Bact.* 79: 816.

[9] Mandels, M. and E. T. Reese. 1957. Induction of cellulase in *Trichoderma viride* as influenced by carbon sources and metals. *J. Bact.* 73: 269.

[10] Mandels, M., F. W. Parrish and E. T. Reese, 1962. Sophorose as an inducer of cellulase in *Trichoderma viride*. J. Bact. 83: 400.

[11] Mandels, M. and D. Sternberg, 1976. Recent advances in cellulase technology. J. Ferment. Technol. 54: 267.

[12] Michel, D., K. A. Gilles, and J. K. Hamilton, 1956. Colorimetric method for determination of sugars and related substances. *Anal. Chem.* 28. 350.

[13] Nisizawa, T., H. Suzuki, M. Nayama and K. Nisizawa, 1971. Inductive formation of cellulase by sophorose in *Trichoderma viride*. *J. Biochem.* 70: 375.

[14] Reese, E. T. 1975. Summary statement of the enzyme system. Ibid, p. 77.

[15] Reese, E. T., J. E. Lola and F. W. Parrish, 1969. Modified substrates and modified products as inducers of carbohydrases. *J. Bact.* 100: 1151.

[16] Sumner, J. B. 1949. Laboratory Experiments in Biological Chemistry. Acad. Press, New York. N. Y. p. 38.

[17] Takasaki, Y., Y. Kosugi and A. Kanbayashi, 1969. Streptomyces glucose isomerase. In Ferment. Adv., (ed. Perlman, D.,) Academic Press, N. Y. p. 561.

[18] Toyama, N. and K. Ogawa, 1975. Sugar production from agricultural woody wastes by saccharification with *Trichoderma viride* cellulase. Biotechnol. Bioengineering Symp., No. 5, p. 225.

Induction and Regulation of Cellulase Formation in Trichoderma

III. The Inductive Formation of Cellulase by Glucose Waste Molasses and Extracts of Pods of *Sophora Japonica L.* *

Molasses from glucose refinery, could strongly induce cellulase formation in washed mycelia of various strains of *Trichoderma pseudokoningii*. After the components of molasses from glucose refinery were seperated by charcol and paper chromatography and tested for inductive activity individually, it was evident that the induction of cellulase formation by the molasses was mainly due to its sophorose impurity, the other impurity gentiobiose had only feeble inductive activity. The extracts *of pods of Sophora japonica L.* had similarly the ability to induce the production of cellulase. The principle and methods for increasing production were proposed on the basis of regulatory mechanism of cellulase formation.

* Y. S. Zhu, C. Tan(Shanghai Institute of Plant Physiology, Academia Sinica)

木霉纤维素酶的诱导形成及其调节

Ⅳ. 高产变异株纤维素酶合成调节的变化——酶活提高原因的初步分析*

提要

拟康氏木霉纤维素酶高产变异株 EA_3－867 和 N_2－78 在合成培养基琼脂平板和马铃薯葡萄糖琼脂平板上菌落明显缩小，生长变慢。但在蛋白胨酵母膏琼脂平板上，小菌落恢复成大菌落，生长速度同野生型菌株一致：EA_3－867 和 N_2－78 在不同液体培养基中纤维素酶活力均显著高于野生型菌株 1096 和木－3。洗涤菌丝体的诱导测定也证明高产变异株的纤维素酶诱导活性有明显的提高。变异株和野生型菌株洗涤菌丝体诱导形成的纤维素酶组分，从聚丙烯酰胺凝胶电泳图谱判断，未见明显差异，仅变异株中代表纤维素酶活性的蛋白带显著加深。高产变异株纤维素酶产量的增加是由于菌株对诱导剂敏感性的增加和对降解物阻遏敏感性的减弱。

前文中指出，拟康氏木霉纤维素酶高产变异株 EA_3－867 和 N_2－78 的固体曲、液体曲和酶制剂的各种纤维素酶活性均显著高于野生型菌株 1096 和木－3（中国科学院植物生理研究所纤维素酶组等，1978），其纤维素酶的诱导形成受诱导和降解物阻遏二个系统的调节（朱雨生等，1978 a、b、c）。本文进一步对变异株和野生型菌株在几种液体培养基中纤维素酶活性，及洗涤菌丝体纤维素酶凝胶电泳作了比较。试验初步证明，高产变异株纤维素酶活性的提高是由于纤维素酶合成的调节系统发生了变化。

一、材料和方法

（1）菌种：拟康氏木霉（*Trichoderma pseudokoningii Rafai*）EA_3－867 和 N_2－78 是分别从野生型菌株 1096（又称 AS 3.3112）和木－3 诱变得到的高产变异株。

（2）试剂：纤维素粉、槐糖、葡萄糖、CMC 等同前文（朱雨生等，1978a）。

（3）液体培养方法。

① 稻草粉培养基成分（%）：稻草粉 1，麸皮 0.5，$(NH4)_2SO_4$ 0.2，KH_2PO_4 0.1，$CaCO_3$ 0.1，葡萄糖母液 1。菌丝接种，30℃旋转式摇床（104 次/分）培养 60 小时。

② 纤维素粉培养基成分（%）：纤维素粉 0.5，$(NH_4)_2SO_4$ 0.14，KH_2PO_4 0.2，尿素

朱雨生，中国科学院上海植物生理研究所

1978 年 5 月 8 日收到。此文发表于 1978 年"植物生理学报"4 卷 2 期 143－151 页。

* 本工作承沈善炯教授鼓励和指教，特此致谢。

0.03，$MgSO_4 \cdot 7H_2O$ 0.03，$CaCl_2$ 0.03，蛋白胨 0.1；每升含微量元素：$FeSO_4 \cdot 7H_2O$ 5 毫克，$MnSO_4 \cdot H_2O$ 1.56 毫克，$ZnCl_2$ 1.67 毫克，$CoCl_2$ 2 毫克，pH 5.3。500 毫升三角瓶中装入 100 毫升培养基，孢子悬浮液接种，30℃左右，往复式摇床(108 次/分)培养。

③ 甘油培养基：上述纤维素粉培养基中纤维素粉改成甘油(0.6%)，其余相同。

(4) 酶活力测定方法。

① CMC 酶活(C_x)：详见前文(朱雨生等，1978a)，除表中特别指出外，一般酶测定温度为 40℃。在规定的反应系统中每产生 0.5 毫克葡萄糖为 1 个活力单位。

② 滤纸崩溃、滤纸糖(F.P.)及棉花糖(C_1)：见前文(中国科学院植物生理研究所纤维素酶组等，1978)。

(5) 糖、蛋白和生长测定：参照前文的方法(朱雨生等，1978a)。

(6) 洗涤菌丝体静止细胞诱导培养方法：详见过去的报道(朱雨生等，1978 b)。在不同菌种洗涤菌丝体纤维素酶凝胶电泳的试验中，500 毫升三角瓶中装入 100 毫升甘油培养基，接入不同菌种孢子悬浮液，每个菌种各接二瓶，培养二天后的洗涤菌丝体合并，于 100 毫升诱导培养基中加 3×10^{-4} mol/L 槐糖，恒温(27℃左右)震荡培养 24 小时，于 3 号细菌漏斗抽滤，滤液冷冻干燥即为胞外酶制剂。菌丝体用 0.01 mol/L，pH 5 磷酸缓冲液多次洗涤后，用超声波发生器(H60025 型)，于冰浴中声击处理 20 分钟(300 mA)，低温离心(3 000 rpm，20 分钟)，所得上清液经冷冻干燥即成胞内酶制剂。

(7) 酶组分的聚丙烯酰胺凝胶电泳分离法：参照前文的方法(朱雨生等，1978a)。每管注样量为粗酶制剂 5 毫克，电泳 2～4 小时(电流为 2 安培/管)，氨基黑染色，用本所电子室自制的凝胶电泳扫描仪绘制蛋白带峰形图。凝胶段酶活分布测定是在未染色的凝胶上(电泳毕)，从前沿蛋白带为起始点，每隔 3 毫米切割一段，于 1 毫升柠檬酸(0.05 mol/L)-磷酸盐(0.1 mol/L)缓冲液(pH 4.8)中研磨浸提，取清液测定 C_1、C_x、FP 和蛋白含量。

二、试验结果

1. 变异株和野生型菌株菌落大小和生长的差异

试验比较了四株菌基在不同培养基平板上菌落大小和菌丝生长的差异(见表 1、图 3 和图 6)。可见，变异株在马铃薯葡萄糖培养基或合成培养基琼脂平板上，菌落明显缩小，但在蛋白胨酵母膏琼脂平板上，与野生型菌株相比，菌落大小相仿。四株菌株在合成培养基和蛋白胨酵母膏培养基中液体培养时菌丝体生长的差异也显示同样的规律，蛋白胨和酵母膏显著促进变异株的生长。

2. 变异株和野生型菌株在不同培养基中纤维素酶活性的比较

试验分别比较了四株菌株在稻草粉培养基(见表 2)、纤维素粉培养基和甘油槐糖培养基中液体培养时纤维素酶活性的差异(见表 3)。可知，变异株的 C_x 酶、滤纸糖和棉花糖等活性均有较明显的提高，尤其在稻草粉培养基中，酶活提高更为明显。

表 4 和表 5 是四株菌株洗涤菌丝体用槐糖诱导的胞外纤维素酶和酶制剂纤维素酶各组分活性的比较，证明了变异株纤维素酶的诱导活性有了显著的增加。由表 2 和表 5 还可看出，变异株胞外可溶性蛋白含量也有所增加。

表1　四株菌株在不同培养基上菌落大小(琼脂平板上)和菌丝体生长(摇瓶培养中)的比较

生长比较 / 培养基 / 菌种	菌落直径/厘米*			菌丝体重量 /(毫克/100毫升)**	
	马铃薯葡萄糖培养基(3天)	合成培养基(4天)	蛋白胨酶母膏培养基(3天)	合成培养基(2天)	蛋白胨酵母膏培养基(2天)
1096	4.8	4.5	6.5	258	308
EA_3 - 867	0.9	2.5	5.5	228	303
木-3	3.5	3.8	5.5	263	315
N_2 - 78	0.8	2.0	5.1	190	314

注：* 马铃薯葡萄糖培养基(%)：马铃薯20。葡萄糖2，琼脂2。

合成培养基(%)：纤维素粉2，$(NH_4)_2SO_4$ 0.14，KH_2PO_4 0.2，尿素0.03，$MgSO_4 \cdot 7H_2O$ 0.03，$CaCl_2$ 0.03，$FeSO_4 \cdot 7H_2O$ 0.000 5，$CoCl_2$ 0.000 2：$ZnCl_2$ 0.000 17，$MnSO_4$ 0.000 16，琼脂2。

蛋白胨酵母膏培养基(%)：蛋白胨0.2，酵母膏0.1，琼脂2。

所试菌种在上述培养基平板上作单菌落培养，30℃，3～4天后测量菌落大小。

** 合成培养基配方同上，但纤维素粉用0.6%甘油代替；蛋白胨酵母膏培养基是在合成培养基中增加蛋白胨0.2%，酵母膏0.1%，孢子悬浮液接种后，于30℃振荡培养2天后，测定菌丝体重量。

表2　四株菌株在稻草粉培养基中(摇瓶)纤维素酶活力的比较*

菌种	C_x /(单位/毫升酶)	滤纸崩溃**	滤纸糖 /(毫克糖/毫升酶)	棉花糖 /(毫克糖/毫升酶)	可溶性蛋白 /(毫克糖/毫升酶)
1096	36.8	⊥$^{-}$	0.72	1.5	0.96
EA_3 - 867	260.0	++	3.62	9.9	1.65
木-3	47.6	⊥	0.88	1.9	1.00
N_2 - 78	576.0	卌$^{+}$	10.60	13.4	2.40

注：* 培养基配方和培养条件见方法部分说明。

** 滤纸膨胀软化为+；滤纸破碎断裂，但仍成形为++；滤纸完全崩溃，呈糊状为+++。⊥代表滤纸崩溃程度比"+"差。

表3　四株菌株在纤维素粉培养基和甘油培养基中纤维素酶活力的比较

培养基 / 酶活 / 菌种	纤维素粉培养基*			甘油培养基**		
	C_x /(单位/毫升酶)	滤纸糖 /(毫克糖/毫升酶)	棉花糖 /(毫克糖/毫升酶)	C_x /(单位/毫升酶)	滤纸糖 /(毫克糖/毫升酶)	棉花糖 /(毫克糖/毫升酶)
1096	56	1.3	4.4	40	0.9	1.8
EA_3 - 867	96	1.8	6.0	140	1.7	3.6
木-3	60	1.3	5.6	68	0.9	2.4
N_2 - 78	108	2.0	8.0	176	2.2	6.0

注：* 培养基见方法部分，32℃，往复式摇床(108次/分)培养7天。

** 培养基见方法部分，32℃，往复式摇床(108次/分)培养2天，加入槐糖(10^{-4} mol/L)，再继续培养5天，测定酶活。

表 4 四株菌株洗涤菌丝体用槐糖诱导形成纤维素酶活力的比较

菌种	C_x/(单位/毫升)
1096	5.6
EA_3 - 867	12.6
木-3	5.2
N_2 - 78	15.4

注：菌丝体重量：1096：12.9 毫克/瓶，EA_3 - 867：11.4 毫克/瓶，木-3：13.1 毫克/瓶，N_2 - 78：9.5 毫克/瓶。槐糖浓度 10^{-4} mol；诱导温度 30℃；诱导时间 24 小时。

表 5 四株菌株洗涤菌丝体用槐糖诱导形成的纤维素酶制剂活力的比较*

菌种	C_x /(单位/毫升)	滤纸糖 /(毫克糖/毫克酶)	棉花糖 /(毫升糖/毫克酶)	可溶性蛋白 /(微克/毫升酶)
1096	6.8	1.4	4.4	50
EA_3 - 867	10.8	1.8	5.0	60
木-3	7.6	1.2	4.6	35
N_2 - 78	22.8	4.5	8.8	70

注：* 四株菌株在甘油培养基中于 27℃分别培养 48 小时的菌丝体（各 200 毫升菌丝培养物）用缓冲液洗涤后，加 3×10^{-4} mol/L 槐糖，再振荡培养 24 小时，胞外酶液冷冻真空干燥，即成胞外酶制剂，均配成 1%浓度测定。

3. 变异株和野生型菌株纤维素酶凝胶电泳的比较

对四株菌株洗涤菌丝体用槐糖诱导形成的纤维素酶胞外酶和胞内酶的聚丙烯酰胺凝胶电泳(2 小时)进行了比较，结果如图 4 和图 5 所示。变异株 EA_3 - 867 和 N_2 - 78 胞外酶的蛋白带，分别与野生型菌株 1096 和木-3 类似，而胞内蛋白带的数目明显减少(从 6 条减少至 3 条)，可能是变异株中非纤维素酶的蛋白带减少了。凝胶段酶活的切割测定，证明胞外酶二条颜色深的蛋白带位置具有纤维素酶活性(棉花糖和 C_x)。图 7 是四株菌株胞外纤维素酶电泳 4 小时进一步分离的凝胶电泳图。可以看出，变异株和野生型菌株胞外酶均清楚地由四条区带组成(Ⅰ～Ⅳ)，总的蛋白带数在洗涤菌丝体用槐糖诱导较为单纯的条件下，远较液体培养或固体培养时(如以稻草粉或纤维素粉为碳源)为少。变异株，尤其是 N_2 - 78 的区带Ⅲ，其次是Ⅳ明显加深。四株菌株胞外酶的凝胶电泳扫描图(见图 1)也证实变异株和野生型菌株的蛋白峰一样，仅变异株中Ⅲ，其次是Ⅳ和Ⅰ高于野生型菌株。凝胶段切割后酶活测定，进一步证明峰Ⅲ主要代表 C_1 酶活性(棉花糖表示)，Ⅳ和Ⅰ主要具有 C_x 酶活性，区带Ⅱ由三条带组成，除了Ⅱ-3 具有少量 C_1 酶活性外，其余均测不到纤维素酶活性(见图 2)。从图 7 和图 2 中可以看到槐糖诱导的胞外蛋白很纯，绝大部分是纤素维酶。

4. 变异株和野生型菌株纤维素酶诱导形成时对槐糖诱导和葡萄糖阻遏敏感性的比较

对四株菌株洗涤菌丝体用不同浓度的槐糖诱导其纤维素酶的形成，发现变异株和野生型菌株对槐糖的敏感性是不同的。变异株 EA_3 - 867 和 N_2 - 78 用 10^{-6} mol/L 槐糖即能诱导纤维素酶的形成，而野生型菌株 1096 和木-3 要用 10^{-4} mol/L 槐糖才能诱导纤维素酶的形成。两者槐糖浓度差异达 100 倍(见表 6)。曾比较了四株菌株对槐糖的利用情况，未见对槐糖的吸收利用有明显的差异。

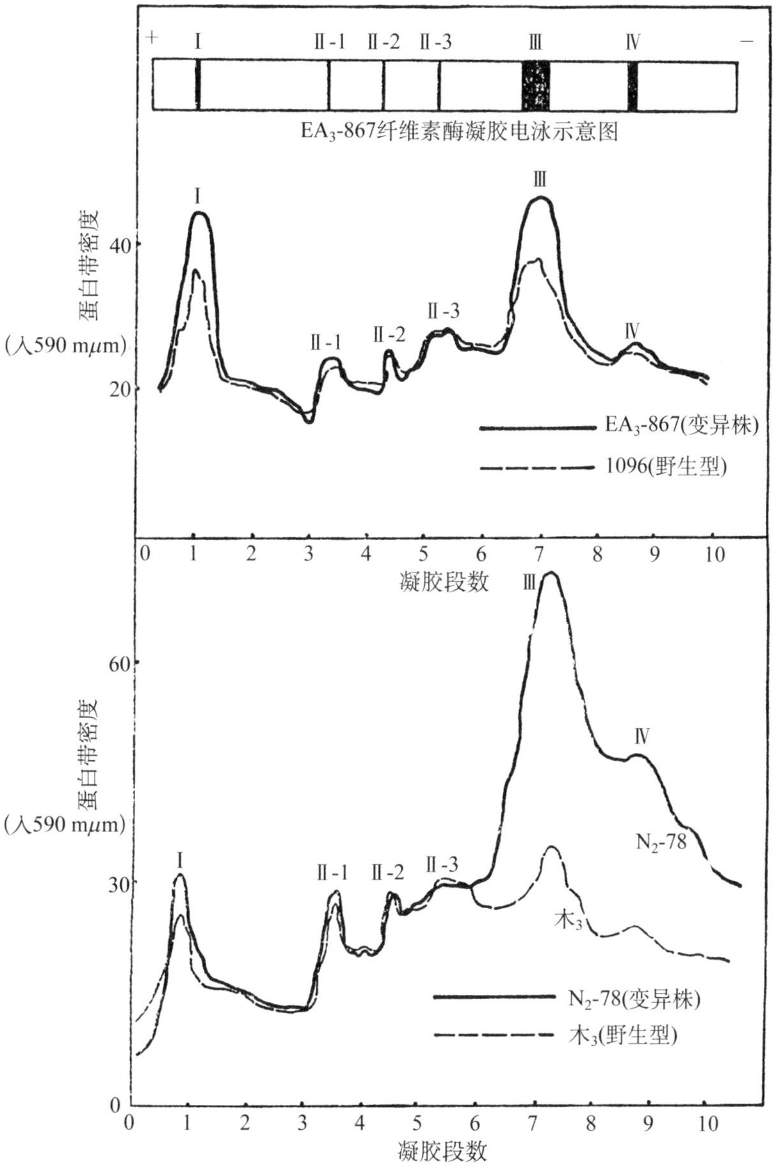

图 1　四株拟康氏木霉菌株洗涤菌丝体用槐糖诱导形成的纤维素酶凝胶电泳的扫描图

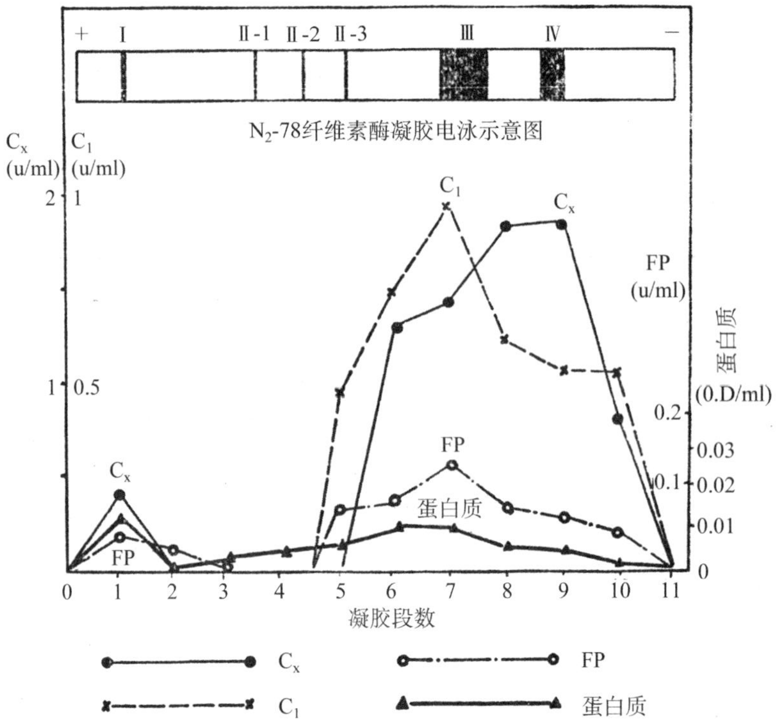

图 2　N_2 - 78 纤维素酶凝胶电泳示意图及其酶活分布

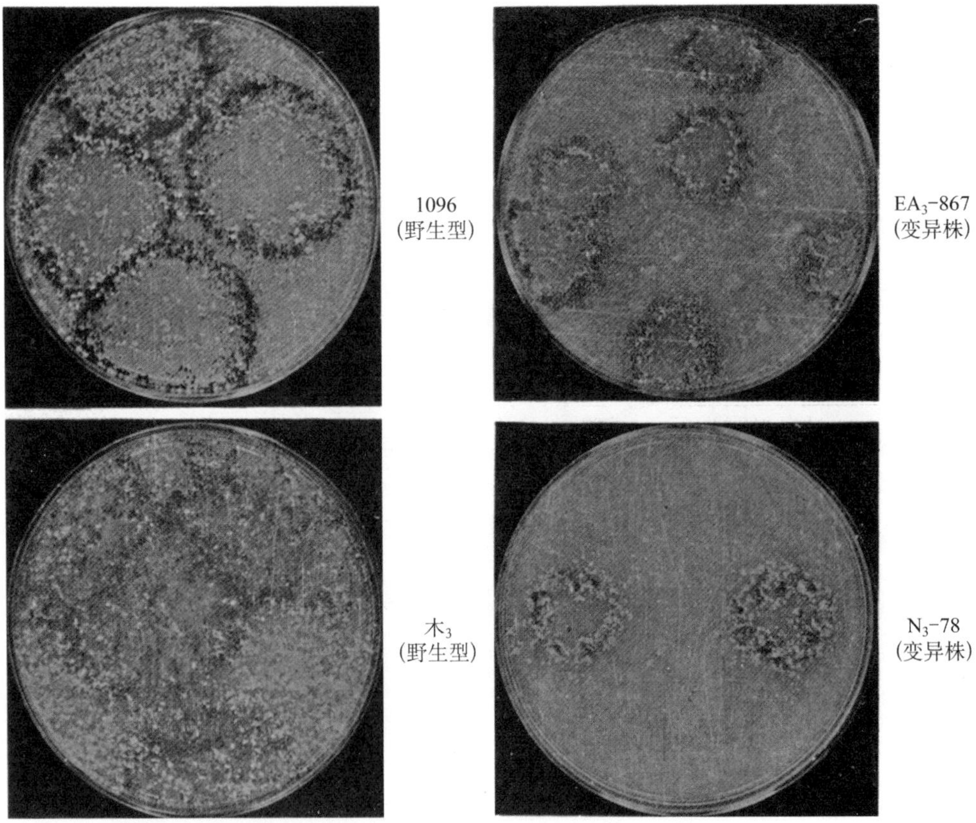

图 3　四株拟康氏木霉菌株在合成培养基琼脂平板上菌落形态、大小的比较(30℃,生长 4 天)

1096、木- 3 为野生型菌株,EA_3 - 867 和 N_2 - 78 分别为纤维素酶高产变异株

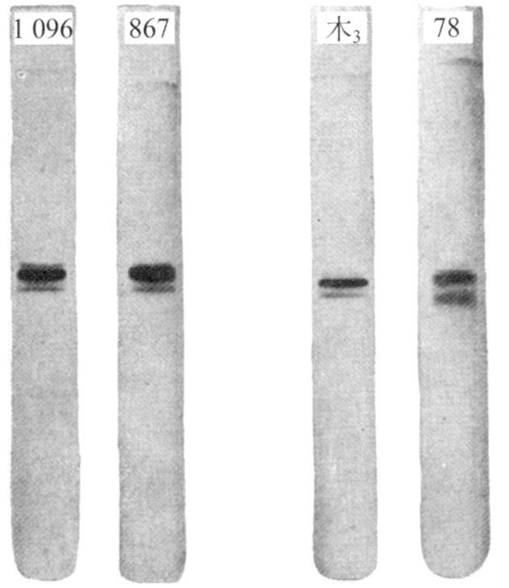

图 4 四株菌株洗涤菌丝体用槐糖诱导形成的纤维素酶(胞外)聚丙烯酰胺凝胶电泳图谱(电泳 2 小时)

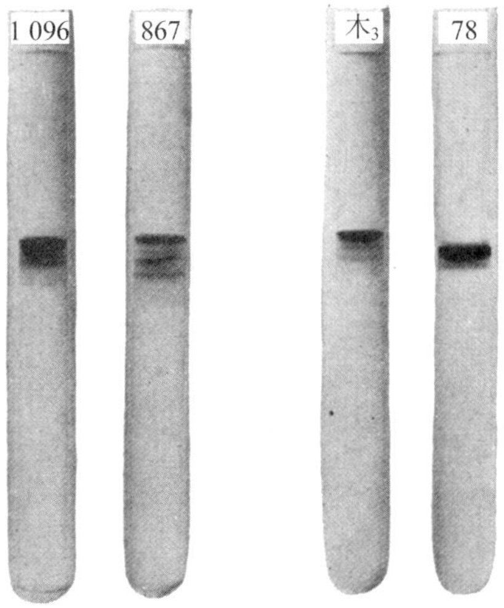

图 5 四株菌株洗涤菌丝体用槐糖诱导形成的胞内酶聚丙烯酰胺凝胶电泳图谱(电泳 2 小时)

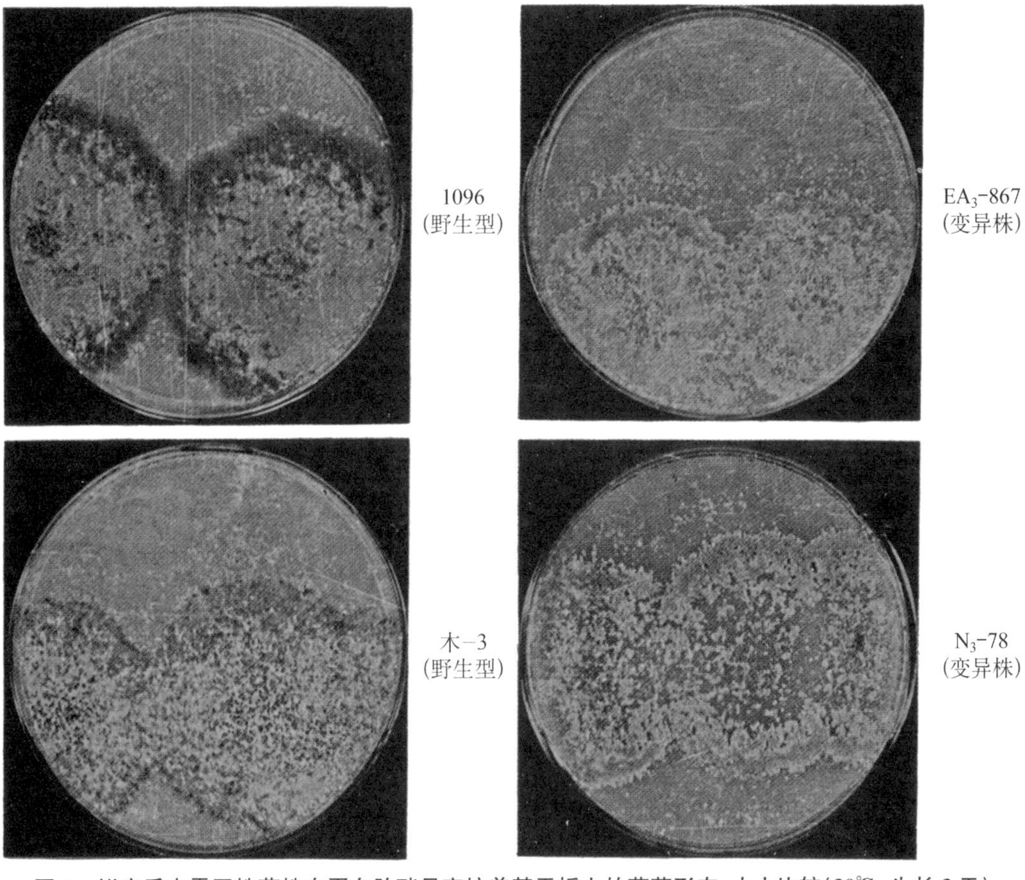

图 6 拟康氏木霉四株菌株在蛋白胨酵母育培养基平板上的菌落形态、大小比较(30℃,生长 3 天)

表 6 四株菌株洗涤菌丝体纤维素酶诱导形成对槐糖敏感性的比较

菌种	槐糖浓度/(mol/L)	C_x/(单位/毫升)		
		4 小时	6 小时	24 小时
1096	10^{-7}	0	0	0
	10^{-6}	0	0	0
	10^{-5}	0	0	0
	10^{-4}	0.5	3.5	10.0
	10^{-3}	2.0	5.9	19.6
EA_3 - 867	10^{-7}	0	0	0
	10^{-6}	0	0	4.8
	10^{-5}	0.5	2.4	8.4
	10^{-4}	1.2	3.2	13.6
	10^{-3}	1.0	3.0	14.4
木- 3	10^{-7}	0	0	0
	10^{-6}	0	0	0
	10^{-5}	0	0	0
	10^{-4}	2.4	5.7	12.0
	10^{-3}	3.0	6.2	23.6
N_2 - 78	10^{-7}	0	0	0
	10^{-6}	0	3.0	9.2
	10^{-5}	1.3	4.6	16.8
	10^{-4}	1.5	5.5	21.2
	10^{-3}	2.1	6.0	27.6

注：菌丝体重量：1096：12.0 毫克/瓶；EA_3 - 867：9.5 毫克/瓶；木- 3：12.7 毫克/瓶；N_2 - 78：8.5 毫克/瓶。
诱导温度 28℃；槐糖浓度 10^{-4} mol/L。

表 7 是四株菌株洗涤菌丝体用槐糖诱导形成纤维素酶对葡萄糖阻遏敏感性的比较，证明了变异株和野生型菌株对葡萄糖阻遏的敏感性也是不同的。变异株 EA_3 - 867 和 N_2 - 78 纤维素酶诱导形成只能被 10^{-2} mol/L 浓度的葡萄糖所阻遏，而野生型菌株 1096 和木- 3 显然对葡萄糖阻遏较为敏感，10^{-3} mol/L 浓度的葡萄糖即能阻止槐糖对洗涤菌丝体纤维素酶的诱导作用，两者浓度相差 10 倍。

若将四株菌株的洗涤菌丝体先同槐糖预保温 17 小时，让纤维素酶充分形成，然后，加入不同浓度的葡萄糖，同样证明变异株同野生型菌株纤维素酶形成对葡萄糖阻遏的敏感性差 10 倍(见表 8)。

表 7　四株菌株洗涤菌丝体纤维素酶诱导形成对葡萄糖阻遏敏感性的比较

菌种	葡萄糖浓度	4 小时		6 小时	
		C_x/(单位/毫升酶)	抑制/%	C_x/(单位/毫升酶)	抑制/%
1096	0	0.70	0	1.8	0
	10^{-2} mol/L	0	100	0	100
	10^{-3} mol/L	0.60	14	1.2	33
EA_3 - 867	0	1.0	0	2.4	0
	10^{-2} mol/L	0	100	0	100
	10^{-3} mol/L	1.0	0	2.4	0
木-3	0	0.6	0	1.0	0
	10^{-2} mol/L	0	100	0	100
	10^{-3} mol/L	0.4	33	0.4	60
N_2 - 78	0	0.8	0	1.4	0
	10^{-2} mol/L	0	100	0	100
	10^{-3} mol/L	0.8	0	1.6	0

注;菌丝体重量 1096：14.7 毫克/瓶;EA_3 - 867：10.7 毫克/瓶;木-3：12.7 毫克/瓶;N_2 - 78 2.8 毫克/瓶;槐糖浓度：10^{-4} mol/L;诱导温度：30℃。

表 8　四株菌株洗涤菌丝体与槐糖预保温后加入不同浓度葡萄糖对纤维素酶诱导形成阻遇作用的比较

菌种	葡萄糖浓度	2 小时		7 小时	
		C_x/(单位/毫升酶)	抑制/%	C_x/(单位/毫升酶)	抑制/%
1096	0	4.0	0	4.6	0
	10^{-2} mol/L	2.4	40	2.8	39
	2×10^{-3} mol/L	2.6	35	3.2	30
	10^{-3} mol/L	3.6	10	4.0	13
EA_3 - 867	0	11.4	0	11.4	0
	10^{-2} mol/L	9.8	14	9.8	22
	2×10^{-3} mol/L	11.0	3	11.0	0
	10^{-3} mol/L	11.4	0	11.4	0
木-3	0	5.6	0	7.0	0
	10^{-2} mol/L	3.8	32	4.4	37
	2×10^{-3} mol/L	4.0	28	5.2	26
	10^{-3} mol/L	5.2	7	6.0	14
N_2 - 78	0	10.4	0	10.4	0
	10^{-2} mol/L	8.8	15	8.8	29

（续表）

菌种	葡萄糖浓度	2 小时		7 小时	
		C_x/(单位/毫升酶)	抑制/%	C_x/(单位/毫升酶)	抑制/%
N_2 - 78	2×10^{-3} mol/L	9.0	13	9.0	1
	10^{-3} mol/L	10.4	0	10.4	1

注：菌丝体重量 1096：13.0 毫克/瓶；EA_3 - 867：8.5 毫克/瓶；木- 3：11.7 毫克/瓶；N_2 - 78：6.0 毫克/瓶。

洗涤菌丝体先与槐糖（10^{-4} mol/L）静止保温（20℃）17 小时，取样测定，作为 0 小时 C_x 酶活（单位/毫升酶），1096 为 2.6，EA_3 - 867 为 8.0，木- 3 为 3.2，N_2 - 78 为 5.6。立即加入不同浓度葡萄糖，于 30℃振荡培养 2 小时、7 小时测定 C_x 活力变化。

三、讨论

原核生物酶合成的遗传控制曾作过广泛的研究（堀内忠郎等，1970），对大肠肝菌乳糖利用的研究导致了操纵子学说的创立（Jacob 等，1961）。但是，真核生物酶合成的调节控制研究较少（Neuberger 等，1975；MacLean，1976）。纤维素酶合成的遗传控制被认为是很重要的研究领域（Pathak 等，1973）。但除了 Myers 等（1966）和 Eberhart 等（1966）在粗糙链霉霉中发现 CMC 酶和纤维二糖酶是受一个调节基因调节，和 Stewart 等（1976）首次报道在纤维单胞菌中分离到抗降解物阻遏的变异株，以及对绿色木霉的纤维素酶缺失变种进行初步的研究外（Nevalainen 等，1978），未在文献中更多地见诸这类报告。

木霉是简单的真核生物之一，其胞外纤维素酶至少包含三个组分：C_1、C_x 和 β-葡萄糖苷酶（纤维二糖酶）。天然纤维素的分解需这三种组分的协同作用（Reese，1975）。由此可设想，分解天然纤维素的操纵子也至少包含三个结构基因。同时，木霉纤维素酶是典型的诱导酶，推测可能存在调节基因和阻遏蛋白。我们从野生型木霉菌株 1096 和木- 3 分别得到的二株高产变异株 EA_3 - 867 和 N_2 - 78，在各种培养条件下，纤维素酶各组分（C_1、C_x 酶）均有大幅度的同步提高现象，同时，槐糖对 C_1、C_x 酶和纤维二糖酶有同步诱导的现象，所以，三个功能相关的酶可能受同一个调节基因控制。目前，关于纤维素酶结构基因和调节基因的遗传分析以及相互的关系，尚一无所知。

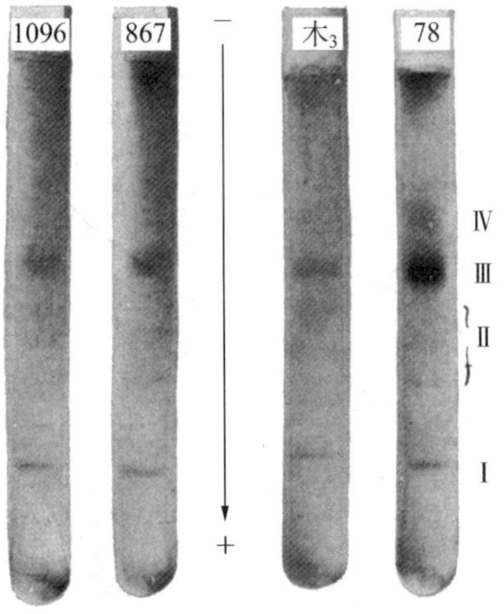

图 7　四株菌株洗涤菌丝体槐糖诱导形成的胞外纤维素酶聚丙烯酰胺凝胶电泳图（电泳 4 小时）

高产变异株在不同培养基中进行液体培养时，纤维素酶活性均高于野生型菌株，而不同菌株洗涤菌丝体的诱导测定又证明了变异株中纤维素酶诱导活性有了很大的提高。但是，纤维素酶的性质（如最适 pH、最适温度）和胞外纤维素酶组分（至少从聚丙烯酰胺凝胶电泳图谱上判断），未见什么明显的差异。所不同者，主要是酶蛋白产量的增加（见图 7 和图 2）。换言之，变异株的结构基因可能未变，而影响酶产量的调节基因发生了变化。丸尾文治（1976）曾发现枯草杆菌淀粉酶的高产突变种也是调节基因发生了突变。木霉的纤维素酶产量调节系统是诱导和降解物阻遏。本试验证实了变异株纤维素酶合成的这二个调节系统确实发生了变化：对诱导剂敏感性增加和对葡萄糖阻遏敏感性的减弱导致纤维素酶产量的增加。

木霉纤维素酶形成对诱导剂敏感性增加的一种可能是阻遏蛋白结构的变化，使之与诱导剂的结合更为紧密，如同乳糖操纵子中 i^t 高产突变一样（堀内忠郎等，1960），以至变异株在比野生型菌株低 100 倍的槐糖浓度下，诱导纤维素酶的形成。变异株 EA_3 - 867 和 N_2 - 78 纤维素酶形成对降解物阻遏敏感性减弱了 10 倍，但仍能为较高浓度的葡萄糖（10^{-2} mol/L）所阻遏，所以属于部分去阻遏的菌株。抗降解物阻遏的突变可以发生在启动子，或者决定 CRP 蛋白或腺苷酸环化酶的基因上，同时，主要代谢途径功能速率的基因突变会间接影响降解物阻遏的程度（Patricia，1976）。从本试验看来，后一种可能是存在的。高产变异株在合成培养基上生长差，在合成培养基琼脂平板上菌落小；但在含蛋白胨和酵母膏的培养基中，生长同野生型菌株一样，在其琼脂平板上小菌落恢复成大菌落。可见，变异株可能缺乏蛋白胨和酵母膏中为生长所必需的成分，减慢了生长速率和降低了代谢速率，从而减弱了降解物阻遏。微生物产物高产变异株往往是小菌落的现象是颇为普遍的，本试验为探讨这类现象的原因也许提供一些线索。

参考文献

[1] 中国科学院植物生理研究所纤维素酶组，上海酒精二厂，1978. 二株高活力纤维分解菌——拟康氏木霉 EA_3 - 867 和 N_2 - 78 的获得及其菌种特性的比较，微生物学报，18(1)：27.

[2] 朱雨生，谭常，1978 a：木霉纤维素酶的诱导形成及其调节，Ⅰ. 槐糖对木霉 EA_3 - 867 纤维素酶形成的诱导作用，微生物学报，18：(即将刊出)

[3] 朱雨生，谭常，1978 b：木霉纤维素酶的诱导形成及其调节，Ⅱ 槐糖对木霉 EA_3 - 867 洗涤菌丝体纤维素酶形成的诱导作用及降解物阻遏现象，植物生理学报，4(1)：1.

[4] 朱雨生，谭常，1978 c：木霉纤维素酶的诱导形成及其调节，Ⅲ. 葡萄糖母液对纤维素酶诱导效应的分析，植物生理学报，4(1)：19.

[5] 堀内忠郎，尾过望，1970，酵素合成の调节，化学增刊，42：61.

[6] 丸尾文治，1976. 酸酵と工业，34：12.

[7] Ebcrhart, B., D. V. Crossa, and L. R. Chase, 1966. β - Glucosidase system of *Neurospora crossa* I. β - Glucosidase and cellulase activities of mutant and wild-type strains. *J. Bact.* 87: 761.

[8] Jacob, F. and J. Monod, 1961. Genetic regulatory mechanisms in the systhesis of proteins. *J. Mol. Biol.* 3: 318.

[9] Maclean, N. 1976. Control of Gene Expression, Academic Press, London, New York, San Francisco.

[10] Myers, MG. and B. Eberhart, 1966. Rcgulation of cellulase and cellobiase in *Neurospora crossa*. *Biochem. Biophys.* Res. Commun. 24: 782.

[11] Neuberger, A. 1975. Enzyme induction, *Basic Life Sciences*, 6: 1. London, Plemm.

[12] Nevalainen, K. M. H. and E. T. Palva, 1978. Production of extracellular enzymes in mutants isolated from *Trichoderma viride* unable to hydrolyse cellulose. *App. (and) Environ. Microbiol*, 35: 1.

[13] Pathak, A. N. and T. K. Ghose, 1973. Cellulase-I, *Process Biochem.* 8: 35.

[14] Patricia, H. C., 1976. Mutant isolation, 2th. *Intern. Symp.* on *Gen. Indus. Microorganisms.* ed. K. D. Macdomald, Academic Press, p. 15.

[15] Reese, E. T., 1975. Summary statement on the enzyme system. *Biotech. Bioeng. Symp.* 5: 77.

[16] Stewart, B. and J. M. Leatherwood, 1976: Derepressed synthesis of cellulomonas, *J. Bact.* 128: 609.

Induction and Regulation of Cellulase Forma Tion in Trichoderma. Ⅳ. Changes in Regu Latory Mechanisms of Cellulase Synthesis of Two Mutants with High Cellulase Yields

——A Preliminary Analysis of the Mechanism of Increase in Cellulase Production*

Two mutants with high cellulase yields of *Trichoderna pseudokiningii Rafai* EA_3 - 867 and N_2 - 78, showed small colonies on agar plates with synthetic medium or potato-glucose medium, and grew slowly, but on agar plates with peptone-yeast extract medium the small colonies converted into colonies as large as wild strains. The cellulase production of EA_3 - 867 and N_2 - 78 on either cellulose or glycerolsophorose media was higher than that of their wild parents 1096 and MO_3, respectively. The inductive activity for cellulase production of washed mycelia of two mutants was also enhanced. The cellulase components, induced by sophorose, of washed mycelia of the mutants and wild strains showed no significant differences, at least in the patterns of polyacrylamine gel electrophoresis. But the protein bands of cellulase of the mutants were considerably deeper in colour than that of their wild parents. It is evident that the increase in cellulase production of both mutants results from the changes in the regulatory mechanism of cellulase systhesis, i. e. , the mutants showed higher responsiveness to inducer and lower susceptibility to catabolite repression than those of their wild parents.

* Y. S. Zhu(Shanghai Institute of Plant Physiology, Academia Sinica)

木霉纤维素酶的诱导形成及其调节

V. 拟康氏木霉(*Trichoderma pseudokoningii* Rafai) N_2－78 洗涤菌丝体纤维素酶诱导形成过程中核酸代谢的变化

拟康氏木霉 N_2－78 洗涤菌丝体在槐糖诱导下合成纤维素酶(C_1 酶、C_x 酶和 β－葡萄糖苷酶)的同时，RNA 含量也发生有规律的变化。采用加核酸酶抑制剂和低温、快速抽提的方法，从槐糖诱导的 N_2－78 洗涤菌丝体中提取总 RNA，经寡聚(dT)－纤维素柱分离，得到含 Poly(A)的 RNA。聚丙烯酰胺凝胶电泳和亲合层析分离证明，经槐糖诱导后，菌丝体中 Poly(A)－RNA 含量增加。

一、引言

槐糖对拟康氏木霉洗涤菌丝体的纤维素酶形成有强力的诱导效应，并且受核酸代谢抑制剂的强烈抑制(朱雨生等，1978b)。本研究从槐糖诱导的 N_2－78 洗涤菌丝体中抽提了 RNA，分离得到 Poly(A)－RNA，比较了槐糖诱导纤维素酶形成过程中 Poly(A)－RNA 含量的变化。

二、材料和方法

1. 菌种

拟康氏木霉 N_2－78。

2. 方法

(1) 洗涤菌丝体纤维素酶诱导方法：参照前文的方法(朱雨生等，1978b)，500 毫升三角瓶中盛 100 毫升洗涤菌丝体，槐糖浓度为 10^{-4} mol/L，诱导温度为 28℃。

(2) 纤维素酶测定方法(C_1 酶、C_x 酶和 β－葡萄糖苷酶)：见前文(朱雨生等，1978a)。

(3) 蛋白质、RNA 测定方法：可溶性蛋白按 Lowry 法(1951)测定。RNA 用甲基间苯二酚试剂测定(Chargaff，1955)。

(4) RNA 抽提分离。

① 抽提：500 毫升左右经槐糖诱导 6 小时的菌丝体，用 3 号细菌漏斗抽滤后，迅速用冷的

吴永强，朱雨生，陈薇，高锦华，费锦鑫，中国科学院上海植物生理研究所

1979 年 2 月 27 日收到。此文发表于 1979 年"植物生理学报"5 卷 4 期 335－341 页。

* 本工作在沈善炯教授指导下完成，还得到中国科学院上海细胞生物所李文裕等同志帮助，特此致谢.

0.01 mol/L Tris-HCl (pH 7.5)洗涤，立即转移到预冷的100毫升0.01 mol/L Tris-HCl (pH 7.5)抽提缓冲液内(内含S. D. S. 0.5%，二乙基焦碳酸钠1%，或肝素0.05%)，菌丝体浓度1%左右(以干重计)，在组织捣碎机中低温捣碎3分钟(1 200转/分)，立即加等体积酚液(苯酚：氯仿：异戊醇=66：33：1)，抽提三次，水相加醋酸钾，使其浓度成2%，再加2.5倍体积-20℃的无水酒精沉淀RNA，用乙醇-0.2 mol/L NaCl (2：1)溶液洗涤沉淀，离心得到总RNA。

② Poly(A)-RNA的分离：基本上按Aviv等(1972)的亲合层析法分离。总RNA样品溶于0.01 mol/L Tris-HCl (pH 7.5)-0.5 mol/L KCl缓冲液中，加入预先用同样缓冲液平衡过的寡聚(dT)-纤维素柱(上海试剂二厂产品)中，洗脱采用三步(升温)法或二步法。三步法是①先用0.01 mol/L Tris-HCl (pH 7.5)-0.5 mol/L KCl (0～2℃)洗脱；②再升温到54℃，用同样缓冲液洗脱；③最后在54℃改用0.01 mol/L Tris-HCl (pH 7.5)溶液洗下含Poly(A)的RNA(上海实验生物所三室等，1977)。二步法的第一步用0.01 mol/L Tris-HCl (pH 7.5)-0.5 mol/L KCl洗脱，而后直接用中性重蒸馏水洗脱Poly(A)-RNA(李文裕等，1978)。

③ RNA电泳：按Loening法(1967)分离，聚丙烯酰胺凝胶浓度为3%，管子6.7×0.56厘米，加样100微克左右。缓冲液含(克/升)：Tris 4.85，醋酸钠2.72，EDTA 0.29，pH 7.4。每管电流为5毫安。冰浴中电泳2小时左右，0.5%甲烯蓝染色，7%醋酸中退色。用CS-900岛津双波长色层扫描仪扫描。

三、试验结果

1. N_2-78洗涤菌丝体被槐糖诱导不同时间纤维素酶、蛋白和核酸含量的变化

N_2-78洗涤菌丝体在槐糖诱导下，经过3小时左右的延迟期，开始合成胞外纤维素酶的各个组分(C_1酶、C_x酶和β-葡萄糖苷酶)，24小时左右达到高峰(见图1)。胞外可溶性蛋白含

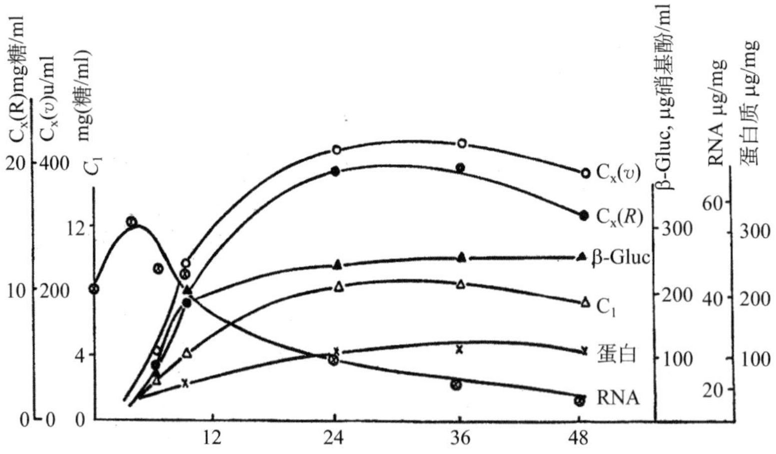

图1 N_2-78洗涤菌丝体被槐糖诱导不同时间胞外纤维素酶、可溶性蛋白和菌体RNA含量的变化，25 ml三角瓶中盛5 ml洗涤菌丝体(7.6 mg干重)，于无甘油的无机盐培养基中震荡诱导，槐糖浓度：10^{-4} mol/L，诱导温度：32℃，震荡频率：108次/分

C_1酶以棉花糖表示；C_x(R)代表还原糖法测定的外切CMC酶活；C_x(ν)代表黏度法测定的内切CMC酶活；β-Gluc代表水解对-硝基苯-β-葡萄糖苷的葡萄糖苷酶活力。

量也有类似的变化，而菌丝体 RNA 含量开始增加，到 3～6 小时后则迅速下降。用超声波破碎 24 小时的菌丝体，比较细胞内外纤维素酶的含量，证明绝大部分纤维素酶均分泌到细胞外（见表 1）。

表 1　N_2－78 洗涤菌丝体被槐糖诱导的纤维素酶各组分和可溶性蛋白在细胞内外的分布

酶分布部位	C_x		C_1 棉花糖 /(mg 糖/ml)	β-葡萄糖苷酶 /(μg 硝基酚/ml)	可溶性蛋白 /(μg/ml)
	还原糖法 /(mg 糖/ml)	黏度法 /(u/ml)			
胞外	20	591	8.2	250	184
胞内	2	85	1.0	50	86

注：诱导时间 24 小时，胞外酶测定是取培养菌丝的滤液，胞内酶测定是收集菌丝，用缓冲液洗涤后，经超声处理（20 KC，10 分钟）的样品中测定。

2. 菌丝体总 RNA 的抽提

经槐糖诱导 6 小时的 N_2－78 洗涤菌丝体，采用加核酸酶抑制剂二乙基焦碳酸钠和低温快速破碎及酚液抽提的方法，得到总 RNA，$A_{260}/A_{280}>2$，得率 2%左右（占干菌体）。图 2 是对

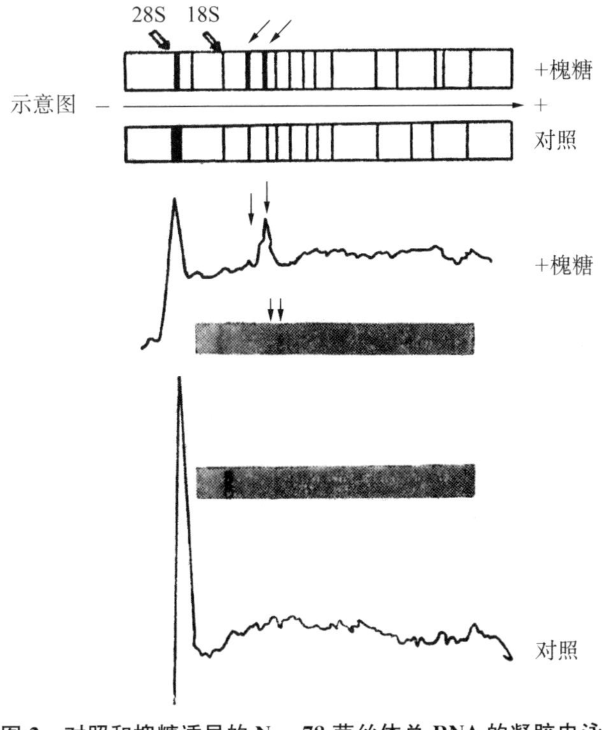

图 2　对照和槐糖诱导的 N_2－78 菌丝体总 RNA 的凝胶电泳图和扫描图

聚丙烯酰胺胶浓 3%，管 6.7×0.56 cm，RNA 样品：300 μg，电流：5 毫安/管，电泳时间：2 小时，甲烯蓝染色，650 mμm 波长扫描，箭头（↓）所示可能是 mRNA 位置。

照和经槐糖诱导的 N_2 - 78 菌丝体总 RNA 的凝胶电泳图和扫描图。图谱上显示 10 多条 RNA 带，在 18 S r-RNA 带右面(2.3 厘米和 2.5 厘米两处)有两条 RNA 色带经槐糖诱导后显著加深。根据其他材料的研究(李文裕，1978；李明，1978；Jacobson 等，1975)，18 S r-RNA 下的区带可能是 mRNA 部分。

3. 菌丝体 Poly(A)- RNA 的分离

100 毫升菌丝体的总 RNA(约 700 A_{260} 单位)溶于 100 毫升 0.01 mol/L Tris-HCl (pH 7.5)-0.5 mol/L KC1 中，经寡聚(dT)-纤维素柱(13×1.2 厘米左右)三步洗脱层析，可以得到三个峰：峰Ⅰ、峰Ⅱ和峰Ⅲ(见图 3)，其中 A_{260}/A_{280} 分别为 2.24、2.14 和 1.46。峰Ⅰ是不吸附在柱上的 RNA，经聚丙烯酰胺凝胶电泳检定，主要是 r-RNA，其含量最高，得率占总 RNA 的 98%以上。峰Ⅱ是吸附不紧密的 RNA。低离子强度下洗脱下来的峰Ⅲ是 Poly(A)- RNA，得率占总 RNA 的 1.06%。

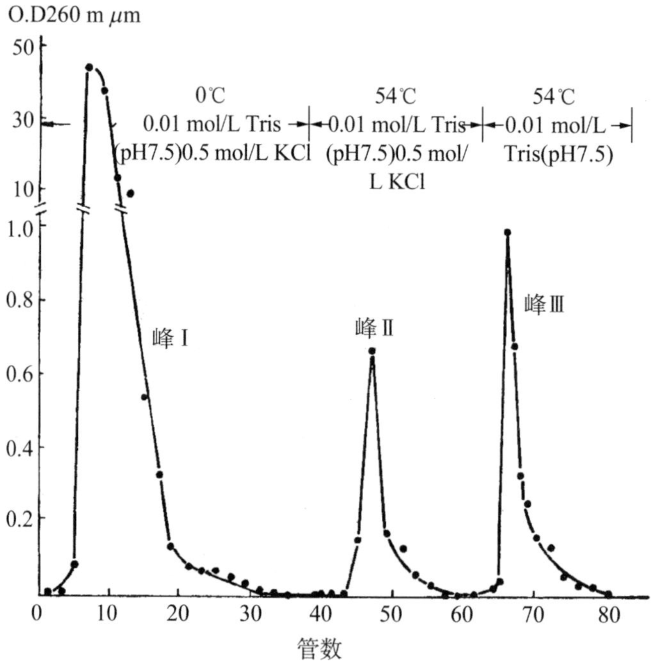

图 3 N_2 - 78 菌丝体经槐糖诱导的 RNA 的寡聚(dT)-纤维素柱层析分离图(三步法)

图 4 是总 RNA 经二步洗脱得到的层析图，有二个峰：峰Ⅰ和峰Ⅱ，A_{260}/A_{280} 分别为 2.30 和 1.63。其中峰Ⅱ按文献报道(李文裕等，1978)是 Poly(A)- RNA，占总 RNA 0.96%。用三步法或二部法洗脱，都表明 Poly(A)- RNA 约占总 RNA 的 1%。

图 3 和图 4 中的 Poly(A)- RNA 经透析、冷冻干燥后，经过聚丙烯酰胺凝胶电泳检定，得到较为单纯的区带(见图 5)，位置与图 2 中的 mRNA 位置接近。

图 6 是对照(不加槐糖)和经槐糖诱导 6 小时的总 RNA 样品的寡聚(dT)-纤维素柱层析图(二步法洗脱)。可见，经槐糖诱导后的 Poly(A)- RNA 含量显著增加。

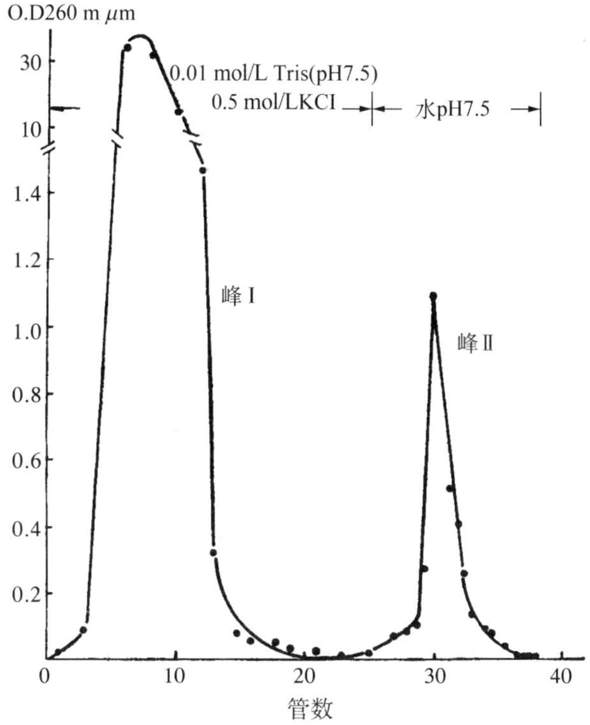

图 4　N_2 –78 菌丝体经槐糖诱导的 RNA 的寡聚(dT)-纤维素柱层析分离图(二步法)

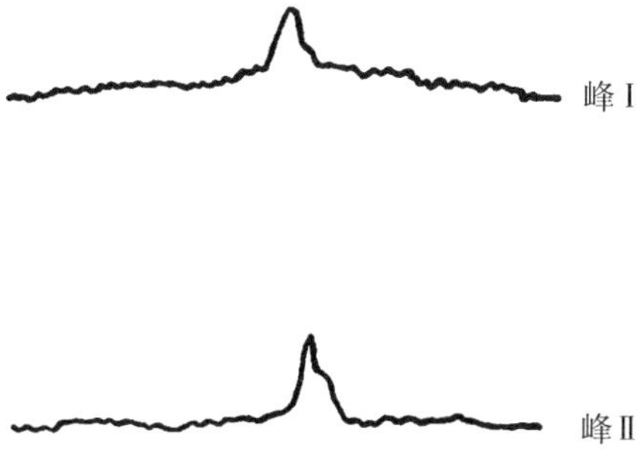

图 5　Poly(A)- RNA 的凝胶电泳扫描图

峰 II 为二步洗脱法得到的 Poly(A)- RNA，峰 III 为三步洗脱法得到的 Poly(A)- RNA。

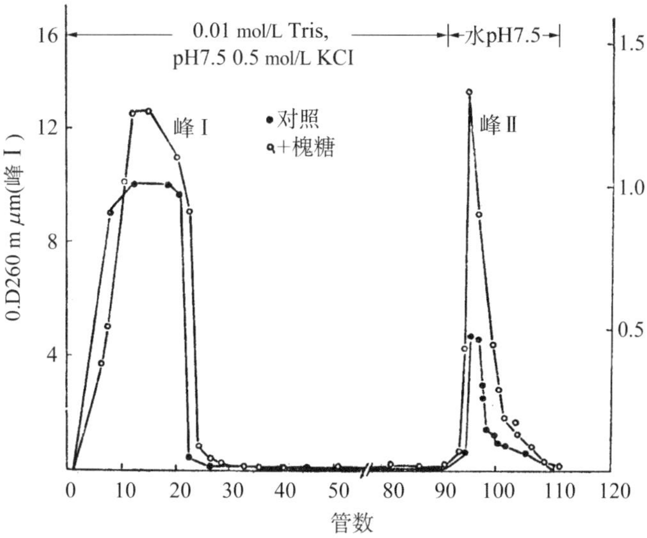

图 6　槐糖诱导和对照总 RNA 经寡聚(dT)-纤维素柱的层析图

四、讨论

拟康氏木霉洗涤菌丝体在槐糖诱导下只经过 3 小时左右的延迟期，就能合成纤维素酶蛋白，并迅速分泌到细胞外。洗去槐糖，纤维素酶合成则停止。核酸代谢抑制剂（5′-氟尿嘧啶、6-氮杂尿嘧啶）和蛋白合成抑制剂（放线酮）对纤维素酶的诱导形成，有强烈抑制作用（朱雨生等，1978b）。所以，核酸代谢在纤维素酶诱导形成中占着核心地位，而其中 mRNA 的调控则是细胞基因表达的重要环节。在高等真核生物中，已证明 mRNA 具有 Poly(A)的结构，且大多是长寿命的（Aviv 等，1972；Lockard 等，1969；Verma 等，1974；李文裕等，1978；李明等，1978；佐野浩，1977），因此可以用寡聚(dT)-纤维素柱的亲合层析法，专一地分离 mRNA。在大多数微生物中，由于 mRNA 寿命很短，难以提取分离。最近报道高等真菌——黏菌中也存在含 Poly(A)的 mRNA（Jacobson 等，1975）。真菌木霉是低等的真核生物，但至今文献中未见有 Poly(A)-RNA 分离的报道。我们从槐糖诱导纤维素酶开始大量合成期的 N_2-78 洗涤菌丝体中，采用加核酸酶抑制剂和低温快速抽提等防止 mRNA 破坏的方法分离核酸，经过寡聚(dT)-纤维素柱分离，也得到含 Poly(A)-RNA 的峰，$A_{260}/A_{280}=1.5$ 左右，得率占总 RNA 的 1%左右，电泳位置与文献记载的 mRNA 位置接近。同时，无论用总 RNA 电泳或寡聚(dT)-纤维素柱亲合层析的方法，均证明槐糖诱导后的 Poly(A)-RNA 含量增加了。此结果与植物中赤霉素诱导大麦糊粉层 α-淀粉酶和ⅠAA 诱导豌豆下胚轴纤维素酶形成时，Poly(A)-RNA 增加的现象是类似的（Ho 等，1974；Verma 等，1975）。槐糖诱导的 Poly(A)-RNA 是否就是纤维素酶的 mRNA，还有待生物活性测定的进一步鉴定。

参考文献

[1] 上海实验生物所三室、上海试剂二厂核酸研究组，1977。oligo(dT)-纤维素与 oligo(u)-纤维素的合成及其应用。生物化学与生物物理进展，1977(4)：8。

[2] 朱雨生、谭常，1978 a. 木霉纤维素酶的诱导形成及其调节Ⅰ. 槐糖对木霉 EA_3-867 纤维素酶形成的诱导作用。微生物学报，18：320。

[3] 朱雨生、谭常，1978 b. 木霉纤维素酶的诱导形成及其调节Ⅱ. 槐糖对木霉 EA_3-867 洗涤菌丝体纤维素酶形成的诱导作用及降解物阻遏现象。植物生理学报，4：1。

[4] 李文裕、王佩瑜、李明、许远钟、刘海湖、韦澄、陈瑞铭、叶秀珍、陆荣华，1978。正常肝信息核糖核酸的翻译和对离体肝癌细胞分化的逆转分化作用。实验生物学报，11：109。

[5] 李明、徐链、李文裕，1978。大鼠肝癌甲胎蛋白 mRNA 在麦胚蛋白合成系统中的翻译（简报）。实验生物学报，11：145。

[6] 石田各香雄、田中信男、渡边力编集，1975。微生物学。朝仓书店。

[7] 佐野浩，1977。真核细胞メッセンジャ-RNAの精制。蛋白质、核酸、酵素，22：1303。

[8] Aviv, H., Leder, P., 1972. Purification of biological active globin messenger RNA by chromatography on oligothymidylic acid-cellulose. *Proc. Nat. Acad. Sci.* USA, 69: 1048.

[9] Chargaff, E., Davison, J. N., 1955. The Nucleic Acid: Chemistry and Biology, Academic press, N. Y. Vol. 1, p. 301.

[10] Ho, D. T-H., Varner, J. E., 1974. Hormonal control of messenger ribonucleic acid metabolism in barley aleurone layers. *Proc. Nat. Acad. Sci.* USA, 71: 4783.

[11] Jacobson, A., Lane, C. D., Alton, T., 1975. Electrophoretic separation of the major species of *slime mold* messenger ribonucleic acid. Microbiology-1975, p. 490, ed. Schlessinger, D., American Society

for Microbiology, Washington, D. C.

[12] Lockard, R. E., Lingrel, J. B., 1969. The synthesis of mouse hemoglobin β - chains in a rabbit reticulocyte cellfree system programmed with mouse reticulocyte 9s RNA. *Biochem. Biophys. Res. Comm.*, 37: 204.

[13] Loening, U. E., 1967. The fraction of high-molecular weight ribonucleic acid by polyacryamide gel electrophoresis. *Biochem.* J., 102: 251.

[14] Lowry, O. M., Rosebrough, N. J., Farr, A. L., Randall, R. J., 1951. Protein measurement with the folinphenol reagent. J. *Biol. Chem.*, 193: 265.

[15] Verma, D. P. S., 1974. Isolation and in vitro translation of soybean leghaemoglobin mRNA. *Nature*, 251: 77.

[16] Verma, D. P. S., Maclachlan, J. A., Byrne, H., Fwings, D., 1975. Regulation and in vitro translation of messenger ribonucleic acid for cellulase from auxin-treated pea epicotyls. *J. Biol. Chem.*, 250: 1019.

Induction and Regulation of Cellulase Formation in Trichoderma

V. The Changes in Nucleic Acid Metabolism of Washed Mycelia of Trichoderma Pseudokoningii Rafai N_2 – 78 During Inductive Formation of Cellulase*

The cellulase production (C_1, C_x, and β – glucosidase) of washed mycelia of *Trichoderma pseudokoningii* Rafai N_2 – 78 was induced by sophorose, meanwhile, the change of nucleic acid content in washed mycelia showed a certain regularities. By the way of addition of inhibiter of nuclease and fast extraction at lower temperature, the total nucleic acids were isolated from washed mycelia of N_2 – 78 treated with sophorose, and from them a Poly (A)– RNA was separated by oligo (dT)– cellulose chromatography. The content of Poly (A)– RNA in mycelia treated with sophorose was shown by affinity chromatography and polyacrylamide gel electrophoresis to be higher than the control.

* Wu Yong-qiang, Zhu Yu-Sheng, Chen wei, Gao Jin-hua, Fei Jin-xin (Shanghai Institute of Plant Physiology, Academia Sinica)

木霉纤维素酶的诱导形成及其调节

Ⅵ. 拟康氏木霉(*Trichoderma pseudokoningii* Rafai) N_2-78 被槐糖诱导形成的纤维素酶组分的分离、纯化及性质

提要

拟康氏木霉洗涤菌丝体被槐糖诱导形成的纤维素酶,经 DEAE-Sephadex A-50 离子交换层析、Sephadex G-100 凝胶过滤和聚丙烯酰胺凝胶电泳等方法,初步分离得到 C_1 酶、C_x 酶和 β-葡萄糖苷酶三个组分,其相对分子质量经聚丙烯酰胺凝胶电泳测定,分别为 67 000、62 000 和 42 000。C_1 酶用聚丙烯酰胺凝胶电泳、醋酸纤维素薄膜电泳、免疫电泳和超离心鉴定,已证明是匀一的组分,为糖蛋白,富含甘氨酸、天门冬氨酸、苏氨酸、丝氨酸和谷氨酸。纤维素酶的三个组分对天然纤维素、微晶纤维素和或磷酸膨胀纤维素的作用存在协同效应。研究了纤维素酶组分的热稳定性及 pH 稳定性。

前文报道了拟康氏木霉纤维素酶的三个组分 C_1、C_x 和 β-葡萄糖苷酶能被槐糖同步诱导(朱雨生等,1978a, 1978b)。本文进一步讨论,拟康氏木霉洗涤菌丝体用槐糖诱导形成的胞外纤维素酶的三个组分,经葡聚糖凝胶过滤、离子交换树脂层析和聚丙烯酰胺凝胶电泳等步骤分离纯化的方法,以及纤维素酶组分间的协同效应。

一、材料与方法

1. 菌种

拟康氏木霉 N_2-78 是自野生型菌株木-3 经过多种物理化学因素诱变得到的高产菌株(中科院上海植生所纤维素酶组等,1978)。

2. 试剂

槐糖是自槐豆荚中提取分离得到的结晶制品(朱雨生等,1978a)。Sephadex G-100、DEAE-SephadexA-50(50—140 目)系瑞典 Pharmacia 公司产品。微晶纤维素是 E. Merck 公司产品。磷酸膨胀纤维素是用微晶纤维素按 Walseth 方法制备(1952)。羧甲基纤维素钠盐(CMC)是上海长虹塑料厂出品,黏度(pM)为 300～600 厘泊。对-硝基苯葡萄糖苷是英国

朱雨生,吴永强,陈薇,高锦华,费锦鑫,中国科学院上海植物生理研究所。此文发表于 1980 年“植物生理学报”6 卷 1 期 1-9 页。

1979 年 2 月 27 日收到。

* 本工作承沈善炯教授指导,宋廷生同志帮助 C_1 酶超离心分析,中国科学院有机化学研究所王思清同志帮助分析 C_1 酶氨基酸成分,在此一并致谢。

Koch-Light Laboratories 公司产品。

3. 方法

(1) 纤维素酶制剂制备方法。5 升三角瓶中装入 1 升液体培养基,成分见前文(朱雨生等,1978a)。28℃振荡培养 48 小时后,抽滤,菌丝体用二倍体积的 0.01 mol/L、pH 5.0 的磷酸缓冲液洗涤后,转入到 1 升缺碳源的诱导培养基中(5 升三角瓶中),内含槐糖(5×10^{-5} mol/L),于 28℃振荡培养 24 小时,抽滤,取滤液,用 80%饱和度的硫酸铵沉淀蛋白,经 Sephadex G-25 脱盐,冷冻干燥,即得槐糖诱导的胞外纤维素酶制剂。

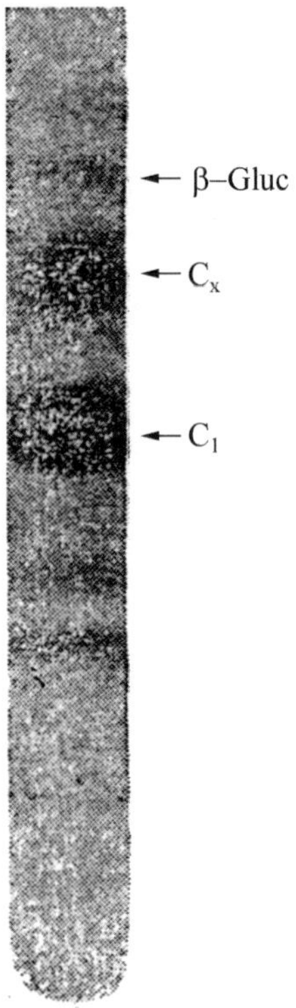

图 1 拟康氏木霉纤维素酶的聚丙烯酰胺凝胶电泳

胶浓 7%;纤维素酶样品 2 mg;电流 2 mA/管;电泳时间 2.5 小时,图中对照管用氨基黑染色。另一样品(测酶活)管依对照管的染色带位置切割,用 1 毫升蒸馏水浸提后,测定酶活力,经鉴定,示 β-Gluc 的带只含 β-Gluc 酶活力(7 000 μg 硝基苯/ml),示 C_x 的带只含 C_x 酶活力(8.8 mg 糖/ml),示 C_1 的带除含 C_1 酶活力(1.9 mg 糖/ml)外,还含有部分 C_x 活力(1.8 mg 糖/ml)。

(2) 纤维素酶活力测定方法。

① C_x 酶(CMC 酶活):a. 还原糖法($C_{x(R)}$)见前文 B 法(中科院上海植生所纤维素酶等,1978),酶活力以毫克糖/毫升酶表示;b. 黏度法($C_{x(v)}$):见前文(同上),酶活力以单位(u)/毫升酶表示。

② β-葡萄糖苷酶(β-Gluc):V0.2 毫升适当稀释的酶液,加 0.1 毫升对-硝基苯葡萄糖苷(0.5%)和 0.1 毫升 0.1 mol/L pH 4.5 的醋酸缓冲液,50℃反应 30 分钟,加 4 毫升 0.1 mol/L Na_2CO_3 于 430 mμm 比色,测定释放的硝基苯。酶活力以每毫升酶液释放的硝基苯微克数(微克硝基苯/毫升酶)表示。

③ C_1 酶(以棉花糖活力表示):见前文(同上)。

④ 微晶纤维素活力:V0.25 毫升酶液,加 0.25 毫升 0.5%的微晶纤维素和 0.5 毫升 0.1 mol/L、pH4.5 的醋酸缓冲液,于 50℃下反应 4 小时,用 DNS 试剂定糖,酶活力以毫克糖/毫升酶表示。

⑤ 磷酸膨胀纤维素活力:V 0.5 毫升酶液,加 0.5 毫升 0.1 mol/L、pH 4.5 的醋酸缓冲液及 4 毫升 0.5%的磷酸膨胀纤维素,于 50℃下反应 2 小时,用 DNS 试剂定糖。酶活力以毫克糖/毫升酶表示。

(3) 电泳。

① 聚丙烯酰胺凝胶电泳:V 参考 Hedrick 等的方法(1968),胶长 0.59 厘米×8 厘米,胶浓 7%,缓冲液为 0.005 mol/L Tris-0.038 mol/L 甘氨酸溶液,pH 8.7,电流 2 毫安/管,氨基黑染色,7%醋酸中退色。在电泳分离纤维素酶时,电泳完毕,参照对照管的染色带位置切割凝胶(见图 1),用蒸馏水浸提酶液,调 pH 至 5.0 左右,对蒸馏水透析,冷冻干燥。纤维素酶分子量测定时,采用 5%、6%、7%和 8%胶浓,以牛血清蛋白(三种聚合态)、过氧化氢酶、胃蛋白酶、淀粉酶、乳酸脱氢酶和肌酸激酶等为对照,测量不同胶浓下蛋白的相对泳动,作图求出各种蛋白的斜率,并对分子量作图。C_1 酶分子量还用 SDS 凝胶电泳法测定(Weber,1969)。

② 醋酸纤维素薄膜电泳。醋酸纤维素薄膜(8 厘米×2 厘米)直线点样后电泳半小时左右,电压 90～110 伏,电流 0.4～0.6 毫安/毫米,氨基黑染色。

③ 免疫电泳:a. 纤维素酶抗血清制备:槐糖诱导的胞外纤维素酶经

聚丙烯酰胺凝胶电泳初步纯化的纤维素酶蛋白(以 C_1 酶为主,含 C_x 酶和 β-葡萄糖苷酶)作抗原,加 Freund (1:1 v/v)佐剂,皮下注射兔子,第一次注射为每头兔子 2 毫克,10 天后第二次注射(0.75 毫克/头),再隔 10 天后作第三次注射(1 毫克/头)。一个月后,取耳静脉血作效价测定,从颈动脉抽血,制备抗血清。b. 电泳方法:琼脂糖(1.2%)溶于 0.02 mol/L、pH 8.6 的巴比妥酸缓冲液中,铺琼脂板 7.5 厘米×2.5 厘米,纤维素酶抗原点样后于巴比妥酸缓冲液中电泳 3 小时,电压 5 伏/厘米,电泳完毕,在琼脂糖板中央开槽,滴入纤维素酶抗血清 50 微升,于室温扩散 24 小时,氮基黑染色,7%醋酸中退色。

(4) 超离心。

用 P hywe 超离心机(东德)测定,纤维素酶 C_1)浓度为 10 毫克/毫升、缓冲液为 0.1 mol/L,pH 4.5 的醋酸缓冲液,转速 48 000 转/分,温度 20℃。

(5) 糖测定。

① C_1 酶结构糖分析(Tomita 等,1974):4 毫克纯化的 C_1 酶,于 2 毫升 1N H_2SO_4 中水解(沸水浴)4 小时,Ba $(OH)_2$ 中和后,溶液通过阳离子树脂 717 和阴离子树脂 732,去离子后的流出液浓缩,用酚-硫酸法定糖(朱雨生等,1978a)。

② 纤维素酶水解产物纸谱分析:见前文(朱雨生等,1978c)。

(6) 氨基酸成分测定。

称取 4.07 毫克纯化的 C_1 酶,于 5.4N HCl 中,110℃水解 24 小时。长短柱分别为 50 和 10 厘米,上样量 0.5 毫克。用日制的日立 KLA-3B 型氨基酸分析仪测定氨基酸成分。

二、试验结果

1. 纤维素酶组分的分离与纯化

称取 100 毫克槐糖诱导的胞外纤维素酶,溶于 10 毫升醋酸-氢氧化钠缓冲液(0.1 mol/L、pH 5.0)中,先上 Sephadex G-75 柱,除了去除一部分杂蛋白外,纤维素酶各组分基本上分不开,于是上 DEAE-Sephadex A-50 柱。

(1) DEAE-Sephadex A-50 分离。DEAE-Sephadex A-50 预先用醋酸-氢氧化钠(0.1 mol/L、pH 5.0)平衡。上样后,用同样的缓冲液洗脱,在 No. 4～No. 14 出现蛋白峰(A),含 C_x 酶和 β-葡萄糖苷酶,但没有 C_1 酶活性。待蛋白含量降为 OD_{280} 值在 0.02 以下时(No. 40),换用 0.1 mol/L、醋酸-氢氧化钠缓冲液和 0.1 mol/L 醋酸-氢氧化钠- 0.5 mol/L NaCl, pH 5.0 的梯度洗脱,在 No. 54～No. 68 之间出现一蛋白主峰(B)(见图 2)。B 峰能分解棉花(C_1 酶),含有少量的 C_x 酶。将 B 峰合并,对水透析 40 小时,冷冻干燥,再用聚丙烯酰胺凝胶电泳进一步纯化(见方法部分)。40 毫克样品,经四根聚丙烯酰胺凝胶电泳分离,约得到 18 毫克较纯的 C_1 酶白色冷冻干燥制剂,此时不含 C_x 和 β-葡萄糖苷酶(见表 1)。故经过 DEAE-Sephadex A-50 和凝胶电泳,首先得到 C_1 酶。

目前文献中报道 C_x 和 β-葡萄糖苷酶分离较困难。本文系用 Sephadex G-100 凝胶过滤和选择性失活的方法。上述 A 峰中的 C_x 和 β-葡萄糖苷酶不能用 SE-Sephadex C-50 分开,故对蒸馏水透析 24 小时后,冷冻干燥,溶于 5 毫升 0.01 mol/L、pH 5.5 的醋酸-醋酸铵缓冲液中,上 Sephadex G-100 柱。

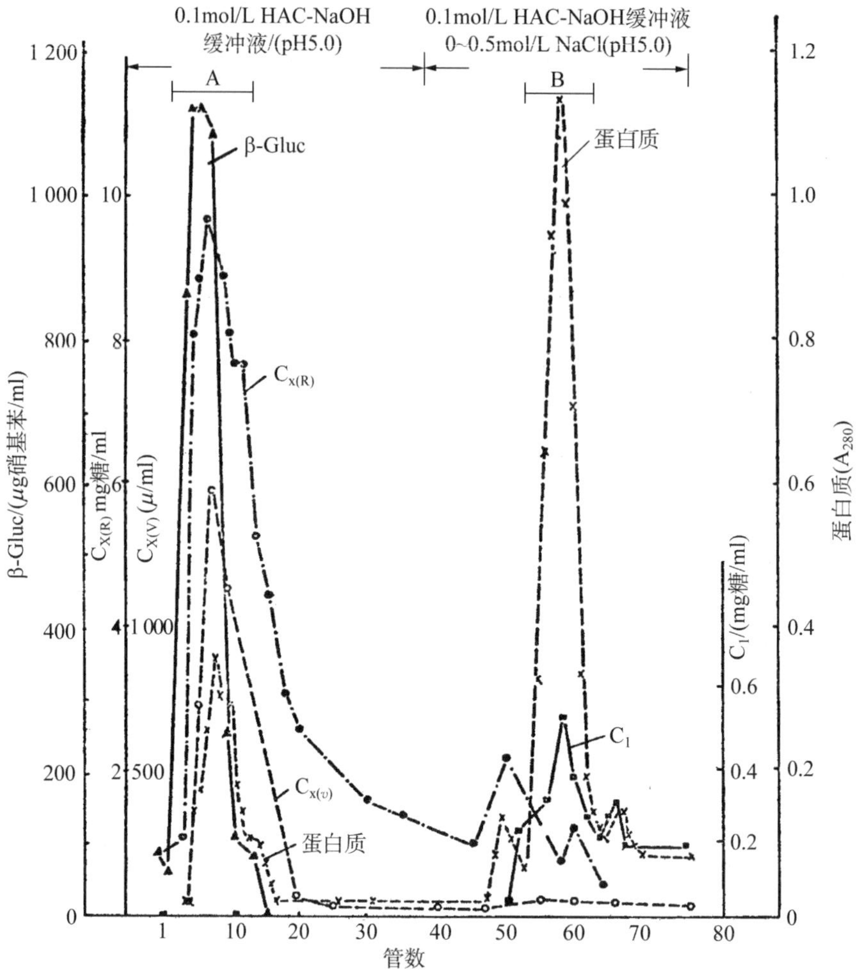

图 2 拟康氏木霉纤维素酶经 DEAE-Sephadex A-50 柱(18.5×1.8 cm)的离子交换层析图

样品上柱后，先用 0.1 mol/L HAc-NaOH 缓冲液(pH 5.0)洗脱，流速 10 ml/小时，分部收集，每管 5.2 ml，到第 40 管后，换用 0.1 mol/L HAc-NaOH 和 0.1 mol/L HAc-NaOH-0.5 mol/L NaCl (pH 5.0)缓冲液的梯度洗脱，流速 20 m/小时，加样后第三管开始收集，每管收集 4.3ml。

表 1 分离的纤维素酶组分 C_1、C_x 和 β-葡萄糖苷酶的底物专一性

作用底物 / 酶组分	棉花 /(mg 糖/ml 酶)	CMC /(mg 糖/ml 酶)	对-硝基苯葡萄糖苷 /(mg 硝基苯/ml 酶)
C_1	0.4	0	0
C_x	0	2.0	0
β-葡萄糖苷酶	0	0	3.2

注：C_1 酶是图 1 中 B 峰经电泳纯化的组分，浓度为 1 mg/ml；C_X 酶为图 2 中的 D 峰部分(0.6 mg/ml)；β-萄葡糖苷酶为图 2 中 C 峰(0.5 mg/ml)部分再经选择性失活得到的部分，加酶量约为 0.1 ml，测定方法见方法部分。

(2) Sephadex G-100 分离 Sephadex G-100 悬于 0.01 mol/L、pH 5.5 的醋酸-醋酸铵缓冲液中，于 90℃浸泡 6 小时，装柱。样品上柱后，用同样的缓冲液洗脱，在 No.13～No.20 之

间出现蛋白主峰(C)(见图 3),含 C_x 和 β-葡萄糖苷酶。但在 No. 20～No. 27 之间肩部(D)只含 C_x 酶,而不含 β-葡萄糖苷酶(见表 1),故通过 Sephadex G-100 分离得到 C_x 酶。

(3) C_1 和 β-葡萄糖苷酶选择性失活的方法 C 峰中的 C_x 和 β-葡萄糖苷酶活性最后通过选择性变性的方法分开。即在 pH 2.5、0.1 mol/L 醋酸缓冲液中,于 45℃处理酶液 24 小时,可使 β-葡萄糖苷酶 100%失活,而绝大部分 C_x 酶被保存(见图 4);相反,在 pH 7.5、0.005 mol/L Tris-0.038 mol/L 甘氨酸缓冲液中,于 45℃处理 8 小时,可使 C_x 酶 100%失活,但保存绝大部分 β-葡萄糖苷酶。此法虽不能去除 C_x 酶蛋白,但可得到不含 C_x 酶活性,只含 β-葡萄糖苷酶活性的组分,以便作下面的重组试验。

纤维素酶三个组分中 C_1 含量最高,得以进一步作纯度鉴定,及研究其性质。

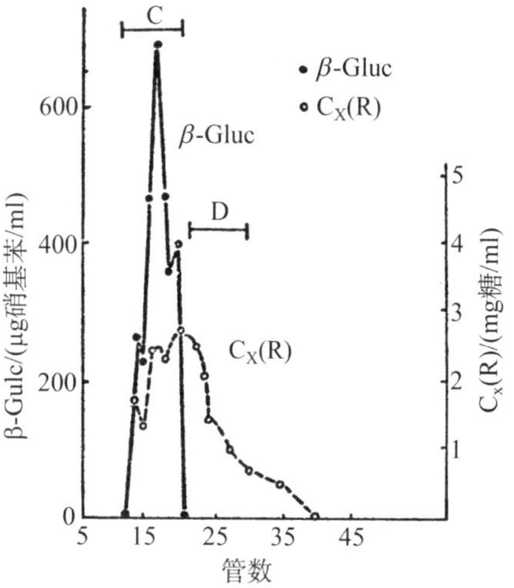

图 3 拟康氏木霉纤维素酶通过 Sephadex G-100(88.5 cm×1.9 cm)的凝胶过滤层析图

流速 10 ml/小时,每管 5 ml。

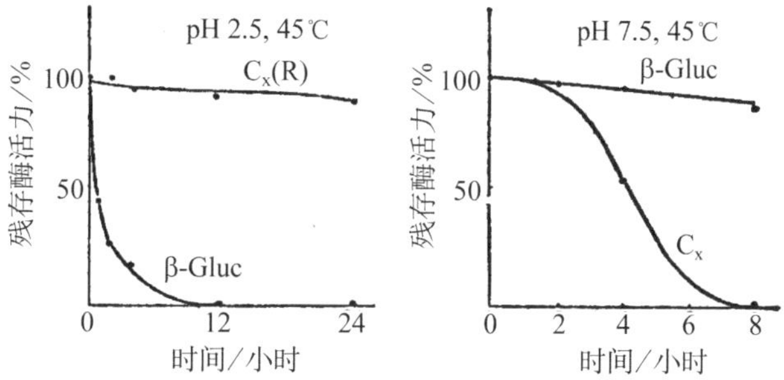

图 4 拟康氏木霉 C_x 酶和 β-葡萄糖苷酶的 pH 稳定性的比较

2. C_1 酶的纯度鉴定

(1) 醋酸纤维素薄膜电泳:C_1 酶成单一蛋白带(见图 5)。

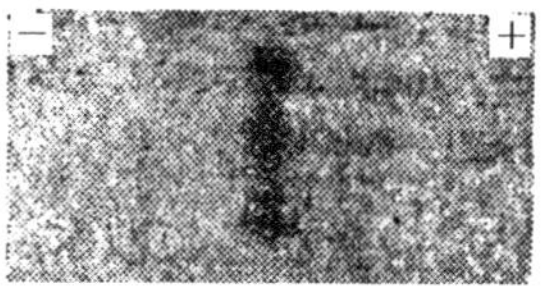

图 5 C_1 酶的醋酸纤维素薄膜电泳

电流 0.5 毫安/厘米,加样量 50 μg,电泳时间 30 分钟。

(2) 聚丙烯酰胺凝胶电泳:C_1 酶为单一蛋白带(见图 6),只能测到 C_1 酶活力。

(3) 免疫电泳:纤维素酶的抗血清与 C_1、C_x 或 β-葡萄糖苷酶均能起沉淀反应,并且纤维素酶三个组分的活力均受抗血清不同程度的抑制(见图 7),证实纤维素酶的抗血清确实有抗体活力。混合酶的免疫电泳呈三条弧线,而 C_1 酶的免疫电泳只出现一条沉淀弧线(见图 8)。

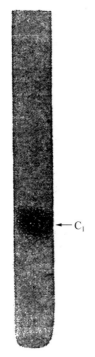

图 6　C_1 酶的聚丙烯酰胺凝胶电泳

加样量 500 μg；胶浓 7%；电流 4 毫安/管；电泳时间 2.5 小时。

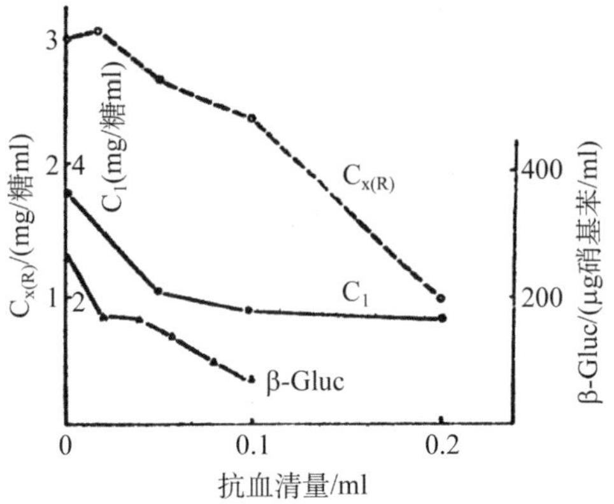

图 7　纤维素酶抗血清对 C_1、C_x 和 β-葡萄糖苷酶的抑制作用

图 8　C_1 酶和纤维素酶混合酶的免疫电泳

琼脂糖浓度 1.2%；电压 5 伏/cm；电泳时间 3 小时。电泳后，中央开槽，滴入纤维素酶抗血清 50 ul。室温下扩散 24 小时，氨基黑染色，干燥保存。

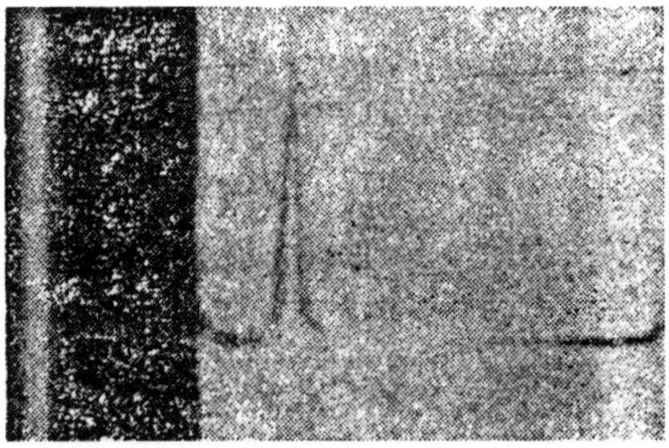

图 9　C_1 酶的超离心图

C_1 酶的浓度 10 mg/ml、0.1 mol/L、pH 4.5 醋酸缓冲液；转速 48 000 转/分。

(4) 超离心：C_1 酶的超离心图谱是呈单一对称峰（见图 9），说明 C_1 酶是超离心纯的。

3. 纤维素酶纯化酶的性质

(1) 相对分子质量。C_1 酶、C_x 酶和 β-葡萄糖苷酶用 Hedrick 等的聚丙烯酰胺凝胶电泳法测定，相对分子质量分别为 67 000、62 000 和 42 000[见图 10(a)]。C_1 酶相对分子质量用 Weber 的 SDS 聚丙烯酰胺凝胶电泳法测定也是 67 000[见图 10(b)]。

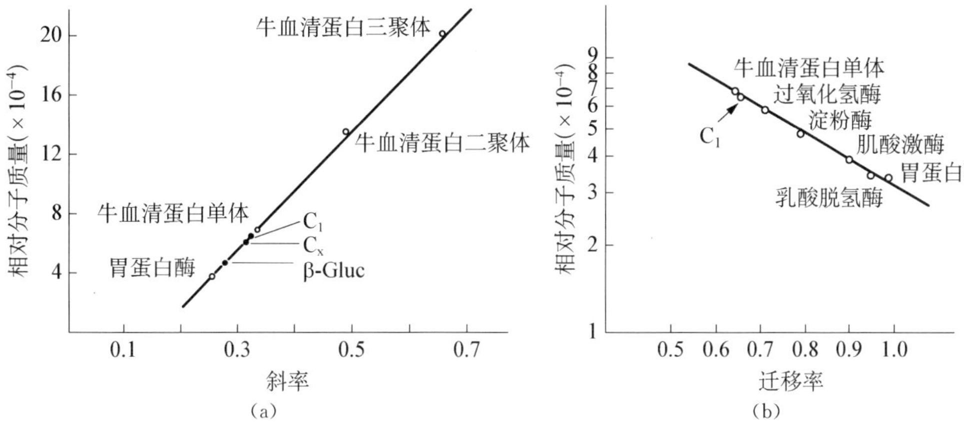

图 10　几种标准蛋白的电泳迁移率与其分子量的关系

图中标准蛋白的相对分子质量如下：牛血清蛋白单体 67 500，牛血清蛋白二聚体 136 000，牛血清蛋白三聚体 204 000，胃蛋白酶 35 000。用 SDS 凝胶电泳测定 C_1 酶相对分子质量，几种标准蛋白的分子量：牛血清蛋白（单体）67 500；过氧化氢酶 58 000；淀粉酶 50 000；肌酸激酶 40 000；乳酸脱氢酶 36 000；胃蛋白酶 35 000。

（2）C_1 酶成分。纯化的 C_1 酶蛋白经硫酸水解，用酚-硫酸法测得含结构糖（1％左右）。氨基酸成分测定结果如表 2 所示。可见，C_1 酶 16 种氨基酸中主要为亲水性氨基酸，如甘氨酸、天门冬氨酸、苏氨酸、丝氨酸和谷氨酸。疏水性氨基酸和碱性氨基酸含量较低。有意思的是测不到胱氨酸和半胱氨酸，而含有半胱磺酸。

表 2　拟康氏木霉 N_2－78 C_1 酶的氨基酸成分

氨基酸成分	克分子含量比/％	氨基酸成分	克分子含量比/％
赖氨酸	3.23	谷氨酸	9.62
组氨酸	0.75	丙氨酸	6.36
精氨酸	1.91	缬氨酸	5.04
半胱磺酸	1.81	亮氨酸	5.68
天门冬氨酸	13.17	异亮氨酸	2.20
苏氨酸	12.08	脯氨酸	5.79
丝氨酸	10.83	酪氨酸	4.08
甘氨酸	14.67	苯丙氨酸	2.77

（3）底物专一性。C_1 酶只分解棉花，不分解 CMC 和对-硝基苯葡萄糖苷。C_x 酶只分解 CMC，不分解棉花和对-硝基苯葡萄糖苷。β-葡萄糖苷酶只分解对-硝基苯葡萄糖苷（见表 1）。

（4）纤维素酶组分间的协同效应。C_1 酶在纤维素酶三个组分中占的蛋白含量最高（见图 1），但纯化的 C_1 酶分解棉花、微晶纤维素和磷酸膨胀纤维素的活力很低，当 C_1 酶与 C_x 酶及 β-葡萄糖苷酶重组后，则分解三种纤维素的能力大大增强（见表 3）。而且，可以看出，主要是 C_1 与 C_x 酶之间表现协同效应。

表 3 C_1、C_x 和 β-葡萄糖苷酶之间的协同效应

酶分组	棉花活力/(mg/糖/ml 酶)	协同倍数	磷酸膨胀纤维素活力/(mg/糖 ml 酶)	协同倍数	微晶纤维素活力/(mg/糖 ml 酶)	协同倍数
C_1	0.3		3.8		1.0	
C_x	0		0		0	
β-葡萄糖苷酶	0		0		0	
C_x+β-葡萄糖苷酶	0		0		0	
C_1+C_x	1.6	3.5 倍	11.0	2.9 倍	1.08	1.8 倍
C_1+C_x+β-葡萄糖苷酶	1.7	5.7 倍	11.2	2.9 倍	1.85	1.9 倍

C_1 酶作用于磷酸膨胀纤维素的主要产物为纤维二糖，尚有少量葡萄糖等(纸谱鉴定)；C_1 酶与 C_x 酶及 β-葡萄糖苷酶重组后分解磷酸膨胀纤维素的主要产物除了纤维二糖外，葡萄糖则增多。

(5) 热稳定性和 pH 稳定性。

① 热稳定性：C_1 酶、C_x 酶和 β-葡萄糖苷酶在 0.1 mol/L、pH 5.0 醋酸缓冲液中，于 60℃ 热处理 10 分钟，失活一半左右，于 100℃处理 10 分钟，C_1 酶全部失活，C_x 酶和 β-葡萄糖苷酶大部分失活。

② pH 稳定性：C_x 酶在微碱性(pH 7.5)中不稳定，而 β-葡萄糖苷酶在酸性介质中(pH 2.5)不稳定(见图 4)。此特点曾被利用来分离 C_x 和 β-葡萄糖苷酶活性。

三、讨论

纤维素酶是复合酶。Reese (1950)提出了 C_1-C_x 酶的作用假说以后，不少研究人员分离了 C_1 酶，并研究其性质(Selby 等，1967；Halliwell 等，1973；Wood，1968；Wood 等，1972)。但是，对于 C_1 酶的作用，争论较多(朱雨生，1978)。近来，不少研究人员认为，纤维素酶分解天然纤维不是像 Reese 所假设的那样，先由 C_1 酶作用，后由 C_x 酶作用，而是 C_x 酶先作用，提供可反应的链端，再由 C_1 酶作用，分解产生纤维二糖。C_x 酶是内切型 β-1,4-葡聚糖酶，而 C_1 酶是外切型 β-1,4-葡聚糖纤维二糖水解酶(Wood 等，1972；Wood，1975；Halliwell 等，1973)。中国科学院微生物研究所纤维素酶组从绿色木霉中分离纯化到 C_1 酶，并对 C_1 酶作用方式持类似的观点(1976)。Tomita (1974)则提出天然纤维素的分解是靠随机性较大的 CMC 酶和随机性较小的微晶纤维素酶协同作用的结果。一般从木霉在复杂的培养基中以天然纤维素(如稻草粉、麸皮等)为碳源作长时间培养所得的滤液中得到的纤维素酶成分较复杂，杂蛋白较多。本工作采用木霉洗涤菌丝体以槐糖诱导的方法得到的胞外纤维素酶成分较单纯(见图 1)，为进一步分离纯化可能带来方便。

拟康氏木霉纤维素酶主要通过 DEAE-Sephadex A-50、Sephadex G-100、凝胶电泳和选择

性失活的方法，初步将 C_1 酶、C_x 酶和 β-葡萄糖苷酶分开。尤其是 C_1 酶经聚丙烯酰胺凝胶电泳和免疫电泳，以及超离心均证明是均一的组分，其含量在三个组分中占绝对优势。氨基酸成分分析证明酸性氨基酸占优势，此结果与 Tomita 等(1974)的报告类似。苏氨酸和丝氨酸含量高可能与多糖水解活性有关。纤维素酶三个组分一旦分开，分解天然纤维素或变性纤维素的能力便大大下降，重组后活力可恢复，相互间(主要是 C_1 与 C_x 酶之间)表现出协同效应，此结果与其他研究者在绿色木霉和康氏木霉中发现的一致(Selby 等，1967；Wood 等，1972)。在我们试验中，根据凝胶电泳和凝胶过滤等看来，C_x 酶可能不止一个成分。最近，Wood 等(1977)报道，采用高分辨力的等电焦聚法等，从康氏木霉的纤维素酶中进一步得到二种 C_1 酶、六种 C_x 酶和二种 β-葡萄糖苷酶，其中只有二种 C_x 酶和 C_1 酶之间有协同效应。本试验中 C_1 酶作用的主要产物是纤维二糖，看来也是支持 C_1 酶基本上是外切型 β-1，4-葡聚糖纤维二糖水解酶的，但由于产物中尚有少量葡萄糖，所以也许是一种随机性较小的外切酶。

参考文献

[1] 中国科学院上海植物生理研究所纤维素酶组、上海酒精二厂，1978. 二株高活力纤维分解菌 EA_3 - 867 和 N_2 - 78 的获得及其特性的比较，微生物学报，18:27 - 38.

[2] 中国科学院微生物研究所纤维素酶组，1976. 绿色木霉纤维素酶系中 C_1 酶的提纯与性质，微生物学报，16:240 - 248.

[3] 朱雨生，1978. 纤维素酶研究进展，生物科学动态，第 6 期，P. 29.

[4] 朱雨生、谭常，1978a. 木霉纤维素酶的诱导形成及其调节，I. 槐糖对木霉 EA_3 - 867 纤维素酶形成的诱导作用，微生物学报，18:320 - 331.

[5] 朱雨生、谭常，1978b. 木霉纤维素酶的诱导形成及其调节，Ⅱ. 槐糖对木霉 EA_3 - 867 洗涤菌丝体纤维素酶形成的诱导作用及降解物阻遏现象，植物生理学报 . 4:1 - 18.

[6] 朱雨生、谭常 . 1978c. 木霉纤维素酶的诱导形成及其调节，Ⅲ. 葡萄糖母液和槐豆荚提取液对纤维素酶诱导效应的分析，植物生理学报，4:19 - 26.

[7] Halliwell, G., Griffin, M., 1973. The nature and mode of acrion of the cellulolytic componcnt C_1 of *Tricho-derma koningii* on native cellulose. Biochern. J, 135:587 - 594.

[8] Hedrick, J. L., Smith, A. J., 1968. Size and charge isomer separation and estimation of molecular weight of protein by disc gel electrophoresis. *Arch. Biochem. Biophys.*, 126:155 - 164.

[9] Reese, E. T., 1950. The biological degradation of soluble cellulose derivatives and its relationship to the mechanism of cellulose hydrolysis. *J. Bacteriol.*, 59:485 - 497.

[10] Selby, K., Maitland, C. C., 1967. The cellulase of *Trichoderma viride* separation of the components involved in the solubilization of cotton. *Biochem. J.* 104:716 - 724.

[11] Tomita, Y., Suzuki, H., Nisizawa, K., 1974.. Further purification and properties of avicelase, a cellulase component of less random type from *Trichoderma viride*. *J. Ferment. technol.* 52:233 - 246.

[12] Walseth, C. S., 1952. Occurrence of cellulases in enzyme preparation from microorganisms. Technical Association of Pulp and Paper Industry (TAPPI), 35:228 - 238.

[13] Weber, K., Osborn, M., 1969. The reliability of molecular weight determination by dodecyl sulfate-polyacrylamide gel electrophoresjs. *J Biol. Chem.*, 224:4406 - 4412.

[14] Wood, T. M., 1968. Cellulolytic enzyme of Trichoderma koningii. Separation of components attacking native cotton. *Biochem. J.*, 109:217 - 227.

[15] Wood, T. M., 1975. Properties and mode of action of cellulases. Biotech. Bioengineering Symp., 5: 111 - 137.

[16] Wood, T, M., McCrae, S. 1., 1972. The purification and properties of the C_1 component of *Trichoderma koningii cellulasa*. *Biochem*. *J*., 128:1183 - 1192.
[17] Wood, T. M., McCrae, S. I., 1977. The mechanism of cellulase action with particular reference to the com-ponent. *In* Bioconversion. of Ccllulosic Substances into Energy Chemicals and Microbiol Protein, ‘Ghose, T. K: (ed). Symp. lIT, p. 111 - 141, New Delhi.

Induction and Regulation of Cellulase Formation in Trichoderma

VI. Separation, Purification and Properties of Cellulase Components of *Trichoderma Pseudokonirngii* Rafai N_2 – 78 Induced by Sophorose*

A cellulase preparation of *Trichoderma pseudokoningii* Rafai N_2 – 78, induced by sophorose, was fractionated by DEAE-Sephadex A-50 and Sephadex-100 column chromatography, and polyacrylamide gel electrophoresis. The components C_1, C_x, and β-glucosidase were obtained, and their molecular weights were estimated to be 67,000, 62,000, and 42,000, respectively. The homogeneity of C_1 component, which is a carbohydrate-protein complex and is rich in glycine, aspartic acid, threonine, serine, and glutamic acid, was verified by polyacrylamide gel and cellulose acetate film electrophoresis, immuno-electrophoresis and ultracentrifugal analysis. The C_1 component showed a strong synergistic action with C_x and β-glucosidase components in the degradation of cotton, avicel and Walthes cellulose. The hydrolysates of Walthes cellulose after incubation with C_1 contained mainly cellobiose and a trace of glucose. The stability of these components to high temperature and different pH has also been studied.

* Zhu Yu-sheng (Chu Yu-shung), Wu Yong-qiang, Chen Wei, Gao Jin-hua, Fei Jin-xin (Shanghai Institute of Plant Physiology', Academia Sinica)

第四篇

光合作用和叶绿体基因表达的光氧调控（1980—1986）

Nicotiana Chloroplast Genome①

I. Chloroplast DNA Diversity

Summary

The chloroplast DNA restriction fragment patterns of several higher plants from different classes, subclasses. orders, and families were examined. Regardless of the restriction enzyme used. There were few similarities among these restriction patterns. Within the genus *Nicotiana*, however, most of the restriction fragments were common to all species examined. The few uncommon fragments remaining were distributable with respect to the taxonomic classification of *Nicotiana* species. Estimates of divergence between selected families, genera, and species were obtained using data gathered from several restriction enzyme patterns.

Introduction

In higher plants, chloroplasc DNA (ct-DNA) was traditionally considered to be highly conserved. This concept arose from the observation that ct-DNA exists in multiple copies per organelle (Herrmann, 1970; Kirk, 1971). The strong conservative force of ct-DNA polyploidy in evolution was supported by the homologous coding information for rRNA, tRNA, and the large subunit of ribulose 1, 5-bisphosphate carboxylase (RuBPCase) as well as the remarkably uniform physicochemical properties of higher plant ct-DNA (Kung, 1977; Bedbrook and Kolodner, 1979). Recently, however, studies of ct-DNA by restriction enzyme (Atchison et al., 1976; Vedel et al., 1976; Frankel et at., 1979) and DNA hybridization techniques (Bisaro and Siegel, 1980) have demonstrated a diversity among chloroplast genomes which was more pronounced as the taxonomic plant groups diverged. Thus ct-DNA restriction patterns of closely related plants would be expected to have many fragment size classes in common. While this was not the case with ct-DNA from *Cucumis* or *Lupinus* species (Atchison et al., 1976), striking similarities in restriction fragment

---

P. R. Rhodes, Y. S. Zhu, and S. D. Kung, Department of Biological Sciences. University of Maryland. Baltimore County (UMBC), Catonsville. MD 21228. USA

此文发表于“Mol. Gel. Genet” 182:106－111.

① *Offprint requests to*: S. D. Kung.

patterns haive been observed with ct-DNA from several wheat (Vedel and Quetier, 1978) and *Nicotiana* (Atchison et al. , 1976) species.

Variations in ct-DNA nucleotide sequence among *Nicotiana* species have been demonstrated for the highly conserved large subunit of RuBPCase (Kung et al. , 1977). As we report here, these differences in the primary structure of RuBPCase large subunit can be correlated with the ct-DNA restriction pattern. Furthermore. variation in the ct-DNA restriction pattern parallel the accumulated geographic. morphological. and cytogenetic evidence upon which the taxonomic classification of *Nicotiana* species is based.

Materials and Methods

Plant Material

Tobacco seedlings were transplanied into 5-inch soil pots in the greenhouse. Plants were allowed to grow for 4 to 5 weeks after transplanting and then covered with black cloth 2 days before the leaves were harvested.

All other plants were grown in vermiculite under low light intensity and leaves or cotyledons were harvested from 5 to 10 day old seedlngs without transplantation.

Purification of Chloroplast

After deveining, 100 to 500 g or freshly harvested 1eaves or cotyledons were homogenized in a metal waring blender with liquid nitrogen (Rhodes and Kung, 1981). When the liquid nitrogen had evaporated and the powder had warmed to -5℃ at room temperature. three weight volumes of isolation buffer containing 0. 33 mol/L sorbitol. 5 mmol/L $MgCl_2$, 50 mmol/L Tris-HCl (pH 8. 0). 0. 1% BSA and 0. 1 mmol/L mercaptoethanol (Walbot, 1977) were added to the powder with rapid mixing. The suspension was filtered through miracloth and centrifuged at 1. 500 xg for 15 min. The crude chloroplast pellets were gently resuspended in the same buffer and puriried through a discontinous silica sol gradient (Walbot, 1977) containing 0. 24 mol/L sucrose throughout.

Isolation of Chloroplast DNA

Purified chloroplasts were lysed in 2% sarkosyl and DNA was isolatcd according to the method of Kolodner and Tewari (1975). The purity of ct-DNA was examined by the denaturation and renaturation procedure used by Kung and Williams (1969).

Restriction Enzyme Analysis of Chloroplast DNA

ct-DNA was digested with EcoRI, HindI, BamHI, SmaI, or Xhol as directed by the supplier, New England Biolabs. The restriction fragments were separated by electrophoresis in a 0. 8% or 1. 5% agarose slab gel according to the procedure or Helling et al. (1974). Ethidium bromide-stained bands were visualized with a model C-62 Chromatovue transilluminator (Ultra-violet Product. Inc.). The EcoRI and HindIII single or double digest fragments of lambda phage DNA were used as molecular weight markers.

Results

The HindIII fragment patterns of ct-DNA from distantly related plants are presented in Figure 1 to illustrate that there was little resemblance between different classes (monocot and dicot), subclasses (spinach and pumpkin) or orders (tobacco and pumpkin). Within families (tobacco and tomato; not shown); however, some similarities began to appear. This variability among ct-DNA restriction patterns is apparent with all restriction enzymes used even when relatively few fragments are generated.

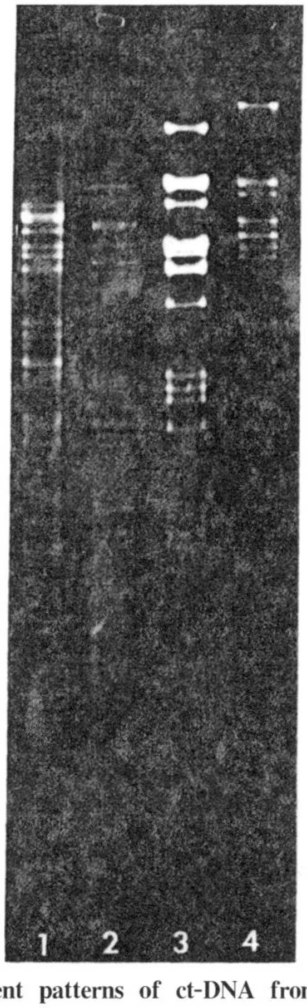

Figure 1 Restriction fragment patterns of ct-DNA from various plants generated by digestion with HindIII.

1—corn; 2—spinach; 3—tobacco; 4—pumpkin. Fragments were separated by electrophoresis for 18 h in 1% agarose gels

In contrast to the heterogenetiy of ct-DNA restriction fragments observed among the more distantly related plants of Figure 1, the ct-DNA prepared from over 20 species of *Nicotiona* exhibit remarkable similartry wth all restriction enzymes used thus far. When relatively few fragments of ct-DNA are generated, many *Nicotiana* species display very similar patterns (see Figure 2). The molecular weights summed from fragments generated by SmaI or XhoI are shown in Table 1 and are in close agreement with the reported value of 9.6×10^7 daltons for *N. tabacum* ct-DNA (Jurgenson and Bourque, 1980). Considerably greater variation among ct-DNA restriction patterns is obtained with EcoRI which generates more than 40 distinct fragments (see Figure 3). Even though the identifiable EcoRI fragments account for only 63% to 72% of the chloroplast genome (see Table 1), the greater frequency of its restriction site increases significantly the fraction of base pairs sampled and the likelihood of observing ct-DNA differences. Thus, EcoRI was chosen for routine analyses of ct-DNA.

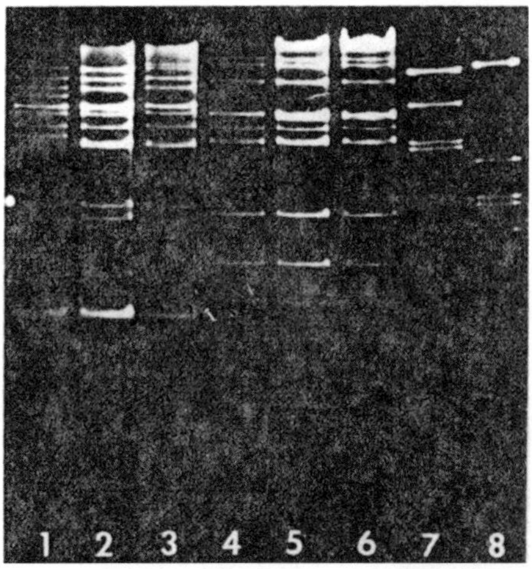

Figure 2 Restriction fragment pattern of ct-DNA from different *Nicotiana* species generated by digestion with XhoI (1—3) and SmaI (4—6). 1 and 4—*N. otophora*; 2 and 5—*N. glauca*; 3 and 6—*N. excelsior*; 7—SmaI fragments of lambda DNA and 8—EcoRI fragments of lambda DNA. Fragments were separated by electrophoresis for 18 h in 0.8% agarose gels

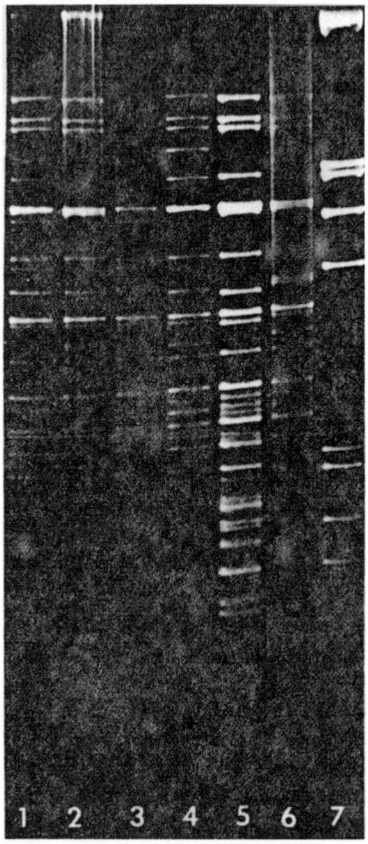

Figure 3 Restriction fragment patterns of ct-DNA from six species of *Nicotiana* generated by digestion with EcoRI. 1—*N. excelsior*; 2—*N. glauca*; 3—*N. gossei*; 4—*N. otophora*; 5—*N. tomentosiformis*; 6—*N. paniculata* and 7—EcoRI and HindⅢ double digest of lambda DNA. Fragments were separated by electrophoresis for 42 h in 1.5% agarose gels

Table 1 Molecular weight of chloroplast DNA from several *Nicotiana* species determined from single and double enzyme digests of ct-DNA. Lambda DNA fragments produced by SmaI, XhoI, BamHI, and EcoRI were used as molecular weight markers. In order to separate the larger fragments for accurate molecular weight determination, photographs were taken of the 0.8% agarose gels arter 20, 40, and 60 h of electrophoresis

Nicotiana Species	Molecular Weight $\times 10^{-6}$ daltons			
	SmaI	*XhoI*	*SmaI*×*XhoI*	*EcoRI*
N. excelsior	92	95	—	73
N. glauca	92	94	—	61
N. tabacum	94	97	96	61
N. tomentosiformis	87	96	—	64
N. otophora	92	95	—	68

Many or the well-separated EcoRI bands are common to all of these *Nicotiana* species; however, certain fragments, most obviously those larger than 2 megadaltons are quite variable (see Figure 3). Figure 4 illustrates a densitometer tracing of the EcoRI fragment pattern of *N. tomentosiformis* ct-DNA. Most or these restriction fragments (1,2,3,5,10,13, and 14 for instance) are found in ct-DNA from every *Nicotiana* species. Amone those conserved fragments. EcoRI fragments 5 and 10 have recently been reported to contain the ribosomal RNA genes (Sugiura and Kusuda, 1979) which have undergone little divergence during evolution (Thomas and Tewari, 1974). Those fragments which are not common to all *Nicotiana* species are listed in Table 2 along with variable EcoRI fragments from several other species. A major portion of these fragments may be considered moderately conserved since they occur in species from several ct-DNA groups (fragments 7 or 9, for example). Others (such as fragments 3a, 6 or 8) occur in only a few of the species examined. Based upon these data, ct-DNA fragments totaling 15 to 25 mcegadaltons can vary among *Nicotiana* species and, or these, at least 13.5 megadalions are frequently present in species from several major ct-DNA groups. Thus much or the *Nicotiana* chloroplast genome is apparently highly conserved.

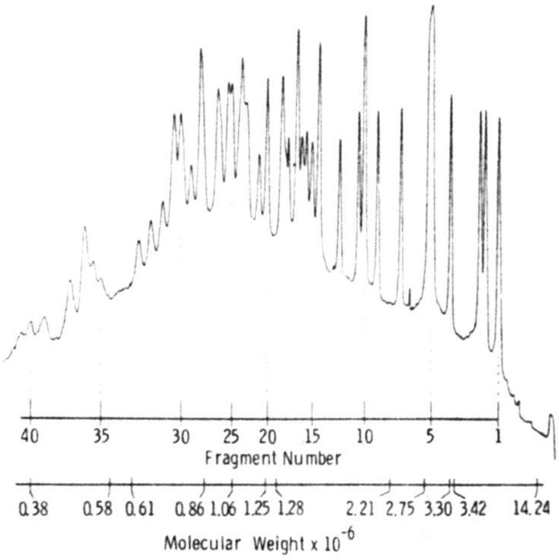

Figure 4 **Microdensitometer tracing of restriction fragment pattern of ct-DNA from *N. tomentosiformis* generated by EcoRI. The fragments were separated in a 1.5% agarose gel slab at 1.25 V/cm for 42 h**

The distribution of variable EcoRI fragments (see Figure 3) was used as a basis for organizing these *Nicotiana* species into groups. Similar groupings were also obtained from analyses of BamHI fragments (see Figure 5(a)) which, like the EcoRI ct-DNA groups, could

Table 2 Size classes of EcoRI fragments which vary among *Nicotiana* species (expressed in megadaltons). Different molecular weights for the same fragment number represent real variations in fragment size for which separate identification numbers have not yet been assigned. Lambda DNA fragments produced by single and double digestion with EcoRI and HindIII were used for molecular weight determinaiions of fragments separated by electrophoresis for 42 h in 1.5% agarose slab gels

Fragment No.	Species from ct-DNA Group						
	I *N. gossei*	*N. excelsior*	II *N. glauca*	III *N. plumbaginifolia*	IVa *N. tomentosiformic* *N. tomentosa*	lVb *No. otophora*	IVc *N. paniculata* *N. knightiana*
3a	—	—	—	—	—	3.75	—
4	3.31	3.31	—	—	3.31	3.31	—
6	2.76	2.76	—	—	—	—	—
7	2.36	2.36	2.36	2.36	2.36	2.36	—
8	2.21	2.21	—	—	—	—	—
8a	—	—	—	—	—	—	2.06
9	2.02	2.02	2.02	2.02	1.96	2.02	2.02
11	1.81	1.81	1.81	1.81	1.79	1.81	1.81
11a	—	1.73	1.73	1.73	—	1.73	1.73
12	1.69	1.69	—	—	—	—	—
15	1.51	1.51	1.51	1.53	1.51	1.51	1.51
16	1.47	—	—	1.47	1.47	1.47	1.47
16a	—	1.45	1.45	—	1.45	—	—
18	1.39	1.39	1.39	1.39	1.36	1.39	1.39
22	1.18	1.18	1.18	1.18	1.15	1.16	1.18

in turn be associated with the isoelectric focusing patterns of the large subunit of RuBPCasc (Chen et al., 1976). As can be seen from Figure 3 and Table 2. EcoRI fragments 6, 8, and 12 are unique to RuBPCase group A (see Table 3). This group includes all *Nicotiana* species indigenous to Australia. Although some diversity in fragment pattern among these species exists in this groups II were designated as ct-DNA group I. The ct-DNA groups II and III, which are analogous to RuBPCase groups B and C (see Table 3), have nearly all of the moderately conserved fragments which occur among the various ct-DNA groups. Even though groups II and III are very similar, separate group assignments were maintained because of the remarkable uniformity of restriction patterns within group III. For example, all species in group III have identical isoelectric focusing of the large subunit of RuBPCase and appear to have identical ct-DNA (see Figure 5(b)) except for *N. langsdorffii*. This species exhibited minor differences in intensity and band position of EcoRI fragments ranging in molecular weight from 0.7 to 1.25×10^6 daltons.

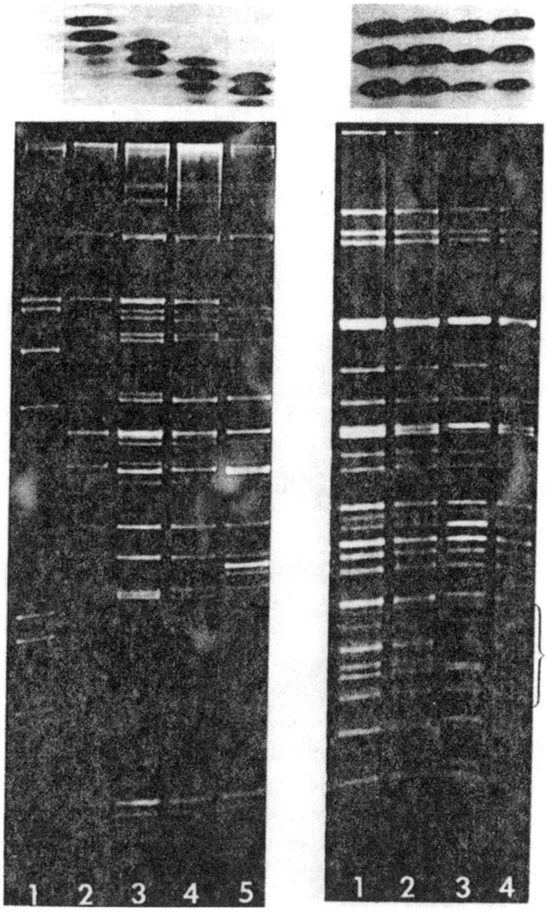

(a) (b)

Figure 5 An association of the isoelectric focusing patterns of RuBPCase large subunits (*top*) with BamHI (a) or EcoRI (b) restriction fragment patterns of ct-DNA from different species of *Nicotiana* (*bottom*).

(a) 1—EcoRI and HindIII double digest of lambda DNA; 2—*N. gossei*; 3—*N. glauca*; 4—*N. tabacum*; 5—*N otophora*. (b) 1—*N. tabacum*; 2—*N. sylvestris*; 3—*N. langsdorffii*; 4—*N. longtflora*. The variable EcoRI fragment region in (b) is marked with brace. The restriction fragments were separated by electrophoresis for 42 h in a 1.5% agarose gel slab

Unlike members of the other ct-DNA groups, the six species examined in RuBPCase group D present a more complex pattern of ct-DNA variation. According to EcoRI fragment patterns, these species form three subgroups: (IVa) *N. setchellii*, *N tomentosa* and *N. tomentosiformis*; (IVb) *N. otophora*; (IVc) *N. knightiana* and *N. paniculata*. Based upon the data in Table 2 *N. otophora*, which shares fragments 4,7,15, and 16 with group IVa and fragments 9,11, 11a, 15,16, and 18 with group IVc, appears to be intermediate between these two groups.

Discussion

The results presented here indicate that the degree of relatedness of restriction patterns of higher plant ct-DNAs reflects their position in the taxonomic scheme. Tobacco and tomato, for example, are classified in the same family and share several similar EcoRI fragments whereas tobacco and corn. which are from different classes (dicot and monocot), show no common feature. This is essentially the same observation obtained by DNA-DNA hybridization studies in which 85% and 15% sequence homologies were estimated between tobacco and tomato and between tobacco and barley, respectively (Bisaro and Siegel, 1980). Similarly some common fragments are generated by HindIII (see Figure 1), EcoRI, BamHI, or Smal (Rhodes and Kung, 1981) from ct-DNA of tobacco and spinach which share 60%—70% sequence homology (Bisaro and Siegel, 1980). If estimates of ct-DNA sequence homology based upon restriction fragment patterns parallel those based upon hybridization data then the restriction fragment evidence may be used for similar comparisons between *Nicotiana* species.

In theory, as long as the restriction enzyme recognition site represents a random distribution of nucleotides and only base substitution is considered as a mechanism for sequence alteration, the proportion of restriction fragments (S) shared by any two ct-DNAs can be used to estimate the fraction of nucleotides that are different (P_o) between the two DNA sequences. Thus in practice, the number of restriction sites on the ct-DNA must be determined for each enzyme included in the study. Unlike SmaI, the observed EcoRI (see Figure 3), BamHI (see Figure 5), and XhoI (see Figure 2) recognition sites agree in number with those predicted from an assumption of random distribution of nucleotides (Upholt, 1977) and thus are suitable for this type of analysis. In order to more accurately estimate the degree of homology, data from EcoRI, BamHI, and Xhol digests of ct-DNA were combined (Gotoh et al. , 1979). The resulting P_o values presented in Table 4 for tobacco, spinach and tobacco, tomato agree quite well with sequence divergence estimates (40% and 15% respectively) obtained from DNA-DNA hybridization studies (Bisaro and Siegel, 1980).

Estimates of sequence divergence between *N. glauca* (Group II) and members of ct-DNA groups I, III, or IVb (see Table 3) confirm that group II ct-DNA is most closely related to group III ct-DNA. This 0.21 % estimate of divergence was similar to that obtained from comparisons among ct-DNA group I species and represents a basal level of mutation for *Nicotiana* ct-DNA (0.63 mutations/300 base pairs). The estimates of divergence and consequently of mutation (3 to 6 mutations/300 base pairs) were higher between ct-DNA groups I : II and II : IVb as well as among groups IVa, IVb, and IVc (not shown). It is worth noting, however, that even our lowest estimate of mutation rate for the entire chloroplast genome is higher than the mutation rate for the large subunit of RuBPCase estimated by Chen et al. (1976) to be 0.27 mutations/100 amino acids. Apparently certain regions of the chloroplast genome may experience fewer mutational events than others.

Table 3 Classification of *Nicotiana* species based on ct-DNA, RuBPCase, and morphology

Species	ct-DNA	Group Based on		
		RuBPCase	Taxonomy	
		Large Subunit	Subgenus	Section
N. excelsior	I	A	Petunioides	Suaveolentes
N. gossei				
N. megalosiphon				
N. suaveolens				
N. glauca	II	B	Rustica	Paniculatae
N. longsdorffii	III	C	Petunioides	Alatae
N. Iongiflora				
N. plumbaginifolia				
N. sylvestris				
N. tabacum	III	C	Tabacum Genuinae	
N. sitchellii	IVa	D	Tabacum Tomentosae	
N. tomentosa				
N. tomentosiformis				
N. otophora	IVb	D'	Tabacum Tomentosae	
N. knightiana	IVc	D'	Rustica	Paniculatae
N. paniculala				

Our grouping of *Nicotiana* species (see Table 3) and estimates of sequence divergence (see Table 4) based upon their ct-DNA restriction patterns are complemented by the taxonomic and biochemical organization of *Nicotiana* species outlined by Goodspeed (1954) and Chen et al. (1976) as presented in Table 4. In ct-DNA group I, for instance, *N. excelsior*, *N. gossei*, and *N. megalosiphon* have ct-DNA patterns similar to that of *N. suaveolens* and were postulated by Goodspeed to be derived from a cross between *N. suaveolens* and *N. fragrans* on the basis of morphological, geologic, and cytologic evidence. These four species have identical large subunits and multipte small subunits of RuBPCase which suggest that they were derived from one maternal and several paternal parents through multiple interspecific hybridizations. The EcoRI fragment pattern characteristic of these ct-DNAs supports this view. In a similar manner. the nearly identical EcoRI fragment patterns and identical isoelectric focusing patterns of RuBPCase large subunit for species in group III (see Figure 5(b)) suggest a single female progenitor for these species. The identity of ct-DNA of *N. tabacum* with *N. sylvestris* and other members in group III was predictabte since *N. sylvestris* is the female progenitor of *N. tabacum* (Gray et al., 1974). However, the low level of sequence divergence (see Table 4) between *N. glauca* and members of ct-DNA group III was unexpected and implies a closer relationship than the assigned taxonomic categories would indicate.

Table 4 Numbers of restriction enzyme cleavage sites which are common or different between pairs of chloroplast DNAs. Calculations of S (fraction of common recognition sites) and P_o (fraction of nucleotide sites for which the two DNA sequences differ from each other) were based upon the equitations of Gotoh et al. (1979)

Enzyme	No. common site/No. different sites				
	N. sylvestris: *spinach*	*N. sylvestris*: *tomato*	*N. excelsior*: *N. glauca*	*N. glauca*: *N. plumbaginifolia*	*N. glauca*: *N. otophora*
EcoRI	2;32	11;30	38;4	37;2	36;6
BamHI	1;35	—	22;3	24;0	19;10
XhoI	3;32	—	17;1	17;0	14;5
S	0.11	0.42	0.95	0.99	0.87
P_o	30%	13%	0.85%	0.21%	2.29%

Also unexpected was the sequence diversity among members of ct-DNA groups IVa and IVb. According to Goodspeed these two groups have a common origin which is separate from that of ct-DNA group IVc. Yet analyses similar to those presented in Table 4 indicate a high degree of divergence (2.67%) between ct-DNA groups IVa and IVb which was nearly equal to that obtained between these groups and IVc. Some of these new fragments may have arisen from addition of a restriction recognition site as was demonstrated for a polymorphic form or *N. debneyi* (Scowcroft, 1979). Perhaps others arose with the loss of a restriction site or through deletion or inversion of DNA segments. But by whatever means, the changes in nucleotide sequence which occur in groups IVa, IVb, and IVc are the most extensive of all *Nicotiana* species examined thus far.

Acknowledgement. This work was supported by NSF grant PCM7812126, NIH grant GM27746-01 and USDA Cooperative agreements 12 - 14 - 1001 - 967 and 12 - 14 - 1001 - 810.

References

[1] Atchison BA, Whitfeld PR, Bottomley W (1976) Comparison of chloroplast DNAs by specific fragmentation with EcoRI endonuclease. Mol Gen Genet 148:263 - 269

[2] Bedbrock JR, Kolodner R (1979) The structure of chloroplast DNA. Ann Rev Plant Physiol 30:593 - 620

[3] Bisaro D, Siegel A (1980) Sequence homology between chloroplast DNA from several higher plants. Plant Physiol 65:234 - 237

[4] Chen K, Johal S, Wildman SG (1976) Role of chloroplast and nuclear DNA genes during evolution of fraction 1 protein. In: Bucher Th, Neupert W, Sebald W, Werner S (eds) Genetics and biogenesis of chloroplasts and mitochondria. Elsevier/Norik Holland Biomedical Press. Amsterdam, The Netherlands, pp 3 - 11

[5] Frankel R, Scowcroft NR, Whitfeld PR (1979) Chloroplast DNA variation in isonuclear male-sterile lines of *Nicotiana*. MoI Gen Genet 169:129 - 135
[6] Goodspeed TH (1954) The Genus *Nicotiana*. Chromica Botanica. Waltham, MA
[7] Gotoh O, Hayashi JI, Yonekawa H. Tagashira Y (1979) An improved method for estimating sequence divergence between related DNAs from changes in restriction endonuclease cleavage sites. J MoI Evol 14:301 - 310
[8] Gray JC, Kung SD, Wildman SG. Sheen SJ (1974) Origin of *Nicotiana tabacum* L. detected by polypeptide composition of Fraction 1 proteins. Nature 252:226 - 227
[9] Helling RB, Goodman HW, Boyer HW (1974) An analysis of endonuclease R EcoRI fragments of DNA from lambdoid bacteriophages and other viruses by agarose-gel electrophoresis. J Virol 14:1235 - 1241
[10] Herrmann RB (1970) Multiple amounts of DNA related to the size of chloroplasts l. An autoradiographic study. Planta 90:80 - 96
[11] Jurgenson JE, Bourque DP (1980) Mapping or rRNA genes in an inverted repeat in *Nicotiana tabacum* chloroplast DNA. Nucleic Acids Res 8:3505 - 3516
[12] Kirk JTO (1971) Chloroplast structure and biosynthesis. Ann Rev Biochem 40:161 - 196
[13] Kolodner R, Tewari KK (1975) The molecular size and conformation of chloroplast DNA from higher plants. Biochim Biophys Acta 402:372 - 390
[14] Kung SD (1977) Expression of chloroplast genomes in higher plants. Ann Rev Plant Physiol 28:401 - 437
[15] Kung SD, Lee CL, Wood DD, Mercarello MM (1977) Evolutionary conservation of chloroplast genes coding for the large subunit of fraction 1 protein. Plant Physiol 60:89 - 94
[16] Kung SD, Williams JP (1969) Chloroplast DNA from broad bean. Biochim Biophys Acta 195:434 - 435
[17] Rhodes PR, Kung SD (1981) Chloroplast DNA isolation: Purity achieved without nuclease digestion. Can J Biochem (submitted) Scowcroft WR (1979) Nucleotide polymorphism in chloroplast DNA of *Nicotiana debneyi*. Theor Appl Genet 55: 133 - 137
[18] Sugiura M, Kusuda J (1979) Molecular cloning of tobacco chloroplast ribosomal RNA genes. MoI Gen Genet 172:137 - 141
[19] Thomas JR, Tewari KK (1974) Conservation of 70S ribosomal RNA genes in the chloroplast DNAs of higher plants. Proc Natl Acad Sci USA 71:3147 - 3157.
[20] Upholt WB (1977) Estimation of DNA sequence divergence from comparison of restriction endonuclease digest. Nucleic Acids Res 4:1257 - 1265
[21] Vedel F. Quetier F. Bayen M (1976) Specific cleavage of chloroplast DNA from higher plants by EcoRI restriction nuclease. Nature 263:440 - 442
[22] Vedel F. Quetier F (1978) Study oF wheat phylogeny by EcoRI analysis of chloroplastic and mitochondrial DNA. Plant Science Lett 13:97 - 102
[23] Walbot V (1977) Use of silica sol step gradients to prepare bundle sheath and mesophyll chloroplasts from *Panicum maximum*. Plant Physiol 60:102 - 108

Communicated by G. Melchers
Received November 17, 1980

Nicotiana Chloroplast Genome*

Ⅱ. Chloroplast DNA Alteration

Summary

The analysis or restriction patterns or chloroplast DNA (ct-DNA) generated by several enzymes reveals that there is a high frequency of ct-DNA alteration in the male sterile but not in the fertile cytoplasma of *N. tabacum*. The altered restriction patterns and maps suggest that recombination of ct-DNA may have occurred. However, other alternatives such as deletions, insertions, and/or point mutations cannot be ruled out.

Introduction

Based on a number of biochemical and functional considerations chloroplast DNA (ct-DNA) of higher plants has long been thought to be highly conserved (Kirk, 1971). This view should be re-examined in light of current evidence. The results obtained from restriction enzyme analysis reveal a wide range of variability (Atchison et al., 1976; Vedel et al., 1976). For instance, in the genus *Nicotiana* ct-DNA exhibits a high degree of diversity within the realm of its common features (Rhodes et al., 1981). The species specific variation in this genus offers an excellent opportunity to study the evolution of *Nicotiana* ct-DNA. One approach is to analyze all ct-DNA in this genus and trace its origin. This will undoubtedly involve a great deal of effort because there are over 60 species. Alternatively, investigation of ct-DNA from various cultivars of the commercial species, *N. tabacum*, may yield more revealing information because the ancestral parents of this species are known (Gray et al., 1974), as well as, the parentage of each of the cultivars.

It has been established that the major force of speciation in *Nicotiana* is interspecific hybridization (Goodspeed, 1954) in which the ct-DNA is always transmitted maternally (Wildman et al., 1973; Frankel et al., 1979). The cytoplasmic male sterilc (CMS) cultivars of *N. tabacum* were also derived from interspecific hybridizations between various

---

S. D. Kung. Y. S. Zhu, K. Chen, G. F. Shen, and V. A. Sisson, Department of Biological Sciences. University of Maryland, Baltimore County. Catonsville. Marvland 21228. USA

此文发表于 1981 年"Mol. Gen. Genet" 183:20 – 24.

* Presem Address: Tobacco Laboratory. SEA, USDA. Beltsville. M D 20705. USA

Offprinr requests to: S. D. Kung

species as the female and *N. tabacum* as the male followed by several generations of backcrossing. This scheme evolved because investigators were only interested in the manipulation of this cultivated species and the reciprocal cross was often difficult or impossibie to makc. Therefore, there is a parallel between the origin of speciation of Nicotiana and a number of *N. tabacum* cultivars. The distinct advantage of using the cultivars is that the parents involved in each cross are known. Any alteration of ct-DNA in the cultivars can be detected and the mechanism involved can be elucidated.

Alteration of ct-DNA was recently reported in CMS lines of *N. tabacum* (Frankel et al., 1979). This study is designed to investigate further ① whether alteration on ct-DNA exists only in the CMS cytoplasm or in the normal cultivars as well. ② the common mechanism of such alteration. ③ the relative position of this altered region on the circular ct-DNA. The cultivars of *N. tabacum* employed in this study covered a wide range of genetic backgrounds including CMS lines of unusual origins. One was obtained from a rare cross (Pittarelli and Stavely, 1975), the only reported case of a direct interspecific cross between these two species in the literature, and the other originated as the result of a spontaneous, cytoplasmic mutation (Berbec, 1974). The analysis of restriction patterns of ct-DNA generated by several enzymes reveals that there is a high frequency of ct-DNA alteration in the male sterile but not in the fertile cytoplasms or *N. tabacum*. The altered restriction patterns and maps suggest that recombination of ct-DNA may have occurred. Other possible alternatives which can cause rearrangements in the chloroplast genome such as deletions and insertions or, point mutations which create or eliminate restriction site sequences cannot be ruled out.

Materials and Methods

Plant Material. Seeds of male sterile *N. tabacum* cultivars were kindly supplied by Drs. S. J. Sheen. University of Kentucky; J. Berbec'. Department of Tobacco Breeding and Cultivation. Pulway. Poland; prof. L. Burk, USDA, Oxford. N. C., and Mr. G. W. Pittarelli, USDA, Beltsville, Md. Tobacco seedlings of approximately 1-inch in height were transplanted into 5-inch soil pots in the greenhouse. Plants were allowed to grow for 4 to 8 weeks depending on the season after transplanting and then covered with black cloth 48 h before the leaves were harvested.

Purification of Chloroplasts. Chloroplasts were purified by the procedure previously described by Rhodes et al. (1981). Freshly harvested leaves were homogenized in a metal Waring blender with liquid nitrogen (Rhodes and Kung, 1981). When the liquid nitrogen had evaporated and the temperature had warmed to -5℃ at room temperatureure, three weight volumes of isolation buffer (Walbot, 1977) were added to the powder with rapid mixing. The suspension was filtered through miracloth and centrifuged at $1,500\times g$ for 15 min. The crude chloroplast pellets were gently resuspended in the same buffer and purified through a discontinuous silica sol gradient (Walbot, 1977) containing 0.24 mol/L sucrose throughout.

Preparation of Chloroplast DNA. Chloroplast DNA was prepared from the purified chloroplasts according to the method of Kolodner and Tewari (1975). The chloroplasts were lysed in 2% sarkosyl and DNA was fractionated by CsCl density gradient centrifugation. The purity of ct-DNA was occassionally checked by the denaturation and renaturation procedure described by Kung and Williams (1969).

Restriction Enzyme Analysis of Chloroplast DNA. Chloroplast DNA was digested with EcoR1, BamHI, SmaI, or SalI as directed by the supplier, New England Biolabs. The restricton fragments were separated by electrophoresis in a 0.8% or 1.5% agarose slab gel according to the procedure of Helling et al. (1974). Ethidium bromide-stained bands were visualized with a Model C-62 Chromatovue transilluminator (Ultra-Violet Product, Inc.). The EcoRI and HindIII single or double digest fragments of lambda phage DNA were used as molecular weight markers.

Results and Discussion

Nicotiana ct-DNAs exhibit distinct common features, as well as, individual variables in their EcoRI restriction patterns (see Figure 1). There are five major regions of restriction fragments plus a few minor bands in region 6, as arranged in order of decreasing molecular weight in Figure 1(a). Each region contains both constant and variable fragments. The frequency of changes and ratio of constant to variable fragments in each region varies from species to species [see Figure 1(a)]. Therefore, the overall ct-DNA restriction pattern in Nicotiana is generally species specific. However, within the species *N. tabacum*, the normal cultivars possess identical patterns [see Figure 1(b)]. This identity was confirmed by using several restriction enzymes (BamHI, SmaI, HindIII, SalI, and EcoRI) in single or double digestion experiments. No variation in restriction patterns of ct-DNA occurs among any of the seven commercial cultivars, Turkish Samsun, Burley 21, NC 95, Ky 14, Md 609, John Williams Broadleaf (JWB), and JWB yellow mutant, analyzed so far. Figure 2 is a densitometer tracing of a typical EcoRI restriction pattern of *N. tabacum* ct-DNA. There are about 40 identifiable bands, many which are doublets or triplets. Bands 8,9,12,16,17,22, 27,29,33, and 39 are bimolar whereas bands 4, 15, and 18 are possibly trimolar. The molecular weights of the EcoRI fragments range from 4.96 to 0.48×10^6 daltons. This pattern is highly reproducible and the corresponding band numbers will be referred to repeatedly throughout this paper.

Chloroplast DNAs from CMS lines of *N. tabacum* constitute the exceptions to this pattern. They either have an altered ct-DNA pattern differing from both parents, or are identical to the female parent of the original cross as expected from the mode of maternal inheritance in Nicotiana (Kung, 1976). Figure 3 represents an EcoRI SalI and SmaI

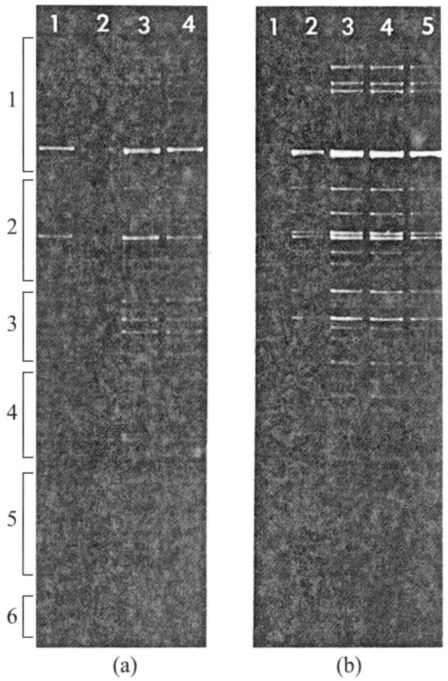

Figure 1 Restriction fragment patterns of ct-DNA from four species of Nicotiana (a) and five cultivars of the species N. tabacum (b) generated by EcoRI.

(a) 1—*N. gossei*; 2—*N. tabacum*; 3—*N. langsdorffii*; 4—*N. otophora* (b) 1—Maryland 609 (double digestion with EcoRI and HindIII); 2—NC 95; 3—Burley 21; 4—Ky 14; 5—Turkish Samsun. Fragments were separated by electrophoresis for 42 h in 1.5% agarose gels as in all the following figures except Figure 4

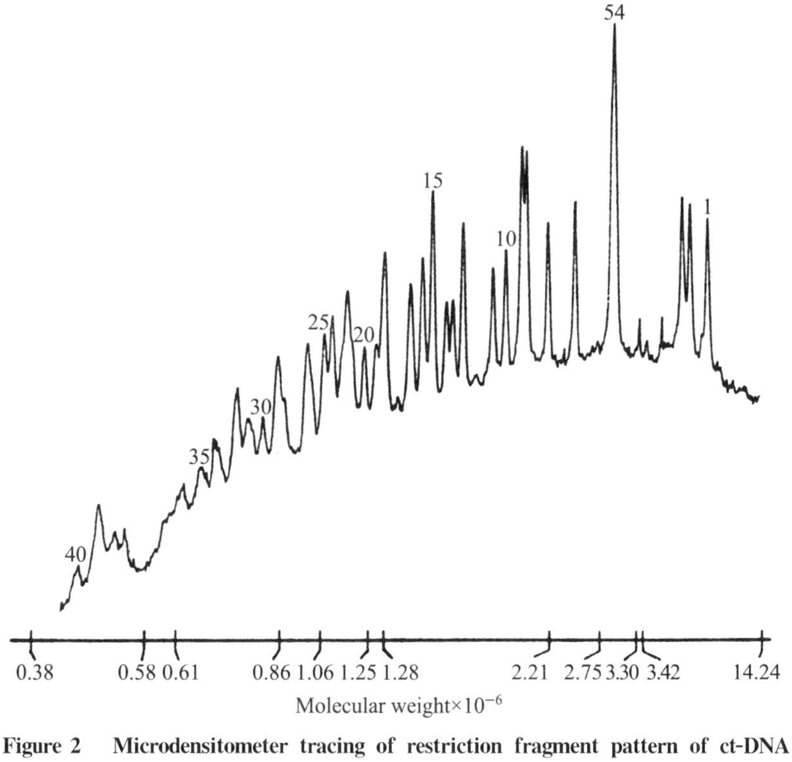

Figure 2 Microdensitometer tracing of restriction fragment pattern of ct-DNA from *N. tabacum* generated by EcoRI fragment numbers are indicated

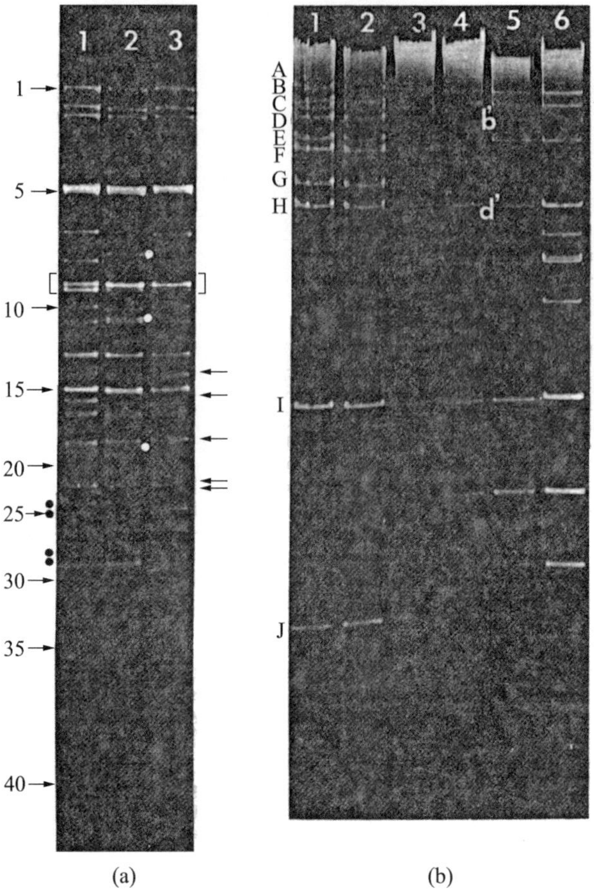

Figure 3 (a) EcoRI restriction fragment patterns of ct-DNA from 1. N. tabacum, 2. CMS line and 3. N. plumbaginifolia. Fragments unique to the CMS line, similar to the female and male parents are marked with white dots, arrows, and black dots respectively. (b) Restriction patterns of ct-DNA generated by SalI (1 - 3) and SmaI (4 - 6) enzymes. Lanes I and 4, N. plumbaginifolia; 2 and 5. CMS line; and 3 and 6 N. tabacum. SalI fragment are marked with capital letters while the SmaI fragments are labelled with lower case latters

restriction patterns of ct-DNAs prepared from a CMS line and its parents. This CMS line was derived from the alloplasmic combination between *N. plumbaginifolia* cytoplasm and *N. tabacum* chromosomes. The overall patterns of ct-DNA from both parents are similar but there are some noticeable differences. In comparison with *N. tabacum* the ct-DNA fragment 9 of *N. plumbaginifolia* is obviously reduced from a doublet to a single band [compare the bracketed regions of lanes 1 and 3 in Figure 3(a)]. The position of fragments 13, 14, 16, 18, 19 and 24 have shifted slightly. Also, a new fragment is present between fragments 16 and 17 and the intensities or fragments 28 and 29 are reversed [see Figure 3 (a), lane 3]. In comparison to the patterns of its parents the EcoRI restriction pattern of ct-DNA of the CMS line is strikingly different [see Figure 3(a), lane 2]. ① It has three unique fragments not present in either parent (marked with white dots); ② fragments 6 and 7 common to both parents have disappeared; ③ there are five fragments which are similar to the male parent (marked with black dots); ④ there are five fragments similar to the female parent (marked with arrows). The presence of new EcoRI fragments and fragments characteristic to both parents in this CMS line indicate recombination or other alterations may have occurred between or within the parental ct-DNAs.

This suggestion is supported by the restriction fragment pattern generated by SmaI [see Figure 3(b)]. When ct-DNA from *N. tabacum*, *N. plumbaginifolia*, and the CMS line are treated with SmaI they exhibit distinct differences. *N. tabacum* has at least 12 SmaI fragments (a—l) similar to that determined by Jurgenson and Bourque (1980) and Jurgenson (1980). According to Jurgeson's numbering scheme the fragments d, f, h, i, j, and l are doublets [see Figure 3(b), lane 6]. In Figure 3(b) fragment f is resolved as two single bands. *N. plumbaginifolia* differs from *N. tabacum* by the presence of bands b' and d' and the absence of bands e, f, g, and a copy of j [see Figure 3(b), lane 4]. As in the case of the EcoRI fragment pattem, the ct-DNA of the CMS line possesses fragments common to both parents. However, the size of fragment e is reduced as compared to that in N. tabacum and there is only a single copy of j as in *N. plumbaginifolia* [see Figure 3(b), lane 5]. Based on the fragment sizes determined by Rhodes and Kung (1981) the combined molecular weight of bands e, f, g, and j (19.3×10^{-6}) is equivilent to that of bands b' and d'. Therefore, it can be deduced that the ct-DNA of the CMS line may represent a recombined molecule. This assumption is supported by the restriction map of *N. tabacum* constructed on the basis of SmaI and SalI fragments in which the fragments e, f, g, and j are sequentially arranged adjacent to one of the inverted repeats on the chloroplast genome (Jurgenson and Bourque, 1980; Jurgenson, 1980). On the basis of the serial order of fragments e, f, g, and j of Jurgenson (1980), the fragment sizes of Rhodes and Kung (1981), and the evidence provided in this study, the possible site of recombination is proposed to be between SmaI fragments f and g of *N. tabacum* and b' and d' of *N. plumbaginifolia*. This proposed site of recombination can account for the observed fragment pattern of the CMS line ct-DNA which contains fragments e and f of the male and fragment d' of the female parents [see Figure 3(b), lane 4]. No CMS lines containing the b' of the female and fragment g of male parent was detected since only one type can survive in a cell (Eberhard, 1980). It should also be noted that no alteration of SalI fragments is observed [see Figure 3(b), lane 1—lane 3].

The second case of ct-DNA alteration was detected from a CMS line originating from a spontaneous, cytoplasmic mutation (Berbec, 1974). As can be noted from Figure 4(a) (compare lane 1 with lane 3) in the altered form the relative proportion of fragments 8 and 9, and 10 and 11 (bracket 1) are reversed. Fragments 14 and 21 (brackets 2 and 3) are reduced in size and a new fragment, 17′ (bracket 3), is present. Here, the ct-DNA of N. tabacum [see Figure 4(a), lane 3] was altered to another type resembling *N. glauca* [see Figure 4(a), lane 2].

A third case is ct-DNA from a CMS line derived from an interspecific cross between *N. glauca* and *N. tabacum*. In this case the EcoRI fragments of ct-DNA of this CMS line follow the expected mode of maternal inheritance. They are identical to those of *N. glauca*. However, when this was further examined by BamHI enzyme, a single extra fragment (marked with dot in Figure 4(b), lane 2) appeared in the CMS line.

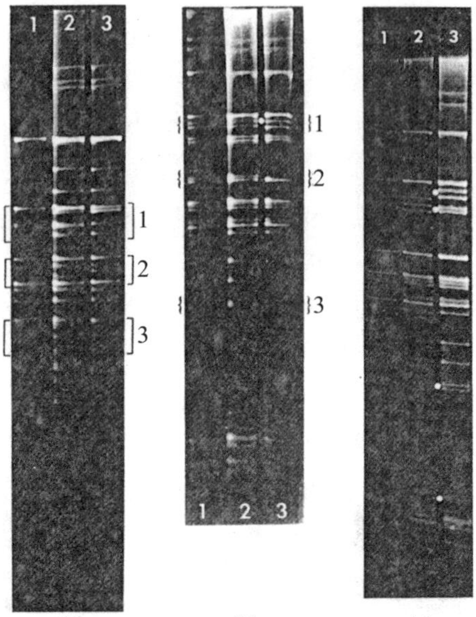

(a) (b) (c)

Figure 4 (a) EcoRI fragment pattern of ct-DNA from a CMS line derived from a cytoplasmic mutation of *N. tabacum*. 1—cytoplasmic mutant; 2—*N. glauca*; 3—*N. tabacum*. The altered fragment pattern of ct-DNA of the cytoplasmic mutant is identical to that of *N. glauca*, which is distinguishable from *N. tabacum* in the marked regions.

(b) BamHI restriction fragment pattern of ct-DNA of the CMS line derived from a cross of *N. glauca* and *N. tabacum*. 1—*N. glauca*; 2—CMS line; 3—*N. tabacum*. Note the ct-DNA of the CMS line resembles that of *N. glauca* in the marked regions and it also differs from *N. glauca* by having an extra fragment (marked with white dot).

(c) BamHI restriction fragment pattern of ct-DNA of the CMS line derived from the cross between *N. repanda* and *N. tabacum*, 1—*N. repanda*; 2—CMS line; 3—*N. tabacum*

Two other CMS lines showed no alteration of ct-DNA. One is the direct interspecific cross of *N. repanda* and *N. tabacum* [see Figure 4(c)]. This was indeed a rare cross since this was the only reported success in the literature (Pittarelli and Stavely, 1975). The analysis of both RuBPcase (data not shown) and ct-DNA confirm the identity of this cross. The other case involves a cross between *N. fragrans* and *N. tabacum* (data not shown). Both cases simply follow the mode of maternal inheritance.

Conclusion

The high degree of diversity of Nicotiana ct-DNA reflects the complex nature of its evolution. Thus, the search for a common mechanism of ct-DNA alteration is both stimulating and challenging. In Nicotiana speciation was primarily achieved by repeated interspecific hybridizations in nature. The study of different cultivars of *N. tabacum*, several of which originated by a similar mechanism, may provide important insight into the evolutionary events from which the current diversity of ct-DNA was derived. The results presented here are very encouraging. There is a high frequency of ct-DNA alteration in the CMS lines. Such alteration seems to be localized in one region (Kung et al., 1981). The pattern of alteration ranges from having a single extra fragment to a more complex pattern which can be best described as recombination. For example, in the more complex case noted above, ct-DNA recombination may have occurred (see Figure 3). Among the SmaI and EcoRI bands in this CMS line (see Figure 3), some are new and not found in either parent, some bands resemble the female and some resemble the male parent, with the rest of the bands common to all three. Several of the new bands are indeed unique and have not been found in any of the 35 species thus far examined. This suggests that ct-DNA recombination may have occurred in this particular hybrid. The restriction map of Jurgenson and Bourque (1980) strongly supports this view. This also, raises some interesting points. ct-DNA must pair in a manner similar to chromosome pairing during meiosis. Although recombination of mitochondrial DNA in somatic hybrids of Nicotiana have been reported (Belliard et al., 1979), caution should still be exercised in interpreting new results. This is the first reported instance that ct-DNA of two different species may have recombined. It is our opinion that these results can be viewed as strong evidence demonstrating that ct-DNA did actually recombine in Nicotiana. However, other alternatives such as deletions, insertions and/or point mutations cannot be ruled out.

The recombination of ct-DNA molecules with each other in a chloroplast has been reported (Kolodner and Tewari, 1979). Recombination events may be essential in the evolution of Nicotiana ct-DNA. Otherwise, by the simple mode of maternal inheritance Nicotiana ct-DNA could not have evolved to such a high degree of diversity (Rhodes et al., 1981). In some higher plants, including Nicotiana, the male parent contributes mostly nuclear materials and very little cytoplasm during fertilization. The number of organelles coming from the male is extremely small, if any, and they are altered and excluded (Vaughn et al., 1980). However, it is still possible there are survivals of male origin. Thus, recombination of ct-DNA from different sources is not totally impossible, but rather, infrequent. The possible recombination of ct-DNA reported here can be tested by protoplast fusion experiments. In fact, this has been done and no sign of such recombination was observed (Belliard et al., 1979). The ct-DNA from the protoplast fusion product of *N. glauca* and *N. langsdorffii* was also analyzed in this laboratory and we found only one or the other parental ct-DNA is present and no recombination occurred (data not shown). It should be pointed out that if ct-DNA recombination occurs only rarely one would not expect to detect it in the few protoplast fusion products examined.

In the second case of CMS line studied here, the ct-DNA was simply mutated to another type. This mutation may involve a simple inversion of a DNA segment containing a restriction site. This suggests mechanisms of mutation involved in chromosomal exchanges can also be applied to organelle DNA. These include translocation, duplication, inversion, and deletion.

Alteration of ct-DNA is not found in all CMS cytoplasms. One possible explanation is that alteration (recombination) occurs only in crosses where the ct-DNA of both parents are quite similar, as in the case of *N. labacum* with *N. plumbaginifolia* (see Figure 3), but not with *N. repanda* or *N. fragrans* which are different. Even with *N. plumbaginifolia* the recombination is a rare event since no alteration was detected in a similar cross reported by Frankel et al. (1979). The biological implication of this phenomenon and its association with CMS is currently under investigation in this laboratory. The results will be important in assessing the role of ct-DNA in cytoplasmic male sterility and will be published elsewhere.

Acknowledgement

We thank Drs. S. J, Sheen. J. Berbec, Prof. L. Burk. and Mr. G. W. Pittarelli for their generosity of supplying the male sterile seed. This work was supported by NSF grant PCM-7812126, NIH grant GM27746-01 and USDA Cooperative agreements 12-14-1001-967 and 12-14-1001-810.

References

[1] Atchison BA, Whitfeld PR, Bottomiey W (1976) Comparison of chloroplast DNAs by specific fragmentation with EcoRI endonuclease. Mol Gen Genet 148:263 - 269

[2] Belliard G, Vedel F. Pelletier G (1979) Mitochondrial recombination in cytoplasmic hybrids of *Nicotiana tabacum* by protoplasmic fusion. Nature 281:401 - 403

[3] Berbec' J, Berbec' A (1976) Growth and development of *Nicotiana tabacum* L. from reconstitution in the cytoplasm of *Nicotiana glauca* Grah Gcnet Pol 17:309 - 318

[4] Berbec' JA (1974) Cytoplasmic male sterile mutation from, *Nicotiana tabacum* L. F. Pflanzenzuechtung 73: 204 - 216

[5] Eberhard WG (1980) Evolutionary consequences of intercellular organelle competition. Q Rev Biol 55: 231 - 249

[6] Frankel R, Scowcroft NR, Whitfeld PR (1979) Chloroplast DNA variation in isonuclear male-sterile lines of *Nicotiana*. Mol Gen Genet 169:129 - 135

[7] Goodspecd TH (1954) The Genus-*Nicotiana*. Chron Bot Waltham. Mass., 636 pp

[8] Gray JC, Kung SD. Wildman SG. Sheen SJ (1974) Origin of *N. tabacum* L. detected by polypeptide composition of fraction 1 protein. Nature 252:226 - 227

[9] Helling RB, Goodman HW. Boyer HW: (1974) An analysis of endonuclease R EcoRI fragments of DNA from lambdoid bacteriophages and other viruses by agarose-gel electrophoresis. J Virol 14:1235 - 1241

[10] Jurgenson JE (1980) *Nicotiana tabacum* chloroplast DNA: structurce and gene content. Ph D. Dissertation. University of Arizona, 149 pp

[11] Jurgenson JE, Bourque DP (1980) Mapping of RNA genes in an inverted region in *N. tabacum* chloroplast DNA. Nucl Acids Res 8:3505 - 3516

[12] Kirk JTO (1971) Chloroplast structure and biosynthesis. Annu Rev Biochem 40:161 - 196

[13] Kolodner R, Tewari KK (1975) The molecular size and conformation of chloroplast DNA from higher plants. Biochim Biophys Acta 402: 372 - 390

[14] Kolodner R, Tewari KK (1979) Inverted repeats in chloroplast DNA from higher plants. Proc Natl Acad Sci USA 76:41 - 45

[15] Kung SD (1976) Tobacco fraction 1 protein: a unique genetic marker. Science 191:429 - 434

[16] Kung SD (1977) Expression of chloroplast genomes in high plants. Annu Rev Plant Physiol 28:401 - 437

[17] Kung SD, Williams JP (1969) DNA from broad bean. Biochim Biophys Acta 195:434 - 445

[18] Kung SD, Zhu SY, Shen GF. Chen K (1981) *Nicotiana* chloroplast genome III. Chloroplast DNA evolution. Theor Appl Genet (in press)

[19] Pittarelli GW, Stavely JR (1975) Direct hybridization of Nictotiana repanda X *N. tabacum*. J Herd 66: 281 - 284

[20] Rhodes PR, Kung SD (1981) Chloroplast DNA isolation: Purity achieved without nuclease digestion. Can J Biochem (submitted)

[21] Rhodes PR, Zhu YS. Kung SD (1981) *Nicotiana* chloroplast genome

[22] I. Chloroplast DNA diversity. Mol Gen Genet 182:106 - 111 Vaughn KC. DeBonte LR. Wilson KG. Schaeffer GW (1980) Organelle alteration as a mechanism for maternal inheritance. Science 208:196 - 198

[23] Vedel F, Quetier F, Bayen M (1976) Specific clcavage of chloroplast DNA from higher plants by EcoRI restriction nuclease. Nature 263:440 - 442

[24] Walbot V (1977) Use of silica sol step gradients to prepare bundle sheath and mesophyll chloroplasts from *Panicum maximum*. Plant Physiol 60:102 - 108

[25] Wildman SG. Lu-Liao C, Wong-Staal F (1973) Maternal inheritance. cytology. and macromolecular composition of defective chloroplasts in variegated mutant of *Nicotiana tabacum*. Planta 113:293 - 312

Communicated by G. Melchers
Received March 23 / June 2, 1981

Nicotiana Chloroplast Genome

III. Chloroplast DNA Evolution*

Summary

Nicotiana chloroplast genomes exhibit a high degree of diversity and a general similarity as revealed by restriction enzyme analysis. This property can be measured accurately by restriction enzymes which generate over 20 fragments. However, the restriction enzymes which generate a small number (about 10) of fragments are extremely useful not only in constructing the restriction maps but also in establishing the sequence of ct-DNA evolution. By using a single enzyme, SmaI, a elimination and sequential gain of its recognition sites during the course of ct-DNA evolution is clearly demonstrated. Thus, a sequence of ct-DNA evolution for many Nicotiana species is formulated. The observed changes are all clustered in one region to form a "hot spot" in the circular molecule of ct-DNA. The mechanisms involved for such alterations are mostly point mutations but inversion and deficiency are also indicated. Since there is a close correlation between the ct-DNA evolution and speciation in Nicotiana a high degree of cooperation and coordination betwen organellar and nuclear genomes is evident.

Key words: *Nicotiana*-Chloroplast DNA-Restriction fragments-Deletion-Evolution

Introduction

Nicotiana chloroplast genomes possess many distinct properties and unique features. The fragment patterns generated by restriction enzymes exhibit a high degree of diversity as well as a general similarity (Rhodes et al., 1981; Kung et al., 1981). For example, the individual restriction pattern is species specific for any given species while the overall configuration is characteristic of the genus Nicotiana (Kung et al., 1981). In many species, even variation of a fragment contains sufficient information to indicate its taxonomic position or to reflect its evolutionary relationship. Therefore, restriction patterns of *Nicotiana* chloroplast DNA (ct-DNA) can be used in this context to identify species in a manner similar

---

* S. D. Kung, Y. S. Zhu, and G. F. Shen, Department of Biological Sciences, University of Maryland, Baltimore County (UMBC) Catonsville, Md. (USA)

此文发表于"Theor. Appl. Genet" 61:73－79.

to the taxonomic, cytogenetic (Goodspeed, 1954) and biochemical (Chen et al., 1976) classifications. This is due mainly to the existence of the unusual taxonomic and evolutionary relationships among *Nicotiana* species. Many of the 64 *Nicotiana* species have originated through interspecific hybridization followed by doubling of the chromosomes (Goodspeed, 1954). In each hybridization the ct-DNA is generally inherited maternally (Wildman et al., 1974) with only infrequent alteration (Kung et al., 1981). Thus, the present-day species with 24 pairs of chromosomes are more advanced than those possessing 12 pairs. Consequently, if a 12-paired chromosome species shares an identical ct-DNA fragment pattern with that of a 24-paired chromosome species a close phylogenetic relationship between them is implied. It is likely that the former is an ancestor (female) of the latter. The best studied case is the origin of *N. tabacum* ($N=24$) in which *N. sylvestris* ($N=12$) is the female parent in the original cross (Gray et al., 1974) and both of these species possess identical ct-DNA (Rhodes et al., 1981) as directed by the mode of maternal inheritance.

This study is designed to investigate the restriction pattern of ct-DNA from many closely related species having different levels of ploidy. From these results the sequence of ct-DNA evolution in *Nicotiana* is formulated. This is accomplished by a simple procedure using a single restriction enzyme without radiolabelling the fragments. In this paper we present our findings and discuss their significance.

Materials and Methods

Plant Material

Tobacco seedlings of approximately 1-inch in height were transplanted into 5-inch soil pots in the greenhouse. Plants were allowed to grow for 4 to 8 weeks depending on the season after transplanting and then covered with black cloth 48 hours before the leaves were harvested.

Purification of Chloroplasts

Chloroplasts were purified by the procedure previously described by Rhodes et al. (1981). Freshly harvested leaves were homogenized in a metal Waring blender with liquid nitrogen (Rhodes and Kung, 1981). When the liquid nitrogen had evaporated and the temperature had warmed to -5℃ at room temperature, three weight volumes of isolation buffer (Walbot, 1977) were added to the powder with rapid mixing. The suspension was filtered through miracloth and centrifuged at $1,500\times g$ for 15 min. The crude chloroplast pellets were gently resuspended in the same buffer and purified through a discontinuous silica sol gradient (Walbot, 1977) containing 0.24 mol/L sucrose throughout.

Preparation of Chloroplast DNA

Chloroplast DNA was prepared from the purified chloroplasts according to the method of Kolodner and Tewari (1975). The chloroplasts were lysed in 2% sarkosyl and DNA was fractionated by CsCl density gradient centrifugation.

Restriction Enzyme Analysis of Chloroplast DNA

Chloroplast DNA was digested with EcoRI, BamHI, HindIII, XhoI, PstI, KpnI, SmaI, or SalI as directed by the supplier, New England Biolabs. The restriction fragments were seperated by electrophoresis in a 0.8% or 1.5% agarose slab gel according to the procedure of Helling et al. (1974). The fragments produced by PstI, KpnI, SmaI, and SalI were analyzed in 0.8% agarose slab gels. Ethidium bromide-stained bands were visualized with a Model C-62 Chromatovue transilluminator (Ultra-Violet Product, Inc.). The EcoRI and HindIII single or double digest fragments of Lambda phage DNA were used as molecular weight markers.

Results and Discussion

Figure 1 illustrates the restriction patterns of *N. tabacum* ct-DNA generated by eight different enzymes. It is evident that EcoRI produces the largest number of fragments among the enzymes used and therefore has the highest resolving power to uncover differences. As represented in Figure 2(a), of the 40 fragments generated by the EcoRI enzyme, the first 10—13 in regions I and II are worth noting. In many species there are 6 fragments in each of these two regions. Fragments 1—3 and 6 in region I are stable and present in all 35 *Nicotiana* species examined so far. In contrast, fragments 4 and 5 are extremely variable.

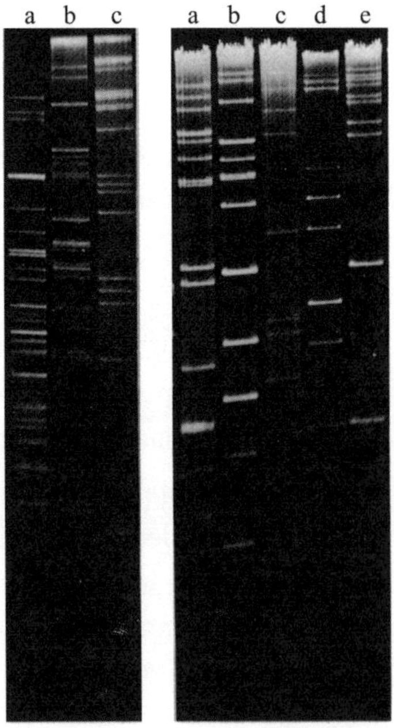

(a) (b)

Figure 1 Restriction fragment patterns of N. tabacum ct-DNA digested with several enzymes

(a): EcoRI a; BamHI b; and Hind III c (b): Xho I a; SmaI b; KpnI c; PstI d; and SalI e. Fragments were seperated by electrophoresis for 42 h in 1.5% (a) and 0.8% (b) agarose gels

One or both can be altered; one or both may be absent. Unlike region I, the six fragments in region II are stable with only a few exceptions. The constant fragments may indicate that a mutation in this region is not likely to survive because of the essential function of the fragment. Indeed, the fragments 6 and 9 are known to contain the 23s and 16s rRNA genes respectively (Sugiura and Kusuda, 1979).

In region III only half of the fragments are constant. The variability increases in the remaining regions (IV—VI). It is such a combination of diversity and similarity in fragment pattern that forms the basis of species specificity as well as the overall identity of *Nicotiana* ct-DNA. The EcoRI patterns in Figure 2(a) contain essential information reflecting many important evolutionary events. For example, the ct-DNA of *N. noctiflora* bears close resemblance to that of *N. gossei* because they are historically related [see Figure 2(a) lane c and d]. The same information exists among the numerous fragments generated by BamHI [see Figure 2(b)] and many other restriction enzymes. Among the seven *Nicotiana* species analyzed in Figure 2(b), ct-DNAs of *N. longiflora*, *N. sylvestris* and *N. tabacum* (lane c, e, and f) have an identical BamHI fragment pattern. Although *N. longiflora* and *N. langsdorffii* ct-DNAs (lanes c and g) bear a close resemblance they are clearly different. *N. glauca* ct-DNA (lane b) is quite similar to that of *N. plumbaginifolia* (not shown) but

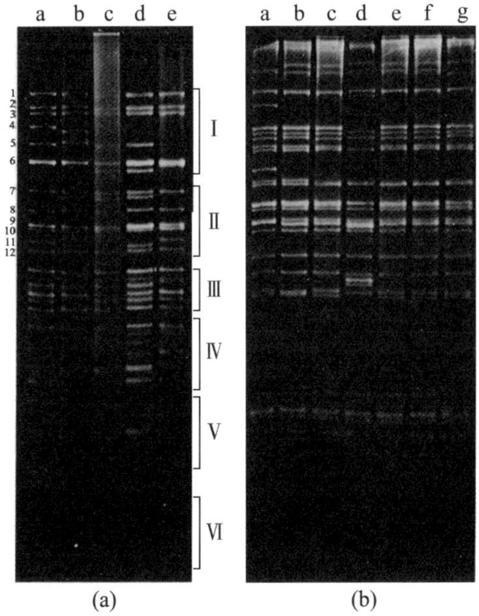

Figure 2 Restriction fragment patterns of ct-DNA from several *Nicotiana* species digested with EcoRI (a) and BamHI (b).

(a) the EcoRI fragments are grouped into 6 regions (I—IV) and the fragments of N. otophora in the first two regions are numbered (1—12). The species are *N. otophora* a; *N. africana* b; *N. noctiflora* c; *N. gossei* d — and *N. tabacum* e (b) *N. gossei* a; *N. glauca* b; *N. otophora* c; *N. longiflora* d; *N. sylvestris* e; *N. tabacum* f and *N. langsdorffii* g. Fragments were seperated by electrophoresis for 42 h in 1.5% agarose gels is distinguishable from that of both *N. gossei* and *N. otphora* (lane a and d). The difference of ct-DNA between *N. gossei* and *N. otophora* is remarkable. They differ in 13 out of the 27 BamHI bands. This matches the extent of variation in their EcoRI fragments in which 18 of the 40 bands are different (see Figure 2(a) lanes a and d). Therefore, both EcoRI and BamHI can be used to measure accurately the degree of variability of ct-DNA in *Nicotiana* species.

In the case of restriction enzymes which generate only a small number of fragments [see Figure 1(b)] SalI and SmaI were selected for this study because they have been frequently used to construct restriction maps of higher plant ct-DNA (Bedbrook and Bogorad, 1976; Jurgenson and Bourque, 1980; Herrmann et al., 1980).

Figure 3(a) is a typical SalI restriction pattern of *Nicotiana* ct-DNA in which there are 9—11 bands. There is no difference in SalI pattern among many *Nicotiana* species and only in four species closely related to *N. tomentosa* are some small fragments combined to form a large one. In these species (*N. tomentosa*, *N. tomentosiformis*, *N. knightiana* and *N. paniculata*) fragments 7 and 10 are replaced by fragment 6 which is now represented as doublet [see Figure 3(a), lane b].

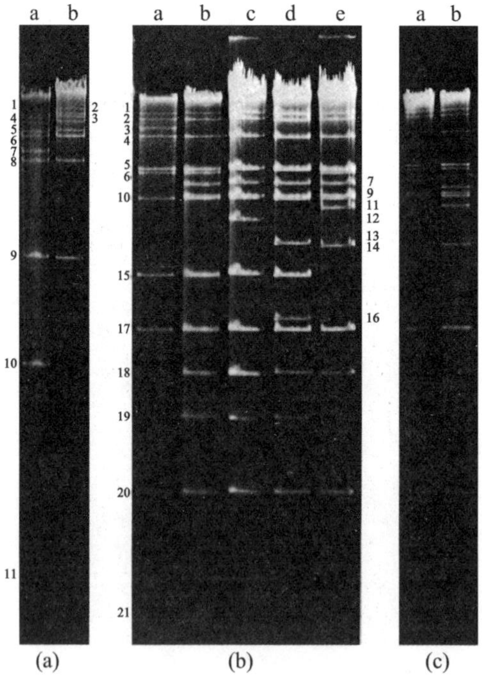

Figure 3 Restriction fragment patterns of ct-DNA from several Nicotiana species digested whith SalI (a) and SmaI (b and c).

In (a) there are a total of 11 SalI fragments in *N. tabacum* a and 9 in *N. tomentosa* b. (b) and (c): there are a combined total of 21 SmaI fragments (as numbered) found in these species. The molecular weights and distribution of each fragment are listed in Table 1. The species are *N. plumbaginifolia* a; *N. langsdorffii* b; *N. sylvestris* c; *N. otophora* d; and *N. tomentosa* e. (c): *N. tomentosiformis* a and *N. paniculata* b. Fragments were seperated by electrophoresis in 0.8% agarose gels for 24 h in (a) and (b), 42 h in (c) in order to achieve better seperation between fragments 9 and 10

In contrast, the SmaI fragment patterns are variable and provide much more revealing information. Figures 3(b) and (c) show the SmaI pattern of ct-DNA from several key *Nicotiana* species on the evolutionary scheme. They represent six unique types of SmaI ct-DNA fragment patterns. In the 22 species examined there are a combined total of 21 distinct SmaI bands. Eight are present in every species and thirteen are variable and scattered (see Table 1). On the average there are 13—15 bands in each species. The fragments 17 and 20 are doublets in all species whereas fragment 15 exists in duplicate in most species but is missing in *N. tomentosa*, *N. tomentosiformis*, *N. knightiana* and *N. paniculata*. Only in *N. tabacum*, *N. sylvestris* and *N. longiflora* are there two copies of fragment 18 (see Table 1). Since the molecular weight of each fragment is known (see Table 1) it is therefore possible to trace the origin and conversion of each variable band in all six species. For example, fragment 3 is unique and found only in *N. plumbaginifolia* [see Figure 3(b), lane a]. It is replaced in *N. langsdorffii* by fragments 7 and 9 [see Figure 3(b), lane b]. The combined molecular weight of fragment 7 and 9(12.0×10^6) is practically the same as that of

Table 1 Restriction fragment size generated by SmaI from six *Nicotina* ct-DNA

Fragment Number	Molecular Weight $\times10^6$	*N. plunbaginifolia*	*N. langsdorffii*	*N. tabacun*[a]	*N. otophora*	*N. tomentose*	*N. paniculata*
1	21	+	+	+	+	+	+
2	17	+	+	+	+	+	+
3	12.5	+	—	—	—	—	—
4	11.5	+	+	+	+	+	+
5	7.2	++	++	++	++	++	++
6	6.9	+	+	—	—	+	+
7	6.3	—	+	+	+	+	—
8	6.0	—	—	—	—	—	+
9	5.7	—	+	+	+	+	+
10	5.6	+	+	+	+	+	+
11	5.1	—	—	—	—	+	+
12	4.9	—	—	+	—	—	—
13	4.1	—	—	—	+	—	—
14	4.0	—	—	—	—	+	+
15	3.5	++	++	++	++	—	—
16	2.6	—	—	—	+	—	—
17	2.4	++	++	++	++	++	++
18	1.95	+	+	++	+	+	+
19	1.55	+	+	+	+	—	—
20	1.15	++	++	++	++	++	++
21	0.61	+	+	+	+	—	—

[a] Data from Rhodes and Kung (1981); ++ = two copies.

fragment 3 (12.5×10^6). Since all the other fragments are constant the conversion from fragment 3 of *N. plumbaginifolia* to 7 and 9 of *N. langsdorffii* is clearly demonstrated. It is thus deduced that ct-DNA of *N. langsdorffii*, a more advanced species, has evolved from *N. plumbaginifolia* by the introduction of a single SmaI site into fragment 3 of *N. plumbaginifolia* [see Figure 3(b), lanes a, b and Figure 4]. Likewise, a gain of a single SmaI site into fragment 6 of *N. langsdorffii* ct-DNA at different positions has generated the ct-DNAs of *N. sylvestris* and of *N. otophora* [see Figure 3(b), lanes c、d and Figure 4]. The two new SmaI fragments generated in each case are fragments 12 and 18 for *N. sylvestris*, 13 and 16 for *N. otophora*. Their combined molecular weights are 6.85×10^6 and 6.7×10^6 which is essentially equal to the original size of 6.9×10^6 for fragment 6 (see Table 1). This

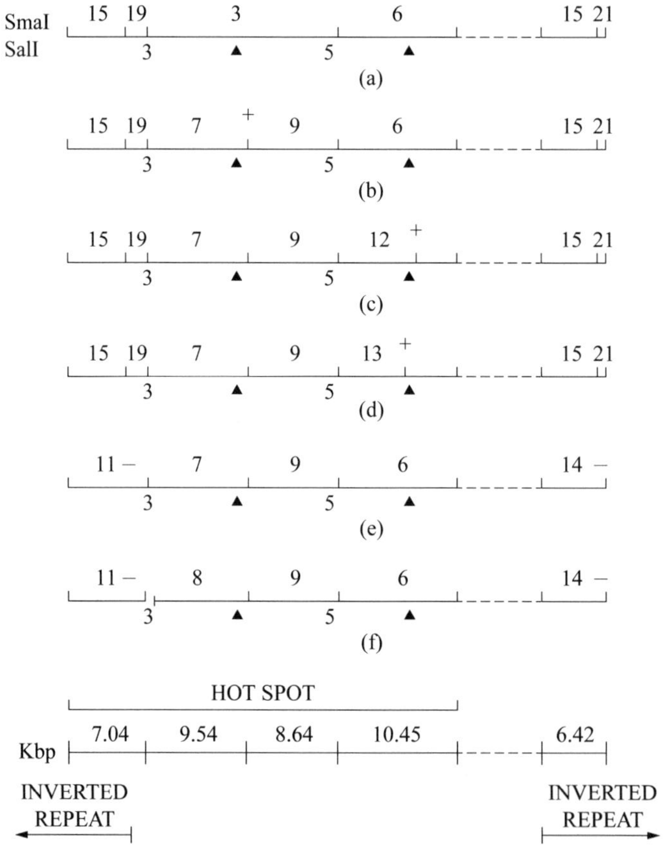

Figure 4 **A diagramatic representation of restriction maps of two segments of *Nicotiana* chloroplast genome showing the locations of gain (+) and elimination (−) of SmaI (1) and SalI (▲) sites. The relative positions of the inverted repeats and the suggested "hot spot" are also indicated; (a) *N. plumbaginifolia* has the unique SmaI fragment 3; (b) an gain of a SmaI site at (+) position converted fragment 3 into fragments 7 and 9 in *N. langsdorffii*; (c) a second gain of SmaI site at (+) position divided fragment 6 into fragments 12 and 18 in *N. sylvestris* or 13 and 16 in *N. otophora*; (d) The elimination of SmaI sites between fragments 15 and 19, 15 and 21 at (−) positions generated fragments 11 and 14. A possible deletion of a segment of SmaI fragment 8 of *N. paniculata*; (f) as compared with that 7 of *N. tomentosa*; (e) is marked by a small breakage. All fragment numbers used here are based on Figure 3 and Table 1**

is confirmed by nick translation and hybridization experiment (data not shown). The alternative interpretation in this case is that initially both *N. sylvestris* and *N. otophora* could have identical SmaI sites but subsequently an inversion occurred in one of them.

In *N. tomentosa*, *N. tomentosiformis*, *N. knightiana* and *N. paniculata* elimination instead of gain of the SmaI site seems to be the mechanism of the evolution of their ct-DNA. As illustrated in Figure 3(b) and (c), the unique fragment 11(5.1×10^6) of *N. tomentosa* appeared which has a combined molecular weight (5.1×10^6) of fragments 15 and 19 not found in this species (see Table 1). The formation of fragment 14(4.0×10^6) resulted from the combination of the missing fragments 15 (second copy, 3.5×10^6) and 21(0.61×10^6). Therefore, the elimination of SmaI sites between fragments 15 and 19, 15 and 21, is clearly demonstrated (see Figure 4). Since all other fragments are equal, the reduction of fragment size from 7 to 8 in *N. paniculata* [see Figure 3(c), lane b and Figure 4] may indicate a deletion. To date, no polymorphism for a restriction site has been observed in this study although such intraspecific variability was reported in *N. debneyi* (Scowcroft, 1979) and in animal mitochondrial genomes (Castora et al., 1980).

From the combination of gain and elimination of SmaI sites it is established that fragments 7 and 9, 12 and 18, 13 and 16, 15 and 19 and 15 and 21 are linked. This is in agreement with the SmaI restriction map of *N. tabacum* ct-DNA (unpublished data) on which fragments 15, 19, 7, 9, 12 and 18 are linked together and arranged in this order. On the basis of this information a sequence of evolution of ct-DNA from several species in *Nicotiana* is proposed. Figure 4 is a diagramatic representation indicating the sequential addition of three SmaI sites and the deletion of SmaI and SalI sites during the course of ct-DNA evolution. Because all of these observed changes are clustered together in one region they may represent a "hot spot" as defined by Benzer (1961). This region or at least the SmaI site seems to mutate much more frequently than others.

Figure 5 is a phylogenetic tree of ct-DNA evolution within this genus. Goodspeed (1954) assummed that the ancestral stocks having 5 pairs of chromosomes had diverged into several branches. It appears the present-day species *N. plumbaginifolia* represents one of them at the 12-paired level. In this proposed scheme of evolution of the *Nicotiana* chloroplast genome, *N. plumbaginifolia* may have served as the ancestral stock from which all the Australian and some of the American species were derived. It is proposed here that ct-DNA of *N. langsdorffii* originated directly from *N. plumbaginifolia*. There is adequate cytogenetic evidence to support this claim (Goodspeed, 1954). *N. langsdorffii* may in turn have given rise to *N. noctiflora*, *N. longiflora* and *N. alata*. The evidence that *N. noctiflora* includes *N. langsdorffii* elements is available (Goodspeed, 1954). At this point of evolution the ancestor of *N. noctiflora* had migrated to Australia and formed the current Australian population. *N. noctiflora* ct-DNA has similar or identical restriction fragment patterns generated by SalI, SmaI, BamHI and EcoRI, to that of *N. langsdorffii* and many Australian species including *N. suaveolens*, *N. gossei*, *N. excelsior*, *N. megalosiphon*

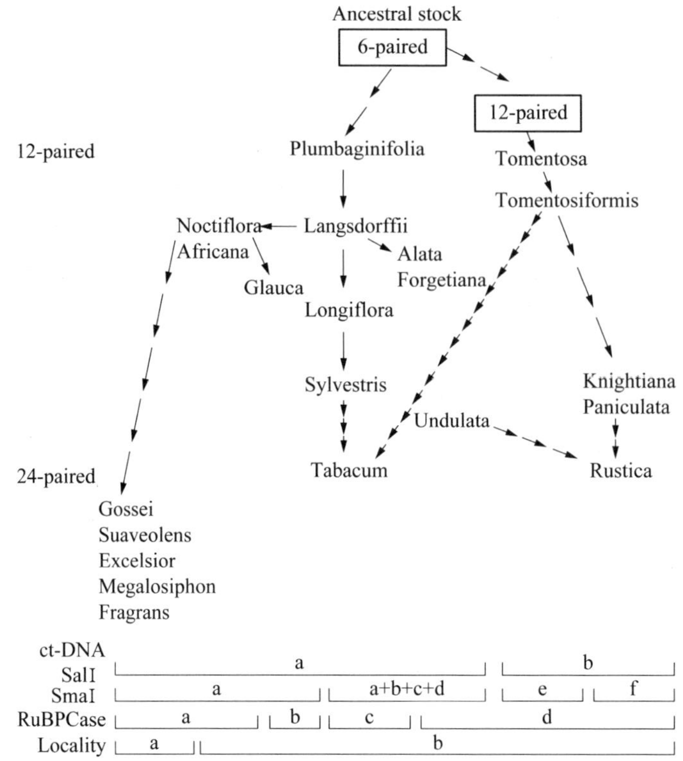

Figure 5 **A phylogenetic tree of ct-DNA evolution in *Nicotiana* based on the restriction pattern of ct-DNA supported by other biochemical (RuBPCase) and distributional data. Arrows indicate the sequence of evolution. *N. tabacum* originated from a cross between *N. sylvestris* and *N. tomentosiformis* whereas *N. rustica* was a product between *N. paniculata* and *N. undulata* (a to f indicates differences)**

Table 2 The relationship of ct-DNA from five closely related species determined by restriction enzyme analysis

Enzymes	Distribution[a]	*N. tabacum*	*N. sylvestris*	*N. longiflora*	*N. langsdorffii*	*N. plumbaginifolia*
SalI	T	11	11	11	11	11
	C		11	11	11	11
	S		0	0	0	0
SmaI	T	14	14	14	14	13
	C		14	14	13	11
	S		0	0	1	2
BamHI	T	27	27	27	27	27
	C		27	27	24	24
	S		0	0	3	3
EcoRI	T	40	40	40	39	38
	C		40	39	31	28
	S		0	1[b]	8	10

[a] The total (T), common (C) (with *N. tabacum*), and species specific (S) fragments,

[b] This difference is so small that it can be considered identical.

and *N. fragrans*. The evidence is therefore very convincing that the ancestor of *N. noctiflora* is indeed the link between these two populations of *Nicotiana* as often suggested (Goodspeed, 1954; Chen et al. , 1976). The restriction enzyme analysis of *N. africana* ct-DNA also indicate that it may have been derived from an ancestor similar to *N. noctiflora*.

The proposed sequence of ct-DNA evolution from *N. plumbaginifolia* to *N. iangsdorffii*, *N. iongiflora*, *N. sylvestris* and *N. tabacum* deserves special treatment. These are very closely related species and all belong to the same section Alatae except *N. tabacum* (Goodspeed, 1954) which is the progeny of a cross between *N. sylvestris* and *N. tomentosiformis* (Gray et al. , 1974). This proposal is based on the restriction analysis of their ct-DNAs but is strongly supported by evidence obtained from taxonomic and biochemical studies (Goodspeed, 1954; Chen et al. , 1976). The recent biochemical evidence is particularly convincing. All five species in this sequence contain the identical isoelectric focusing pattern of the large subunit of RuBPCase, a product of the chloroplast genome (Chen et al. , 1976). Their differential relationship is clearly illustrated in Table 2 which strongly supports this proposed sequence.

Chloroplast DNAs from *N. tomentosa*, *N. tomentosiformis*, *N. knightiana* and *N. paniculata* are clearly separated from all the other species studied here. Their immediate ancestor is different from but probably parallel to *N. plumbaginifolia*. However, the close relationship of this group is highly supported by other evidence (Goodspeed, 1954). Furthermore, the evidence obtained from amino acid analysis and peptide mapping of the large subunit of RuBPCase (Kung et al. , 1976) and the current analysis of restriction fragments of ct-DNA [see Figure 3(b)、(c) and Figure 4] further divide these species into two subgroups; *N. tomentosa* and *N. tomentosiformis* for one and *N. knightiana* and *N. paniculata* for the other (see Figure 5). This is in agreement with the results obtained previously (Rhodes et al. , 1981).

It is interesting to note that in this proposed scheme of evolution (see Figure 5) there is a close correlation between the restriction pattern of ct-DNA and the isoelectric focusing pattern of the large subunit of RuBPCase. It is well established that the large subunit of RuBPCase is coded by ct-DNA (Kung, 1976). Whether such a correlation bears any functional relationship is currently under investigation.

Conclusions

Several remarkable features are evident from this study. First, a single restriction enzyme (SmaI) can uncover sufficient variation of ct-DNA among *Nicotiana* species to construct a sequence of its evolution. It is a simple procedure and does not need to radioactively label the fragments or a second enzyme digestion. Second, a sequential gain, as well as elimination, of the SmaI site is a unique feature of ct-DNA evolution in *Nicotiana*. Third, the great majority of mutations are clustered in one region to form a "hot spot".

Fourth, most of the variability is due to point mutation but deletion and inversion are also implicated. However, recombination has been suggested but only as a rare event (Kung et al., 1981). Finally, there is a close correlation between the ct-DNA evolution and speciation in *Nicotiana* indicating a high degree of cooperation between organelle and nuclear genomes in evolution as well as in expression (Kung, 1977).

Nicotiana species studied here cover a wide range with respect to their biochemistry, cytogenetics, morphology and distribution. They include many key species having 12 pairs and 24 pairs of chromosomes indigenous to both America and Australia. Particular attention was focused on the ancestral linkage between those two continents as well as the parentage and descendant relationship during the course of ct-DNA evolution. Results obtained from the restriction fragment analysis of ct-DNA clearly demonstrate that the ancestor of *N. noctiflora* is the bridging species between the Western Hemisphere and Australian populations.

Results presented here also demonstrate that restriction fragment analysis of ct-DNA is a powerful tool and can provide important information in the study of evolution that exceeds many other means available today (Brown et al., 1979). This is demonstrated by the formulation of a sequence of ct-DNA evolution involving several key Nicotiana species (see Figure 4). Moreover, by selecting proper restriction enzymes such as SmaI to treat several *Nicotiana* species a restriction map of ct-DNA can be constructed in a manner similar to using several enzymes to digest a single type of organelle DNA.

Acknowledgement

This work was supported by NIH grant GM 27746 - 01 and U. S. D. A. cooperative agreement 58 - 3244 - 0 - 157 from the tobacco Laboratory. Suggestions offered by Dr. B. P. Brandley during the preparation of this manuscript are deeply appreciated.

References

[1] Bedbrook, J. R.; Bogorad, L. (1976): Endonuclease recognition sites mapped on zea mays chloroplast DNA. Proc. Natl. Acad. Sci. (USA) 73, 4309 - 4313

[2] Benzer, S. (1961): On the topography of the genetic fine structure. Proc. Natl. Acad. Sci. (USA) 47, 403 - 416

[3] Brown, W. M.; George, M. Jr.; Wilson, A. C. (1979): Rapid evolution of animal mitochrondrial DNA. Proc. Natl. Acad. Sci. (USA) 76, 1967 - 1971

[4] Castora, F. J.; Arnheim, N.; Simpson, M. V. (1980): Mitochondrial DNA polymorphism: evidence that variants detected by restriction enzymes differs in nucleotide sequence rather than in methylation. Proc. Natl. Acad. Sci. (USA) 77, 6415 - 6419

[5] Chen, K.; Johal, S.; Wildman, S. G. (1976): Role of chloroplast and nuclear DNA genes during evolution of fraction 1 protein. In: Genetics and Biogenesis of Chloroplasts and Mitochondria (eds. Bucher, T.; Neupert, W.; Sebald, W.; Werner, S.), pp 3 - 11. Amsterdam: Elsevier North Holland Biomed. Press

[6] Goodspeed, T. H. (1954): The genus Nicotiana. pp. 283 - 314 Waltham, Mass.: Chronica Botanica
[7] Gray, J. C.; Kung, S. D.; Wildman, S. G.; Sheen, S. J. (1974): Origin of Nicotiana tabacum L. detected by polypeptide composition of Fraction 1 proteins. Nature 252, 226 - 227
[8] Helling, R. B.; Goodman, H. W.; Boyer, H. W. (1974): An analysis of endonuclease R EcoRI fragments of DNA from lambdoid bacteriophages and other viruses by agarose-gel electrophoresis. J. Virology 14, 1235 - 1241
[9] Herrmann, R. G.; Whitfeld, P. R.; Bottomley, W. (1980): Construction of Sal I/PsT I restriction map of spinach chloroplast DNA using Low-gelling-temperature-agarose electrophoresis. Gene 8, 179 - 191
[10] Jurgenson, J. E. (1980): Nicotiana tabacum chloroplast DNA: structure and gene content, Ph D. Dissertation, Univ. Arizona
[11] Jurgenson, J. E.; Bourque, D. P. (1981): Mapping of rRNA genes in an inverted repeat *in Nicotiana tabacum* chloroplast DNA. Nucleic Acids Res. 8, 3505 - 3516
[12] Kolodner, R.; Tewari, K. K. (1975): The molecular size and conformation of chloroplast DNA from higher plants. Biochim. Biophys. Acta 402, 372 - 390
[13] Kung, S. D.: (1976). Tobacco fraction 1 protein: a unique genetic marker. Science 191, 429 - 434
[14] Kung, S. D. (1977): Expression of chloroplast genomes in higher plants. Ann. Rev. Plant Physiol. 28, 401 - 437
[15] Kung, S. D.; Lee, C. L.; Wood, D. D.; Moscarello, M. M. (1977): Evolutionary conservation of chloroplast genes coding for the large subunit of fraction 1 protein. Plant Physiol. 60, 89 - 94
[16] Kung, S. D.; Zhu, Y. S.; Chen K.; Shen, G. F.; Sisson V. (1981): Nicotiana chloroplast genome II. Chloroplast DNA alteration. Mol. Gen. Genet. 183, 20 - 24
[17] Rhodes, P. R.; Zhu, Y. S.; Kung, S. D. (1981): Nicotiana chloroplast genome I. chloroplast DNA diversity. MoL. Gen. Genet. 182, 106 - 111
[18] Rhodes, P. R.; Kung, S. D. (1981): Chloroplast DNA isolation: Purity achieved without nuclease digestion. Can. J. Biochem. (in press)
[19] Scowcroft, W. R. (1979): Nucleotide polymorphism in chloroplast DNA of Nicotiana debneyi, Theor. Appl. Genet. 55, 133 - 137
[20] Sugiura, M.; Kusuda, J. (1979): Molecular cloning of tobacco chloroplast ribosomal RNA genes. Mol. Gen. Genet. 172, 137 - 141
[21] Wildman, S. G.; Lu-Liao, C.; Wong-Staal, F. (1973): Maternal inheritance, cytology, and macromolecular composition of defective chloroplasts in variegated mutant of Nicotiana tabacum. Planta 113, 293 - 312
[22] Walbot, V. (1977): Use of silica sol step gradients to prepare bundle sheath and mesophyll chloroplasts from *Panicum maximum*. Plant Physiol. 60, 102 - 108

Received May 8, 1981
Communicated by D. von Wettstein

Dr. S. D. Kung
Y. S. Zhu
G. F. Shen
Department of Biological Sciences
University of Maryland
Baltimore County
Catonsville, Md. 21 228 (USA)

Nicotiana Chloroplast Genome

Ⅴ. Construction, Mapping and Expression of Clone Library of *N. otophora* Chloroplast DNA

Summary

A total of 27 out of 28 BamHI restriction fragments of *N. otophora* ct-DNA have been stably cloned into plasmid pBR322. Each cloned fragment was identified by its electrophoretic mobility. By using this cloned library and a second restriction enzyme, SmaI, a physical map of *N. otophora* ct-DNA, was also constructed. The genes for rRNA and the LS of RuBPCase have been localized in the clone and marked on the map. No unusual structural feature of Nicotiana ct-DNA was detected in this species. Attempts to perform the in vivo expression of the cloned fragments employing the plasmid-directed protein synthesis system in maxicells met with some success. It is clearly demonstrated that certain cloned ct-DNA fragments could direct the synthesis of chloroplast polypeptides in E. coli.

Introduction

The structure and function of higher plant chloroplast genomes have been studied extensively in recent years. It is well established that chloroplast DNA (ct-DNA) exists in circular form and has an average contour length of 50 μm corresponding to a molecular weight of 10^8 daltons (Bedbrook and Kolodner, 1979). This circular molecule is divided into four regions: two inverted repeats separated by a small and a large single-copy region (Bedbrook and Kolodner, 1979). The inverted repeats contain the genes for ribosomal and transfer RNAs while the single-copy region contains the genes for various proteins and tRNAs (Steinmetz et al., 1978). Potentially, ct-DNA has a coding capacity for up to 150 polypeptides of 4×10^4 daltons each (Kung, 1977). To date only a few proteins coded by ct-DNA and synthesized in isolated chloroplasts have been identified (Ellis, 1981). They are, for example, the large subunit (LS) of ribulose 1,5-bisphosphate carboxylase-oxygenase

Y. S. Zhu, E. J. Duvall, P. S. Lovett, and S. D. Kung, Department of Biological Sciences, University of Maryland Baltimore County, Catonsville, MD 21228, USA

Offprint requests to: S. D. Kung.

此文发表于 1982 年“Mol. Gen. Genet” 187:61 - 66.

(RuBPCase) (Kung, 1976), several subunits of coupling factor (Nelson et al., 1980), cytochrome f (Doherty and Gray, 1979), elongation factor G and T (Ciferri et al., 1979), cytochrome b_{559} (Zielinski and Price, 1980), Dicyclohexylcarbodiimide-binding protein (Doherty and Gray, 1980) and apoprotein of chlorophyll-protein complex 1 (Zielinski and Price, 1980).

The site of coding information for the polypeptides mentioned above was determined by genetic or biochemical approaches. The direct method for testing the coding capacity of chloroplast genomewould be to clone the entire chloroplast genome and then test for the expression of each clone. To achieve this we have cloned the entire *Nicotiana* otophora chloroplast genome in E. coli and tested each cloned ct-DNA fragment for expression in E. coli cells. In this communication we report the construction of a clone library and present a restriction map of *N. otophora* chloroplast genome including the localization of genes for rRNA and the LS of RuBPCase. Finally, we demonstrate that certain cloned ct-DNA fragments direct the synthesis of chloroplast DNA-directed polypeptides in E. coli.

Materials and Methods

Plant and Bacterial Strains. N. otophora was grown in a greenhouse for 2—3 months. The leaves were harvested for preparation of ct-DNA. Escherichia coli strains HB101 and CSR603 (recA1, uvrA6, phr-1) were used as the host in transformation experiments. Plasmid pBR322 was cloning vector.

Isolation of Chloroplast and Plasmid DNA. N. otophora ct-DNA was isolated as described previously (Rhodes and Kung, 1981). Plasmid DNA was isolated and purified by centrifugation in CsCl/ethidium bromide equilibrium gradients (Kolodner and Tewari, 1975).

Cloning of ct-DNA in E. coli. The "shotgun" approach was employed in this study. 10 μg of *N. otophora* ct-DNA and 2 μg of pBR322 DNA were digested with BamHI, then ligated according to the procedure of Bolivar and Backman (1979). The total ligation mixture (50 μl) was used to transform 0.1 ml of competent HB101 or CSR603 cells (Pagert and Ehrlich, 1979). The ampicillin resistant transformants were screened for tetracycline sensitivity. Tetracycline sensitive transformants were lysed and plasmid DNA was partially purified. After digestion with BamHI, each was analyzed for their ct-DNA inserts on 1% horizontal agarose slab gel using BamHI digests of *N. otophora* ct-DNA and pBR322 DNA as markers.

Cloning of Specific ct-DNA Fragments in E. coli. N. otophora ct-DNA was digested with BamHI and electrophoresed on 0.8% low-gelling-temperature-agarose according to the method of Herrman et al. (1980). Specific fragments of interest were cut from the gel, melted at 65℃, and isolated from agarose by phenol extraction and ethanol precipitation. The purified ct-DNA fragment was ligated to plasmid pBR322 pretreated with BamHI, then used to transform 0.1 ml of competent HB101 cells.

Isolation of Probes for r-RNA and RuBPCase LS Genes. EcoRI-generated fragment 4 (E4), and 8 (E8) and 29 (E29) from *N. tabacum* ct-DNA were isolated from low-gelling-temperature-agarose gel. The fragments E8, E4 and E29 corresponding to 16S, 23S and SS r-RNA genes respectively (Sugiura and Kusuda, 1979; Fluhr and Edelman, 1981) were used as probes for r-RNA genes. Spinach RuBPCase LS gene probe was isolated from the clone PJEA 4 digested with BamHI and AvaI (Erion et al., 1981; Weissbach, personal communication). Then, this 2.0 kb ct-DNA fragment released from chimeric plasmid pBR322 after digestion contains whole RuBPCase LS gene.

Nick Translation and Hybridization. ct-DNA fragments for probing r-RNA and RuBPCase LS genes were labelled with [α-^{32}P] dATP by the nick-translation reaction (Maniatis et al., 1975) and hybridized to BamHI or SmaI fragments of *N. otophora* ct-DNA blotted on nitrocellulose paper (Southern, 1975). The localization of genes for 16S, 23S, and 5S r-RNA and RuBPCase LS on the *N. otophora* ct-DNA was determined by autoradiography.

In vivo Protein Synthesis. All the BamHI fragments in cloned HB101 were subcloned in maxicell CSR 603. These subclones were individually examined for their ability to direct the synthesis of new polypeptides in CSR 603. After irradiation with UV light (0.5 $JM^{-2}S^{-1}$) the cells were fed with ^{35}S-methionine. The newly synthesized polypeptides were identified by SDS polyacrylamide gel followed by autoradiography (Sancar et al., 1979).

Results

Cloning and Construction of N. otophora Chloroplast Genome Library

The restriction enzyme BamHI was used to construct a genome library of *N. otophora* ct-DNA because it has a single site within the tetracycline resistance gene of the plasmid pBR322 (Bolviar et al., 1977). BamHI generated 28 fragments of *N. otophora* ct-DNA. In comparison, SmaI generated fewer fragments and EcoRI generated more fragments (see Figure 1).

Initially, a mixture of *N. otophora* ct-DNA and plasmid pBR322 was digested with BamHI, ligated and introduced into E. coli (see materials and methods). The transformed E. coli cells were first selected on ampicillin plates and then screened for tetracycline sensitivity by dual plating on ampicillin and on tetracycline plates. The frequency of transformation is about 0.5%. Moreover, the frequency of insertion is much higher, a total of 207 recombinants (Amp^r, Tc^s) were obtained from 2002 colonies (Amp^r). After digestion with BamHI, the chimeric plasmids carrying ct-DNA inserts of various sizes in the recombinants were further analyzed electrophoretically and compared with plasmid pBR322 and BamHI treated *N. otophora* ct-DNA. Most of the BamHI ct-DNA fragments (23/28) have been cloned in E. coli HB101. Plasmid pBR322 can carry up to four different BamHI ct-DNA fragments. The largest ct-DNA fragment inserted is twice the size of plasmid pBR322.

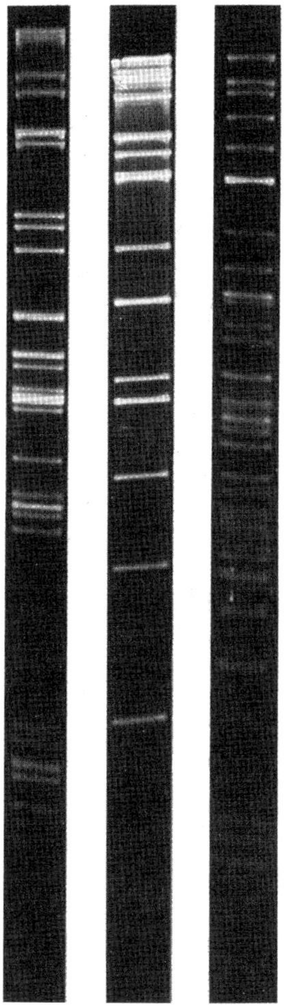

Figure 1 Restriction fragment patterns of ct-DNA from *N. otophora* generated by digestion with BamHI (a), SmaI (b) and EcoRI (c)

Fragments larger than this, such as BamHI ct-DNA fragments 1,2 and 3, seem to encounter difficulty in being inserted into plasmid pBR322. Therefore those fragments have not been successfully cloned by using the shot gun method with plasmid pBR322 as vehicle in this manner. Alternatively, the BamHI fragments 1,2,3 of *N. otophora* ct-DNA were separated and recovered from the low-gelling-temperature-agarose gel, then ligated to pBR322 and introduced into E. coli individually. Thus, except BamHI fragment 8, a clone library of the entire *N. otophora* chloroplast genome has been constructed (see Figure 2).

Mapping of SmaI and BamHI Restriction Sites Map of N. otophora Chloroplast DNA

The basic strategy in constructing the SmaI and BamHI restriction sites map is to take advantage of the available information on several SmaI sites mapped on *N. tabacum* (Jurgenson, 1980). The SmaI restriction sites on *N. otophora* ct-DNA are identical to that of *N. tabacum* with only one exception (Kung et al., 1982). The sequential order of SmaI

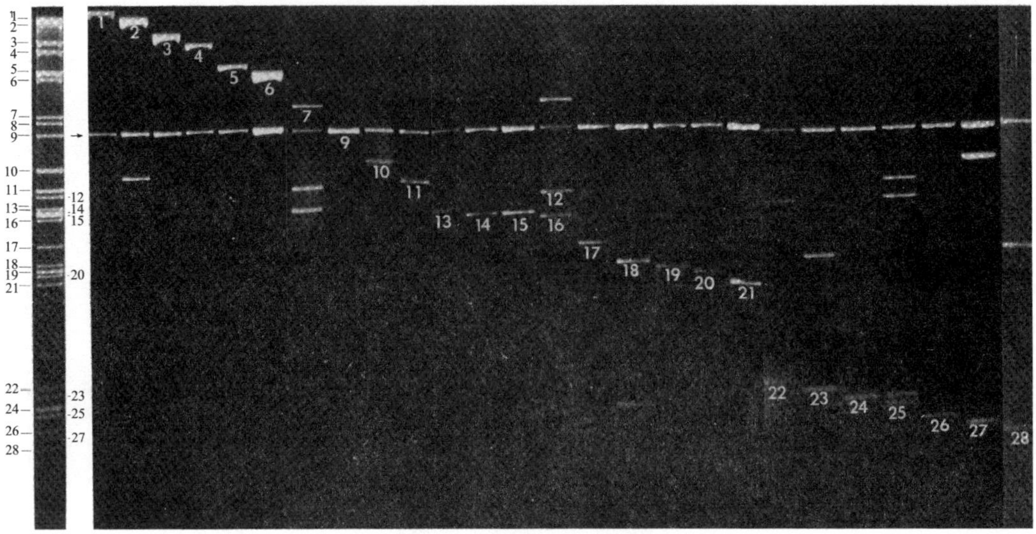

Figure 2 Clone library of *N. otophora* chloroplast genome containing 28 BamHI fragments (a). All the BamHI fragments (except # 8) are cloned in pBR322 (b). In several cases there are more than one insert per plasmid. Each cloned fragment was numbered and identified electrophoretically on a 1.0% agarose gel using BamHI fragments of ct-DNA (a) as standards. Plasmid pBR322 is marked by arrow

fragments on the left half of the large single-copy region (see Figure 3) was arranged according to the information provided by SmaI digestion of ct-DNA from several Nicotiana species (Kung et al., 1981). This region consists of five SmaI fragments in the order (counter-clockwise) of-13 - 5 - 6 - 8 - 10 -. It is 24.0 kb in length accounting for approximately 15% of the total genome. The order of SmaI fragments in the other half of the large single-copy region was aligned as-3 - 7′- 1 - 12′- 15 - based on the information obtained from hybridization of SmaI and SalI fragments. Sall restriction map for *N. tabacum* ct-DNA of Seyer et al. (1981) was used as a standard. The SmaI restriction sites of the inverted repeats (- 9 - 4 - 11 - 14 -) (- 9′- 4′- 11′- 14′-) and the small single-copy region (- 2 -) were essentially adapted from the map of Jurgenson and Bourque (1980). Thus, the complete SmaI restriction map was generated for *N. otophora*. Based on this map, the BamHI restriction sites were also mapped in the following manner: ① primary digestion: ct-DNA was digested with SmaI and BamHI individually or in combination (see Table 1); ② secondary digestion: each SmaI or BamHI fragment was cut from low-gelling-temperature-agarose gel, melted and digested with a second enzyme BamHI or SmaI respectively; ③ DNA hybridization. The sequential order of some very small BamHI fragments in a large SmaI fragment was determined by further restriction enzyme digestion using XhoI, BglI or PvuII and their known restriction sites map of *N. tabacum* (Seyer et al., 1981). The ct-DNA of *N. otophora* is 160.5 kb in length (see Table 1).

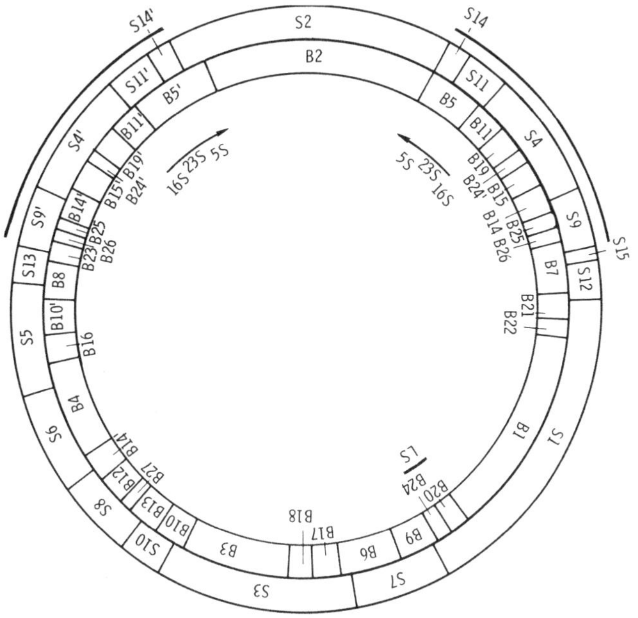

Figure 3 The physical map of BamHI and SmaI restriction sites of *N. otophora* chloroplast genome. The location of the genes for 5S, 16S and 23S rRNA and the large subunit (LS) of RuBPCase are marked. The inverted repeats are indicated by double lines

Table 1 Sizes (kb) and stoichiometries (+) of primary and secondary fragments of *N. otophora* ct-DNA generated by digestion with SmaI, BamHI and SmaI/BamHI

Primary fragments #	BamHI	SmaI	Secondary fragments #	SmaI/ BamHI
1	21.5	28	1	21.5
2	21	25.8	2	20.0
3	10.5	17.4	3	10.5
4	9	11.2++	4	7.0
5	7.0++	9.5	5	5.0
6	6.8	8.6	6	3.6
7	5.0	8.5	7	3.5
8	4.8	6.5	8	3.3++
9	4.4	5.3++	9	3.0
10	3.5++	4.2	10	2.8++
11	3.2	4.0++	11	2.75
12	3.0	3.3	12	2.7

(continued)

Primary fragments #	BamHI	SmaI	Secondary fragments #	SmaI/ BamHI
13	2.9	2.6	13	2.6++
14	2.87+++	1.95++	14	2.55++
15	2.80++	1.2	15	2.5
16	2.75		16	2.3
17	2.8		17	2.2++
18	2.3		18	2.16
19	2.2++		19	2.0++
20	2.16		20	1.95++
21	2.1		21	1.9
22	1.4		22	1.8
23	1.36		23	1.72++
24	1.32+++		24	1.5
25	1.3++		25	1.39
26	1.2++		26	1.36++++
27	1.15		27	1.32+++
28	1.12		28	1.3++
			29	1.2++
			30	1.15
			31	1.12
			32	1.10

Localization of the Genes for Ribosomal RNA and the LS of RuBPCase

The 16S, 23S and 5S rRNA genes of *N. tabacum* have been identified to reside on EcoRI fragments 8,4 and 29 respectively (Sugiura and Kusuda, 1979; Fluhr and Edelman, 1981). These fragments were used as probes to localize rRNA genes in the cloned BamHI fragments of *N. otophora* ct-DNA. Figure 4 shows that the ^{32}P-labeled EcoRI fragment 8 (E8) of *N. tabacum* ct-DNA containing the 16S rRNA gene hybridized specifically with BamHI fragment 11 (B11) and SmaI 4 (S4) of *N. otophora* ct-DNA. Similarily, the EcoRI fragment 4 (E4) containing the 23S rRNA gene hybridized with BamHI fragment 5 (B5),11 (B11) and SmaI fragment 11 (S11) (data not shown). Thus, the B11 and B5 of *N. otophora* ct-DNA contain the 16S and 23S rRNA genes. Likewise, the EcoRI fragment 29 (E29) containing 5S rRNA gene hybridized with B5 and S14. As in the case of both EcoRI (E8, E4 and E29) and SmaI (S4, S11 and S14) fragments, B11 and B5 are also doublets

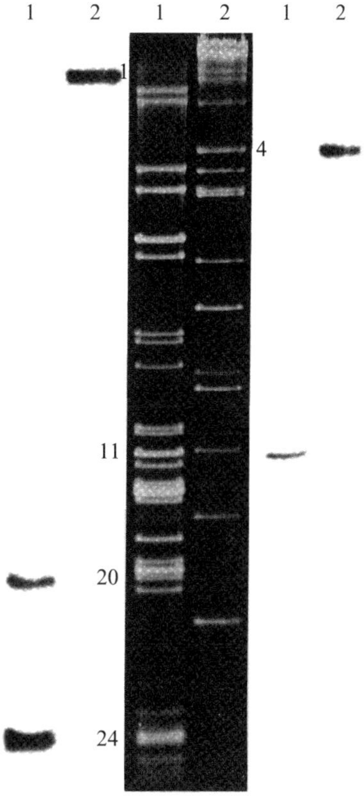

Figure 4 Localization of the genes for 16S rRNA and the large subunit (LS) of RuBPCase on the BamHI and SmaI fragments of *N. otophora* chloroplast genome. The probe for the LS is a hybrid plasmid from the clone PJEA 4 which consists of pBR322 and a 2.0 kb of spinach chloroplast DNA fragment containing the LS. After nick-translation the probe was hybridized to the BamHI and SmaI fragments. The autoradiograph showed two labeled BamHI [(a)1] and one SmaI [(a)2] fragments, corresponding to BamHI fragments 20 and 24, and SmaI fragment 1. The probe for 16S rRNA was EcoRI fragment 8 of *N. tabacum* ct-DNA. Both BamHI fragment 11 and SmaI fragment 4 were labeled [(b) 1 and 2, (c) 1 and 2]

(see Table 1) meeting the requirement as being part of the inverted repeats (see Figure 3).

In a similar manner, the BamHI fragments containing the gene for LS of RuBPCase was also identified using spinach RuBPCase LS gene as probe. The LS gene of *N. otophora* ct-DNA containing a single BamHI restriction site is located on fragments B20 and B24 as determined by hybridization (see Figure 4). These fragments correspond to SmaI fragments 1 and EcoRI 4 fragment (see Figure 1 and Figure 3).

Expression of Nicotiana Chloroplast Genes in vivo

The plasmid-directed protein synthesis system was employed to test whether the cloned ct-DNA can be expressed in vivo in maxicells CSR 603. Since UV light inhibits chromosomal DNA replication, all the proteins synthesized in the UV irradiated maxicells are plasmid-directed. Thus, the expression of any ct-DNA inserts on the plasmid can be detected. [^{35}S] methionine was incorporated into almost every protein synthesized by E. coli (CSR 603) without UV treatment [see Figure 5(a)] as compared to Figure 5(b). However, the polypeptide synthesis directed by the plasmid after UV irradiation is quite evident as revealed by autoradiography [see Figure 5(c)]. Figure 5(c) illustrates the synthesis of a major protein by plasmid pBR322 alone. In this case a 31 kD protein [see Figure 5(c)] was made which is most likely the β-lactamase of plasmid pBR322 origin (Sancar et al., 1979). When many cloned fragments of ct-DNA were tested for in vivo expression in this manner plasmid pBR322 carrying ct-DNA inserts of BamHI 7,12, and 16 (in clone # 17) produced three additional products of approximately 43,29, and 13.5 kD [see Figure 5(c)]. Thus, the in vivo expression of chloroplast genes in E. coliis clearly demonstrated. Furthermore, in seven clones where different inserts were ligated to pBR322, a 2 - 3 fold enrichment of the β-lactamase synthesis was also observed [see Figure 5(c)].

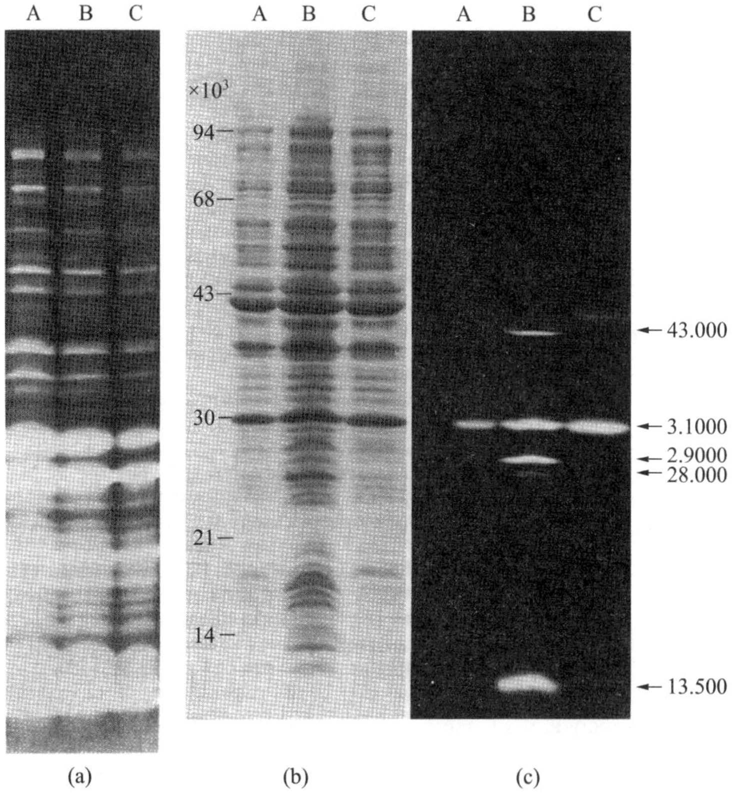

Figure 5 Labeling of plasmidcoded polypeptides in UV-irradiated "maxicell" CSR603.

(a) Autoradiography of labeled polypeptides of maxicells on SDS polyacrylamide gel before UV-irradiation (control) (b) The patterns of maxicell polypeptides on SDS-polyarcylamide gel stained with Bromphenol Blue (c) Autoradiography of (b) showing plasmid-coded polypeptides

In Figure 5(a)(b)(c), A CSR 603 carrying pBR322. B Clone # 17 in which plasmid pBR322 carries N. otophora ct-DNA Bam fragment 7,12,14. C Clone # 8 indicating Enrichment of the β-lactamose synthesis by specific ct-DNA inserts in pBR322. Numerals indicate marker of molecular weight (daltons)

Discussion

The construction of clone library of nuclear or organellar genomes is an important operation in explorating the potentials of genetic manipulation of plants. Only recently a library of the French bean nuclear genome has been established (Sun et al., 1981). Organellar genes, on the other hand, have been cloned individually (Bedbrook et al., 1977; Coen et al., 1977) prior to the construction of any comprehensive chloroplast genome libraries (Palmer and Thompson, 1981). This is the first clone library of *Nicotiana* chloroplast genome. *Nicotiana* chloroplast genome has been the subject of intensive investigations in recent years (Fluhr and Edelman, 1981; Jurgenson, 1980; Kung et al., 1981; Seyer et al., 1981). However, there is no report on chloroplast genome library in this genus. In this study, *N. otophora*, a closely related species of *N. tabacum*, was selected to construct a clone library of *Nicotiana* chloroplast genome. A total of 27 out of 28 BamHI fragments of *N. otophora* chloroplast DNA were ligated into plasmid pBR322 and cloned in E. coli cells. The inability to clone a specific BamHI fragment 8(4.8 kb) after many futile attempts is frustrating. Palmer and Thompson (1981) experienced a similar difficulty in their attempt to clone pea chloroplast genome. They attributed this failure to some unknown feature of its sequence which may interfere with the cloning process or efficiency or even to affect the survival.

This clone library is used as the base for constructing a physical map of *N. otophora* chloroplast genome. The sequential order of BamHI fragments was aligned by using secondary restriction digestion. Thus, a restriction map of the SmaI and BamHI was completed. The genes for rRNA and the LS of RuBPCase have been identified in the clone as well as localized on individual fragments generated by various enzymes. No unusual structural feature was detected in this species.

The plasmid-directed protein synthesis system is employed for the in vivo expression experiments. In this maxicell system the β-lactamase is synthesized under the direction of plasmid pBR322 after UV treatment. However the product coded for by the tetracycline resistant gene on this plasmid is apparently not synthesized. A similar result was also reported by others (Erion et al., 1981). In the case of many cloned ct-DNA fragments tested, only three labeled polypeptides (representing 1% of ct-genome) can be detected. The identity of these polypeptides have not yet been determined. Nevertheless, the in vivo expression of chloroplast genes in E. coli usingthe plasmid-directed protein synthesis system is clearly demonstrated.

Acknowledgements

This work was supported by NH grant CM27746 - 01 and U. S. Department of

Agriculture Cooperative Agreement 58 - 32U4 - 0 - 157 from the tobacco laboratory to S. O. K. and NSF grant PCM 78 - 05755 to P. S. L.

References

[1] Bedbrook JR, Kolodner R (1979) The structure of chloroplast DNA. Annu Rev Plant Physiol 30:593 - 620

[2] Bolivar F, Backman K (1979) Plasmids of *E. coli* as cloning vectors. Methods Enzymol 68:245 - 267

[3] Ciferri O, DiPasquale G, Tiboni O (1979) Chloroplast elongation factors are synthesized in the chloroplast. Eur J Biochem 102:331 - 335

[4] Dagert M, Ehrlich DS (1979) Prolonged incubation in calcium chloride improves the competence of *E. coli* cells. Gene 6: 23 - 28 Doherty A, Gray JC (1979) Synthesis of cytochrome f by isolated pea chloroplasts. Eur J Biochem 98:87 - 92

[5] Doherty A, Gray JC (1980) Synthesis of a dicyclohexylcarbodiimide binding proteolipid by isolated pea chloroplasts. Eur J Biochem 108:131 - 136

[6] Ellis J (1981) Chloroplast proteins: synthesis, transport and assembly. Annu Rev Plant Physiol 32: 111 - 137

[7] Erion JL, Toinowski J, Weissbach H, Brot N (1981) Cloning, mapping, and in vitro transcription-translation of the gene for the large subunit of RuBPCase from spinach chloroplasts. Proc Natl Acad Sci USA 78:3459 - 3463

[8] Fluhr R, Edelman M (1981) Physical mapping of Nicotiana tabacum chloroplast DNA. Mol Gen Genet 181:484 - 490

[9] Herrmann RG, Whitfeld PR, Bottomley W (1980) Construction of a SalI/PstI restriction map of spinach chloroplast DNA using low-gelling-temperature-agarose electrophoresis. Gene8:179 - 191

[10] Jurgenson JE, Bourque DP (1980) Mapping of RNA genes in an inverted region in N. tabacum chloroplast DNA. Nucl Acids Res 8:3505 - 3516

[11] Jurgenson JE (1980) Nicotiana tabacum chloroplast DNA: Structure and gene content. PhD Dissertation, Univ Arizona

[12] Kolodner R, Tewari KK (1975) The molecular size and conformation of the chloroplast DNA from higher plants. BiochemBiophys Acta 402:372 - 390

[13] Kung SD (1976) Tobacco fraction I protein: a unique genetic marker. Science 191:429 - 434

[14] Kung SD (1977) The expression of chloroplast genomes in higherplants. Annu Rev Plant Physiol 28: 401 - 437

[15] Kung SD, Zhu YS, Shen GF (1982) Nicotiana chloroplast genome. III. Chloroplast DNA evolution. Theor Appl Genet 61:73 - 79

[16] Maniatis T, Jeffrey A, Kleid DG (1975) Nucleotide sequence of rightward operator of phage. Proc Natl Acad Sci USA 72:1184 - 1188

[17] Nelson N, Nelson H, Schatz G (1980) Biosynthesis and assembly of the proton-translocating adenosine triphosphatase complex from chloroplasts. Proc Natl Acad Sci USA 77:1361 - 1364

[18] Palmer JD, Thompson WF (1981) Clone banks of the mung bean, pea and spinach chloroplast genomes. Gene 15:21 - 26

[19] Rambach A, Hogness DS (1977) Translation of *Drosophila melanogaster* sequences in *E. coli*. Proc Natl Acad Sci USA 74:5041 - 5045

[20] Rhodes PR, Kung SD (1981) Chloroplast DNA isolation: purity achieved without nuclease digestion. Can J Biochem 59: 911 - 915

[21] Sancar A, Hack AM, Rupp WD (1979) Simple method for identification of plasmid-coded proteins. J

Bacteriol 137:692 - 693

[22] Seyer P, Kowallick KV, Herrmann RG (1981) A physical map of Nicotiana tabacum plastid DNA including the location of structural genes for ribosomal RNAs and the large subunit of Ribulose Bisphosphate Carboxylase/Oxygenase. Curr Genet 3:189 - 204

[23] Steinmetz A, Mubumbila M, Keller M, Burkard G, Weil JH (1978) Mapping of tRNA genes on the circular DNA molecule of *spinacia Oleracea* chloroplasts in chloroplast development. Akoyunoglou G, Argyroudi-Akoyunoglou (eds) Elsevier/North Holland, Amsterdam New York Oxford, pp 573 - 580

[24] Sugiura M, Kusuda J (1979) Molecular cloning of tobacco chloroplast ribosomal RNAgenes. Mol Gen Genet 172:137 - 141

[25] Southern EM (1975) Detection of specific sequences among DNA fragments separated by gel electrophoresis. J Mol Biol 98: 503 - 517

[26] Sun SM, Slightom JL, Hall TC (1981) Intervening sequences in a plant gene-comparison of the partial sequence of cDNA and genomic DNA of French bean phaseolin. Nature 289:37 - 41

[27] Zielinski RE, Price CA (1980) Synthesis of thylakoid membrane proteins by chloroplasts isolated from spinach. J Cell Biol 85:435 - 445

Communicated by G. Melchers

Received May 18, 1982

Nicotiana chloroplast genome

7. Expression in *E. coli* and *B. subtilis* of tobacco and Chlamydomonas chloroplast DNA sequences coding for the large subunit of RuBP carboxylase

Summary. RuBPCase. the enzyme responsible for carboxylation and oxidation of RuBP in a wide variety of photosynthetic organisms, is the major protein found in the chloroplast. Here we present the first evidence for direct expression in *E. coli* and *B. subtilis* of tobacco and *Chlamydomonas* ct-DNA sequences coding for the LS of RuBPCase as demonstrated by a simple in situ immunoassay.

Key words: *Nicotiana* — *Chlamydomonas* — Chloroplast DNA — RuBPCase — *E. coli* — *B. subtilis*

Ribulose-1, 5-bisphosphate carboxylase (RuBPCase) (Ec 4.1.1.39), the enzyme responsible for both CO_2 fixation and evolution in a wide variety of photosynthetic organisms, is the major protein found in the chloroplast (Kung 1977). In higher plants it occurs as an oligomer of eight large subunits (LS) (MW 53,000) and eight small subunits (SS) (MW 12,000 - 15,000) (Jensen and Bahr 1977). The LS, containing the catalytic sites, are coded by the chloroplast genome (Chan and Wildman 1972) and are synthesized in the chloroplast (Gray and Kekwick 1973). The SS. whose function remains uncertain, are coded by the nuclear genome and synthesized in the cytoplasm as a precursor of higher molecular weight (MW 20,000) that is cleaved and transported into the chloroplast (Chua and Schmidt 1978; Highfield and Schmidt 1978). This enzyme provides an attractive model for studying nucleus-chloroplast cooperation and gene regulation in plants. Genes for the LS of RuBPCase from maize (Coen et al. 1977), spinach (Bottomley and Whitfeld 1978; Erion et al. 1981) and *C. reinhardii* (Malnoë et al. 1979) have been expressed in vitro in coupled

---

Present address*: Department of Cellular and Developmental Biology, Harvard University. 16 Divinity Avenue, Cambridge. MA 02138. USA

Y. S. Zhu*, P. S. Lovett, D. M. Williams and S. D. Kung,本文发表于 1984 年“Theor Appl Genet. 67:333 - 336.”

University of Maryland, Baltimore County, Department of Biological Science. 5401 Wilkens Avenue, Catonsville, MD 21228, USA

Received May 20. 1983; Accepted July 28, 1983

Communicated by D. von Wettstein

transcription-translation systems, and maize and wheat LS genes have been expressed in vivo in *E. coli* (Gatenby et al. 1981). Here we present the first evidence for direct expression in *E. coli* and *B. subtilis* of chloroplast DNA (ct-DNA) sequences from tobacco and *Chlamydomonas* coding for the LS of RuBPCase using a simple in situ immunoassay.

A clone library of *Nicotiana otophora* ct-DNA has been constructed and mapped with respect to *Bam*HI and *Sma*I sites (Zhu et al. 1982). Unique restriction fragments generated by cleavage of the ct-DNA with *Sma*I. *Sal*I. *Hind*III or *Eco*RI contain the intact LS gene suitable for expression studies, whereas *Bam*HI and *Pst*I cut within the gene. This was determined by using the spinach LS gene as a probe previously inserted into plasmid pBR322 (pJEA4) by Erion et al. (1981). Figure 1A shows the identification of the tobacco LS gene by the hybridization of ^{32}P-labeled pJEA4 to *N. otophora* ct-DNA digested with *Sma*I. *Sal*I, *Pst*I, *Hind*III, *Eco*RI and *Bam*HI. The fragments Sal 6 (14 kb) and Hind 2 (11 kb) containing the intact LS gene were isolated from total *N. otophora* ct-DNA and cloned in *E. coli* using pBR325 and pBR322 as the respective vectors (see Figure 1 B). Cloned fragments Sal 6 (in plasmid PRCZ2) and Hind 2 (in plasmid PRCZ1) hybridized with ^{32}P-labeled spinach LS gene probe, confirming that both clones contain the LS gene, although these fragments are considerably larger than the LS gene itself (1.3 – 1.6 kb).

The LS gene from *Chlamydomonas reinhardii* was used for cloning in *Bacillus*. This LS gene was previously cloned in plasmid pBR322 (pLM401) by L. Metz (personal communication, see Figure 2 A). Plasmid PRCZ3 (see Figure 2 B) was constructed by substituting the 0.7 kb *Eco*RI-*Hind*III region of the *B. subtilis* expression plasmid pPL608 (Williams et al. 1981) with the *Eco*RI-*Hind*III fragment (1.15 kb and 0.75 kb) spanning the anterior region of the *Chlamydomonas* LS gene. This substitution is clearly illustrated with the dotted lines in Figure 2 between A and B. Two contiguous *Hind*III fragments (0.75 kb and 1.25 kb) of pLM401 (see Figure 2 A), which contain the major posterior region of the LS gene, were inserted into the *Hind*III site of plasmid pPL608, generating pRCZ4. This insertion is also marked with the dotted lines in Figure 2 between A and C. Both recombinant plasmids were transformed into *B. subtilis* strain IS53 with selection for the neomycin-resistance trait specified by pPL608.

For detection of the LS gene product a simple in situ immunoassay was employed (Anderson et al. 1980). The clones CSR603 (PRCZ1) and CSR603 (PRCZ2) containing the *N. otophora* LS gene were inoculated and incubated on a N-Z bottom agarose plate overnight at 37℃. CSR603 cells containing pBR322 and pBR325 were used as controls. After incubation, the cells in the plates were lysed with chloroform vapor and lysozyme. A channel was made in the center of the plate into which was added antiserum against the LS of RuBPCase. Incubation was continued for another day at room temperature. A sharp immunoprecipitation line was formed between CSR603 (PRCZ1) and the central channel demonstrating that this clone has produced the LS polypeptide (see Figure 3 A). *B. subtilis* cells carrying PRCZ3 and PRCZ4 and control cells containing the vector pPL608 were similarly

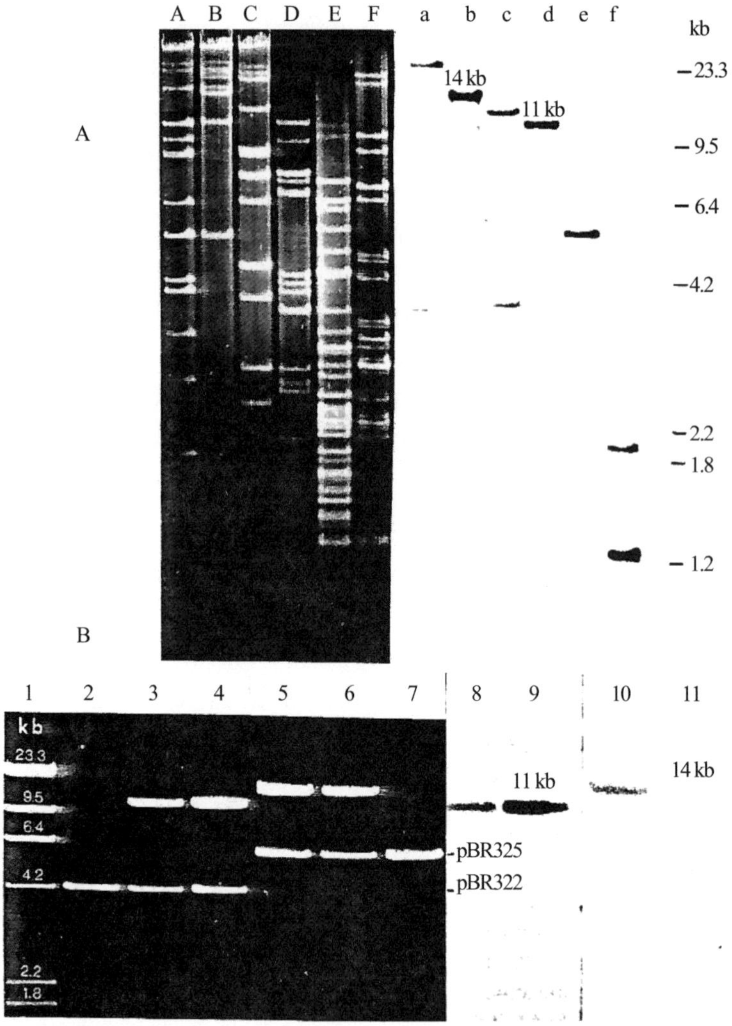

Figure 1A **Restriction patterns (*left*) and hybridization to ^{32}P-labelled spinach LS gene (*right*) of *N. otophora* ct-DNA. 10 μg of purified ct-DNA (Rhodes and Kung 1981; Kolodner and Tewari 1975) were digested with *Sma*I, *Sal*I, *Pst*I *Hind*III. *Eco*RI and *Bam*HI and subjected to electrophoresis in a 1% agarose gel (*channels* A→F, respectively). DNA fragments in the gel were transferred by blotting to nitrocellulose filter paper (Smith and Summers 1980). The plasmid pJEA4 DNA was prepared according to Lovett and Keggins (1979), and labelled with [α-^{32}P] dATP by the nick-translation reaction (Maniatis et al. 1975). Hybridization and autoradiography were made according to Southern (1975). On the *right* (*a*→*f*) with size marker is an autoradiograph of the gel shown on the *left*, after hybridization whth the spinach LS gene. B Cloning of *Hind 2* and *Sal 6* restriction fragments of *N. otophora* ct-DNA in *E. coli* and hybridization to ^{32}P-labelled spinach LS gene. 10 μg of *N. otophora* ct-DNA was digested with *Hind*III or *Sal*I, and electrophoresed through 0.8% low-gelling-temperature agarose according to Herrmann et al. (1980). The purified ct-DNA fragment *Hind 2* was ligated to *Hind*III-digested pBR322, and fragment *Sal 6* was ligated to *Sal*I-digested pBR325 according to Bolivar and Backman (1979). The hybrid plasmids were transformed into CSR603 or HB101 cells as described by Dagart and Ehrlich (1979). Amp^r Tc^s recombinants were selected and the ct-DNA inserts were analysed. *Channels 1, 2, 3,* and *4* contain *Hind*III digests of *1* λDNA; *2* pBR322; *3* and *4* PRCZ1 (*Hind 2* fragment in pBR322 cloned in HB101 and CSR603 respectively). *Channels 5, 6,* and *7* contain *Sal*I digests of *5* and *6*, PRCZ2 (*Sal 6* fragment in pBR325 cloned in HB101 and CSR603, respectively); *7* pBR325. *Channels 8, 9, 10* and *11* are autoradiograms demonstrating the hybridization of ^{32}P-labelled spinach LS gene to gel *channels 3, 4, 5* and *6***

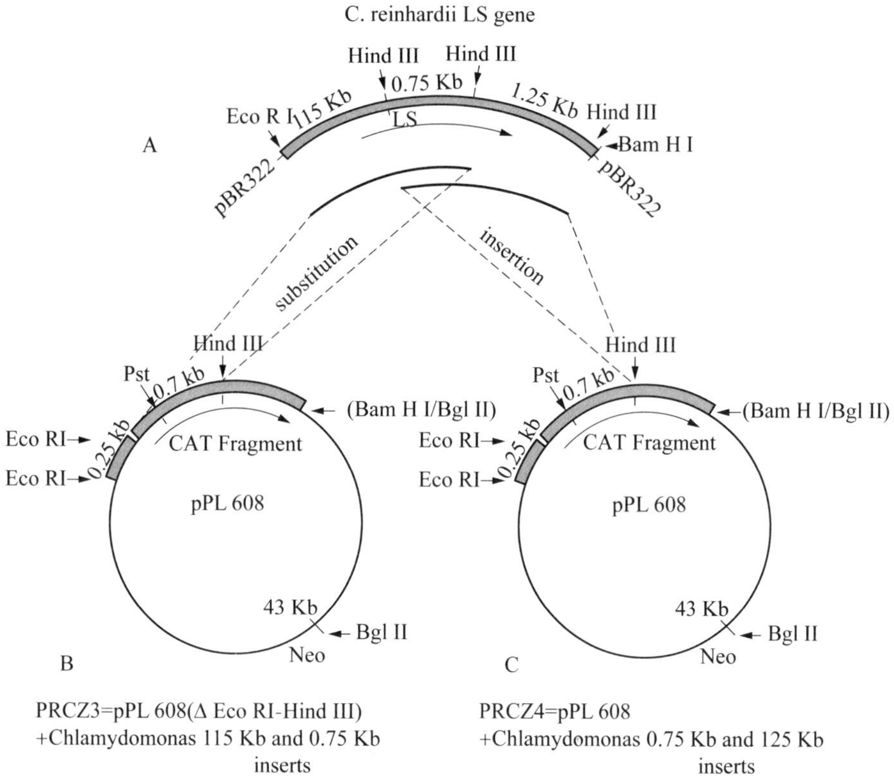

Figure 2A–C **Cloning of the *C. reinhardii* LS gene in *B. subtilis*. A *C. reinhardii* ct-DNA fragment (3. 2 kb) was inserted into *Eco*RI and *Bam*HI sites of pBR322 and cloned in HB101 as recombinant plasmid pLM401 by L. Metz. It contains the LS gene as marked by the *arrow* indicating the direction of transcription according to Metz's data. The central region of this gene is located in a 0. 75 kb *Hind*III fragment. The 5' and 3' ends are located within the 1. 15 kb *Eco*RI-*Hind*III fragment and 1. 25 kb *Hind*III fragment respectively. B Construction of PRCZ3: 10 μg of pPL608 DNA was digested with *Hind*III for 2 h at 37℃, then partially digested with *Eco*RI for 5 min at room temperature. A plasmid fragment containing the 4. 3 kb *Eco*RI-*Hind*III region plus the 0. 25 kb *Eco*RI promoter fragment as marked by the split bar, was separated and isolated by low-gelling-temperature agarose gel electrophoresis. The anterior portion of the *Chlamydomonas* LS gene (1. 15 kb and 0. 75 kb) in Figure 2 A was obtained by partial *Hind*III digestion of the *Eco*RI linear form of PLM401. The deleted form of pPL608 and anterior portion of the *Chlamydomonas* LS gene were ligated and introduced into *B. subtilis* IS53 as described by Lovett and Keggins (1979). C Construction of PRCZ4: Plasmids PLM401 and pPL608 were digested with *Hind*III and ligated, and transformed into *B. subtilis* IS53. Neo^r, Cm^s recombinants were selected and the inserts were analysed. A clone PRCZ4 containing 0. 75 kb and 1. 25 kb *Hind*III fragments was obtained**

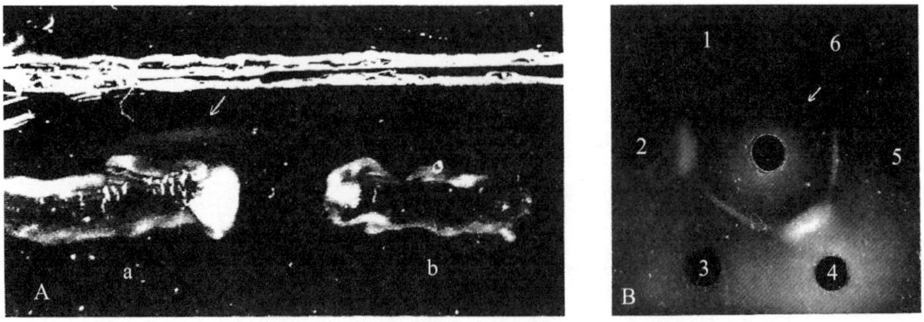

Figure 3A **Detection of the *N. otophora* LS polypeptide synthesis in *E. coli* by in situ immunoprecipitation reaction. Clones containing different ct-DNA inserts were cultured on N-Z bottom agarose plate containing ampicillin (20 μg/ml) in a 50 mm petri dish. After lysis and washing, a channel was made in the center of the plate and 100 μl of an antiserum against LS was loaded. Overnight incubation at room temperature allowed development of a zone of immunoprecipitation (marked by *arrow*) around CSR603 (PRCZ1) (*a*), whereas control cells (*b*) containing plasmid pBR322 exhibited no reaction. B Detection of the synthesis of LS polypeptide of *N. otophora*, *N. tabacum*, spinach and *C. reinhardii* in *E. coli* or *B. subtilis* by Ouchterlony double diffusion. Five ml of cultures were grown in L broth at 37℃ for 4 h. Cell pellets were washed with TES buffer (Lovett and Keggins 1979) and lysed by boiling in 0. 1 ml of 2% SDS for 2 min. After removal of free SDS by acetone treatment, twenty μl of lysates were added to the outer holes. Twenty μl of antiserum against LS was loaded into the central hole. After incubation for one day at room temperature, the immunoprecipitates were observed in all clones containing LS gene except in control cells.**

1. **PRCZ1: Hind 2 (*N. otophora*) /pBR322/CSR603**
2. **PRCZ2: Sal 6 (*N. otophora*) /pBR325/CSR603**
3. **pLM401: *Chlamydomonas* LS gene/ pBR322/HB101**
4. **pJEA4: Spinach LS gene/pBR322/HB101**
5. **pHPE4: *Eco* R4 (*N. tabacum* BR325/HB101**
6. **PRCZ3: anterior *Chlamyclomonas* LS gene/pPL608/IS53**

cultured on N-Z bottom agarose plate, lysed with chloroform vapor and lysozyme, and overlayered with molten agarose mixed with the antiserum against LS. Incubation continued for one week at 4℃. Colonies harboring both recombinant plasmids showed an immunoprecipitation ring. whereas the control cells exhibited no immunoreaction (data not shown). This result was confirmed by immunoprecipitation of extracts of these clones with antiserum against LS in an Ouchterlony double diffusion test (Figure 3B). Figure 3 B shows expression in *B. subtilis* of anterior portion of the LS gene from *Chlamydomonas* and in *E. coli* of the LS genes from spinach, tobacco, and *Chlamydomonas*. As it is expected, the smaller incomplete polypeptide made by the anterior portion of the LS gene from *Chlamydomonas* (well 6) migrated faster as compared to the complete LS polypeptide made by the entire LS genes from spinach, tobacco, and *Chlamydomonas* (wells 1 – 5) in *E. coli*. The sizes of the complete LS of tobacco and spinach are similar (52 – 53,00 daltons, Erion et al. 1981). Based on the intensity of immunoprecipitation line of a standard, the amount of LS produced in each cell was approximated to be 0. 1 pg.

The results presented here demonstrate direct expression in *E. coli* and *B. subtilis* of the LS genes from tobacco and *Chlamydomonas*. Previous studies demonstrated that expression of the maize LS gene in *E. coli* was initiated within its own promoter (Gatenby 1981). Our results also indicate that the expression in *E. coli* of tobacco LS gene was directed by its promoter, since the LS gene and its promoter are located in the middle of the Hind 2 fragment (11 kb). Furthermore, this promoter is able to initiate the expression in *E. coli* of the galactokinase gene on the promoter-less plasmid pK01 (C. M. Lin and X. F. Kong, unpublished data).

This is the first demonstration of expression of a plant gene in *Bacillus*. *Hind*III fragments (0. 75 and 1. 25 kb) cloned in PRCZ4 represent a large posterior part of the LS gene of *C. reinhardii* presumably lacking its own promoter. Its expression in *Bacillus*. which is stimulated by chloramphenicol (0. 1 μg/ml) (data not shown), is apparently initiated by a strong promoter located whthin the 0. 25 kb *Eco*RI fragment in pPL608 (Williams et al. 1981) (Figure 2C). The unique *Hind*III site in pPL608 is located within a structural gene specifying a chloramphenicol-inducible, chloramphenicol acetyltransferase (CAT). Based on the data of Williams et al. (1981) it is expected that the LS polypeptide in PRCZ4 is synthesized as a hybrid with the amino terminal portion of chloramphenicol acetyltransferase. Transcription of the LS gene in the clone PRCZ3, which contains the anterior portion of the LS gene (Figure 2B) is either initiated within its own promoter or within the plasmid promoter.

Acknowledgements. The authors are grateful to Drs. L. Metz and H. Weissbach for the generous gifts of their clones pLM401 and pJEA4 respectively. This investigation was supported by NIH grant CM22746 - 01, U. S. Dept of Agriculture cooperative agreement 58 - 3204 - 0 - 157 from the tobacco laboratory and NSF grants PCM 78 - 05755 and PCM 82 - 02701.

References

[1] Anderson D. Shapio L. Skalka AM (1980) In situ Immunoassays for translation products. Methods Enzymol 68: 428 - 437

[2] Bolivar F. Backman K (1979) Plasmids of *E. coli* as cloning vectors. Methods Enzymol 68: 245 - 267

[3] Bottomley W. Whitfeld PR (1978) The products of cell-free transcription and translation of total spinach chloroplast DNA In: Akoyunoglou G (ed) Chloroplast development. Elsevier, Amsterdam, pp 657 - 660

[4] Chan PH, Wildman SG (1972) Chloroplast DNA codes for the primary of the large subunit of fraction I protein. Biochim Biophys Acta 227: 677 - 680

[5] Chua NH, Schmidt GW (1978) Post-translational transport into intact chloroplast of a precursor to the small subunit of ribulose-1, 5-bisphosphate carboxylase. Proe Natl Acad Sci USA 75: 6110 - 6114

[6] Coen DM, Bedbrook JR, Bogorad L. Rich A (1977) Maize chloroplast DNA fragment encoding the large subunit of ribulose bisphosphate carboxylase. Proc Natl Acad Sci USA 74: 5487 - 5491

[7] Dalgart M, Ehrlich SD (1979) Prolonged incubation in calcium chloride improves the competence of *E. coli* cells. Gene 6: 23 - 28

[8] Erion JL, Tarnowski J, Weissbach 11, Brot N (1981) Cloning, mapping, and in vitro transcription-translation of the gene for the large subunit of ribulose-1, 5-bisphosphate carboxylase from spinach

chloroplasts. Proc Natl Acad Sci USA 78: 3459 - 3463
[9] Gatenby AA, Castleton JA, Saul MW (1981) Expression in *E. coli* of maize and wheat chloroplast genes for large subunit of ribulose bisphosphate carboxylase, Nature 291: 117 - 121
[10] Gray JC, Kekwick RGO (1973) Synthesis of the small subunit of ribulose-1, 5-diphosphate carboxylase on cytoplasmic ribosomes from green leaves, FEBS Lett 38: 67 - 69
[11] Herrmann RG, Whitfeld PR, Bottomley W (1980) Construction of a SalI/PstI restriction map of spinach chloroplast DNA using low-gelling temperature-agarose electrophoresis, Gene 8: 179 - 191
[12] Highfield PE, Schmidt GW (1978) Synthesis and transport of the small subunit of chloroplast ribulose bisphosphate carboxylase, Nature 271: 420 - 425
[13] Jensen KG, Bahr JT (1977) Ribulose-1, 5-bisphosphate carboxylase-oxygenase, Annu Rev Plant Physiol 28: 379 - 478
[14] Kolodner R, Tewari KK (1975) The molecular size and conformation of the chloroplast DNA from higher plants. Biochim Biophys Acta 402: 372 - 390
[15] Kung SD (1977) Expression of chloroplast genomes in higher plants. Annu Rev Plant Physiol 28: 401 - 437
[16] Lovett PS, Keggins KM (1979) *Bacillus subtilis* as a host for molecular cloning. Methods Enzymol 68: 342 - 357
[17] Malnoë P, Rochaix JD, Chua NH, Spahr PF (1979) Characterization of the gene and messenger RNA of the large subunit of ribulose-1, 5-diphosphate carboxylase in *Chlamydomonas reinhardii*. J Mol Biol 133: 417 - 434
[18] Maniatis T, Jeffrey A, Kleid DG (1975) Nuclcotide sequence of right-ward operator of phage. Proc Natl Acad Sci USA 72: 1184 - 1188
[19] Rhodes PG, Kung SD (1981) Chloroplast DNA isolation: purity achieved without nuclease digestion. Can J Biochem 59: 911 - 915
[20] Smith GE, Summers MD (1980) The bidirectional transfer of DNA and RNA to nitrocellulose or diazobenzyloxymethylpaper. Anal Biochem 109: 123 - 129
[21] Southern EM (1975) Detection of specific sequences among DNA fragments separated by gel clectrophoresis. J Mol Biol 98: 503 - 517
[22] Williams DM, Schoner RG, Duvall EJ, Preis LH, Lovett PS (1981) Expression of *Escherichia coli* trp genes and the mouse dihydrofolate reductase gene cloned in *Bacillus subtilis*, Gene 16: 199 - 206
[23] Zhu YS, Duvall EJ, Lovett PS, Kung SD (1982) *Nicotiana* chloroplast genome, V. Construction, mapping and expression of clone library of *N. otophora* chloroplast DNA. Mol Gen Genet 187: 61 - 66

Light-Induced Transformation of Amyloplasts into Chloroplasts in Potato Tubers①②

Abstract

The transformation of amyloplast into chloroplasts in potato (Solanum tuberosum L.) tuber tissue can be induced by light. Excised potato tuber discs illuminated with white light of 3,000 lux began to synthesize chlorophyll after a lag period of 1 day, and continued to synthesize chlorophyll for 3 weeks. In this paper we present evidence, based on ultracentrifugal sedimentation and immunoprecipitation, that the light mediated synthesis of Ribulose-1,5-bisphosphate carboxylase began 1 day after illumination with white light.

When illuminated the chloroplasts isolated from light-grown potato tuber tissue incorporated [^{35}S]methionine into polypeptides, one of which has been identified as the large subunit of Ribulose-1, 5-bisphosphate carboxylase. These chloroplasts are functional as determined by O_2 evolution in the Hill reaction.

A great deal of data is now available on the biochemistry of the development of chloroplasts from etioplasts[6]. Much less is known about the development of chloroplasts from amyloplasts, a starch storing organelle, which exists in storage tissue, roots, and some callus. It is well known that potatoes turn green on exposure to light for several days. Not only does the study of the organelle transformation and light-regulated gene expression stimulate theoretical interest, but the greening of potato also has some practical importance, since the greening of the potato is accompanied by the formation of the poisonous alkaloid, solanine[5]. The morphological and ultrastructural changes in the development of chloroplasts from amyloplasts were described in several electron microscopic studies[3,11]. In

---

Received for publication November 9,1983 and in revised form January 19,1984

Yu S. Zhu, Denise L. Merkle-Lehman, Shain D. Kung, Department of Biological Sciences, University of Maryland Baltimore County, Catonsville, Maryland 21228

此文发表于 1984 年"Genet Plant Physiol" 75:142 - 145.

① Supported by National Institutes of Health grant CM22746 - 01 and United States Department of Agriculture agreement 58 - 3204 - 0 - 157.

② Abbreviations: RuBPCase, ribulose-1,5-bisphosphate carboxylase; LS, large subunit.

this communication we present some biochemical evidence to show the function of chloroplasts from potato tuber based on the biosynthesis of Chl, RuBPCase[2], and Hill activity.

Materials and Methods

Preparation, Culture, and Light Treatment of Potato Tuber Discs. Centennial and Katahdin, two varieties of potato tubers supplied by the United States Department of Agriculture were used in this study. The tubers were peeled and sterilized with Amphyl (National Laboratories, Lehn and Fink Industrial Products Division of Sterling Drug Inc., New Jersey). Discs ($10 \times 2\ mm^2$) were made with a sterile cork borer and a gel slicer. These discs, after washing in sterile water, were placed on 1% agar plates in sterile Petri dishes (9 cm), 40 discs per dish. The discs were illuminated at room temperature by fluorescent tubes with intensity of 3,000 lux.

Chlorophyll and RuBPCase Determination. Chl was determined according to Arnon[2]. RuBPCase was detected with a Model E analytical ultracentrifuge, as previously described[9]. Schlieren pictures were taken at 44,770 rpm, 10 min after attaining this speed. An Ouchterlony double diffusion test was employed to determine the RuBPCase specifically. The antiserum was prepared against RuBPCase from tobacco[15].

Light-Driven Protein Synthesis in Chloroplasts. Ten g of green potato tuber discs which had been exposed to white light for 7 d were homogenized in a Waring Blendor with cold isolation buffer (sucrose 0.35 mol/L, Hepes-NaCl 25 mmol/L, EDTA 2 mmol/L, isoascorbateNa 2 mmol/L, pH 7.6) and filtered through 2 layers of Miracloth. The resultant filtrate was centrifuged at 30g for 1 min to remove starch granules, and the supernatant was then centrifuged at 2,500 g for 1 min. The pellet was resuspended in 1 ml of KCl suspension (KCl 0.2 mol/L, Tricine-KOH 66 mmol/L, $MgCl_2$ 6.6 mmol/L). The chloroplast suspension was transferred to a Petri dish (3 cm in diameter) and incubated with [^{35}S] methionine (10 μCi) at room temperature under white light (4,000 lux). During a 2 h incubation, the incorporation of [^{35}S]methionine into proteins was measured. After the 2 h incubation the radioactive polypeptides were separated on an SDS-polyacrylamide gradient gel (8%—15%), followed by fluorography as described by Blair and Ellis[4].

O_2 Evolution Assay. The ability of chloroplasts from potato tuber discs to evolve O_2 was measured in the Hill reaction according to the procedure of Marsho et al.[12]. The incubation system consisted of ferricyanide 1.7 mmol/L, methylamine 3 mmol/L, glyceraldehyde 10 mmol/L, and chloroplasts corresponding to 15 μg of ChI in a total volume of 0.6 ml. The reaction was initiated by irradiation with red light (22.4 mW/cm^2). The O_2 evolution was measured and recorded with a polarograph.

Results

Chlorophyll Synthesis in Light-Illuminated Potato Tuber. Our experiments showed that the greening of potato tuber depended on varieties, storage temperature, light intensity, and wavelength of light. Out of more than ten varieties of potatoes tested, two varieties, Centennial and Katahdin, which turn green more quickly under light, were selected in this study. Storage of potato tuber below 4℃ retarded or inhibited the transformation of amyloplasts into chloroplasts. Blue light was most effective in inducing greening of potato tubers. No red light stimulation was observed.

Potato tuber discs illuminated with white light began to green with a lag period of about 1 d. The greening continued for 3 weeks after culture (see Figure 1). After illumination for 3 weeks, the potato tuber tissue contained 10 μg Chl/g of fresh tissue, about one hundredth of the Chl content in normal leaves. This is primarily due to fewer chloroplasts per cell. The chloroplasts were not distributed uniformly in discs and were probably linked to some specific ultrastructure in the cortex. Microscopic observation revealed that the amyloplasts were comprised of two types: large (55×80 μm) and small (5—20 μm). Most of them were stainable with I_2 - KI. Upon exposure of the discs to light the small amyloplasts only turned greenish.

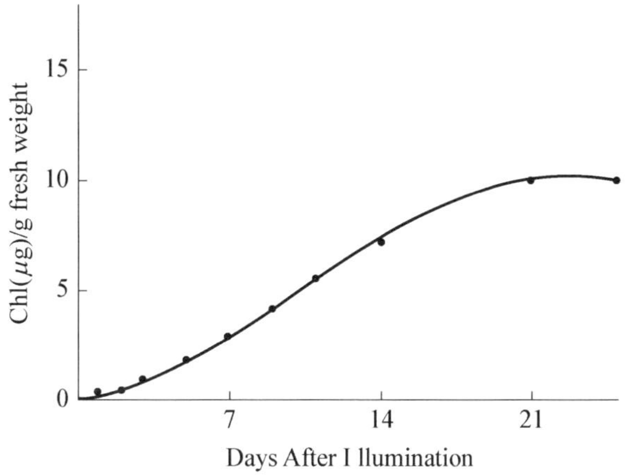

Figure 1 **Time course of the synthesis of chlorophyll in potato tuber discs during continuous illumination with white light**

RuBPCase Biosynthesis in Potato Tuber during Greening. Since RuBPCase is the most abundant soluble protein in the chloroplasts[10], it can be used as an important biochemical marker of chloroplasts. The de novo synthesis of RuBPCase was clearly demonstrated by ultracentrifugal sedimentation (data not shown) and specific immunoprecipitation (see Figure 2). The Schlieren pattern of extracts from normal leaves consists of four peaks representing 80S cytoplasmic ribosomes, 70S chloroplast ribosomes, 18S Fraction I protein (RuBPCase), and 4 to 6S Fraction II proteins[8,9]. The extract from potato tuber stored in the dark lacked the peak of RuBPCase, whereas a small peak was observed in discs exposed to 3 d of light, indicating the light initiated the synthesis of RuBPCase (data not shown).

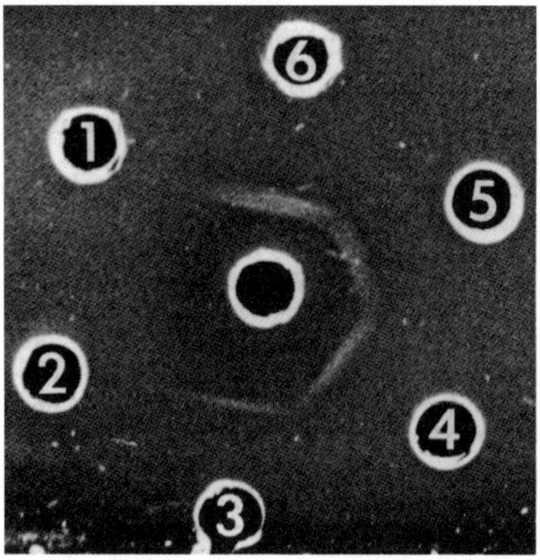

Figure 2 Biosynthesis of RuBPCase in potato tuber tissue as assayed by Ouchterlony double diffusion 0 h (1), 6 h (2), 12 h (3), 24 h (4), 48 h (5), and 72 h (6) after illumination. The central well contained antibody to LS. One g of potato tuber discs was homogenized with 0.1 ml of Tris 80 mmol/L, $MgCl_2$ 20 mmol/L, KCl 40 mmol/L, pH 8.5. After centrifugation at 12,000 g for 10 min, 20 μl of the supernatant were used for assay of RuBPCase

The biosynthesis of RuBPCase was also demonstrated by an Ouchterlony double diffusion assay (see Figure 2). The light-mediated synthesis of RuBPCase started at 1 d after illumination and increased during greening. The RuBPCase content was estimated to be ~6 μg/g fresh potato tuber tissue, whereas the RuBPCase content of a typical green leaf is 5 to 10 mg/g fresh tissue[10].

Further evidence for the de novo synthesis of RuBPCase in light-treated potato tuber was obtained from the experiment on light-driven protein synthesis in chloroplasts. The chloroplasts, isolated from potato tuber discs after illumination for 7 d, exhibited a higher activity of protein synthesis, as demonstrated by the incorporation of [^{35}S]methionine into proteins (see Figure 3). In contrast to light-driven protein synthesis in chloroplasts, the chloroplasts in the dark and especially amyloplasts, either in the light or dark, exhibited a very low protein synthesis activity. A number of radioactive polypeptides synthesized in the chloroplasts in the light were recognized on SDS-polyacrylamide gels followed by fluorography (see Figure 4). It was observed that some chloroplast polypeptides (mol wt 39,000, 50,000, 52,000, 55,000, 94,000, 96,000) were synthesized more readily in the light than in the dark. One of these polypeptides co-migrated with purified unlabeled LS of RuBPCase (mol wt 52,000), and was identified as the LS, based on this and results presented in Figure 2. There is also one strongly light-initiated polypeptide (mol wt 64,000) which does not appear in the dark.

O_2 Evolution of Chloroplasts from Light-Induced Potato Tubers. Upon illumination with red light, the chloroplasts, isolated from light-treated potato tuber discs after 3 weeks of light treatment, evolved O_2 using ferricyanide as an electron acceptor. This Hill reaction

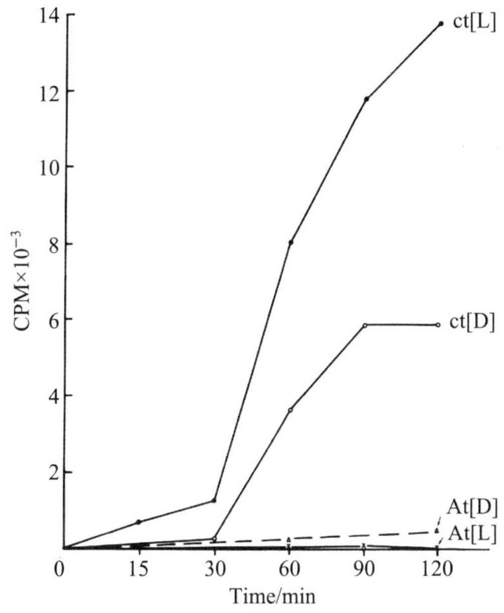

Figure 3 **Incorporation of [^{35}S]methionine into proteins in the light-driven chloroplast protein synthesis system. 5-μl samples were added to 20% TCA containing 10 μg/ml unlabeled methionine. Proteins were precipitated with 5% TCA containing 10 μg/ml unlabeled methionine, and counted with a Mark I scintillation counter. ct, chloroplasts; At, amyloplasts (control); L, light-incubated; D, dark-incubated.**

activity is comparably high (177 μmol O_2/mg Chl · h), indicating that an active electron transport reaction took place in this chloroplast preparation.

Discussion

The cells of potato tuber contain a large number of amyloplasts, whose function is to accumulate and store starch in the form of reserve starch granules. It is of interest that upon exposure to light, the amyloplasts are transformed into chloroplasts, which have a different function. The changes in ultrastructure of potato tuber amyloplasts during greening were investigated by electron microscope. It was shown that the main developmental features were elongation of vesicles into thylakoids, the differentiation of grana and the appearance of ribosomes in the stroma[3], although the ultrastructure is generally less well developed compared to that for normal leaves. On the other hand, there is very little information on biochemical alterations in potato tuber during greening. To confirm the transformation of amyloplasts into chloroplasts, this study provides some biochemical evidence: ① the synthesis of photosynthetic pigments; ② the synthesis of RuBPCase and other proteins; ③ Hill reaction activity.

Before illumination the potato tubers contain no Chl or Pchl, but do contain carotenoid which increases during greening[1]. Very low light intensity (400 lux) was required to

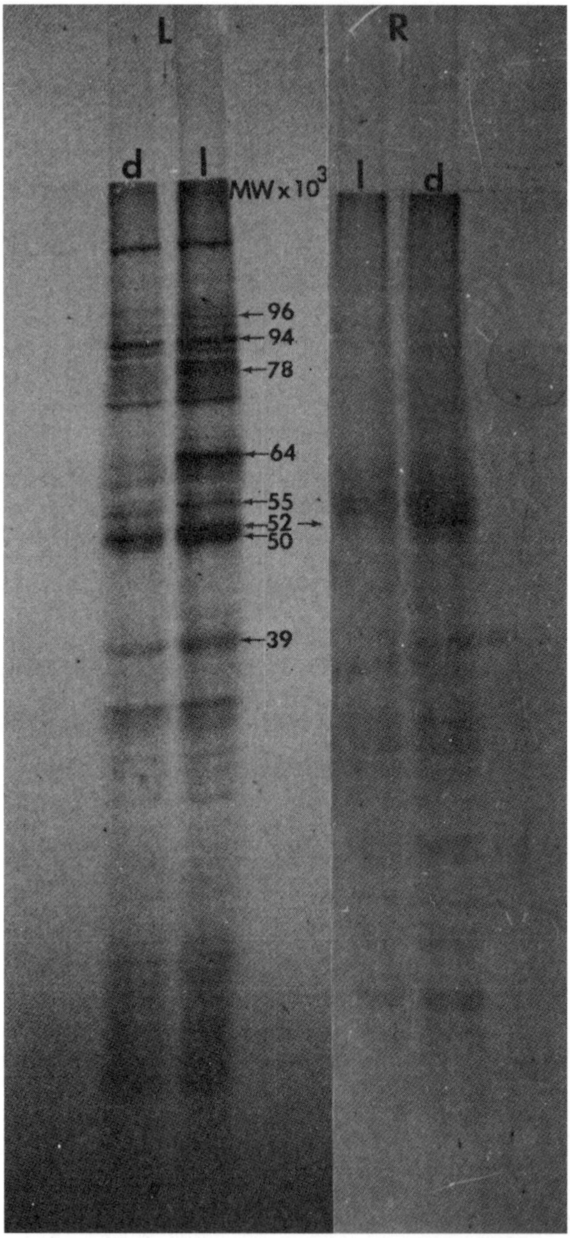

Figure 4 SDS-PAGE of light- (1) and dark- (d) initiated polypeptides synthesized in chloroplasts isolated from light-treated potato tuber discs. (R) photograph of stained gel. (L) fluorograph of (R). Arrows indicate the mol wt of light-stimulated polypeptides and band corresponding to LS. Althhough the dark-incubated sample was more concentrated than the light-incubated sample (R), the fluorograph (L) indicates there was more incorporation of [^{35}S]methionine into the light-induced polypeptides than into the dark-initiated polypeptides, demonstrating that more polypeptides were synthesized in the light than in the dark. The samples were incubated with [^{35}S]methionine for 120 min

initiate greening. The light may penetrate into the potato tuber discs and evoke the development of chloroplasts from amyloplasts. During the development of chloroplasts from amyloplasts, assembly of Chl into the newly synthesized membranes occurs. Cold storage may cause the breakage of membranes[13]; however, some studies suggested that amyloplast membranes remain intact in cold storage[14]. It was frequently observed that the chloroplasts were formed as streaks in the potato tuber discs, suggesting a special structure is linked to the chloroplast development.

The Schlieren pattern of the extracts from light-induced potato tuber tissue as well as its immunoprecipitation reaction with antiserum to RuBPCase demonstrated that RuBPCase, an important enzyme in the photosynthetic carbon cycle, was de novo synthesized during greening. This result was further confirmed by the active light-dependent protein synthesis in the isolated chloroplasts. The biosynthesis of RuBPCase during greening suggests the involvement and operation of CO_2 assimilation in the chloroplasts. Furthermore, the photosynthetic function of electron transport and O_2 evolution in isolated chloroplasts from potato tuber was shown by the high Hill reaction activity.

The transformation of amyloplasts into chloroplasts is absolutely light-dependent. The fact that red light did not stimulate this transformation indicates that phytochrome may not participate in this regulation. This coincides with the result that potato tuber does not have phytochrome[7]. What is the photoreceptor in this light-induced organelle transformation? How does light turn on the genes for the development of chloroplasts, and turn off the genes for the development of amyloplasts? Undoubtedly, this system provides an attractive and challenging model for investigations into the molecular mechanisms underlying the photoregulation of development and gene expression.

Acknowledgments

We thank Dr. Ray Webb of United States Department of Agriculture, Beltsville, Maryland for providing us with different varieties of potato, and F. J. Xi for her help in determination of Hill activity.

LITERATURE CITED

[1] ANSTIS PJP, DH NORTHCOTE 1973 Development of chloroplasts from amyloplasts in potato tuber discs. New Phytol 72:449 - 463

[2] ARNON DI 1949 Copper enzymes in isolated chloroplasts. Polyphenoloxidase in Beta vulgaris. Plant Physiol 24:1 - 15

[3] BADENHUIZEN NP, R SALEMA 1976 Observations of the development of chloroamyloplasts. Rev Biol (Lisb) 6:139 - 155

[4] BLAIR GE, RJ ELLIS 1973 Protein synthesis in chloroplasts 1. Light-driven synthesis of the large subunit of fraction I protein by isolated pea chloroplasts. Biochim Biophys Acta 319:223 - 234

[5] FORSYTH AA 1954 British Poisonous Plants. Her Majesty's Stationery office, London

[6] KIRK JTO, RAE TILNEY-BASSETT 1978 The plastids. Growth and Differentiation of Plastids. Part

I. Formation of the Chloroplast during Greening of the Enolated Plant. Elsevier/North-Holland Biomedical Press, Amsterdam, The Netherlands, pp 720 - 773

[7] KOUKKARI WL, WS HILLMAN 1966 Phytochrome levels assayed by in vivospectrophotometry in modified underground stems and storage roots. Physiol Plant 19:1073 - 1078

[8] KUNG SD 1977 Expression of chloroplast genomes in higher plants. Annu Rev Plant Physiol 28:401 - 437

[9] KUNG SD, PR RHODES 1981 Hormonal effects on the biosynthesis of tobacco RuBPCase in vitro. Beitr Tabakforsch Int 11:44 - 49

[10] KUNG SD, TC Tso 1978 Tobacco as a potential food source and smoke material: soluble protein content, extraction, and amino acid composition. J Food Sci 43:1844 - 1852

[11] LOBOV UP, PI BONDAR 1977 The RNA of potato tuber amyloplasts. Fiziol Rast 24:318 - 322

[12] MARSHO TV, PM SOKOLOVE, RB MACKAY 1980 Regulation of photosynthetic electron transport in intact spinach chloroplasts. Plant Physiol 65:703 - 706

[13] OHAD I, I FREIDBERG, Z NEEMAN, M SCHRAMM 1971 Biogenesis and degradation of starch I. The fate of the amyloplast membrane during maturation and storage of potato tubers. Plant Physiol 47: 465 - 477

[14] WETZSTEIN HY, C STERLING 1978 Integrity of amyloplast membranes in stored potato tubers. Z Pflanzenphysiol Bd 90S: 373 - 378

[15] ZHU YS, PS LOVETT, DM WILLIAMS, SD KUNG 1983 Nicotiana chloroplast genome 7 expression in E. coli and B. subtilis of tobacco and Chlamydomonas chloroplast DNA sequences coding for the large subunit of RuBP carboxylase. Theor Appl Genet 67:333 - 336

Phytochrome Control of Levels of mRNA Complementary to Plastid and Nuclear Genes of Maize*

Abstract

The involvement of phytochrome in the control of the levels of RNA transcribed from maize plastid and nuclear genes was examined. The effects of illumination with red light, farred light, or red light followed by farred light on relative amounts of RNAs complementary to maize plastid genes for the large subunit of ribulose bisphosphate carboxylase (RuBPCase); the 32-kDa thylakoid membrane triazine herbicide binding B protein of photosystem II; the α, β and $\in$ subunits of CF_1; subunit III (proton-translocating) of CF_0; the reaction center proteins A1 and A2 of photosystem I; two other light-induced genes for membrane proteins of photosystem II (ORFs 353 and 473); and one gene for an unidentified membrane protein (UORF 443) were measured by hybridization of labeled DNA probes to samples of leaf RNA. Transcripts of two nuclear-encoded genes, the genes for the small subunit of RuBPCase and the light-harvesting chlorophyll a/b binding protein, were studied in the same way. The levels of RNA complementary to all of these light induced genes were significantly increased within 3 to 6 hours after brief illumination with red light. The stimulatory effects of red light were largely reversed by subsequent illumination with farred light. It is concluded that phytochrome controls increases in the levels of mRNAs complementary to certain plastid and nuclear genes in dark-grown maize

---

* Received for publication December 26, 1984 and in revised form June 3, 1985. 此文发表于"Plant Physiol" 79:371 – 376.

① Yu S. Zhu, Present address: Deparlment of Microbiology, University of Illinois at Urbana-Champaign, Urbana, IL 61801.

② Shain D. Kung, Permanent address: Department of Biological Sciences, University of Maryland Baltimore County, Catonsville, MD 21228.

③ Lawrence Bogorad

Department of Cellular and Developmemal Biology, The Biological Laboratories, Harvard University, Cambridge, Massachusetts 02138

Various segments of this research were supported in part by research grants from the National Science Foundation, the Competitive Research Grants Office of the United States Department of Agriculture, and the National Institute of General Medical Sciences (L. B.). It was also aided by the Maria Moors Cabot Foundation for Botanical Research of Harvard University.

seedlings.

During light-induced development of etioplasts into chloroplasts, morphological, structural and biochemical changes occur, including the accumulation of Chl, the formation of thylakoid membrane proteins and lipids, increases in levels of some enzymes, an increase in chloroplast RNA polymerase activity, and increases in various RNA species[1,5,6,10,14,17,25,28]. The first major plastid DNA sequence whose transcript level was shown to increase during greening was photogene 32, the gene for a 32-kDa thylakoid membrane polypeptide in maize[3,17]. This polypeptide has been shown to bind triazine herbicides and appears to be the B protein of PSII; thus, the gene is designated ps2B[23]. (It has also been designated psbA[18].) Photo-regulation of ps2B transcription has also been reported in other plants[12,20,28]. Recently, a number of other photoregulated genes have been mapped on the maize chloroplast chromosome and characterized in this laboratory (Castroveijo K M T, Muskavitch E T, Krebbers H, Roy D R, Russell L, Bogorad, unpublished data)[8,25]. These include, in addition to ps2B: rcL—encoding the large subunit of ribulose biphosphate carboxylase[19]; cf1 A B and E—encoding the α, β, and ∈ subunits, respectively of CF_1[4] (Rodermel S, Bogorad L, unpublished data) the coupling factor 1 for photosynthetic phosphorylation[25]; cfoIII-encoding subunit III of CF_0[25]; pslA1 and ps1A2— encoding polypeptides of P700—the reaction center of PSI[13,25]; ORFs 473, 353—encoding PSII proteins of 473 and 353 amino acids, respectively-and UORF 443-encoding an unidentified membrane polypeptide (Krebbers E T, Muskavitch K M, Orr E A, Schantz R, Castroveijo M and Bogorad L, unpublished data). In addition, expression of nuclear genes rcS (encoding the small subunit of ribulose biphosphate carboxylase) and/or cab (encoding the light harvesting Chl a/b binding protein) have been demonstrated to be induced by light in barley, Lemna, pea, bean, and maize[2,16,25,28,29]. For convenience, the term 'photogene' is used to designate a gene whose transcript level increases significantly during light-induced development.

The major photomorphogenic system of plants is driven by the R and FR reversible phytochrome receptor[11,21,22,27]. In some cases, a blue light receptor may be involved in photomorphogenesis[21,24]. Phytochrome is a bile pigment-protein complex that has two photointerconvertible forms: Pr and Pfr with absorption maxima at 660 and 730 nm, respectively[11]. Photoconversion of Pr to Pfr by R potentiates a diverse array of morphogenic responses that are not exhibited if the Pfr is promptly reconverted to Pr by illumination with FR.

To understand the molecular mechanisms that govern light-dependent changes during greening, it is necessary to identify the photoreceptor (or photoreceptors) that regulate expression of maize photogenes. It has been shown by others[2,16,20,26,28,29] that RNA levels of rcS, cab, ps2B, and rcL are affected by phytochrome in at least one plant species. The results presented here demonstrate that the levels of RNAs complementary to these and several other sets of photoregulated plastid and nuclear genes of maize are controlled by phytochrome.

Materials and Methods

Growth of Plants

Maize seeds (FR9 cms X FR 37; Illinois Foundation Seeds, Inc.) were soaked overnight in running water① and sown in moist Vermiculite. Seedlings were grown in narrow plastic trays ($36\times7\times5$ cm^3) in the dark at 25℃ for 8 to 10 d.

Light Treatment

Monochromatic light was obtained from Airequipt 125 500 W projectors equipped with Baird-Atomic interference filters. The R source had a peak transmittance at 660 nm and an energy fluence rate of 0.9 W/m^2 at the level of plants. For the 725 nm FR source, this value was 0.7 W/m^2.

Extraction of RNA

About 10 g of leaves were harvested from the seedlings under a dim green safe light and ground in a mortar with liquid N_2. The frozen powder was suspended in 25 ml of hot lysis buffer (0.2 mol/L sodium borate, 30 mmol/L EGTA, 2% (w/v) sodium triisopropylnaphthalenesulfonate, pH 9.0) and was extracted with phenol/chloroform by the procedure previously described[9]. After repeated precipitation with alcohol and centrifugation, the RNA pellet was resuspended in 2 mol/L LiCl. The final pellet was dissolved in distilled H_2O and analyzed spectrophotometrically and by agarose gel electrophoresis.

Isolation of Chloroplast and Plasmid DNA

cpDNA was prepared as described previously[9]. Plasmid DNAs were isolated and purified by cesium chloride centrifugation. The designations of the cloned plasmids used for probing transcripts of plastid and nuclear genes of maize are listed in Table 1 together with the sizes of the cloned inserts and the genes encoded.

Radioactive Labeling of DNA

Plasmid DNA for hybridizations to membrane-bound RNA samples was labeled with (α-^{32}P) dATP by the nick-translation reaction. Labeled plasmid DNA was separated from unincorporated nucleotides by chromatography on Biogel P-60.

---

① Abbreviations: CF_1, coupling factor for photosynthetic phosphorylation; cpDNA, plastid DNA; ORF, open reading frame; UORF, unidentified open reading frame; R, red light; FR, far-red light; kb, kilobases; kbp, kilobase pairs.

Dot Blots

Aliquots of RNA were denatured by heating to 60℃ in 7.5% formaldehyde, 10 mmol/L Na-phosphate (pH 6.8), for 30 min, according to the procedure of White and Bancroft[30]. For analysis by hybridization with radioactive DNA probes, series of RNA containing dots were formed on strips of Gene Screen Hybridization Transfer membrane (New England Nuclear). Each 5-mm diameter dot was formed by applying 5 μg RNA in 10 μl to the Gene Screen Membrane presoaked in 2XSSC (1XSSC = 0.15 mol/L NaCl, 0.015 mol/L Na-citrate) by gentle capillary action using a stack of four layers of wet 3 MM-Whatman paper and dry paper toweling (about 2 cm high) underneath the Gene Screen. Dot blots were air-dried and membranes were baked in a vacuum at 80℃ for 4 h.

Northern Blots

Total leaf RNA was glyoxylated in 1 mol/L glyoxal, 10 mmol/L Na-phosphate (pH 6.8) at 50℃ for 1 h and separated electrophoretically on 1.2% agarose gel using TBE (89 mmol/L Tris, 89 mmol/L boric acid, 2.5 mmol/L EDTA, pH 8.3) as running buffer[9]. The denatured RNA in 2XSSC was transferred to Gene Screen by blotting. These 'Northem' blots were dried and baked in the same manner as the Gene Screen dot blots.

Hybridization

Gene Screen sheets carrying RNA samples as dot blots and Northern blots were presoaked in hybridization buffer (50% [v/v] formamide, 3XSSC, 0.1% [w/v]SDS, 1X Denhardt's solution, 100 μg/ml calf thymus DNA) at 37℃ for 4 h, then hybridized with nick-translated ^{32}P-labeled plasmid DNA (10^6—10^7 cpm) in 10 ml of fresh hybridization buffer at 42℃ for 16 h[9]. After hybridization, the Gene Screen membranes were washed extensively in 2XSSC, 0.5% (w/v) SDS, at 65℃ for 1.5 h, then in 1XSSC, 0.1% SDS (w/v), at 65℃ for another 1.5 h. In experiments using heterologous plasmid probes for rcS and cab from pea and barley respectively, less formamide (10%) and lower temperatures were used for hybridization (37℃) and washing (42℃) for better hybridization.

The washed and dried dot and Northern blots were exposed to X-ray films at −70℃ using intensifying screens for autoradiography. To assess mRNA quantitatively in the dot hybridization, several exposures of the films were scanned with a Zeineh Soft Laser Scanning Densitometer with a peak area integrator. This allowed us to estimate the level of an individual mRNA relative to total RNA which is mainly rRNA. A linear relationship was found between the scanner readings and the RNA on the Gene Screen in amounts of 5 μg and below hybridized with ^{32}P-labeled DNA (about 10^6 cpm) for a strip of 10 RNA dots. Therefore, aliquots containing identical amounts of total RNA (5 μg), monitored both by spectrophotometry (A at 260 nm) and RNA gel electrophoresis (mainly cytoplasmic 25S, 18S rRNA, and plastid 16S, 23S rRNA) were spotted onto Gene Screen.

Results

Increase in Maize Plastid Photogene RNAs under Continuous or Intermittent Illumination

The time courses of increases in the levels of RNAs complementary to photoregulated plastid genes in response to continuous or brief intermittent illumination were determined in the following experiment conducted to establish conditions for illumination with monochromatic light. Eight dold dark-grown maize plants were exposed to continuous white light; a single 1-min white light pulse ('1'); or two 1-min white light pulses separated by a 3-h interval of darkness ('2'). Plants were harvested 3, 6, or 9 h after initiation of illumination (see Figure 1, bottom right). The changes in the level of transcripts of rcL, ps2B, ps1A1, ps1A2, and maize plastid Bam fragment 3 containing cflA and cfOlII were measured by dot hybridization and are shown as time-course curves as show in Figure 1. The results indicate that not only continuous white light but also a single white light pulse (1 min) is sufficient to bring about an increase in the level of RNAs complementary to various

Table 1 Plasmids Used for Probing Transcripts of Plastid and Nuclear Genes of Maize

Gene Location	Plasmid	Insert	Size (bp)	Gene (s) Encoded[a]		References
				[19]	[18]	
Plastid[b]	pZmc461	Pst-Pst (from Bam9)	560	*rcL*	*rbcL*	[19]
	pZmc427	BamHI-EcoRI (from Bam8)	2,000	*ps2B*	*psbA*	[7],[25]
	pZmc545	Bam2	9,800	ORFs353,473		[c][19]
	pZmc527	Bam3	8,700	*cf1A* *cf0III*	*atpA* *atpH*	[19],[25]
	pZmc415	Bam24	1,450	*cf1A*	*atpA*	[19],[25]
	pZR4876	Sal (from Eco e)	2,700	*cf1BE*	*atpBE*	[19]
	pZmc556	Bam17	2,500	*ps1A1*		[13],[19],[25]
	pZmc404	Bam21′	1,670	*ps1A2*		[13],[19],[25]
	pZmc569	Bam20	1,700	UROF443		[d][19]
	pZmc518	Bam13,17′	3,100,2,500	*r16*	*16S rDNA*	[19]
Nucleus	pSSU160	HindIII (pea)	160	*rcS*		[4]
	pAB96	Pst (barley)	850	*cab*		[16]

ⓐ Two parallel nomenclatures given are presented in references [18] and [19]. ⓑ Plastid probes were cloned maize plastid DNA sequences as designated by Larrinua *et al.* [19] and listed in the "Inserts" column. ⓒ Krebbers E T, Muskavitch K M T, Bogorad L, unpublished data. ⓓ Schantz R, Orr E A, Bogorad L, unpublished data.

photoregulated genes that is detectable after 3 h of dark incubation following the light pulse. However, the levels of complementary RNAs under continuous light increased more than under light pulses, and two light pulses separated by 3 h induced higher levels of complementary RNAs than a single light pulse. Brief illumination with R resulted in similar kinetics for the appearance of RNAs (data not shown). A program of briefillumination, 3 h of darkness, brief illumination, and harvest after 3 h of additional darkness was used in subsequent experiments to study the involvement of phytochrome. This experimental sequence resulted in changes in RNA pool sizes that were adequate to measure while it reduced the probability of secondary effects that might occur when using continuous illumination or repeated brief illumination over periods of days rather than hours.

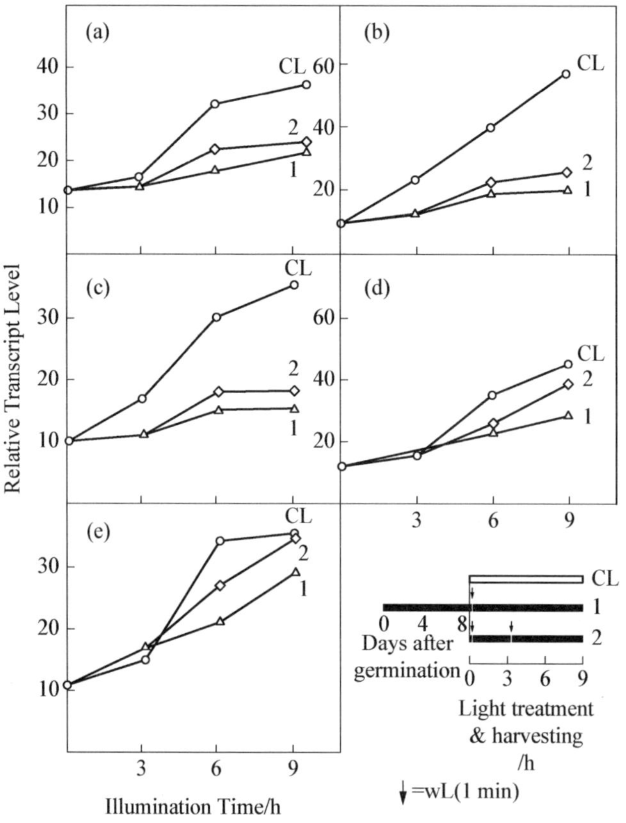

Figure 1 Levels of maize photogene transcripts after illumination of etiolated seedlings with pulses of white light. Eight d-old dark-grown plants were treated with (line CL) continuous white light; one (line 1) or two (line 2) 1-min pulses of white light (WL) at 0 and 3 h as marked by arrows. The treatment program is shown in the diagram at the lower right-hand corner. Total RNA was isolated at 0, 3, 6, 9, h and hybridized with plasmid probes for rcL (a), ps2B (b), cf1A, and cfOIII included in plastid chromosome fragments Bam 3 (c), plA1 (d), and ps1A2 (e). The transcript levels were measured by dot hybridization and expressed as scanning units. The numbers are the averages of three measurements for a given RNA. The variations in values were from 4 to 15% ofthe average

Phytochrome Control of the Levels of Plastid and Nuclear Photogene RNAs

The involvement of phytochrome in controlling photogene expression was tested by briefly exposing etiolated maize plants to R (1 min) or to FR (5 min) or to R (1 min) followed by R (5 min) and measuring the relative levels of complementary RNA [see Figure 2(a)]. Morphological changes in plants exposed to different light treatments were checked roughly to verify that the illumination conditions used elicited classical phytochrome-mediated responses in etiolated maize seedlings. After all the plants were moved to continuous white light for 24 h directly from darkness or exposed to two R pulses each followed by 3 h darkness, those seedlings preilluminated with R pulses showed, more rapid Chl accumulation than seedlings not preilluminated with R pulses.

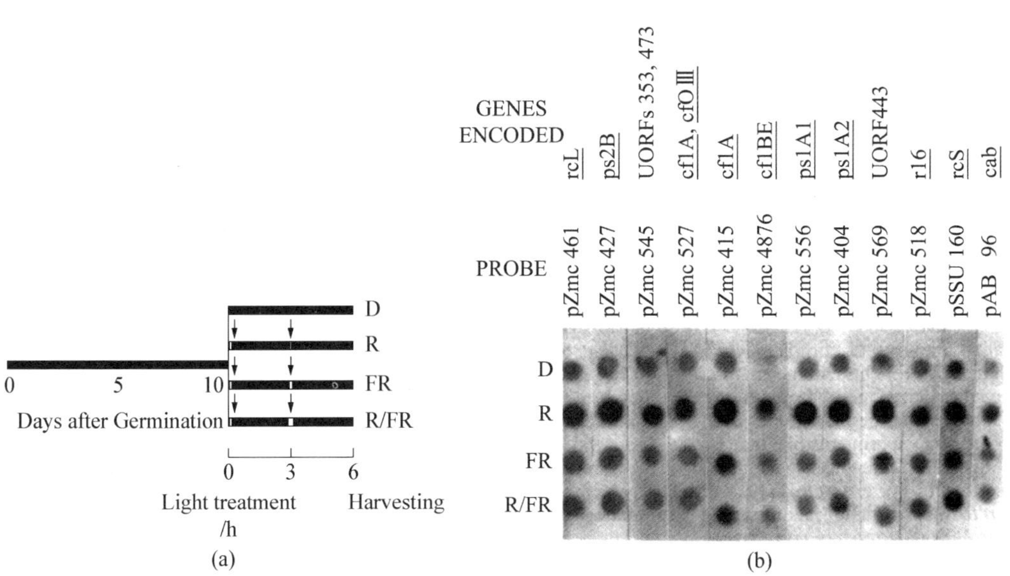

Figure 2 (a) Scheme of R and FR treatments. Maize plants were grown in the dark for 10 d, then illuminated with R (1 min), FR (5 min), or R/ FR (1 min/5 min) at 0 and 3 h, as indicated by the arrows. The leaves were harvested at 6 h. Dark bars represent periods in darkness. (b) Dot hybridization of maize RNA with cloned probes for plastid and nuclear genes. Total RNAs were extracted from dark-grown leaves or leaves illuminated with R, FR, or R/FR, as shown in (a). Equal amounts of RNA (5 μg) were denatured and spotted onto Gene Screen, hybridized with ^{32}P-labeled probes, and exposed to X-ray film, as described in "Materials and Methods." The scanning densitometer values of films from similar expenments are shown in Table 2.

Dot Blots

Figure 2(b) shows dot hybridization analyses of maize leaf mRNA from different light treatments using ^{32}P-labeled cloned probes. For each previously identified photogene, the relative amounts of complementary RNAs were dramatically enhanced by R. This enhancement effect was largely reversed by immediate subsequent illumination with FR.

Table 2 lists the magnitudes of the light-induced changes in RNAs from experiments like that for which dot blots as shown in Figure 2(b). Differences of about 10% to 20%, were seen between experiments. For example, the amount of RNA complementary to rcL doubled after R illumination compared to the dark control. This is roughly comparable to the 2.4-fold increase in RNA complementary to rcS. R-induced increases in RNA pools complementary to photogenes ranged up to about 3.5 to 3.8-fold. Exposure to FR alone generally had no effect; the increases of about 20% (UORFs 353, 473, and rcS) after illumination with FR are probably not significant. Illumination with FR for 5 min immediately following each 1-min exposure to R blocked the effect of R almost completely except for ps2B, cf1BE, rcS, and cab; in these cases, reversal was incomplete. The levels of maize plastid rRNAs exhibit a transitory increase of approximately 10% to 20% after about 3 h of continuous illumination of dark-grown seedlings (Davey J E, Bogorad L, unpublished data); the slight apparent increase in transcripts of r16 (the gene for 16S rRNA) here tends to confirm the earlier observation.

Table 2 Effects of R and FR Light on the Abundance of Transcripts of Light-Induced Plastid and Nuclear Genes of Maize

Plasmids	Probes for	mRNA Levels (Scanner Units, %)[a]			
		D[b]	R	FR	R/FR
pZmc461	*rcL*	100	205	100	105
pZmc427	*ps2B*	100	280	105	135
pZmc545	ORF 353,473	100	325	125	115
pZmc527	*cf1A*, cf0III	100	320	90	110
pZmc415	*cflA*	100	370	100	110
pIR4876	*cf1B*, E	100	300	130	170
pZmc556	*ps1A1*	100	340	90	120
pZmc404	*ps1A2*	100	345	100	120
pZmc 569	UORF 443	100	340	95	90
pZmc 518	*r16*	100	140	110	110
pSSu 160	*rcS*	100	240	125	190
pAB 96	*cab*	100	270	130	190

ⓐEach value is the average of duplicate or triplicate measurements of each RNA in different hybridizations in two or three experiments with dark control as 100%. The variation is from 5 to 20%. ⓑD, dark; R, red light (1 min); FR, far-red light (5 min); R/FR, red light (1 min) followed by far-red light (5 min). For detailed scheme of light treatment [see Figure 2(a)].

Northern Blots

Northern blot analyses were carried out for mRNAs of some photogenes; these confirmed the results of the dot blot assays. Equal amounts of total leaf RNA were separated electrophoretically by size on agarose gels. The denatured RNA transcripts were transferred to Gene Screen and hybridized with ^{32}P-labeled nick translated DNA probes for transcripts of rcL, ps2B, r16, ps1A1, and ps1A2. The results of autoradiography as shown in Figure 3. The major transcripts ofrcL (1. 6 plus 1. 8 kb) and ps2B (1. 2 kb) increased remarkably upon R irradiation. This increase was reversed if FR treatment followed exposure to R. The same is true for ps1A1 and ps1A2 whose major transcripts are 3. 3 kb, 1. 7 kb and 3. 7 kb, 2. 2 kb, 1. 75 kb, respectively. However, r16 showed a relatively stable transcript (1. 54 kb) level after illumination with R.

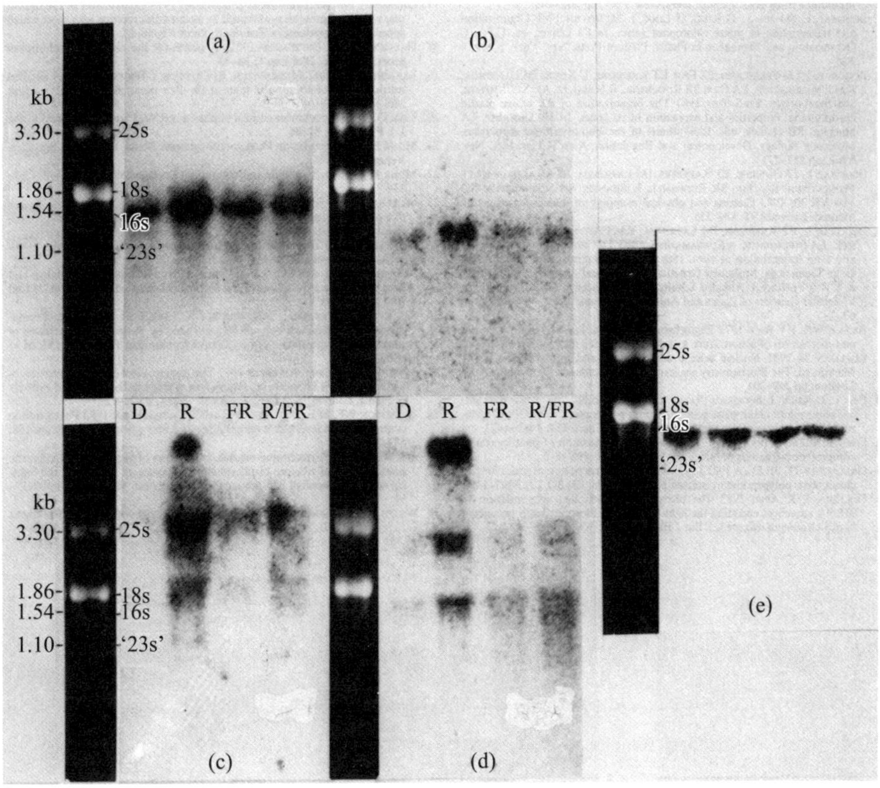

Figure 3 **Northem hybridization of maize RNA with cloned labeled probes for plastid genes. Equal aliquots (15 μg) of glyoxylated RNA from dark-grown (d) maize leaves or from leaves of plants subjected to treatment with R, FR, or R/FR were fractionated electrophoretically on 1. 2% agarose gels, transferred to Gene Screen, and hybridized with ^{32}P-labeled probes (see Table 1) for rcL (a), ps2B (b), ps1A1(c), ps1A2(d), r16(e). Four treatments (DR, FR, R/FR) have the same patterns and same intensities of rRNA onthe gels. The ethidium bromide-stained gels are shown in the left-hand panels corresponding to each probe. The locations of cytoplasmic 25S and 18S rRNAs, plastid 16S rRNA, and '23S' rRNA breakdown products which are used as mol wt markers are indicated on the gels. The sizes of RNAs are indicated by numerals (kb)**

Discussion

The 139-kbp maize plastid chromosome is comprised of two large inverted repeats of 22 kbp each plus two unique regions. Thus, the chromosome contains 117 kbp of unique sequences[19]. Photogenes occupy about 19% of the unique sequences and are distributed among at least six different transcription units (Muskavitch K M T, Krebbers E T, Bogorad L, unpublished data)[25]. These transcription units could not be distinguished by the kinetics of increase in complementary RNAs upon illumination of dark-grown plants but showed three distinctive patterns of change on prolonged continuous illumination[25]. In this investigation, we have focused on the question of whether or not phytochrome is involved in regulating the initial increases of RNA levels for these maize photogenes upon illumination of dark-grown seedlings.

Many photoregulated developmental phenomena in plants are mediated by phytochrome. Although it is not clear yet exactly how and at which level it acts, Mohr[21] proposed that phytochrome might regulate "differential gene activation and gene repression." This hypothesis has been supported by the observations that the levels of RNA transcripts of rcS, cab (in barley, Lemna, pea and bean), ps2B (in mustard, pea and bean), and rcL (in pea and bean) are affected by phytochrome[2,16,20,26,28,29]. We have used somewhat different illumination procedures and come to the same conclusions for etiolated maize seedlings with regard to these as well as other genes.

We have shown here, by dot blot and Northern hybridization procedures, that increases in the levels of transcripts complementary to maize photogenes, ps2B, rcL, pslA1, cflA, cfOIII, and cf1BE as well as the photogenes UORF 443, ORF 353, and ORF 473 observed after illumination of dark-grown seedlings (Muskavitch K M T, Krebbers E T, Bogorad L, unpublished data) are regulated via phytochrome. The increases are observed after brief illumination with R but the effect of illumination by R is completely or largely reversed by FR, as is characteristic of phytochrome-mediated photomorphogenesis. However, the possibility of involvementof photoreceptors other than phytochrome cannot be ruled out; the synthesis of some plant proteins is responsive to blue light[24]. Rodermel and Bogorad[25] have shown that, in continuously illuminated dark-grown seedlings, the levels of transcripts of several maize photogenes drop after reaching maxima after 10 to 20 h in the light. The results described in this paper do not address the mechanism responsible for these drops.

Many developmental and metabolic changes are triggered by illumination of etiolated seedlings. For example, some limited photosynthetic activity begins within a few minutes after commencement of illumination. The optimal experimental arrangement is to induce a change large enough to measure with confidence in the parameter under study while eliciting no, or only the smallest possible, changes in other, possibly interacting, systems. We have chosen experimental conditions close to the minimum needed to elicit clearly measurable changes in RNA pool sizes in order to minimize possible effects of secondary changes on RNA metabolism. Earlier work showed that illumination of etiolated maize seedlings with white light for 1 min initiated plastid maturation effects that were completed over a period of about 3 h in darkness, and could be reinforced by an additional brief exposure to light[14].

In the present experiments, we have measured only RNA pool sizes because of technical limitations that do not permit us to determine rates of transcription in vivo. The size of each pool reflects the sum of synthesis and destruction. It seems unlikely that changes in pool sizes observed here result from constant rates of synthesis of photogene RNAs coupled with photoinduced changes in RNA lifetimes without postulating rises and falls in the activities of a number of ribonucleases that act on specific transcnpts. First, RNA pools for most maize plastid genes do not change significantly during light-induced development of etioplasts to chloroplasts[25]. Second, transcripts of at least one region of the maize plastid chromosome decrease, rather than increase, in amount after illumination (Lukens J, Bogorad L, unpublished data). Thus, it seems most reasonable to conclude that phytochrome, acting somewhere in the cell, regulates one or more events leading to increases in the rates of transcription of maize plastid photogenes. This suggestion is also supported by Gallagher and Ellis's finding, using isolated nuclei from dark and light-grown pea leaves, that light increases the transcription of the genes cab and rcs[15].

Acknowledgmems

We thank S. R. Rodermel for his technical help, and K. M. T. Muskavitch, L. Fish, and E. A. Orr for providing plasmids pZmc 545, pZmc 404, and pZR 4876, respectively. We also thank J. Bedbrook and K. Apel for providing plasmids pSSu 160 and pAB 96, respectively.

Note Added in Proof—UORF 353 has been shown to code for the D-2 protein of PS II and UORF 473 to be the gene for the 44 kDa Ch1 protein of the PS

II reaction center (E. T. Krebbers, K. M. T. Muskavitch, M. Castroveijo, H. Roy, D. R. Russell, L. Bogorad, unpublished data).

Literature cited

[1] AKOYUNOGLOU G, JH ARGYROUDI-AKOYUNOGLOU 1978 Chloroplast Development. Elsevier/North Holland Biomedical Press, Amsterdam

[2] APEL K 1979 Phytochrome-induced appearance of mRNA activity for the apoprotein of the light-harvesting chlorophyll a/b protein of barley (*Hordeum vulgare L.*). Eur J Biochem 97:183 - 185

[3] BEDBROOK JR, G LINK, DM COEN, L BOGORAD, A RICH 1978 Maize plastid gene expressed during photoregulated development. Proc Natl Acad Sci USA 75:3060 - 3064

[4] BEDBROOK JR, S SMITH, RJ ELLIS 1980 Molecular cloning and sequenang ofcDNA encoding the precursor to the small subunit of chloroplast ribulose 1,5-bisphosphate carboxylase. Nature (Lond.) 287:692 - 697

[5] BOGORAD L 1967 The role of cytoplasmic units. Control mechanisms in plastid development. Dev Biol Suppl 1:1 - 31

[6] BOGORAD L 1975 Eukaryotic intracellular relationships. In A Tzagoloff, ed, Membrane Biogenesis. Plenum Press, New York, pp 201 - 246

[7] BOGORAD L, SO JOLLY, G KIDD, G LINK, L MCINTOSH 1980 Organization and transcription of maize chloroplast genes. In CJ Leaver, ed, Genomè Organization and Expression in Plants. Plenum Press, New York, pp 291 - 304

[8] BOGORAD L, LD CROSSLAND, LE FISH, ET KREBBERS, U KUCK, IM LARRINUA, KMT MUSKAVITCH, EA ORR, SR RODERMEL, R SCHANTZ, AA STEINMENTZ, SM STIRDIVANT, YU-S ZHU 1983 The organization of the maize plastidchromosome. Properties and expression of its genes. In JP Thornber, LAStaehlin, RB Hallick, eds, Biosynthesis of the Photosynthetic Apparatus, Molecular Biology, Development and Regulation. Alan R Liss, Inc. , New York, pp 257 - 272

[9] BOGORAD L, EJ GUBBINS, ET KREBBERS, IM LARRINUA, BJ MULLIGAN KMT MUSKAVITCH, EA ORR, SR RODERMEL, R SHANTZ, AA STEINMENTZ, GD Vos YK YE 1983 Cloning and physical mapping of maize plastid genes. Methods Enzymol 97:524 - 555

[10] BOGORAD L, ET KREBBERS, IM LARRINUA, KMT MUSKAVITCH, SR RODER-MEL, AA STEINMENTZ, A SUBRAMANIAN 1983 The struaure of maize genes and their transcription in vitro. 15th Miami Symposium on Advances in Gene Technology: Molecular Genetics of Plants and Animals. In K Dowdney, RW Voellmy, F Amad, J Schultz, eds, Advances in Gene Technology: Molecular Genetics of Plants and Animals. Academic Press, New York, pp 63 - 79

[11] BRIGGS WR, HV RICE 1972 Phytochrome: chemical and physical properties and mechanism ofaction. Annu Rev Plant Physiol 23:293 - 436

[12] EDELMAN M 1981 Nucleic acids of chloroplasts and mitochondria. In A Marcus, ed, The Biochemistry of Plants, Vol 6. Academic Press, New York, London, pp 249 - 301

[13] FISH L, U KUCK, L BOGORAD 1985 Two partially homologous adjacent lightinducible maize chloroplast genes encoding polypeptides of the P700 chlorophyll a-protein complex photosystem I. J Biol Chem 260:1413 - 1421

[14] FORGER III JM, L BOGORAD 1973 Steps in the acquisition of photosynthetic competence by plastids of maize. Plant Physiol 52:491 - 497

[15] GALLAGHER TF, JR ELLIS 1982 Light-stimulated transcription of genes for two chloroplasts polypeptides in isolated pea leaf nuclei. EMBO J 1:1493 - 1498

[16] GOLLMER I, K APEL 1983 The phytochrome-controlled accumulation of mRNA sequences encoding the light-harvesting chlorophyll a/b protein of barley (*Hordeum vulgare L*). Eur J Biochem 133:309 - 313

[17] GREBANIER AE, KE STEINBACK, L BOGORAD 1979 Comparison of the molecular weights of proteins synthesized by isolated chloroplasts with those which appear during greening in Zea mays. Plant Physiol 63:436 - 439

[18] HALLICK RB, W BOTTOMLEY 1983 Proposals for the naming of chloroplast genes. Plant Mol Biol Rep 1:38 - 43

[19] LARRINUA IM, KMT MUSKAVITCH, EJ GUBBINS, L BOGORAD 1983 A detailed restriction endonuclease site map of the Zea mays plastid genome. Plant Mol Biol 2:129 - 140

[20] LINK G 1982 Phytochrome control of plastid mRNA in mustard (Sinapis alba L.). Planta 154:81 - 86

[21] MOHR H 1972 Lectures on Photomorphogenesis. Berlin, New York, Springer-Verlag, pp 48 - 59

[22] MOHR H 1977 Phytochrome and chloroplast development. Endeavor 1:107 - 114

[23] MULLIGAN B, N SCHULTE S, L CHEN, L BOGORAD 1984 Nucleotide sequence of a multiple copy gene for the B protein of photosystem II of a cyanobacterium. Proc Natl Acad Sci USA 81:2693 - 2697

[24] RICHTER G 1984 Blue light control of the level of two plastid mRNAs in cultured plant cells. Plant MoL Biol 3:271 - 276

[25] RODERMEL SR, L BOGORAD 1985 Maize plastid photogenes: mapping and photoregulation of transcript levels during light-induced development. J Cell Biol 100:463 - 467
[26] SASAKI Y, T SAKIHAMA, T KAMIKUBO, K SHINOZAKI 1983 Phytochromemediated regulation of two mRNAs encoded by nuclei and chloroplasts of ribulose-1,5-bisphosphate carboxylase/oxygenase. Eur J Biochem 133:617 - 620
[27] SMITH H, EE BILLETT, AB GILES 1977 The photocontrol of gene expression in higher plants. In H Smith, ed, Regulation of Enzyme Synthesis and Activity in Plants. London, New York, Academic Press, pp 93 - 127
[28] THOMPSON WF, M EVERETT, NO POLANS, RA JORGENSEN 1983 Phytochrome control of RNA levels in developing pea and mung-bean leaves. Planta 158:487 - 500
[29] TOBIN EM 1981 Phytochrome-mediated regulation of messenger RNAs for the small subunit of ribulose bisphosphate carboxylase of maize and the lightharvesting chlorophyll a/b protein in Lemna gibba. Plant MoL Biol 1:35 - 51
[30] WHITE BA, FC BANCRoFT 1982 Cytoplasmic dot hybridization. J Biol Chem 257:8569 - 8572

The Organization of the Maize Plastid Chromosome: Properties and Expression of Its Genes*

The maize chloroplast chromosome, in common with many other chloroplast DNAs, is a circle comprised of two large inverted repeated segments separated by two stretches of unique sequences. A number of genes have been mapped on the chromosome and have been sequenced to reveal features of the chromosome and of plastid genes. These genes have a mixture of prokaryotic and eukaryotic features.

Two maize plastid developmental programs in which the abundance of mRNAs vary are discussed. In differentiation of bundle sheath and mesophyll cells, the gene for the large subunit of ribulose bisphosphate carboxylase is transcribed little if at all in mesophyll cells but abundantly in bundle sheath cells. In the development of chloroplasts from etioplasts in response to illumination, extensive analyses have shown that five regions of the chromosome totaling about 19% of the total unique sequences are represented by increased amounts of RNA starting within 2 h after dark-grown seedlings are illuminated and continuing for 10—20 h. For three of the five photogene-containing regions, very large transcripts are seen together with a number of classes of smaller transcripts. The relationships between the larger and smaller transcripts is not known. In some cases, eg. the gene for large subunit of

---

* Lawrence Bogorad, Lyle D. Crossland, Leonard E. Fish, Enno T. Krebbers, Ulrich Kück, lgnacio M. Larrinua, Karen M. T. Muskavitch, Elizabeth A. Orr, Steven R. Rodermel, Rudi Schantz, Andre A. Steinmetz, Steven M. Stirdivant, and Yu-Sheng Zhu
Department of Cellular and Developmental Biology, Harvard University, Cambridge, Massachusetts 02138

此文发表于 1984 年"ICN-UCLA Symposium". "Biosynthesis of the Photosynthetic Apparatus: Molecular Biology, Development and Regulation" Alan R. Liss, Ind., N. Y., 14: 257 - 272

Authors' present addresses: Enno T. Krebbcr's present address is Max-Planck-Institur für Züchtungsforschung. Abtcilung Saedler. D-5000 Koln 30. Federal Republic of Germany; Ulrich Kück's present address is lehrstuhl für Allgemeine Botanik. Rhur-Universirat Bochum. postfach 10 21 48. D-4630 Bochum, Federal Republic of Germany; Ignacio M. Larrinua's present address is Lilly Research Laboratories MC625. 307 East McCarty Street, Indianapolis. Indiana 46285; Karen M. T. Muskavitch's present address is Department of Biology. Indiana University. Jordan Hall. Bloomiington, Indiana 47405; Rudi Schantz's present address js Universite Louis Pasteur. Laboratoire Dc Physiologie Vegegale 28. Rue Goethc. 67083. Strasbourg CEDEX. Francc; Andre A. Stcinmetz's present address is IBMC. 15 rue Renc Descartes, F-67084. Strasbourg. CEDEX. France; Yu-Shung. Zhu's present address is Department of Microbiology. University of Illinois at Urbana-Champaign, 131 Burrill Hall. 407 South Goodwin Avenuc. Urbana. Illinois 61801.

Rcccivcd Scptcmbcr 16. 1983.

Key words: maize plastids, transcription, Shine-Dalgarno sequences, promoters ribulose bisphosphate carboxylase, the distribution of transcript sizes differs at various stages of greening.

One approach to analyzing mechanisms of regulation of transcription has been the study of the action of the maize plastid DNA-dependent RNA polymerase in vitro. This system preferentially transcribes certain chloroplast genes over others from supercoiled chimeric plasmids comprised of maize plastid DNA sequences cloned into bacterial plasmid vehicles, and promises to be useful for studying DNA sequences that may regulate expression as well as polypeptides that may interact with templates or the polymerase to alter the array of transcripts in the plastid during stages in a developmental program.

It has been known since early this century that greening of plants, ie, developments of plastids, is affected by some genes transmitted in a Mendelian manner, but there are in addition uniparentally transmitted genetic factors that regulate greening. The latter genes were said to be in the "plastome", the former in the genome (ie, the nuclear genome). The bulk of this report deals with properties of the plastid genetic material, primarily of Zea mays, discovered during the past few years using recombinant DNA technology and the techniques of molecular biology.

In the early 1960s ways that three nuclear genes affect plastid development were determined. A mutation at the *yellow stripe* I (ysl) locus of maize results in, as the gene designation indicates, yellow-striped plants. The lesion is in the uptake of iron by roots[1]. The phenotype matches symptoms of moderate iron deficiency in maize. The effects of a recessive mutation in barley *xantha* 23 can be overcome by administration of leucine to the seedling, and barley mutant *albina* 7 becomes phenotypically normal when aspartic acid is administered[2]. These few examples establish the principle that chloroplast development depends upon the adequacy of the culture medium, ie, the composition of the cytoplasm in which it grows, and that starvation retards plastid development-mutations that reduce the levels of plastid nutrients in the cell below acceptable limits, and block normal plastid development. It is useful to remember that the plastids of barley *xantha* 23, barley *albina* 7 and ysl, are different from one another and yet all are capable of developing to maturity when the plant is supplied with the appropriate material. The differences in appearance of the plastids suggests that starvation for one nutrient does not have the same effect as starvation for another and that the production of certain plastid elements may be affected differently from others by the absence of a particular essential nutrient.

Combinations of transmission genetics and analyses of the gene products have been used to locate genes for chloroplast ribosomal proteins in Chlamydomonas and of the genes for the targe and small subunits of RUBISCO (ribulose bisphosphate carboxylase). The results revealed an unexpected intimacy between nuclear and chloroplast gene products in multimeric components of the plastid. Some mutations to erythromycin-resistance in Chlamydomonas reinhardi were transmitted in a Mendalian manner and judged to be in nuclear genes; one was transmitted uniparentally as expected for an organefle gene[3,4]. The nuclear mutation was shown to occur at the locus ery M-I of nuclear linkage group XI and this locus was shown to code for protein 6 of the large subunit of the chloroplast ribosome[5,6]. Related data showed that some chloroplast ribosomal proteins of Chlamydomonas were products of the nuclear genome and others of the plastid genome. Wildman and co-workers[7,8] found, as a result of crosses between various species of Nicotiana and analyses of the subunits of RUBISCO, that the gene for the small subunit was transmitted according to the rules of Mendel and thus was judged to be a nuclear gene, while the gene for the large subunit was transmitted uniparentally and thus was judged to be a plastid gene. Realization that genes for polypeptides of multimeric plastid components are dispersed among the separate compartmentalized genomes of the eukaryotic cell called for serious consideration and analyses of possible mechanisms for distribution of genes among multiple genomes during the evolution of eukaryotic cells[9-11] and for the study of mechanisms for integration of the expression of dispersed genes.

One long-term goal of the study of the molecular developmental biology of plastids is to understand mechanisms for the integration of expression of plastid and nuclear genes for subunits of multimeric plastid components. Another goal of this work is to learn details of the structures of proteins of the photosynthetic apparatus and their functional relationships to one another. To proceed toward either of these goals. it is necessary to understand the structures of plastid genes and how their expression is regulated as well as the structures of nuclear genes for plastid components and how their expression is regulated.

The opportunity to study the molecular developmental biology of plastids followed from the firm identification of DNA and ribosomes in plastids in the early 1960s and the emergence of recombinant DNA technology in the mid-1970s. Physical mapping and cloning of chloroplast genes could be carried out with any eukaryotic green plantand it has been done with many by now. The particular case of Zea mays. with which the present report is mainly concerned, serves as an example. As outlined below, genes for stable and abundant RNAs such as rRNAs and tRNAs that can be purified relatively easily have been mapped by various molecular hybridization techniques. On the other hand, a major approach employed to locate genes coding for proteins has been to express a segment of cloned plastid DNA (cloned in E coli) in vitro using a transcription-translation system and identifying the protein immunologically or by other means[12,13]. By now enough plastid genes have been mapped and sequenced to reveal a number of features of plastid chromosome and gene strucrure as well as to provide well-characterized templates for studying the regulation of transcription in vivo and in vitro.

The Organization of the Plastid Chromosome

The first restriction maps of plastid DNAs showed the chloroplast chromosomes of maize[14,15] and of spinach[16] to be made up of two large inverted repeated segments separated by two regions of unique sequences. A number of plastid chloroplast chromosomes have been analyzed since then by electron microscopy[17] as well as by restriction endonuclease restriction site mapping. To date only those of pea and broad bean have been found to lack a pair of inverted repeated segments[18,19] and the chloroplast chromosome of Euglena gracilis is unique among those studied to date in containing three comparatively small repeated segments plus part of a repeaccd sequence that occurs still another time[20-22].

The inverted repeated segments in maize were shown to contain copies of genes for 16S, 23S, and 5S rRNAs[14,15]. These genes were mapped in detail by Southern[23] hybridization, heteroduplex mapping, and by R-Looping[24]. Besides establishing that the inverted repeats are 22 kbp (kilobase pairs) in length and showing the order of these genes on the chromosome to be 16,23, and 5S rRNA, it was determined that the genes for the two larger RNAs are separated by an intervening sequence of about 2,100 bp. This intervening region was subsequently shown to contain genes for alanine and isoleucine tRNAs[25]. The large inverted repeats of maize also contain a number of tRNAs that lie outside of the rRNA coding regions[26-28] as well as at least one UORF (unidentified open reading frame).

The large repeated sequences in other plastid DNAs and the smaller tandemly repeated DNA sequences in Euglena have now been shown to also contain rRNA genes although these genes have generally not been mapped as precisely in the other higher plant chromosomes as they have been in maize, Chlamydomonas or Euglena (In maize, besides the mapping described above, the regions containing the 16 and 23S rDNAs plus the intervening sequence have been sequenced by Kossel and coworkers[25,29,30].).

Genes for plastid tRNAs have been assigned to restriction fragments of spinach[31] and maize[28]. Although fewer tRNA genes have been identified than are expected to be present (from consideration of codon usage and comparisons with other cases), it seems likely that plastid chromosomes contain genes for a complete set. It is interesting that the distribution of tRNA genes between the inverted repeated and the unique sequence regions is different in maize and spinach chloroplast chromosomes. The $tRNA^{His}$ gene of maize is near a border inside each of the inverted repeats[26,28]; in the spinach chromosome there is one copy of this gene and it is in the large unique region, ie, outside of the inverted repeats, but near the edge of one of its inverted repeats[31]. A gene for leucine tRNA is in the unique region of maize but is in the inverted repeats of spinach. In a formal sense, rearrangements of this sort can be imagined to occur, at least in part, by movements of the tRNA genes into or out of the inverted repeats or by migration of the ends of the inverted repeats. It seems unlikely that the ends of both inverted repeats would migrate to precisely corresponding positions on the chromosome or that tRNA genes would simultaneously insert at exactly the same points in both of the inverted repeats; thus one must imagine that each event occurred one time (for each change) and was then copied into the second repeat or one of the repeats was lost. Any DNA sequence included between the borders of an inverted repeat is repeated. The genetic mechanism for repetition is not known but it seems that plastids can survive without it, as is seen in peas and broad beans. The rearrangement of large segments of plastid chromosome in relation to phylogeny has been addressed[19]. The relative locations of tRNA genes may provide useful information for phylogenetic analyses at different levels.

Features of Plastid Genes

Maize plastid genes for the large subunit of ribulose bisphosphate carboxylase (rc-L), the beta and epsilon subunits of the coupling factor CF_i (cfl-BE), the ribosonial protein S4 (rp-S4), subunit III of CF_0 (cfo-III), the alpha subunit of CF_i (cfl-A), photogene 32[32] or ps2B, ie, the gene for the the B protein of photosystem 2, plus seven UORFs have been sequenced to date [33 - 35; S. R. Rodermel and L. Bogorad. unpublished; U. Kück. L. Fish. and L. Bogorad, unpublished; E. T. Krebbers, K. M. T. Muskavitch. and L. Bogorad. unpublished; A. Steinmetz and L. Bogorad. unpublished; L. McIntosh and L. Bogorad. unpublished.] In all, DNA sequences have been determined for 11 polypeptides containing an aggregate of more than 5,000 amino acids. In addition, genes for 12 tRNAs have been sequenced [25, 27, 36, 37; E. T. Krebbers, A. Steinmetz, and L. Bogorad, unpublished].

Conserved sequences upstream of transcription start sites for a number of plastid genes from maize and other plants resemble prokaryotic promoters. A consensus sequence is[37]:

$$\overset{\text{“}-35\text{”}}{{}^{\mathrm{A}}_{\mathrm{g}}\mathrm{TT}^{\mathrm{G}}_{\mathrm{c}}{}^{\mathrm{A}}_{\mathrm{c}}\mathrm{N}^{\mathrm{a}}_{\mathrm{t}}}\cdots\cdots 15\text{–}20\ \text{nucleotides}\cdots\cdots \overset{\text{“}-10\text{”}}{\mathrm{T}^{\mathrm{AAG}}_{\mathrm{tta}}}\mathrm{AT}.$$

Variations on this consensus sequence are considerable and nested pairs of such conserved sequences occur upstream of the coding regions of a few genes. At the present time, however, there is no direct experimental evidence that these conserved sequences serve as recognition sites for plastid DNA-dependent RNA polymerase.

Another feature observed in this first plastid gene to be sequenced-maize rc-L[33] is the presence of a short nucleotide sequence complementary to nucleotides at the 3′ end of maize 16S rRNA[29]. This sequence is analogous to the Shine-Dalgarno sequence of E. coli messenger RNAs; it occurs a few nucleotides upstream of the translation start site. Such sequences (given as they would appear in the transcription product of the gene) that have been seen in plastid genes of maize and other plants are shown in Table 1. These sequences vary in the number of nucleotides complementary to 16S rRNA as well as in the distance between the complementary sequence and the 5′ end of the initiating methionine translation codon. There is no homologous plastid translation system available at present to determine whether these differences significantly influence the relative translation frequency of different mRNAs. There is one case in which such a sequence has not been recognized at the expected location [41].

Table 1 Putative 16S rRNA Binding Sites of Chloroplast Genes

Gcnc	Organism	Sequence		Ref
16S RNA	Z mays	3′-UUUCCUCCACUAGG		
rpS4	Z mays	AAAAAUAAA	AAAGGAG	UCUUC AUG [35]
rpS19	N tab	AAAAAAAU	AGGAG	UAAGCUU GUG [38]
rcL	Z mays	UGAGUUGUAG	GGAGG	GACUU AUG [33]
rcL	N tab	UGAGUUGUAG	GGAGG	GAUUU AUG [39]
rcL	S oler.	GAAGUUGUAG	GGAGG	GACUU AUG [40]
cfIB	Z mays	UUGAAAAUUAU	GUGAU	AAUU AUG [34]
cfIE	Z mays	GAGGAGAGCAAAUUG	AAG	AA AUG [34]
cflE	S oler	AUGGAGAGCAAAUU-	AAAG	AA AUG [41]
cf0III	Wheat	UGACAC	GAGG	AACUCACACC AUG [42]
cf0III	Z mays	ACAC	GAGG	AACUCAUCAUAA AUG [a]
cfIA	N tab.	AAAAUAAUAAG	AAAG	AAUCC AUG [43]
ps2B	S oler	AUAAACC	AAG	AUUUUUACC AUG [44]
ps2B	N deb.	AUAAACC	AAG	AUUUUUACC AUG [44]
1. 6kb	Z mays	AUGAAAA	AAAGGAG	UAAGCU GUG [26]
UORF	Z mays	AAUUGAAU	AGGAGG	AUCACU AUG [b]
UORF	Z mays	GGCAGCUC	AGGA	UCAGCCUC AUG [b]
UORF	Z mays	UCCGGAAAG	AGGAGG	ACUUA AUG [c]
UORF	Z mays	AGGAGGAUUUG	AAAGG	CAUU AUG [d]
UORF	N tab	AGUAGU	AGGAGG	CAAACCUU AUG [38]

[a] S. R. Rodermel and L. Bogorad [unpublished]; [b] E. T. Krebbcrs and L. Bogorad [unpublished]; [c] U. Kück. L. Fish, and L. Bogorad [unpublished]; [d] L;. Fish. U. Kück, and L. Bogorad [unpublished].

Unlike mitochondria, chloroplast genes use the universal code for amino acids. All of the codons possible in the universal code are employed in one protein or another that has been sequenced to date. The data in Table 2 show codon synonym usage distribution for some amino acids of three maize and one tobacco plastid genes[33-35,38]. In the bottom section of the table are the synonyms most highly preferred in the plastid genes but the bias is not the same in different plastid genes (eg, Tyr/UAU in rp-S4 and cfl-B vs rc-L). Differences between ribosomal protein genes rp-S4 and rp-S19 may be due to differences in the organisms from which the genes were sequenced, maize and tobacco, respectively.

Two kinds of overlaps of coding regions have been observed in the maize plastid chromosome. A gene for a 1. 6-kb transcript overlaps divergently the gene for $tRNA^{His}$[26]. Another overlap has been observed in the coding sequences for the beta and epsilon subunics of CF_1[34]. In the latter case, the last A of the triplet coding for lysine becomes, together with the U and G of the 3′ terminal codon of the mRNA for the beta subunit of CF_1, part of

Table 2 Synonym Usage in Chloroplast Genes*

Codon		rc-S4	rp-S19	rc-L	cf1-B
UUC	Phe	25	0	43	22
CUG	Leu	7	0	3	4
AUC	IIe	15	20	37	35
CCG	Pro	12	0	19	15
CAC	His	33	17	27	0
CAG	GIn	33	0	23	18
AAC	Asn	29	17	54	28
CGU	Arg	43	50	32	29
GAA	Glu	67	75	71	73
UUU	Phe	75	100	57	78
UAU	Tyr	100	100	53	92
CAU	His	100	83	73	100
GAU	Asp	86	100	79	62
AAA	Lys	83	90	69	50
Reference		[35]	[38]	[33]	[34]
Organism		Z. m.	N. t.	Z. m.	N. t.

* Percentage of use to code for the amino acid.

Z. m. =Zea mays; N. t. =Nicotiana tabaccum: rp-S4: gene for ribosomat protein S4 (35); rp-S19: gene for ribosomal protein; S19: rc-L: gene for the large subunit of ribulose bisphosphate carboxylasc; cfl-B: gene for the β subunit of CF_1.

the translation initiation codon for the epsilon subunit.

As described above, in the polycistronic message for the beta and epsilon subunits of CF_1 the coding regions overlap with a one nucleotide frameshift. There are other cases in the maize plastid chromosome in which two genes are transcribed in a single RNA molecule although the genes themselves are well separated from one another. One example is the single transcript containing messages for cfo-III and cf1-A [S. R. Rodermel and L. Bogorad, unpublished].

Maize chloroplast genes for alanine, isoleucine, leucine, and valine tRNAs (tA-UGG, tI-GAU, tL-UAA, and tV-UAC) contain introns [25,36; E. T. Krebbers, A. Steinmetz, and L. Bogorad, unpublished]. The introns range in size from 468 to 949 bp and, unlike nuclear genes for tRNAs in yeast, the introns are not always at the same position in the tRNA gene. For example, the intron in the tL splits the anticodon while in the other three genes the introns are within the anticodon loop but at different locations.

In summary, we have a fairly clear idea of consistent (but not necessarily universal) features of plastid genes but the number of cases studied is comparatively small and thus we may have a rather narrow view of the range of variation that occurs. These features are the presence of conserved sequences resembling prokaryotic promoters associated with each gene or each transcript; the presence of a Shine Dalgarno-like sequence upstream of the translation start site of most genes; and the use of the universal nucleotide code for amino acids. The more variable features include divergent overlaps of coding regions and frameshift overlaps in the same strand; polycistronic as well as monocistronic messages, and introns.

Molecular Developmental Biology of Plastids

Maize is a C4 plant. Thus, there are two distinctive developmental programs for the principal photosynthetic cells of its leaves. Mesophyll cells lack ribulose bisphosphate carboxylase but fix carbon dioxide with a different enzyme system. Bundle sheath cells fix CO_2 using ribulose bisphosphate carboxylase but these cells are relatively poor in many of the components of photosystem II. There are a number of other differences as well.

The genes for ribulose bisphophate carboxylase and the beta subunit of CF_1 lie adjacent to one another on the maize chloroplast chromosome but they are transcribed divergently[45] (see Figure 1). We have used cloned fragments from within these two genes to determine whether the levels of their transcripts are the same in bundle sheach and mesophyll cells. Mesophyll protoplasts of maize can be separated from strands of bundle sheath cells by enzymatic digestion of mesophyll cell walls, filtration, etc. RNA isolated from preparations of the two cell types was labeled and hybridized to gene sequences. Transcripts of cfl-B are abundant in plastids of both cell types but transcripts of rc-L are abundant only in bundle sheath cells. Barring strikingly different life times for these two mRNAs in the two cell types, these data indicate that the expression of these genes is regulated at transcription[46,47]. Preparations of mesophyll cells are estimated to be contaminated by about 3% of bundle sheath cells[46]. A small amount of rc-L mRNA is found in mesophyll cell preparations; this may be contamination or there may be a small amount of transcription of rc-L in mesophyll cells. The mechanisms for control of expression of these two genes in mesophyll vs bundle sheath cells are obviously of considerably interest.

Photoregulated maturation of plastids is another interesting developmental program. In general, angiosperm seedlings grown in darkness on seed reserves are pale yellow. These etiolated plants contain plastids (etioplasts) that are arrested at a characteristic stage of development. Etioplasts lack chlorphyll and many of the proteins and lipids required for photosynthetic function. Upon illumination, or soon afterward, chlorophyll accumulates together with new proteins and lipids; low levels of photosynthetic competence can be detected within a few hours after illumination of seedlings although complete maturation of the chloroplast takes considerably more time.

Chloroplast DNA has been hybridized with RNA from etioplasts or RNA from chloroplasts to determine what fraction of the chromosome is transcribed in illuminated

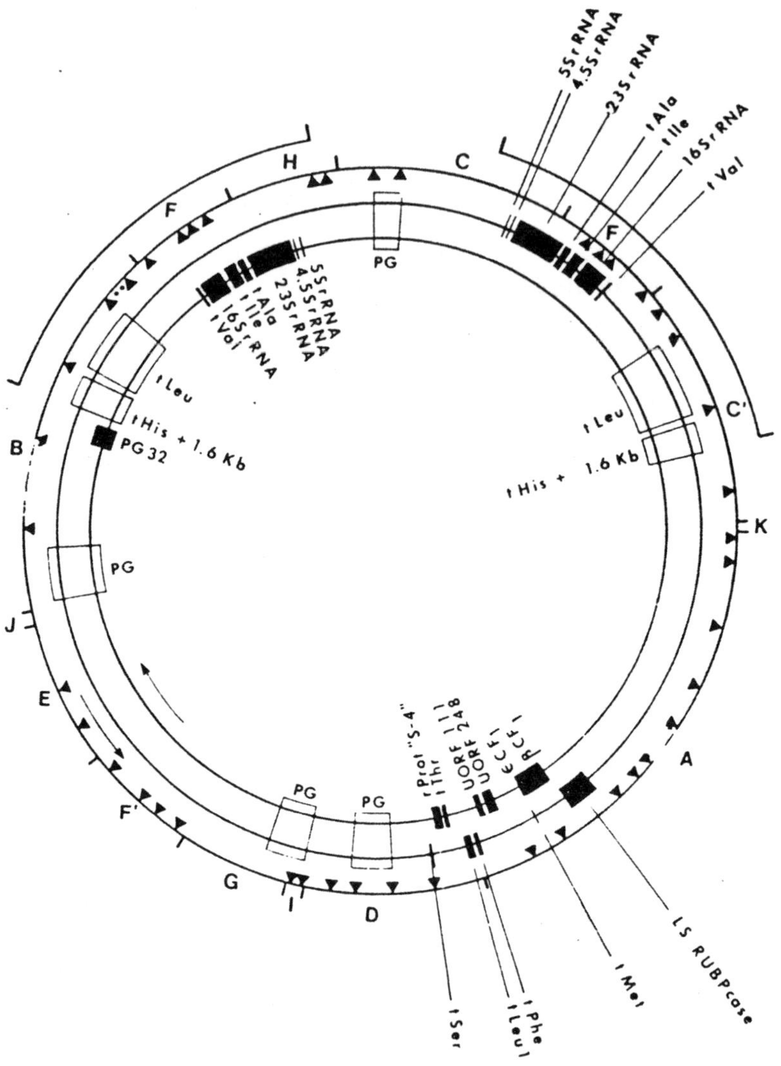

Figure 1 A map of the I40-kbp maize plastid chromosome. The two outermost arcs show the positions of the large inverted repeated sequences that contain ribosomal RNAs (r16, r23, r4.5, r5); the introncontaining tRNA genes for alanine and isoleucine are in the spacer between the genes for the 16 and 23S rRNAs. Other tRNA genes of the inverted repeats that have been sequenced are tV-GAC, tL-CAA, and tH-GUG. The rilled triangles inside of the outer circle of this diagram show the sites of recognition by BamHI. The lines on the outside of the outer circle indicare SalI recognition sites. Filled boxes in the inner circle show positions of genes that are transcribed in a clockwise manner and those on the middle full circle are transcribed in a counterclockwise direction. UORF designates unidentified open rceading frames. Segments of the chromosome besides PG32 represented by increased numbars by transcripts after dark-grown plants have been illuminated (ie. photogane regions) are shown by open boxes labeled "PG." In addition, pools of transcripts of the genes for the large subunit of ribulosc bisphosphatc carboxylase and the β and E subunits of CF_1 increase upon illumination of dark-grown seedling

plants but not dark-grown seedlings. About 14% of the maize plastid chromosome that is transcribed in chloroplasts was found to be untranscribed in etioplasts [L. Haff and L.

Bogorad, unpublished]. In these and some other experiments, RNA pools are being measured and thus it is more strictly correct to state that etioplast RNA lacks transcripts of about 14% of the chromosome that are present in the chloroplast RNA population. Considering other limitations of the experiments, it is more appropriate to state that transcripts of about 14% of the chromosome are either absent or are present in much lower amounts in etioplast than in chloroplast RNA.

The first gene whose transcript level was found to be raised upon illumination of seedlings, ie, a "photogene" was identified by a combination of techniques. These included comparison of the hybridization of radioactive RNAs from etioplasts and chloroplasts against fragments of chloroplast DNA generated by a restriction endonuclease and translation of these RNAs in vitro to identify differences in their polypeptide products[32]. These approaches were combined with analyses of proteins made by isolated chloroplasts and by intact plants at various stages during light induced development[48,49]. From all of these studies it was concluded that the primary translation product of this maize photogene is a 34. 5 kDa (kilodalton) thylakoid membrane protein that is processed to 32 kDa after association with the membrane. The gene was consequently designated photogene 32. Isolated maize plastids can synthesize the 34. 5-kDa primary translation product (or, at least, can complete the synthesis of started polypeptidc chains) but have little capacity for processing it to its mature 32-kDa form[49]. Photogene 32 was mapped to a portion of Bam fragment 8 of the maize plastid chromosome. When intact green leaves are supplied with ^{35}S-methionine, the first membrane protein to be labeled is the product of photogene 32. Since there seems to be a relatively small amount of this protein in membranes, despite its rapid labelling, it seemed reasonable to conclude that the protein must turn over relatively rapidly. The photogene 32 product has been shown subsequently to bind azido-atrazine[51] and has been judged to function in electron transport at the Q_B site in the photosynthetic electron transport chain.

We have recently gone on to map additional maize plastid photogenes—ie, genes for which more transcripts are present during light-induced development than in darkness-by improved methods for detecting them by hybridization against chloroplast DNA fragments. Furthermore, we have used cloned fragments of the maize chloroplast chromosome rather than fragments of total plastid DNA as sources of DNA for localizing of photogenes on the map of the chloroplast chromosome. The latter has two advantages: ① confusion is avoided-two or more sequences of a single size class can lie in a single band after agarose gel electrophoresis of total plastid DNA digested with a restriction enzyme, but the likelihood of this happening is sharply reduced when smaller portions of the chromosome are used; ② a single DNA fragment can be further analyzed in detail in hybridization experiments. Three well-separated regions of the maize chloroplast chromosome have been shown to harbor "highly responsive" photogenes. In addition, a "minor photogene" regions, sequences for which transcript levet changes are smaller than for the "major photogene regions" has been found at a fourth location in the small unique DNA region of the chromosome (see Figure 1). Thus, there is not a single photogene "operon" in the maize chloroplast chromosome but there are several regions from which transcription is stimulated by illumination of dark-grown seedlings. These photoresponsive regions comprise about 19% of the unique DNA (counting only one inverted repeat) of the maize chloroplast chromosome. This number is amazingly close to the roughly 14% estimated at from saturation hybridization data.

One objective of mapping photogenes is to open the way to studying the mechanisms of photoregulation of transcription. Now that we know that there is not a single photogene operon, we must ask whether all of the plastid photogenes of maize are of a single expression class despite this. Is transcription of all photogenes stimulated to the same extent? Is the interval between initiation of illumination and increase in the transcript pool size the same for all photogene regions? Is transcription of all photogenes affected in the same manner when plants thar have been illuminated are returned to darkness?

These questions were addressed by using cloned fragments of photogenes to measure the amounts of complementary RNA present in seedlings at different stages of light-induced development using either the dot blot[51] or Northern hybridization methods. We learned from the dot blot hybridization experiments that there are differences in the kinetics of induction of transcription of the various photoregulated regions. We have also found that under our conditions transcript pools of all photogenes reach maxima at between 10 and 44 h after the commencement of illumination and then, eicher in light or darkness, the size of each pool drops. The singular exception to this is the level of transcripts of photogene 32.

To study these problems, RNA was isolated from dark-grown maize seedlings and from seedlings of the same growth lot illuminated for 2,10,20,40 or 68 h: some samples were returned to darkness after 20 h of illumination. In other experiments, harvests were made after dark-grown seedlings had been illuminated for 2,4,6, or 10 h or for 3,6,12,18,24, or 36 h. Aliquots of RNA samples were affixed to Gene Screen filters (New England Nuclear. Waltham. Massachusetts). Appropriate cloned DNA fragments were nick-translated to provide ^{32}P-labeled probes to measure the abundance of specific RNA species in each sample using the dot blot hybridization procedure. In other experiments, the Northern procedure was used; ie, RNA was denatured and subjected to agarose gel electrophoresis, the RNAs were transferred to Gene Screen, and nick-translated probes were used to identify bands of complementary RNA. In the latter case, the sizes of transcripts could also be established.

Careful and sensitive analyses demonstrated that transcripts of photogenes were present in etiolated plants but that these increased in abundance during illumination. For example, RNA complementary to Bam fragments 17 and 21 increased about four fold in abundance during the first 10 h of illumination. RNA from Bam fragments 3 and 24 increased to about the same extent but only after 20 h of continuous illumination. One striking observation is that transcripts of three of the four photogene regions studied in detail decrease in abundance starting after ten or twenty h of continuous illumination of the dark-grown seedlings. By 44 or 68 h of illumination the RNA transcribed from these regions was only about as abundant as the same species in etiolated plants. The steady-state level of transcripts for rc-L, for example, was slightly below that in etiolated plants after the seedlings had been illuminated continuously for 44 h. An exception to this pattern is the transcript pool of photogene 32. When seedlings were returned to darkness for 20 h and maintained there for another 48 h, the pool sizes for photogene 32 mRNA remained about constant; the sizes of these pools continued to increase until 68 h (the termination of the experiment) when seedlings were maintained in the light.

We conclude that not all photogenes are the same expression class. They differ with regard to ① whether mRNA pool sizes increase throughout the first 68 h of illumination, as is the case for photogene 32, or ② whether a peak is reached at 10 or 20 h after illumination and pool sizes decline after that. The latter class can be further subdivided depending upon the time at which the maximum size of the pool is reached under our illumination conditions.

Another striking set of observacions is that, except for cfl-BE, for which only a single size of transcript is seen in maize, each photogene region is represented by from two to 13 transcripts of different size classes.

The simplest case of multiple-sized transcripts is of the gene for the large subunit of RUBISCO. The most abundant transcript of rc-L in etiolated plants is 1. 6 kb, but a trace of a 1. 8 kb RNA is visible in Northern hybridizations. After only 2 h of illumination, the 1. 8-kb transcript is prominent and after 20 h it comprises about one-third of this RNA complemencary to rc-L. But by 44 h the level of the 1. 8-kb transcript is down to perhaps 20% of its maximum level and 24 h later its contribution to the total is very small (The "1. 8-kb" transcript is longer at its 5′ end than the 1. 6-kb transcript.).

More complex patterns of RNAs are seen for the three photogene regions we have studied that yield polycistronic mRNAs. For each of these regions there is a transcript of about 5 kb that includes messages for at least two different genes. In addition, these regions, or parts of them, are represented by a larger number of transcripts of different sizes. In one case, 13 different transcript size classes have been identified as complmentary to various regions across a stretch of the plastid chromosome of about 5—6 kbp. These transcripts, like those of rc-L, are not present equimolar amounts at each stage of light-induced development. Some of the smaller transcripts differ from the largest one representing that region at the 5′ end and other less than maximum-sized transcripts are shorter at their 3′ ends. We do not know whether the 5′ differences result from the use of multiple transcription initiacion sites, purposeful processing, or degradation. Neither do we know whether the 3′ differences can be attributed to two or more possible different termination sites, purposeful processing, or degradation. Differences in abundance of some of the multiple transcripts from regions that contain more than one gene can be imagined to reflect a mechanism for controlling the relative abundance of translatable messages for different genes within a single operon. Whether this is true remains to be determined.

One photogene operon contains coding sequences for cf1-A and cfo-III (We suspect that there are other genes in this region that are transcribed into the same 5 kb RNA but the sequencing has not been completed.). Another maize photogene region contains fused coding sequences for the beta and epsilon subunits of coupling factor CF_1[34]. Changes in RNA pool size in response to illumination and development, including the decreases seen after 20 h in the light, are parallel for the two DNA sequences that code for polypeptide components of the CF_0-CF_1 complex, ie, cfl-BE and cfl-A plus cfo-lll. The conserved sequence resembling a prokaryotic promorer 5′ to the transcribed portion of cfl-BE is known but that for the cfl-A plus cf0-1ll is not yet known. It will be interesting to see if there are common sequences that serve as part of the expression controlling mechanisms and, in general, to understand how expression of these two operons is coordinated.

The observations on photogene transcription in vivo outlined above lead to some questions. What are the mechanisms for bringing about increases in the sizes of pools of photogene transcripts? What determines the maximum sizes and the time at which a pool reaches that size? What signal or system leads to a reduction in pool size? What determines the steady-state pool size in mature plastids? Some aspects of these questions are now reasonably well-defined in outline: dark-grown maize seedlings have low levels of transcripts of every light-inducible gene we have studied; increases in the sizes of transcript pools for all of the photogenes can be detected within about 2 h after illumination-there may be a single underlying mechanism (e. g. , a change in RNA polymerase activity) for increasing transcription of all photogenes or there may be a number of signals all of which appear at the same time, assuming that the rate of RNA turnover is unaltered; photoregulated transcription may involve alternate promoters or terminators or changes in RNA processing or degradation; photogenes and photogene operons differ from one another with regard to the time at which pool sizes reach a maximum and whether they subsequently remain constant or decrease in darkness or in the light-decreases in sizes of the pools may be under negative or positive control. Repression can be imagined via a feedback loop as might be found already in etioplasts lacking products of the nuclear-cytoplasmic system and again after the gene products have accumulated (in mature plastids) or, as a positive (i. e. , derepression or enhancement) control model. The level of accumulated transcripts might be related to sizes of pools of nuclear gene specific products, eg, components of mnultimeric systems in the plastid. We have no idea of how the nuclear-cytoplasmic system might sense the completion of plastid components. Nor do we know whether the cytoplasmic products are produced all the time or are turned over after plastids mature.

Other related problems raised by the observations described here have been mentioned earlier. How are the difterent sizes of transcripts produced from a single transcription unit-transcription start or stop site differences, processing, degradation? Is there any functional significance underlying this phenomenon? Some answers will come from analyses of the transcripts themselves. Other answers will come from research on transcription and processing.

In Vitro Transcription of Chloroplast Genes

It has been known for some time [52, 53; Haff and Bogorad, unpublished] that illumination of dark-grown maize seedlings quickly results in the production of more plastid RNAs as well as increased activity of the plastid DNA-dependent RNA polymerase. As noted earlier, some species of transcripts increase considerably in abundance during light-induced development. In the other developmental situation we have examined, ie, differentiation of mesophyll and bundle sheath cells in maize, as also described above, the rc-L gene is strikingly differently expressed in the two cell types[47,48]. In order to explore the mechanisms underlying such apparent transcriptional regulation of gene expression in plastids, a number of years ago we started to study details of the structure and properties of maize plastid DNA-depedent RNA polymerase[26,47,52,54-56].

This enzyme was solubilized from thylakoids of sucrose gradient-purified plastids[54-56]. It has been shown to be different in properties and structure from the type II nuclear DNA-dependent RNA polymerase of maize[57-59]. Although purified preparations of nuclear polymerase II and the maize plastid polymerase have some polypeptides of about the same size in common, these are, in fact, different[59] polypeptides as deduced from fingerprints[59]. Tewari and Goel[60] have recently found that purified preparations of pea plastid DNA-dependent RNA polymerase contain polypeptides of the same sizes as purified preparations of the maize enzyme[55,58,61].

When characterized cloned maize plastid sequences first became available[12,13,24,33,45,62,63], it became possible to study their transcription by the plastid enzyme as a step toward developing an in vitro system capable of mimicking transcription of chloroplast genes as it occurs in vivo, ie, to experimentally test mechanisms of developmental regulation of gene transcription in maize plastids by synthesis of the system in vitro.

The first experiments toward this end[56] were directed at determining whether partially purified maize plastid polymerase can preferentially transcribe maize plastid DNA sequences in chimeric plasmids replicated in E. coli. Maize plastid RNA polymerase solubilized from the membranes of sucrose gradient-purified plastids was chromatographed on DEAE-cellulose and eluted with a gradient of KCl. The bulk of the enzyme activity was eluted wich about 0.2 mol/L KCI. When tested with a chimeric plasmid template containing two cloned maize plastid DNA sequences—one of which included rRNA genes—the enzyme transcribed both the bacterial plasmid sequences and the cloned chloroplast sequences about equally, judging from Southern hybridization experiments. A fraction that eluted with about 0.5 mol/L KCl, when added to the polymerase activity elutcd with 0.2 mol/L KCl, stimulated the transcription of the plasmid DNA template, but the most striking effect was preferential transcription of some chloroplast sequences over others and over the bacterial vehicle DNA.

These results wre extended using as a template a relatively small fragment of plastid DNA cloned in the plasmid pMB9. The relative transcription of chloroplast and plasmid sequences was assessed quantitatively. The DEAE-cellulose-purifed RNA polymerase transcribed the chloroplast and vehicle sequences about equally. However, when the 27. 5-kDa polypeptide (the S factor) that was purifed from the 0. 5 mol/L KCl eluate of the DEAE column was included in the reaction mixture, transcription of the cloned chloroplast sequences was about an order of magnitude greater than transcription of the vehicle DNA. Another notable feature of this system is a requirement for a supercoiled DNA template. If the template was relaxed with E. coli topoisomerase most of the preferential transcription of plastid DNA sequences was lost. The latter observation also confirmed an assessment carried out by Southern hybridization on another fragment of chloroplast DNA. We concluded from these experiments that preferential transcription of chloroplast DNA sequences required the polymerase, the S factor polypeptide, and a supercoiled DNA template[57]. In the next set of experiments we showed that the 5′ termini of in vivo and in vitro transcripts of the gene for $tRNA^{His}$ are the same[26].

The earliest experiments[56], as noted, indicated that not all chloroplast DNA sequences are transcribed equally in vitro by the maize polymerase plus S factor system. This question was examined quantitatively in a particularly attractive situation. The gene for the large subunit of ribulose bisphosphate carboxylase and a gene for a 2. 2-kb transcript, later identified as cfl-BE that codes for the beta and epsilon subunits of the coupling factor CF_1, are separated from one another by about 100,300 bp and are transcribed divergently. Maize plastid fragment Bam 9 contains all of rc-L and about 1. 4 kbp of cfl-BE. This is an interesting pair to study because the two are differentially expressed in mesophyll and bundle sheath cells of maize[46,47].

First, it was shown[47] using a BglII subfragment of maize plastid DNA fragment Bam 9 cloned in the bacterial vector RSF1030 that even without the S factor chloroplast DNA sequences were transcribed almost nine times more actively than the RSF-1030 DNA but in the presence of the S factor hybridization to RSF-1030 was not detectable above the background level. When transcription of rc-L vs cf1-BE was studied, the ratio of transcription was about 1. 1 in the absence of the S factor and 3. 2 in its presence. Thus, the earlier observation regarding differential transcription of parts of chloroplast DNA cloned in E. coli[56] was confirmed (Another observation, related to the topography of the template, was made in the latter sets of experiments[47].). It was found that using two different sized fragments of linear chloroplast DNA as templates for RNA polymerase "run-off" experiments, RNAs of the expected sizes could not be found, leading us to conclude that supercoiled or circular forms of template might be required for initiation at specific sites. The latter has been confirmed in more recent experiments [L. D. Crossland and L. Bogorad, unpublished].

In all, the experiments wich cloned maize chloroplast DNA sequences and maize chloroplast RNA polymerase in vitro encourage us to use this system to study developmentally regulated transcription.

COMMENTS

The focus in this discussion has been primarily on knowledge gained about plastid chromosomes and genes from analyses of the maize chloroplast chromosome and its genes. Other work on plastid molecular biology, including the sequencing of genes, has been carried out in a number of laboratories with chloroplast DNA of spinach, Nicotiana, pea, wheat, Chlamydomonas, Euglena, and other species. It is beyond the scope of this limited discussion to include all of these items although some data from such studies appear in Table 1.

Currently available information on plastid chromosomes and genes will guide future analyses of chloroplast chromosomes and investigations of mechanisms for the control of plastid gene expression and the integration of expression of plastid and nuclear genes. Furthermore, the protein products of several of the genes that have been cloned and sequenced are of great interest for understanding mechanisms of photosynthesis; this knowledge remains to be exploited to broaden our understanding of this process.

Acknowledgments

Various parts of the research described here were supported by grants from the National Science Foundation, the Competitive Grants Office of the United States Department of Agriculture, and the National Institute of General Medical Sciences. It was also supported in part by the Maria Moors Cabot Foundation of Harvard University.

REFERENCES

[1] Bell WD, Bogorad L, Mclllrath WJ: Botan Gaz 124:1,1962.
[2] Walles B: In Goodwin TW (ed): "Biochemistry of Chloroplasts." New York: Academic Press, 1967. Vol II, pp 633 - 654.
[3] Mets LJ, Bogorad L: Science 174:707,1971.
[4] Mets L. Bogorad L: Proc Natl Acad Sci USA 69:3779,1972.
[5] Davidson JN, Hanson MR. Bogorad L: Mol Gen Gencl, 132:119,1974.
[6] Hanson MR, Davidson JM, Mets LJ, Bogorad L: Mol Gen Gent 132:105,1974.
[7] Chan P. Wildman SG: Biochim Biophys Acta 277:677,1972.
[8] Kawashima N, Wildman SG: Biochim Biophys Acta 262:42,1972.
[9] Bogorad L: Science 188:891,1975.
[10] Bogorad L, Bedbrook JR. Davidson JN, Hanson M, Kolodner R: Brookhaven Symp Biol 29:1,1977.
[11] Bugorad L: In Schiff JA (ed): "Origin of Chloroplast." Amsterdam: Elsevier/North-Holland Inc. 1982. pp 277 - 295.

[12] Coen DM, Bedbrook JR. Bogorad L, Rich A: Proc Natl Acad Sci USA 74:5487. 1977.
[13] Bedbrook JR, Coen DM. Beaton AR, Bogorad L, Rich A: J Biol Chem 254:905,1979.
[14] Bedbrook JR, Bogorad L: In Bucher T, Neupert W, Sebald W, Werner S (eds): "Genetics and Biogenesis of Chloroplasts and Mitochondria." Amsterdam: North-Holland Publishing Company, 1976. pp 369 - 374.
[15] Bedbrook JR, Bogorad L: Proc Natl Acad Sci USA 73:4309. 1976.
[16] Hermann RG, Bohnert H-J, Driesel A. Hobom G: In Bucher T, Neupert W, Sebald W, Werner S (eds): "Genetics and Biogenesis of Chloroplasts and Mitochondria." Amsterdam: North-Holland Publishing Company. 1976, pp 351 - 359.
[17] Kolodner R, Tewari KK: Proc Natl Acad Sci USA 76:41,1979.
[18] Koller B. Dfius H: Mol Gen Genet 178:261,1980.
[19] Palmer JD, Thompson WF: Cell 29:537,1982.
[20] Gray PW, Hallick RB: Biochemistry 18:281,1978.
[21] Jenni B, Sturz E: Eur J Biochem 88:127,1978.
[22] Jenm B, Stutz E: FEBS Lett 102:95,1979.
[23] Southern EM: J Mol Biol 98:503,1975.
[24] Bedbrook JR, Kolodner R. Bogorad L: Cell 11:739,1977.
[25] Koch W, Edward K. Kossel H: Cell 25:303,1981.
[26] Schwarz Z, Jolly SO. Steinmetz AA. Bogorad L: Proc Natl Acad Sci USA 78:3423,1981.
[27] Schwarz Z, Kossel H. Schwarz E. Bogorad L: Proc Natl Acad Sci USA 78:4748,1981.
[28] Selden RF. Steinmetz A. Mclntosh L. Bogorad L. Burkard G. Mubumbila M, Kuntz M. Crouse EJ. weil JH: Plant Mol Biol 2:141,1983.
[29] Schwarz Z, Kossel H: Nature 283:739,1980.
[30] Edwards K, Kossel H: Nucleic Acids Res 9:2853,1981.
[31] Driesel AJ, Crouse EJ, Gordon K, Bohnert HJ, Herrmann RG, Steinmetz A, Mubumbila M, Keller M, Burkard G, Weil JH: Gene 6:285,1979.
[32] Bedbrook JR, Link G. Coen DM, Bogorad L, Rich A: Proc Natl Acad Sci USA 75:3060,1978.
[33] Mclntosh L, Poulsen C, Bogorad L: Nature 288:556,1980.
[34] Krebbers ET, Larrinua IM, Mclntosh L, Bogorad L: Nucleic Acids Res 10:4985,1982.
[35] Subramanian AR, Steinmetz A, Bogorad L: Nucleic Acids Res 15:5277,1983.
[36] Steinmetz AA, Gubbins EJ, Bogorad L: Nucleic Acids Res 10:3027,1982.
[37] Steinmetz AA, Krebbers ET, Schwarz Z, Bogorad L: J Biol Chem 258:5503. 1983.
[38] Sugita M, Sugiura M: Nucleic Acids Res 11:1913,1983.
[39] Shinozaki K, Sugiura M: Nucleic Acids Res 10:4923,1982.
[40] Zurawski G, Perrot B, Bottomley W, Whitfeld PR: Nucleic Acids Res 9:3251,1981.
[41] Zurawski G, Bottomley W, Whitfeld PR: Proc Natl Acad Sci USA 79:6260,1982.
[42] Howe CJ, Auffret AD, Doherty A, Bowman CM, Dyer TA, Gray JC: Proc Natl Acad Sci USA 79:6903,1982.
[43] Deno H, Shinozaki K, Sugiura M: Nucleic Acids Res 11:2185,1983.
[44] Zurawski G, Bohnert HJ, Whitfeld PR, Bottomley W: Proc Natl Acad Sci USA 79:7699,1982.
[45] Link G, Bogorad L: Proc Natl Sci USA 77:1832,1980.
[46] Link G, Coen DM, Bogorad L: Cell 15:725,1978.
[47] Jolly SO, Mclntosh L. Link G, Bogorad L: Proc Natl Acad Sci USA 78:6821,1981.
[48] Grebanier AE, Coen DM, Rich A, Bogorad L: J Cell Biol 78:734,1978.
[49] Grebanier AE, Steinback KE, Bogorad L: Plant Physiol 63:436,1979.

[50] Steinback KE, Mclntosh L, Bogorad L, Arntzen CJ: Proc Natl Acad Sci USA 78:7463,1981.
[51] Kafatos FC, Jones WC, Efstratiadis A: Nucleic Acids Res 7:1541,1979.
[52] Bogorad L: Dev Biol [Suppl] 1:1,1967.
[53] Apel K, Bogorad L: Eur J Biochem 67:615,1976.
[54] Bottomley W, Smith HJ, Bogorad L: Proc Natl Acad Sci USA 68:2412,1971.
[55] Smith HJ, Bogorad L: Proc Natl Acad Sci USA 71:4839,1974.
[56] Jolly SO, Bogorad L: Proc Natl Acad Sci USA 77:822,1980.
[57] Strain JC, Mullinix KP. Bogorad L: Proc Natl Acad Sci USA 68:2647,1971.
[58] Mullinix KP, Strain GC, Bogorad L: Proc Natl Acad Sci USA 70:2386,1973.
[59] Kidd GH, Bogorad L: Proc Natl Acad Sci USA 76:4980,1979.
[60] Tewari KK, Goel A: Biochemistry 22:2142,1983.
[61] Kidd GH, Bogorad L: Biochim Biophys Acta 609:14,1980.
[62] Bogorad L, Jolly SO. Link G, Mclntosh L, Poulsen C, Schwarz Z, Steinmetz AA: In Bucher T, Sebald W, Weis H (eds): "Biological Chemistry of Organelle Formation." Berlin: Springer, 1980, pp 87 - 96.
[63] Bogorad L, Jolly SO, Kidd G, Link G, Mclntosh L: In Leaver C (ed): "Genome Organization and Expression in Plants." London: Plenum, 1980, pp 291 - 304.

Effects of Light, Oxygen, and Substrates on Steady-State Levels of mRNA Coding for Ribulose-1, 5-Bisphosphate Carboxylase and Light-Harvesting and Reaction Center Polypeptides in Rhodopseudomonas sphaeroides*

The mRNA levels specific for ribulose-1,5-bisphosphate carboxylase, light-harvesting I polypeptides α and β, and reaction center polypeptides L and M were assayed by use of a series of DNA probes specific for each cognate mRNA. Both the steady-state amounts and sizes of the specific mRNAs were measured as a function of the light intensity incident to the culture, the presence or absence of oxygen, and the type of substrate present in the growth medium. Northern hybridization revealed at least two and possibly three transcripts for ribulose-1,5-bisphosphate carboxylase. The cellular level of mRNA specific for ribulose-1,5-bisphosphate carboxylase increased in consort with enzyme activity as a function of both light intensity and reducing state of the substrate. Neither mRNA nor enzyme activity was detectable in aerobically grown cells. For the light-harvesting I and reaction center polypeptides there exist two transcripts, the larger of which appears to be a polycistronic mRNA possessing information for all four polypeptides and a smaller transcript specific for only the α and β polypeptides of the light-harvesting I complex. The regulation of each of these mRNAs was affected by light and oxygen, but was not significantly affected by the oxidation-reduction state of the substrate.

The process of photosynthesis in photosynthetic organisms, whether plants, algae, or procaryotes, is strongly affected by environmental factors. Among these fators, light is certainly one of the most important. The regulation of photosynthetic genes by light has been well established in green plants, algae, blue-green bacteria, and photosynthetic bacteria[1,7,25], but the molecular basis for this regulation is largely unknown.

The purple nonsulfur photosynthetic bacterium Rhodopseudomonas sphaeroides provides an attractive system for studying the regulation of the activity of photosynthetic genes by light, O_2,

* Yu Sheng Zhu, Samuel Kaplan, Department of Microbiology, University of Illinois at Urbana-Champaign, Urbana, Illinois 61801 Received 26 December 1984/Accepted 11 March 1985
此文发表于 1985 年“J. Bacteriol” 162:925 – 932.

and other environmental parameters. Upon the removal of oxygen, the cell develops an extensive intracellular membrane that comprises the photosynthetic apparatus[5,6,11,12,16,17]. Further, the amount of intracellular membrane per cell and the composition of the intracellular membrane are functions of the light intensity[11,16,17]. Therefore, the synthesis and composition of the intracellular membrane are under at least two different levels of control, the pO_2 and the light intensity.

In addition to the photosynthetic membrane components, ribulose-1, 5-bisphosphate (RuBP) carboxylase (RuBPCase) in *R. sphaeroides*, a key enzyme of the Calvin cycle and a major soluble protein in the cytoplasm[21], represents another model system for studying the regulation of photosynthetic gene expression. Two structurally and immunologically distinct forms of RuBPCase, forms I and II, exist in *R. sphaeroides*[27]. Form I is composed of eight large and eight small subunits and resembles the eucaryotic enzyme in its kinetic parameters[27]. Form II, which lacks small subunits, is reported to be a hexamer of large subunits and resembles the procaryotic dimer enzyme as typified by Rhodospirillum rubrum[13,14,21]. The cellular levels of both forms of the enzyme are dependent on light, oxygen, and type of carbon corresponding author, substrate[29]. The light-and oxygen-dependent regulation of the synthesis of soluble and photosynthetic membrane proteins of *R. sphaeroides* is also clearly demonstrated by recent experiments involving shifts in these environmental parameters[5,7].

With the advent of recombinant DNA techniques, DNA segments defining many of these important membrane and soluble proteins have become available. The authors of a recent paper have begun to examine specific mRNA levels in a related organism, *Rhodopseudomonas capsulata*[8]. The availability of specific DNA probes defining the genetic regions for the RuBPCase form Ⅱ[22] and for both the lightharvesting polypeptides (LH-1) (P. Kiley and S. Kaplan, unpublished data), and the reaction center polypeptides (RC-L and RC-M)[30,31] of *R. sphaeroides* has made it possible to qualitatively and quantitatively probe specific mRNAs by hybridization with these unique cloned fragments. The study presented here demonstrates the identification and regulation in the levels of mRNA coding for RuBPCase, LH-1, RC-L, and RC-M polypeptides in *R. sphaeroides* grown under defined steady-state conditions of light, oxygen, and carbon source.

Materials and Methods

Bacterial strains, media, and growth conditions

R. sphaeroides wild-type strain 2. 4. 1 was used throughout. The cells were grown on modified Sistrom medium A[26] supplemented with 0. 4% butyrate and 0. 1% $NaHCO_3$ as the carbon source unless otherwise noted. The cells were fully adapted to the butyrate medium before inoculation. Those cells undergoing photoheterotrophic growth were grown at 30℃ on the butyrate medium in completely filled flat screwcap bottles (100 ml) normally at near-saturating light intensity (10 W/m^2), provided by a bank of lumiline lamps (General Electric Co.). When light intensity was to be a variable, the cells were also grown under conditions designated low light intensity at 3 W/m^2 in comparison with 10 W/m^2, designated as the high intensity light. The light energy was measured with a Yellow Springs-Kettering model 6.5A radiometer through a coming colored-glass filter (CS No. 7 - 69; 620 to 1,100 nm). Dark-shift experiments were performed by wrapping the growth vessels from high light-grown cells with aluminum foil for 1 h at 30℃. Partial aerobic growth was maintained by chemoheterotrophic culturing of the cells (100 ml in a 500-ml flask) on a gyratory shaker at 30℃. In some cases the culture (100 ml) was sparged with a gas mixture of O_2-N_2-CO_2 (25 : 74 : 1) to obtain high O_2 partial pressure; such conditions are designated full aerobic growth[5]. In addition to butyrate, malate and succinate (0.4%) were also used as carbon sources in certain experiments. Cell growth was routinely monitored turbidimetrically with a Klett-Summerson colorimeter (Klett Manufacturing Co., Inc.) equipped with a No. 66 filter. One Klett unit is equivalent to a cell density of approximately 10^7 cells per ml. Chemoheterotrophically grown cells and photoheterotrophically grown cells were harvested at the cell densities indicated below.

RuBPCase, protein, and bacteriochlorophyll assay

RuBPCase activity was assayed with toluene-treated cells based on the method of Tabita et al.[28], with some minor modifications. Samples (20 ml) of cells were harvested, washed twice in 10 ml of cold Tris (50 mmol/L)-EDTA (10 mmol/L) buffer (pH 8.0), and then suspended in 2 ml of the same buffer. A 1-ml amount of cold toluene was added and mixed, and the cells were kept on ice for 10 min. After centrifugation at 5,000 rpm for 10 min, the cells were washed with Tris-EDTA buffer and suspended in 1 ml of the same buffer. Reactions were performed in scintillation minivials with rubber stoppers and contained the following: 0.05 ml of 50 mmol/L $MgCl_2$, 50 mmol/L N-2-hydroxyethylpiperazine-N′-2-ethanesulfonic acid (pH 7.2), 0.05 ml of 100 mmol/L $NaH^{14}CO_3$ (specific activity, 300 to 1,000 cpm/nmol), 0.05 ml of toluene-treated cell suspension, 0.025 ml of 10 mmol/L RuBP, and 0.075 ml of water. All components minus the RuBP were mixed and preincubated for 10 min at 33℃ to activate the enzyme. Prewarmed RuBP was added to initiate the reaction. The assay was performed at 33℃ for 3 to 5 min and stopped with the injection of 0.025 ml of 6 N HCl. The reaction mixture was dried down at 80℃ under N_2 in a fume hood, allowed to cool, and suspended in 0.1 ml of water. The samples were then counted in Tritosol[22]. One unit of RuBPCase is defined as the amount of enzyme needed to carboxylate 1 μmol of RuBP in 1 min at 23℃.

Protein concentrations were determined by a modified Lowry procedure[19]. The total bacteriochlorophyll (Bchl) content of the cells was estimated by absorbancy measurements at 775 nm of acetone-methanol (7 : 2, vol/vol) extracts with an extinction value of 75 mmol/L[9].

Extraction of RNA

A 100-ml sample of fresh cell culture was rapidly poured onto 100 ml of crushed, frozen extraction buffer containing 80 mmol/L Tris-chloride (pH 7.5), 10 mmol/L $MgCl_2$, 10 mmol/L β-mercaptoethanol (freshly added), 25 mmol/L NaN_3, and chloramphenicol (200 μg/ml), rapidly mixed, and centrifuged at 7,000 rpm for 3 min. The cells were lysed by adding 5 ml of lysis buffer containing 0.5% sodium dodecyl sulfate, 10 mmol/L EDTA, 2 mmol/L O-phenanthroline, 0.2 mg of heparin per ml, and 0.25 mg of proteinase K per ml and then frozen in a dry ice-ethanol bath and thawed at 37℃ for three successive cycles. The cell lysates were incubated at 37℃ for 20 min and then extracted with hot (65℃) phenol (redistilled)-chloroform (1 : 1, vol/vol, three or four times until a clear aqueous phase was obtained. DNA and RNA were precipitated by adding 2 volumes of ethanol to the aqueous phase and dissolved in 5 ml of 50 mmol/L Tris-chloride-5 mmol/L $MgCl_2$ (pH 7.8). DNA was digested by incubation with RNase-free DNase I (50 μg/ml) at 37℃ for 20 min. DNase I was removed by phenol extraction, and the RNA was finally collected by ethanol precipitation. About 1 mg of RNA was obtained from 100 ml of cells. The ratio of absorbancy at 260 nm to that at 280 nm was 1.8 to 2.0. RNA was used either for dot blots (5 μg per dot) after denaturing with formaldehyde or for Northern blot (15 μg per lane) hybridization after glyoxylation and agarose gel electrophoresis.

Isolation of chromosomal DNA

The bulk DNA of *R. sphaeroides* was isolated basically by the procedure of Saito and Miura[23]. The total bulk DNA was digested with *Bam*HI and separated on an agarose gel.

Radioactive labeling of RNA and DNA. Total *R. sphaeroides* RNA was radiolabeled in vitro by phosphorylating the 5′-OH groups produced by partial alkaline hydrolysis of RNA with ^{32}P from [γ-^{32}P]ATP with T4 polynucleotide kinase by the method of Bogorad et al.[4]. RNA samples for Southern hybridization were adjusted to equal specific radioactivity (10^6 cpm/μg). The plasmid DNA or DNA fragment (see Table 1) used for dot blot and Northern hybridizations were labeled with [α-^{32}P]dCTP by nick translation by the protocol of Bethesda Research Laboratories, Gaithersburg, Md. The labeled DNA and [α-^{32}P] dCTP were separated with a spun column of Sephadex G-50 by the method of Maniatis et al.[18].

Table 1 Plasmids used for probing photosynthetic genes of *R. sphaeroides*

Plasmid	Vector	Insert	Size/kb	Gene encoded	Reference
pLI10	pAS621	*Bam*HI fragment of *R. sphaeroides* DNA	3.0	RuBPCase Ⅱ	[22]
C-2-C	M13mp10WRF	SalI-*Bam*HI fragment	1.2	RuBPCase II	Unpublished ⓐ
pJW1	pBR322	*Bam*HI fragment of *R. sphaeroides* DNA	12.5	LH-1, RC-L, RC-M	[30]
		KpnI-PstI fragment	1.2	LH-1, portion of RC-L	Unpublished ⓑ

(continued)

Plasmid	Vector	Insert	Size/kb	Gene encoded	Reference
pJW1	pBR322	AluI-AluI fragment	0.51	LH-1	Unpublished[b]
		PvuII-SalI fragment	0.46	RC-L	Unpublished[b]
		XhoI-NruI fragment	0.37	RC-M	Unpublished[b]
pUI135	pBR322	SalI fragment of pJW1	7.3	LH-1, RC-L, portion of pJW1 vector	Unpublished[c]
pUI128	pBR322	SalI fragment of pJW1	2.35	RC-M, portion of RC-L	Unpublished[c]

[a] p. Hallenbeck and S. Kaplan.
[b] P. Kiley and S. Kaplan.
[c] S. Kuhl and S. Kaplan.

Dot blots and Northern and Southern hybridizations

Dot blots and Northern and Southern hybridizations were performed as previously described[4]. The plasmids used as probes for hybridization are listed in Table 1. To quantitatively assess mRNA in either the dot blots or Northern hybridizations, the films were scanned with a GS300 transmittance reflectance scanning densitometer (Hoefer Scientific Instruments) linked to a recording integrator (Waters M 730 data module). For quantitative comparisons, the dots on the filters were sometimes cut out and counted in 2,5-diphenyloxazole-1, 4-bis-(5-phenyloxazolyl) benzene cocktail with a Beckman liquid scintillation counter.

Chemicals

The DNA from plasmids pLI10, C-2-C, pUI135, and pUI128 were generous gifts of P. Hallenbeck and S. Kuhl, respectively. The DNA from plasmid pJW1 and cloned KpnI-PstI, AluI, Pvull-SalI, and XhoI-NruI fragments derived from pJWl were kindly supplied by P. Kiley. [γ-^{32}P]ATP was made by T. J. Donohue by the Johnson and Walseth technique[15]. RuBP was obtained from Sigma Chemical Co., St. Louis, Mo. $NaH^{14}Co_3$ (53 mCi/mmol) was obtained from New England Nuclear Corp., Boston, Mass. Agarose, restriction enzymes, and a nick translation reagent kit were purchased from Bethesda Research Laboratories. Gene Screen (New England Nuclear) was used for all hybridizations. Proteinase K was purchased from Boehringer Mannheim Biochemicals, Indianapolis, Ind. RNase-free DNase I was obtained from Worthington Diagnostics, Freehold, N. J. Phenol was redistilled, and all other chemicals were reagent grade.

Results

Effects of light, oxygen, and substrates on RuBPCase and Bchl. RuBPCase is a major soluble cytoplasmic enzyme characteristic of photosynthetic growth and was chosen for study. Further, Bchl levels were used as a measure of photosynthetic development, since the level of Bchl is a measure of intracellular membrane content and composition[16]. Four different growth conditions were used to study the effects of light and O_2 on RuBPCase and Bchl. ① In the high-light conditions, *R. sphaeroides* cells were photosynthetically grown under near-saturating high intensity light (10 W/m^2). ② In the dark conditions, the cells were grown under high light intensity and were shifted to the dark for 1 h. ③ In the low-light conditions, cells were photosynthetically grown under low-intensity light (3 W/m^2). ④ In partial O_2, the cells were aerobically cultured on a shaker. Because cells growing under these different growth conditions had different growth rates, an effort was made to harvest the cells at comparable cell densities (from 110 to 125 Klett units) unless otherwise noted. After the cells were harvested, the RuBPCase activity and Bchl content were determined. Both RuBPCase and Bchl were affected by light and O_2 (see Table 2). The shift of cells from high-intensity light to the dark resulted in a 25% reduction in RuBPCase activity, but only a 10% reduction in the level of Bchl, after 1 h. Under steady-state low-intensity light, the RuBPCase activity was approximately 50% of its high-light level, whereas the level of Bchl was approximately 80% greater than its high-light level, as expected. The partially aerobically grown cells had negligible RuBPCase activity, whereas the Bchl content, although low, was detectable. At 110 to 125 Klett units on a rotary shaker the cells were beginning to go through a low-O_2 phase followed by the induction of the photosynthetic apparatus (see below).

Table 2 Effects of light and O_2 on the Bchl content and RuBPCase activity of *R. sphaeroides*[a]

Growth condition[b]	Bchl		RuBPCase	
	μg/mg of protein	%	U/mg of protein	%
HL	3.87	100	0.127	100
D	3.52	90	0.095	75
LL	6.90	178	0.071	75
O_2	0.48	12	0.001	0.7

[a] Cells were grown on Sistrom medium containing 0.4% butyrate and 0.1% sodium bicarbonate.
[b] HL, high-intensity light (10 W/m^2); D, shift from high-intensity light to dark for 1 h; LL, low-intensity light (3 W/m^2); O_2, partially aerobic growth.

The cellular levels of RuBPCase and Bchl were also compared among the cells growing on the different substrates, namely, butyrate, malate, and succinate (see Table 3). The cells were grown at high light and harvested during the late log phase (between 200 and 220 Klett units). Under these conditions the cells are undergoing severe shading; however, such conditions are useful in seeing maximal RuBPCase levels. The levels of Bchl showed only slight, but reproducible, differences on the different substrates, whereas the RuBPCase activity showed a marked decrease when the cells were grown on malate (20-fold) and succinate (8-fold) relative to butyrate-grown cells. The relative order of RuBPCase levels (see Table 3) is the reverse of the relative growth rates.

Table 3 Effects of carbon substrates on Bchl content and RuBPCase activity of *R. sphaeroides*

Substrate	Bchl		RuBPCase	
	μg/mg of protein	%	U/mg of protein	%
Butyrate	8.4	100	0.141	100
Malate	7.2	85	0.007	5
Succinate	7.9	94	0.019	13

Light-, O_2-, and substrate-regulated levels of mRNA coding for RuBPCase, LH-1, RC-L, and RC-M

Since RuBPCase levels and Bchl content are regulated by light, O_2, and substrates, these parameters were varied to learn more about the nature of gene regulation in *R. sphaeroides*. The RNA isolated from cells grown under the different conditions was hybridized with ^{32}P-labeled DNA probes for genes specifying RuBPCase, LH-1, RC-L, and RC-M, and semiquantitative measurements were made by dot hybridization. The Northern hybridization procedure affords additional information about the numbers and sizes as well as the amounts of the transcripts.

Dot hybridization

Figure 1 presents the results obtained from dot hybridizations of DNA probes against bulk RNA isolated from cells grown under the conditions designated. The quantitative measurements obtained from scintillation counting of the filters are summarized in Tables 4 and Table 5. The quantitative results were similar when obtained by densitometric scans of the dot blots. The C-2-C clone contains a 1.0-kilobase (kb) SalI-*Bam*HI fragment of pLI10 cloned into M13 phage (P. Hallenback, personal communication). This 1.0-kb fragment contains most of the gene for RuBPCase. The KpnI-PstI fragment of pJW1 contains the LH-1 gene as well as a portion of RC-L, and the subclones pUI135 and pUI128 are the probes for LH-1 plus RC-L and for RC-M genes, respectively. It can be determined from the data in Figure 1(a) and Table 4 that the shift of cells from high light to dark for 1 h resulted in the degradation of greater than 70% of the mRNA for RuBPCase. Similarly, cells grown under steady-state low light intensity (3 W/m^2) had a 20% reduction in the level of mRNA specifying RuBPCase when compared with cells growing at high light. Under aerobic growth conditions, no mRNA specifying RuBPCase was detectable. Although mRNA specifying LH-1, RC-L, and RC-M responded to light and O_2, several major differences were observed

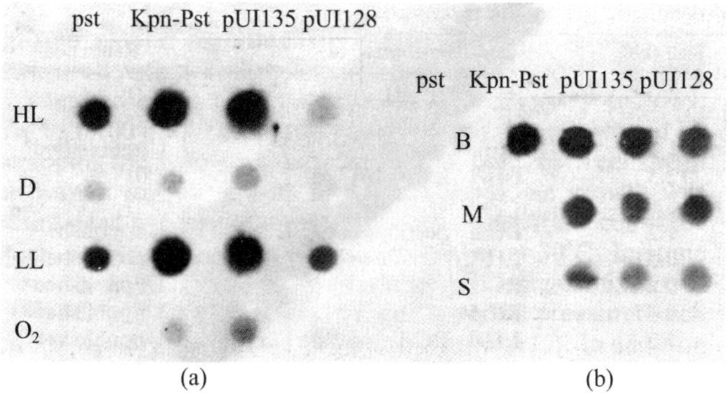

Figure 1 **Dot blot hybridization of *R. sphaeroides* RNA with cloned probes for RuBPCase, LH-1, RC-L, and RC-M. Cell cultures were grown and RNA was isolated as described in the text. Equal amounts of RNA (5 μg) after denaturation were spotted onto Gene Screen, hybridized with ^{32}P-labeled probes, and exposed to X-ray film. Each probe employed is listed above the figures and described in Table 1, and the culture conditions are listed to the left of the figures. Abbreviations: HL, high-intensity light; D, shift from highintensity light to dark for 1 h; LL, low-intensity light; O_2, partially aerobic growth; B, butyrate; M, malate; S, succinate**

Table 4 Effects of light and O_2 on levels of mRNA coding for RuBPCase, LH-1, RC-L, and RC-M of *R. sphaeroides*[a]

Growth condition[b]	Relative amt/%			
	RuBPCase	LH-1	LH-1 plus RC-L	RC-M
HL	100	100	100	100
D	34	30	28	25
LL	80	183	152	147
O_2	<1	25	28	25

[a] The relative amounts of mRNA were measured by dot hydridization, quantified by scintillation counting of the filters, and expressed as the percentages of mRNA level under high-intensity light (set as 100). Only the results within a column can be compared, not those among columns.

[b] See footnote b of Table 2.

Table 5 Effects of carbon substrates on levels of mRNA coding for RuBPCase, LH-1, RC-L, and RC-M of *R. sphaeroides*[a]

Substrate	Relative amt/%			
	RuBPCase	LH-1	LH-1 plusRC-L	RC-M
Butyrate	100	100	100	100
Malate	6	78	64	86
Succinate	14	58	63	79

[a] See footnote a of Table 4.

when we compared these results with those observed for RuBPCase. First, the decreased light intensity used for growth from 10 to 3 W/m^2 resulted in an increased level of mRNA specifying these membrane proteins; second, cells grown under partial aerobiosis still had a measurable level of mRNA (30% to 40%). This is not unusual, since detectable levels of Bchl were present. However, under high pO_2, LH-1 mRNAs were still present despite the absence of Bchl. On the other hand, a very different pattern of responses was observed when the same mRNA levels were measured in cells grown on different substrates (see Figure 1 (b), Table 5). The cells grown on butyrate synthesized substantially higher levels of mRNA for RuBPCase as compared with that synthesized when cells were grown on either malate (16-fold) or succinate (7-fold). However, relatively small differences were observed in the levels of mRNA for the specific membrane proteins when cells were grown on the different carbon substrates.

Northern hybridization

Figure 2 shows a Northern hybridization of isolated RNA with DNA probes for the RuBPCase, LH-1, RC-L, and RC-M. The RNA species were extracted from *R. sphaeroides* cells grown on butyrate under different conditions of light and O_2 as described above. After glyoxylation, the RNA was separated by agarose gel electrophoresis. Two strong bands of rRNA, 16S and 14S, can be seen on the agarose gel (see Figure 2(a)(b), right lanes). The ratio of 16S rRNA to 14S rRNA is roughly 2.1 : 1, indicating that there has been little or no degradation of rRNA during isolation[20]. The RNA was transferred to Gene Screen and hybridized with ^{32}P-labeled nick translated C-2-C DNA. Three transcripts (2.3, 2.0, and 1.4 kb) were found when fragment C-2-C was used as the probe (see Figure 2(a)). Each of these hybridization complexes was very stable, resisting melting out up to a wash temperature of slightly over 90℃. The ratio of 2.3-, 2.0-, and 1.4-kb transcripts is 1 : 4 : 2.5. Each of the three transcripts is coordinately light and oxygen regulated. Further, the amount of transcript apparent in low light-grown cells relative to high light-grown cells is precisely what one would expect from the dot blot results (see Figure 1). When a 0.91-kb internal PstI fragment of the RuBPCase gene was used as a specific probe, the same three transcripts were obtained (data not shown). No transcripts for RuBPCase were detected in aerobically grown cells. Even more interesting are the transcripts for the LH-1 polypeptides. The probe for LH-1 is the 1.2-kb KpnI-PstI fragment of pJW1 containing the LH-1 genes as well as a portion of the RC-L gene. Two major transcripts (2.6 and 0.8 kb) and two minor transcripts (2.0 and 1.45 kb) are seen in Figure 2(b), panel 1. Under high-intensity light with the KpnI-PstI probe, the ratios of the major transcripts (2.6 and 0.8 kb) to one another were approximately equal. However, after 1 h in the dark or in the presence of O_2 the amount of larger transcript (2.6 kb) decreased, but the concentration of the small transcript (0.8 kb) changed only slightly; therefore, the ratio of the 0.8-kb transcript to the 2.6-kb transcript reached a value of 12, suggesting a greater rate of turn over of the 2.6-kb

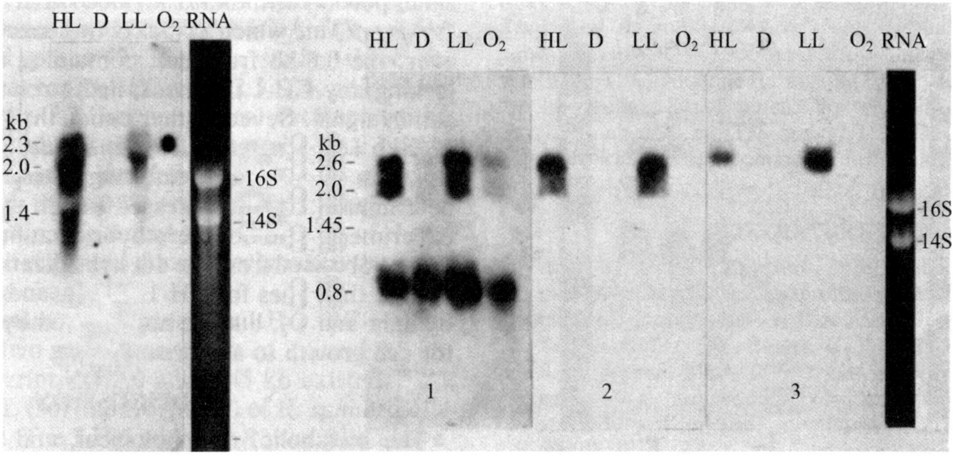

Figure 2 **Northern hybridization of *R. sphaeroides* with cloned probes for RuBPCase, LH-1, RC-M, and RC-L. Samples of glyoxylated RNA from *R. sphaeroides* cells, grown photosynthetically on the buryrate medium under the conditions described in the legend to Figure 1, were fractionated on 1.0% agarose gel (rightmost lanes of (a) and (b)), transferred to Gene Screen, and hybridized with ^{32}P-labeled probes for RuBPCase (a) and LH-1 (b), panel 1, RC-L (b), panel 2, and RC-M (b), panel 3. The probes for RuBPCase and LH-1 are listed in Table 1. The cloned fragment of PvuII-SalI (0.46 kb) and XhoI-NruI (0.37 kb) were used as the probes specific for RC-L and RC-M, respectively. Both are the internal portions of their respective genes**

transcript relative to the 0.8-kb transcript. Under low-intensity light the ratio of the small transcript to the large transcript increased to a value of 1.9.

Northern hybridizations were performed with probes specific for RC-L [see Figure 2(b), panel 2] and RC-M [see Figure 2(b), panel 3], which indicated that these genes were expressed on the large transcript (2.6 kb), but were absent from the small transcript (0.8 kb). Since these four structural genes coding for LH-1, RC-L, and RC-M are closely linked on the chromosome ([31]; P. Kiley and S. Kaplan, unpublished data), it seems reasonable to assume that the larger transcript represents a single polycistronic transcript specifying the four polypeptides: LH-1 α and β, RC-L, and RC-M. The LH-1 polypeptides appear to be specified by the small transcript. Unlike the response observed when the C-2-C probe for RuBPCase was used, no significant changes were found in the relative abundance of the two major transcripts for LH-1 when the cells were grown on the different substrates (data not shown).

Because the KpnI-PstI probe contains, in addition to the LH-I genes, a portion of the distal RC-L gene, the observed ratio of the small transcript (0.8 kb) to the large transcript (2.6 kb) (1 : 1 to 2 : 1) is not an accurate measure of the true steady-state ratio of each of these transcripts, since the RC-L and RC-M coding regions reside only on the large transcript. Therefore, an AluI DNA fragment containing only the information for the LH-1 polypeptides as well as a small amount of proximal DNA was used for Northern hybridization

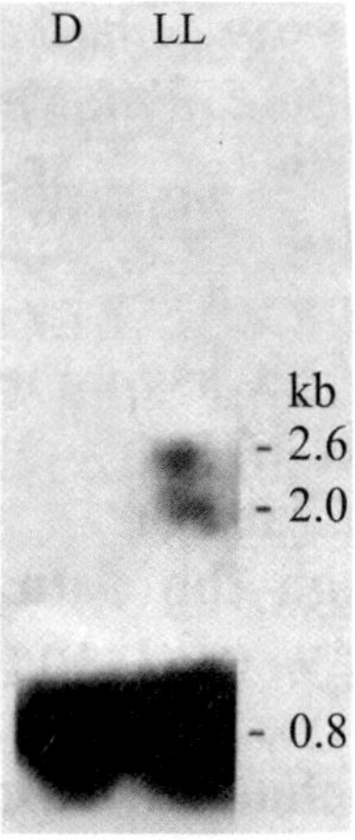

Figure 3 Northern hybridization of *R. sphaeroides* with a specific probe for LH-1. The hybridization conditions were the same as in Figure 1, except a 0.51-kb AluI-AluI fragment of pJW1 was used as a specific probe for LH-1

analysis, and the relative levels of the small and large transcripts were determined (see Figure 3). From the densitometer tracings of Figure 3, the steady-state ratio of the small transcript to the large transcript was 10 : 1. From five independent determinations of this kind, the average ratio for the small to the large transcript was 11.2 : 1, ranging from 7 : 1 to 18 : 1.

Southern hybridization

The results from the dot blots and Northern hybridization experiments were further confirmed by Southern hybridization. Figure 4 shows the results of Southern hybridization of a *Bam*HI digest of plasmid pLI10 DNA with ^{32}P-labeled bulk RNA, as described above. Plasmid pLI10 is a pBR322 derivative containing a trp promoter and having an insert of a 3.0-kb fragment of *R. sphaeroides* DNA containing the gene coding for RuBPCase II [see Table 1, Figure 4(c)]. The results, employing different conditions of light and O_2 are shown in Figure 4(a); the right lane is an agarose gel on which the 4.9-kb vector and the 3.0-kb *Bam*HI insert were separated after electrophoresis. The bulk RNA was homologous only to the 3.0-kb *Bam*HI insert. The hybridization intensity, therefore, represents primarily the relative amount of mRNA coding for RuBPCase II. Cells grown under high-intensity light had the highest concentration of mRNA for RuBPCase; these levels were greatly reduced when cells were grown under low-intensity light or incubated in the dark. There was no observable hybridization of RNA derived from aerobically grown cells.

Figure 4(b) demonstrates the changes in the levels of mRNA for RuBPCase II when the cells were grown on different substrates. Butyrate-grown cells resulted in the highest mRNA level for RuBPCase when compared with the results obtained for cells grown on succinate and malate. The relative mRNA concentrations resulting from growth on the three substrates agreed well with the values determined from the dot hybridization experiments.

Figure 5 shows the Southern hybridization with plasmid pJW1 DNA and *R. sphaeroides* derived RNA. The pJW1 DNA was digested with SalI, KpnI, NruI, and Hinc Ⅱ. Approximately 13 electrophoretic bands were obtained, as shown on the agarose gel on the right lanes of Figure 5(a) and (b). A partial map of the restriction sites located on pJW1 is presented in Figure 5(c). The 0.98-kb HincII-KpnI, 1.06-kb KpnI-SalI, and 0.6-kb SalI-NruI fragments contain genes coding for LH-1, RC-L, and RC-M, respectively (P. Kiley and S. Kaplan, unpublished data). The autoradiographs demonstrated that the LH-1 genes (0.98 kb) had the strongest hybridization signal. The 1.06-kb fragment had a rather strong hybridization signal, probably due to the fact that a small portion of the 3′ end of the LH-1 genes extended into this fragment, which contains the entire RC-L gene. However, the 0.6-

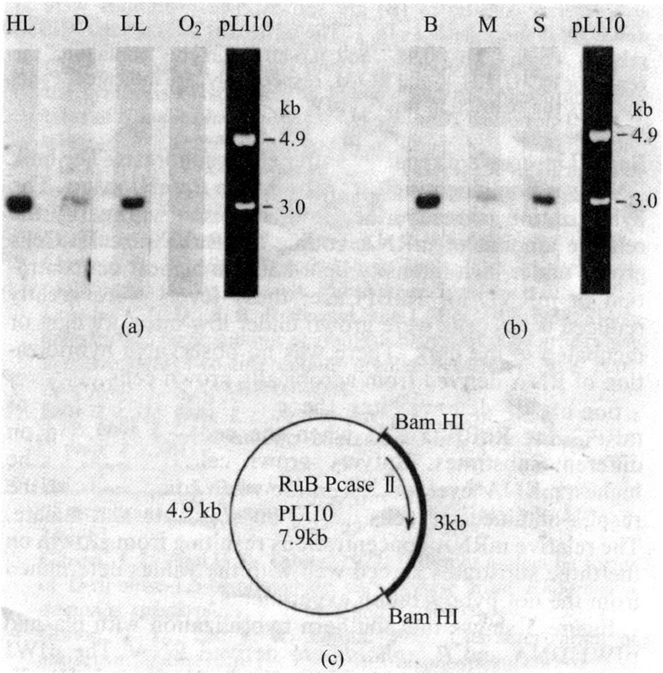

Figure 4 Southern hydridization of pLI10 *Bam*HI digest with ^{32}P-labeled RNA. The pLI10 DNA was disgested with *Bam*HI, separated on a 1.0% agarose gel, transferred to Gene Screen, and hybridized with ^{32}P-labeled RNA from *R. sphaeroides* cells. The cells were grown on butyrate medium under different lighting conditions (a) or on different substrates (b) as described in the legend to Figure 1. The rightmost lanes of (a) and (b) are agarose gels, and the remaining lanes are autoradiograms of these gels hybridized with the indicated probes, where the 4.9-kb vector and 3.0-kb *Bam*HI insert are indicated. Panel (c) shows the restriction map of pLI10 indicating the location of the gene coding for form II RuBPCase

kb fragment, containing the RC-M gene and lacking any LH-1 fragment, had a relatively weak hybridization signal. Several other bands that were located on the 12.0-kb DNA insert of *R. sphaeroides* also hybridized with RNA to various degrees, suggesting that other genetic determinants were expressed under the conditions of this experiment. The Southern hybridization results verified the results obtained from the dot hybridization experiments, that is, that the genes for LH-1, RC-L, and RC-M are regulated by light and O_2, but are not affected by the substrates used for cell growth to any great degree.

Discussion

The metabolic, morphological, and biophysical changes induced by the imposition of different environmental regimens on *R. sphaeroides* have been well studied[11,16,17]; nevertheless, information specifying the molecular events governing such changes with respect to gene regulation is limited. The present investigation with carefully defined probes

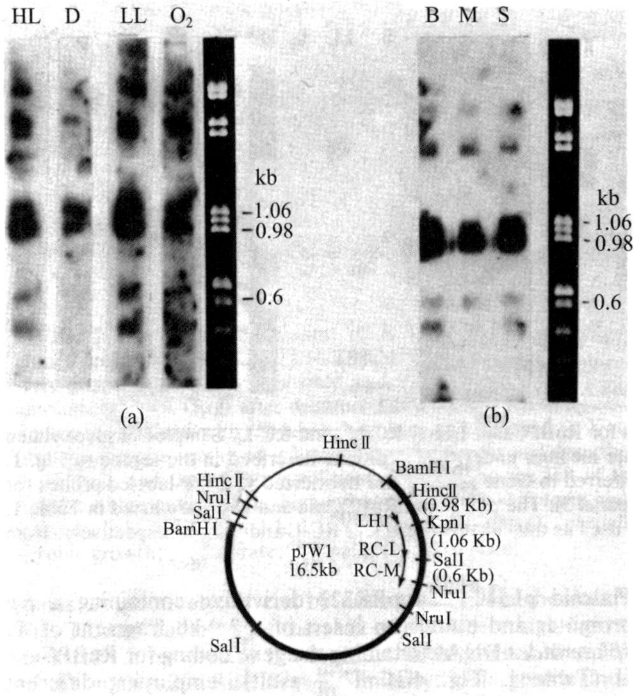

Figure 5 **Southern hybridization of pJW1 with ^{32}P-labeled RNA. The pJW1 DNA was digested with Hincll, KpnI, SalI, and NruI and blotted onto Gene Screen after fractionation on an agarose gel. The hybridization results employing different light and O_2 treatments (a) and different substrates (b) are shown. The conditions were as described in the legend to Figure 1. The rightmost lanes are the agarose gels on which 1.06-, 0.98-, and 0.6-kb fragments containing the genes for RC-L, LH-1, and RC-M, respectively, are indicated. Panel C shows the restriction map of pJW1**

and methods provides substantial, detailed information about the regulation of mRNA coding for several photosynthetic membrane proteins as well as RuBPCase in cells subjected to a variety of cultural conditions.

The hybridization experiments reported here demonstrate the importance of light as a trigger in the regulation of photosynthetic gene expression. Cells grown under low-light conditions revealed significantly reduced levels of mRNA for RuBPCase. However, under similar conditions the level of LH-1-specific mRNA was shown to be very high, as might be anticipated because more photosynthetic membrane is produced to compensate for the decreased light intensity and because there is a greater density of photosynthetic units per unit of membrane[32]. Based on the total radioactivity added for hybridization, the specific radioactivity of the RNA (3.6×10^7 cpm/μg of RNA), and the amount of radioactivity shown to hybridize, we calculated that the mRNA specific for LH-1 was about 0.006% of total RNA, but 0.2% of the total mRNA.

Growth in the presence of O_2 completely abolished the expression of the gene for

RuBPCase, whereas a considerable level of mRNA specifying LH-1 was still synthesized in the presence of O_2, although no Bchl was detected under high O_2 partial pressure. On the other hand, the amounts of mRNA fro RC-L and RC-M significantly decreased (see Figure 2 (b)), indicating that the mRNAs for RC-L and RC-M are very unstable.

As might have been anticipated, the substrates used to support photosynthetic growth had a major influence on the mRNA levels for RuBPCase. Butyrate has been shown to be the best source of reducing power and shows a very high level of RuBPCase in *R. sphaeroides*[27]. The high level of RuBPCase can be accounted for by the increased transcriptional activity of the RuBPCase gene, which shows a 17-fold increase in cognate mRNA accompanied by a 20-fold increase in RuBPCase activity on butyrate as compared with that on malate. Although the data presented here cannot exclude the possibility of translational regulation of RuBPCase expression, the data are consistent with a transcriptional control mechanism. On the other hand, the substrates had relatively small effect on the expression of the genes for LH-1, RC-L, and RC-M.

The Northern hybridization experiments revealed that multiple transcripts may exist for both RuBPCase and LH-1. A single gene for the large subunit of RuBPCase from maize yields two differentially regulated mRNAs with sizes of 1. 6 and 1. 8 kb[10]. More complex patterns of RNA are seen for other photoregulated genes[3]. In this study, three *R. sphaeroides* transcripts for RuBPCase of 2. 3, 2. 0, and 1. 4 kb were found, with either C-2-C or an internal Pst fragment of the RuBPCase gene (data not shown) as a probe. The 1. 4-kb transcript was the minimal size for the structural gene of RuBPCase. However, the 2. 0-kb transcript was the major transcript observed. We also found that the levels of these three transcripts were regulated coordinately when the light and O_2 pressure were changed.

Northern hybridization of the genes for LH-1, RC-L, and RC-M revealed that two major transcripts of 2. 6 and 0. 8 kb and two minor transcripts of 2. 0 and 1. 45 kb existed. The genes coding for RC-L[30] and RC-M[31] of *R. sphaeroides* have been sequenced. The LH-1 genes in *R. capsulata* have been sequenced and code for two polypeptides, B870α and B870β. In *R. sphaeroides*, the LH-1 genes have approximately 350 base pairs, with a spacer of about 150 base pairs, and are linked to the RC-L and RC-M genes (P. Kiley and S. Kaplan, unpublished data). Although these four genes may have separate start codons, they probably constitute the same operon[33]. Since the 2. 6-kb transcript is long enough to code for LH-1, RC-L, and RC-M, it probably gives rise to a polycistronic mRNA expressing at least these four genes. The 0. 8-kb transcript is long enough to code for LH-1 and may be a single transcript of only the LH-1 genes. This assumption is confirmed by the results of the Northern hybridization of RNA with probes specific for RC-L and RC-M (see Figure 2 (b))[2,3], because only the 2. 6-kb transcript, but not the 0. 8-kb transcript, hybridized under these conditions. If we assume that each of these transcripts is initiated from the same site and ignore the mechanism by which the smaller as opposed to the larger is generated, then this may be the basis for the existence of a relatively fixed, 15 : 1 stoichiometry between

LH-1 and the RC polypeptides[7]. The average ratio of the small transcript (0. 8 kb) to the large transcript (2. 6 kb) (11. 2 : 1) is within reasonable error identical to the fixed ratio observed for the gene products of the LH-1, RC-L, and RC-M genes. Therefore, the relative steady-state abundance of these two transcripts may be the sole basis for the stoichiometry of the gene products. The exact mechanism, initiation, termination, or selective degradation leading to the steady-state abundance of each of these transcripts in *R. sphaeroides* remains to be determined. In a paper that appeared recently, Belasco et al. [2] made essentially identical observations for the presence and relative abundance of two mRNA species for the LH-1, RC-L, and RC-M genes in *R. capsulata*. These authors concluded that selective degradation from the 3′ end of the large transcript is responsible for the steady-state level of the small and large transcripts. The 2. 6-kb transcript greatly diminished when the cells were shifted from light to dark or when cells were grown aerobically, whereas the concentration of the 0. 8-kb transcript was maintained at almost the same levels under identical conditions (see Figure 2). The different stability of two transcripts may be involved in the regulation of the LH-1, RC-L and RC-M genes in *R. sphaeroides* and, as suggested by Belasco et al. [2], for *R. capsulata*. This result will be described in more detail in a separate paper.

Acknowledgments

This work was supported by Public Health Service grants GM15590 and GM31667 from the National Institutes of Health and National Science Foundation grant PCM 8317682 to S. K.

We are most grateful to J. Williams for providing plasmid pJW1, to P. Kiley for providing cloned KpnI-PstI, AluI, PvuII-SalI, and XhoI-NruI fragment DNA, and to P. Hallenbeck for providing plasmid pLI10, C-2-C DNA, and the internal Pst DNA fragment. We thank T. J. Donohue for making [γ-^{32}P] ATP and W. D. Sheperd for his technical assistance. We especially thank S. Kuhl for providing plasmid pUI135 and pUI128 DNAs as well as for his helpful discussion and comments during the writing of this paper.

References

[1] Apel, K., and K. Koppstech. 1978. Light-induced appearance of mRNA coding for the apoprotein of the light-harvesting chlorophyll a/b protein. Eur. J. Biochem. 85:581 - 588.

[2] Belasco, J. G., J. T. Beatty, C. W. Adams, A. von Gabain, and S. N. Cohen. 1985. Differential expression of photosynthesis genes in *R. capsulata* results from segmental differences in stability within the polycistronic rxc A transcript. Cell 40:171 - 181.

[3] Bogorad, L., L. D. Crossland, L. E. Fish, E. T. Krebbers, U. Kuck, I. M. Larrinua, K. M. T. Muskavitch, E. A. Orr, S. R. Rodermel, R. Schantz, A. A. Steinmetz, S. M. Stirdivant, and Yu. S. Zhu. 1983. The organization of the maize plastid chromosome properties and expresion of its genes. In UCLA Symp. Mol. Cell. Biol. New Series 14:257 - 272.

[4] Bogorad, L., E. J. Gubbins, E. T. Krebbs, I. M. Larrinua, B. J. Mulligan, K. M. T. Muskavitch, E.

A. Orr, S. R. Rodermel, R. Shantz, A. A. Steinmetz, G. D. Vos, and Y. K. Ye. 1983. Cloning and physical mapping of maize plastid genes. Methods Enzymol. 97:524 - 555.

[5] Chory, J., T. J. Donohue, A. R. Varga, L. A. Staehelin, and S. Kaplan. 1984. Induction of the photosynthetic membranes of *Rhodopseudomonas sphaeroides*: biochemical and morphological studies. J. Bacteriol. 159:540 - 554.

[6] Chory, J., and S. Kaplan. 1982. The in vitro transcription-translation of DNA and RNA templates by extracts of *Rhodopseudomonas sphaeroides*. J. Biol. Chem. 257:15110 - 15121.

[7] Chory, J., and S. Kaplan. 1983. Light-dependent regulation of the synthesis of soluble and intracytoplasmic membrane proteins of *Rhodopseudomonas sphaeroides*. J. Bacteriol. 153:465 - 474.

[8] Clark, W. G., E. Davidson, and B. L. Marrs. 1984. Variation of levels of mRNA coding for antenna and reaction center polypeptides in Rhodopseudomonas capsulata in response to changes in oxygen concentration. J. Bacteriol. 157:945 - 948.

[9] Clayton, R. K. 1966. Spectroscopic analysis of bacteriochlorophylls *in vitro* and *in vivo*. Photochem. Photobiol. 5:669 - 677.

[10] Crossland, L. D., S. R. Rodermel, and L. Bogorad. 1984. Single gene for the large subunit of ribulose bisphosphate carboxylase in maize yields two differentially regulated mRNAs. Proc. Natl. Acad. Sci. U. S. A. 81:4060 - 4064.

[11] Drews, G. 1978. Structure and development of the membrane system of photosynthetic bacteria. Curr. Top Bioenerget. 8:161 - 207.

[12] Drews, G., and J. Oelze. 1981. Organization and Differentiation of membranes of phototropic bacteria. Adv. Microb. Physiol. 22:1 - 81.

[13] Gibson, J. L., and F. R. Tabita. 1977. Different molecular forms of D-ribulose-1, 5-bisphosphate carboxylase from *Rhodopseudomonas sphaeroides*. J. Biol. Chem. 252:943 - 949.

[14] Gibson, J. L., and F. R. Tabita. 1977. Characterization of antiserum against form II ribulose 1, 5-bisphosphate carboxylase from *Rhodopseudomonas sphaeroides*. J. Bacteriol. 131:1020 - 1022.

[15] Johnson, R. A., and T. F. Walseth. 1979. The enzymatic preparation of [α-^{32}P]ATP, [α-^{32}P]GTP, [α-^{32}P]cAMP, and [^{32}P]cGMP, and their use in the assay of adenylate and guanylate cyclases and cyclic nucleotide phosphodiesterases. Adv. Cyclic Nucleotide Res. 10:135 - 167.

[16] Kaplan, S. 1978. Control and kinetics of photosynthetic membrane development, p. 809 - 839. In R. K. Clayton and W. R. Sistrom (ed.), The photosynthetic bacteria. Plenum Publishing Corp., New York.

[17] Kaplan, S., and C. J. Arntzen. 1982. Photosynthetic membrane structure and function, p. 65 - 151. In Govindjee (ed.), Photosynthesis in energy conversion by plants and bacteria, vol. Academic Press, Inc., New York.

[18] Maniatis, T., E. F. Fritsch, and J. Sambrook. 1982. Molecular cloning, p. 466. Cold Spring Harbor Laboratory, Cold Spring Harbor, N. Y.

[19] Markwell, M. A. K., M. H. Suzanne, L. L. Bieber, and N. E. Tolbert. 1963. A modification of the Lowry procedure to simplify protein determination in membrane and in protein samples. Anal. Biochem. 87:206 - 210.

[20] Marrs, B., and S. Kaplan. 1970. 23S Precursor ribosomal RNA of *Rhodopseudomonas sphaeroides*. J. Mol. Biol. 49:297 - 317.

[21] Miziorko, H. M., and G. H. Lorimer. 1983. Ribulose-1, 5-bisphosphate carboxylase/oxygenase. Annu. Rev. Biochem. 521:507 - 537.

[22] Muller, E. D., J. Chory, and S. Kaplan. 1984. Cloning and expression of the ribulose-1, 5-bisphosphate carboxylase gene of *Rhodopseudomonas sphaeroides*. J. Bacteriol. 161:469 - 472.

[23] Saito, H., and K. Miura. 1963. Preparation of transforming deoxyribonucleic acid by phenol

treatment. Biochim. Biophys. Acta 72:619 - 629.

[24] Sauer, L., and L. A. Austin. 1978. Bacteriochlorophyll-protein complex from the light-harvesting antenna of photosynthetic bacteria. Biochemistry 17:2011 - 2019.

[25] Shepherd, H. J., S. G. Ledoight, and S. Howell. 1983. Regulation of light-harvesting chlorophyll-binding protein (LHCP) mRNA accumulation during the cell cyle in Chlamydomonas reinhardii. Cell 32:99 - 107.

[26] Sistrom, W. R. 1977. Transfer of chromosomal genes mediated by plasmid R68.45 in *Rhodopseudomonas sphaeroides*. J. Bacteriol. 131:526 - 532.

[27] Tabita, F. R. 1981. Molecular regulation of carbon dioxide assimilation in autotrophic microorganisms, p. 70 - 82. In H. Dalton (ed.), Microbial growth on Cl compounds. Heydon and Sons, London.

[28] Tabita, F. R., P. Caruso, and W. Whitman. 1978. Facile assay of enzymes unique to the Calvin cycle in intact cells, with special reference to ribulose-1,5-bisphosphate carboxylase. Anal. Biochem. 84: 462 - 472.

[29] Tabita, F. R., L. E. Sarles, R. G. Quirey, Jr. Weaver, and F. E. Waddill. 1983. Molecular regulation, mechanism, and enzymology of autotrophic carbon dixoide fixation, p. 148 - 154. In Microbiology-1983. American Society for Microbiology, Washington, D. C.

[30] Williams, J. C., L. A. Steiner, G. Feher, and M. I. Simon. 1984. Primary structure of the L subunit of the reaction center from *Rhodopseudomonas sphaeroides*. Proc. Natl. Acad. Sci. U. S. A. 81:7303 - 7307.

[31] Williams, J. C., L. A. Steiner, R. C. Ogden, M. I. Simon, and G. Feher. 1983. Primary structure of the M subunit of the reaction center from *Rhodopseudomonas sphaeroides*. Proc. Natl. Acad. Sci. U. S. A. 80:6505 - 6509.

[32] Yen, G. S. L., B. D. Cain, and S. Kaplan. 1984. Cell-cycle-specific biosynthesis of the photosynthetic membrane of *Rhodopseudomonas sphaeroides*. Structural implication. Biochim. Biophys. Acta 777:41 - 55.

[33] Youvan, D. C., M. Alberti, H. Begusch, E. J. Bylina, and J. E. Hearst. 1984. Reaction center and light-harvesting I genes from Rhodopseudomonas capsulata. Proc. Natl. Acad. Sci. U. S. A. 81:189 - 192.

[34] Yu, K. L., B. Hohn, H. Falk, and G. Drews. 1982. Molecular cloning of the ribosomal RNA genes of the photosynthetic bacterium Rhodopseudomonas capsulata. Mol. Gen. Genet. 188:392 - 398.

Origin of the mRNA Stoichiometry of the *puf* Operon in Rhodobacter sphaeroides*

(Received for publication, September 16,1985)

The LH-I structural genes are located 5′ of the RC-L and -M structural genes on what has been designated as the *puf* operon of *Rhodobacter sphaeroides*. Analysis of *puf* operon expression in *R. sphaeroides* by Northern hybridization with probes specific for individual structural genes has identified two transcripts encoded by this operon. The large (2.6 kilobase pairs (kb)) transcript contains sequences for all four polypeptides of the *puf* operon, whereas the small (0.5 kb) transcript, which is more abundant (10—15-fold) than the large transcript under photosynthetic growth, is homologous only to the two LH-I structural genes. Transcription of the *puf* operon during photosynthetic growth under saturating light conditions is increased approximately 3-fold relative to growth in the presence of oxygen while the relative ratio of these two transcripts is independent of the incident light intensity. Analysis of the turnover of the two transcripts ($t_{1/2}$ of 9 and 20 min for the large and small transcripts, respectively) indicates that 5′ processing is the initial step in the degradation of the large transcript and that the molar excess of the small transcript cannot be accounted for by differences in the rates of turnover of these two mRNA species. Analysis of the 5′ ends of the 2.6- and 0.5-kb transcripts, their relative abundance, and stabilities indicates that these two transcripts have different 5′-ends corresponding to 75 and 104 base pairs upstream from the start of the LH-I β structural gene, respectively. Northern hybridization analysis with specific synthetic deoxyoligonucleotide probes confirmed that the two transcripts differ by 29 bases at their 5′-ends, suggesting that differential transcript

---

* Yu Sheng Zhu, Patricia J. Kiley, Timothy J. Donohue, Samuel Kaplank, From the Department of Microbiology, University of Illinois at Urbana-Champaign, Urbana, Illinois 61801

此文发表于 1986 年"J. Biol. Chem." 261:10366-10374.

This work was supported by Grants GM15590 and GM31667 from the National Institutes of Health to S. Kaplan. This work was presented at the 5th International Meeting on Photosynthetic Procaryotes held in Grindelwald, Switzerland, in 1985. It was at this meeting that the nomenclature employed here for this operon was decided upon. The genus name of this bacterium has recently been changed to Rhodobacter from Rhodopseudomonas[7]. The costs of publication of this article were defrayed in part by the payment of page charges. This article must therefore be hereby marked "advertisement" in accordance with 18 U.S.C. Section 1734 solely to indicate this fact.

‡ Predoctoral fellow supported by United States Public Health Service Training Grant GM07283 from the National Institutes of Health.

initiation may be involved in regulating the relative levels of these two mRNA species in vivo, although we cannot rule out complex mechanisms of post-transcriptional processing.

In *Rhodobacter sphaeroides*, a purple nonsulfur photosynthetic bacterium, the two light-harvesting complexes constitute the major pigment-protein components found in the photosynthetic membrane[1]. They are composed of the light-harvesting I (B875, designated LH-I) and the light-harvesting II (B800 – 850, designated LH-II) bacteriochlorophyll-protein complexes. A third spectral complex, designated the reaction center (RC) bacteriochlorophyll-protein complex, is composed of equimolar amounts of the reaction center polypeptides L, M, and H (designated RC-L, RC-M, and RC-H). The LH-I complex accepts light energy from a pool of peripherally arranged LH-II complexes (which represent the primary source of photon capture) and transfers this excitation energy to a centrally located RC complex where the primary photochemical reaction occurs[2]. The LH-I complexes are found in the photosynthetic membrane in a fixed molar excess (approximately 12-fold) over the RC complex[3]. However, each of the two polypeptides within an LH-I complex is present in a 1 : 1 stoichiometry. The LH-I complex together with the RC complex have been termed the fixed photosynthetic unit, PUF[4]. Recently, it has been found that the structural genes for the two polypeptide subunits of LH-I (β, α) and the RC-L and RC-M polypeptides are closely linked in a single operon in *R. sphaeroides*[5,6], ① similar to that reported for Rhodobacter capsulatus[8], and which we designate the *puf* operon. The structural genes for RC-L (*puf* L)[6], RC-M (*puf* M)[5], and LH-I (*puf*B, A)[1] have been sequenced, and the two LH-I genes (*puf*B, A) are located proximal to the distal *puf* L, M genes. Since the polypeptides comprising an individual pigment-protein complex derived from the *puf* operon are synthesized and inserted into the photosynthetic membrane[1-3,9] in unique but different stoichiometries in both *R. sphaeroides* and *R. capsulatus*, elucidating the mechanism by which the coordinate but differential expression of these genes is accomplished of primary importance and interest.

We have previously reported[10] that two transcripts are derived from the *puf* operon: the larger and less abundant transcript is a polycistronic mRNA possessing information for all four polypeptides; the smaller but more abundant transcript is specific for only the LH-I β and α apoproteins. The ratio of these two transcripts reflects the ratio of the LH-I and RC complexes found in the photosynthetic membrane. In this study we have extended our analysis of the stability, regulation by light and oxygen, and 5′-ends of the stable mRNA species encoded by the *puf* operon of *R. sphaeroides*. Similar studies were recently performed by Belasco et al.[11] for *R. capsulatus*, and they concluded that the large transcript, following selective degradation from the 3′-end, is the precursor of the smaller transcript.

The present work demonstrates that the two transcripts of the *puf* operon in *R.*

---

① P. J. Kiley, T. J. Donohue, and S. Kaplan, manuscript in preparation.

sphaeroides have different 5′ termini; the 5′ terminus of the large (2. 6 kb①) and small (0. 5 kb) transcripts have been precisely mapped to 75 and 104 nucleotides, respectively, upstream of the start codon of the *puf*B gene. Our data also show that differences in the stability of these two transcripts may exert secondary post-transcriptional control over the stoichiometry of the LH-I and RC complexes within the photosynthetic membrane of *R. sphaeroides*. Finally, we discuss these results in light of differential transcription initiation versus post-transcriptional modifications of primary transcripts as possible mechanisms for achieving the stoichiometry of the small and large *puf* operon transcripts.

Experimental Procedures

Bacterial Strains and Cell Growth

R. sphaeroides 2. 4. 1 or 2. 4. 1 (pUIIA2) was used where indicated in this study. Strain 2. 4. 1 (pUIIA2) contains plasmid pUIIA2 which has an *R. sphaeroides* promoter and translational regulatory regions fused to the lac operon described by Casadaban et al. [12] which is contained in the broad host range plasmid pUI108[13]. This strain was employed for the studies described, because measurement of β-galactosidase activity and lac-specific mRNA represented reliable internal controls. The β- galactosidase activity was measured as described previously[12]. Control experiments (some of which are contained in this paper) have shown no differences in the expression of the *puf* operon in this strain relative to that observed for *R. sphaeroides* 2. 4. 1. The cells were grown on Sistrom's medium A containing 0. 4% butyrate and 0. 1% $NaHCO_3$ as the carbon source unless otherwise stated. For experiments employing *R. sphaeroides* (pUI1A2) this medium was supplemented with streptomycin at 25 μg/ml. Photoheterotrophically grown cells were cultured to a cell density of 5×10^8 cells/ml at 30℃ under different light intensities: 10 watts/m^2 was designated high intensity light (HL) and 3 watts/m^2 was designated low intensity light (LL), as previously described[10]. The generation times of *R. sphaeroides* 2. 4. 1 (pUI1A2) are 5. 5 and 12 h in the presence of high and low light, respectively. Dark treatment was performed by wrapping the growth vessels from high light-grown cells with aluminum foil for 1 h at 30℃. Aerobic growth (5-h generation time) for strain 2. 4. 1 (pUI1A2) was maintained by sparging with a gas mixture of O_2 : N_2 : CO_2(25 : 74 : 1).

In experiments determining the decay of *puf*-specific mRNA, cells were grown on succinate (0. 4%) containing medium sparged with a gas mixture of N_2 : CO_2(98 : 2) to a cell density of 2×10^9 cells/ml at high intensity light. Where indicated, transcription initiation was inhibited by the addition of rifampicin (200 μg/ml), and cells were harvested for RNA isolation at the times indicated. The decay of the *puf*-specific mRNA was also measured following either a shift of photoheterotrophically grown cells to dark conditions or

① The abbreviations used are: kb, kilobase pair (s); Pipes, 1,4-piper-azinediethanesulfonic acid.

by changing the gas mixture from N_2 : CO_2 (98 : 2) to O_2 : N_2 : CO_2 (25 : 74 : 1). Cells were withdrawn for RNA isolation at various time intervals as indicated in the text. Zero time samples were those withdrawn from the culture just prior to imposition of the new growth conditions.

Extraction of RNA, Dot Blots, and Northern Hybridization

R. sphaeroides RNA was isolated and extracted from cells as described previously[10]. Dot blots and Northern hybridizations with nick-translated probes were performed according to the previously described procedures[14]. Dot blots were used for quantitative assessment of cellular mRNA levels in response to light and O_2. The Northern hybridization provided additional information on the number, sizes, and the relative amount of the individual *puf* transcripts. Quantitation was performed by scintillation counting of the filters or densitometer scans of autoradiograms well within the linear response of the X-ray film. The levels of specific cellular mRNA species determined by dot blots and Northern hybridizations were in agreement.

Northern hybridizations with ^{32}P-labeled deoxyoligonucleotide probes were performed as follows. The hybridization buffer was essentially that described by Wallace et al. [15] and consisted of 90 mmol/L Tris-Cl (pH 7.5), 0.9 mol/L NaCl, 6 mmol/L EDTA, 0.5% sodium dodecyl sulfate, 10% dextran sulfate, 5×Denhardt's, 100 μg/ml Escherichia coli tRNA, and 1 mmol/L ATP to minimize background hybridization from unincorporated [γ-^{32}P] ATP used to label the deoxyoligonucleotide probes. The preheated (55℃) hybridization buffer was passed through a 0.45 μm nitrocellulose filter prior to use to reduce background hybridization. GeneScreen filters were prepared for hybridization by a 1-h incubation with hybridization buffer at 55℃ followed by at least a 1-h incubation at room temperature (23℃) prior to the addition of the labeled deoxyoligonucleotide probes. Hybridization was conducted at room temperature for 12—16 h followed by washing of the filter (2 washes of 5 min each) at the indicated temperature in 6×SSC. While still moist, the washed filters were covered with plastic wrap for exposure to X-ray film; the use of moist filters facilitates the removal of nonspecifically bound probes in subsequent higher stringency washes at elevated temperatures.

S-1 Nuclease Protection Analysis

S-1 nuclease protection analysis was performed essentially as described by Mullet et al. [16]. The source of plasmid DNA for purification of restriction fragments was a series of specific subclones of plasmid pJW1[15] in the plasmid pUC19. [1] Restriction fragments were eluted from 5% polyacrylamide gels and purified by chromatography over DE52 prior to ethanol precipitation[17]. Individual restriction fragments were dephosphorylated with bacterial alkaline phosphatase prior to 5′-end labeling with [γ-^{32}P] ATP and T_4 polynucleotide kinase[17]. The [^{32}P]DNA was denatured, and single complementary strands were separated on 5% polyacrylamide gels[17]. For hybridization, the purified 5′-end-labeled single-stranded [^{32}P] DNA (5,000—20,000 cpm/assay) and carrier yeast tRNA (10 μg/

assay) were resuspended in 80% formamide (10 μl/assay), boiled for 5 min, and then added to a dry pellet containing *R. sphaeroides* RNA (50 μg/assay) and S-l hybrdization buffer to give a final concentration of 400 mmol/L NaCl, 40 mmol/L Pipes-NaOH, pH 6.4, 1 mmol/L EDTA. Hybridization was performed at 37℃ for 4 h. The sample was then diluted 1 : 10 with a mixture containing 250 mmol/L NaCl, 30 mmol/L NaOAc (pH 4.6), 1 mmol/L $ZnSO_4$, 20 μg/ml denatured salmon sperm DNA, and 25 units of S-l nuclease were added. The mixture was incubated at 30℃ for 40 min. After phenol chloroform extraction, the S-l-protected nucleic acids were analyzed on either a 5% polyacrylamide gel or a 6% polyacrylamide, 8.3 mol/L urea DNA-sequencing gel relative to Hinfl-or AluI-generated pBR322 molecular weight standards[17]. For precise mapping of 5′-ends, DNA sequence ladders were generated by Maxam and Gilbert sequencing of the appropriate end-labeled restriction fragment[17].

Primer Extension Assay

The indicated 5′-end-labeled double stranded primer [^{32}P]DNA was denatured; the strands were separated and hybridized to *R. sphaeroides* RNA in the same manner as for the S-1 protection assay. The primer extension assay was performed using the procedure of Mullet et al.[16]. The reaction proceeded in 20 μl of 50 mmol/L Tris-HCl, pH 8.3, 70 mmol/L KCl, 5 mmol/L $MgCl_2$, 10 mmol/L dithiothreitol, 0.1 mg/ml bovine serum albumin, 1 mmol/L of each dNTP, and 20 units of avian myeloblastosis virus reverse transcriptase at 37℃ for 60 min. After phenol extraction and ethanol precipitation, the nucleic acids were analyzed on polyacrylamide gels by the same procedure used for the S-l protection analysis.

Materials

With the exception of avian myeloblastosis virus reverse transcriptase (Seikagaku America, Inc., St. Petersburg, F L), all restriction endonucleases or DNA-modifying enzymes were the products of either Bethesda Research Laboratories or New England Biolabs (Beverly, M A). GeneScreen and $^{32}PO_4$ (carrier free) were from New England Nuclear, and [α-^{32}P]dCTP (8000 ci/mmol) was from Amersham Corp. Dextran sulfate was obtained from Pharmacia Fine Chemicals, and nitrocellulose was supplied by Schleicher and Schuell. With the exception of phenol, which was distilled before use, all other chemicals were of reagent grade and were used without further purification. Synthetic deoxyoligonucleotides were synthesized at the Biotechnology Center on the campus of the University of Illinois at Urbana-Champaign on an Applied Biosystems model 380A DNA synthesizer.

Results

Effects of Light and Oxygen on the Levels of mRNAs Specifying LH-I, RC-L, and RC-M

The cellular levels of mRNA specifying the LH-I, RC-L, and RC-M polypeptides in *R. sphaeroides* 2.4.1 (pUI1A2) were significantly affected by light and oxygen as revealed by dot blot and Northern hybridization analysis. These results are similar to those reported earlier[10] for *R. sphaeroides* 2. 4. 1 lacking plasmid (pUI1A2). Table 1 shows a quantitative estimate of the mRNA levels obtained by dot hybridization in cells grown under different physiological conditions. For these experiments an AluI-AluI (see Figure 1, fragment a, 0. 51 kb), PυuII-SalI (see Figure 1, fragment b, 0. 46 kb), or XhoI-NruI (see Figure 1, fragment c, 0. 37 kb) restriction fragment was used as a specific probe for the LH-I, RC-L, and RC-M structural genes, respectively. Dark treatment of high light-grown cells for 1 h resulted in a 35% reduction in the level of mRNA specific for the LH-I polypeptides and a 90% reduction in the level of *puf*L and *puf*M specific mRNA, suggesting a differential stability of these two transcripts. Steady-state photosynthetic growth at low light intensity (3 watts/m^2) resulted in an increased level of the mRNA specifying these intracytoplasmic membrane proteins (3. 5-fold for LH-I; 2. 5 - 2. 8-fold for RC-M and RC-L) when compared to growth at 10 watts/m^2. These data parallel the whole cell specific Bchl content of these cells which was used to approximate the level of intracytoplasmic membrane (see Table 1). Cells grown under steady-state aerobic conditions contained a detectable level of mRNA specific for LH-I (31% of that found at high light photosynthetic growth), whereas the level of mRNA specific for *puf*L and *puf*M decreased nearly 20-fold relative to their steady-state levels in high light-grown cells. The levels of mRNA specifying lacZ from *R. sphaeroides* 2. 4. 1 (pUI1A2) under different growth conditions were also measured using phage mpllw replicative form DNA as a probe for the lacZ gene. Under either high light or aerobic growth conditions *R. sphaeroides* 2. 4. 1 (pUI1A2) had nearly identical growth rates, LacZ-specific mRNA, and β-galactosidase activity (data not shown). When the growth rate was reduced

Table 1 Effects of light and O_2 on bacteriochlorophyll content and mRNA levels specifying LH-I, RC-L, and RC-M polypeptides of *R. sphaeroides* 2. 4. 1 (pUI1A2)

Growth condition[a]	Bchl[b]		Relative mRNA amount[c]		
			LH-I	RC-L	RC-M
	μg/mg protein	%		%	
HL	6. 5	100	100	100	100
D	6. 0	0	9265	10	9
LL	29. 4	452	358	284	258
O_2	0	0	31	6	7

[a] HL, cells were photosynthetically grown on butyrate medium using high intensity light (10 watts/m^2). D, the HL-grown cells were shifted to darkness for 1 h. LL, cells were photosynthetically grown using low intensity light (3 watts/m^2). O_2, cells were grown aerobically (25% O_2).

[b] The total bacteriochlorophyll (Bchl) content of cells was estimated by absorbance measurements at 775 nm of acetone/methanol (7 : 2, v/v) extracts of whole cells assuming a millimolar extinction coefficient of 75[18]. Protein concentrations were assayed by a modified Lowry procedure[19].

[c] The relative amounts of mRNA were measured by dot hybridization and by scintillation counting of the filters and are expressed as the percentages of the mRNA level under high intensity light (HL) (set as 100). Because different restriction fragment probes were employed, only the results within a column can be compared, not those between columns.

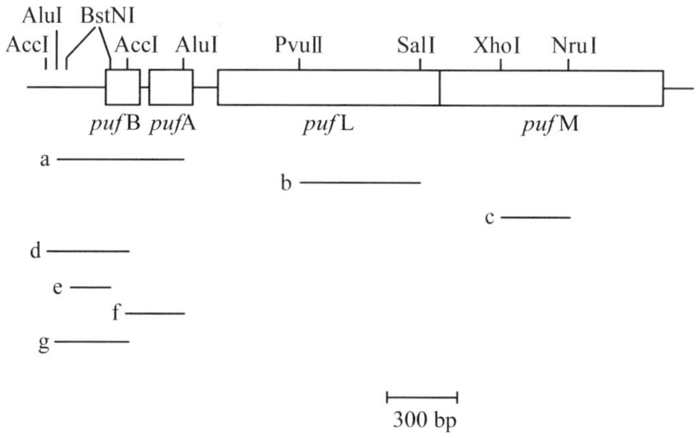

Figure 1 Restriction map of the *puf* operom. The genes coding for the β and α subunits of LH-I, RC-L, and RC-M polypeptides in *R. sphaeroides* lie in a single operon. Individual restriction sites used to generate the a (AluI), b (PvuII-SalII), c (XhoI-NruI), d (AccI), e (BstNI), and f (AccI-AluI) probes used for hybridization are indicated. The AccI, AluI, and BstNI sites depicted within the 5′-end of the *puf* operon are shown for the purposes of illustration; these do not represent the sole sites for these restriction endonucleases within this operon. bp, base pairs

over 50% by growth in the presence of low light the Lac-specific mRNA level declined by a factor of 4, and the level of β-galactosidase activity was reduced by a factor of 3 (data not shown).

As shown in Figure 2, the Northern hybridizations using the specific probes, a, b, and c referred to above, against RNA isolated from *R. sphaeroides* 2. 4. 1 (pUI1A2). Two major RNA species (2. 6 and 0. 5 kb) were observed when the AluI fragment was used as a specific probe for the LH-I structural genes. The large *puf* operon transcript hybridized with the ^{32}P-labeled probes specific for LH-I, RC-L, and RC-M structural genes, whereas the small transcript only hybridized with the probe for the LH-I structural genes. Therefore, the large transcript represents a polycistronic mRNA specifying the *puf*B, -A, -L and -M polypeptides, and the small transcript encodes only the *puf*B and -A polypeptides. The ratio of the small to the large transcript under either high or low light conditions shown in Figure 2 was approximately 10 : 1. Both the experiments reported in this work and in a previous study[10] have documented the molar excess of the small *puf* operon transcript relative to the large mRNA with the range from numerous independent determinations on steady-state photosynthetic cells varying from approximately 7 to 18 to 1. The small *puf* operon transcript was even more abundant under nonphotosynthetic conditions, i. e. darkness (ratio > 50 : 1) or oxygen (ratio 20 : 1). Besides these two major transcripts, several minor transcripts (2. 0, 1. 4, 1. 1 kb) were observed only when hybridization was with the probes specific for either *puf*L and *puf*M. Therefore, this analysis suggests that these minor transcripts might represent specific degradation products of the large transcript (see Figure 2).

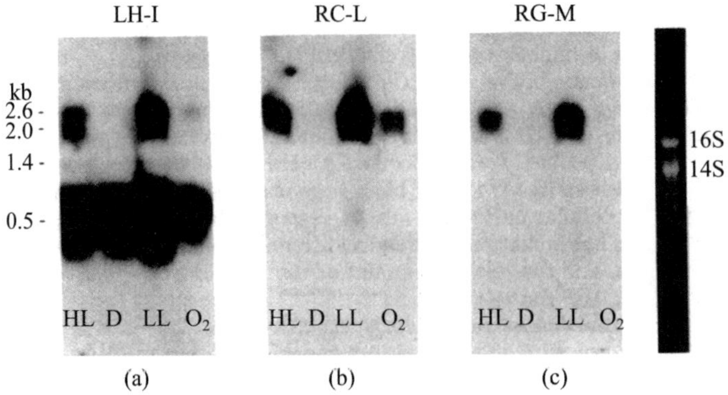

Figure 2 Northern hybridization of RNA from *R. sphaeroides* (pUI1A2) with probes specific for the LH-I, RC-L, and RC-M polypeptides structural genes. RNA was isolated from *R. sphaeroides* cells grown photosynthetically on butyrate medium under the conditions described in the footnotes to Table 1. After glyoxylation, approximately 30 μg of bulk RNA was separated on a 1.0% agarose gel and transferred to GeneScreen. Hybridization with the specific probes for the LH-I, RC-L, and RC-M polypeptide structural genes listed in Figure 1 was performed as described under "Experimental Procedures." The rightmost lane is an ethidium bromide-stained agarose gel showing the 14 and 16 S rRNA of *R. sphaeroides*[32]

Differential Decay of the Large and Small *puf* Operon Transcripts

In order to obtain a quantitative estimate of the kinetics of decay of the 2.6- and 0.5 -kb mRNA species, the decay rates of these transcipts in high light-grown cells of *R. sphaeroides* 2.4.1 (pUI1A2) were studied following inhibition of transcription using rifmpicin. Samples were taken at various time intervals for the assay of specific mRNA levels by dot blots and Northern hybridizations employing the specific prodes described above and in Figure 2. The results as shown in Figure 3(a) where it can be seen that the decay of these transcripts followed exponential kinetics after the addition of rifampicin. The half-life ($t_{1/2}$) of the mRNA sequences specifying the LH-1 structural genes was 20 min, whereas the corresponding value for the sequences corresponding to *puf*L and *puf*M as 9 min. Similar decay rates for the individual mRNA sequences requences by the two major transcripts were also obtained by analysis of Northern hybridizations (data not shown). Because the LH-I-sequences contained within the small transcript are approximately 15-fold more abundant than homologous sequences encoded on the large transcript in high light-grown cells (see above), the values reported for the half-life of the LH-I specific sequence have not been corrected for the decay of the large transcript. However, such a correction would be minimal and would not significantly alter the calculated results for mRNA half-lives.

The kinetics of decay of these transcripts were different when the high light-grown cells were shifted to darkness. The levesls of the small and large transcripts specific for LH-I, RC-L, and RC-M polypeptides increased for the first 5 min following a shift to the dark and then decreased at different rates. In general, the turnover of these transcripts was slower under such conditions relative to the rifampicin-treated cells [see Figure 3(c)]. The initial increase in transcription of this operon may reflect a derepression in the rate of transcription of this operon due to light limitation, but it is clear that the small transcript is more stable than the large transcript under these conditions. Because of the physiologic complexity of the light to dark shift we cannot draw any further major conclusions.

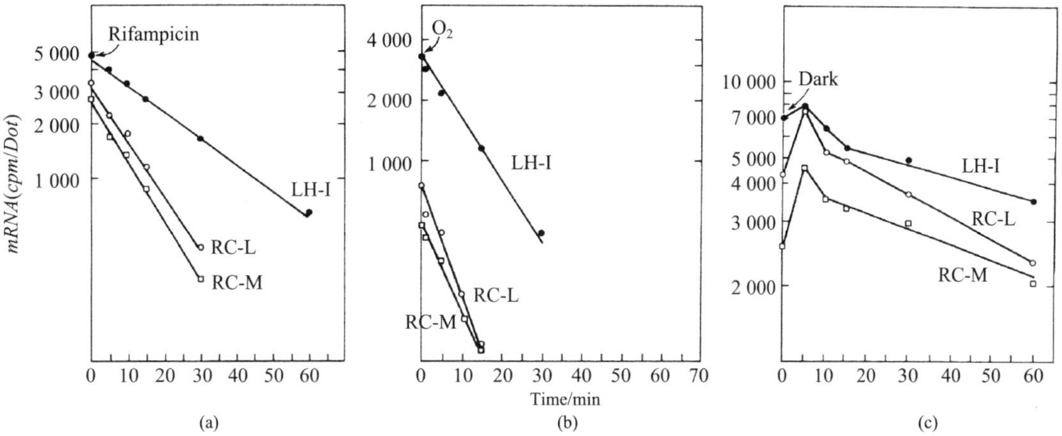

Figure 3 Decay of mRNA specifying the LH-I, RC-L, and RC-M polypeptide structural genes of the *puf* operon of *R. sphaeroides* (pUI1A2) Cells were grown photoheterotrophically on succinate medium[10]. The mRNA decay in photoheterotrophically grown cells after addition of rifampicin (a), shifting cells to aerobic growth conditions (b), or to darkness (c) was measured by dot hybridization. The 0 min point represents samples harvested just prior to imposition of the treatment. The 5-, 10-, 15-, 30-, and 60-min points represent samples removed following imposition of treatment. The amounts of mRNA were assessed by scintillation counting of the filter and expressed in cpm/dot. For details see under "Experimental procedures"

Following a shift from photoheterotrophic to aerobic conditions the half-lives of the transcripts (as determined by dot hybridization) homologous to the LH-I, RC-L, and RC-M. Similar results were also obtained from Northern hybridization analysis (see Figure 4). It can also be seen that the large (2.6 kb) transcript was rapidly degraded after exposure to O_2 and was present only at very low levels even after 10 min [see Figure 4(b) and (c)]. In contrast the level of the small transcript decreased more slowly and was clearly detectable 30 min following the shift to aerobic conditions [see Figure 4(a)]. An additional important feature of the results of the Northern hybridization was that significant amounts of discrete-sized transcripts (2.0, 1.8, 1.4, and 1.1 kb) were observed when the X-ray films were exposed for long periods [see Figure 4(b) and (c)]. All of these intrermediate-sized transcripts hybridized only with probes specific of the RC-L and RC-M structural genes. Further, when the specific RC-L and RC-M probes were used in Northern hybridization [see Figure 4(b) and (c)] many other hybridizing species, resulting in a smear, were observed between the 2.6- and 1.1-kb regions. One possible interpretation of these resulted from processing of the large transcipt at specific sites located toward the 5′-end of the transcript; the smear may be caused by extensive degradation of the large transcript, and the gaps between the individual degradation products may represent some RNA sequences which were relatively resistant to ribonuclease attack. On the other hand, when the LH-I sepecific probe was employed [see Figure 4(a)] the loss of the large transcript appeared to be more rapid, and fewer intermediates were apparent. These data coupled with the kinetic data in Figure 3 and the previously discussed data in Figure 2 suggest that the breakdown of the large transcript proceeds from the 5′-to the 3′-end of the mRNA.

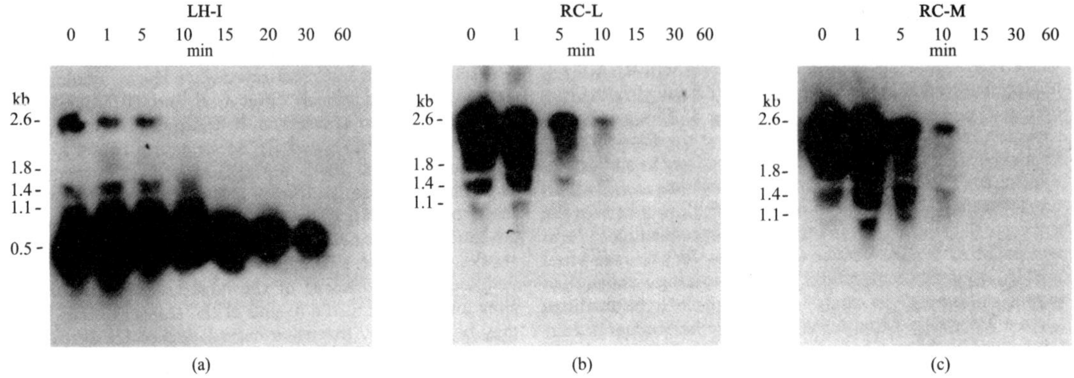

Figure 4 **Northern hybridization with DNA probes specific for the LH-I (a), RC-L (b), and RC-M (c) polypeptides to RNA from a photosynthetic culture of *R. sphaeroides* (pUI1A2) after a shift to aerobic conditions. Conditions for growth, RNA extraction, and hybridization were the same as described in the legend to Figure 3. The transcripts with different sizes (2.4,1.8,1.4,1.1,0.5 kb) and changes in their amounts following the shift (0,1,5,10,15,30,60 min) are indicated**

Determination of the 5′-Ends of the 2.6- and 0.5-kb *puf* Operon Transcripts by S-1 and Primer Extension Analysis

The 5′ termini of the two mRNAs were mapped by S-1 protection and primer extension analysis. Several different single-stranded DNA fragments with a 5′-labeled end within the *puf*B or A structural gene and a 3′-end upstream of the *puf*B structural gene were chosen as probes for hybridization with bulk RNA. The results are shown in Figure 5. The ^{32}P-labeled single-stranded AccI (see Figure 1, fragment d, 350 bp), AluI (see Figure 1, fragment a, 500 bp), and Bst NI (see Figure 1, fragment e, 190 bp) fragments were hybridized with RNA isolated from high light-grown cells and treated with S-1 nuclease. In all cases two distinct S-1-protected DNA fragments were observed in addition to the uppermost renatured DNA probe band. The sizes of the three pairs of S-1-protected fragments were 390 and 360 bp [see Figure 5(a)], 180 and 150 bp [see Figure 5(b)], and 120 and 90 bp [see Figure 5(c) a], respectively, when using the AluI, AccI, and BstNI restriction fragments. The larger S-1-protected fragments were always in excess of the smaller fragments (approximately (12—15) : 1), and the difference in sizes between the protected fragments was always approximately 30 bp. When S-1 protection experiments were performed on bulk RNA derived from cells undergoing a photosynthetic to aerobic shift it is clear that the larger

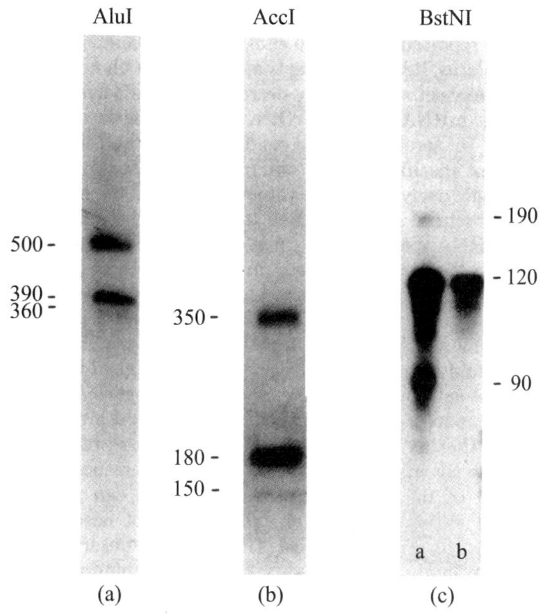

Figure 5 Analysis of the 5′-end of the two LH-I transcripts of *R. sphaeroides* (pUI1A2) by S-1 mapping. Purified AluI (see Figure 1, fragment a), AccI (see Figure 1, fragment d), and BstNI (see Figure 1, fragment e) restriction fragments were 5′-end labeled with [γ-^{32}P]ATP and T_4 polynucleotide kinase. The coding strand of the DNA was isolated, hybridized with RNA from photosynthetically [(a)、(b)、(c)a] or aerobically (c)b grown cells, treated with S-1 nuclease, and the S-1-protected fragments were electrophoresed on 5% polyacrylamide gels, as described under "Experimental Procedures". The fragments of 500, 350, and 190 bp in length represent renatured AluI, AccI, and BstNI DNA fragments, respectively. In addition, two S-1-protected fragments of different molar abundance (390, 360 for AluI; 180, 150 for AccI; 120, 90 for BstNI) were observed in all three S-1 analyses. Based on the known DNA sequence for the *puf* operon, the 5′-ends of each transcript can be mapped to approximately 100 and 70 nucleotides upstream of the start codon of the *puf*B gene from this analysis

fragment is more stable than the smaller one [see Figure 5(c)b and Figure 6(a)] with the ratio of the large to small S-1-protected fragment at time zero being 17 : 1, at 5 min following the shift to aerobic conditions the ratio of the larger fragment to the smaller fragment was 20 : 1; at 15 min, 30 : 1; and at 60 min, 35 : 1. Based on the fact the three pairs of protected fragments (e. g. larger fragments corresponding to 390, 180, and 120 bp versus smaller ones corresponding to 360, 150, and 90 bp) identified in S-1 protection analysis share common properties (i. e. similar relative ratios and stabilities) to the two transcripts of the *puf* operon (0.5 and 2.6 kb) identified in Northern blots, these results map the 5′-termini of the small (0.5 kb) and large (2.6 kb) transcripts to approximately 100 and 70 nucleotides upstream of the start codon for the *puf*B structural gene, respectively.

A 212-bp AccI-AluI restriction fragment (see Figure 1, fragment f) within the *puf*B and -A structural genes was used to identify the 5′-end of the *puf* operon transcripts by primer extension analysis. After labeling the 5′-end of this fragment with [γ-^{32}P]ATP, the

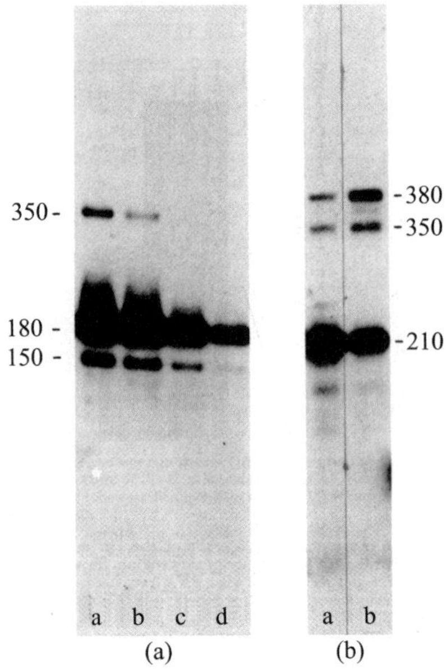

Figure 6 Analysis of the 5′-ends of the two transcripts from photosynthetic cells of *R. sphaeroides* (pUI1A2) after a shift to aerobic conditions by S-1 protection (a) and primer extension (b) assays. The S-1analysis using an AccI restriction fragment (see Figure 1, fragment d) in A was the same as shown in Figure 5. The RNA was isolated from cells exposed to high oxygen at different time intervals (0,5,15, 60 min as represented by (a)a, (a)b, (a)c, (a)d) as in shown Figure 4. A more abundant and relatively stable band (180 nucleotides) and a less abundant and relatively unstable band (150 nucleotides) were observed on 5% polyacrylamide gels. The primer extension was performed using a AccI-AluI fragment (see Figure 1, fragment f) as a primer as described under "Experimental Procedures". The extended cDNA using RNA isolated from photosynthetic cells (0 min) ((b)a) and from cells 60 min after exposure to oxygen ((b)b) were 380 and 350 nucleotides in length. The band at 210 nucleotides in length was renatured primer itself

single-stranded DNA was separated on a 5% polyacrylamide gel and hybridized with RNA isolated from cells grown either under steady-state photosynthetic conditions or from photosynthetic cells following a shift to aerobic conditions for 1 h. Two primer-extended fragments of approximately 380 and 350 bp were observed either on 5% polyacrylamide gels [see Figure 6(b), a and b] or on 6% polyacrylamide, 8.3 mol/L urea gels (data not shown) when using RNA from either culture. The sizes of these two primer-extended fragments were as expected from the results of S-1 mapping, so that the results of both the S-1 and primer extension analyses mapped the 5′-termini for the small and large *puf* operon transcripts to approximately 100 and 70 nucleotides upstream of the start of the *puf*B structural gene, respectively.

In order to more precisely determine the 5′-ends for the *puf* operon transcripts, the S-1-protected fragments obtained using the AccI restriction fragment (see Figure 1, fragment d) were analyzed on a 6% polyacrylamide, 8.3 mol/L urea gel alongside a Maxam and Gilbert sequence ladder of the AtuI-AccI restriction fragment (see Figure 1, fragment g). The restriction fragment used for Maxam and Gilbert sequencing was obtained by labeling the 5′-

ends of the AccI restriction fragment (see Figure 1, fragment d) followed by digestion with AluI and purification of the large labeled AluI-AccI restriction fragment from a 5% polyacrylamide gel. Figure 7 shows the results of this analysis with the two 174- and 144-base pair S-1-protected fragments analyzed alongside the sequenced restriction fragment. This experiment, when analyzed together with the known DNA sequence of this region,[1] precisely maps the 5′ termini for the small and large *puf* operon transcripts to 104 and 75 nucleotides upstream of the start of the *puf*B structural gene.

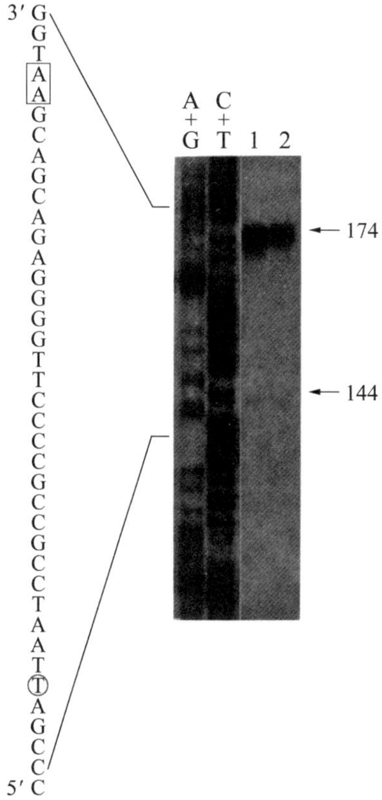

Figure 7 Alignment of the 5′-ends of the *puf* operon transcripts with the DNA sequence of this region. Shown are the S-1-protected fragments obtained when RNA from steady-state photosynthetic cells (lane 1) or photosynthetic cells 1 h after a shift to aerobic conditions (lane 2) were hybridized to an end-labeled AccI single-stranded restriction fragment (see Figure 1, fragment d) prior to S-1nuclease treatment as described under "Experimental Procedures". These samples were analyzed on a 6% polyacrylamide, 8.3 mol/L urea gel alongside a Maxam and Gilbert sequence ladder containing the noncoding strand of the AluI-AccI restriction fragment (see Figure 1, fragment g) generated as described under "Results"

Since the DNA sequence of this region of the *puf* operon has been obtained by independent methods[1] we have shown only the G plus A and C plus T chemical reactions which allow positioning of the transcription initiation sites to 104 and 75 nucleotides upstream of the start of the *puf*B structural gene as shown. The corresponding DNA sequence of the noncoding strand is illustrated on the left with the transcript initiation sites highlighted with a circle or a box. The experiments performed to date have not allowed precise resolution as to which of the two possible A residues initiates transcription for the large transcript, so both nucleotides are highlighted. However, for the purposes of further discussion and clarity in the text we have assumed it to be the second base which is 104 nucleotides upstream of the initiation of the *puf*B structural gene

Differential Hybridization of the *puf* Operon Transcripts with Specific Deoxyoligonucleotide Probes

The previously presented analysis of the 5′-ends of the small and large *puf* operon transcripts, their relative abundance and stabilities in vivo, strongly suggests that the small and large *puf* operon transcripts have different 5′ termini upstream of the *puf*B structural gene (104 and 75 nucleotides 5′, respectively). In order to test this hypothesis directly, two specific deoxyoligonucleotides were used as hybridization probes against bulk *R. sphaeroides* DNA. The LHI-4 deoxyoligonucleotide was synthesized to be complementary to a region within the mRNA coding for the *puf*B structural gene while the LHI-5 deoxyoligonucleotide was complementary to the mRNA from 78 to 96 nucleotides upstream of the *puf*B initiation codon. The results obtained when these deoxyoligonucleotides as well as an LH-I structural gene probe were used in Northern blots against bulk *R. sphaeroides* 2.4.1 RNA from cells grown under different physiological conditions are shown in Figure 8. As expected, both the LH-I-specific and the LHI-4 probe hybridized to both the small and large *puf* operon transcripts under high stringency washing conditions (75℃ for the LH-I-specific probe, 50℃ for the LHI-4 probe). In contrast, the LHI-5 probe hybridized to mRNA corresponding only to the small *puf* operon transcript under high stringency (65℃) washing conditions. Therefore, the results of this experiment when considered together with the previously presented data unambiguously show that the large *puf* operon transcript has a 5′ terminus 75 nucleotides upstream of the start of the *puf*B structural gene with the transcript encompassing the four *puf* operon structural genes. In contrast, the small *puf* operon transcript contains a 5′ terminus 104 nucleotides 5′ of the *puf*B initiation codon and extends just beyond the *puf*B and A genes but does not include the L and M genes. The available information on the factors which may influence the origins of these two 5′-termini in vivo, the termination of the small *puf* operon transcript prior to the *puf*L structural gene, and the ramifications of the differential use of two transcripts are summarized under "Discussion".

Discussion

The LH-I and RC complexes from *R. sphaeroides* are highly organized within the photosynthetic membrane for efficient transfer of light energy[1,2,20]. The expression of the genes coding for the polypeptides of these complexes is also highly regulated with respect to growth conditions such as light and oxygen as demonstrated in this study. Light was shown to regulate the *puf* operon in *R. sphaeroides*, with high intensity light partly repressing the transcription of the *puf* operon approximately 3-fold. In addition, overall expression of the *puf* operon was repressed an additional 3-fold by oxygen (see Table 1, Figure 2).

A striking aspect of *puf* operon expression is that the LH-I (β and α), RC-L, and RC-M polypeptides are synthesized in vivo at widely different levels. Several possibilities may explain this observation; multiple promoters, attenuation control, or different translation efficiencies have been observed to regulate differential gene expression within individual operons of other procaryotic systems[21-24].

Previous[10] and present studies have shown that there are two mRNA species homologous to the structural genes for the LH-I complex polypeptides of *R. sphaeroides*. They were reported in the previous study as 2.6 and 0.8 kb in length based on their mobility on agarose gels. According to more accurate measurements on polyacrylamide gels performed in this work and DNA sequence determinations for this region[1] the small transcript is 0.5 kb in length. The larger and less abundant mRNA (2.6 kb) is a polycistronic mRNA homologous to all four structural genes (*puf*B, -A, -L and -M) whereas the smaller more abundant mRNA (0.5 kb) is also polycistronic but specific only for the *puf*B and A polypeptides, with the ratio of the smaller to the larger mRNA ranging from 7—18 to 1 ([10] this study) with an average value of approximately 12 : 1. The ratio of the two transcripts approximates the ratio of LH-I to RC complexes found within the photosynthetic membrane. Similar results were recently reported in *R. capsulatus*[11], and Belasco et al.[11] concluded that the ratio of the small to large transcript encoding the LH-I and RC genes in the analogous rxcA (*puf*) locus of *R. capsulatus* results predominantly from the selective degradation of a specific segment of the initial large transcript at the 3′-end of the molecule resulting in a stabilized small transcript.

Our previous[10] and present studies also show different stabilities for the two transcripts of the *puf* operon in *R. sphaeroides*. The half-lives of the small and large transcripts were 20 and 9 min, respectively, after inhibition of transcription initiation with rifampicin. These results are very similar to those reported by Belasco et al.[11] and Dierstein[25] for *R. capsulatus*. It is interesting that exposure of photosynthetic cells to oxygen significantly decreased the half-lives of both of these mRNA species. When photosynthetic cells were exposed to aerobic growth conditions, the half-life of the transcript specific for *puf*L and *puf*M decreased to 5 and 6.5 min, respectively, while the value for the $t_{1/2}$ of the mRNA for LH-I structural genes declined to 10 min. The rapid loss of these mRNA species under these conditions is not surprising based on the abrupt cessation in the insertion of photosynthetic membrane components when photosynthetic cells are subjected to aerobic conditions[26]. An accelerated degradation of the mRNA specifying photosynthetic membrane proteins would be an efficient regulatory mechanism for photosynthetic cells when they are exposed to oxygen. At present we have no information as to how the transcripts are degraded and whether oxygen, as a signal for aerobic growth, directly or indirectly affects ribonuclease activities responsible for turnover of these transcripts. However, we can propose a degradation pathway for the large transcript based on the existence of intermediate size mRNA species observed in the Northern hybridization studies and the differential effect of O_2 on mRNA degradation. It would appear that processing of the large mRNA (2.6 kb) occurs from the 5′-region. In addition, there appear to be regions within the 2.6-kb transcript which are more resistant to exonuclease degradation. Although the experiments reported here were not designed to specifically follow the degradation pathway of the mRNA, the observed results are consistent with the above interpretation.

The differential degradation of the two transcripts of the *puf* operon undoubtedly results in different steady-state concentrations of the mRNA specific for LH-I and RC-L and M structural genes. However, the different ratios of these two mRNA species (10—20-fold) cannot be explained solely by differences in mRNA stability (2-fold). Our ability to demonstrate two different 5′-ends, with the small transcript being 29 bases longer at the 5′-end than the large transcript, would preclude the large transcript (as defined here) being a precursor of the small transcript (again as defined here). Furthermore, it appears that breakdown of the large transcript proceeds in a 5′ to 3′ direction, which would further mitigate against it being a precursor of the small transcript as suggested for *R. capsulatus*[11]. The simplest interpretation of these data is that two different metabolically stable transcription initiation sites exist, the frequency of use of each being dependent upon upstream promoter affinities for RNA polymerase. Further, the selection of one initiation site relative to the other would result in transcription of the complete operon while selection of the other would result in termination of transcription between *puf*B, A and *puf*L, M. The secondary structure of the *puf* operon mRNA between *puf* A and *puf* L would be in accord with such a hypothesis[1]. Additionally, we have been unable to detect any mRNA sequences within 500 bp upstream to the PstI site of the large and small transcripts reported here (see lane 5, Figure 8).

It is also possible that transcription of the *puf* operon initiates at only one site, and post-transcriptional modification(s) results in degradation at the 5′-end of some fraction of the primary transcripts, resulting in the two 5′-ends observed here. These post-transcriptional modifications would, in some unknown manner, be involved in the formation of the two transcripts (2. 6 and 0. 5 kb) in very precise stoichiometries. Clearly, if this is the route by which these two transcripts are derived and the observed stoichiometry achieved, the mechanism(s) of post-transcriptional processing of the primary *puf* transcript of *R. sphaeroides* is far more complex than that envisioned for *R. capsulatus*[11].

If differential transcription initiation gives rise to the large and small transcripts, then the question of the factors regulating termination of the two *puf* operon transcripts at the molecular level need to be investigated. The data presented in this manuscript show that the small *puf* operon transcript, if it were to initiate 104 nucleotides upstream of the start of the *puf*B structural gene, terminates prior to the start of the *puf* L structural gene. DNA sequence analysis of the intercistronic region between *puf*A and -L reveals the existence of two possible stem-loop structures which, by analogy to other well studied procaryotic systems, would be classified as factor-dependent transcription terminators[27,28]. Whether either or both of these structures function in termination of the small *puf* operon transcript remains to be determined. Conversely, the results obtained would indicate that transcription of the large *puf* operon mRNA, which has a 5′ terminus 75 base pairs upstream of the *puf*B

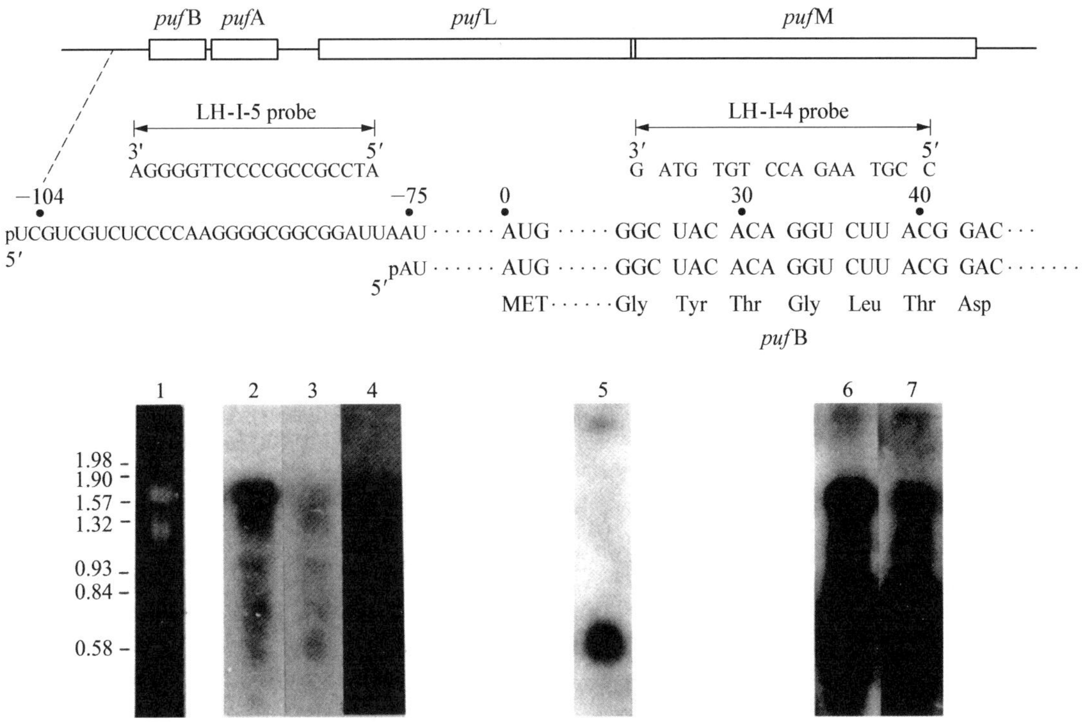

Figure 8 Selective Northern hybridization of the small and large *puf* operon transcripts with specific deoxyoligonucleotide probes

The top of the figure illustrates the structure of the *R. sphaerides puf* operon, the sequence of the 5′-ends of the two transcripts encoded by this operon, and the specific deoxyoligonucleotide probes LHI-4 and LHI-5 are also shown. Bulk RNA (30 μg) from *R. sphaeroides* 2. 4. 1 cells grown photosynthetically in the presence of low light resolved on a 1. 2% agarose gel was used for all samples. Shown in lane 1 is the ethidium bromide fluorescent material from a typical bulk RNA sample along with the migration of bacteriophage λ DNA molecular weight standards. Lanes 2 and 3 show the results obtained after a low (room temperature) and high (65℃) stringency wash using the LH-I-5 probe, respectively. Lane 4 is the same as lane 3 except for a longer exposure to demonstrate the absence of the 2. 6-kb transcript. Lane 5 shows the results obtained using a PstI-XorII restriction fragment probe for the LH-I genes. The PstI site is approximately 600 bp upstream of the start of *puf*B, and the XorII site is within *puf*B. Lanes 6 and 7 show the two *puf* operon transcripts visualized after a low (room temperature) and high (50℃) stringency wash using the *puf*B-specific LHI-4 probe, respectively. The high stringency wash temperature for each deoxyoligonucleotide probe was based on the theoretical melting temperature (52℃ for LH-I-4, 66℃ for LH-I-5) of these molecules. Theoretical T_m were calculated assuming 2℃ of melting temperature for each A-U base pair and 4℃ for each G-C base pair[15].

structural gene, continues past any possible regulatory signals involved in terminating transcription of the small *puf* operon mRNA. No information is currently available on the existence of any transcription termination structures 3′ of the *puf*M structural gene. In order to more fully understand the factors regulating *puf* operon expression in *R. sphaeroides*, experiments probing the molecular mechanisms which control transcription initiation and/or termination in this bacterium need to be undertaken. In addition, the existence of protein factors which may interact either directly or indirectly with either or both of the *puf* operon transcripts to aid in transcription termination[29,30] or antitermination[29,31] remains to be determined. Finally, we may conclude that the promoter structure for the *R.*

sphaeroides puf operon could be complex. The possible existence of two precise and differentially regulated sites for transcription initiation of this operon would undoubtedly be determined by selective recognition or binding of RNA polymerase to different domains within the *puf* operon promoter.

Acknowledgment — We wish to thank Dr. B. Orozco for his advice on the S-1 nuclease and primer extension analysis.

References

[1] Kaplan, S. , and Arntzen, C. J. (1982) in *Photosynthesis Energy Conversion by Plants and Bacteria* (Govindjee, ed) Vol. 1, pp. 65 - 151. Academic Press, Orlando, FL
[2] Drews, G. , Peters, J. , and Dierstein, R. (1983) *Ann. Microbiol. (Paris)* 134B, 151 - 158
[3] Chory, J. , and Kaplan, S. (1983) *J. Bacteriol.* 153,465 - 474
[4] Kaplan, S. , Cain, B. D. , Donohue, T. J. , Shepherd, W. D. , and Yen, G. S. L. (1983) *J. Cell. Biochem.* 22,15 - 29
[5] Williams, J. C. , Steiner, L. A. , Feher, G. , and Simon, M. 1. (1984) Proc. *Natl. Acad. Sci.* U. S. A. 81,7303 - 7307
[6] Williams, J. C. , Steiner, L. A. , Ogden, R. C. , Simon, M. I. , and Feher, G. (1983) *Proc. Natl. Acad. Sci. U. S. A.* 80,6505 - 6509
[7] Imhoff, J. F. , Trüper, H. G. , and Pfennig, N. (1984) *Int. J. Syst. Bacteriol.* 34,340 - 343
[8] Youvan, D. C. , Bylina, E. J. , Alberti, M. , Begusch, H. , and Hearst, J. E. (1984) *Cell* 37,949 - 957
[9] Drews, G. (1978) *Curr. Top. Bioenerg.* 8,161 - 207
[10] Zhu, Y. S. , and Kaplan, S. (1985) *J. Bacteriol.* 162,925 - 932
[11] Belasco, J. G. , Beatty, J. T. , Adams, C. W. , Gabain, A. V. , and Cohen, S. N. (1985) *Cell* 40,171 - 181
[12] Casadaban, M. J. , Martinez-Arias, A. , Shapira, S. K. , and Chowy, J. (1983) *Methods Enzymol.* 100,298 - 308
[13] Nano, F. E. , Shepherd, W. D. , Watkins, M. M. , Kuhl, S. A. , and Kaplan, S. (1985) *Gene* 34,219 - 226
[14] Zhu, Y. S. , Kung, S. D. , and Bogorad, 1. (1985) *Plant Physiol.* 79,371 - 376
[15] Wallace, R. B. , Johnson, M. J. , Hirose, T. , Miyake, T. , Kawashima, E. H. , and Itakura, K. (1981) *Nucleic Acids* Res. 9,879 - 893
[16] Mullet, J. E. , Orozco, E. M. , and Chua, N-H. (1985) Plant Mol. *Biol.* 4,39 - 54
[17] Maxam, A. M. , and Gilbert, W. (1980) *Methods Enzymol.* 65,499 - 560
[18] Clayton, R. K. (1966) *Photochem. Photobiol.* 5,669 - 677
[19] Markwell, M. A. K. , Suzanne, M. H. , Bieber, L. L. , and Tolbert, N. E. (1963) *Anal. Biochem.* 87, 206 - 210
[20] Youvan, D. C. , and Ismail, S. (1985) *Proc. Natl. Acad. Sci.* U. S. A. 82,58 - 62
[21] Kennell, D. , and Riezman, H. (1977) *J. Mol. Biol.* 114,1 - 21
[22] Barry, G. , Squires, C. L. , and Squires, C. (1980) *Proc. Natl. Acad. Sci.* U. S. A. 77,3331 - 3335
[23] Valentin-Hansen, p. , Hammer, K. , Larsen, J. E. L. , and Svendsen, 1. (1984) *Nucleic Acids. Res.* 12,5211 - 5224
[24] McCarthy, J. E. G. , Schairer, H. U. , and Sebald, W. (1985) *EMBO J.* 4,519 - 526
[25] Dierstein, R. (1984) *Eur. J. Biochem.* 138,509 - 518
[26] Shephierd, W. D. , and Kaplan, S. (1983) *J. Bacteriol.* 156,1322 - 1331

[27] Bauer, C. E., Carey, J., Kasper, L. M., Lynn, S. P., Waechter, D. A., and Gardner, J. F. (1983) *in Gene Function in Procaryotes* (Beckwith, J., Davies, J., and Gallant, J. A., eds) pp. 65 - 89, Cold Spring Harbor Laboratory, Cold Spring Harbor, NY

[28] Holmes, W. M., Platt, T., and Rosenberg, M. (1983) *Cell* 32, 1029 - 1032

[29] Platt, T., and Bear, D. G. (1983) in *Gene Function in Procaryotes* (Beckwith, J., Davies, J., and Gallant, J. A., eds) pp. 123 - 161, Cold Spring Harbor Laboratory, Cold Spring Harbor, NY

[30] Friedman, D. I., Schauer, A. T., Mashni, E. J., Olson, E. R., and Baumann, M. F. (1983) in *Microbiology-1983* (Schlessinger, D., ed) pp. 39 - 42, American Society for Microbiology, New York

[31] Gottesman, M. E., Adhya, S., and Das, A. (1980) *J. Mol. Biol.* 140, 57 - 75

[32] Marrs, B., and Kaplan, S. (1970) *J. Mol. Biol.* 49, 297 - 317

Regulation of Expression of Genes for Light-Harvesting Antenna Proteins LH-I and LH-II; Reaction Center Polypeptides RC-L, RC-M, and RC-H; and Enzymes of Bacteriochlorophyll and Carotenoid Biosynthesis in Rhodobacter Capsulatus* by Light and Oxygen①

Abstract

RNA levels were measured by blot hybridization to study the coordinate and differential expression of *Rhodobacter capsulatus* genes for light-harvesting I antenna proteins LH-I and LH-II; reaction center (RC) polypeptides L, M, and H; and bacteriochlorophyll and carotenoid biosynthesis in response to light and O_2. The genes for LH-II α and β subunits only have one transcript, 0.5 kilobase (kb) long, whereas the genes for LH-I have two transcripts (0.5 and 2.6 kb). The small transcript (0.5 kb) is the mRNA only for LH-I β and α polypeptides, whereas the large transcript (2.6 kb) codes for RC-L, RC-M, and the β and α polypeptides of LH-I, as well as the product of an unknown open reading frame designated ORF C2397. These five genes thus comprise a single operon (designated the *puf* operon). The mRNA specifying the LH-II polypeptides is more abundant, more sensitive to changes in O_2 concentration, and shows a variation over a wider range than that of the mRNA for LH-I, indicating that the genes for LH-II and LH-I/RC are regulated independently. The gene for RC-H (*puh*A) has at least two transcripts (1.2 and 1.4 kb) that initiate within ORF F1696 and respond differentially to light intensity. The expression

---

Yu Sheng Zhu and John E. Hearst. Division of Chemical Biodynamics, Lawrence Berkeley Laboratory, and Department of Chemistry, University of California, Berkeley, CA 94720 Communicated by Melvin Calvin, June 20,1986

此文发表于 1986 年"Proc. Natl. Acad. Sci. USA." 83:7613 - 7617.

① The publication costs of this article were defrayed in part by page charge payment. This article must therefore be hereby marked "advertisement" in accordance with 18 U. S. C. § 1734 solely to indicate this fact.

Abbreviations: LH-I and LH-II, light-harvesting proteins I and II; RC, reaction center; Bchl, bacteriochlorophyll; Crt, carotenoid (s); ORF, open reading frame; kb, kilobase (s).

* The genus name of this bacterium has been changed from Rhodopseudomonas to Rhodobacter[1].

关键词:photosynthetic bacteria/Rhodopseudomonas capsulata/*puf* operon/polycistronic mRNA/photooxidative damage of the genes coding for RC-L, RC-M, and RC-H is coordinately regulated by light intensity and O_2 concentration. An increase in light intensity causes a decrease in the expression of the genes for LH-I, LH-II, and RC proteins. The genes coding for the enzymes in the bacteriochlorophyll biosynthetic pathways are regulated by light intensity and O_2 in a manner similar to that of the genes for LH and RC proteins. The *crt* genes coding for the enzymes in carotenoid biosynthetic pathways, however, are regulated in an opposite fashion: high light intensity results in increased expression of *crt* genes. These results are interpreted based on the protective function of carotenoids under high light intensity in the presence of O_2.

The purple nonsulfur photosynthetic bacterium *Rhodobacter capsulatus* [formerly *Rhodopseudomonas capsulata*[1]] provides an attractive model for studying the regulation of photosynthetic genes by light, O_2, and other environmental factors. When the concentration of O_2 is lowered, the cell develops an extensive intracytoplasmic membrane and induces the biosynthesis of light-harvesting LH-I (B870) and LH-II (B800-B850) and reaction center (RC) proteins, as well as bacteriochlorophyll (Bchl) and carotenoids (Crt)[2-4]. Once assembled, the photochemically active complexes harvest and convert light energy into chemical energy[5,6]. The size and structure of the photosynthetic apparatus are also influenced by light intensity[2-4,6].

The genes for the β (*puf*B) and α (*puf*A) subunits of the LH-I complex, as well as the L (*puf*L) and M (*puf*M) subunits of the RC complex, are located in the *Bam*HI-C-*Eco*RI-B fragment of the R' plasmid pRPS404 isolated from *R. capsulatus*[7]. They are cotranscribed in vivo both in *R. capsulatus*[8] and in *Rhodobacter sphaeroides*[9]. The gene for the H subunit (*puh*A) of the RC lies in the *Bam*HI-F fragment of R'[10], whereas the genes for the β (*puc*B) and α (*puc*A) subunits of LH-II are outside this photosynthetic gene cluster[11]. The structural genes for LH-I, LH-II, RC-L, RC-M, and RC-H in *R. capsulatus* have been sequenced[7,10,11]. The genes for Bchl and Crt biosynthesis have been mapped by genetic methods[12,13] and transposon mutagenesis[14,15].

It has been shown that lowering the O_2 concentration in cultures of *R. capsulatus* and *R. sphaeroides* increases the levels of mRNA for LH-I, RC[9,16], and LH-II polypeptides[17], while the amounts of RNA for pigment-biosynthesis enzymes show only a small increase when the O_2 concentration is reduced[13,16]. According to one set of experiments, Crt biosynthesis is not directly regulated by O_2[18]. More recently, a sequential and independent expression of LH-I and LH-II genes has been shown in *R. capsulatus* during ICM development[19].

In this paper we report a systematic investigation of mRNAs of photosynthetic genes in *R. capsulatus* under different growth conditions, using dot blot hybridization as well as blot hybridization of electrophoretically fractionated RNA. The results show a coordinate and differential regulation of the genes for LH (I, Ⅱ) and RC (L, M, H) polypeptides and pigment (Bchl, Crt) biosynthesis by light intensity and O_2 concentration.

Materials and Methods

Bacterial Strains, Media, and Growth Conditions

R. capsulatus rifampicin-resistant strain SB1003 was used. The cells were grown on malate-minimal RCV medium[20]. Those cells growing photoheterotrophically were cultured at 30℃ in 170-ml screw-cap bottles filled completely to the top. Light was provided by a bank of lumiline lamps at high light intensity (30 W/m^2) or low light intensity (6 W/m^2). Darkshift experiments were performed by wrapping the growth vessels from high light-grown cells with aluminum foil and incubating them for 1 h at 30℃. Highly aerobic growth (designated as high O_2) was maintained by chemoheterotrophic culturing of the cells (80 ml of medium in a 1,000-ml Erlenmeyer flask) on a gyratory shaker (200 rpm). Semiaerobic growth (designated as low O_2) was carried out in vessels filled to 80% capacity with growth medium on a gyratory shaker (200 rpm). The cells at mid-log phase ($(4-7)\times 10^8$ cells per ml) were harvested for measurement of absorption by Bchl and chromatophores and for RNA extraction.

Preparation of Plasmid, M13 Phage, and Oligonucleotide Probes

The plasmid, M13 phage, and oligonucleotide probes for various photosynthetic genes of *R. capsulatus* are listed in Table 1 and Figure 1. M13 phage DNA was extracted according to the procedure described in the 1982 Bethesda Research Laboratories Cloning and Dideoxy Sequencing Instruction Manual. The nucleotide sequences of inserts in individual recombinant M13 phages were determined and confirmed by dideoxy sequencing[22]. Two oligodeoxynucleotides containing 15 bases complementary to the sequence encoding the β subunit of LH-II and to the open reading frame (ORF) C2397 (see Figure 1 and Table 1) were synthesized on a Biosearch Sam One DNA synthesizer using phosphite triester chemistry and were purified by thin-layer chromatography[23].

Table 1 M13 phage, plasmids, or oligonucleotides used as probes

Probe	Insert*	Gene or gene product	Ref.
M13†			
T319	*Bam*HI-C 422 - 342	LH-I(α)	[10]
T330	1392 - 1094	RC-L	[10]
S331	1704 - 1607	RC-M	[10]
T210	3647 - 3413	RC-H	[10]
S316	3038 - 2927	C2814	[10]
T214	*Bam*HI-F 581 - 467	NR‡	[27]
S216	1202 - 1045	F1025	[10]
H227	3062 - 2974	F1696	[10]

(continued)

Probe	Insert*	Gene or gene product	Ref.
T224	3377 - 2531	RC-H, F1696	[10]
F3981	*Bam*HI-K 133 - 1	F3981	[10]
Plasmids			
pFL205	*Bam*HI-E	bch J, -G, -D	[13,15]
pFL120	*Bam*HI-D	*bch*H, -K, —F	[13,15]
pFL227	*Bam*HI-J	*crt*E, -F	
pFL103	*Bam*HI-H	*crt*B, -I, -A; bchl 13,15	
pFL104	*Bam*HI-G	*crt*C, -D	[13,15]
pFL268	*Bam*HI-M	*crt*E	[13,15]
Oligonucleotides			
3′ TACTGACTGCTATTT 5′		LH-II (β)	[11]
3′ AGGTACAAGCTGTTT 5′		C2397	[10]

* Numbers (bp) correspond to positions shown in Figure 1.
† Single-stranded probes synthesized from M13 DNA were complementary to known ORFs, based on DNA sequence analysis.
‡ Nitrogenase reductase.

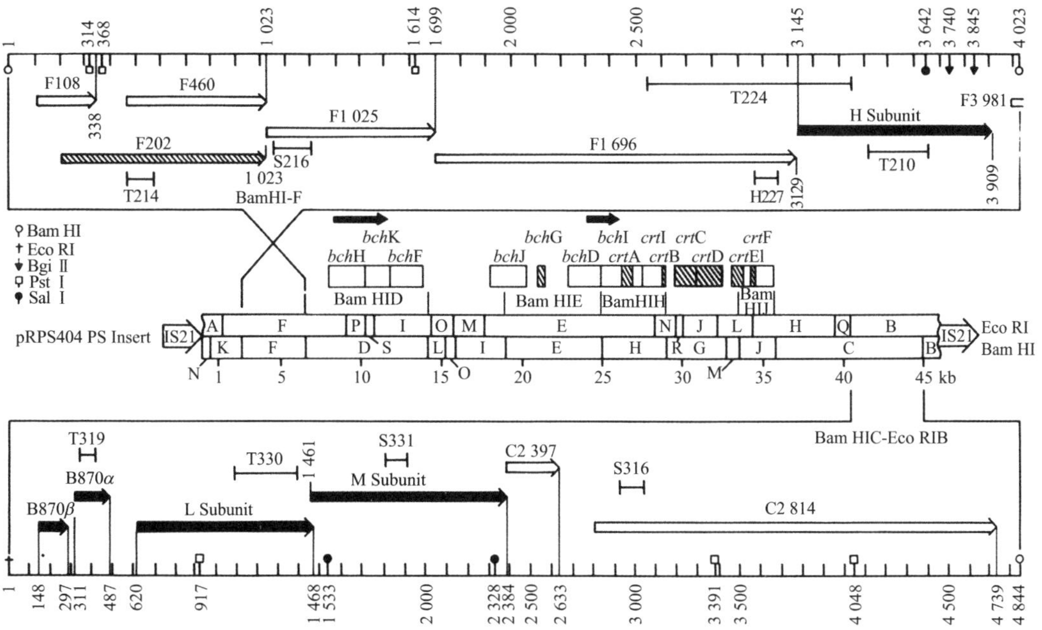

Figure 1 **Photosynthetic gene cluster organization and location of hybridization probes. *Bam*HI and *Eco*RI restriction sites are shown within the 46-kb photosynthetic gene cluster from *R. capsulatus* carried by the R' plasmid pRPS404[10,15]. The *Bam*HI-F and *Bam*HI-C-*Eco*RI-B fragments are magnified and shown at top and bottom. The *Bam*HI-E and -D fragments, containing the genes for Bchl biosynthetic enzymes, and the *Bam*HI-J, -H, -G, and -M fragments, containing the genes for Crt biosynthetic enzymes, are represented above pRPS404. Solid arrows indicate the coding regions and the direction of transcription. Open arrows represent putative protein-coding sequences (ORFs). The small bars above or below these arrows indicate the hybridization probes for specific genes (see Table 1)**

RNA Isolation and Blot Hybridization

Total RNA was extracted as described[9]. After the RNA was denatured in formaldehyde, it was fractionated by electrophoresis in 6% formaldehyde/1.2% agarose gels[24] and transferred to GeneScreen membranes (New England Nuclear)[25]. The plasmid DNA was labeled with [α-^{32}P]dCTP by nick-translation. The M13 phage DNA was labeled with [α-^{32}P]dATP by extension of the annealed sequence primer with DNA polymerase I Klenow fragment at 15℃ for 1 h, using 50 μCi (1 Ci = 37 GBq) of [α-^{32}P]dATP in 20 μl of 10 mmol/L Tris. HCl/1 mmol/L $MgCl_2$/60 mmol/L NaCl/1 mmol/L dithiothreitol/0.2 mmol/L dCTP/0.2 mmol/L dUTP/0.2 mmol/L dGTP, pH 7.9. The oligonucleotide probes were 5′-end-labeled by the polynucleotide kinase reaction[26]. Dot blot and electropherogram-replica hybridizations were as described[25]. To quantitate hybridization, autoradiograms were scanned with a GS 300 transmittance-reflectance scanning densitometer (Hoefer, San Francisco).

Results

Genes for LH-I and LH-II

A study on absorption spectra of pigments and chromatophores indicated that the synthesis of Bchl is reduced by O_2 under steady-state condition and the synthesis of the LH-II complex is more strongly repressed by O_2 than the LH-I complex (data not shown). This result is supported by measurements of the mRNAs for LH-I and LH-II by dot hybridization (see Figure 2, Table 2). The level of mRNA for LH-II was 30 times less in aerobically grown cells as compared to high-light-grown cells. We conclude that the level of mRNA for LH-II is more sensitive to O_2 concentration than that of the LH-I mRNA. This conclusion was further confirmed by blot hybridization of electrophoretically fractionated total RNA. Only one transcript [0.5 kilobase (kb)] for LH-II was found [see Figure 3(a)]. Its level was greatly reduced when cells were grown under high O_2 conditions. The genes for LH-I have two major transcripts: 0.5 and 2.6 kb [see Figure 3(b)]. The 0.5-kb transcript is only complementary to the genes for LH-I β and α subunits, whereas the 2.6-kb transcript is homologous to the five structural genes coding for LH-I (β, α), RC (L, M) and C2397, as shown by hybridization with specific probes for LH-I (see Figure 3(b)), RC-L [see Figure 3(c)], and RC-M and C2397 (data not shown). The level of the 2.6-kb mRNA appears more sensitive to changes in O_2 compared to the 0.5-kb mRNA [see Figure 3(b) and (c)].

Genes for RC L, M, and H Polypeptides

Dot blot (see Figure 2, Table 2) and electropherogram-replica [see Figure 3(c) and (d)] hybridizations indicated highly coordinate regulation among the genes for the subunits L, M, and H of the RC complex. As shown by dot hybridization, when the high-light-grown cells were shifted to the dark for 1 h, about 70% of the mRNAs for RC-L, RC-M, and RC-H

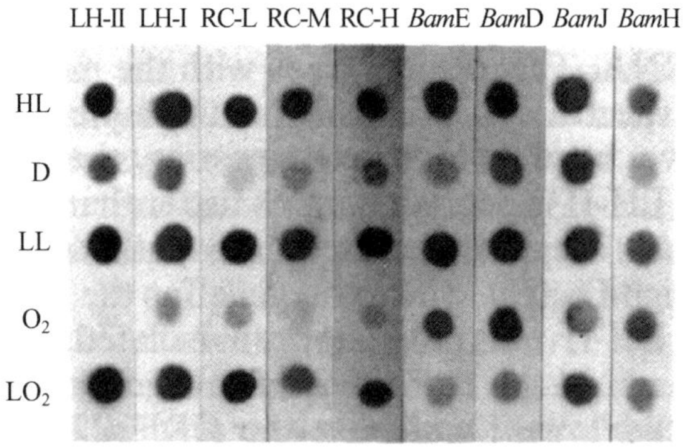

Figure 2 Dot blot hybridization of *R. capsulatus* RNA with cloned probes for LH (I, II), RC (L, M, H) and Bchl (*Bam*E and -D) Crt (*Bam*J and -H) biosynthetic enzymes. Cells were grown under various growth conditions as indicated at left (HL, high light; D, shift from HL to dark for 1 h; LL, low light; O_2, highly aerobic growth; LO_2, semiaerobic growth), and RNA was isolated as described in the text. Equal amounts of RNA (5 μg) after denaturation were spotted onto GeneScreen membranes, hybridized with ^{32}P-labeled probes, and exposed to X-ray film. Each probe is listed above the dot blots and described in Table 1

Table 2 Effects of light and O_2 on mRNA levels

Gene or gene product	mRNA level, %*				
	High light	Dark (1 h)	Low light	High O_2	Low O_2
LH-II	100	35	182	3	113
LH-I	100	51	173	28	110
RC-L	100	25	130	27	100
RC-M	100	28	139	32	96
RC-H	100	28	176	30	106
*bch*J, -G, -D (*Bam*HI-E)	100	25	124	76	36
*bch*H, -K, -F (*Bam*HI-D)	100	36	119	140	30
*crt*E, -F (*Bam*HI-J)	100	62	80	30	60
*crt*A, -I, -B; bchI (*Bam*HI-H)	100	31	80	105	29
*crt*C, -D (*Bam*HI-G)	100	55	82	59	34
*crt*E (*Bam*HI-M)	100	69	76	72	69

* The relative amounts of mRNA were measured by dot hybridization, quantified by densitometry of X-ray film, and expressed as the percentages of mRNA level under high light (set as 100).

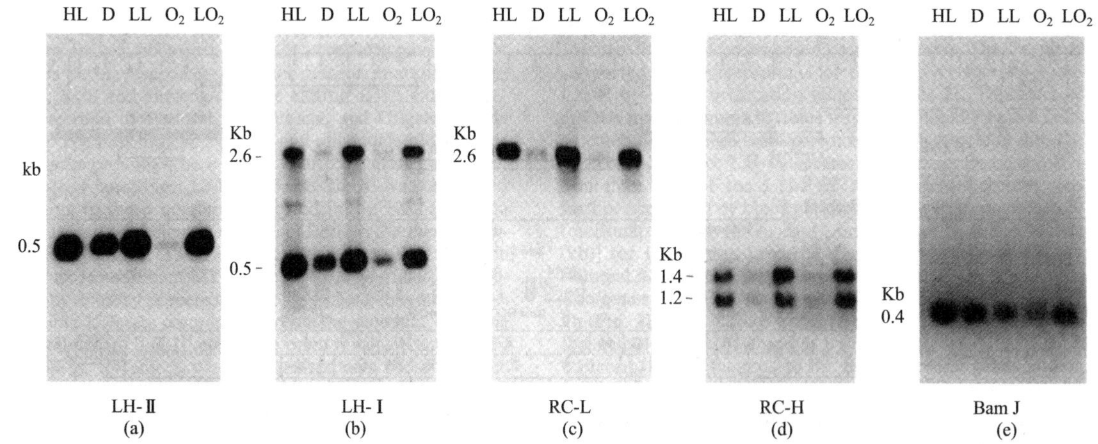

Figure 3 Blot hybridization of electrophoretically fractionated total RNA from *R. capsulatus* with cloned probes for LH-I, LH-II, RC-L, RC-H, and Crt biosynthetic enzymes. RNAs from *R. capsulatus* cells grown under various growth conditions (HL, D, LL, O_2, LO_2; see legend to Figure 2) were denatured in formaldehyde, fractionated in 1.2% agarose gel, transferred to nitrocellulose (a) or GeneScreen [(b)—(e)], and hybridized with ^{32}P-labeled probes for LH-II (a), LH-I (b), RC-L (c), RC-H (d), and Crt (e) (listed in Table 1)

were degraded. Under steady-state low light, a 30%—76% increase in mRNA levels was observed. This increase agrees with the increase in mRNAs for LH-I and LH-II, and it is expected because more photosynthetic membrane is produced to compensate for the decreased light intensity. Steady-state aerobic cells (high O_2) showed a 70% reduction in mRNA compared to high-light-grown cells, whereas the low-O_2 condition had no noticeable effect on the RC-L, RC-M, and RC-H mRNA levels. This result agrees with hybridizations of electrophoretically fractionated RNA (see Figure 3). We found that the gene for RC-H had two major transcripts, of 1.2 and 1.4 kb. They hybridized with the probes for the internal region of the gene for RC-H (T210) and with upstream regions coding for F1696 (H227) (see Figure 1). There was no hybridization with the downstream segment encoding F3981in *Bam*HI-K (data not shown). Therefore, we assume that the mRNA for RC-H initiates within the coding region designated ORF F1696. Although the amounts of the 1.2- and 1.4-kb transcripts were greatly reduced in dark or high-aeration growth conditions, they were regulated differently by light intensity. Under high light, the 1.2-kb transcript is more abundant than the 1.4-kb transcript, with a 1.2-kb/1.4-kb ratio of 2.5; under low light the 1.4-kb transcript predominates, with a 1.2-kb/1.4-kb ratio of 0.5.

Genes for Biosynthesis of Bchl and Crt

The genes for pigment biosynthesis were transcribed at lower levels than those for LH and RC subunits, but they were responsive to light intensity and O_2 concentration, as measured by dot blot analysis. Although the changes in Bchl content in response to light and O_2 roughly corresponded to the levels of the LH and RC mRNAs (data not shown), the changes in the levels of mRNA for Bchl and Crt biosynthetic enzymes had the following features (see Figure 2, Table 2). ① Under high light intensity, the amounts of mRNAs of bch genes (*bch*J, -G, -D, -H, -K, and -F) located in the *Bam*HI-E and *Bam*HI-D fragments showed a slight but reproducible decrease, whereas the amounts of mRNAs of *crt* genes (*crt*C, -D, -A, -I, and -B) carried by the *Bam*HI-H, *Bam*HI-G, and *Bam*HI-M fragments showed an increase, compared to that under low light. ② Under steady-state low O_2 the levels of all mRNAs for pigment biosynthetic enzymes declined.

A small and abundant 0.4-kb transcript hybridized with labeled *Bam*HI-J fragment, which has been mapped to contain *crt*F and *crt*E[13,15] [see Figure 3(e)]. Densitometry revealed that the amounts of the 0.4-kb transcript increased about 3-fold when cells were grown under high light relative to that under low light.

In addition to the genes for LH-I β and α subunits and RC L and M polypeptides, which are clustered in the *Bam*HI-C-*Eco*RI-B fragment, and the gene for the RC H polypeptide, which is in *Bam*HI-F fragment of the R' plasmid, DNA sequence data show that the *Bam*HI-C-*Eco*RI-B fragment contains ORFs C2397 and C2814 and the *Bam*HI-F fragment contains ORFs F108, F460, F1025, and F1696 and an ORF homologous to the gene for nitrogenase reductase[7,27]. A series of M13 phage DNA probes (see Table 1) specific for those regions was used to determine which regions are transcribed under the described photoheterotrophic or chemoheterotrophic growth conditions. Only a transcript for ORF C2397 was detected by dot blot or electropherogramreplica hybridizations (data not shown). The mRNA homologous to the gene for C2397 was shown to have the same size (2.6 kb) and same response to light intensity and O_2 concentration as the mRNAs for RC-L and RC-M, when a 15-mer oligonucleotide was used as a specific probe (see Table 1) for the gene for C2397 (data not shown).

Discussion

Recently a great deal of attention has been paid to the regulation of photosynthetic genes in *R. capsulatus* and *R. sphaeroides*[8,9,13,16-19]. In the present study we have used an improved method for RNA isolation[9] and have probed transcripts with DNA fragments specific to particular genes wherever possible. Hybridization of electrophoretically resolved RNAs, in combination with dot blot analysis, was used to measure mRNA quantity and sizes in order to obtain specific information about the regulation of the photosynthetic genes cluster in *R. capsutatus* in response to light and O_2.

Recent studies of physical structure show that the RC subunits L, M, and H comprise a single complex in the intracytoplasmic membrane where the photochemical reaction takes place. Each RC is surrounded by 12—18 LH-I (β and α subunits) complexes. The more abundant LH-II (β and α subunits) complexes, located peripherally to LH-I, harvest light energy and transfer exciton energy to LH-I and then to the RC[28]. Bchl and Crt, absorbing light across the visible and near-infrared region, bind to LH-II, LH-I, and RC[6]. The location of genes coding for these polypeptides parallels the structural and functional association of these polypeptides[7,11]. The structural, functional, and genetic features of these complexes define the patterns of regulation of the genes coding for these polypeptides. Earlier studies on proteins indicated that LH-II and LH-I/RC complexes were sequentially accumulated during intracytoplasmic membrane differentiation of *R. capsulatus* and that they were regulated independently by light and O_2[6]. Here we provide evidence for the independent regulation of the genes for LH-I and LH-II, based on the results from a study of mRNA levels. We found that the level of LH-II mRNA is more sensitive to change in O_2 concentration, and that it varies more than that of LH-I mRNA. This result agrees with the measurement of absorption spectra of protein-pigment complexes in vivo. In addition, we have evidence suggesting that the mRNA specific for LH-II is degraded much faster than that of LH-I, when the cells are shifted from photosynthetic growth conditions to high O_2[21]. Since the gene for the third (14-kDa) subunit of LH-II has not yet been isolated, we have no information about the regulation of this gene.

In this study, we found that the ORF C2397 was transcribed, and the level of its mRNA changed in response to light and O_2 in a similar way to that of the genes for RC-L and RC-M. Since ORF C2397, 236 base pairs long, is located immediately 3′ to the gene for RC-M and there is a hairpin structure after the ORF C2397[10], we suggest that the five structural genes for LH-I β and α subunits, RC-L, RC-M, and C2397 comprise a single operon and that the transcription of this operon stops after the gene for C2397. Whether the gene for C2397 codes for a structural protein, or it functions in the assembly of the LH and RC, or it has some other functional role is not known. We assume that it must have some function related to the LH and RC. Although the genes for LH-I (β, α), RC (L, M, H), and C2397 generally showed similar regulation by light and O_2, some differential regulation was observed. The 2.6-kb mRNA for LH-I (β, α), RC (L, M), and C2397 is more responsive to the change in O_2 concentration than the 0.5-kb mRNA. Such differential expression of these two transcripts was also observed in *R. sphaeroides*[9].

A close correlation between pigment (Bchl, Crt) and LH/RC biosynthesis has been reported in *R. capsulatus* (3,4,6). We found a general coordinate regulation between genes for Bchl biosynthesis and the genes for LH and RC in response to light and O_2. Some difference was observed, however. The mRNAs of *bch*H, -K, and -F (in *Bam*HI-D) and *bch*J, -D, and -G (in *Bam*HI-E) genes increased less than mRNAs for LH and RC when the light intensity decreased. The levels of mRNAs of these bch genes decreased not only when the cells were shifted from photosynthetic growth condition to high O_2 (data not shown) but also under steady-state low O_2 condition (see Figure 2, Table 2), indicating that the genes for Bchl biosynthesis respond more stringently to O_2 concentration. The unexpected results were that under steady-state high O_2, the level of mRNA homologous to the *Bam*HI-D fragment increased. An O_2-shift experiment also showed the induction of an O_2-activated gene (data not shown). We believe that there is an O_2-activated gene (s) located in the *Bam*HI-D fragment.

It is of interest that the genes for Crt biosynthesis are regulated differently from the other photosynthetic genes in response to light intensity and O_2. We have found that under steady-state high O_2, the level of mRNA from the *Bam*HI-H fragment was relatively high (see Figure 2, Table 2), and it increased when the cells were shifted from anaerobic to aerobic conditions[21]. We have evidence suggesting that *crt*A in the *Bam*HI-H fragment is activated by O_2. This gene is responsible for the oxidation of spheroidene to spheroidenone. Such regulation in response to O_2 may be related to the function of Crt. It has been long known that Crt has two functions: harvesting light and protecting cells including the photosynthetic apparatus from photooxidative damage, which only occurs in the presence of both high light and O_2[6]. Activation of *crt*A and other Crt genes by O_2 may be part of the protective mechanism by which Crt scavenges O_2 radicals in the cell[21]. Here we have also found another type of protective mechanism in response to high light. That is, in contrast to the decrease in the levels of mRNA for LH, RC, and Bchl biosynthetic enzymes, an increase in light intensity raised the levels of mRNA from a number of *crt* genes located in the *Bam*HI-M, *Bam*HI-G, and *Bam*HI-H fragments. Although the increase in transcription of *crt* genes in response to high light seems to be a plausible protective mechanism, we do not rule out the possibility that regulation may also occur posttranscriptionally, including the activation of Crt biosynthetic enzymes by light and O_2.

The 0.4-kb transcript from the *Bam*HI-J fragment also showed a response to high light, in the opposite fashion to the mRNAs for LH and RC [see Figure 4(e)]. It did not hybridize to either regions genetically mapped as *crt*F or *crt*E (G. A. Armstrong and J. E. H., unpublished data). The response of this transcript to O_2 is very similar to that of mRNAs for LH and RC but is very different from that of mRNAs for Crt biosynthesis. On the other hand, transcript level increases in response to high light as do the mRNAs for Crt biosynthesis, except to a greater extent, but it is very different from that of mRNAs for LH and RC in this respect. The nature and function of this small transcript remain to be characterized.

Multiple transcripts for RC-H (1.4 and 1.2 kb) were detected. The 1.2- and 1.4-kb transcripts of the RC-H gene, which is 764 base pairs long[10], are probably initiated from the middle of ORF F1696. We do not know whether they have different initiation sites or whether the 1.2-kb transcript is the product of processing of the 1.4-kb transcript. The relatively long 5′ noncoding region (0.4—0.6 kb) may contain important regulatory sequences.

Assay for transcripts from all of the ORF putative genes near the LH and RC gene clusters in the *Bam*HI-C-*Eco*RI-B and *Bam*HI-F fragments resulted in the detection of mRNA from only C2397. We suggest that the putative genes in both fragments are either not expressed under the growth conditions used or that their mRNAs, if present, are below the limits of detection by our methods.

We thank D. Cook, F. Leach, G. Armstrong, and N. Stover for helpful discussions;

F. Leach for kindly providing plasmids pFL103, —120, —205, and —227; and M. Alberti for technical assistance with phage DNA preparation and DNA sequencing. This work was supported by the Office of Basic Energy Sciences, Biological Energy Research Division of the Department of Energy under Contract DE-ACO30 - 76SFOOO98, by Department of Agriculture Grant 84CRCR11485, and by NationalInstitutes of Health Grant GM30786. The Biosearch Sam One was leased with Program Development Funds from the Director of Lawrence Berkeley Laboratory.

References

[1] Imhoff, J. F., Truper, H. G. & Pfenning, N. (1984) *Int. J. Syst. Bacteriol.* 34,340 - 343.

[2] Drews, G. (1978) *Curr. Top. Bioenerg.* 8,132 - 136.

[3] Ohad, I. & Drews, G. (1982) in *Photosynthesis* Ⅱ: *Development, Carbon Metabolism and Plant Productivity*, ed. Govindjee (Academic, New York), pp. 89 - 140.

[4] Dierstein, R., Tadros, M. & Drews, G. (1984) *FEBS Lett.* 24,219 - 223.

[5] Cramer, W. A. & Crofts, A. R. (1982) in *Photosynthesis I: Energy Conversion by Plants and Bacteria*, ed. Govindjee (Academic, New York), pp. 427 - 433.

[6] Drews, G. & Oelge, J. (1981) Adv. *Microbiol. Physiol.* 22,1 - 92.

[7] Youvan, D. C., Alberti, M., Begusch, H., Bylina, E. J. & Hearst, J. E. (1984) *Proc. Natl. Acad. Sci. USA* 81,189 - 192.

[8] Belasco, J. G., Beatty, J. T., Adams, C. W., Von Gabain, A. & Cohen, S. N. (1985) *Cell* 40,171 - 181.

[9] Zhu, Y. S. & Kaplan, S. (1985) *J. Bacteriol.* 162,925 - 932.

[10] Youvan, D. C., Bylina, E. J., Alberti, M., Begusch, H. & Hearst, J. E. (1984) *Cell* 37,949 - 957.

[11] Youvan, D. C. & Ismail, S. (1985) Proc. *Natl. Acad. Sci.* USA 82,58 - 62.

[12] Yen, H. C. & Marrs, B. (1976) J. Bacteriol. 126,619 - 629.

[13] Biel, A. J. & Marrs, B. L. (1983) *J. Bacteriol.* **156**,686 - 694.

[14] Youvan, D. C., Elder, J. T., Sandlin, D. E., Zsebo, K., Alder, D. P., Panopoulos, N. J., Marrs, B. L. & Hearst, J. E. (1982) *J. Mol. Biol.* **162**,17 - 41.

[15] Zsebo, K. M. & Hearst, J. E. (1984) *Cell* 37,937 - 947.

[16] Clark, W. G., Davidson, E. & Marrs, B. L. (1984) J. *Bacteriol.* 157,945 - 948.

[17] Klug, G., Kaufmann, N. & Drews, G. (1984) *FEBS Lett.* 177,61 - 65.

[18] Biel, A. L. & Marrs, B. L. (1985) *J. Bacteriol.* 162,1320 - 1321.

[19] Klug, G., Kaufmann, N. & Drews, G. (1985) *Proc. Natl. Acad. Sci. USA* 82,6485 - 6489.

[20] Weaver, P. F., Wall, J. D. & Gest, H. (1975) *Arch. Microbiol.* 105,207 - 216.

[21] Zhu, Y. S., Cook, D., Leach, F., Armstrong, G., Alberti, M. & Hearst, J. E. (1986) *J. Bacteriol.*, in press.

[22] Sanger, F., Nicklen, S. & Coulson, A. R. (1977) *Proc. Natl. Acad. Sci. USA* 74,5463 - 5467.

[23] Urbina, G. A., Sather, G. M., Liu, W. -C., Gillen, M. F., Duck, P. D., Bender, R. & Ogilvie, K. K. (1981) *Science* 214,270 - 274.

[24] Maniatis, T., Fritsch, E. F. & Sambrook, J. (1982) *Molecular Cloning: A Laboratory Manual* (Cold Spring Harbor Laboratory, Cold Spring Harbor, NY), pp. 202 - 203.

[25] Zhu, Y. S., Kung, S. D. & Bogorad, L. (1985) *Plant Physiol.* 79,371 - 376.

[26] Zoller, M. J. & Smith, M. (1983) *Methods Enzymol.* 100,468 - 500.

[27] Hearst, J. E., Alberti, M. & Doolittle, R. F. (1985) *Cell* 40,219 - 220.

[28] Drews, G. (1985) *Microbiol. Rev.* 49,59 - 70.

Oxygen-Regulated mRNAs for Light-Harvesting and Reaction Center Complexes and for Bacteriochlorophyll and Carotenoid Biosynthesis in Rhodobacter capsulatus during the Shift from Anaerobic to Aerobic Growth*

The stability and regulation by oxygen of mRNAs for the photosynthetic apparatus in Rhodobacter capsulatus have been studied by using proflavin to inhibit transcription and by shifting cells from anaerobic to aerobic conditions. The results from the inhibition experiments show that the mRNA for the light-harvesting LH-II polypeptides (β, α) is more stable than that for the light-harvesting LH-I polypeptides (β, α) during anaerobic growth, whereas the mRNAs for the reaction center polypeptides L (RC-L), M (RC-M), and H (RC-H) are less stable than both the LH-I and LH-II mRNAs. When photosynthetic cells are shifted from anaerobic to aerobic conditions, an immediate decrease in the levels of mRNA for the LH-I, LH-II, RC-L, RC-M, and RC-H proteins was observed. The level of mRNA for the LH-II proteins, however, is more sensitive to oxygen and is reduced faster than the level of mRNA for the LH-I proteins. These results suggest that oxygen represses the expression of genes coding for the light-harvesting antenna and reaction center complexes and may selectively accelerate the degradation of mRNA for the LH-II proteins. The mRNAs for several enzymes in the bacteriochlorophyll biosynthetic pathway are regulated by oxygen in a similar manner. The mRNAs for carotenoid biosynthetic enzymes, however, are regulated by oxygen in a different way. We have found that the amounts of mRNAs for carotenoid biosynthetic enzymes, relative to the amounts of mRNAs for LH and RC, increased during the shift from anaerobic to aerobic conditions. We have particularly shown that although the expression of most photosynthetic genes in *R. capsulatus* is repressed by oxygen, the *crt*A gene, located in the *Bam*HI H fragment of the R' plasmid pRPS404 and

---

* Yu Sheng Zhu, David N. Cook, Francesca Leach, Gregory A. Armstrong, Marie Alberti, and John E. Hearst
Division of Chemical Biodynamics, Lawrence Berkeley Laboratory, and Department of Chemistry, University of California, Berkeley, California 94720
Received 23 June 1986/Accepted 9 September 1986
Corresponding author.
此文发表于 1986 年"J. Bacteriol." 168:1180 - 1188.

responsible for the oxidation of spheroidene to spheroidenone, responds to oxygen in an opposite fashion. This enzymatic oxidation may protect the photosynthetic apparatus from photooxidative damage.

Rhodobacter capsulatus[19], which normally inhabits muddy lake bottoms and sewage lagoons, is a gram-negative, facultative, photosynthetic bacterium. It can generate metabolic energy by either aerobic respiration or anaerobic photosynthesis depending on the oxygen concentration and light intensity in its environment. The photosynthetic apparatus is composed of pigment-protein complexes for the conversion of light to chemical energy, including a shortwavelength light-harvesting antenna, LH-II (also denoted B800－850 by its near-infrared absorption maxima), a long wavelength light-harvesting antenna, LH-I (also denoted B870), and a reaction center complex consisting of three polypeptides (RC-L, RC-M, and RC-H). Each of these complexes binds bacteriochlorophyll and carotenoid pigments. The photosynthetic apparatus contains additional electron transport proteins to complete the cyclic pathway of electron transport. A decrease in oxygen concentration induces formation of the photosynthetic apparatus, which is localized within invaginations of the cellular membrane known as the intracytoplasmic membrane[16,20,26]. High levels of oxygen inhibit the formation of the photosynthetic apparatus, and under these conditions the cells use oxygen as a terminal electron acceptor for energy generation through aerobic respiration[16,17]. Oxygen concentration thus is one of the primary environmental factors governing the switch between alternative growth modes in *R. capsulatus*. However, the mechanism whereby oxygen controls photosynthetic gene expression is unknown.

The effects of oxygen on various photosynthetic functions have been well documented. Early studies by Cohen-Bazire et al.[12] demonstrated that oxygen represses bacteriochlorophyll biosynthesis in purple nonsulfur photosyn thetic bacteria. The control of bacteriochlorophyll formation by both oxygen and light has also been described[1,2,27]. The transcription of several genes for bacteriochlorophyll biosynthesis in *R. capsulatus* has been shown to be regulated by oxygen[4]. In addition, it has been stated that the carotenoid biosynthesis is not directly regulated by oxygen[5]. Recently, we[39-41] and others[3,8,21] have shown that the levels of mRNA for LH-I, LH-II, and RC complexes in *Rhodobacter sphaeroides* or *R. capsulatus* are regulated by oxygen. Furthermore, we have found that the levels of mRNAs for several genes coding for bacteriochlorophyll and carotenoid biosynthesis are also regulated by oxygen[39]. In this paper, we have conducted a shift experiment from anaerobic photosynthetic conditions to aerobic conditions to analyze the effects of oxygen on the stabilities of transcripts for LH and RC proteins and transcripts for bacteriochlorophyll and carotenoid biosynthetic enzymes. The results presented in this communication demonstrate that oxygen immediately represses the expression of the genes coding for LH-II, LH-I, RC-L, RC-M, and RC-H proteins and bacteriochlorophyll biosynthetic enyzmes. Oxygen also significantly stimulated the degradation of the mRNA for the LH-II proteins.

Furthermore, we have found that the *crt*A gene responsible for the oxidation of spheroidene to spheroidenone in *R. capsulatus* is oxygen activated.

Material and Methods

Bacterial Strains and Culture Conditions

R. capsulatus SB1003 (Rif^r)[24] was cultured on malate-minimal RCV medium[38]. Cells growing photosynthetically (designated PS cells) were anaerobically cultured at 30℃ in completely filled screw-cap bottles (170 ml) at a light intensity of (30 W/m^2) provided by a bank of lumiline lamps in an incubation chamber. The shift experiment from anaerobic (photosynthetic) to aerobic conditions was performed by abruptly shifting the PS cells to a 2-liter Erlenmeyer flask on a shaker and culturing those cells at the same light intensity and the same temperature with vigorous shaking (300 rpm) to achieve high aeration. In experiments determining the decay of specific mRNAs, cells were cultured photosynthetically to a density of 5×10^8 cells per ml (0.95 mg of protein per ml; 8.5 μg of bacteriochlorophyll per mg of protein). Transcription was then inhibited by the addition of proflavin (100 μg/ml). Samples of 10 ml of cells were harvested just before adding the inhibitor to give a 0-min time point. After the cells were quickly mixed with proflavin and covered with a layer of sterile paraffin oil to exclude oxygen, the remaining cells were cultured under the same conditions and thereafter harvested by pipetting 10 ml each at various time intervals (5, 10, 20, or 40 min or 1, 2, 3, 6, 9, 12, or 24 h) for RNA extraction. The *crt*A transposon mutant, KZR9A12, used for determining the *crt*A regulation has been previously described[43].

Pigment Extraction and Chromatophore Preparation

Total cellular pigments (bacteriochlorophyll and carotenoids) of 10 ml of cells were extracted with and dissolved in 10 ml of acetone-methanol (7 : 2, vol/vol) for spectroscopic analysis as previously described[9]. For the chromatophore absorption study 5 ml of photosynthetically or aerobically grown cells was harvested and suspended in 1 ml of Tris (5 mmol/L, pH 8.0)-EDTA (5 mmol/L) buffer as described previously[39]. The absorption spectra of the pigments in acetone-methanol and of chromatophores in Tris-EDTA buffer were measured in a Cary 118 and a Varian 2300 spectrophotometer, respectively, based on total cell number or amounts of protein. Protein concentrations were assayed by a modified Lowry method[23].

Extraction and Thin-Layer Chromatography of Carotenoids

Cells were harvested and extracted with acetone-methanoi as above under N_2 and in dim light. After solvents were evaporated with a stream of N_2, the pigments were redissolved in petroleum ether and spotted onto an analytical silica-gel plate for thin-layer chromatography. The solvent system for chromatography was petroleum ether-acetone (9 : 1).

Plasmid, M13 Phage, and Synthetic Oligonucleotide Probes

The plasmid, M13 phage, and oligonucleotide probes for varous genes coding for LH and RC proteins and for bacteriochlorophyll and carotenoid biosynthetic enzymes were prepared as previously described[39] and are listed as shown in Table 1 and Figure 1.

Table 1 Phage M13, plasmids, and oligonucleotides used for probing photosynthetic genes of R. capsulata

Probe	Insert (segment [bp])	Size (bp)	Gene encoded	Reference
M13 phage[a]				
T319	*Bam*HI-C 422 - 342	80	LH-I(α)	[36]
T330	1,392 - 1,094	298	RC-L	[36]
S331	1,704 - 1,607	97	RC-M	[36]
T210	3,647 - 3,413	234	RC-H	[36]
Plasmid				
pFL205	*Bam*HI-E	6,400	*bch*J, *bch*G, *bch*D	[31,34,43]
pFL120	*Bam*HI-D	7,000	*bch*H, *bch*K, *bch*F	[31,34,43]
pFL227	*Bam*HI-J	2,200	*crt*F, *crt*E	[31,34,43]
pFL103	*Bam*HI-H	3,200	*crt*A, *crt*B, *crt*I, *bch*I	[31,34,43]
pFL104	*Bam*HI-G	3,900	*crt*C, *crt*D	[31,34,43]
pFL268	*Bam*HI-M	870	*crt*E	[31,34,43]
pRC1	Cosmid clone[b]		23S, 16S, 5S rRNA	[35]
Oligonucleotide 3′		15	LH-II	[37]
TACTGACTGCTATTT 5′				

[a] Single-stranded probes synthesized from M13 DNA were complementary to known open reading frames based on DNA sequence analysis.

[b] Size, 35 to 45 kb.

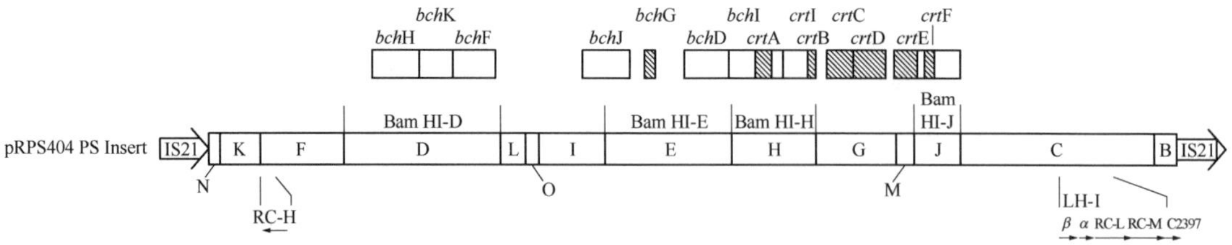

Figure 1 Photosynthetic gene cluster map. *Bam*HI restriction sites are shown within a 46-kb photosynthetic gene cluster from *R. capsulatus* carried by the R' plasmid pRP404[24,36,41]. The genes coding for LH-I (β and α), RC-L, and RC-M subunits and an open reading frame C2397, are located in the *Bam*HI C fragment, whereas the gene coding for the RC-H subunit resides in the *Bam*HI F fragment. The arrows indicate the direction of transcription. The *Bam*HI fragments between these two genetic loci contain a number of bch genes (in *Bam*HI-D and -E) and *crt* genes (in *Bam*HI-H, -G, -M, and -J). The shaded areas indicate the genes determined by cluster point mutations conferring the same phenotype. The blank areas represent genes determined by transposon mutagenesis and complementation

RNA Extraction, Dot Blots, and Northern Hybridization

Total cellular RNA was extracted as previously described[40], except that a minipreparation was developed by reducing the cell volume to 10 ml to handle large numbers of samples more easily. The yields of total RNA per 10 ml of cells was about 300 μg, enough for dot blots and Northern hybridization analysis. The labeling of plasmid DNA, phage DNA, oligonucleotides, and all dot blot and Northern hybridizations were as previously described[39,42].

Chemicals

[γ-^{32}P]dATP, [α-^{32}P]dATP, and [α-^{32}P]dCTP were purchased from Amersham Corp., Arlington Heights, Il. Agarose and a nick translation reagent kit were obtained from Bethesda Research Laboratories, Gaithersburg, Md.; Gene Screen membrane and nitrocellulose filters were products of New England Nuclear Corp., Boston, Mass. Proteinase K was purchased from Boehringer Mannheim Biochemicals, Indianapolis, Ind. DNase I was obtained from Worthington Diagnostics, Freehold, N.J. Macoloid used for purifying DNase I was kindly supplied as a gift by NL Chemicals, Hightstown, N.J. Phenol was obtained from Sigma Co., St. Louis, Mo., and redistilled. All other chemicals were reagent grade.

Results

Absorption Spectra of Pigments and Chromatophores under Different Oxygen Concentrations

To study the effects of oxygen concentration on the synthesis of pigments (bacteriochlorophyll and carotenoids) and on light-harvesting proteins (LH-I and LH-II), *R. capsulatus* SB1003 was grown photosynthetically for 22 h to reach a density of 5×10^8 cells per ml. These logarithmically growing photosynthetic cells were designated PS. Samples of PS cells were immediately poured into two Erlenmeyer flasks covered with two layers of gauze; one was shifted to low oxygen (medium/flask ratio, 8 : 10, vol/vol) designated as LO_2, and the other was shifted to high oxygen (medium/flask ratio, 0.8 : 10, vol/vol) designated as O_2. Both were continuously cultured with vigorous shaking for 24 h. Steady-state high aerobic cells were obtained by inoculating oxygen-adapted cells into fresh RCV medium (medium/flask ratio, 0.8 : 10, vol/voL and culturing them aerobically in a shaker for 18 h to reach a similar density (5×10^8 cells per ml) for comparison. After 24 h, cells shifted from photosynthetic to LO_2 conditions has grown only slightly because of a long period of adaptation, whereas cells shifted to O_2 doubled in cell density after adaption to the O_2 condition. The aeration of cells is directly correlated with their pigmentation: photosynthetic growth yields cells which are yellowish brown, LO_2 conditions result in a purplish-redculture, O_2 conditions yield yellowish-pink cells, and steadystate aerobic growtn gives rise to yellowish cells. The difference in colors of cell cultures results from the difference in the relative content and composition of carotenoids and bacteriochlorophyll. This is shown by the absorption spectra of the extracted pigments from the four types of cell cultures (see Figure 2). The pigments from PS cells show absorption peaks typical of bacteriochlorophyll at 360, 600, and 770 nm and peaks typical of carotenoids at 427, 453, and 484 nm. Bacteriochlorophyll content gradually diminished after the cells were shifted to LO_2 or O_2 and was present only in trace amounts in steady-state aerobic cells. The decrease in amount of bacteriochlorophyll was correlated to oxygen concentration. The carotenoid content was also reduced about 50% when the cells were shifted to the LO_2 or O_2 conditions for 24 h. Carotenoids were present in steady-state aerobic cells (O_2) at about 25% of the level in PS cells based on the area under the absorption bands between 400 and 500 nm. However, the amount of carotenoid relative to bacteriochlorophyll increased under high oxygen.

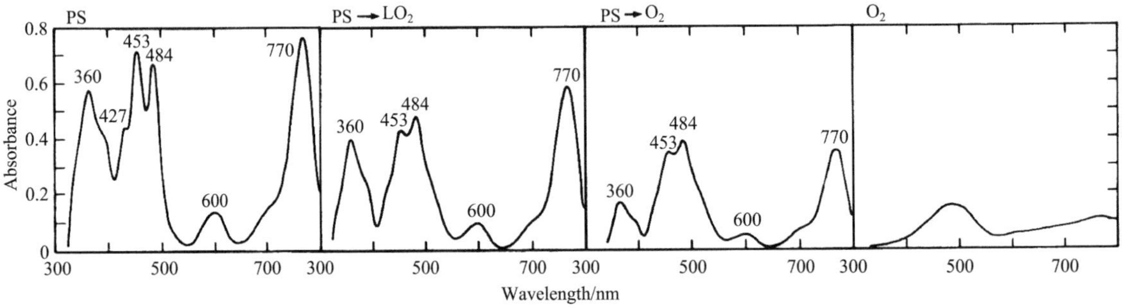

Figure 2 Absorption spectra of pigments of *R. capsulatus* under different oxygen conditions. Cells of *R. capsulatus* were grown under anaerobic photosynthetic conditions (PS) to the midlog phase and then shifted to low O_2 (PS →LO_2) or high O_2 (PS →O_2) for 24 h. Steady-state aerobic cells (O_2) were obtained by inoculating oxygen-adapted cells and culturing them at high oxygen for 1 day. The pigments from the four types of cell cultures (5×10^8 cells per ml) were extracted with acetone-methanol (7 : 2). The visible absorption spectra were determined by using a Cary 118 spectrophotometer and normalized based on the total cell numbers (see Materials and Methods)

A more detailed analysis of the change in carotenoid content after the oxygen shift reveals that the amounts of carotenoid increased during the first hour by about 10% and then decreased over 24 h (data not shown). It is interesting to note that the composition of carotenoids altered during the shift, as indicated by changes in the shape of the absorption peaks (see Figure 2). The PS cells had a major absorption peak at 453 nm, a second peak at 484 nm, and a minor peak at 427 nm, all due to carotenoids. After the cells were exposed to oxygen (LO_2 or O_2), the peak at 427 nm disappeared, and the peak at 484 nm predominated, although all peaks became more rounded. Those three peaks were not separated in steady-state aerobic cells. The changes in the absorption peaks of carotenoids at 427, 453, and 484 nm were detectable even 10 min after the cells were exposed to oxygen (data not shown) and gradually became more significant with time (from 0 to 24 h) (see Figure 3). Such changes in absorption by carotenoids indicate that the composition of carotenoids underwent a significant alteration when the PS cells were exposed to oxygen. This conclusion was verified directly by thin-layer chromatography (data not shown). Based on published thin-layer chromatography and mass spectroscopy analyses[28], we have found that the major carotenoid of the PS cells was spheroidene. After the PS cells were shifted to LO_2 or high O_2 conditions, spheroidene was converted to spheroidenone. This conversion was correlated with oxygen concentration. The steady-state aerobic cells had much less total carotenoids; spheroidenone was the major carotenoid under these conditions. In agreement with the pigment absorption spectra, we have also observed by thin-layer chromatography that the amount of carotenoid relative to bacteriochlorophyll increased after the cells were shifted to high oxygen.

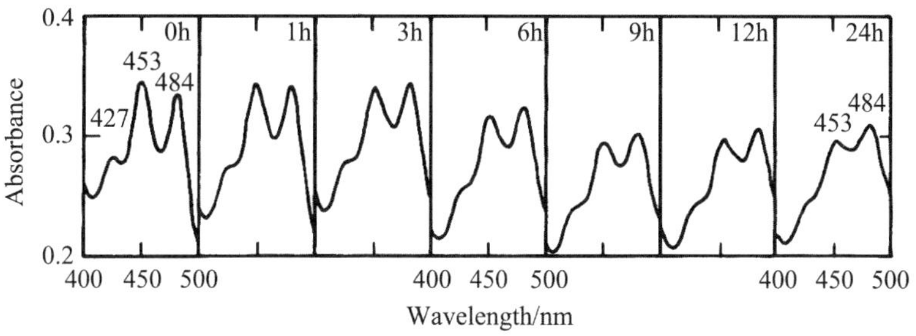

Figure 3 Absorption spectra of carotenoids of *R. capsulatus* after the shift of cells from photosynthetic to O_2 conditions. PS cell cultures were shifted to high O_2 as in Figure 2 and then harvested at different time points as indicated. Only the absorption peaks of carotenoids at 427, 453, and 484 nm are presented

These results were also demonstrated by measurement of absorption spectra of chromatophores from cell lysates (see Figure 4). The chromatophores had absorption peaks at 375, 510, 800, and 855 nm due primarily to bacteriochlorophyll complexed to light-harvesting proteins and at 450, 480, and 510 nm due to carotenoids. After the PS cells were

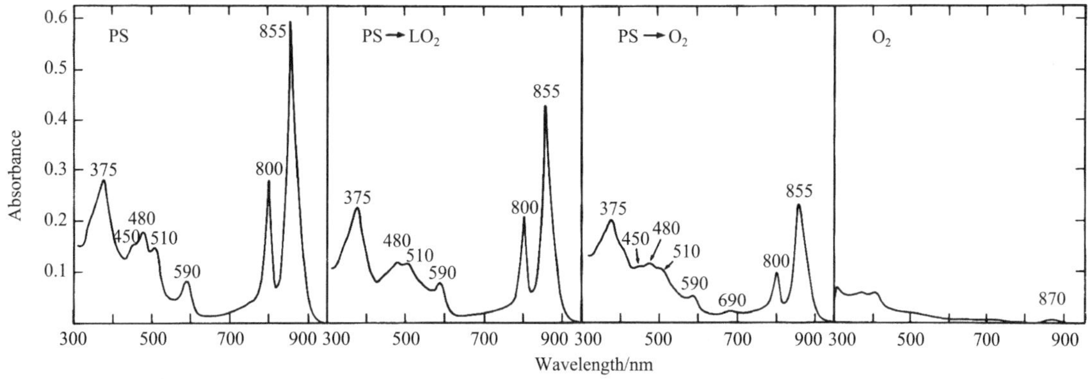

Figure 4 In vivo absorption spectra of chromatophores of *R. capsulatus* under different oxygen conditions. Chromatophores were isolated from cells cultured under the different oxygen conditions as described in the legend to Figure 2 and in Materials and Methods and were then subjected to infrared-visible absorption spectroscopy as previously described[38]

shifted to aerobic conditions, the absorption by carotenoids in chromatophores was reduced and the shape ofthe peaks was altered as for the extracted pigments. The amount of carotenoid relative to LH-I and LH-II, however, was found to increase in response to high oxygen. LH-I LH-II complexes, as measured by bacteriochlorophyll absorption peaks at 870 and 800 to 850 nm, respectively, also diminished after shifting to L02 or 02, and they were absent in steadystate aerobic cells except for a minor peak attributable to LH-I at 870 nm. The peak at 855 nm is composed of bacteriochlorophyll absorption from both the LH-I and the LH-II complexes. The decrease in the LH-I and LH-II complexes in response to oxygen may result from the decreased biosynthesis and dilution of the existing LH pool by cell division.

Decay of rRNA and mRNAs for LH-I and LH-II Polypeptides and the Effects of Oxygen

One possible mechanism for regulating gene expression in response to oxygen would be to increase the decay rate of certain mRNAs after a shift from anaerobic conditions. To test this hypothesis, we measured the change in the levels of RNA under various conditions. As a control, we measured the half-lives of selected mRNAs after the addition of the transcriptional inhibitor proflavin to the cell culture by using Northern hybridization (see Figure 5) and dot-blot analysis. The genes coding for the LH-II proteins had one transcript (0. 5 kilobase (kb)), whereas the genes coding for the LH-I proteins had two transcripts: a 0. 5-kb transcript coding for the LH-I proteins only and a 2. 6-kb transcript coding for LH-I, RC-L, and RC-M proteins and an as yet unidentified open reading frame[39] (see Figure 5).

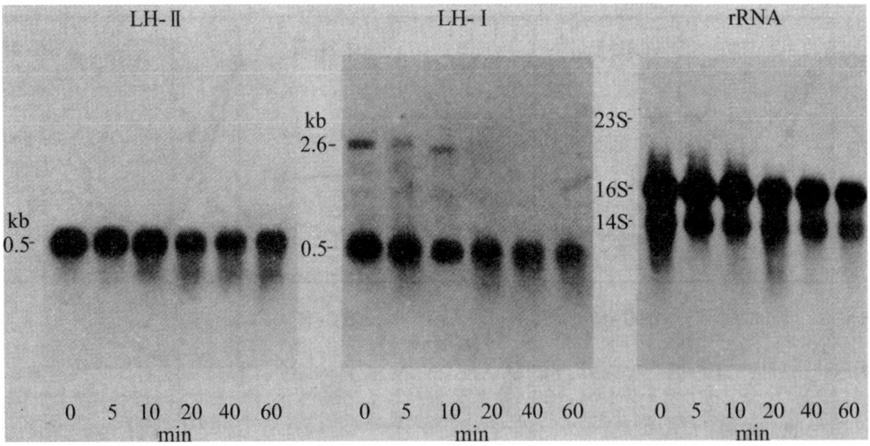

Figure 5 **Northern hybridization with probes specific for the LH-II, LH-I polypeptides, and rRNAs to RNA from a photosynthetic culture of *R. capsulatus* after treatment with proflavin. Cells were grown photosynthetically on RCV medium for 1 day and then treated with proflavin (100 μg/ml) as described in Materials and Methods. The 0 point represents RNA samples harvested just before addition of proflavin. The points at 5, 10, 20, 40, and 60 min represent samples removed after proflavin treatment. Total RNA was extracted using a mini preparation. Equal amounts of RNA were fractionated on an agarose gel and then transferred to nitrocellulose (LH-II) or Gene Screen (LH-I, rRNA) and hybridized with ^{32}P-labeled probes for the LH-II, LH-I, and rRNA as previously described[41]. The probes are listed in Table 1 and Figure 1**

The half-life of the 0. 5-kb LH-II mRNA was 24 min, whereas the half-lives of the 0. 5-kb LH-I mRNA and the 2. 6-kb polycistronic mRNA were 19 and 10 min, respectively. These results suggest that, under steady-state photosynthetic growth conditions in the presence of proflavin, the mRNA for the LH-II is more stable than the mRNA for the LH-I, and that the 0. 5-kb mRNA for the LH-I proteins alone is more stable than the 2. 6-kb mRNA which also codes for the RC-L and RC-M. For the purposes of comparison, we measured the half-life of rRNAs which have the 23S, 16S, 14S, and SS species (see Figure 5). Normally the 23S rRNA is unstable and is specifically degraded to the 16S and 14S rRNA during extraction[25]. The half-life of total rRNA was about 24 min, as measured by dot hybridization with pRCl containing 23S, 16S, and 5S rRNA as a probe[35].

We also examined the stability of mRNAs after shifting from photosynthetic to O_2 growth. There were no significant changes in the amounts of rRNA after this shift (16S, 14S) (see Figure 6). However, the mRNAs for the LH-II and LH-I decreased markedly. The mRNA for LH-II decreased with a half-life of 17 min, whereas the mRNA for LH-I decreased with a half-life of 23 min. Therefore, the level of mRNA for the LH-II is more sensitive to oxygen than is that for the LH-I (see Figure 6).

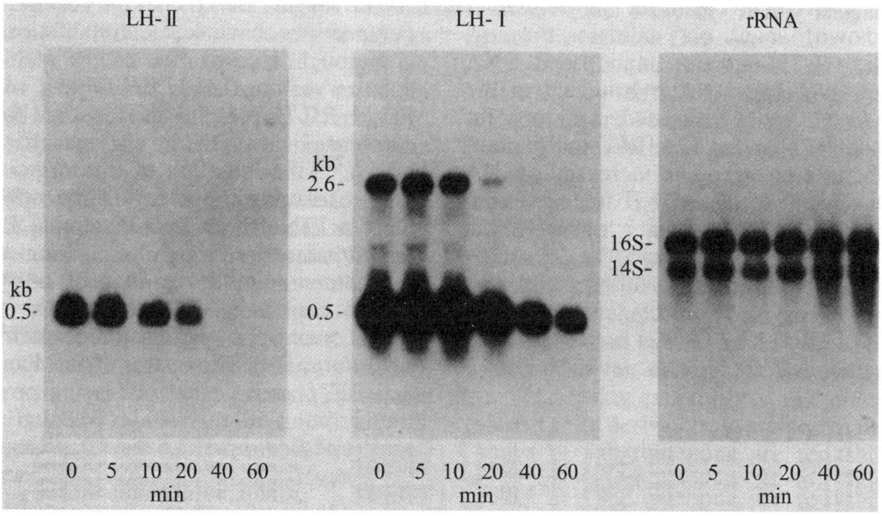

Figure 6 **Northern hybridization with DNA probes specific for the LH-II and LH-I polypeptides and rRNAs to RNA from a photosynthetic culture of *R. capsulatus* after a shift to O_2. Photosynthetic cells were shifted to O_2 growth conditions as described in Materials and Methods. RNA extraction and hybridization were as described in the legend to Figure 5**

Decay of mRNAs for RC-L, RC-M, and RC-H and the Effect of Oxygen

Quantitative measurements of mRNA levels for RC-L, RC-M, and RC-H by dot-blot analysis after cells were treated with proflavin or shifted to oxygen as shown in Figure 7. The results reveal that the levels of mRNAs for RC-L, RC-M, and RC-H responded

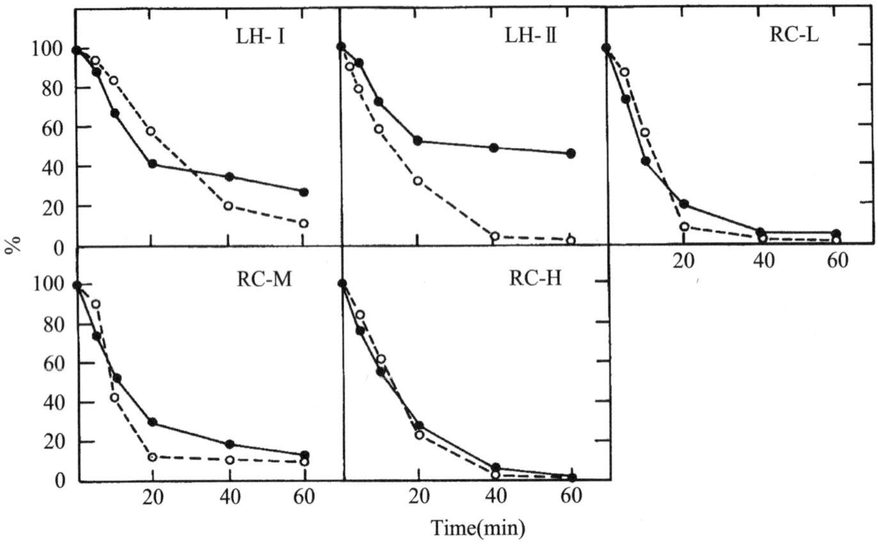

Figure 7 **Time course of mRNA levels for LH-I, LH-II, RC-L, RC-M, and RC-H. Photosynthetic cells were treated with proflavin which inhibits the transcription or exposed to oxygen, and RNA was isolated as in Figure 5 and Figure 6. Symbols: •, proflavin treatment; ○, O_2 shift. Dot blots were made and hybridized with ^{32}P-labeled probes as described in the text. The amounts of mRNA were assessed by densitometry of X-ray film. The mRNA are expressed with the zero time point chosen as 100%**

coordinately to both the transcriptional inhibitor and to the O_2 shift. The messages had similar half-lives of about 10 min in the presence of proflavin and in the O_2 shift experiment.

Decay of mRNAs for Bacteriochlorophyll and Carotenoid Biosynthesis and Oxygen Regulation

Particular *Bam*HI restriction fragments ofthe R' plasmid pRPS404 have been mapped genetically and shown to contain predominantly *crt* (in *Bam*HI-H, -M, -G, -J) or *bch* (in *Bam*HI-E, -D) genes (see Figure 1)[31,41]. We used these DNA fragments as probes for dot hybridization to determine how the levels of mRNA for pigment biosynthetic enzymes change during the O_2 shift, and we observed a broad range of responses. The kinetics of mRNA decay was similar to that for the LH and RC proteins when the fragments *Bam*HI-E (*bch*D, *bch*G, and *bch*J) and *Bam*HI-J (*crt*F, *crt*E, and an unidentified, heavily transcribed gene)[39] were used as probes (see Figure 8). In contrast, the level of mRNAs from the *Bam*HI-D fragment (*bch*F, *bch*K, and *bch*H) decreased during the first 20 min after the shift and then increased in the next 40 min (see Figure 8). In general, the levels of mRNA from fragments containing *crt* genes increased relative to that of mRNA for RC and LH during the shift from anaerobic to aerobic conditions. An increase in total mRNA level was found for the *Bam*HI M fragment (data not shown), which only contains the *crt*E gene based on genetic[4,31] and our unpublished DNA sequencing data. The level of mRNA hybridizing to the *Bam*HI G fragment (*crt*C, *crt*D) decreased, but to a far smaller extent than the mRNAs for LH, RC, and bacteriochlorophyll biosynthetic enzymes. The increased mRNA level was most noticeable for the *Bam*HI H fragment containing the *crt*A gene; it reached a maximum of about two fold 1 h after the shift (see Figure 8). Although the *Bam*HI H fragment has been mapped to contain other genes, we attribute this response mainly to the *crt*A gene, since the significant stimulation of mRNA by O_2 was not observed in the *crt* transposon mutant KZR9A12 (data not shown).

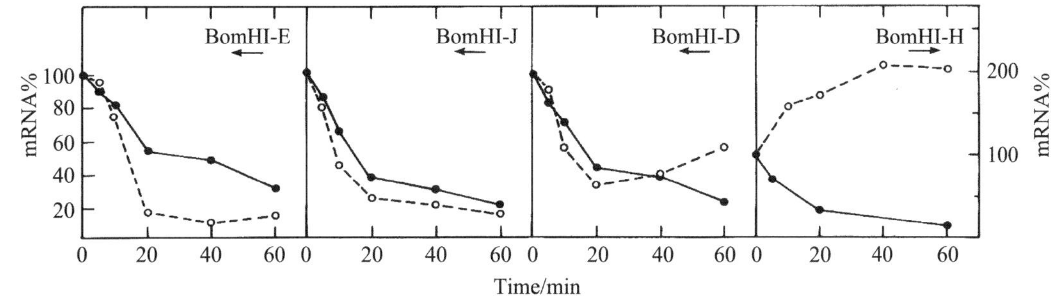

Figure 8 Time course of mRNA levels for bacteriochlorophyll and carotenoid biosynthetic enzymes in photosynthetic cells. Photosynthetic cells were subjected to proflavin and O_2 treatments as in Figure 7. The probes used for dot hybridization were pBR325 plasmids containing the *Bam*HI -E, -J, -D, and -H fragments. These clones and the genes harbored in these clones are listed in Table 1. The arrows underneath *Bam*HI refer to the vertical axis for each set of measurements

Discussion

R. capsulatus and other facultative anaerobes such as *Salmonella typhimurium* and *Escherichia coli* regulate gene expression for anaerobic and aerobic metabolism depending on environmental conditions, but the events controlling these processes are not well understood. Buck and Ames[6] suggested that tRNA modification, which mediates attenuation of transcription, may play a role in this control in *Salmonella* species. Yamamoto and Droffner[33] found that activity of topoisomerase I, associated with relaxation of chromosomal DNA in Salmonella species, is necessary for the expression of genes required for aerobic growth, whereas activity of gyrase, associated with supercoiling of chromosomal DNA, is necessary for expression of genes required for anaerobic growth. In this study we have examined the levels and stability of mRNAs from a number of photosynthetic genes and investigated the effects of oxygen on the transcription and the turnover of mRNAs.

Our previous studies on oxygen and light effects[39] and others[3] showed that two transcripts which code for pigment-binding proteins, a short 0.5-kb mRNA from LH-I and a longer 2.6-kb mRNA for LH-I, RC-L and RC-M polypeptides, have different stabilities. In the present study a thorough investigation of the stabilities of mRNAs for light-harvesting (LH-I, LH-II) and reaction center (RC-L, RC-M, RC-H) proteins and for some bacteriochlorophyll and carotenoid biosynthetic enzymes has been conducted by inhibiting the initiation of transcription with proflavin and then measuring the decay of the mRNA by means of dot blots and Northern hybridizations. Proflavin is known to inhibit transcription by intercalating with DNA template, a mechanism similar to that of actinomycin D[7,32]. Actinomycin D has been shown to block transcription of the genes coding for photosynthetic membrane proteins in *R. capsulatus*[13]. The results of half-lives of mRNAs for LH-I and RC proteins measured using proflavin as inhibitor of transcription in this study well agree with the previous results of *R. capsulatus* and *R. sphaeroides* with rifampin or actinomycin D as an inhibitor of transcription[3,13,40]. The half-lives of mRNAs for LH-II and LH-I were measured as 24 and 19 min, respectively; the half-lives of mRNAs for RC proteins were about 10 min. Therefore, the LH-II mRNA is more stable than the LH-I mRNA, and the LH-I mRNA (0.5 kb) has greater stability than the RC mRNA under photosynthetic conditions. Comparison of the abundance of the LH-I and LH-II mRNAs by using specific oligonucleotide probes of the same length indicates that the LH-H mRNA is more abundant than the LH-I mRNA (data not shown), and the ratio of RC mRNA to LH-I mRNA is about 1 : 10 to 1 : 20([3,39,40]; this study). The stoichiometry of the RC, LH-I, and LH-II polypeptides (RC-LH-I-LH-II, 1 : 10 to 20 : 0 to 80) in the membrane[15] can well be explained by the difference in abundance of the respective mRNAs. The difference in the abundance of mRNAs, however, cannot be accounted for only by the difference in the mRNA decay rates (half-life RC-LH-I-LH-II, about 1 : 2 : 2.5). The data thus suggest that the differential rates of transcription of the genes for these pigment-binding proteins must account for the difference in the abundance of their respective mRNAs and ultimately for the stoichiometry of the assembled protein complexes.

Kinetic studies of mRNA decay demonstrate that the levels of mRNA for LH, RC, and bacteriochlorophyll biosynthesis were immediately reduced after the PS cells were shifted to aerobic conditions. In contrast, the contents of LH, RC complexes, and bacteriochlorophyll and cell growth did not show significant changes during a corresponding short term (1 h) of oxygen shift (data not shown), indicating that the level of mRNAs is more sensitive to oxygen compared with their proteins. It is likely that oxygen inhibits transcription somehow and then, in certain cases, stimulates the degradation of mRNAs. Our data for this argument are twofold. First, for the majority of genes studied, including those for LH-I, RC-L, RC-M, and RC-H, the decreases in mRNA levels upon exposure to oxygen are similar to the kinetics observed when proflavin was added to photosynthetic cells. This argues that oxygen probably inhibits the transcription of these genes. Second, in the case of LH-II transcripts, the kinetics of the decrease in mRNA was significantly faster than that observed upon addition of proflavin based on both Northern (see Figure 6) and dot hybridization (see Figure 7). In this case, oxygen probably both inhibits transcription and accelerates the degradation of LH-II mRNA, although the mechanism is unknown. There are several possibilities, namely, that either the LH-II mRNA becomes more susceptible than the LH-I mRNA to RNase attack in the presence of oxygen, or that the LH-II mRNA and LH-I mRNAs are degraded by two types of RNase whose activities have a different sensitivity to oxygen. Alternatively, it also might be possible that oxygen strongly blocks the translation process for LH-II, resulting in release of mRNA from the ribosome where it is attached and increasing its susceptibility to nuclease cleavage.

The mRNAs for bacteriochlorophyll and carotenoid biosynthesis show a number of different responses to oxygen. For the *Bam*HI -E and -J fragments, the response to oxygen was similar to the LH and RC. The *Bam*HI -J fragment contains a heavily transcribed unidentified gene as well as *crt*F and *crt*E, which have previously been mapped[31,34,43]. The size of the unidentified transcript is 0.4 kb ([39]; unpublished data). The response to oxygen of this fragment is due mainly to this small transcript and is similar to that of the genes for RC and LH polypeptides. The nature and function of this unidentified gene remain to be characterized. For the *Bam*HI -D probe, the level of mRNA decreased during the first 20 min after the shift and then increased. From this result we assume that there is an oxygen-induced gene (s) located in the *Bam*HI -D fragment which is highly transcribed under steady-state aerobic conditions[39]. When the PS cells were shifted to oxygen, the genes for bacteriochlorophyll biosynthetic enzymes (*bch*F, *bch*K, *bch*H) were repressed, and their mRNAs were degraded at the early stage of aeration, and then unknown oxygenactivated gene (s) in the *Bam*HI -D fragment was induced. Since the *Bam*HI D fragment is 7 kb long, it is large enough to accomodate other genes in addition to *bch*F, *bch*K, and *bch*H, which have been only roughly mapped by genetic complementation[31,34].

For the *Bam*HI -H fragment, the level of mRNA increased after the O_2 shift. The response to oxygen in this fragment, which occurs in an opposite fashion compared with most other photosynthetic genes, is mainly due to the *crt*A gene. This conclusion is supported by the result that a transposon insertion in *crt*A abolishes this response. The *crt*A gene, which probably codes for an oxygenase responsible for the oxidation of spheroidene to spheroidenone[28], has been mapped to this restriction fragment. Earlier studies[14,29] show that when anaerobic cultures of *R. sphaeroides*, a closely related species, are aerated, the carotenoid pigment spheroidene is enzymatically converted to spheroidenone. In ^{18}O-labeling experiments, the added keto group has been shown to be derived from molecular oxygen[30]. Absorption peaks of spheroidenone are red shifted by about 30 nm compared with spheroidene[18,22]. Our spectroscopic studies of the pigments (see Figure 2 and Figure 3) reveal that the carotenoid composition rapidly changed when the PS cells were shifted to oxygen, based on absorption shifts to longer wavelengths between 400 and 500 nm. We have confirmed this result by thin-layer chromatography analysis. Under the photosynthetic conditions, the major carotenoid was found to be spheroidene; under aerobic conditions the major carotenoid was found to be spheroidenone. It has long been understood that carotenoids have two functions in photosynthetic bacteria. They serve both to harvest light energy[10] and to protect the photosynthetic apparatus from photooxidative damage which occurs in the presence of light and oxygen[11]. Here we suggest that the increase in expression of the *crt*A gene in the *Bam*HI -H fragment as well as other *crt* genes by oxygen represents one type of protection by carotenoids from the potentially damaging effects of oxygen. The oxidation of spheroidene to spheroidenone, catalyzed by an oxygen-activated oxygenase which is coded by the *crt*A gene, may scavenge oxygen which can cause damage to the photosynthetic apparatus.

Acknowledgments

We thank G. Drews for kindly providing plasmid pRCl and N. Stover and D. O' Brien for discussion.

This work was supported by the Office of Basic Energy Sciences, Biologial Energy Research Division of the U. S. Department of Energy under contract DE-ACO30 - 76SF00098, by U. S. Department of Agriculture grant 84 CRCR 11485, and by Public Health Service grant GM 30786 from the National Institutes of Health. The Biosearch Sam One was leased with Program Development Funds from the Director of Lawrence Berkeley Laboratory.

References

[1] Aagaard, J., and W. R. Sistrom. 1972. Control of synthesis of reaction center bacteriochlorophyll in photosynthetic bacteria. Photobiology 15:209 - 225.

[2] Arnheim, K., and J. Oelze. 1983. Differences in the control of bacteriochlorophyll formation by light and oxygen. Arch. Microbiol. 135:299 - 304.

[3] Belasco, J. G., J. T. Beatty, C. W. Adams, A. V. Gabain, and S. N. Cohen. 1985. Differential expression of photosynthetic genes in *Rhodopseudomonas capsulata* results from segmental differences in stability within the polycistronic rxcA transcript. Cell 40:171 - 181.

[4] Biel, A. J., and B. L. Marrs. 1983. Transcriptional regulation of several genes for bacteriochlorophyll biosynthesis in Rhodopseudomonas capsulata in response to oxygen. J. Bacteriol. 156:686 - 694.

[5] Biel, A. J., and B. L. Marrs. 1985. Oxygen does not directly regulate carotenoid biosynthesis in Rhodopseudomonas capsulata. J. Bacteriol. 162:1320 - 1321.

[6] Buck, M., and B. N. Ames. 1984. A modified nucleotide in tRNA as a possible regulator of aerobiosis: synthesis of cis-2-methylthioribosylzeatin in the tRNA of Salmonella. Cell 36:525 - 531.

[7] Chamberlin, M. J. 1974. Bacterial DNA-dependent RNA polymerase, p. 373. In P. D. Boyer (ed.), The Enzyme, vol. 10. Academic Press, Inc., New York.

[8] Clark, W. G., E. Davidson, and B. L. Marrs. 1984. Variation of levels of mRNA coding for antenna and reaction center polypeptides in Rhodopseudomonas capsulata in response to changes in oxygen concentration. J. Bacteriol. 157:945 - 948.

[9] Clayton, R. K. 1966. Spectroscopic analysis of bacteriochlorophylls in vitro and in vivo. Photochem. Photobiol. 5:669 - 677.

[10] Cogdell, R. J. 1978. Carotenoids in photosynthesis. Philos. Trans. R. Soc. Lond. Ser. B. 284:569 - 579.

[11] Codgell, R. J., M. F. Hipkins, W. MacDonald, and T. G. Truscott. 1981. Energy transfer between the carotenoid and the bacteriochlorophyll within the B-800 - 850 light-harvesting pigment-protein complex of *Rhodopseudomonas sphaeroides*. Biochim. Biophys. Acta 634:191 - 202.

[12] Cohen-Bazire, G., W. R. Sistrom, and R. Y. Stanier. 1957. Kinetic studies of pigment synthesis by non-sulfur purple bacteria. J. Cell Comp. Physiol. 49:25-68.

[13] Dierstein, R. 1984. Synthesis of pigment-binding protein in toluene-treated Rhodopseudomonas capsulata and in cell-free systems. Eur. J. Biochem. 138:509-518.

[14] Dorothea, S. H. 1985. Carotenoids in photosynthesis. I. Loca-tion in photosynthesic membranes and light-harvesting function. Biochim. Biophys. Acta 811:325-355.

[15] Drews, G. 1985. Structure and functional organization of lightharvesting complexes and photochemical reaction centers in membranes of phototropic bacteria. Microbiol. Rev. 49:59-70.

[16] Drews, G., and J. Oelze. 1981. Organization and differentiation of membranes of phototropic bacteria. Adv. Microb. Physiol. 22:1-81.

[17] Drews, G., J. Peters, and R. Dierstein. 1983. Molecular organization and biosynthesis of pigment-protein complexes of Rhodopseudomonas capsulata. Ann. Microbiol. (Paris) 134:151-158.

[18] Goodwin, T. W. 1980. The biochemistry of carotenoids, vol. 1. Plants. Chapman & Hall, Ltd., New York.

[19] Imhoff, J. F., H. G. Truper, and N. Pfenning. 1984. Rearrange-ment of the species and genera of the phototrophic purple nonsulfur bacteria. Int. J. Syst. Bacteriol. 34:340-343.

[20] Kaplan, S., and C. J. Arntzen. 1982. Photosynthetic membrane structure and function, p. 65-151. In Govindjee (ed.), Photosynthesis in energy conversion by plants and bacteria, vol. 1. Academic Press, Inc., New York.

[21] Klug, G. T., N. Kaufmann, and G. Drews. 1985. Gene expression of pigment-binding proteins of the bacterial photosynthetic apparatus: transcription and assembly in the membrane of Rhodopseudomonas capsulata. Proc. Natl. Acad. Sci. USA 82:6485-6489.

[22] Manwaring, J., E. H. Evans, G. Britton, and D. R. Schneider. 1950. The identification of desmethylspheroidenone as a major carotenoid in aerobic cultures of Rhodopseudomonas capsulata. FEBS Lett. 110:47-49.

[23] Markwell, M. A. K., M. H. Suzanne, L. L. Bieber, and N. E. Tolbert. 1963. A modification of the Lowry procedure to simplify protein determination in membrane and in protein samples. Anal. Biochem. 87:206-210.

[24] Marrs, B. L. 1981. Mobilization of the genes for photosynthesis from Rhodopseudomonas capsulata by a promiscuous plasmid. J. Bacteriol. 146:1003-1012.

[25] Marrs, B. L., and S. Kaplan. 1970. 23S precursor ribosomal RNA of *Rhodopseudomonas sphaeroides*. J. Mol. Biol. 49:297-317.

[26] Miller, K. R. 1985. The photosynthetic membrane: prokaryoticand eukaryotic cells. Endeavour New Ser. 9:175-182.

[27] Oelze, J., and K. Arnheim. 1983. Control of bacteriochlorophyll formation by oxygen and light in Phodopseudomonas sphaeroides. FEMS Microbiol. Lett. 19:197-199.

[28] Scolnik, P. A., M. A. Walker, and B. L. Marrs. 1980. Biosynthesis of carotenoids derived from neurosporene in Rhodopseudomonas capsulata. J. Biol. Chem. 255:2427-2432.

[29] Schneour, E. A. 1962. Carotenoid pigment conversion in *Rhodopseudomonas sphaeroides*. Biochim. Biophys. Acta 62:534-540.

[30] Shneour, E. A. 1962. The source of oxygen in *Rhodopseudomonas sphaeroides* carotenoid pigment conversion. Biochim. Biophys. Acta 65:510-511.

[31] Taylor, D. P., S. N. Cohen, W. G. Clark, and B. L. Marrs. 1983. Alignment of genetic and restriction maps of the photosynthesis region of the Rhodopseudomonas capsulata chromosome by a conjugation-mediated marker rescue technique. J. Bacteriol. 154:580-590.

[32] Waring, M. J. 1965. The effects of antimicrobial agents on ribonucleic acid polymerase. Mol. Pharmacol. 1:1 - 13.

[33] Yamamoto, N. , and M. L. Droffner. 1985. Mechanisms determining aerobic or anaerobic growth in the facultative anaerobic Salmonella typhimurium. Proc. Natl. Acad. Sci. USA 82:2077 - 2081.

[34] Yen, H. -C. , and B. Marrs. 1976. Map of genes for carotenoid and bacteriochlorophyll biosynthesis in Rhodopseudomonas capsulata. J. Bacteriol. 126:619 - 629.

[35] Yu, P. -L. , B. Hohn, H. Falk, and G. Drews. 1982. Molecular cloning of the ribosomal RNA genes of the photosynthetic bacterium Rhodopseudomonas capsulata. Mol. Gen. Genet. 188:392 - 398.

[36] Youvan, D. C. , E. J. Bylina, M. Alberti, H. Begusch, and J. E. Hearst. 1984. Nucleotide and deduced polypeptide sequences of the photosynthetic reaction-center, B870 antenna and flanking polypeptides from Rhodopseudomonas capsulata. Proc. Natl. Acad. Sci. USA 81:189 - 192.

[37] Youvan, D. C. , and S. Ismail. 1985. Light-harvesting II (B800-B850) complex structural genes from Rhodopseudomonas capsulata. Proc. Natl. Acad. Sci. USA 82:58 - 62.

[38] Weaver, P. F. , J. D. Wall, and J. Gest. 1975. Characterization of Rhodopseudomonas capsulata. Arch. Microbiol. 105:207 - 216.

[39] Zhu, Y. S. , and J. E. Hearst. 1986. Regulation of expression of genes for light harvesting (LH-I, LH-II), reaction center (RC-L, M, H), bacteriochlorophyll and carotenoid biosynthesis inRhodobacter capsulalus by light and oxygen. Proc. Natl. Acad. Sci. USA 38:7613 - 7617.

[40] Zhu, Y. S. , and S. Kaplan. 1985. Effects of light, oxygen and substrates on steady-state levels of mRNA coding for ribulase-1, 5-bisphosphate carboxylase and light-harvesting and reaction center polypeptides in *Rhodopseudomonas sphaeroides*. J. Bacteriol. 162:925 - 932.

[41] Zhu, Y. S. , P. J. Kiley, T. J. Donohue, and S. Kaplan. 1986. Origin of the mRNA stoichiometry of the *puf* operon in Rhodobacter sphaeroides. J. Biol. Chem. 261:10366 - 10374.

[42] Zhu, Y. S. , S. D. Kung, and L. Bogorad. 1985. Phytochrome control of levels of mRNA complementary to plastid and nuclear genes of maize. Plant Physiol. 79:371 - 376.

[43] Zsebo, K. M. , and J. E. Hearst. 1984. Genetic-physical mapping of a photosynthetic gene cluster from R. capsulata. Cell 37:937 - 947.

Oxygen and Light Regulation of Expression of Genes for Light Harvesting (LH-I, LH-II), Reaction Center (RC-L, RC-M, RC-H), Pigment Biosynthesis and a Transcriptional Role in the Protective. Function of Carotenoids in *Rhodobacter Capsulatus* *

Abstract

The coordinate and differential expression of genes for light harvesting I (LH-I) and II (LH-II) antenna, reaction center (RC) polypeptides L, M, and H, bacteriochlorophyll (Bchl) and carotenoid (Crt) biosynthesis in response to light and O_2 in Rhodobacter capsulatus has been studied by measuring the mRNA levels using dot blot and Northern hybridizations. The genes for LH-II only have one transcript (0. 5 kb), while the genes for LH-I have two transcripts (0. 5 and 2. 6 kb). The small transcript (0. 5 kb) is the mRNA only for LH-I (β, α) polypeptides, whereas the large transcript (2. 6 kb) codes for RC-L, RC-M, and the β and α polypeptides of LH-I, as well as an unknown ORF C2397. These five genes thus comprise a single operon (designated the *puf* operon). The mRNA specifying the LH-II polypeptides is more abundant, and more sensitive to changes in O_2 concentration and shows a variation over a wider range than that of the mRNA for the LH-I, indicating that the genes for the LH-II and LH-I/RC are independently regulated. The gene for RC-H (*puh*A) has at least two transcripts (1. 2 and 1. 4 kb) which initiate within ORF F1696 and respond differentially to light intensity. The genes coding for RC-L, RC-M and RC-H are coordinately regulated by light intensity and O_2 concentration. We have shown that an increase in light intensity and O_2 concentration causes a decrease in the expression of the genes for LH-I, LH-II, RC proteins, and Bchl biosynthetic enzymes. The genes coding for

---

* Yu Sheng Zhu and John E. Hearst, Division of Chemical Biodynamics, Lawrence Berkeley Laboratory and Department of Chemistry, University of Califomia, Berkeley, California 94720

Biggens, J. (ed.), Progress in Photosynthesis Research, Vol. IV. ISBN 90 247 3453 3 © 1987 Martinus Nijhoff Publishers, Dordrecht. Printed in the Netherlands.

此文发表于 1987 年"Progress in Photosynthesis Research." 4:717 – 720. Biggens(ed.) Martinus Nijhoff Publishers, Dordrecht, printed in the Netherland.

the enzymes in Crt biosynthetic pathways, however, are regulated in an opposite fashion: high light and high O_2 result in increased expression of *crt* genes. This increased transcription of *crt* genes is interpreted based on the protective function of Crt against photooxidative damage. An interaction and interdependence of the genes for LH, RC proteins, Bchl and Crt biosynthetic enzymes were also observed using various mutants.

The purple nonsulfur photosynthetic bacterium, *R. capsulatus*, provides an attractive model for studying the regulation of photosynthetic genes by light, O_2 and other environmental factors. When the concentration of oxygen is lowered, the cell develops an ICM and induces the biosynthesis of LH-I (B870), LH-II (B800-B850), and RC proteins, as well as Bchl and Crt[1-3]. Once assembled, the photochemically active complexes harvest and convert light energy into chemical energy.

The genes for the β(*puf*B) and α (*puf*A) subunits of LH-I complex, as well as the L (*puf*L) and M (*puf*M) subunits of RC complex, are located in the *Bam*C-*Eco*B fragment of the R-prime plasmid pRPS404 isolated from *R. capsulatus*[4]. They are co-transcribed in vivo both in *R. capsulatus*[5] and *Rhodobacter sphaeroides*[6,7]. The gene for the H (*puh*A) subunit of the RC lies in the *Bam*F fragment of the R prime[8], whereas the genes for the β (*puc*B) and α (*puc*A) subunits of the LH-II are outside of this photosynthetic gene cluster[9]. The structural genes for LH-I, LH-II and RC-L, RC-M, RC-H in *R. capsulatus* have been sequenced[4,8,9]. The genes for Bchl and Crt biosynthesis have been mapped by genetic methods[10] and transposon mutagenesis[11].

In this paper we report a systematic investigation of mRNAs of photosynthetic genes in *R. capsulatus* under different growth conditions using dot and Northen hybridizations. The results show a coordinate and differential regulation of the genes for LH (I, II), RC (L, M, H) and pigment (Bchl, Crt) biosynthesis by light intensity and O_2 concentration.

Materials and Methods

Bacterial strains, media and growth conditions *R. capsulatus* Rif^R strain SB1003 was used throughout this study. Mutants KZR9A12 (*crt*A-), KZR9F4 (*crt*I-), UAR87 (RC-H-) and SB1007 (bchC-) were used where indicated. Light intensities are denoted as follows: high light (HL), low light (LL), dark (D); O_2 conditions are denoted: high O_2(O_2), low O_2(LO_2). Cell culture conditions are previously described [12].

Absorption spectroscopy of pigments. See [13].

Preparation of plasmid, M13 phage and oligonucleotide probes. See [12].

RNA isolation, dot blots and Northern hybridization. See [12].

Results and Discussion

Effects of light and O_2 on the mRNA levels of genes for LH, RC proteins, Bchl and Crt

biosynthesis. Light intensity and O_2 concentration significantly effect the levels of mRNA for LH, RC and pigment biosynthesis (see Table 1). It has been shown that: ① A shift to dark causes the decreased expression of photosynthetic genes; ② An increase in light intensity results in decreased expression of the genes for LH, RC and Bchl biosynthesis, but it results in increased expression of *crt* genes; ③ O_2 greatly reduced the expression of genes for LH, and RC; however, the relative amounts of mRNAs from *crt* genes rise; ④ The mRNA for LH-II is more sensitive to O_2 than that of the mRNA for LH-I, indicating that the genes for LH-II and LH-I/RC are independently regulated. These findings are supported by Northern hybridization (see Figure 1). The gene for LH-II has one transcript (0.5 kb), whereas the gene for LH-I has two transcripts: 0.5 kb and 2.6 kb. The 0.5 kb transcript is more abundant and is only complementary to the genes for LH-I (β, α), while the 2.6 kb transcript hybridizes to five genes coding for LH-I (β, α), RC-L (see Figure 1), RC-M and ORF 02397 (data not shown). The 2.6 kb transcript is more sensitive to O_2 than the 0.5 kb transcript (see Figure 1). The gene for RC-H has two transcripts, 1.2 and 1.4 kb. They initiated within ORF F1696, and they are regulated differently by light intensity.

Table 1 Effects of light and O_2 on the mRNA levels of genes for light harvesting, reaction center, bacteriochlorophyll and carotenoid biosynthesis in *R. capsulatus*

Genes	mRNA/%*			
	HL	D	LL	O_2
LH-II	55	19	100	2
LH-I	59	30	100	17
RC-L	77	19	100	21
RC-M	72	20	100	23
RC-H	57	16	100	17
BchJ, G, D (*Bam*E)	81	20	100	62
BchH, K, F (*Bam*D)	84	30	100	118
*crt*A, I, B, BchI (*Bam*H)	125	39	100	131
*crt*C, D (*Bam*G)	132	67	100	72
*crt*E (*Bam*M)	132	91	100	95

Decay rate of mRNA from photosynthetic genes and O_2 regulation. In order to study the effects of O_2 on the expression of photosynthetic genes in more detail, we have conducted an O_2 shift experiment. *R. capsulatus* strain SB1003 was grown photosynthetically (see Figure 2, PS), then shifted to LO_2 (medium/flask=8/10, v/v) (see Figure 2, PS→ LO_2), or to high 02 (medium/flask=0.8/10, v/v) (see Figure 2, PS-O_2). The absorption spectra of pigments shows that the Bchl content (peaks at 360, 600 and 770 nm) decreases during the shift. A more detailed time course reveals that the total Crt content (absorption between 400 and 500 nm)

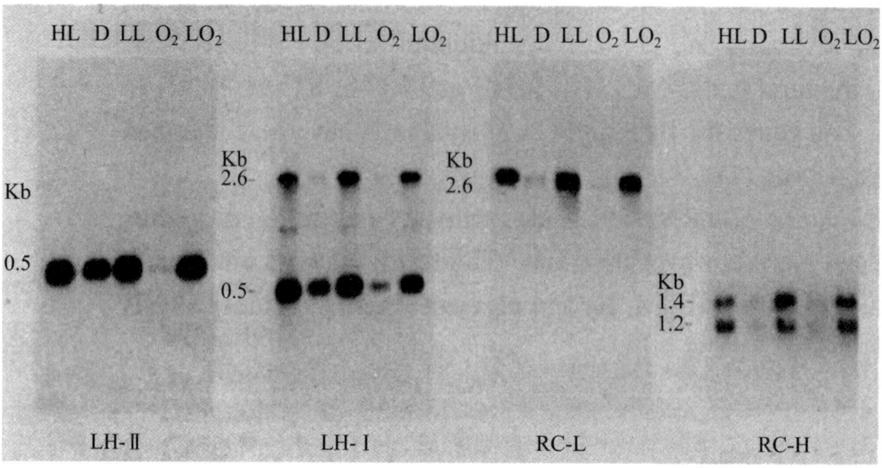

Figure 1 Northern hybridization of ***R. capsulatus*** with cloned probes for LH-I, LH-II, RC-L, and RC-H. RNAs from ***R. capsulatus*** cells grown on different growth conditions (HL, D, LL, O_2, LO_2) were denatured in formaldehyde, fractionated on 1.2% agarose gel, transferred to Gene Screen, and hybridized with ^{32}P-labeled probes for LH-II, LH-I, RC-L, and RC-H

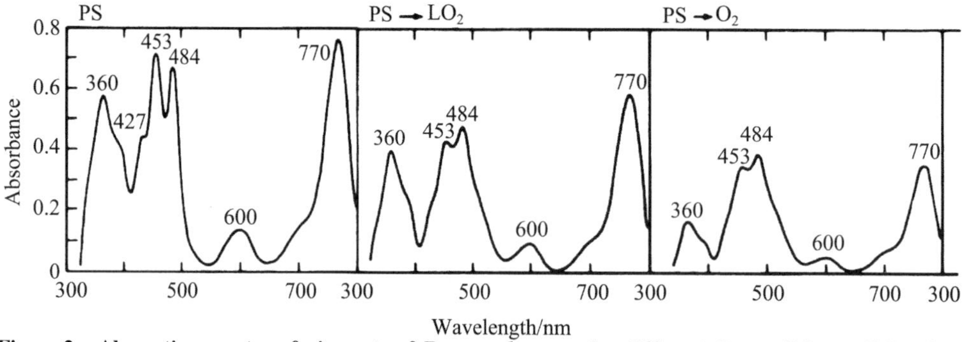

Figure 2 Absorption spectra of pigments of ***R. capsulatus*** under different O_2 conditions. Cells of ***R. capsulatus*** were grown under anaerobic photosynthetic conditions (PS) to mid-log phase, then shifted to low O_2 (PS→LO_2) for 24 h. The pigments were extracted with acetone: methanol (7 : 2). The visible absorption spectra were determined using a Cary 118 spectrophotometer

increases slightly in the first 3 hours, then decreases (data not shown). The change in relative ratio of A453 and A484 is due to the oxidation of spheroidene, a major Crt in PS cell, to spheroidenone, a major Crt in aerobic cells, as shown by TLC analysis (data not shown). This oxidation is catalyzed by a metal-containing oxygenase[14], encoded by *crt*A located in *Bam*H fragment[11]. Figure 3 shows the decay rate of mRNAs and the effects of O_2. When an inhibitor of transcription, proflavin, is added to cell cultures, the levels of mRNA from all photosynthetic genes decrease. The half-lives of mRNAs for LH-II, LH-I and RC (L, M, H) are 24, 19 and 10 min respectively. O_2 probably turns off the transcription of genes for LH (I, II), RC (L, M, H) and Bchl biosynthesis, resulting in a

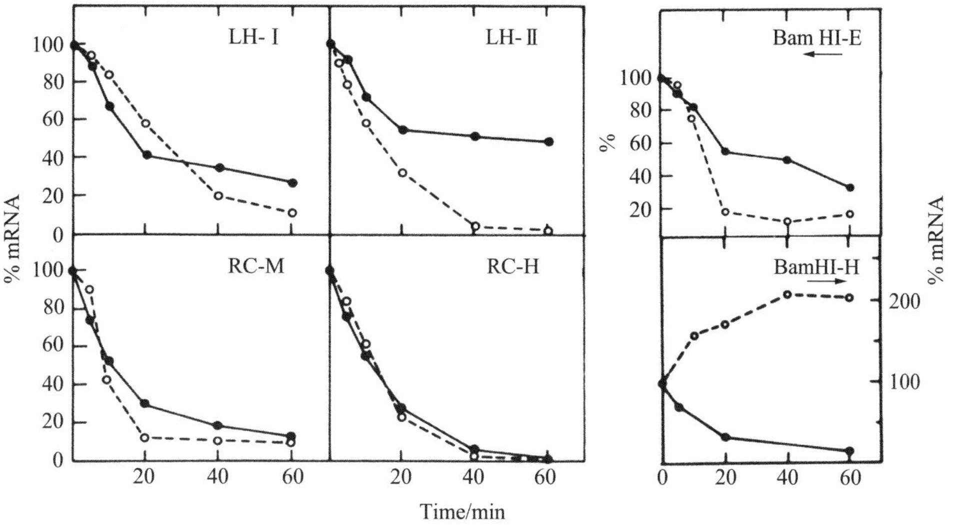

Figure 3 **Decay rate of mRNA levels for LH-I, LH-II, RC-M, RC-H, and for bacteriochlorophyll (*Bam*E) and carotenoid (*Bam*H) biosynthetic enzymes. Photosynthetic ceUs were treated with proflavin (solid line) which inhibits the initiation of transcription, or exposed to O_2 (dotted line). Cells were harvested at various time intervals and RNAs were extracted and measured by dot blot hybridization**

decrease in their mRNAs. O_2 appears also to stimulate the degradation of mRNA for LH-II. It is interesting to note that the level of mRNA in *Bam* H fragment containing *crt*A, I, B and bchI increases two fold after O_2 shift Such an opposite response to O_2 is mainly due to *crt*A, because a Tn5. 7 insertion in *crt*A (in KZR9A12) abolishes this response (data not shown). Crt has been long known to have two functions: harvesting light and protecting cell from photooxidative killing[15]. Plants and bacteria have a variety of mechanisms for such protection[16]. Carotenoids, as important constituents of chloroplast membranes, react with singlet O_2 and therefore protect chlorophyll and membranes against damage[16]. This is the first report that such protection can be regulated at the level of gene expression under high light and O_2.

Cooperation and interdependence of photosynthetic genes. The photosynthetic genes for LH, RC and pigment biosynthesis are interdependent. Mutation in each of the genes for RC-H (UAR87)[17], Crt (9F4) or Bchl (MB1007) biosynthesis results in significantly reduced amounts of mRNAs from the rest of the genes (see Figure 4). A longer exposure of the autoradiograph (data not shown) shows that although the mRNA levels in these mutants are very low, they are detectable. A close link exists between genes for LH-II and Crt biosynthesis. A comparison of the decay rate of mRNAs for LH-I, LH-II, RC-L, RC-M and RC-H after treatment with proflavin as transcription inhibitor reveals that these mRNAs in the mutants are at least stable as in wild-type. Therefore, the reduction in mRNA levels in these mutants is due to transcriptionsl control. We suggest that proper and accurate assembly of the photosynthetic compontents in ICM is necessary for transcription. Absence

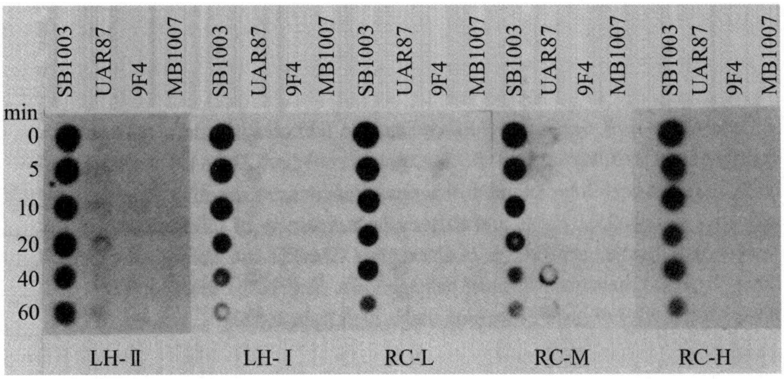

Figure 4 Decay of mRNAs for LH-I, LH-II, RC-L, RC-M and RC-H in wild type and mutants after proflavin treatment

SB1003 Wild type ps$^+$
UAR 87 RC-H-ps-
KZR9F4 CrtI-ps$^+$
MB1007 bchC-ps-

of one of these components causes failure of assembly of the photosynthetic apparatus. Accumulation of these components (either proteins or pigments) may tum off transcription. Futhermore, unassembled proteins may become more susceptible to endogenous proteinase.

References

[1] Drews, G. (1978) Curr. Top. Bioenerg. 8,132 - 136.

[2] Ohad, I. & Drews, G. (1982) in Photosynthesis II: Development, carbon metabolism and plant productivity, ed. Govindjee (Academic Press, New York), pp. 89 - 140.

[3] Dierstein, R., Tadros, M. & Drews, G. (1984) FEBS Lett. 24,219 - 223.

[4] Youvan, D. C., Alberti, M., Begusch, H., Bylina, E. J. & Hearst, J. E. (1984) Proc. Natl. Acad. Sci. USA 81,189 - 192.

[5] Belasco, J. G., Beatty, J. T., Adams, C. W., Von Gabain, A. & Cohen, S. N. (1985) Cell 40,171 - 181.

[6] Zhu, Y. S. & Kaplan, S. (1985) J. Bacteriol. 162,925 - 932.

[7] Zhu, Y. S., Killey, P. J., Donohue, T. J. and Kaplan, S. (1986) J. Biol. Chem., (in press).

[8] Youvan, D. C., Bylina, E. J., Alberti, M., Begusch, H., & Hearst J. E. (1984) Cell 37,949 - 957.

[9] Youvan, D. C. & Ismail, S. (1985) Proc. Natl. Acad. Sci. USA 82,58 - 62.

[10] Biel, A. L. & Marrs, B. L. (1983) J. Bacteriol. 156,686 - 694.

[11] Zsebo, K. M. & Hearst, J. E. (1984) Cell 37,937 - 947.

[12] Zhu, Y. S. & Hearst, J. E. (1986) Proc. Natl. Acad. Sci. USA, (in press).

[13] Zhu, Y. S., Cook, D., Leach F., Armstrong, G., Alberti, M. & Hearst, J. E. 0986) J. Bacteriol. (submitted).

[14] Shneour, E. A. (1962) J. Bio. Chem. 62,534 - 540.

[15] Cogdell, R. J. (1978) Philos. Trans. R. Soc. Lond. Ser. B. 248,569 - 579.

[16] Halliwell, B. (1984) Chloroplast metabolism. pp. 180 - 206. Claremdon Press, Oxford.

[17] O'Brien, P. A., Stover, N. & Heast, J. E. (1986) J. Bacteriol. (submitted).

Transcription of Oxygen-Regulated Photosynthetic Genes Requires DNA Gyrase in *Rhodobacter Capsulatus*

(Supercoiling/DNA Conformation/Oxygen Regulation)①②

Abstract

The regulation of the photosynthetic genes by DNA supercoiling in *Rhodobacter capsulatus* has been studied by using gyrase inhibitors in vivo and by measurement of mRNA levels of more than a dozen genes. The results demonstrate that the levels of mRNA for light-harvesting (I, II) and reaction center (L, M, H) proteins, bacteriochlorophyll biosynthetic enzymes, ribulose-bisphosphate carboxylase (EC 4.1.1.39), and the mRNAs from the open reading frames Q and R decreased immediately and dramatically upon addition of novobiocin and coumermycin. In contrast, the mRNAs for carotenoid biosynthetic enzymes, the cytochrome bc_1 complex, and constitutively expressed mRNA under aerobic conditions for light-harvesting I and for reaction center (L, M) proteins are less sensitive to the inhibitors. In accordance with these results, the biosynthesis of bacteriochlorophyll is markedly repressed by gyrase inhibitors novobiocin, coumermycin, nalidixic acid, and oxolinic acid either under anaerobic conditions or during a shift from aerobic to anaerobic conditions. The synthesis of light-harvesting (I, II) bacteriochlorophyll complexes is also inhibited by novobiocin and coumermycin. The kinetics of specific mRNA changes and the differential sensitivity of anaerobic and aerobic genes to the gyrase inhibitors strongly suggest that DNA supercoiling is involved in the differential expression of photosynthetic genes in response to the level of oxygen in *R. capsulatus*.

Yu Sheng Zhu, John E Hearst

Division of Chemical Biodynamics, Lawrence Berkeley Laboratory, and Department of Chemistry, University of California, Berkeley, CA 94720

Communicated by Martin Gellert, February 22, 1988 (received for review November 9, 1987)

① The publication costs of this article were defrayed in part by page charge payment. This article must therefore be hereby marked "advertisement" in accordance with 18 U. S. C. § 1734 solely to indicate this fact.

② Abbreviations: LH, light harvesting; RC, reaction center; BChl, bacteriochlorophyll; Crt, carotenoid; ORF, open reading frame; RuP_2Case, ribulose-bisphosphate carboxylase.

此文发表于 1988 年"Proc. Natl. Acad. Sci. USA" 85:4209－4213.

Rhodobacter capsulatus is a facultative purple nonsulfur Gram-negative bacterium capable of growing either aerobically in the dark (chemotrophic) or anaerobically in the light (phototrophic), primarily depending on the oxygen tension in the environment[1]. In the presence of oxygen, cells grow aerobically with an active respiratory electron transport chain as an energy source. Lowering the oxygen tension induces anaerobic growth with a photosynthetic electron transport chain producing energy. In the latter case the photosynthetic intracytoplasmic membrane composed of pigment-protein complexes including two light-harvesting (LH) antennae (I, II) and three reaction center (RC) polypeptides (L, M, H) develops. The pigments bacteriochlorophyll (BChl) and carotenoid (Crt) bound to the LH (I, II) polypeptides capture photons and funnel this energy into the RC. In the special membrane protein environment of the RC containing BChl, Crt, bacteriopheophytin, quinone, and iron, the charge separation occurs and the electron transport begins. The energetic photoelectron drives a cyclic series of redox complexes consisting of quinones and cytochromes (bc_1, c_2) in the membrane. The resultant proton gradient across the membrane drives ATP synthesis by way of an ATPase. The switch of the growth modes by oxygen is the result of activation and inactivation of a large number of genes responsible for different metabolic pathways[2].

Earlier studies have indicated that the biosynthesis of Bchl in purple nonsulfur bacteria is inhibited by oxygen[3]. Several enzymes in the BChl biosynthetic pathway, such as δ-aminolevulinate synthase, magnesium protoporphyrin methyltransferase[4], and magnesium protoporphyrin chelatase[5], appear to be repressed by oxygen. The transcription of a few genes in the Bchl biosynthetic pathway has been demonstrated to be regulated by oxygen using lac fusion[6]. The synthesis of pigment binding proteins involved in LH and the photochemical reaction in the photosynthetic membranes is also repressed by oxygen[7]. Recently, a number of studies[8-15] has shown that the levels of mRNA for LH (I, II) and RC (L, M, H) complexes in *R. capsulatus* or *Rhodobacter sphaeroides* are reduced in response to oxygen. Although most photosynthetic genes are repressed by oxygen, the Crt genes, particularly the *crt*A gene, are activated by oxygen, perhaps for the protection of the cells from photooxidation[9]. It has been reported that an open reading frame (ORF) (designated as Q) upstream from the *puf* operon may function as a regulatory gene to control the *puf* operon and Bchl biosynthesis in response to oxygen[16]. Recently, the same authors have modified their hypothesis suggesting that the Q gene may code for a Bchl binding protein①. We have found that two ORFs exist in the *Eco*Q fragment of *R. capsulatus*. The additional ORF [1473 base pairs (bp)]designated as R is located upstream from the Q gene (225 bp) with a space of 1 bp (M. Alberti and J. E. H., unpublished data). The R gene in *R.*

① Bauer, C. E. & Marrs, B. L., Symposium on Molecular Biology of Photosynthetic Procaryotes, June 8 – 10, 1987, University of Wisconsin-Madison, p. 2 (abstr.).

sphaeroides seems to regulate the insertion of LH-I in the membrane①. Furthermore, we have found that the expression of both putative genes is also regulated by oxygen and light (see below). This raises the further question: What is the fundamental mechanism for oxygen regulation?

Since most of the photosynthetic genes are coordinately expressed in response to the depletion of oxygen, there might be a common mechanism for their coordinate expression. Here we present in vivo evidence using the gyrase inhibitors novobiocin and coumermycin that the expression of the photosynthetic genes in *R. capsulatus* in response to oxygen is mediated by the supercoiling of DNA.

Materials and Methods

Bacteria and Anaerobic and Aerobic Growth Conditions

R. capsulatus SB1003 was grown either photoheterotrophically or chemoheterotrophically as described[9,10]. The photo-heterotrophically grown cells were cultured at 30℃ and at a light intensity of 30 W/m^2 in 15-ml screw-cap tubes. The tubes were filled completely to the top with malate-minimal RCV medium[17]. The doubling time of cells under these conditions is about 3 h. The gyrase inhibitors novobiocin, coumermycin, nalidixic acid, and oxolinic acid were added when the cells were grown to midlogarithmic phase $(5—7)\times 10^8$ cells per ml). The cells were harvested at various time points and used for pigment, chromatophore, and RNA preparation[9,10].

For the experiments involving a shift from aerobic to photosynthetic growth, the aerobic growth was maintained by chemoheterotrophic culture (10 ml of medium in a 120-ml Erlenmeyer flask) on a Gyratory shaker (200 rpm). Induction of the photosynthetic genes was performed by pouring the midlogarithmic-phase aerobic cells into 15-ml screw-cap tubes filled to the top and by culturing them under the anaerobic photosynthetic conditions. Novobiocin or coumermycin was immediately added to cell cultures after the shift from the aerobic to the anaerobic conditions. The cells were harvested after 3 h for pigment and RNA extraction.

Assay of Pigments and Chromatophores

Photosynthetic pigments were extracted with acetone/methanol (7 : 2). The total Bchl content of the cells was estimated by absorbancy measurements at 775 nm[10]. Chromatophores were prepared from the cell lysate[10]. The absorption spectra of the pigments and the chromatophores were measured as described[9,10].

---

① Kaplan, S., Symposium on Molecular Biology of Photosynthetic Procaryotes, June 8 - 10, 1987, University of Wisconsin-Madison, p. 1 (abstr.).

RNA Isolation and Blot Hybridization

Total RNA was extracted as described[8]. After purification, the RNA was fractionated by electrophoresis and transferred to the mem-branes. Procedures for dot blots and RNA transfer hybrid-izations and the preparation of hybridization probes (plasmid DNA, M13 phage DNA, and synthetic oligonucleotides) for various photosynthetic genes have been described[9,10]. To quantify hybridization, either the autoradiograms were scanned with a GTS 300 transmittance-reflectance scanning densitometer (Hoefer, San Francisco) or the filters were assayed for radioactivity in 2,5-diphenyloxazole-1,4-bis (5- phenyloxazolyl) benzene cocktail with a Packard model 3385 liquid scintillation counter.

Results

Differential Inhibition by Novobiocin and Coumermycin of the mRNAs for LH and RC Subunits, Bchl and Crt Biosyn-thetic Enzymes, Ribulose-Bisphosphate Carboxylase (RuP_2-Case; EC 4.1.1.39), and Cytochromes

The superhelicity of DNA in bacteria is determined by a dynamic balance between gyrase, which supercoils the DNA, and topoisomerase I, which relaxes the DNA. Thus, inhibition of gyrase leads to a decrease in negative superhelicity[18]. If the expression of the photosynthetic genes in *R. capsulatus* is dependent on the supercoiling of the DNA, inhibitors of gyrase would repress transcription and reduce the levels of these gene products. We have measured the mRNA levels in vivo of various genes in *R. capsulatus* cells with or without novobiocin and coumermycin by dot or RNA transfer hybridization using probes for specific genes. To determine whether or not the inhibition is specific to anaerobic genes for photosynthesis, we have also examined the effect of gyrase inhibitors on the levels of mRNA for the cytochrome bc_1 complex. The cytochrome bc_1 consists of an FeS protein, a cytochrome b, and a cytochrome c_1; the genes coding for these three subunits are on the *fbc* operon[19]. Figure 1(c) shows that the *fbc* operon is expressed in anaerobic and aerobic conditions.

Dot hybridization (see Figure 1) shows that the levels of the various mRNAs were constant in the absence of the drugs over the course of the 3-h incubation. The levels of various mRNAs, however, changed in different patterns after drug treatment. The response of the various RNA species to gyrase inhibition roughly falls into one of two broad classes: gyrase-dependent, oxygen-regulated RNAs; and gyrase-less dependent, constitutive RNAs. The first group includes the genes for the structural components of the photosynthetic apparatus (LH-I, LH-II, RC), BChl biosynthesis, and CO_2 fixation (RuP_2Case). These mRNAs are immediately and dramatically reduced after the addition of gyrase inhibitors as shown in Figure 1(a). The patterns of mRNA for LH-II, RC-M, and RC-H (data not shown) were similar to LH-I and RC-L, respectively. The half-lives of mRNAs for LH proteins following addition of the gyrase inhibitor were ≈ 20 min, and those for RC proteins were about 10 min,

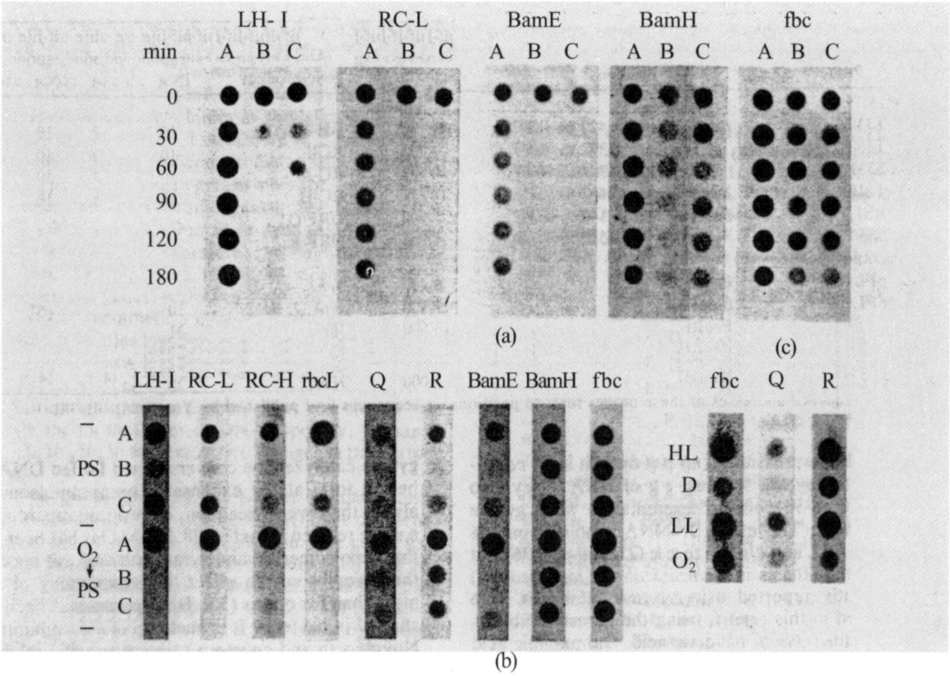

Figure 1 Dot blot hybridization of *R. capsulatus* RNA with gene probes for LH and RC subunits, Bchl and Crt biosynthetic enzymes, RuP_2Case, the Q and the R products, and cytochrome bc_1 complex. (a) Time course of RNA levels in the presence of inhibitors. Novobiocin (50 μg/ml) (marked as C) and coumermycin (50 μg/mg) (marked as B) were added to photosynthetically growing cells. A represents RNA from the photosynthetic cells without the inhibitor treatment. The total RNA was extracted from an equal number of cells at various times (0, 30, 60, 90, 120, 180 min) as indicated on the left. The RNA was denatured in formaldehyde, blotted onto a GeneScreen membrane, hybridized with ^{32}P-labeled DNA probes, and exposed to X-ray film as described[8]. The probes are marked on the top of each panel. (b) Dot hybridization. The top three rows represent the RNA from photosynthetically growing cells (marked as PS). The RNA was isolated from the cells 15 min after addition of coumermycin and novobiocin. The bottom three rows represent the RNA isolated from the cells 3 h after a shift from aerobic to anaerobic conditions (marked as O_2→ PS). A, control; B, coumermycin (50 μg/ml); C, novobiocin (50 μg/ml). Most probes have been described[9,10]. The probe for Q is an M13 clone, ESL1, containing an *Eco*RI-SalI fragment in the ORF Q from 142 to 342 bp 5′ to the *puf*B gene. The probe for R is an M13 clone, S357, containing an intenal Sau3A insert within the ORF R from 539 to 740 bp 5′ to the *puf*B gene; it may partially overlap with the Q mRNA, based on S1 nuclease analysis. * ESL1 and S357 are within the *Eco*RI-Q fragment of pRPS404. The probe for the *rbc*L gene is a plasmid, pLI10, carrying a 0.3-kilobase (kb) heterologous DNA insert coding for form II RuP_2Case of *R. sphaeroides*[20]. The same size of major transcript of rcbL and its same response to oxygen and light were observed in *R. capsulatus* (unpublished data) as in *R. sphaeroides*[8]. The probe for fbc genes is a plasmid, pRSFI, containing an insert coding for the entire fbc operon of *R. capsulatus*; only one 2.9-kb transcript from this operon was found[19]. (c) Dot hybridization of mRNAs from fbc and the Q and the R genes. The mRNA was isolated from cells grown under different oxygen and light conditions as described[10]. The anaerobic photosynthetic conditions are designated as HL (high light), D (dark), and LL (low light). The aerobic condition is designated as O_2

a result similar to the half-lives observed following addition of proflavin as an inhibitor of transcription[9]. (See last column, Table 1.) The levels of mRNA from bcJ, E, D in *Bam*HI-E fragment (see Figure 1(a)) and mRNA from *rbc*L and ORF Q, R (not shown) were also significantly reduced by the inhibitors within 30 min.

Table 1 M13 phage, plasmids, or oligonucleotide used as probes

Probe	Positions*		Size/bp	Gene or gene product	Ref.	Half-life of corresponding mRNA, min
M13						
T319	*Bam*HI-C	422 to 342	80	LH-I(α)	[10]	19
T330		1392 to 1094	298	RC-L	[10]	10
S331		1704 to 1607	97	RC-M	[10]	10
T210	*Bam*HI-F	3647 to 3413	234	RC-H	[10]	10
ESL1	*Bam*HI-C	−142 to −342	200	ORF Q		
S357		−539 to −740	201	ORF R		
Plasmid						
pFL205	*Bam*HI-E		6,400	*bch*J, -G, -D	[10]	
pFL103	*Bam*Hl-H		3,200	*crt*A, -I; bchI	[10]	5†
pRSF1	*Eco*RI		10,000	fbc	[21]	
pLI10	*Bam*HI		3,000	rcbL	[22]	
Oligonucleotide			15-mer	LH-II	[10]	24

* The sequence addresses of these probes refer to positions on sequences first published by Youvan et al. [34].
†Specific for *crt*A.

The second class of genes that are less sensitive to gyrase inhibitors includes genes for cytochromes and genes for Crt biosynthetic enzymes typified by *crt*A, I in the *Bam*HI-H fragment [see Figure 1(a)] and CrtE in the *Bam*HI-M fragment (not shown). The levels of mRNA from these genes and from the *fbc* genes decreased at a slower rate and to smaller extents, reaching half of their initial values after ≈3 h [see Figure 1(a)].

The reduction of the levels of mRNA from the genes coding for LH-I, RC-L, and RC-H and from the genes *rbc*L, Q, R, and *bch* (in *Bam*HI-E) under photosynthetic conditions can be detected by dot hybridization in as short a time as 15 min following the inhibitor treatment, as shown in another set of experiments [see top three rows of Figure 1(b), marked as PS]. The sensitivity of different classes of genes to the inhibitors has also been tested in the shift from aerobic to photosyn thetic conditions. The cells were grown aerobically and then shifted to anaerobic photosynthetic conditions in the presence or the absence of novobiocin and coumermycin. RNA was extracted 3 h after the shift. The formation of the photosynthetic membrane is induced in the cells during the transition period characterized by no growth at the early stage. The measurement of mRNA has also revealed the differential sensitivity of different classes of genes to the gyrase inhibitors as shown in the bottom three rows of Figure 1(b), marked as ($O_2 \rightarrow$ PS). The effects of oxygen and light intensity on the mRNA levels from the Q and R genes are shown in Figure 1(c).

The decrease in levels of mRNA for LH-I and RC (L, M) has been observed by using RNA transfer [see Figure 2(a)] and dot [see Figure 2(b)] hybridization even when the novobiocin concentration is reduced to 5 and 10 μg/ml. However, the basal mRNA levels of the same genes from aerobic cells, which is much lower than that in photosynthetic cells, are not affected by novobiocin [see Figure 2(b)].

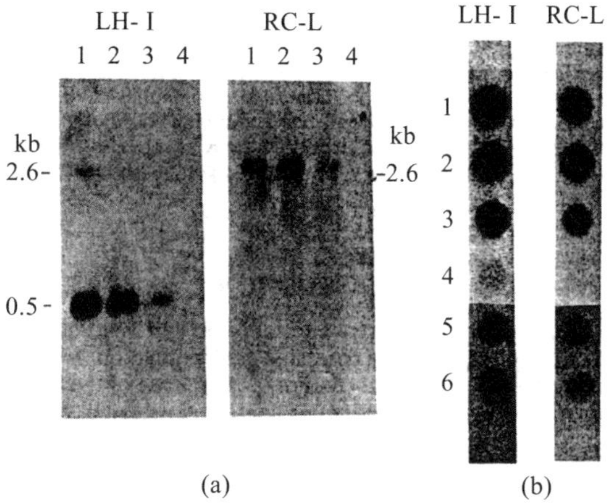

Figure 2 **RNA transfer (a) and dot (b) hybridization of *R. capsulatus* RNA with probes for LH-I and RC-L. (a) Lanes 1, 2, 3, and 4, concentrations of novobiocin at 0, 5, 10, and 50 μg/ml, respectively, under anaerobic photosynthetic conditions. Total RNA was extracted from same number of cells 30 min after drug treatment, fractionated on agarose gel, transferred to GeneScreen, and hybrid-ized with probes for LH-I and RC-L as in Figure 1[10]. The 0. 5-kb and 2. 6-kb transcripts code for LH-I (β and α) proteins, whereas the 2. 6-kb transcript also codes for RC (L, M) proteins. (b) Dots in rows 1, 2, 3, and 4, hybridizations to mRNA samples equivalent in source and conditions as lanes 1, 2, 3, and 4 in (a). Dots in rows 5 and 6, concentrations of novobiocin at 0 and 10 μg/ml, respectively, under aerobic conditions**

Effects of Novobiocin and Coumermycin on the Biosynthesis of Bchl and LH-I and LH-II Complexes

Since the transcription of various photosynthetic genes including the genes involved in Bchl biosynthesis and the genes for LH proteins is inhibited by novobiocin and coumermycin, the biosynthesis of Bchl and LH complexes must also be inhibited by the drugs, but over a longer time scale. We have examined Bchl content in the presence of the gyrase inhibitors and have shown that the accumulation of Bchl in photoheterotrophically grown cells is strongly inhibited by novobiocin and coumermycin (data not shown). The Bchl accumulation is inhibited by novobiocin and coumermycin at concentrations as low as 5 μg/ml. Photosynthetic growth was gradually inhibited as the concentration of the inhibitors was increased from 5 μg/ml to 100 μg/ml, although viability of the cells was not affected up to 150 μg/ml (see also Ref. [21]). The time course of the inhibition of Bchl accumulation indicates that a significant inhibition is observable 30 min after addition of the drugs.

The inhibition of Bchl biosynthesis by the drugs has also been observed in the cells shifted from aerobic to anaerobic photosynthetic conditions. Such a shift is described in the former section. We could detect no growth in our cultures based on optical density measurements at 680 nm during the first 3 h of the induction period with or without the drugs. The Bchl is absent under aerobic conditions. However, the synthesis of Bchl is initiated by the shift to anaerobic conditions in the absence of the drugs and fails to initiate when novobiocin or coumermycin is present. The Bchl content in the control cells was estimated as 1.0 μg/mg of protein after a 3-h induction, whereas Bchl in either novo-biocin- or coumermycin-treated cells was not detectable.

The spectra of pigment extracts and chromatophores (data not shown) have also demonstrated that the accumulation of LH-I, -II complexes as well as Bchl is markedly inhibited by novobiocin and coumermycin.

To further verify the involvement of DNA supercoiling in Bchl accumulation, we have tested two additional inhibitors of gyrase: nalidixic acid (50 μg/ml) and oxolinic acid (50 μg/ml). A similar inhibition of Bchl accumulation by these drugs was observed (not shown).

Discussion

R. capsulatus carries genes for aerobic and anaerobic metabolic modes, whose expression and regulation are determined by environmental oxygen tension. One of the features of this regulation is the coordinated expression of many groups of genes essential for LH and photochemical reactions[8-12,14,15] upon removal of oxygen. It is reasonable to suggest that a common mechanism for oxygen-specific regulation exists that is superimposed on the individual regulation of some of these genes. Although several hypotheses, such as redox potential control[22] or aporepressor-corepressor interactions[23], have been proposed as mediators of the environmental signal, the exact mechanism of oxygen regulation has not been clarified. Recent results in Gramnegative bacteria, however, have shed some light on this question[21,24]. Yamamoto and Droffner[24] have hypothesized that the expression of genes required for anaerobic growth in *Salmonella typhimurium* depends on the supercoiling of chromosomal DNA, whereas the expression of genes required for aerobic growth depends on the relaxation of DNA. Kranz and Haselkorn[21] have shown by lac fusion that the anaerobic expression of the nif operon in *R. capsulatus* is controlled by the supercoiling of DNA. They also noticed a decrease in cellular pigmentation after gyrase inhibitor treatment. The effect of DNA topology on the transcription of *rbc*L and atpB in maize[25] and pea[26] and on hydrogenase synthesis in *Bradyrhizobium japonica*[27] has been recently reported using gyrase inhibitors. The results presented in this report, using the gyrase inhibitors novobiocin, coumermycin, nalidixic acid, and oxolinic acid in vivo, indicate that the regulation by oxygen of the mRNA levels of genes for the LH peptides, RC proteins, and RuP_2Case in *R. capsulatus* is mediated by the supercoiling of DNA.

Superhelical tension may be involved in the regulation of expression of numerous prokaryotic and eukaryotic genes[24-28]. The topology of DNA is mainly determined by two enzymes, topoisomerase I and gyrase, in bacteria. DNA gyrase catalyzes the conversion of relaxed DNA to a superhelical form at the expense of ATP; topoisomerase I catalyzes the reverse reaction, converting supercoiled DNA to a more relaxed form[18,28,29]. There has been speculation that topoisomerase and gyrase might act at specific sites on the chromosome to effect the supercoiling of some genes more than for others[30]. DNA gyrase has been extensively studied in bacteria. It is made up of two subunits, A and B. Novobiocin and coumermycin are specific inhibitors of the bacterial gyrase B subunit, whereas nalidixic acid and oxolinic acid are inhibitors for the A subunit[18,28,29]. We have shown that the ATP-dependent gyrase activity in a crude cell extract from *R. capsulatus* is strongly inhibited by these drugs. The measurement of a decrease in psoralen photoreactivity with the DNA in *Bam*HI-C, F and K fragments upon treatment of cells with novobiocin has provided direct evidence that novobiocin is effecting the superhelicity of genomic DNA in vivo (D. Cook, G. Armstrong, and J. E. H. , to be published elsewhere). It is therefore our position that novobiocin and other gyrase inhibitors can be used as useful tools in the study of the superhelicity of the chromosome.

In this study we have shown that novobiocin and coumermycin have strong inhibitory effects on the levels of mRNA for Bchl biosynthesis, LH, and RC complexes and on their respective mRNAs. The levels of mRNA for LH and RC, however, were not reduced in two novobiocin-resistant mutants, RC232 and RC233, following novobiocin treatment (data not shown). We have also demonstrated a significant inhibition of mRNA from *rbc*L and the ORFs Q and R. This inhibition occurs on a time scale of minutes and is observed only for oxygen-sensitive genes. The concentrations of the inhibitors used (50 μg/ml) in this study do not irreversibly damage *R. capsulatus* cells since cell growth is recovered after washing away the novobiocin. The reduction of mRNA for LH-I and RC (L, M) has been observed after treatment with novobiocin even at concentrations as low as 5 or 10 μg/ml when the cell growth undergoes only a slight inhibition. We, therefore, believe that the inhibition of transcription upon addition of the antibiotics is a specific consequence of the selective repression of transcription of a large number of photosynthetic genes.

As we reported earlier[9], the half-lives of mRNAs for LH-I, LH-II, and RC (L, M, H) in *R. capsulatus* are 19,24, and 10 min, respectively, when proflavin is used as a specific transcription inhibitor. The half-lives of respective mRNAs in the presence of novobiocin or coumermycin measured in this study agree with these values. The similarity of the kinetics of in vivo mRNA levels of this class of genes in response to the gyrase inhibitors and to proflavin provides strong evidence that the primary effect of the gyrase inhibitors is on the transcription. A second class of genes, whose mRNAs have more or less similar half-lives as the first class does upon addition of proflavin, displays vastly different kinetic response to gyrase inhibitors in comparison to proflavin (see Table 1). This differential sensitivity of the anaerobic genes tested and constitutive genes to the gyrase inhibitors supports the hypothesis that the gyrase inhibitors selectively alter the transcription of specific genes, presumably by changing the DNA conformation at or near the respective gene. Two classes of genes, therefore, can be distinguished based on their response to gyrase inhibitors: strongly gyrasedependent anaerobic genes, including those for LH and RC complexes, Bchl biosynthetic enzymes, RuP_2Case, and the Q and R genes; and genes less dependent on gyrase, such as genes for Crt biosynthesis, cytochromes, and the constitutive genes for LH-I and RC complexes that are expressed under aerobic conditions. Thus, we propose that the change in oxygen levels induces changes in DNA topology, perhaps locally, which, in turn, lead to selective changes in transcription of genes.

It is interesting to note that the oxygen-regulated and constitutive mRNAs for LH-I and RC (L, M) have different sensitivities to novobiocin and coumermycin. The genes for LH-I (β and a subunits) and RC (L and M subunits) plus an ORF, C2397, comprise a single operon designated as *puf* operon[10,11]. Two transcripts (0.5 and 2.6 kb) originate from the *puf* operon[10,11,13]. The 0.5-kb transcript codes for LH-I (β and a) polypeptides, whereas the 2.6-kb transcript codes for LH-I (β and α) as well as RC (L, M) and C2397. Two *puf* promoters, oxygen-regulated and constitu-tive, have been postulated recently in *R. caspulatus* on the basis of lac fusions. A simple model would be that the different promoters have different sensitivities to DNA topology. The alternative possibility would be a more complex interaction between the promoter, some upstream DNA regulatory sequence[31], or regulatory protein and gyrase. The different sensitivities of promoters to DNA topology has been demonstrated in other organisms[28,31,32]. The *Escherichia coli* promoters for DNA gyrase subunits[32] and the Chlamydomonas chloroplast promoter for a dark-induced transcript A[31] are even stimulated by novobiocin. Recently, Menzel and Gellert have found that a DNA sequence 20 bp long is responsible for DNA relaxation-stimulated transcription of the E. coli gyrase gene[33]. The mechanism by which supercoiling of DNA may lead to enhanced transcription during photosynthetic growth in *R. capsulatus* is unknown.

We thank J. Willison for kindly providing the mutants RC223 and RC233, S. Kaplan for providing his clone pLIl0 as probe for *rbc*L, and N. Gabellini for providing us with her clone pRSFl for probing fbc. We thank M. Gellert, N. Cozzarelli, and J. Wang for their valuable comments and D. Cook, G. Armstrong, M. Alberti, F. Leach, and D. Burke-Aguero for their help during writing. Most of this work was supported by the Office of Basic Energy Sciences, Biological Energy Research Division of the Department of Energy under Contract DE-ACO30 - 76SF00098. The probes for the Crt genes and BChl genes were obtained from a project supported by National Institutes of Health Grant GM30786. The Biosearch Sam One was leased with Program Development Funds from the Director of Lawrence Berkeley Laboratory.

References

[1] Drews, G. & Oelze, J. (1981) *Adv. Microbiol. Physiol.* **22**, 1 - 92.
[2] Zhu, Y. S. & Hearst, J. E. (1987) in *Plant Biotechnology*, eds. Kung, S. D. & Artzen, C. J.

(Butterworth, London), in press.
[3] Cohen-Bazire, B., Sistrom, W. R. & Stanier, R. Y. (1951) *J. Cell. Comp. Physiol.* **49**, 25 - 68.
[4] Lascelles, J. (1975) *Ann. N. Y. Acad. Sci.* **244**, 334 - 347.
[5] Gorchein, A. (1973) *Biochem. J.* 134, 833 - 845.
[6] Biel, A. J. & Marrs, B. L. (1983) *J. Bacteriol.* **156**, 686 - 694.
[7] Chory, J. & Kaplan, S. (1982) *J. Biol. Chem.* **257**, 15110 - 15121.
[8] Zhu, Y. S. & Kaplan, S. (1985) *J. Bacteriol.* **162**, 925 - 932.
[9] Zhu, Y. S., Cook, D., Leach, F., Armstrong, G., Alberti, M. & Hearst, J. E. (1986) *J. Bacteriol.* **168**, 1180 - 1188.
[10] Zhu, Y. S. & Hearst, J. E. (1986) *Proc. Natl. Acad. Sci. USA* **83**, 7613 - 7617.
[11] Zhu, Y. S., Kiley, P. J., Donohue, T. J. & Kaplan, S. (1986) J. Biol. Chem. **261**, 10366 - 10374.
[12] Zhu, Y. S. & Hearst, J. E. (1987) in *Proceedings of the Seventh International Congress of Photosynthesis*, ed. Biggins, J. (Nijhoff, Dordrecht, The Netherlands), Vol. 4, pp. 717 - 720.
[13] Belasco, J. G., Beatty, J. T., Adams, C. W., Von Gabain, A. & Cohen, S. N. (1985) *Cell* **40**, 171 - 181.
[14] Clark, W. G., Davidson, E. & Marrs, B. L. (1984) *J. Bacteriol.* **157**, 945 - 948.
[15] Klug, G. T., Kaufmann, N. & Drews, G. (1985) *Proc. Natl. Acad. Sci.* USA **82**, 6485 - 6489.
[16] Bauer, C. E., Eleuterio, M., Young, D. A. & Marrs, B. L. (1987) in *Proceedings of the Seventh International Congress of Phorosynthesis*, ed. Biggins, J. (Nijhoff, Dordrecht, The Netherlands), Vol. 4, pp. 669 - 705.
[17] Weaver, P. F., Wall, J. D. & Gest, J. (1975) *Arch. Microbiol.* **105**, 207 - 216.
[18] Wang, J. (1985) *Annu. Rev. Biochem.* **54**, 665 - 697.
[19] Gabellini, N., Hamisch, U., McCarthy, J. E., Hauska, G. & Sebald, W. (1985) *EMBO J.* **2**, 549 - 553.
[20] Muller, E. D., Chory, J. & Kaplan, S. (1984) *J. Bacteriol.* **161**, 469 - 472.
[21] Kranz, R. G. & Haselkom, R. (1986) *Proc. Natl. Acad. Sci. USA* **83**, 6805 - 6809.
[22] Marrs, B. & Gest, H. (1973) *J. Bacteriol.* **114**, 1052 - 1057.
[23] Kaplan, S. (1978) in *The Photosynthetic Bacteria*, eds. Clayton, P. K. & Sistrom, W. R. (Plenum, New York), pp. 809 - 839.
[24] Yamamoto, N. & Droffner, M. L. (1985) *Proc. Natl. Acad. Sci. USA* **82**, 2077 - 2081.
[25] Stirdivant, S. M., Crossland, L. D. & Bogorad, L. (1985) Proc. *Natl. Acad. Sci.* USA **82**, 4886 - 4890.
[26] Lam, E. & Chua, N. -H. (1987) *Plant Mol.* Biol. **8**, 415 - 424.
[27] Novak, P. D. & Maier, R. J. (1987) *J. Bacteriol.* **169**, 2708 - 2712.
[28] Gellert, M. (1981) *Annu. Rev. Biochem.* **50**, 879 - 910.
[29] Otter, R. & Cozzarelli, N. R. (1983) *Methods Enzymol.* **100**, 171 - 180.
[30] Smith, G. R. (1981) *Cell* **24**, 599 - 600.
[31] Thompson, R. J. & Mosig, G. (1987) *Cell* **48**, 281 - 287.
[32] Menzel, R. J. & Gellert, M. (1983) *Cell* **34**, 105 - 113.
[33] Menzel, R. J. & Gellert, M. (1987) *Proc. Natl. Acad. Sci.* USA **84**, 4185 - 4189.
[34] Youvan, D. C., Bylina, E. J., Alberti, M., Begusch, H. & Hearst, J. E. (1984) *Cell* **37**, 949 - 957.

第五篇

植物转基因和生物防治（1987—1988）

Separation of Protein Crystals from Spores of *Bacillus thuringiensis* by Ludox Gradient Centrifugation*

A method is described for the purification of *Bacillus thuringiensis* protein crystals by Ludox gradient centrifugation. This method is simple, inexpensive, fast, and efficient compared with other techniques. It has been successfully used to purify and characterize the protein crystals from several *B. thuringiensis* strains.

We are developing transgenic plants containing toxin genes from various strains of *Bacillus thuringiensis*. Each strain synthesizes an intracellular, parasporal protein crystal, which becomes toxic through proteolytic activation after ingestion by insects[9]. Strains which produce toxins specific for *Lepidoptera*, *Diptera*, or *Coleoptera spp.* are known. Assays of expression of the transferred genes are based on specific antibody reactions and specific insecticidal activities of the toxin proteins. These assays require pure, biologically active toxin crystals as standards and as antigens. Several methods for purification of toxin crystals or toxin proteins of *B. thuringiensis* have been described. These include germination of spores and dissolution of the crystals[3], extraction by biphasic systems by using an organic solvent or high-molecular-weight polymers[5], isopycnic centrifugation in CsCl[4], and gradient centrifugation in NaBr[1]. More recently, a method involving density gradient centrifugation in Renografin (E. R. Squibb & Sons, Princeton, N. J.) has been widely used[11]. However, multiple centrifugations through Renografin are needed to achieve acceptable purity. Yields are low, and Renografin is expensive.

We have found that centrifugation through step gradients of Ludox is a very effective method for purifying toxin crystals. Ludox is an aqueous colloidal silica produced by Du Pont Co., Wilmington, Del., in industrial quantities. It has been used to purify organelles[10]. The Ludox method is simpler, quicker, more efficient, and less expensive than the

---

* Yu Sheng Zhu, Allan Brookes, Ken Carlson, Philip Filner, Sungene Technologies Corporation, San Jose, California 95131-1818
Received 31 October 1988/Accepted 28 January 1989
Corresponding author.
此文发表于 1989 年"Appl. Envir. Microbiol." 55:1279-1281.

Renografin method. Crystals and spores clump less and the band of crystals is sharper in Ludox gradients than in other types of gradients. Consequently, crystals are separated more easily and more completely from spores and debris with fewer centrifugations.

Complete cell lysis to crystals, spores, and cell debris is critical for good separation of the crystals in Ludox gradients. Protein crystals were purified from four subspecies of B. thuringiensis: *B. thuringiensis* subsp. *kurstaki* HD-1, *B. thuringiensis* subsp. *kurstaki* HD-73, and *B. thuringiensis* subsp. *aizawai* IC1[6], which were obtained from the *Bacillus* Genetic Stock Center, Ohio State University; and *B. thuringiensis* sj, which was isolated at Sungene. For all except sj, the cells were grown in 400 ml of CHES (0.5% Casamino Acids, [Difco Laboratories, Detroit, Mich.], 0.4% yeast extract, 0.2% glucose, 0.5% NaCl, 0.01% $MgSO_4$, 0.05% $CaCl_2$) in a 2-liter flask on a reciprocal shaker (200 rpm) at 28℃ overnight. To induce sporulation, the cells were harvested by centrifugation, thoroughly suspended in 400 ml of water, and incubated for another 4 days. For sj, the cells were grown overnight on PYWE (5% peptone, 0.1% yeast extract, 0.5% NaCl (pH 7.5)) and then transferred to NYSM (0.8% nutrient broth, 0.5% yeast extract, 0.5% salt (0.14 mol/L $CaCl_2$, 0.12 mol/L $MgCl_2$, 0.01 mol/L $MnCl_2$) for 4 days[7]. These procedures resulted in virtually complete cell lysis[8].

The lysed cultures were centrifuged at 3,000 rpm (1,600×g) for 10 min. Each pellet was suspended in 200 ml of 1 mol/L NaCl and vigorously shaken to produce foam, which was enriched in spores. After the foam was removed with a spatula, the remaining suspension was centrifuged at 3,000 rpm for 10 min, and the pellet was suspended in 50 ml of H_2O. The crude crystal and spore suspension was disaggregated by sonication at 100 W for 30 s (Braunsonic 2000; Melsungene AG). Aliquots were layered onto Ludox gradients.

Du Pont kindly provided a sample of Ludox HS-40 which was sufficient for this study. Ludox HS-40 is 40% (wt/wt) sodium silica with a density of 1.295 g/ml and pH 9.7. This sol irreversibly precipitates below pH 7, at salt concentrations above 0.1 N, or upon freezing. The pH of 100 ml of Ludox HS-40 was adjusted at room temperature to 8.0 with about 10 ml of 1 mol/L Tris hydrochloride (pH 2.5), with rapid stirring. Desired Ludox concentrations were then made by diluting the pH-adjusted Ludox HS-40 (defined as 100% solution) with distilled water. These Ludox solutions were stable for weeks at 4℃. Typical Ludox gradients were made by layering 5 ml of 40% (vol/vol) Ludox on 5 ml of 50% (vol/vol) Ludox in 30-ml Corex tubes (Coming Glass Works, Coming, N. Y.), all at room temperature. Usually, 2 to 5 ml of the crude suspension of crystals and spores described above was layered onto the Ludox gradient. The gradient was centrifuged in a swinging bucket rotor (no. JS-13; Beckman Instruments, Inc., Fullerton, Calif.) in a centrifuge (no. J2 - 21; Beckman) at 8,000 rpm (10,000 ×g) at 4℃ for 1 h. Large-scale Ludox gradients with similar geometry should work as well, but we did not attempt this because we lacked a suitable rotor. To optimize separations, we sometimes varied the Ludox concentrations between 30% and 70%, depending on the *B. thuringiensis* strain.

Excellent separation of HD-73 or sj crystals was obtained in the 30-ml tubes under the conditions described, with the spores at the bottom of the tube, the crystals at the interface between the Ludox layers, and the debris mostly at the interface between the crude suspension layer and the upper layer of Ludox. For HD-1 and IC1, the positions of crystals and spores were reversed. The sharp band of crystals could easily be collected with a pipette. To estimate purity, we stained the crystals red with safranin and the spores green with malachite green[2]. A single Ludox gradient centrifugation sometimes produced apparently pure crystals by this criterion (about 5 to 10 mg, containing less than 0.1% spores, from 100 ml of culture) (see Figure 1).

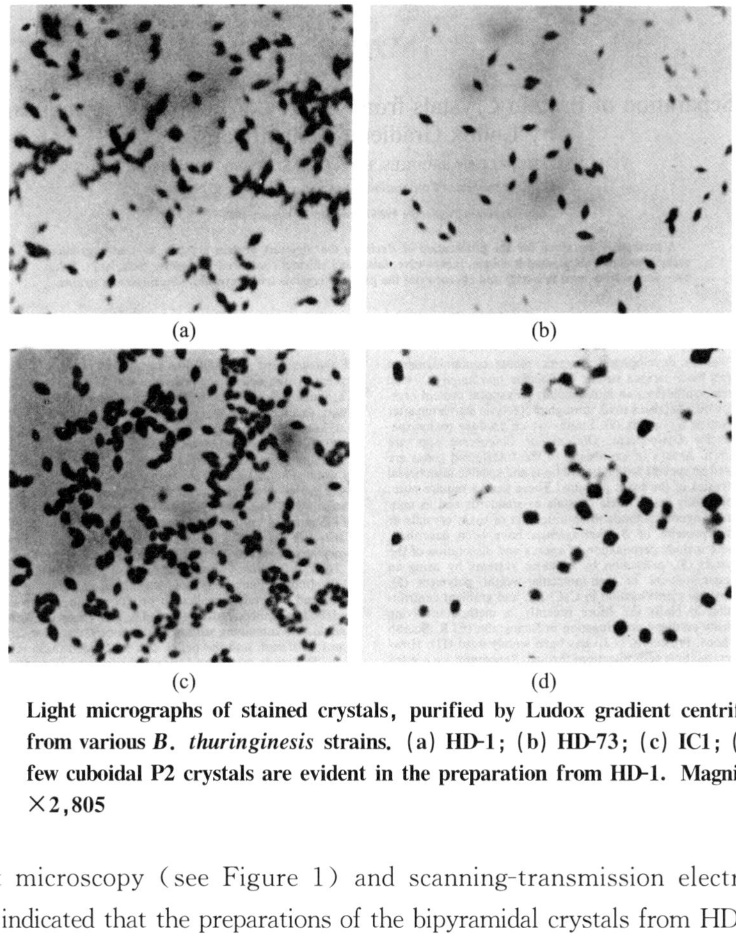

Figure 1 Light micrographs of stained crystals, purified by Ludox gradient centrifugation, from various *B. thuringinesis* strains. (a) HD-1; (b) HD-73; (c) IC1; (d) sj. A few cuboidal P2 crystals are evident in the preparation from HD-1. Magnification, ×2,805

Both light microscopy (see Figure 1) and scanning-transmission electron microscopy (see Figure 2) indicated that the preparations of the bipyramidal crystals from HD-73 and IC1 and the flat, thomboid crystals from sj were free of contaminating particles and were morphologically pure. In addition to the bipyramidal P1 crystals, the HD-1 cultures produced small amounts of cuboidal P2 crystals [see Figure 1(a)]. All of these crystal types were stable in Ludox.

The crystal preparations were also biochemically pure, judging from sodium dodecyl sulfate-polyacrylamide gel electrophoresis (see Figure 3) and their immunochemical properties. The predominance of the 130-kilodalton (kDa) polypeptide in the HD-73 and IC1 crystals and the 130-kDa (Pl) and 60-kDa (P2) polypeptides in HD-1 crystals indicates that negligible

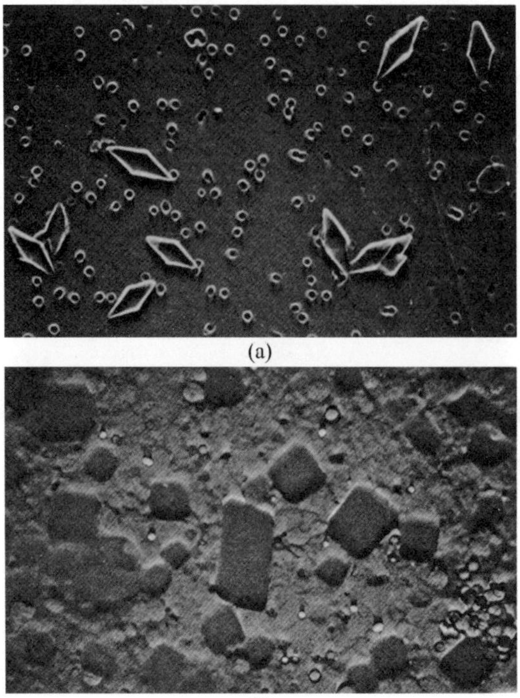

(a)

(b)

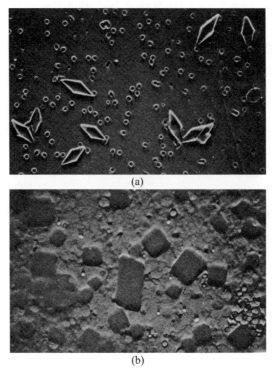

Figure 2 Electron micrographs of B. thuringinesis crystals

(a) HD-1 (scanning electron micrograph) (b) sj scanning-transmission electron micrograph. Magnification, ×8,300. Bar, 3 μm. Scanningtransmission electron microscopy makes images from both the elastically and inelastically forward-scattered electrons. The result is a combination image of the surface and the interior of the sample. The small circular objects are pores in the supporting film

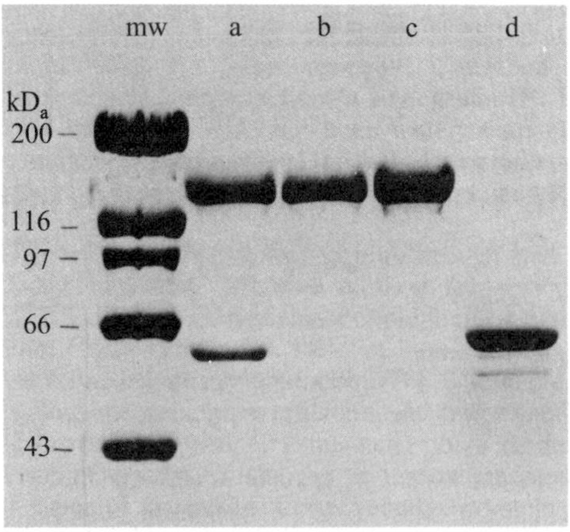

Figure 3 Coomassie blue-stained sodium dodecyl sulfate —10% polyacrylamide gel after electrophoresis of crytal proteins from various B. thuringinesis strains. (a) HD-1; (b) HD-73; (c) IC1; (d) sj. Lane mw, Molecular weight standards. To ensure solubilization of the P2 component (60 kDa) of HD-1, we dissolved the crystals by boiling them for 5 min in 10 mmol/L Tris hydrochloride-10 mmol/L dithiothreitol-8 mol/L urea (pH 8.5)

proteolysis occurred during purification. The crystals from sj had a major polypeptide of 64 kDa and a minor one of 58 kDa. Antibodies against each of the crystal preparations have been raised in rabbits. Ouchterlony tests, enzyme-linked immunosorbent assays, and Western immunoblots showed that the protein in each crystal preparation specifically immunoreacted with the corresponding antibodies.

The purified crystals retained their activities as substrates for insect gut proteases and as toxins to insects. The crystal proteins from HD-73 and sj were processed to polypeptides of the sizes expected for the active toxins, when purified crystals were mixed with gut juice from tobacco hornworms (*Lepidoptera*) or gut extracts from Tenebrio molitor mealworms (*Coleoptera*), respectively. Crystals purified from HD-1 and HD-73 were highly toxic to tobacco hornworm larvae (50% lethal dose, 0.01 μg per third-instar larva), and crystals purified from sj were toxic to mealworm larvae (50% lethal dose, 4 μg per second-instar larva). At 4 μg per larva, HD-1 cyrstals were not toxic to mealworms and sj crystals were not toxic to hornworms.

We thank B. Clark for his electron microscopy work.

References

[1] Ang, B. J., and K. W. Nickerson. 1978. Purification of the protein crystal from *Bacillus thuringiensis* by zonal gradient centrifugation. Appl. Environ. Microbiol. 36:625 - 626.

[2] Bartholemew, J. W. 1980. Stains for microorganisms in smears, p. 375 - 440. In G. Clark (ed.), Staining procedures. The Williams & Wilkins Co., Baltimore.

[3] Cooksey, K. E. 1971. The protein crystal toxin of *Bacillus thuringiensis*: biochemistry and mode of action, p. 247 - 274. In H. D. Burges and N. W. Hussey (ed.), Biological control of insects and mites. Academic Press, Inc. (London), Ltd., London.

[4] Fast, P. G. 1972. The δ-endotoxin of *Bacillus thuringiensis*. III. A rapid method for separating parasporal bodies from spores. J. Invertebr. Pathol 20:139 - 140.

[5] Goodman, N. S., R. J. Gottfried, and M. H. Rogoff. 1967. Biphasic system for separation of spores and crystals of *Bacillus thuringiensis*. J. Bacteriol. 94:485.

[6] Haider, M. Z., E. S. Ward, and D. J. Ellar. 1987. Cloning and heterologous expression of an insecticidal-endotoxin gene from *Bacillus thuringiensis* var. aizawai IC1 toxic to both lepidoptera and diptera. Gene 52:285 - 290.

[7] Herrnstadt, C., G. G. Soares, E. R. Wikox, and D. L. Edwards. 1986. A new strain of *Bacillus thuringiensis* with activity against coleopteran insects. Bio/Technology 4:305 - 308.

[8] Mahillion, J., and J. Delcour. 1984. A convenient procedure for the preparation of highly purified parasporal crystals of *Bacillus thuringiensis*. J. Microbiol. Methods 3:69 - 75.

[9] Nickerson, K. W. 1980. Structure and function of *Bacillus thuringiensis* protein crystal. Biotechnol. Bioeng. 22:1305 - 1333.

[10] Schmitt, J. M., and R. G. Herrmann. 1980. Fractionation of cell organelles in silica sol gradients. Methods Cell Biol. 15:179 - 209.

[11] Sharpe, E. S., K. W. Nickerson, L. A. Bulla, Jr., and J. N. Aronson. 1975. Separation of spores and parasporal crystals of *Bacillus thuringiensis* in gradients of certain X-ray contrasting agents. Appl. Microbiol. 30:1052 - 1053.

第六篇

医药生物工程药物制造和分子诊断

（1989—2006）

More False-Positive Problems*

SIR—Sarkar and Sommer have reported in Scientific Correspondence[1] the use of ultraviolet irradiation as a sterilization technique for the control of contamination in the polymerase chain reaction (PCR), but their procedure does not deal with all the problems. First, they introduce true target DNA and *Taq* DNA polymerase after irradiation, which will produce the same level of sporadic false-positive PCR signals even though all the contaminating carryover molecules in the other PCR reagents are sterilized. Second, their conclusion that the oligonucleotide primers for the PCR retain their full functional integrity after irradiation is not warranted. Primer damage which leads to a compromise in signal sensitivity can be evaluated only when the PCR amplification is limited so that the concentration of product is maintained well below PCR plateau concentrations. Finally, the critical nature of the size and sequence specificity of the PCR product being inactivated was not addressed. Here, we elaborate only on the last point (see Figure 1).

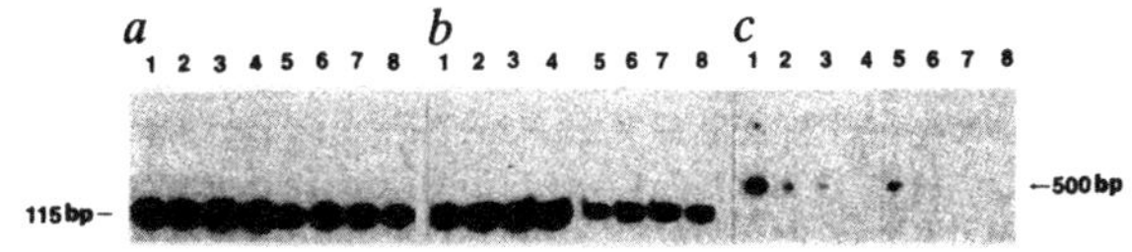

Figure 1　Effect of PCR product length on UV sterilization of PCR products. Two PCR products were examined: a 115 mer (Ou *et al.*[6]) and a 500 mer (kit N801－0055, Perkin-Elmer Cetus). PCR templates were irradiated for 0 min (lanes 1 and 5); 5 min (lanes 2 and 6); 20 min (lanes 3 and 7); 30 min (lanes 4 and 8), amplified for 30 cycles and analysed by PAGE and autoradiography. In panels *a* (115 mer template), *b* (plasmid template) and *c* (λ DNA template). Lanes 1－4 and 5－8 correspond to 10^5 and 10^3 copies of PCR template molecules, respectively. Detection sensitivity with our PCR procedure is at least 100 copies. Experimental details available from authors upon request

Pyrimidine dimers undoubtedly contribute to ultraviolet-induced sterilization by functioning as termination sites during the extension reactions of the PCR procedure. Only a fraction of all pyrimidines within a DNA strand form dimers and, as dimers are both made

* George D. Cimino, Kenneth Metchette, Stephen T. Isaacs, Yu Sheng Zhu, HRI Research Inc, Berkeley, California 94710, USA
此文发表于1990年"Nature" 345:773－774.

and broken by ultraviolet exposure, this fraction establishes a steady-state level which varies with the irradiation wavelength, the type of pyrimidine dimer and the nucleotide sequences next to the dimer site[2, 3]. At steady-state, an upper limit for the number of all dimer defects in a long, irradiated DNA molecule is less than 0.065 defects per base pair[4]. Non-dimer photodamage (for example non-cyclobutane-type pyrimidine adducts, thymine glycols, interstrand and intrastrand DNA-DNA crosslinks and DNA strand breaks) can also be termination sites for *Taq* DNA polymerase. The number of these sites should be at least equivalent to the number of dimer sites[5].

If these defects (0.13 total defects per base pair) are randomly distributed throughout a DNA molecule, a 500-base-long oligonucleotide would have an average of 32 damaged sites. In a population of 10^5 DNA strands containing an average of 32 termination sites per strand, there is a minute probability that any strand is without at least one termination site. By comparison, 100-base-long PCR products will have an average of only 6 modified sites per strand at this same defect density. In a population of 10^5 molecules with just six average termination sites per strand, a statistically large number of molecules will contain no modified sites and therefore are not sterilized.

Furthermore, to be effective for PCR sterilization, the photoinduced defects should be in the sequence region bounded by the 3′ ends of the PCR primers. These considerations predict that ultraviolet irradiation would be effective for long but not for short PCR products. Our experiments demonstrate that this is the case.

We irradiated targets with either 254, 300 or a combination of 254 and 300-nm bulbs. After irradiation, the remaining PCR reaction components were added to each tube and amplification was performed. As shown in the Figure 1, a 115 mer PCR product is efficiently generated from as little as 10^3 target molecules that had been irradiated with a combined 254 and 300 nm exposure independent of whether the target is a synthetic 115 mer or an 11-kb plasmid. Irradiation of linerized λ DNA from which a 500 mer PCR product is made results in 10^3 copies of the λ being sterilized with the combined 254/300 nm procedure. But 10^5 copies of the λ-DNA target show an exposure dependent sterilization. Only 30 minutes of exposure is sufficient to reduce the PCR signal below the sensitivity level of the gel assay.

These data imply that caution is needed when applying direct ultraviolet irradiation as a sterilization process in a preamplification mode. Both the length of the PCR product and its internal sequence must be considered. The 115 mer we used was chosen because it has a very high fraction of adjacent thymines between the prime regions of the PCR product. But the distribution of thymines is such that most of the potential pyrimidine dimers will occur on one strand of the product. This type of distribution must also be taken into account when considering ultraviolet sterilization techniques.

References

[1] Sarkar, G. & Sommer, S. *Nature* **343**, 27 (1990).

[2] Setlow, R. B. & Setlow, J. K. *Proc. natn. Acad. Sci. U.S.A.* **48**, 1250 - 1257 (1962).
[3] Gordon, L. K. & Haseltine, W. A. *Rad Res*, **89**, 99 - 112 (1982).
[4] Rahn, R. O. in *Photophysiology* (ed. Grise. A. C.) 231 - 255 (Academic. New York, 1973).
[5] Rahn, R. O. *Photochem. Photobiol. Rev.* **4**, 267 - 330 (1979).
[6] Ou, C-Y. *et al. Science* **239**, 295 - 297 (1988).

The Use of Exonuclease III for Polymerase Chain Reaction Sterilization*

The carryover of previously amplified sequences (amplicons) into new PCR reactions is a serious problem. Recently Furrer *et al*. reported a 'pre-PCR' sterilization technique using DNase I or restriction enzymes for contamination control[1]. Unfortunately, these methods require the reaction tube to be re-opened to add target DNA and *Taq* polymerase *following* the enzymatic treatment (providing an opportunity for subsequent contamination), and correspondingly, do not address possible carryover contamination from these materials. Furthermore, our results indicate that the endonucleolytic activity of DNase I severely degrades primers.

To overcome these problems, we have developed two alternative protocols for pre-PCR sterilization which utilize exonuclease III (exo III). Exo III catalyzes the sequential cleavage of 5′ mononucleotides from the 3′ hydroxyl end of duplex DNA. In Protocol I, double-stranded target DNA and all PCR reagents (including Taq polymerase) are incubated with exo III (30℃/30 min) followed by the heat inactivation of exo III (95℃/5 min). Protocol I is based on the size difference between the carryover amplicon and the target DNA, with the much smaller amplicon being more readily digested by the exo III treatment. In Protocol II, the target DNA is first denatured in the presence of the PCR reaction components (100℃/5 min) then quick-chilled on ice. Exo III and *Taq* polymerase are then added and the mixture incubated (30℃/30 min) followed by heating (95℃/5 min). Protocol II works because exo III does not degrade single stranded target DNA. Following denaturation, the DNA consists primarily of hybrids between the primers and 1) any contaminating amplicon and 2) the genomic target DNA. Upon treatment with exo III, the 3′ ends of the amplicons, and the primers which are hybridized to either amplicon or genomic DNA, are degraded; however, the genomic DNA remains largely intact. Since excess primer is used, the small quantity degraded by exo III is negligible. With both protocols, the PCR tubes remain closed between sterilization and the end of the PCR. 30℃ is used for the exo III digestion since *Taq*

---

* Yu Sheng Zhu, Stephen T. Isaacs, George D. Cimino and John E. Hearst[1]
HRI Research, Inc., Concord, CA 94520 and [1]University of California, Berkeley, CA 94720, USA
Submitted January 28, 1991
To whom correspondence should be addressed
此文发表于 1991 年“Nucleic Acids Research” 19:2511 - 2512.

polymerase is essentially inactive at this temperature. Reciprocally, 95℃ inactivates exo III but not *Taq* polymerase. By selective activation and inactivation, both enzymes can be present during the procedure.

Figure 1(a) shows that while 5×10^5 copies of the HIV amplicon were completely degraded by the exo III treatment (lane 2), the same sequence remained amplifiable when integrated into the genomic DNA of HIV-infected cells (lanes 4/6). Figure 1(b) shows the results using a linearized plasmid target. Again, 5×10^5 amplicon copies were degraded by the exo III treatment (lane 4), while the plasmid target (lane 2) remained amplifiable (signal slightly reduced).

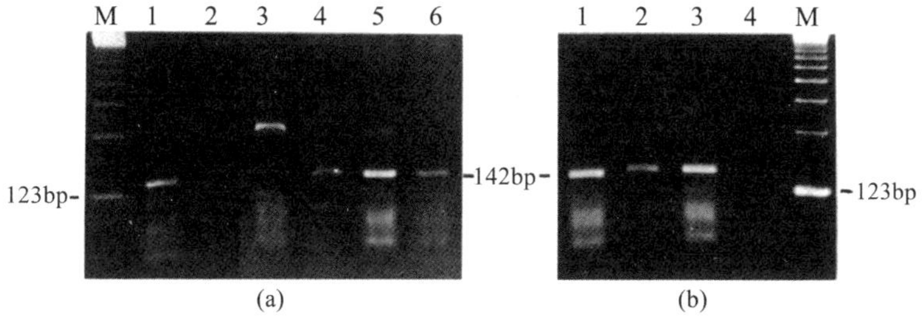

Figure 1 Sterilization with Protocol II. Amplifications were performed for 30 cycles with primer pair SK145/SK431 which provides a 142 base-pair amplicon[2]. *Cellular* target was obtained from HIV-infected H9 cells (provided by Dr. C. Hanson). The infected cells were lysed with proteinase K (120 μg/ml; 55℃/1 h) in 10 mmol/L Tris (pH 8.0), 1 mmol/L EDTA, 0.5% Tween 20 and 0.5% NP-40 followed by amplification of the lysates. *Plasmid target* was pBKBH10S (a gift from Dr. J. Rossi), which contains most of the HIV genome. *Amplicon target* was provided by amplification of the plasmid. PCR products were analysed on a 3% Nusieve/1 %agarose gel

(a) Results with cellular target. Lanes 1/2: amplicon target (5×10^5 copies); lanes 3/4: cellular target (20 HIV infected H9 cells in 10^4 uninfected H9 cells); lanes 5/6: amplicon (5×10^5 copies) plus cellular target. Lanes 1/3/5 were controls; lanes 2/4/6 were treated with 10 units of exo III (BRL). 'M' is the standard 123 base pair ladder (b) Results with plasmid target. Lanes 1/2: amplicon (5×10^5 copies) plus plasmid target (10^3 copies); lanes 3/4: amplicon target (5×10^5 copies). Lanes 1/3 were controls; lanes 2/4 were treated with 10 units of exo III

The advantage of the exo III sterilization is that it destroys both amplicon and primer dimer carryover. Furthermore, the reaction tubes are not opened until the end of PCR. The success of this method does, however, depend on the quality of exo III (no contamination with endonuclease) and on the integrity of the target DNA (high single strand molecular weight).

References

[1] Furrer, B. *et al.* (1990) *Nature* 346, 324.
[2] Kellogg, D. E. and Kwok, S. (1990) in *PCR Protocol* (eds. Innis, M. A. *et al.*) Academic Press, pp. 337 - 355.

Pilot Study of Topical Dinitrochlorobenzene (DNCB) in Human Immunodeficiency Virus Infection*

1 Summary

Dendritic cells, the primary antigen presenting cells of the human immune system, are heavily infected with human immunodeficiency virus (HIV) in patients with the acquired immunodeficiency syndrome (AIDS). Dinitrochlorobenzene (DNCB) is a contact sensitizing agent that acts as a potent immune modulator of dendritic cells. In this pilot study, we examined the safety and efficacy of topical DNCB application in patients with early HIV disease. Topical DNCB was well tolerated by these patients, with an adverse reaction rate of 10%. $CD4^+$ T-cell counts remained stable with repeated DNCB use. In contrast, $CD8^+$ T-cell counts and natural killer cells increased significantly following DNCB sensitization. This increase in $CD8^+$ T-cell and natural killer cell subsets was accompanied by a decrease in HIV replication, as measured by serum HIV RNA levels. Based on this pilot study, we conclude that topical DNCB is safe in early HIV disease and may decrease viral load via a systemic effect on dendritic cells, $CD8^+$ T-cells and natural killer cells. These results require confirmation in larger controlled trials.

2 Introduction

Dendritic cells and epidermal Langerhans cells are blood and tissue mononuclear cells

---

* Raphael B. Stricker[a], Yu Sheng Zhu[a], Blaine F. Elswood[b], Cecilio Dumlao[a], Joanna Van Elk[a], Timothy G. Berger[b], Jordan Tappero[b], William L. Epstein[b] and Dobri D. Kiprov[a]

[a]*Division of Immunotherapy, Department of Medicine, California Pacific Medical Center*; and [b]*Department of Dermatology, University of California School of Medicine, San Francisco, CA, USA*

Received 3 June 1992; revision received 4 August 1992; acccpted 15 January 1993

Immunology Letters, 36(1993) 1 - 6

0165 - 24781931 $6.00

IMLET 01927

此文发表于 1993 年"Immunology Letter" 36:1 - 6.

Key words: Dinitrochlorobenzene; HIV; AIDS; Dendritic cell; Langerhans cell; Antigen presenting cell

Correspondence to: Raphael B. Stricker, M. D., California Pacific Medical Center, California Campus, P. O. Box 7999, San Francisco, CA 94120, USA.

that function as the primary antigen presenting cells of the human immune system[1-5]. Depletion and dysfunction of these cells have been demonstrated in early stages of the acquired immunodeficiency syndrome (AIDS)[6-10]. Recently, dendritic cells have been shown to be heavily infected (up to 25%) with human immunodeficiency virus (HIV) in patients with AIDS[9,10]. Thus, dendritic cells serve as a major reservoir for HIV, and dendritic cell abnormalities contribute significantly to the immune dysfunction seen in AIDS.

1-Chloro-2,4-dinitrobenzene (DNCB) is an organic compound that is used in color photography. Like other contact sensitizing agents, DNCB is a potent modulator of dendritic cell function in vitro, and it has been used to stimulate cellular immune function in vivo[1,11,12]. Because topical application of this compound has a systemic effect on the cells that form the major reservoir for HIV[1], we organized a pilot study to examine the safety and immunologic efficacy of topical DNCB use in HIV infection.

3 Study Design

3.1 Study Subjects

Subjects were enrolled in the study after informed consent was obtained. Inclusion criteria were as follows: HIV seropositivity with $CD4^+$ T-cell counts between 100—600 cells/ml; no prior use of DNCB; either no current use of antiretroviral therapy, or treatment with a stable dose of zidovudine (AZT), dideoxyinosine (ddI) or dideoxycytidine (ddC) for at least six months; and no use of any other experimental medication. Patients with an AIDS diagnosis (significant opportunistic infections or malignancy) were excluded. All subjects had routine follow-up with their primary physicians during the study, and blood test results were shared with the primary physician.

3.2 DNCB Protocol

The protocol was designed as an observational study of topical DNCB application by the study subjects. DNCB was supplied in an acetone solution in concentrations of 10%, 2%, 0.2% and 0.02% (Healing Alternatives Foundation, San Francisco, CA). Once subjects were enrolled in the study, baseline blood samples were drawn for lymphocyte subset testing, and serum was banked at −70℃. Each subject then underwent patch testing with a 2% DNCB solution applied to a 5 mm skin site. The patch test was used to rule out prior sensitization to DNCB. If the patch test was negative, each subject then proceeded to the sensitization step by applying the 10% DNCB solution to a 2.5 cm area of skin, usually on the forearm. Reactivity was achieved when contact dermatitis occurred at the site within 72 h of DNCB application. If reactivity did not occur with the original 10% solution, subjects repeated this dose at weekly intervals until sensitization occurred.

Once reactivity was demonstrated, subjects began weekly application of 2% DNCB to a 2.5 cm site located on the forearms or upper arms. With repeated application, a 'reflare' (reddening or itching at the original application site) was seen. This response confirmed systemic sensitization to the DNCB. Subjects then decreased the weekly dose to the lowest amount that elicited a localized contact dermatitis.

Subjects kept a log of all symptoms during DNCB use. Clinical evaluation was performed every two months by the protocol coordinator.

3.3 Lymphocyte Subset Analysis

Peripheral blood lymphocyte subsets were analyzed using monoclonal antibodies (Coulter Corporation, Hialeah, FL) and two-color flow cytometry, as previously described[13]. The subsets that were analyzed included helper/inducer T-cells ($CD4^+$), suppressor/cytotoxic T-cells ($CD8^+$), cytotoxic T-cells ($CD8^+ CD57^+$) and natural killer cells ($CD56^+$). Blood samples were drawn every two months during the study, and lymphocyte subset analysis was performed within 24 h of blood drawing.

3.4 Serum HIV RNA Quantitation

Quantitation of HIV RNA in banked serum samples was performed using reverse transcriptase (RT) and the polymerase chain reaction (PCR), as previously described[14,15]. In brief, total RNA was extracted from serum using RNazol (Biotecx Laboratories, Houston, TX). HIV RNA was reverse transcribed using random hexamers, amplified by PCR, and then detected by oligonucleotide hybridization[15]. Using a standard curve of synthetic HIV RNA, the amount of HIV RNA in serum samples could be quantitated over a linear range from 50 to 1,000 input copies.

Serum samples were analyzed for HIV RNA prior to DNCB sensitization and then two months after sensitization. In addition, HIV RNA quantitation was performed in select cases every two months over the course of the study.

3.5 Statistical Analysis

The data was not normally distributed. Therefore, the results were analyzed by nonparametric methods using the Wilcoxon Signed Rank Test.

4 Results

The clinical characteristics of the study subjects are shown in Table 1. Twenty homosexual men were enrolled in the study. Sixteen were asymptomatic, while four had lymphadenopathy. Nine patients were on a stable dose of AZT, while eleven patients were on no medication. One subject was on a stable dose of ddI and one was on a stable dose of ddC in addition to AZT. $CD4^+$ T-cell counts ranged from 170/μl to 560/μl prior to entry into the study.

Table 1 Clinical characteristics of study subjects

Total	20
Asymptomatic	16
Lymphadenopathy	4
Medications	
None	11
AZT	9
AZT+ddI	1
AZT+ddC	1
Mean $CD4^+$ T-cell counts	353/μl
Range	170—560/μl

All patients underwent DNCB sensitization, and all were sensitized by the 10% solution. Two subjects experienced significant reactions to the initial DNCB application despite negative patch tests. These reactions included burning, itching and erythema at the application site (see Figure 1). The reactions subsided within 48 h with local care (moisturizing lotion or cream). No subject had a systemic adverse reaction to the DNCB, although most subjects experienced some initial discomfort at the application site. Each subject experienced a 'reflare' within 3—6 weeks after initiation of the 2% DNCB applications.

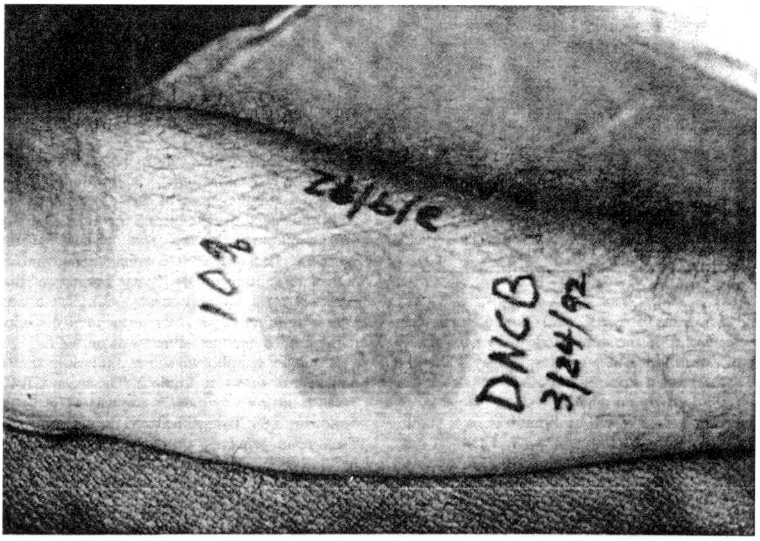

Figure 1 Pronounced contact dermatitis induced by topical DNCB application

Sixteen subjects were followed for 3—20 months after DNCB sensitization. The mean follow-up was seven months. None of the subjects had progression of HIV disease during this time. One patient withdrew from the study because of the 'cosmetic effects' of DNCB application. Another patient discontinued DNCB use because of discouragement by his private physician. These subjects were followed for six months after discontinuing DNCB. Neither subject was on AZT.

Results of lymphocyte subset analysis in the 16 patients are shown in Table 2. The mean $CD4^+$ T-cell count prior to DNCB use was 347 cells/μl. Following DNCB sensitization, the mean $CD4^+$ T-cell count remained stable at 343 cells/μl (P=N. S.). Subjects on AZT had a slight increase in $CD4^+$ T-cells, while subjects not on AZT had a slight decrease. These differences were not significant, however. In contrast to the $CD4^+$ T-cell counts, $CD8^+$ T-cell counts showed a significant increase following DNCB sensitization, from a mean of 904 cells/μl to a mean of 1,068 cells/μl ($P=0.006$). This increase was seen in patients regardless of AZT treatment (see Table 2). The increase in $CD8^+$ T-cells persisted for at least six months during continuous DNCB use (data not shown). Analysis of the cytotoxic T-cell subset ($CD8^+CD57^+$) showed no change with DNCB sensitization (see Table 2). In contrast, the natural killer cell subset ($CD56^+$) increased significantly with DNCB use, from a mean of 95 cells/μl to a mean of 119 cells/μl ($P=0.037$). This increase was independent of AZT treatment (data not shown).

Table 2 Effect of topical DNCB application on lymphocyte subsets

Lymphocyte Subsets (Cells/μl)	N	Pre-DNCB (Mean±SEM)	Post-DNCB(Mean±SEM)
$CD4^+$ (Helper/Inducer)	16	347±30[a]	343±31[a]
-On AZT	7	260±31[a]	273±43[a]
-No AZT	9	414±32[a]	397±35[a]
$CD8^+$ (Suppressor/Cytotoxic)	16	904±112[b]	1,068±106[b]
-On AZT	7	643±148[c]	836±140[c]
-No AZT	9	1,107±133[d]	1,249±130[d]
$CD8^+CD57^+$ (Cytotoxic)	16	229±41[a]	244±37[a]
$CD56^+$ (Natural Killer)	16	95±12[e]	119±15[e]

[a]P=N. S.; [b]P=0.006; [c]P=0.031; [d]P=0.046; [e]p=0.037.

N, number of patients; N. S., not significant. Normal ranges: $CD4^+$ T-cells, 400—1,800/μl; $CD8^+$ T-cells, 300—1,200/μl; $CD8^+CD57^+$ T-cells, 80—200/μl; $CD56^+$ natural killer cells, 150—400/μl. SEM, standard error of the mean.

Quantitation of serum HIV RNA in 11 subjects during DNCB use is shown in Table 3. In the two patients who discontinued DNCB application, the mean HIV RNA level increased from 2,650 copies/ml to 8,800 copies/ml. In contrast, the nine patients who continued to use DNCB showed a significant decrease in HIV RNA levels, from 42,911 copies/ml to 16,111 copies/ml ($P=0.02$). Patients on AZT had a slight increase in HIV RNA copies, while patients not taking AZT showed a threefold decrease in HIV RA copy numbers ($P=0.017$). Subjects with initial HIV RNA copy numbers >10,000/ml showed a greater response to DNCB than subjects with initial HIV RNA copy numbers <10,000/ml (see Table 3).

Table 3 Effect of topical DNCB application on HIV RNA levels in serum

Patients	N	Serum HIV RNA (Copies/ml)	
		Pre-DNCB (Mean±SEM)	Post-DNCB (Mean±SEM)
Discontinued DNCB[a]	2	2,650±1,750[b]	8,800±800[b]
Continued DNCB	9	42,911±17,041[c]	16,111±5,424[c]
-On AZT	2	4,800±400[b]	5,000±3,600[b]
-No AZT	7	53,000±20,221[d]	19,286±6,500[d]
RNA <10,000	5	4,220±894[b]	6,840±1,444[b]
RNA >10,000	6	61,733±22,007[e]	21,400±7,273[e]

[a]Not on AZT; [b]P=N/A; [c]P=0.02; [d]P=0.017; [e]P=0.014.
N, number of patients; N/A, not applicable due to small sample size. SEM, standard error of the mean.

5 Discussion

Although HIV infection is associated with progressive cellular immune dysfunction, the mechanism of this dysfunction is poorly understood[16]. $CD4^+$ T-cells are depleted during the course of HIV disease, but less than 0.2% of these cells become infected with HIV[17]. Recently it has been shown that up to 25% of dendritic cells, the primary antigen presenting cells of the immune system, are infected with HIV[8-10]. Infection of these cells leads to dysfunction and loss of antigen presentation, which may be the most significant factor in the development of cellular immunodeficiency in AIDS[10]. Thus, a treatment modality that promotes normal antigen presentation in HIV disease is desirable.

DNCB exerts a significant immunomodulatory effect on antigen presenting cells in vitro and in vivo[1,4,11]. The effect of topically applied DNCB appears to be systemic in nature, with cellular immune responses appearing at distant sites following DNCB application[1,12,18]. Dendritic cells exposed to DNCB in vivo and in vitro stimulate $CD8^+$ T-cells and natural killer cells[19,20]. Expansion of these lymphocyte populations in response to DNCB may be extremely useful in vivo, since HIV suppressor cells may be activated as part of this response, and viral replication may thus be controlled[21,22]. A concern about DNCB use is that $CD4^+$ T-cells would also be activated, leading to increased viral infection of these cells (the so-called 'food for the virus' theory) with a fall in $CD4^+$ T-cell counts and propagation of the disease[23,24].

Our pilot study of topical DNCB use in patients with early-stage HIV disease demonstrates that DNCB is safe in these patients. The local complication rate from application of the compound was 10% (2/20 subjects), and this type of complication was easily treated with local measures. Extensive allergic contact dermatitis has been reported with excessive DNCB application[25], but this problem was not seen when DNCB was used according to the study protocol. Although each patient in our study experienced a 'reflare' (systemic response) with the DNCB regimen that was used, the optimum DNCB treatment dose in individual subjects has not been determined.

A significant finding of our study was that DNCB application did not result in depletion of $CD4^+$ T-cells. This observation appears to invalidate the 'food for the virus' theory of increased $CD4^+$ T-cell destruction due to DNCB. More important, DNCB sensitization was associated with a significant rise in $CD8^+$ T-cells and natural killer cells (see Table 2), presumably via dendritic cell modulation. Although the quantitative expansion of these lymphocyte populations persisted over time, the qualitative antiviral properties of these $CD8^+$ T-cell and natural killer subsets were not defined. Further study of HIV suppressor subsets in DNCB-treated patients is now in progress.

DNCB application had a significant effect on viral replication, as shown by serum HIV RNA levels (see Table 3). These levels decreased with chronic DNCB use, while they appeared to increase in patients who discontinued DNCB. Patients with higher viral loads (>10,000 copies/ml) had a greater response to DNCB, while patients with lower amounts of circulating HIV RNA (<10,000 copies/ml) maintained these levels over time (see Table 3). These results are consistent with the expansion of $CD8^+$ T-cell and natural killer cell subsets that may control viral replication in response to DNCB.

In summary, topical DNCB application appears to be safe in patients with early HIV disease. Use of DNCB does not cause $CD4^+$ T-cell depletion, and it is associated with an increase in $CD8^+$ T-cells and natural killer cells. This effect is accompanied by a decrease in viral replication, as measured by serum HIV RNA levels. The results of this pilot study require confirmation in larger controlled clinical trials, and the effect of DNCB on HIV suppressor cells merits further evaluation. Testing of this inexpensive, readily available compound in patients with more advanced HIV disease, as well as in seronegative atrisk subjects[26], is also warranted.

Acknowledgements

The authors are indebted to Drs. Stella Knight, Steve Macatonia, Valerie Ng, Brian Hjelle, Judy Wilber and Luc Montagnier for helpful discussion. We are grateful to Joe Brewer and Project Inform for their support, and we also thank Steven Koontz and Kevin Cox for expert technical assistance.

Supported in part by Research Grants 89.024C and 90.041C from California Pacific Medical Center and the Elizabeth Reed Taylor Foundation.

References

[1] Stricker. R. B., Elswood, B. F. and Abrams. D. I. (1991) Immunol. Lett. 29,191.
[2] Knight, S. C. (1989) Res. Immunol. 140,907.

[3] Sprecher, E. and Becker. Y. (1988) Arch. Virol. 103,1.
[4] Marks, J. G., Zaino, R. J., Bressler, M. F. and Williams. J. V. (1987) Int. J. Dermatol. 26,354.
[5] Bhardwaj, N., Friedman, S. M., Cole, B. C. and Nisanian, A. J. (1992) J. Exp. Med. 175. 267.
[6] Belsito, D. V., Sanchez, M. R., Baer, R. L., Valentine, F. and Thorbecke, G. J. (1984) N. Engl. J. Med. 118,481.
[7] Tschachler, E., Groh, V., Popovic, M., Mann, D., Konrad, K., Safai, B., Eron, L., Veronese, F., Wolff, K., Stingl, G. and Wantzin, G. L. (1987) J. Invest. Dermatol. 88,233.
[8] Macatonia, S. E., Patterson, S. and Knight, S. C. (1989) Immunology 67,285.
[9] Hughes, R. A., Macatonia, S. E., Rowe, I. F., Keat, A. C. S. and Knight, S. C. (1990) Br. J. Rheumatol. 29,166.
[10] Macatonia, S. E., Lau, R., Patterson, S., Pinching, A. J. and Knight, S. C. (1990) Immunol. 71,38.
[11] Hanau, D., Fabre, M., Schmitt, D. A., Lepoittevin, J. P., Stampf, J. L., Grosshans, E., Benezra, C. and Cazenave, J. P. (1989) J. Invest. Dermatol. 92,689.
[12] Lee, S., Cho, C. K. and Chun, S. I. (1984) Int. J. Dermatol. 9,624.
[13] Kiprov, D. D., Busch, D. F., Simpson, D. M., Morand, P. R., Tardelli, G. P., Gullett, J. H., Lippert, R. and Mielke, H. (1984) in: Acquired Immune Deficiency Syndrome (M. S. Gottlieb, J. S. Groopman, Eds.) pp. 299 - 308, Liss, New YorK.
[14] Holodniy, M., Katzenstein, D. A., Israelski, D. M. and Merigan, T. C. (1991) J. Clin. Invest. 88,1755.
[15] Zhu, Y. S., Stricker. R. B., Gong, Y., Kiprov, D. D., Isaacs, S. and Cimino, G. (1992) Proceedings of the VIII International Conference on AIDS, Amsterdam.
[16] Edelman, A. S. and Zolla-Pazner, S. (1989) FASEB J. 3,22.
[17] Ho, D. D., Moudgil, T. and Alam, M. (1989) N. Engl. J. Med. 321,1621.
[18] Mills, L. B. (1986) J. Am. Acad. Dermatol. 14,1089.
[19] McKinney, E. C. and Streilen, J. W. (1989) J. Immunol. 143,1560.
[20] Young, J. W. and Steinman, R. M. (1990) J. Exp. Med. 171,1315.
[21] Walker, C. M., Moody, D. J., Stites, D. P. and Levy, J. A. (1986) Science 234,1563.
[22] Mackewicz, C. E., Ortega, H. W. and Levy, J. A. (1991) J. Clin. Invest. 87,1462.
[23] Langhoff, E., McElrath, J., Bos, H. J., Pruett, J., GranelliPiperno, A. and Steinman, R. M. (1989) J. Clin. Invest. 84,1637.
[24] Langhoff, E., Terwilliger, E. F., Bos, H. J., Kalland, K. H., Poznansky, M. C., Bacon, O. M. L. and Haseltine, W. A. (1991) Proc. Nati. Acad. Sci. USA 88,7998.
[25] Reitmeijer, C. A. M. and Cohn, D. L. (1988) Arch. Dermatol. 124,490.
[26] Marion, S. A., Schechter, M. T., Weaver, M. S., McLeod, W. A., Boyko, WJ., Willoughby, B., Douglas, B., Craib, K. J. and O'Shaughnessy, M. (1989) J. AIDS 2,178.

Quantitative Analysis of HIV-1 RNA in Plasma Preparations*

Abstract

HIV-1 RNA extraction methodology, stability and cellular location in plasma were studied by quantitative analysis using reverse transcriptase (RT) and polymerase chain reaction (PCR). HIV-1 RNA as intact virus was stable in plasma at room temperature for at least for 24 h, or stable in RNAzol (Tel-Test, Inc. Texas) at −70℃ for at least 6 months. The HIV-1 RNA PCR signal did not decline significantly after freezing and thawing of the virus in plasma or in RNAzol. To assess the effect of plasma constituents from different individuals upon quantitative PCR, identical copy numbers of HIV LAI were spiked into plasma from 9 different, normal individuals. PCR detection of HIV-1 RNA did not show any significant variation in quantitative signals. Additionally, platelet-rich plasma from three seropositive subjects was fractionated into a platelet-free plasma fraction and a platelet pellet fraction. The quantitative analysis of HIV-1 RNA in these fractions, and in the corresponding peripheral blood lymphocytes (PBLs) from each patient, demonstrated that the majority of the HIV-1 RNA was distributed in the plasma, and the HIV-1 RNA in the plasma of these patients seemed not to be strongly platelet associated.

Keywords: HIV; PCR; RT-PCR; Quantitation; Plasma; Viral load

1 Introduction

Human immunodeficiency virus type 1 (HIV-1) is the etiological agent of the acquired immunodeficiency syndrome (AIDS). The accurate and quantitative determination of HIV-1 in the human host is essential for the diagnosis of HIV-1 infection and for the assessment of

* Yu Sheng Zhu, Yu Gong, George D. Cimino

HRI, Research Inc., 2341 Stanwell Dr., Concord, CA 94520, USA

Accepted 5 October 1994

Corresponding author. Present address: Receptron, Inc., 835 Mande Ave, Mountain View, CA 94039, USA.

此文发表于 1995 年"J. Virol. Methods" 52:287-299.

the efficacy of both antiviral therapy and therapeutic vaccine intervention. In addition, the evaluation of the copy number of HIV-1 RNA sequences in specimens collected for clinical trials requires extensive handling, shipping and storage. An early study showed that cell-free infectious virus in plasma is an indicator of HIV-1 replication (Zagury et al., 1985). Previous studies with quantitative plasma culture indicated that plasma viremia is a useful marker of disease progression and is a potential marker for evaluating the response to antiviral chemotherapy (Ho et al., 1989; Coombset al., 1989). Viral culture, although sensitive, is cumbersome, time-consuming and subject to selection of viral strains. The polymerase chain reaction (PCR), which specifically amplifies HIV nucleic acids from a clinical specimen, provides a useful tool for the direct detection of HIV-1 sequences (Ou et al., 1988, 1990; Hewlett et al., 1988; Rogers et al., 1989; Schnittman et al., 1989; Genesca et al., 1990; Jackson et al., 1990; Bagnarelli et al., 1991; Daar et al., 1991; Holodniy et al., 1991; Ottmann et al., 1991; Michael et al., 1992; Saksela et al., 1994). A sensitive assay was developed to measure quantitatively small amounts of HIV-1 RNA directly from plasma. Ten copies of HIV-1 RNA can be detected by RT-PCR amplification and solution oligonucleotide hybridization detection. However, for quantitative results, the most satisfactory range is 50—1,000 copies. With this assay, HIV RNA is extracted with guanidinium thiocyanate and quantitated relative to a full length synthetic HIV-1 cRNA standard. The inter-assay variation among identical replicates is within a two-fold difference. Using this assay, we have evaluated the stability of HIV-1 in plasma, the effect of plasma constituents from different individuals on the assay, and the distribution of HIV RNA sequences in whole blood of seropositive patients.

2 Materials and Methods

2.1 Preparation of Plasma, Platelets, and Peripheral Blood Lymphocytes

Whole blood (10 ml) was collected in acid citrate dextrose tubes. A platelet-rich plasma preparation (PRP) was prepared by centrifugation of the whole blood at 500 g for 15 min. The PRP fraction was further centrifuged at 6,000 g for 30 min and separated into a supernatant fraction (PFP or platelet-free portion) and a platelet pellet (PP). The PP was resuspended to 1/10 of the original plasma volume with TE buffer. HIV-1 RNAs from all three fractions (PRP, PFP and PP) were extracted with RNAzol (9 : 1, v/v).

PBLs from seropositive patients were prepared from whole blood by standard Ficoll-Hypaque separation. The PBLs were washed in PBS twice, and total RNA was extracted with RNAzol. Normal PBLs were prepared by an alternate method to evaluate the potential of HIV DNA in plasma to contribute to the RT-PCR signal of the guanidinium extracted nucleic acids. One milliliter of fresh blood from a seronegative individual was mixed with 2 ml of Isoton II and 20 μl of Zap-Oglobin II lysing agent (Coulter Diagnostics, Hialeah, FL) at room temperature for 2 min, and centrifuged in a microfuge for 2 min. The white cell pellet was washed with Isoton II and mixed with 800 μl of RNAzol that was adjusted either to pH 4.0 or 8.5 as described below. Then a standard, synthetic HIV-1 cRNA (10^5 copies) and a standard HIV-1 DNA (10^5 copies; linearized pBKBH10S) were spiked into this mixture.

2.2 Extraction of HIV Nucleic Acids with RNAzol

HIV-1 nucleic acids were extracted with RNAzol A, a commercial product of guanidinium thiocyanate, according to the manufacturer's manual, with minor modifications. First, it was found that the addition of tRNA as a carrier RNA during extraction substantially improved the consistency of the results, particularly when low copy numbers of HIV-1 RNA were assessed. Second, a single isopropanol precipitation was sufficient to obtain an RNA preparation suitable for the quantitative RT-PCR assay. And third, to simultaneously extract both HIV RNA and HIV DNA, the pH of RNAzol was adjusted from acidic (pH 4.0) to alkaline (pH 8.5) (Fisher and Favreau, 1991). Briefly, 100 μl of plasma was mixed with 900 μl of RNAzol and 2 μg of tRNA. Eighty microliters of chloroform was added, vortexed vigorously, kept on ice for 15 min, and then spun in a microfuge at 4℃ for 10 min. The top aqueous phase was collected, and 600 μl of isopropanol was added, mixed and kept at −20℃ for 40 min. After centrifugation at 4℃ for 10 min, the RNA pellet was washed once with 75% ethanol, and then again with 100% ethanol. The pellet was air dried and resuspended in 100 μl of diethylpyrocarbonate-treated H_2O.

For the simultaneous extraction of both HIV-1 RNA and HIV-1 DNA, the same procedure was carried out, except that the RNAzol was adjusted to pH 8.5 before mixing with plasma. The pH of the aqueous phase after chloroform extraction was readjusted to pH 4—5 with a 2 mol/L acetate buffer (pH 4.0). This modified procedure ensured that the aqueous phase contained both RNA and DNA.

2.3 In Vitro Synthesis of HIV-1 RNA

A plasmid pBKBH10S (kindly supplied by Dr. J. Rossi) was used as a template for HIV-1 RNA synthesis. pBKBHlOS consists of a complete HIV DNA sequence (8,927 bp) minus the LTR that is positioned downstream from a T7 RNA polymerase promoter sequence. The HIV-1 RNA standard was synthesized by in vitro transcription in a reaction containing 40 mmol/L Tris-HCl, pH 8.0, 8 mmol/L $MgCl_2$, 2 mmol/L spermidine, 50 mmol/L NaCl, 1 μg DNA template, 400 μmol/L rNTPs, 25 units of RNasin and 10 units of T7 RNA polymerase. After incubation at 37℃ for 30 min, the synthesized HIV-1 RNA was extracted with phenol and precipitated with ethanol. The purity and integrity of the synthetic HIV-1 RNA was determined by Northern analysis. The HIV-1 RNA was quantitated by measuring the absorbance at 260 nm. A dilution series of known copies of HIV-1 RNA was made to generate an HIV-1 RNA standard curve of copy number vs PCR product.

2.4 Reverse Transcription (RT) and PCR

Total extracted RNA from 100 μl of plasma, as a precipitate with 2 μg of carrier tRNA, was dissolved in dH_2O. Aliquots of the solubilized RNA, corresponding to 1 to 100 μl of the original plasma. were reverse transcribed into cDNA in a total 20 μl reaction mixture containing 50 mmol/L Tris-HCl, pH 8.3, 150 mmol/L KCl, 10 mmol/L $MgCl_2$, 10 mmol/L DTT, 200 μmol/L dNTPs, 0.1 μmol/L random hexamer primers, 20 units of RNasin and 5 units of AMV reverse transcriptase (Seikagaku America Inc., St. Petersburg, FL). After incubation, first at room temperature for 10 min, then at 37℃ for 30 min, the reaction was terminated by heating at 95℃ for 5 min and quick-cooling on ice. PCR was carried in a 30 μl reaction containing 50 mmol/L Tris-HCl, pH 8.5, 50 mmol/L KCl, 3.3 mmol/L $MgCl_2$, 200 μmol/L dNTPs, 200 μg/ml gelatin, 0.5 μmol/L primer pairs SK38/SK39 (Ou et al., 1988), 1 unit of Taq polymerase (Perkin-Elmer, Norwalk, CT) and 10 μl of the RT product. The amplification protocol consisted of 30 cycles of 95℃ for 30 s, 55℃ for 30 s, and 72℃ for 1 min in a Perkin-Elmer-Cetus thermocycler. Samples were run in duplicate, with negative and positive controls included in all experiments. For quantitation purposes, the HIV RNA was diluted so that the RT-PCR signal was within the range of HIV-1 RNA standard curve. Control experiments that lacked the reverse transcriptase indicated that no HIV DNA was detectable with the RNAzol plasma extracts from seropositive patients.

2.5 Detection and Quantitation of PCR Products

The PCR products were detected by solution oligomer hybridization. Ten microliters of PCR product and 3.3 μl of ^{32}P-labeled SK19 probe ((3—5)$\times10^5$ cpm; 3,000 Ci/mmol/L) in 10 mmol/L EDTA, 15 mmol/L NaCl were heated to 95℃ for 5 min, then hybridized at 55℃ for 15 min. Five microliters of a dye containing 0.01% Bromophenol blue, 0.01% Xylene cyanole FF, 50% glycerol, 0.125 mol/L EDTA, pH 8.0 and 0.1% SDS was added. Aliquots of 5 μl were loaded on a 12% polyacrylamide minigel and run at 300 V for 30 min. The gel was exposed to X-ray film with intensifying screens at −80℃ for 0.5 or 1 h. The bands on the gel corresponding to the HIV-1 gag sequence delineated by SK38/SK39(115 bp) were quantitated either directly with an Ambis Radioanalytic Imaging System (expressed as counts) or by cutting and counting the bands in a scintillation counter (expressed as cpm).

3 Results

A simple RT-PCR assay was developed for use with plasma extracts to evaluate the stability and distribution of HIV in plasma preparations. The sensitivity of the RT-PCR assay as shown in Figure 1. While HIV-1 RNA copies as low as 10 copies were detected, the quantitative range that is normally used is 50—1,000 copies with the synthetic HIV-1 RNA standards. Although 2,000 copies can be measured with the assay, the standard curve becomes flat above this level of target molecules, indicating PCR plateau (data not shown). The variability of the RT-PCR assay for HIV-1 quantitation was assessed by analyzing 15 replicates of 100, 500 or 1,000 copies of the RNA standard. The intra-assay coefficient of variation was found to be 23,16, and 6% for 100,500, and 1,000 copies, respectively. The assay variation in a plasma background was determined with 10 replicates of 1,000 copies of HIV-1 RNA spiked into 100 μl of plasma from a seronegative individual. After RNAzol extraction and RT-PCR, the intra-assay coefficient of variation was found to be 22%. The inter-assay difference was found to be less than two-fold when the same samples were extracted, amplified and quantitated at different times.

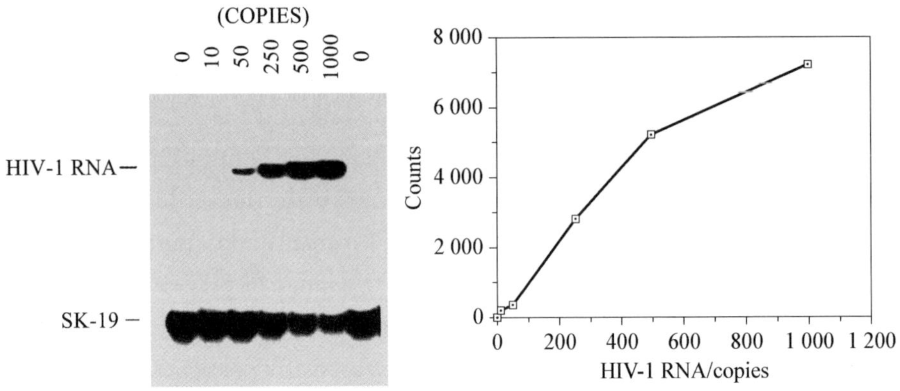

Figure 1 Standard curve of HIV-1 RNA. The left panel is the autoradiograph of the amplification products of different copy numbers of HIV-1 RNA (0,10,50,250,500,1,000). The right panel is the standard curve of HIV-1 RNA (counts vs copy numbers), as determined with the Ambis Imager System

The quantitative assay described here utilizes platelet rich plasma for the assessment of viral load. Guanidinium extraction is commonly used for extraction of RNA from plasma. However, PRP is contaminated with lymphocytes to a level of about 10^3—10^4 per 100 μl, depending upon the preparative procedure. Since HIV DNA from infected cells would contribute to the RT-PCR signal if it were simultaneously extracted with the genomic RNA from HIV particles, we evaluated the level of DNA contamination thatwould be tolerated with the extraction procedure. Both RNAzol A and RNAzol B, two similar commercial products, were evaluated as potential extraction reagents. 10^5 copies of HIV-1 synthetic cRNA and 10^5 copies of linearized pBKBHlOS were simultaneously spiked into RNAzol extracts of PBL preparations (10^6 cells). RNA was extracted with acidic RNAzol (pH 4.0) as usual. As a control, both RNA and DNA were simultaneously extracted with RNAzol that had been adjusted to pH 8.5. Nucleic acids from 100 μl aliquots of both the aqueous phase and the interface of the phase separated RNAzol extracts were analyzed by reverse transcription and PCR amplification [see Figure 2(a)]. Under these conditions both RNA and DNA are amplified. The DNA contribution to the PCR signal was assessed with the same aliquots by amplifying without reverse transcriptase [see Figure 2 (b)]. The autoradiograph of Figure 2 shows that the HIV-1 plasmid DNA could be detected in the aqueous phase when the pH was 8.5. HIV DNA could not be detected in the aqueous region when the extraction occurred under acidic conditions (pH 4.0). HIV DNA was, however,

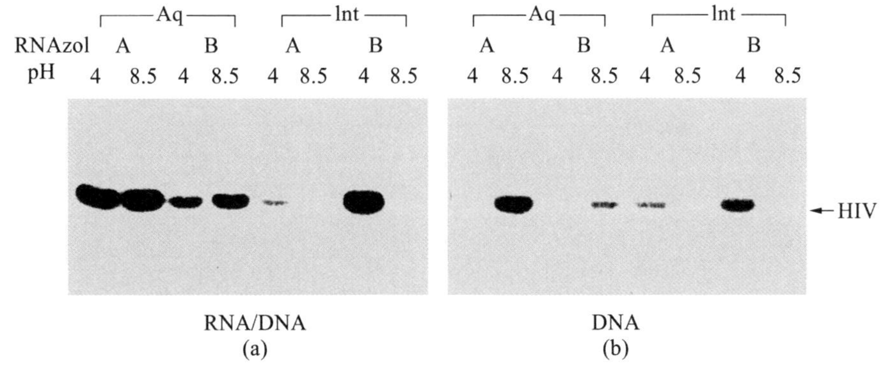

Figure 2 Autoradiograph of the pH effect upon the distribution of HIV-1 DNA and RNA between the aqueous phase (Aq) and the interface (Int) of RNAzol extraction reagents (A or B; two commercial products of TEL-TEST, Inc.). The samples shown in the left panel (labeled RNA/DNA) were reverse transcribed, then PCR amplified. The PCR signals represent both RNA and DNA contributions. The samples shown in the right panel (labeled DNA) were only PCR amplified. These PCR signals are due to DNA targets exclusively, and indicate the DNA target contribution to the PCR signals of the corresponding lanes in the right panel

measurable from material obtained at the interface. These results demonstrate that HIV RNA is selectively isolated from the aqueous phase when acidic RNAzol is used. This is the preferred preparative procedure for the quantitative analysis of the molecular copy number of virus in plasmaor serum samples. Small amounts of contaminating DNA in these samples will not contribute to the RT-PCR signal since they are left in the interface. With the RT-PCR assay described here, 2×10^4 HIV DNA copies in the presence of nucleic acids extracted from 2×10^5 PBLs did not contribute to the PCR signal. Extreme care should used to avoid the interface region during extraction, since the potential for contamination with HIV DNA sequences from originating from cellular material is high.

In order to study the stability of HIV in plasma and to determine the HIV-1 RNA molecular copies of cultured virus, plasma from a normal individual was prepared and spiked with 1 $TCID_{50}$/ml or 100 $TCID_{50}$/ml of HIV LAl. These samples were kept either at room temperature for 0, 1, 4, or 24 h, or were subjected to repeated freezing and thawing manipulations. The HIV-1 RNA from 1 $TCID_{50}$/ml and 100 $TCID_{50}$/ml samples were extracted with acidic RNAzol A, and diluted to 1 ∶ 2 and 1 ∶ 200, respectively, for the RT-PCR assay. This dilution was required to bring the PCR signal within the range that yields quantitative data. The results are shown in Figure 3. One $TCID_{50}$/milliliter of viral culture was estimated to be equivalent to 2,000 copies of the cRNA standard. HIV-1 RNA extracted from the plasma containing 100 $TCID_{50}$/ml of HIV LAI at 1 ∶ 200 dilution resulted in a RT-PCR signal similar to that from the extracted HIV-1 RNA from the plasma with 1 $TCID_{50}$/ml of HIV LAl at a 1 ∶ 2 dilution. These results suggest that the extraction efficiency of HIV RNA is the same for a high-titer sample (200,000 molecular copies/ml) and for a low-titer sample (2,000 molecular copies/ml). Figure 3 also shows that intact HIV is stable in the

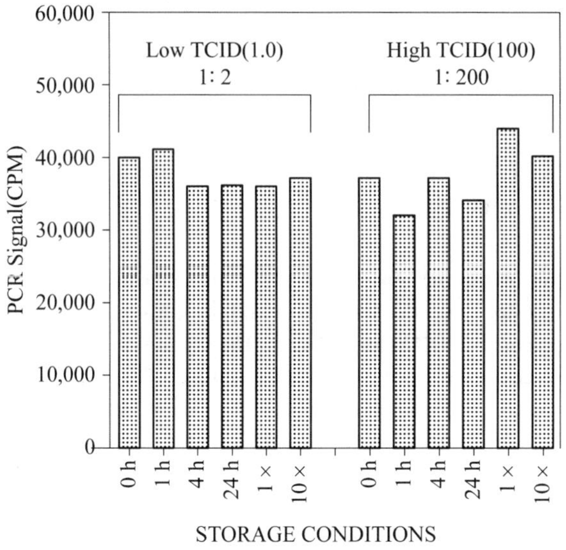

(a)

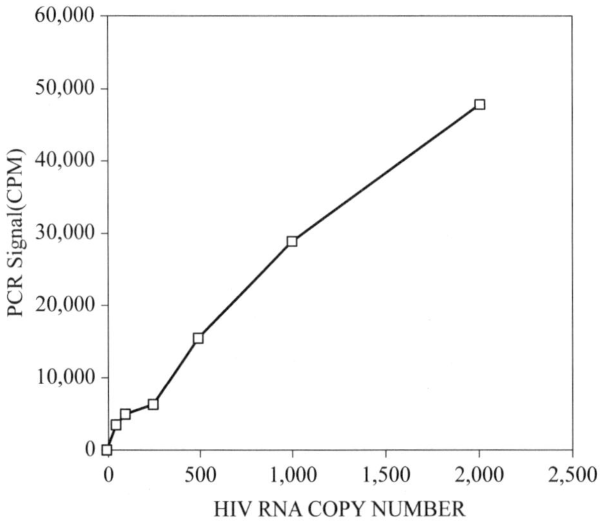

(b)

Figure 3 Stability of HIV-1 in plasma (a). HIV-1 (1 or 100 $TCID_{50}$/ml) was spiked into 100 μl aliquots of plasma. These samples were extracted at either 0, 1, 4 or 24 h after storing at room temperature, or after freezing and thawing once (1×) or 10 times (10×). For RT-PCR amplification, the RNA extracts from the low (1.0) and high (100) $TCID_{50}$/ml samples were diluted 1 : 2 and 1 : 200, respectively, and then aliquotcd into the RT assay. Amplified products were separated on polyacrylamide gel, excised, and quantitated by scintillation counting. As shown with the accompanying dilution series of the standard RNA (b), the dilutions used in this experiment resulted in RT-PCR signals that were within the quantitative range of the assay

plasma at room temperature for at least 24 h. Additionally, a single freeze/thaw had no effect on the RT-PCR signal, although 10 cycles of freeze/thaw shows a slight loss of signal. Similar results with HIV-1 seropositive individuals are presented as shown in Figure 4. The plasma from individuals (a) and (b) were frozen and thawed once (1FT) or 10 times (10FT). The HIV-1 RT-PCR signal was not affected by the repeated treatment. In addition, we found that HIV-1 RNA was stable in frozen RNAzol at least for 6 months or when subjected to repeated freezing and thawing manipulations in RNAzol (data not shown).

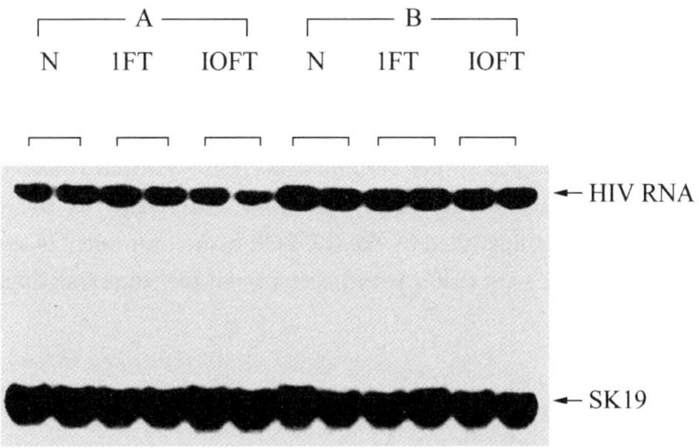

Figure 4 **Stability of HIV-1 RNA in RNAzol. Aliquots of 100 μl of plasma from two HIV-1 seropositive subjects (A and B) were mixed with RNAzol. The RNAzol/plasma mixtures were immediately extracted (N), frozen and thawed once (1FT), or frozen and thawed 10 times (10FT). Duplicate samples were run for each condition. The position of the HIV-specific PCR band (HIV RNA) and the position of the oligonucleotide probe (SK19) are indicated**

Next, the effects of plasma constituents on the yield of an RT-PCR assay signal were investigated, with samples containing identical viral copy numbers. Plasmas were prepared from 9 different seronegative individuals. These plasmas were each spiked to 0.5 and 1 $TCID_{50}$/ml of HIV LAI. HIV-1 RNAs from each sample (100 μl) were extracted, amplified and quantitated. The results (see Figure 5) show that the plasmas from different individuals have a small effect upon the RT-PCR assay. Although patient 3 displayed abnormally high signals with both the low and high spiked samples, upon repeat (2×) analysis with this patient's plasma, the PCR signals were comparable to those of the other volunteers. By quantitation, the average of the 1 $TCID_{50}$/ml samples was determined to be $1{,}800 \pm 730$ copies. The 0.5 $TCID_{50}$/ml fractions were found to contain 780 ± 150 copies. Duplicate plasma samples from volunteer 9 were identically spiked with virus and quantitated in duplicate to assess the reproducibility in a single individual. As shown, the results are consistent. Although the various plasmas resulted in a spread in the intra-assay variation, in all cases, the two-fold differences in each individual plasma were discernible. The high/low molecular copy ratios for the group of samples was 2.30 ± 0.96. These data show that within a single individual, two-fold differences in the RT-PCR signal are statistically

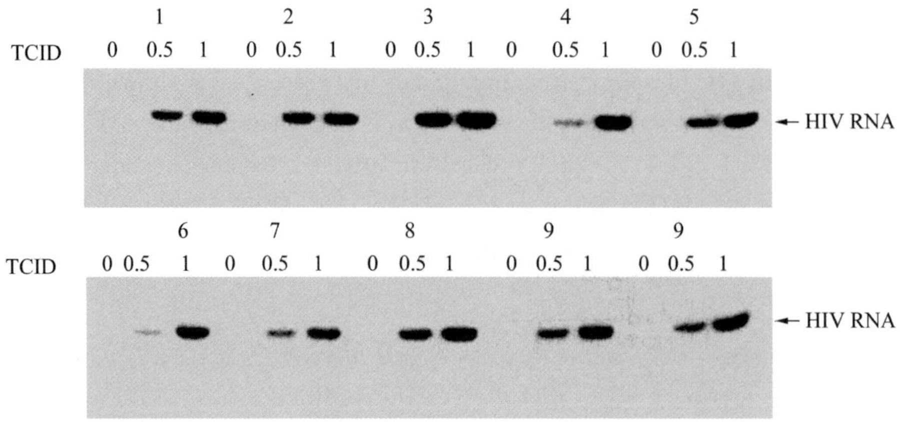

Figure 5 **The effect of various plasmas from different individuals on the HIV-1 RNA assay signal. Intact virus was spiked into plasma from 9 normal volunteers at either 0, 0.5 or 1.0 $TCID_{50}$/ml. In all cases, two-fold differences in the HIV-1 RNA copy numbers can be distinguished by the RT-PCR assay. Sample 9 is shown in duplicate to illustrate the assay's precision with plasma from the same individual**

significant (2.4-times the coefficient of variation for the average of the ratios of the molecular titers of each individual plasma).

Our target sample for the analysis of HIV copy number is PRP. To determine the fraction of free HIV vs platelet-associated HIV in the PRP preparations, the PRP fraction was further separated into a PFP portion and a PP as described in Materials and Methods. HIV-1 RNA signals from same amount of the PRP and PFP, after RT-PCR, were compared to that from the PP. The quantitative results, summarized as shown in Table 1, show that the PRP and PFP have similar concentrations of HIV-1 RNA, indicating that plasma HIV-1 may not be strongly platelet-associated. Furthermore, the majority of HIV-1 RNA was found in the PFP, whereas only 0.5%—3.0% of HIV-1 RNA was found in the PP. The

Table 1 Distribution of HIV-1 RNA among plasma, platelets, and PBLs

Patient number	HIV-1 RNA in plasma/(copies/ml)	HIV-1 RNA in platelet-free plasma/(copies/ml)	HIV-1 RNA associated with platelets/(copies/ml)	HIV-1 RNA associated with PBLs/(copies/ml)
1	110,000	100,000	500	5,000
2	40,000	35,000	1,000	120
3	32,000	30,000	1,000	7,000

The platelet-associated RNA copy number is reported as the number of HIV RNA molecules found associated with the platelets contained in 1 ml of platelet-rich plasma. The PBL-associated HIV is reported as the number of HIV RNA molecules found associated with the PBLs that are present in 1 ml of whole blood. The PBL counts per ml for patients 1, 2, and 3 were 1×10^6, 1.1×10^6, and 1.0×10^6, respectively.

viral load in PBLs from these same patients was also analyzed by RT-PCR, and compared to the HIV copy number in the plasma fractions. It was found that the number of HIV-1 RNA copies in the plasma fractions were very much higher than the number of HIV molecules associated with the PBLs in these patients.

4 Discussion

The evaluation of the effectiveness of antiviral therapy requires an accurate and quantitative measurement of HIV-1 from clinical samples. Basic information on sample handling, processing and storage are required for a credible interpretation of the quantitative measurements. PCR is a powerful tool for assessing the presence of small amounts of nucleic acids. The application of this technique has yielded a great deal of information on the pathogenesis of HIV and the efficacy of anti-HIV drugs (Ou et al., 1988,1990; Kellogg et al., 1990; Holodniy et al., 1991; Schnittman et al., 1991; Michaelet al., 1992; Semple et al., 1993). However, the use of PCR for routine quantitative analysis has been questioned due to the exponential nature of the amplification process and the effects of small quantities of PCR inhibitors that may be present in clinical samples (Gilliland et al., 1990; Stieger et al., 1991; Piatak, et al., 1993a, b; Siebert et al., 1993). A minor change in the amplification efficiency and kinetics will result in a large change in the magnitude of a PCR signal. In addition, the attempted measurement of high copy numbers of targets that yield signals in the PCR plateau range fails to quantitate nucleic acids.

A simple PCR protocol was evaluated for quantitation and relied upon carefull aboratory practice to control the variability of the amplification process. This approach is based upon: ① using and optimizing an RNAzol extraction procedure in the presence of added carrier tRNA which immediately denatures proteins, including any RNases to ensure high yield of RNA; ② a full-length synthetic HIV-1 RNA is used as a quantitative standard and amplified concurrently with a series of patient samples under the same conditions; ③ the HIV-1 RNA extract from a patient specimen is diluted so that quantitation occurs within a valid range of PCR amplification, which is between 50 and 1,000 copies for the assay described here. This assay, however, is able to quantitatively detect 10 copies of HIV-1 RNA (see Figure 1) and 1 copy of HIV-1 DNA (data not shown). Although the level of reproducibility of simple PCR protocols described in the literature spans a large range from two-to ten-fold, as described by Ferre (1992), the variability was controlled to within two-fold under cautious laboratory practice.

The application of the RT-PCR assay described now shows that acidic RNAzol exclusively isolates HIV RNA from plasma preparations in the presence of expected DNA contamination from lymphocytes. Using this reagent, we observed that intact HIV is stable in plasma at room temperature for at least 24 h. Apparently, the envelope and coat proteins of the virus protect the RNA genome from RNases in plasma. In contrast, we have observed that purified HIV-1 RNA is extremely sensitive to plasma RNases. One thousand copies of purified HIV-1 RNA are completely degraded in a 1-min exposure to the plasma (data not shown). Intact HIV is also resistant to the multiple freezing and thawing treatments. This observation is in agreement with another report (Shizuko et al., 1990). The stability of HIV-1 RNA as intact virus in plasma and the stability of isolated HIV-1 RNA in RNAzol, as measured by RT-PCR of a short segment of the RNA genome, indicate that sample handling and storage are not an impediment for long-term clinical studies that are required for evaluating therapeutic vaccines and anti-HIV compounds.

In this study, the distribution of HIV RNA was also evaluated within platelet-free plasma, isolated platelets and peripheral blood lymphocytes from 3 patients with relatively high plasma loads of HIV. Ho et al. (1989) previously reported that the HIV-1 in plasma is platelet associated in many HIV-infected patients, possibly on the outside of the cell and perhaps bound to the Fc receptor. The 3 patients evaluated in this study had plasma viral loads ranging from 32,000 copies/ml to 110,000 copies/ml. We observedthat the platelet-associated HIV fraction (PP) amounts to 500—1,000 molecular copies/ml of plasma, corresponding to only 0.5%—3.1% of the total number of copies of HIV-1 in the plasma. Since the typical human platelet count is 10^8/ml, the results indicate that the number of platelets containing a single virus is in the order of 1 cell per 10^5 platelets. These results imply that RT-PCR analysis of PRP preparations measures predominately free HIV viral particles, as opposed to cell associated viral particles. These results are in agreement with the observations of Piatak et al. (1993c), but are not supported by the observations of Lee et al. (1993). Although our results indicate that as mall fraction of the viral RNA is platelet associated, a different fraction may be determined with culture measurements. Since the laboratory-grown virus has a 2,000 : 1 ratio of molecular titer (as measured by RT-PCR) to infectious titer, it is possible that the population of infectious virus has a higher affinity for the platelets than the non-infectious, packaged viral particles. Such a differential distribution of virus will result in a higher proportion of the HIV being associated with the platelets when measured by infectious culture assays.

The peripheral blood cells from the 3 patients displayed a range of HIV RNA from 120 copies per 10^6 PBLs to 7,000 copies per 10^6 PBLs. These values are in agreement with other observations (Michael et al., 1992). On a per milliliter of blood basis, thisrange of PBL-associated virus corresponds to a wide fraction of the total viral load. The HIV-1 RNA copies in the PBLs was found to range from 0.3% of the plasma load to 21% of the plasma load. The implication of such a change of distribution of HIV-1 between the plasma and cellular fractions in the pathogenesis of AIDS is certainly worthy of further investigation. Recently, Saksela et al. (1994) correlated disease progression with an increase of HIV RNA in PBLs.

While our efforts have concentrated on a simple PCR procedure for quantitation of HIV, several groups have developed competitive PCR procedures for quantitation of HIV DNA and RNA. These protocols are based on quantitation of the relative amounts of PCR products from the target sequence and a competitor of known copy number in a co-amplification system. The attractive feature of the competitive approach is to eliminate the variability of both RT and PCR. However, some drawbacks of this method may limit its application (Ferre, 1993). A plasmid with a deletion must be constructed and cloned for each target. The competitor RNA, which is designed to have a different size and partial sequence homology relative to the true target genome, may have a secondary structure that differs from that of the true target RNA. Therefore, the efficiency and kinetics of amplification of the target and the competitor must be established for each system, particularly in the PCR plateau range. Furthermore, since a competitor of at least 5 different concentrations must be introduced to each specimen to be analyzed, the work load is greatly magnified and may limit the application of the competitive approach for routine use in the clinical laboratory.

With the assay described here, 0.1 ml of plasma is routinely used in a simple protocol that reproducibly differentiates two-fold differences in HIV copy number within the range of 50—1,000 copies input into the assay. The competitive approach of Piatak et al. (1993a, b) describes a sensitivity of 200 HIV-1 RNA copies from 0.5 and 2.8 ml of plasma.

The issue of whether the HIV DNA copy number in the PBLs and the HIV RNA copy number in these same cells track each other during progression of the disease has not yet been resolved. This, in part, has been limited by the inability to simultaneously isolate both RNA and DNA in a quantitative manner. Our preliminary observations with basic RNazol indicate that quantitative recovery of both HIV RNA and HIV DNA can be performed simultaneously under conditions where the RNA remains intact. Application of this method will permit a single extraction process to quantitatively yield both RNA and DNA that can be assessed by quantitative PCR.

Acknowledgements

We wish to acknowledge K. Steimer, F. Sinangil, and M. Urdea for helpful discussions and suggestions.

References

[1] Aoki, S., Yarchoan, R., Thomas, R. V. et al. (1990) Quantitative analysis of HIV-1 proviral DNA in peripheral blood mononuclear cells from patients with AIDS or ARC: decrease of proviral DNA content following treatment with 2′,3′-dideoxyinosine (ddI). AIDS Res. Hum. Retroviruses. 6, 1331 - 1339.

[2] Bagnarelli, P., Menzo, S., Manzin, A., Varaldo, P. E., Montroni, M., Giacca, M. and Clementi, M. (1991) Detection of HIV type 1 transcript in peripheral blood lymphocytes by the polymerase chain reaction. J. Virol. Methods 32, 31 - 39.

[3] Chirgwin, A. E., Przybyla, A. E., MacDonald, R. J. and Rutter, WJ. (1979) Biochemistry 18, 5294 - 5299.

[4] Chomczynski, P. and Sacchi, N. (1987) Anal. Biochem. 162, 156 - 159.

[5] Coombs, R. W., Collier, J. P., Allain, B. et al. (1989) Plasma viremia in human immunodeficiency

virus infection. New Engl. J. Med. 321,1626 - 1631.

[6] Daar, E. S., Moudgil, T., Meyer, R. D. et al. (1991) Transient high levels of viremia with primary human immunodeficiency virus type 1 infection. New Engl. J. Med. 324,961 - 964.

[7] Ferre, F. (1992) Quantitative or semi-quantitative PCR: reality versus myth. PCR Methods Appl. 1,1 - 9.

[8] Fisher, J. and Favreau, M. (1991) Plasmid purification by phenol extraction from guanidinium thiocynate solution: development of an automated protocol. Anal. Biochem. 194,309 - 315.

[9] Genesca, J., Wang, R. Y.-H., Alter, H. J. et al. (1990) Clinical correlation and genetic polymorphism of the human immunodeficiency virus proviral DNA obtained after polymerase chain reaction amplification. J. Infect. Dis. 162,1025 - 1030.

[10] Hart, C., Schochetman, G., Spira, T. et al. (1988) Direct detection of HIV RNA expression in seropositive subjects. Lancet 2,596 - 599.

[11] Hewlett, I. K., Gregg, R. A., Ou, C.-Y. et al. (1988) Detection in plasma of HIV-1 specific DNA and RNA by polymerase chain reaction before and after seroconversion. J. Clin. Immunoassay 11,161 - 164.

[12] Ho, D. (1991) Quantitative distribution of HIV-1 in blood. 1991 Workshop on Viral Quantitation in HIV Infection. Session 1: HIV Quantitation Using Cell Culture. Paris, June 13 - 14. John Libbey Eurotext.

[13] Ho, D., Moudgil, T. and Alam, M. (1989) Quantitation of human immunodeficiency virus type 1 in the blood of infected persons. New Engl. J. Med. 321,1621 - 1625.

[14] Holodniy, M., Katzenstein, D. A., Sengupta, S. et al. (1991) Detection and quantification of HIV RNA in patient serum by use of the polymerase chain reaction. J. Infect. Dis. 163,862 - 866.

[15] Kellogg, D. E., Sninsky, J. J. and Kwok, S. (1990) Quantitation of HIV-1 proviral DNA relative to cellular DNA by the polymerase chain reaction. Anal. Biochem. 189,202 - 208.

[16] Lee, T., Stromberg, R. R., Henrad, D. and Busch, M. P. (1993) Effect of platelet-associated virus on assays of HIV-1 in plasma. Science 262,1585.

[17] Jackson, J. B., Kwok, S. Y., Sninsky, J. J. et al. (1990) Human immunodeficiency virus type 1 detected in all seropositive symptomatic and asymptomatic individuals. J. Clin. Microbiol. 28,16 - 19.

[18] Maniatis, T., Fritsch, E. F. and Sambrook, J. (Eds.) (1982) Molecular Cloning. Cold Spring Harbor Laboratory, Cold Spring Harbor, NY, pp. 188 - 209.

[19] Michael, N. L., Vahey, M., Burke, D. S. et al. (1992) Viral DNA and mRNA expression correlate with the stage of HIV type 1 infection in disease. J. Virol. 66,310 - 316.

[20] Ottmann, M., Innocenti, P., Thenadcy, M. et al. (1991) The polymerase chain reaction for the detection of HIV-1 genomic RNA in plasma from infected individuals. J. Virol. Methods 31,273 - 284.

[21] Ou, C.-Y., Kwok, S., Mitchel, S. W. et al. (1988) DNA amplification for direct detection of HIV-1 in DNA of peripheral blood mononuclear cells. Science 239,295 - 297.

[22] Ou, C.-Y., McDonough, S. H., Cabanas, D. et al. (1990) Rapid and quantitative detection of enzymatically amplified HIV-1 DNA using chemiluminescent oligonucleotide probes. AIDS Res. Hum. Retroviruses 6,1323 - 1327.

[23] Piatak, M., Jr., Luk, K.-C., Williams, B. et al. (1993a) Quantitative competitive polymerase chain reaction for accurate quantitation of HIV DNA and RNA species. Biotechniques 14,70 - 80.

[24] Piatak, M., Jr., Saag, M. S., Yang, S. J. et al. (1993b) High level of HIV-1 in plasma during all stages of infection determined by competitive PCR. Science 259,1749 - 1754.

[25] Piatak, M., Shaw, G., Yang, L. et al. (1993c) Effect of platelet-associated virus on assays of HIV-1 in plasma. Science 262,1585 - 1586.

[26] Rappolee, D. A. (1989) Novel method for studying mRNA phenotypes in single or small numbers of cells. J. Cell Biochem. 39,1 - 11.

[27] Rogers, M. , Ou, C. -Y. , Rayfield, M. et al. (1989) Polymerase chain reaction for early detection of HIV proviral sequence in infants born to seropositive mothers. New Engl. J. Med. 320,1649 - 1654.

[28] Saksela, K. , Stevens, C. , Rubinstein, P. and Baltimore, D. (1994) Human immunodeficiency virus type 1 mRNA expression in peripherial blood cells predicts disease progression independently of the numbers of CD4+lymphocytes. Proc. Natl. Acad. Sci. U. S. A. 91,1104 - 1108.

[29] Scadden, D. T, Wang, Z. and Groopman, J. F. (1992) Quantitation of plasma human immunodeficiency virus type 1 RNA by competitive polymerase chain reaction. J. Infect. Dis. 165,1119 - 1123.

[30] Schnittman, S. M. , Psallidopoulos, H. C. , Lane, L. et al. (1989) The reservoir for HIV-1 in human peripheral blood is a T cell that maintains expression of CD4. Science 245,305 - 308.

[31] Semple, M. G. , Kaye, S. , Loveday, C. et al. (1993) HIV-1 plasma viremia quantification: a non-culture measurement needed for therapeutic trials. J. Virol. Methods. 41,167 - 180.

[32] Sheppard, H. W. , Ascher, M. S. , McRae, B. et al. (1991) The initial immune response to HIV and immune system activation determine the outcome of HIV disease. J. AIDS 4,704 - 712.

[33] Siebert, P. D. and Larrick, J. W. (1993) PCR mimics: competitive DNA fragments for use as internal standards in quantitative PCR. Biotechnique. 14,244 - 249.

[34] Simmonds, P. , Balfe, P. , Peuthrer, J. F. et al. (1991) Human immunodeficiency virus-infected individuals contain provirus in small numbers of peripheral mononuclear cells and at low copy numbers. J. Viol. 64,864 - 872.

[35] Zagury, D. , Fouchard, M. , Vol, J. C. et al. (1985) Detection of infectious HTLV-III/LAV virus in cell-free plasma from AIDS patients. Lancet 2,505 - 506.

A Simplified Method for Quantitation of Human Immunodeficiency Virus Type 1 (HIV1) RNA in Plasma: Clinical Correlates①

Summary

Human immunodeficiency virus type 1 (HIV1) RNA was quantitated in the plasma of HIV1-seropositive patients using a simplified guanidinum-based extraction technique, reverse transcriptase (RT) and the polymerase chain reaction (PCR). Plasma samples were obtained from 15 HIV1-seronegative individuals and 38 HIV1-seropositive patients. Following the extraction of RNA from plasma using "RNAzol", HIV1 RNA was reverse-transcribed using random hexamers, amplified by PCR and then detected by solution oligonucleotide hybridization, of the 15 HIV1-seronegative individuals, 14 were negative for HIV1 RNA by RT-PCR. One high-risk patient who was HTLV-1-seropositive but HIV1-seronegative was found to be positive for HIV1 RNA by RT-PCR. All 38 HIV1-seropositive patients were positive for HIV1 RNA by this technique. The HIV1 RNA levels in plasma varied from 800 to 500,000 copies/ml. patients with advanced clinical disease tended to have HIV1 RNA levels above 25,000 copies/ml. In patients studied serially, an increase in plasma HIV1 RAN correlated with a progressive decline in $CD4^+$ T cells and a deteriorating clinical course. The simplified quantitative RT-PCR assay for HIV1 RNA provides a useful tool for the evaluation and management of HIV disease.

Key-words: AIDS, ARC, HIV, RNA, RT, PCR; Quantitation, Clinical correlates, CD4 counts, AZT.

R. B. Stricker [1,2,3] (*), Y. S. Zhu[3] and B. F. Elswood[3]

[1] HemaCare Corporation, San Francisco, CA, [2] Department of Medicine, California Pacific Medical Center, and [3] North American Biomedical Technologies Corporation, San Francisco, CA (USA)

此文发表于 1995 年"Res. Virol" 146:151 - 158.

① Submitted November 8,1994, accepted December 28,1994.

(*)Address all correspondence to: Raphael B. stricker, HemaCare Corporation, 450 sutter street, suite 1504, San Francisco, CA 94108, USA.

ARC=AIDS-related complex; AZT=azidothymidine (zidovudine); ddC=dideoxycytidine; ddI=dideoxyinosine; DEPC=diethylpyrocarbonate; DNCB=dinitrochlorobenzene; LTR=long terminal repeat; PCR=polymerase chain reaction; RT=reverse transcriptase; *Taq*=*Thermus aquaticus*.

Introduction

The polymerase chain reaction (PCR) has been used as a qualitative method to detect human immunodeficiency virus type 1 (HIV1) DNA in peripheral blood mononuclear cells and plasma (Ou et al, 1988; Schnittman et al., 1990; Genesca et al., 1990; Michael et al., 1992).

Recently the same technique coupled with cDNA synthesis by reverse transcriptase (RT) has been empolyed to amplify HIV1 RNA (Hart et al., 1988; Holodniy et al., 1991a; Michael et al., 1992). Since HIV1 is a positive-stranded RNA virus, quantitative measurement of plasma HIV1 RNA should directly reflect productive viral replication and HIV disease activtiy (Hart et al., 1988; Holodniy et al., 1991a). Using a simplified guanidinium-based extraction technique, we have been able to quantitate HIV1 RNA in blood samples. Here we describe a clinical study of HIV1 RNA quantitation in plasma from patients with AIDS and AIDS-related complex (ARC) using the RT-PCR method.

Patients and Methods

Patients

Plasma samples were obtained from 53 subjects at California Pacific Medical Center. Fifteen individuals were HIV1-seronegative, while thirty-eight individuals were HIV1-seropositive (5 asymptomatic, 13 ARC and 20 AIDS). The ARC and AIDS patients were classified according to the revised criteria from the Centers for Disease Control (CDC) (CDC, 1987). Thirty of these patients had been on stable doses of zidovudine (AZT) for at least six months at the time of plasma sampling. In addition to AZT, one patient was taking dideoxyinosine (ddI). one was on dideoxycytidine (ddC) and one was taking prednisone. Plasma samples were obtained from blood collected in EDTA by centrifugation at 800 g for 15 min at room temperature. The samples were frozen at −60℃ until use. In addition, serial plasma samples were obtained from five patients (3 ARC and 2 AIDS) who were on stable doses of AZT. The samples were obtained every 3—6 months over 1—3 years. $CD4^+$ T-cell counts were measured in eash patient using flow cytometry, as previously described (Kiprov et al., 1984).

Synthesis of HIV1 RNA Standard

For quantitation, a full-length HIV1 RNA was synthesized *in vitro* by T7 RNA polymerase (Stratagene, La Jolla, CA) using pBKBH10S DNA as a template. The template pBKBH10S, kindly provided by Dr. John Rossi (NIH, Bethesda, MD), contains full-length genomec HIV1 (9 kb) less the LTR cloned downstream of the T7 promoter of plasmid pBluescript II-KS(+) (Stratagene). The size and integrity of synthesized HIV1 cRNA was confirmed by Northern analysis, and HIV1 cRNA was quantitated by OD_{206} measurement. A series of dilutions of HIV1 RNA were made and used as standards for comparison in each RT-PCR experiment.

HIV1 RNA Extraction

Extractions were performed on 100 μl aliquots of plasma. Total RNA was extracted from plasma using "RNAzol" (Biotecx, Houston, TX), an acidic guanidinium-thiocyanate-based reagent (Chomcynski and Sacchi, 1987), according to the manufacturer's instructions, with the following modifications: 2 μg of yeast tRNA was added to the plasma at the beginning of each extraction, and only one isopropanol precipitation of the total RNA was performed. Total RNA was dissolved and stored in DEPC-H_2O at −70℃. RNA samples stored in this manner have been shown to be stable for up to six months (Zhu et al., 1995).

Reverse Transcription and PCR

Reverse transcription and PCR were performed as follows. Total RNA from 100 μl of plasma was incubated in 20 μl containing 50 mmol/L Tris-HCl pH 8.3, 150 mmol/L KCl. 10 mmol/L $MgCl_2$ 10 mmol/L DTT, 200 μmol/L dNTPs, 0.1 μmol/L random hexamer, 20 units "RNasin" (Promega Corp., Madison, WI) and 5 units RT (Seikagaku America, Inc., St. Petersburg, FL) at 37℃ for 30 min. After heat heat inactivation of the reaction, the cDNA was amplified by PCR. Ten microliters of cDNA was amplified in 30 μl containing 50 mmol/L Tris-HCl pH 8.5, 50 mmol/L KCl, 2.5 mmol/L $MgCl_2$, 200 μg/ml gelatin, 0.5 μmol/L primer pairs SK38/39 and 1 unit of Taq polymerase (perkin-Elmer Cetus, Norwalk, CT). PCR was carried out for 30 cycles in a DNA thermal cycler (perkin-Elmer Cetus) with the following program: 95℃ for 30 s, 55℃ for 30 s, 72℃ for 60 s followed by a 10-min extension at 72℃ at the end. All samples were run in duplicate. Negative controls (reagents alone) and positive controls (standard HIV1 RNA) were included for each set of amplification reactions.

Detection of PCR Products

The amplified HIV1 RNA *gag* region delineated by the primer pairs SK38/39 was detected by solution olingonucleotide hybridization. The PCR product (10 μl) was hybridized with ^{32}P-la-belled SK19 probe ((3—5)$\times 10^5$ cpm, 10^7 com/pmol) at 55℃ for 15 min. The hybridized *gag* sequence was separated by electrophoresis in a native 12% polyacrylamide gel and then exposed to X-ray film with intensifying screens for one hour. The bands corresponding to the HIV1 sequence (115 bp) were either cut and counted in a scintillation counter or quantitated with an "Ambis Radioanalytic Imaging System"(Ambis Systems, San Diego, CA).

Results

The sensitivity of the RT-PCR procedure was first tested using the HIV1 cRNA standards. Different concentrations of HIV1 cRNA (1 to 10,000 molecules per test) were reverse-transcribed into cDNA using RT, followed by PCR. Using this procedure, the lower limit of detection of HIV1 RNA was 10 copies and the linear range of detection was between 50 and 1,000 copies. We also found that identical quantitative data were obtained for the standards when they were amplified alone or in the presence of plasma extract. Under the acidic conditions of the RNAzol extraction method, RNA is isolated exclusively, and contaminating DNA was not detected in our samples. The loss of RNA in the extraction step was small, with yields consistently greater than 80%. The intra-assay coefficient of variation was found to be 16% (Zhu et al., 1995).

Figure 1 illustrates the analysis of HIV1 RNA gag sequences from 30 of the 53 subjects. The patient population included 2 seronegative individuals and 28 seropositive patients diagnosed as asymptomatic, ARC or AIDS. Duplicate plasma samples were prepared and analysed for each individual, and the duplicates consistently yielded similar results with less than 2-fold variation between samples. While one HIV1-seronegative individual (No. 29) was negative for HIV1 RNA even when the autoradiograph was prepared by overnight exposure, the other HIV1-seronegative individual (No. 7) was positive for HIV1 RNA by RT-PCR (7,000 copies/ml). This high-risk subject is additionally HTLV-I-seropositive and is the active sexual partner of an HIV-seropositive individual. All 28 seropositive patients had positive RT-PCR results for HIV1 RNA. The HIV1 RNA copy numbers measured by RT-PCR for the seropositive patients varied from 800 copies/ml to 500,000 copies/ml. As illustrated in Figure 1, the band intensities of HIV1 RNA from six patients (Nos. 3,4,10, 18,22 and 24) were out of the RT-PCR quantitative range. These plasma samples were further diluted to the proper concentrations so that accurate quantitation could be determined.

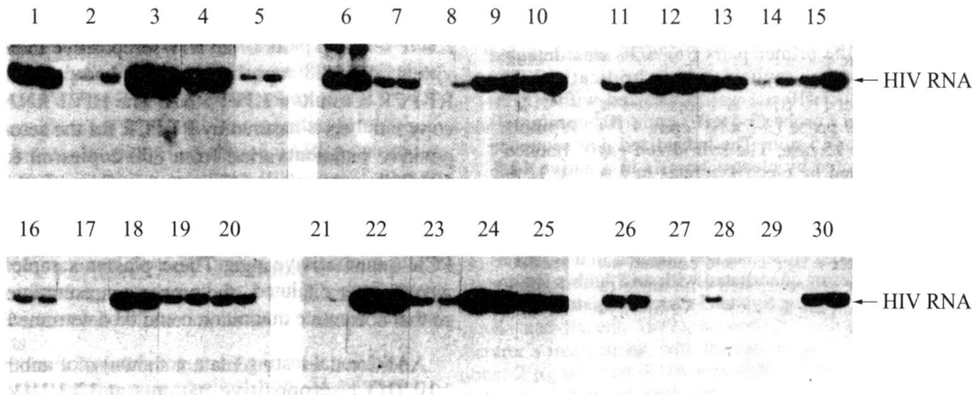

Figure 1 Autoradiographs of HIV1 RNA PCR products obtained from plasma of 30 patients

HIV1 RNA was prepared as described in "Patients and Methods". Samples were run in duplicate

Additional testing (data not shown) of another 10 HIV1-seropositive patients and 13 HIV1 seronegative individuals yielded similar results. All 10 seropositive patients were found to be positive for HIV1 RNA by RT-PCR. Conversely, all 13 HIVl-seronegative individuals were shown to be negative for HIV1 RNA after overnight autoradiograph exposure. However, in one high-risk seronegative individual with a history of multiple blood transfusions and hepatitis B virus infection, the RT-PCR assay yielded an extremely faint signal. Follow-up of this patient is continuing at present.

In order to investigate the relationship between plasma HIV1 RNA concentration and disease status, HIV1 RNA copy numbers were plotted against $CD4^+$ T-cell counts. The results are shown in Figure 2(a). Overall, asymptomatic and AIDS patients showed a good correlation between HIV1 RNA copies and $CD4^+$ T-cell counts, while ARC patients were more variable. An increase in HIV1 RNA copies correlated with a decrease in $CD4^+$ T-cell counts in all AIDS patients except for nos. 16, 23 and 28 from Figure 1. In general, ARC and AIDS patients had higher HIV1 RNA levels compared with asymptomatic individuals (usually above 25,000 copies/ml). HIV1 RNA was also measured in serial plasma samples from five patients over 1—3 years, as shown in Figure 2(b). Three patients had ARC, while two patients had AIDS. All patients were on stable doses of AZT at the time of plasma sampling, and one patient (MC) was also taking ddC. In two of the ARC patients (DE and RA), a rise in $CD4^+$ T-cell counts was associated with a greater than 2-fold fall in HIV1 RNA copies. Conversely, in the AIDS patients (MO and MC), a decline in $CD4^+$ T-cells corresponded to a greater than 2-fold rise in HIV1 RNA. The third ARC patient (YE) had fluctuating levels of $CD4^+$ T-cells that corresponded to fluctuating HIV1 RNA levels.

The relationship between HIV1 RNA levels and AZT treatment is shown in Table 1. The mean HIV1 RNA level in AZT-treated patients was 64,174 copies/ml, while the mean level in patients on no therapy was 17,200 copies/ml. These levels parallel the $CD4^+$ T-cell counts shown in Figure 2(a), with several exceptions (see below).

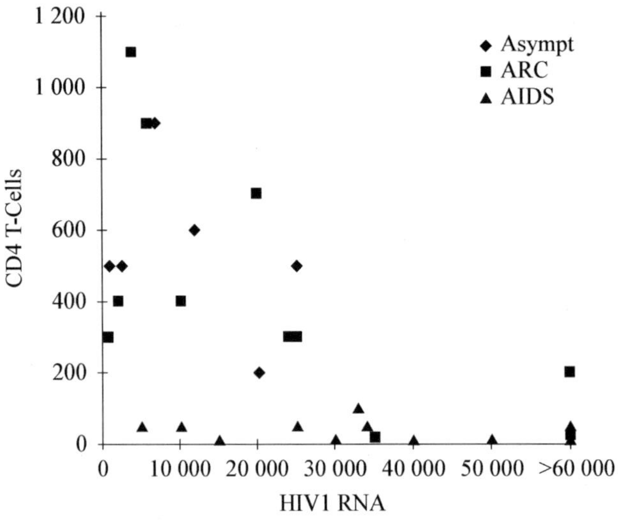

(a)

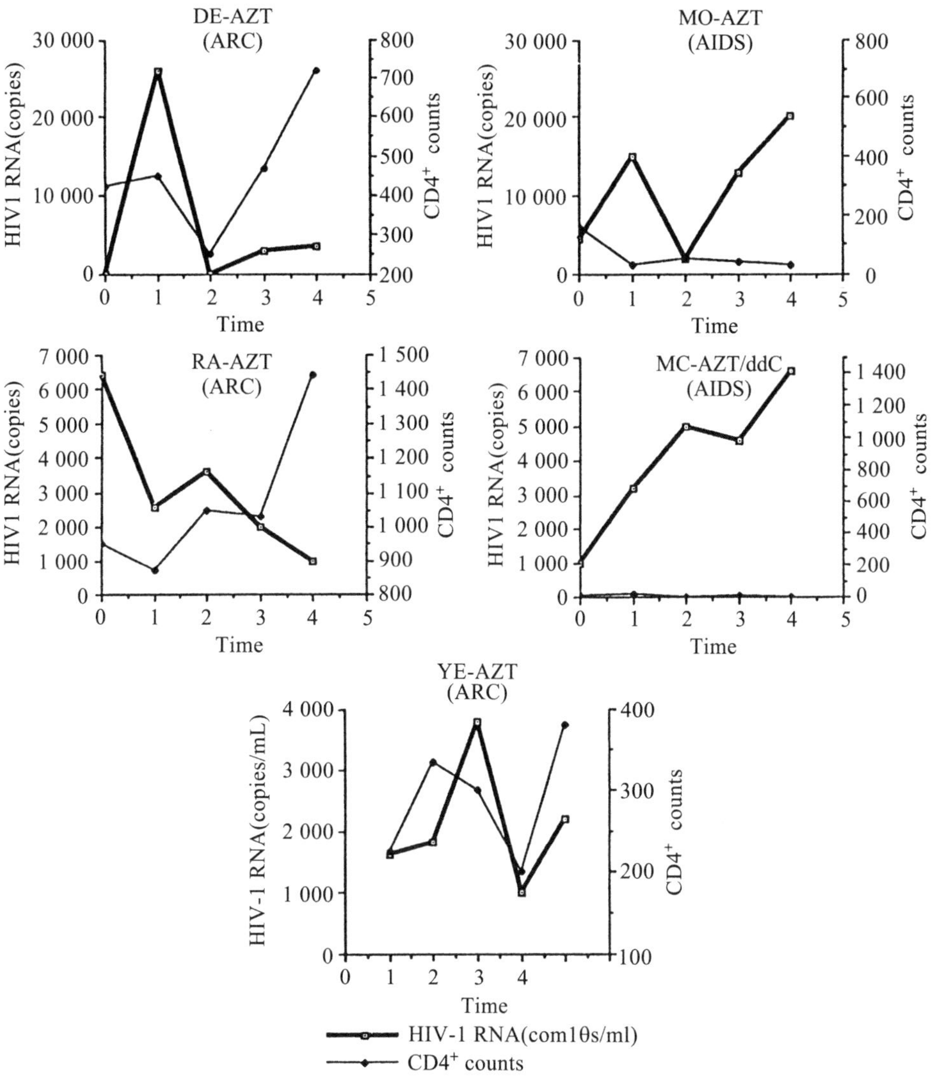

(b)

Figure 2 **(a) Correlation between plasma HIV1 RNA levels (copies/ml), $CD4^+$ T-cell counts (cells/mm^3) and disease status in HIV1-seropositive patients**
Asympt = asymptomatic HIV1 seropositive; ARC, AIDS-related complex
(b) HIV1 RNA levels in serial plasma samples from patients with ARC and AIDS. Time points represent 3—6-month intervals

Discussion

Our study was designed to measure HIV1 RNA in routinely processed clinical samples stored under conventional conditions. The combination of RNAzol plasma extraction and RT-PCR yields a reliable assay for HIV1 RNA, with results obtainable in less than two days. The assay is responsive to 2-fold changes in RNA copy number when this number is maintained within the linear range of the amplification protocol. The effect of this assay on recovery and stability of HIV1 RNA is the subject of a separate publication (Zhu et al., 1995). Here we have investigated the clinical correlates of this technique in patients with HIV1 infection.

Using cumbersome culture methods, investigators have found that plasma viraemia is detectable in all clinical stages of HIV infection (Ho et al., 1989; Coombs et al., 1989). Holodniy et al. (1991a) utilized a PCR method to quantitate viral load in serum. The authors reported that viral RNA copy numbers from HIV1-seropositive patients ranged from 500 to 250,000/ml. In our study, measurements of HIV1 RNA copies in plasma from patients at different stages of infection ranged from 800 to 500,000/ml (see Figure 1). These results are similar to the findings in serum. In an effort to decrease the variability of RNA PCR, quantitative HIV1 RNA assays have recently been developed using the competitive PCR technique (Scadden et al., 1992; Menzo et al., 1992; Piatak et al., 1993). Our results suggest, however, that the additional steps used in the competitive assay may not be necessary when a reliable RNA extraction method and a rigorous amplification protocol are employed. Further comparison of these techniques should be informative.

We examined the relationship between HIV1 RNA levels and clinical status, $CD4^+$ T-cell counts and AZT therapy in our patients. We found that plasma HIV1 RNA levels correlated roughly with the clinical stage of HIV disease, with levels above 25,000/ml indicating advanced disease [see Figure 2(a)]. To a lesser extent, HIV1 RNA levels also correlated with $CD4^+$ T-cell counts. Although these lymphocyte counts are considered to be useful surrogate markers in HIV disease, $CD4^+$ T-cell numbers may fluctuate for a variety of reasons during the course of HIV infection (Amadori et al., 1992). Nevertheless, in serial samples from individual patients, the HIV1 RNA level paralleled the $CD4^+$ T-cell count in each case [see Figure 2(b)]. Thus quantitation of HIV1 RNA appeared to provide a consistent indicator of disease activity in individual cases.

We also found that patients treated with AZT tended to have higher levels of HIV1 RNA than patients on no therapy (see Table 1). This difference probably reflects the more advanced clinical stage of patients on antiretroviral therapy. Previ ous culture studies have suggested that AZT may reduce the viral load in HIV-infected patients (Ho et al., 1989; Coombs et al., 1989). In our serial and cross-sectional analyses, patients with more advanced HIV disease did not appear to respond significantly to AZT, as judged by direct measurement of viral load [see Figure 2(b) and Table 1]. However, our study was not designed to measure the in vivo effect of antiviral therapy, and no conclusions can be drawn from the RNA measurements in AZT-treated patients.

It is of interest that every patient whose plasma had HIV1 RNA copies above 25,000/ml had ARC or AIDS. However, patients whose plasma HIV1 RNA copies were below this level were not necessarily at an early stage of HIV infection. One AIDS patient (No. 23) who received topical treatment with the immune modulator DNCB had a relatively low level of

Table 1 Relationship between plasma HIV1 RNA levels and AZT treatment

Treatment and disease status	No. of patients	HIV1 RNA/(copies/ml)	
		Mean	Range
AZT	19	64,174	800—500,000
asymptomatic	1	2,500	—
ARC	8	47,850	800—200,000
AIDS	10	85,700	10,000—500,000
No AZT:	10	17,200	1,000—33,000
asymptomatic	5	13,000	1,000—25,000
ARC	3	23,000	20,000—25,000
AIDS	2	19,000	5,000—33,000

HIV1 RNA (5,000 copies/ml). A follow-up study has confirmed this initial observation of a decreased viral load in DNCB-treated patients (Stricker et al., 1993). Two other AIDS patients (Nos. 16 and 28) also had relatively low HIV1 RNA levels. One of these patients had recently started ddI therapy, which may have resulted in a decreased viral load (Holodniy et al., 1991b).

In conclusion, direct HIV1 RNA quantitation using RNAzol extraction and RT-PCR provides an accurate and reliable marker for HIV disease activity. This simplified assay should prove useful for monitoring the efficacy of various treatment modalities in HIV infection.

Acknowledgements

Supported in part by Research Grants 89.024C and 90.041C from California Pacific Medical Center and the Elizabeth Reed Taylor Foundation.

The authors thank Drs. Judy Wilber, Valerie Ng, Giuseppe Pantaleo and Anthony Fauci for helpful discussion. We are indebted to Cecilio Dumlao and Joanna Van Elk for sample collection and processing, and we are grateful to Steven Koontz and Lori Sanchez for expert technical assistance.

Méthode simple de dosage de l'ARN plasmatique du VIH1 et correspondances cliniques

Une méthode utilisant une technique d'extraction (guanidinium) simplifiée pour la PCR et la RT (transcriptase réverse) a permis de quantifier l' ARN du virus de l'immunodéficience humaine de type 1 ou HIV1 dans le plasma de 15 sujets séronégatifs et dans 38 cas séropositifs qui se sont avérés positifs pour cette recherche de l' ARN par la RT-

PCR, à des taux allant de 800 à 500,000 copies/ml, ces taux se situant autour de 25,000/ml chez les malades en phase clinique évolutive. Un accroissement de l'ARN correspond à un déclin numérique des lymphocytes $CD4^+$ et à une aggravation clinique. Sur les 15 sujets séronégatifs, 14 se sont avérés négatifs pour la recherche d'ARN par la RT-PCR. Un sujet à risques, séropositif pour le HTLV-I et séronégatif pour le VIH1 (HIV1) s'est montré positif pour l'ARN viral déterminé par la RT-PCR. Cette méthode quantitative et simple est effectivement un outil appréciable dans l'analyse bio-clinique des effets du VIH.

Mots-clés: SIDA, ARC, VIH, ARN, RT, PCR; Détection quantitative, Correspondances cliniques, Numération des CD4, AZT.

References

[1] Amadori, A., De Silvestro, G., Zamarchi, R., Veronese, M. L., Mazza, M. R., Schiavo, G., Panozzo, M., De Rossi, A., Ometto, L., Mous, J., Barelli, A., Borri, A., Salmaso, L. & Chieco-Bianchi, L. (1992), CD4 epitope masking by gp120/anti-gp120 antibody complexes. *J. Immunol.*, 148, 2709 - 2716.

[2] *Centers for Disease Control* (1987), *Revision of the CDC surveillance case definition for acquired immunodeficiency syndrome.* MMWR, 36 (*suppl. ls*), 3S-15S.

[3] *Chomcynski, P. & Sacchi, N.* (1987), *Single-step method of RNA isolation by acid guanidinium thiocyanatephenol-chloroform extraction.* Anal. Biochem., 162, 156 - 159.

[4] Coombs, R. W., Collier, J. P., Allain, B., Nikora, B., Leuther. M., Gjerset, G. F. & Corey, L. (1989), Plasma viremia in human immunodeficiency virus infection. *N. Engl. J. Med.*, 321, 1626 - 1630.

[5] *Genesca, J., Wang. R. Y.-H., Alter, H. J. & Shih, J. W.* (1990), *Clinical correlation and genetic polymorphism of the human immunodeficiency virus proviral DNA obtained after polymerase chain reaction amplification.* J. Infect. Dis., 162, 1025 - 1030.

[6] Hart, C., Spira. T., Moore, J., Sninski, J., Schochetman, G., Lifson, J., Galphin, J. & Ou, C.-Y. (1988), Direct detection of HIV RNA expression in seropositive subjects. *Lancet*, 2. 596 - 599.

[7] Ho, D. D., Moudgil, T. & Alam, M. (1989), Quantitation of human immunodeficiency virus type 1 in the blood of infected persons. *N. Engl. J. Med.*, 321, 1621 - 1625.

[8] *Holodniy, M., Katzenstein, D. A., Sengupta, S., Wang, A. M., Casipit, C., Schwartz, D. H., Konrad, M., Groves. E. & Merigan, T. C.* (1991*a*), *Detection and quantification of HIV RNA in patient serum by use of the polymerase chain reaction.* J. Infect. Dis., 163, 862 - 866.

[9] Holodniy, M., Katzenstein, D. A., Israelski, D. M. & Merigan, T. C. (1991b), Reduction in plasma human immunodeficiency virus ribonucleic acid after dideoxynucleoside therapy as determined by the polymerase chain reaction. *J. Clin. Invest.*, 88, 1755 - 1759.

[10] *Kiprov, D. D., Busch, D. F., Simpson, D. M., Morand, P. R., Tardelli, G. P., Gullett, J. H., Lippert, R. & Mielke, H.* (1984), *Antilymphocyte serum factors in patients with acquired immunodeficiency syndrome, in "Acquired immune deficiency syndrome" (Gottlieb M. S., Groopman J. S. (eds.)) pp.* 299 - 308. *New York, Alan R. Liss, Inc.*

[11] *Menzo, S., Bagnarelli, P., Giacca, M., Manzin, A., Varaldo, P. E. & Clementi, M.* (1992), *Absolute quan titation of viremia in human immunodeficiency virus infection by competitive reverse transcription and polymerase chain reaction.* J. Clin. Microbiol., 30, 1752 - 1757.

[12] Michael, N. L., Vahey, M., Burke, D. S. & Redfield, R. (1992), Viral DNA and mRNA expression correlate with the stage of HIV type 1 infection in disease. J. Virol., 66, 310 - 316.

[13] Ou, C.-Y., Kwok, S., Mitchell, S. W., Mack, D. H., Sninski, J. J., Krebs, J. W., Feorino, P., Warfield, D. & Schochetman, G. (1988), DNA amplification for direct detection of HIV-1 in DNA of peripheral blood mononuclear cells. Science, 239,295 - 297.

[14] Piatak, M., Saag, M. S., Yang, L. C., Clark, S. J., Kappas, J. C., Luk, K. C., Hahn, B. H., Shaw, G. M. & Lifson, J. D. (1993), High levels of HIV-1 in plasma during all stages of infection determined by competitive PCR. Science, 259,1749 - 1754.

[15] Scadden, D. T., Wang, Z. & Groopman, J. E. (1992), Quantitation of plasma human immunodeficiency virus type 1 RNA by competitive polymerase chain reaction. J. Infect. Dis., 165,1119 - 1123.

[16] Schnittman, S. M., Greenhouse, M. C., Psallidopoulos, M., Baseler, M., Salzman, N. P., Fauci, A. S. & Lane, H. C. (1990), Increasing viral burden in $CD4^+$ T-cells from

[17] patients with human immunodeficiency virus (HIV) infection reflects rapidly progressive immunosuppression of clinical disease. Ann. Intem. Med., 113,438 - 443.

[18] Stricker, R. B., Zhu, Y. S., Elswood, B. F., Van Elk, J., Dumlao, C., Berger, T. G., Tappero, J., Epstein, W. F. & Kiprov, D. D. (1993), Pilot study of dinitrochlorobenzene (DNCB) in human immunodeficiency virus infection. Immunol. Lett., 36,1 - 6.

[19] Zhu, Y. S., Gong, Y. & Cimino, G. D. (1995), Quantitative analysis of HIV-1 RNA in plasma preparations. J. Virol. Methods (in press).

Quantitative Restriction Fragment Length Polymorphism: A Procedure for Quantitation of Diphtheria Toxin Gene CRM197 Allele*

Here we present an assay for quantitation of a particular gene allele in DNA mixtures by means of restriction fragment length polymorphism (RFLP) in combination with polymerase chain reaction (PCR). We applied the quantitative RFLP principle for estimation of the relative amount of diphtheria toxin gene CRM197 allele in *Corynebacterium diphtheriae* culture DNA samples. The procedure is based on PCRmediated generation of an artificial AluI restriction site specifically with the CRM197 DNA template. After AluI digestion of the PCR product and polyacrylamide gel electrophoresis of the restriction fragments, the percentage of CRM197 template in the initial DNA sample was determined by scanning a gel negative. The method was shown to give a linear response when applied to template mixtures containing different amounts of CRM197 reference template. For samples where non-CRM197 DNA was detected by AluI RFLP, we designed a further allele-specific PCR assay to determine whether the non-CRM197 template portion was the wild-type toxin gene allele.

In the biopharmaceutical industry, it is important to evaluate the genetic stability of a recombinant protein producing strain because of a concern that mutational alterations may occur in critical DNA sequences during cultivation. This is particularly the case for cultures of *Corynebacterium diphtheriae* which produce the inactive mutant form of diphtheria toxin protein, CRM197.

Diphtheria toxin protein, secreted by pathogenic strains of *C. diphtheriae*, is the causative agent for diphtheria in humans[1]. This 58-kDa protein is able to kill eukaryotic cells by blocking cellular protein syn thesis[2-5]. Toxin protein is encoded by the tox gene of *Corynebacteriophage β* integrated into the chromosome of lysogenic *Corynebacterium*[6,7].

A number of tox mutants which produce altered toxin proteins, designated CRMs (serologically cross-reacting materials), have been isolated by mutagenic treatment of

---

* Elena A. Pushnova (To whom correspondence should be addressed. Fax: 510 - 923 - 4116.) and Yu Sheng Zhu
Chiron Corporation, 4560 Horton Street, Emeryville, California 94608 - 2916
Received January 14, 1998
此文发表于 1998 年"Anal. Biochem." 260: 24 - 29.

Corynebacteriophage β with nitrosoguanidine. CRMs are either completely non-toxic or of greatly reduced toxicity, yet still cross-react with diphtheria antitoxin[8-10]. From the point of view of active immunization against diphtheria, the most interesting of the CRMs that have been isolated is CRM197. Due to a G → A transition in the tox gene leading to a GLY→ GLU substitution at position 52, CRM197 protein is enzymatically inactive, but appears to be serologically indistinguishable from wild-type toxin[2,8,9,11]. In addition to its use for immunization against diphtheria, CRM197 protein is successfully used as an adjuvant protein carrier for carbohydrate antigens[1,12].

Considering the possibility of CRM197 reversion to wild-type during cultivation of bacteria, it is very important to be able to monitor the genetic stability of CRM197 mutation during the process of protein production. The purpose of the present study was to develop a simple assay to quantitate the relative amount of the CRM197 mutant allele in *Corynebacterium* genomic DNA samples, and to distinguish between CRM197 and wild-type tox gene alleles.

For quantitation of the CRM197 mutant allele, a portion of the toxin gene was polymerase chain reaction (PCR)①-amplified with primers generating an artificial AluI restriction site specifically with the CRM197 DNA template. The resulting PCR product was digested with AluI, and restriction fragments were separated in a polyacrylamide gel. The percentage of CRM197 template in the initial DNA sample was determined by scanning the gel negative. Since the quantitative RFLP assay did not allow to detect the wildtype tox gene allele specifically, a complementary allele-specific PCR assay was designed to determine the presence of the wild-type allele.

Materials and Methods

Strains and Media

C. diphtheriae lysogens carrying integrated *Corynebacteriophage* β toxin gene wild-type allele (strain ATCC No. 13812) and CRM197 mutant allele (strain CRM-Ref., C7 (β197)$^{tox-}$ double lysogen)[13], were used in the present study. Bacteria were grown on CY medium as described[13].

Genomic DNA Isolation and Quantitation

Bacterial genomic DNA was extracted using the Invitrogen Easy DNA kit according to the manufacturer's protocol. DNA concentration in the samples was estimated using PicoGreen dsDNA quantitation reagent (Molecular Probes) according to the manufacturer's protocol, with *Escherichia coli strain* B genomic DNA (Sigma) as a standard.

① Abbreviations used: PCR, polymerase chain reaction; RFLP, restriction fragment length polymorphism.

AluI RFLP Quantitative Assay

A125-bp toxin gene fragment which includes the site of CRM197 mutation was PCR-amplified with oligonucleotide primers: Alu197-F, 5′-GACTAAACCTGGTTATGTAGCTTCC -3′ (forward primer); and Alu197-R, 5′-CCGCAGCGTCGTATTTATTGTCGGTACTATAAAGC-3′ (reverse primer).

PCR was carried out in a total reaction volume of 100 μL containing 1×PCR buffer, 5.5 mmol/L $MgCl_2$, 200 μmol/L each dNTP, 10 pmol each primer, 0.2 ng template genomic DNA, and 5 units *Taq* DNA polymerase (Boehringer Mannheim). The PCR mixtures were overlayed with 3 drops of Nujol mineral oil. After predenaturation for 5 min at 95℃, 30 PCR cycles were carried out in Perkin-Elmer Cetus DNA Thermal Cycler 480, each cycle including denaturation for 30 s, 94℃; annealing 30 s, 48℃; and extension 30 s, 72℃; followed by a final extension for 3 min at 72℃.

Three microliters of the PCR product was digested with 10 units AluI (New England Biolabs or Boehringer Mannheim) in a total volume of 10 μl overlayed with a drop of Nujol mineral oil to prevent evaporation. Digestions were carried out to completion overnight at 37℃.

AluI digests (10 μl) were analyzed by 15% polyacrylamide gel (acrylamide: Bis, 19 : 1) electrophoresis in 0.5×TBE buffer (45 mmol/L Tris-borate, 1 mmol/L EDTA, pH 8.3) using Bio-Rad Mini Protean II electrophoresis system. Gels (7 cm long, 10 cm wide, 1 mm thick) were run at 200 V until the bromophenol blue reached approximately 1 cm from the bottom of the gel. After obtaining the photographs of the ethidium bromide-stained gels with Polaroid positive/negative film No. 55, the negatives were scanned using the Personal Densitometer SI (Molecular Dynamics).

The percentage of CRM197 template in the initial DNA sample was calculated using the formula:

$$\%\text{CRM197 template} = \frac{(\text{peak 71 bp area} + \text{peak 34 bp area})}{(\text{peak 105 bp area} + \text{peak 71 bp area} + \text{peak 34 bp area})} \times 100\%.$$

Allele-Specific PCR Assay

DNA samples were subjected to PCR amplification in reaction volumes of 50 μL containing 1×PCR buffer, 5 mmol/L $MgCl_2$, 200 μmol/L each dNTP, and 0.1 ng template genomic DNA. Each mixture was overlayed with 2 drops of Nujol mineral oil. Samples were preheated in the thermal cycler for 10 min at 95℃. As soon as the block temperature reached 95℃, 2.5 units *Taq* DNA polymerase (Boehringer Mannheim) and 5 pmol each of four primers (BTOX5P1, 5′-CGGTGTGGTACACCTGATCTGGTCCGGTTC-3′; JET1372, 5′-GATAAACTCTTCCGTTCCGACTTGC-3′; 197M, 5′-GAAATTATGACGATGATTGGA-AAGA-3′; 197N, 5′-TATTTATTGTCGGTACTATAAAACC-3′ were added to the tubes under the mineral oil. Following preheating, 30 PCR cycles were carried out, each cycle including denaturation for 1 min, 94℃; annealing 1 min, 65℃; and extension 1 min, 72℃. PCR products were finally extended for 7 min at 72℃. Five microliters of each PCR product was analyzed by 2% agarose gel electrophoresis in 0.5×TBE.

Results

Principle of AluI RFLP Assay

In the AluI RFLP assay, a 125-bp tox gene fragment including the site of CRM197 point mutation was PCR-amplified with oligonucleotide primers Alu197-F and Alu197-R. These primers were designed to generate artificial AGCT AluI restriction sites in the PCR product. Introduction of an artificial restriction site by site-directed mutagenesis for RFLP assay purposes is often applied when the normal or mutant sequence, such as CRM197, does not form any known endonuclease restriction site[14]. Amplification with Alu197-F will generate a control AluI site at the 5′-end of the PCR product. Amplification with Alu197-R will generate the AluI site only with the CRM197 template, but not with wild-type or with a template other than CRM197 (see Figure 1). After AluI digestion of the PCR product, the

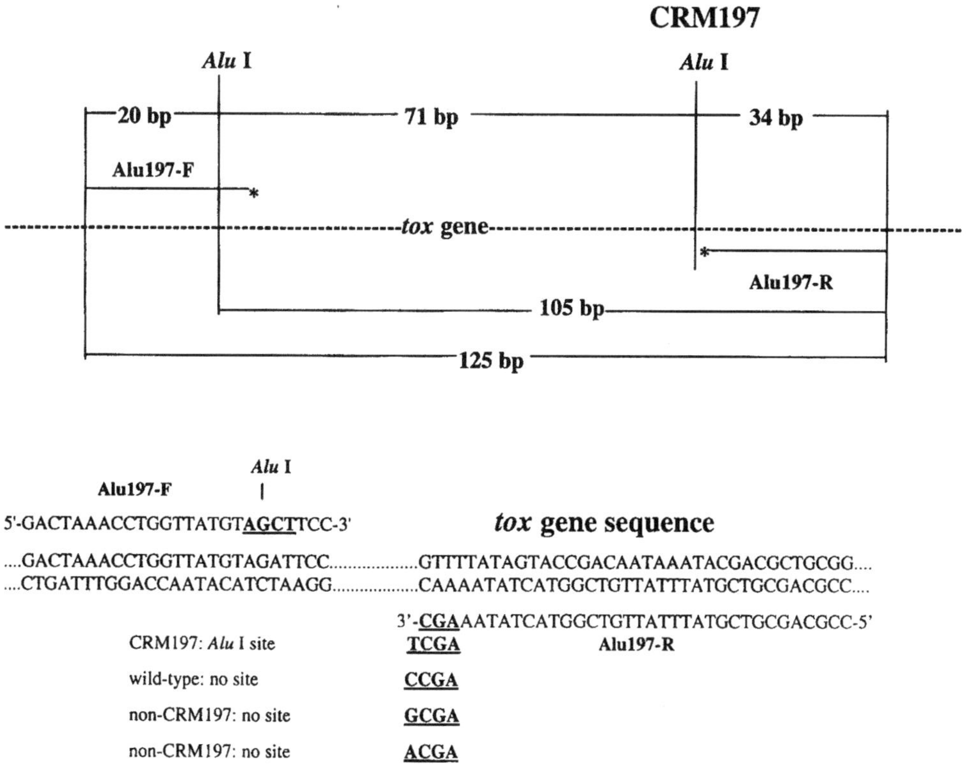

Figure 1 Principle of AluI RFLP approach: restriction map of 125-bp PCR product and sequences generated with different templates

expected RFLP patterns in polyacrylamide gel electrophoresis are 105 + 20 bp (control fragment) for the non-CRM197 DNA template, and 71+34+20 bp (control fragment) for the CRM197 DNA template (see Figure 1).

CRM197 Quantitation in Template Mixtures by AluI RFLP

To determine whether the AluI RFLP approach would accurately estimate the relative amount of CRM197 toxin gene mutant allele in DNA samples, template mixtures containing different percentages of wild-type and CRM197 mutant reference DNAs were subjected to AluI RFLP analysis.

Typical gel patterns in AluI RFLP assay as shown in Figure 2. The scan data, summarized in Table 1, indicate high reproducibility of the AluI RFLP assay in four independent experiments. The percentage of (71 + 34 bp) AluI fragments is directly proportional to the percentage of CRM197 in the template mixture, as demonstrated by the linear regression curve in Figure 2(b) ($R^2=0.997$). The detection limit of the assay is 5% of the particular template DNA in the sample (Figure 2(a)). With lower template amounts, the allele-specific restriction fragments were barely detectable on the gel.

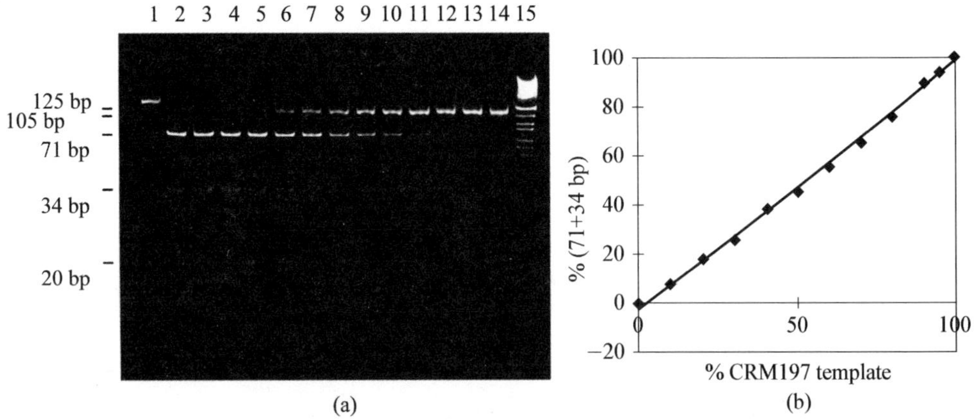

Figure 2 CRM197 quantitation in CRM197 + wild-type template mixtures by AluI RFLP

(a) AluI restriction fragments separated in 15% polyacrylamide gel. Lane 1: nondigested 125 bp PCR product. Lane 2 — Lane 14: AluI fragments of the PCR products generated with 100%, 99%, 95%, 90%, 80%, 70%, 60%, 50%, 40%, 30%, 20%, 10%, and 0% of CRM197-Ref. in the template mixture. Lane 15: Molecular weight marker V (Boehringer Mannheim) (b) Linear regression curve representing correlation between percentage CRM197 in the template mixture and the percentage of (71+34 bp) AluI fragments

Principle of Allele-Specific PCR Assay

In the allele-specific PCR assay for distinguishing between CRM197 and wild-type tox gene alleles, genomic DNA is PCR-amplified with four oligonucleotide primers: BTOX5P1 (forward), JET1372 (reverse), 197M (forward), and 197N (reverse). BTOX5P1 and JET1372 are the flanking primers, while the 197M and 197N primers are specific for the

Table 1 Precision of the AluI RFLP assay

% CRM197-Ref. DNA in template mixture[a]	%(71+34 bp)[b]					%CV
	Experiment				Average	
	1	2	3	4		
0	0	0	0	0	0	0.0
10	5	10	7	9	8	28.6
20	13	20	17	20	18	19.0
30	22	28	26	27	26	10.2
40	35	39	39	39	38	5.3
50	44	48	44	46	46	4.2
60	55	58	55	56	56	2.5
70	66	64	65	66	65	1.5
80	75	76	78	75	76	1.9
90	91	90	90	88	90	1.4
95	96	92	94	94	94	1.7
100	100	100	100	100	100	0.0

[a] Template samples obtained by mixing wild-type and CRM-Ref. DNAs, both at 0.1 ng/μL concentration. The samples were subjected to AluI RFLP assay. The details of the assay are as previously described under Materials and Methods, AluI RFLP quantitative assay section. The percentage (71 + 34 bp) was calculated after scanning of the negative of polyacrylamide gel, in which the AluI digests of PCR products were separated.

[b] Calculated as: $\frac{\text{(peak 71 bp area + peak 34 bp area)}}{\text{(peak 105 bp area + peak 71 bp area + peak 34 bp area)}} \times 100\%$.

CRM197 mutant and the wild-type alleles, respectively, with each matching allele-specific nucleotide at its 3′-end (see Figure 3). Three PCR products were expected with the above set of four primers: a 740-bp control product with BTOX5P1 and JET1372; a 242-bp CRM197-specific product with 197M and JET1372 showing the presence of CRM197 template; and a 547-bp wild-type-specific product with BTOX5P1 and 197N showing the presence of wild-type template (see Figure 3).

Detection of CRM197 and wt Alleles by Allele-Specific PCR

PCR product patterns in allele-specific PCR analysis of the template mixtures, which contain different percentages ofwild-type and CRM197 mutant (CRM-Ref., C7(β197)$^{tox-}$ double lysogen[13]) DNAs, are presented as shown in Figure 4. The estimated sensitivity of the allele-specific assay, with the PCR conditions used, is about 5% of the particular template DNA in the sample. With lower template amounts, the allele-specific PCR products were barely detectable on the gel.

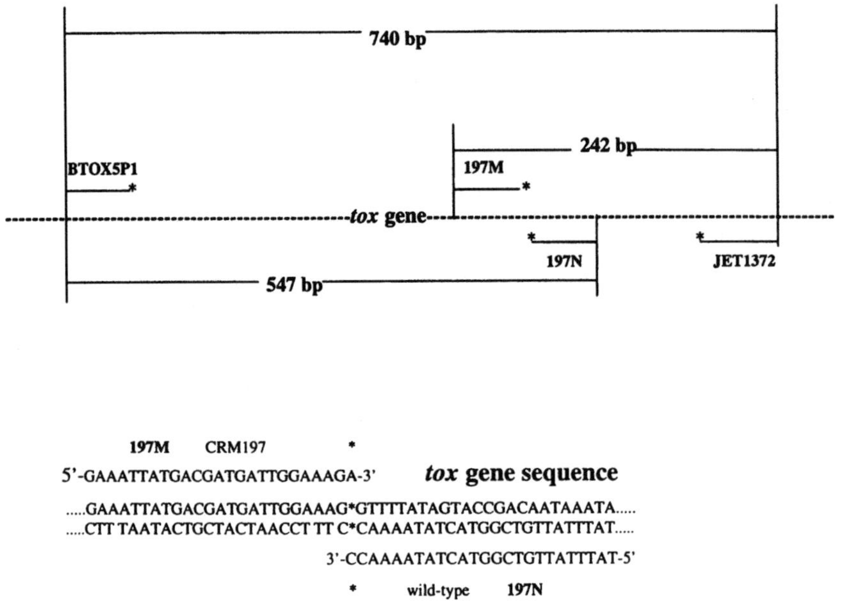

Figure 3 Principle of allele-specific PCR approach: regions of amplification and sequences of allele-specific primers

Discussion

We have demonstrated that our simple and reproducible AluI RFLP assay in combination with allele specific PCR can be successfully used for the accurate quantitation of the relative amount of CRM197 tox gene allele in bacterial DNA samples, as well as for detection of greater than 5% wild-type contamination.

In addition to the RFLP technique, there are several alternative approaches that are currently being used for diagnostics of point mutations, and these could potentially be applied in CRM197 diagnostics. However, these alternative approaches have certain disadvantages when compared to RFLP. Some of those, such as allele-specific oligonucleotide hybridization, oligonucleotide ligation assay, and direct sequencing, appear to be more laborious than our assay, and often require radioactive labeling, which presents safety problems. Others, such as single strand conformational polymorphism, denaturing gradient gel electrophoresis, and mismatch cleavage, have certain limitations with respect to particular bases or mismatches[15-17].

The most popular approach used so far for estimation of allele concentration in DNA samples is quantitative PCR with allele-specific primers[18]. However, the reliability of quantitation with this procedure may be significantly affected by unequal efficiency of amplification with different allele-specific primers. In contrast, our *Alu*I RFLP quantitation assay uses a pair of primers that are not allele-specific. Therefore, both CRM197 and non-

CRM197 allele templates are amplified at random, and with equal efficiency. As a result, the initial allele ratio in the template DNA is correctly represented in the PCR product, and this is the basic advantage of the quantitative RFLP design.

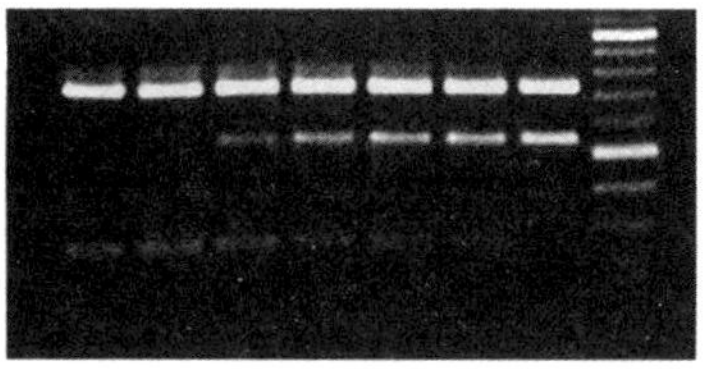

Figure 4 **PCR product patterns with CRM197 + wild-type DNA mixtures in template diagnostics by allele-specific PCR (2% agarose gel. Lane 1—Lane 7: 100%, 99%, 95%, 90%, 50%, 10%, and 0% of CRM197-Ref. in the template mixture. Lane 8: Molecular weight marker XIV (Boehringer Mannheim)**

We have complemented the AluI RFLP assay by an allele-specific PCR assay. Each assay has different advantages, and both work well in combination. The AluI RFLP assay is quantitative, while the allele-specific PCR is not. On the other hand, the allele-specific PCR assay can differentiate both wild-type and CRM197 tox gene alleles, while the AluI RFLP assay is specific for CRM197 allele only. Both assays have same sensitivity, i. e., both are able to detect a minimum of 5% particular allele in the template sample.

The PCR-RFLP procedure described in this article is the first DNA assay applied to the CRM197 toxin protein. So far, the toxicity of CRM197 has been evaluated by using the lethality test in guinea pigs, the cytotoxic activity assay, and the ADP-ribosyltransferase enzymatic assay[19]. We believe that our PCR-RFLP procedure could become an important initial control step for estimation of the genetic purity of bacterial culture prior to manufacturing of CRM197 protein. However, the final protein toxicity assays are also required to ensure vaccine safety.

Acknowledgments

We thank Drs. Michael Innis, Leo Lin, Michael Geier, and Rino Rappuoli for discussion and critical comments, Carmen Arnold and Biqi Gao for their technical assistance, and Titania Buchholdt and Heatherbell Fong for their help in preparation of the manuscript.

References

[1] Pappenheimer, A. M., Jr. (1984) in Bacterial Vaccines (Germanier, R., Ed.), pp. 1 - 36, Academic Press, San Diego.

[2] Uchida, T. (1982) in Molecular Action of Toxins and Viruses (Cohen, P., and Van Heyningen, S., Eds.), pp. 1 - 31, Elsevier, New York.

[3] Honjo, T., Nishizuka, U., Hayashi, O., and Kato, 1. (1968) J. Biol. Chem. 243, 3553 - 3555.

[4] Gill, D. M., Pappenheimer, A. M., Jr., Brown, R., and Kurnick, J. J. (1969) J. Exp. Med. 129, 1-21.
[5] Choe, S., Bennett, M. J., Fujii, G., Curmi, P. M. G., Kantardjieff, K. A., Collier, R. J., and Eisenberg, D. (1992) Nature 357, 216-222.
[6] Greenfield, L., Bjorn, M. J., Horn, G., Fong, D., Buck, G. A., Collier, R. J., and Kaplan, D. A. (1983) Proc. Natl. Acad. Sci. USA 80, 6853-6857.
[7] Murphy, J. R., Kaczorek, M., Leong, D., Delpeyroux, F., Coleman, K., Chencier, N., Boquet, P., Streeck, R. E., and Tiollais, P. (1984) in Bacterial Protein Toxins (Alouf, J. E., Fehrenbach, F. J., Freer, J. H., and Jeljaszewicz, J., Eds.), pp. 29-38, Academic Press, San Diego.
[8] Uchida, T., Pappenheimer, A. M., Jr., and Creany, R. (1973) J. Biol. Chem. 248, 3838-3844.
[9] Uchida, T., Pappenheimer, A. M., Jr., and Harper, A. A. (1973) J. Biol. Chem. 218, 3845-3850.
[10] Laird, W., and Groman, N. (1976) J. Virol. 19, 220-227.
[11] Giannini, G., Rappuoli, R., and Ratti, G. (1984) Nucleic Acids Res. 12, 4063-4069.
[12] Anderson, P., Pichichero, M. E., and Insel, R. A. (1985) J. Clin. Invest. 76, 52-59.
[13] Rappuoli, R. (1983) Appl. Environ. Microbiol. 46, 560-564.
[14] Haliassos, A., Chomel, J. C., Tesson, L., Baudis, M., Kruh, J., Kaplan, J. C., and Kitzis, A. (1989) Nucleic Acids Res. 17, 3606.
[15] Rossiter, B. J. F., and Caskey, C. T. (1994) in The Polymerase Chain Reaction (Mullis, K. B., Ferre, F., and Gibbs, R. A., Eds.), pp. 395-405, Birkhauser, Boston.
[16] Yap, E. P. H., and McGee, J. O'D. (1994) in PCR Technology: Current Innovations (Griffin, H. G., and Griffin, A. M., Eds.), pp. 107-120, CRC Press, Boca Raton, FL.
[17] Landegren, U., Kaiser, R., and Hood, L. (1990) in PCR Protocols (Innis, M. A., Gelfand, D. H., Sninsky, J. J., and White, T. J., Eds.), pp. 92-98, Academic Press, San Diego.
[18] Diaco, R. (1995) in PCR Strategies (Innis, M. A., Gelfand, D. H., and Sninsky, J. J., Eds.), pp. 84-108, Academic Press, San Diego.
[19] Carroll, S. F., and Collier, R. J. (1988) Methods Exzymol. 165, 218-225.

An Easy and Accurate Agarose Gel Assay for Quantitation of Bacterial Plasmid Copy Numbers*

An assay for quantitation of plasmid copy numbers in bacterial cell cultures has been developed and validated. The method combines isolation of total bacterial DNA (including both plasmid and genomic DNA), running a series of two-fold dilutions of total DNA in an agarose gel followed by ethidium bromide staining, and subsequent scanning of the gel picture negatives. We have developed a novel set of rules for integration of the scan data that allows us to achieve high assay precision, accuracy, and sensitivity. The assay validation results were as follows: intra-and interassay precision with %CV of 8.2—9.9 and 7.1—9.8, respectively; ruggedness with %CV of 9.3—17.5; spike recovery of 80—102; and sensitivity of 1 plasmid copy per genome.

Key Words: plasmid copy number; bacteria; DNA; agarose gel; quantitation.

Recombinant proteins produced in bacteria are commonly used as therapeutics and vaccines. Heterologous genes, encoding for recombinant proteins, are generally introduced and maintained in bacterial cells as part of the recombinant plasmid, or expression vector. Accordingly, the general strategy for achieving high yield of recombinant proteins is to select for bacterial strains, bacterial vectors, and growth conditions that confer high plasmid copy number[1,2]. In reality, however, many proteins appear to be toxic to the host bacteria when overproduced. Several studies have shown that a significant increase in expression vector copy number may be associated with decline in the growth rate[3-7], as well as decline in the efficiency of expression of cloned heterologous genes[5-8]. To achieve a high yield of each particular recombinant protein, an optimal copy number must be found (for review see [9,10]).

The quality of cell banks for recombinant protein production is also critical. A common goal of process development in biopharmaceutical production is to establish and document the cell line stability. One of the approaches to genetic stability testing of the cell banks is the

---

* Elena A. Pushnova, Michael Geier, and Yu Sheng Zhu
Chiron Corporation, 4560 Horton Street, Emeryville, California 94608-2916
Received February 22, 2000
此文发表于2000年"Anal. Biochem." 284:70-76.

plasmid copy number assay.

For these reasons, it is of imperative importance to develop reliable and accurate methods for quantitation of plasmid copy numbers. Here we present a plasmid copy number quantitation assay that is an accurate, sensitive, and reproducible assay, and is also easy to perform. The assay can be used for estimation of the plasmid copy numbers in bacterial strains under different growth conditions, and at the different stages of the cell culture growth. The assay is also suitable for detecting any changes in plasmid copy number during the storage of the cell culture.

Materials and Methods

Reagents

TE (pH 8.0) and TBE buffers as well as 6× gel loading dye III (with bromophenol blue only, no xylene cyanol) were prepared as described[11]. Lysozyme powder, proteinase K solution, and RNase, DNase-free solution, were obtained from Boehringer Mannheim. λ DNN/HindIII molecular weight marker (MWM) was purchased from Gibco BRL. Agarose SeaKem GOLD (FMC) was used for electrophoresis. Note that it is critical to use this particular brand of agarose to achieve sufficient separation of genomic and plasmid DNAs.

Strains and Media

Cell banks of *Escherichia coli* transformants A and B, as well as fermentation cell line F, carried plasmid pIL-2 (4.56 kb in size) with integrated IL-2 (interleukin-2) coding sequence. R2Bl growth medium contained 20 g tryptone, 10 g yeast extract, 10 g NaCl, and 1 mg vitamin B_1 per 1 liter medium.

General Description of the Procedure

The sequence of steps comprising the plasmid copy number assay procedure is summarized in Figure 1.

Cell Preparation

E. coli host cells were cultured by inoculating 0.5 ml of overnight-grown fresh cells, or cells freshly thawed from a frozen cell bank, into 25 ml of R2Bl medium. The cells were allowed to grow in an incubator shaker (180 rpm) at 37℃ overnight (18 to 24 h) to the stationary phase. The optical density (OD_{680}) of the culture was measured at harvest (measurement at 1∶50 dilution). The culture volume corresponding to a total 4 OD_{680}, equal to approximately 1.5×10^9 cells (e.g., 400 μl of cell culture at 10 OD_{680}/ml), was transferred to a microcentrifuge tube and spun briefiy (30 s at 10,000 rpm, 4℃). The supernatant was discarded, and the resulting cell pellet was used for bacterial DNA isolation.

Grow and harvest *E. coli*

↓

Lyse the cells with lysozyme

↓

Digest proteins with proteinase K

↓

Digest RNA with RNase

↓

2-fold sequential dilutions of resulting total DNA prep

↓

Run agarose gel

↓

Scan the negatives of gel pictures

↓

Determine gDNA and pDNA peak areas

↓

$$\text{PCN per genome} = \frac{\text{Size of chromosomal DNA (bp)} \times \text{Amount of plasmid DNA}}{\text{Size of plasmid DNA (bp)} \times \text{Amount of genomic DNA}}$$

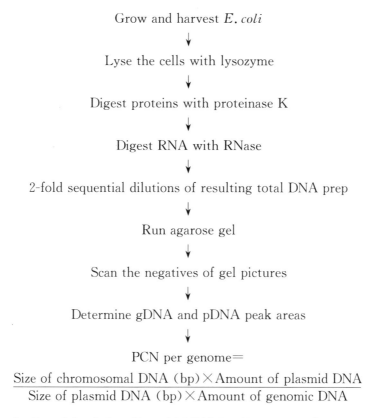

Figure 1 General description of bacterial PCN (plasmid copy number) assay procedure

If needed, bacterial cell pellets or cell cultures can be kept at −20 to −80℃ for at least 1 year prior to DNA isolation (frozen cultures should be thawed, and then spun as described above, before further processing). Alternatively, a short-term storage of freshly harvested bacterial cell culture is possible for at least 5 days at 4℃. We have found that storing the cells as mentioned above will not affect the final plasmid copy number.

Bacterial DNA Isolation and Purification

Bacterial cell pellets equal to 4 OD_{680} were resuspended in 400 μl of 50 mmol/L Tris/50 mmol/L EDTA, pH 8.0, by gentle vortex. The cells were lysed by the addition of 8 μl of 50 mg/ml lysozyme solution in TE, with subsequent incubation at 37℃ for 30 min. Following cell lysis, 4 μl of 10% SDS and 8 μl of 14—15 mg/ml proteinase K solution were added to each tube, and the cellular proteins were digested at 50℃ for 30 min. Proteinase K was then heat-inactivated at 75℃ for 10 min. Next, RNA was removed by the addition of 2 μl of 10 mg/ml RNase solution, followed by incubation at 37℃ for 30 min. DNA was subsequently extracted by adding 425 μl of phenol: chloroform: isoamyl alcohol (25 : 24 : 1), vortexing the tube vigorously and intermittently for 20 s, and sedimenting the tube for 5 min at 10,000 rpm (10,000 g), 4℃. The upper aqueous phase was carefully transferred to a fresh microcentrifuge tube using a wide-opening pipet tip (a regular tip cut about 0.5 cm from the end with a clean razor blade). DNA was further extracted by adding 400 μl chloroform, vortexing the tube vigorously and intermittently for 20 s, and sedimenting the tube for 5 min at 10,000 rpm (10,000 g), 4℃. The upper aqueous phase was carefully transferred to a fresh microcentrifuge tube using a wide-opening pipet tip. The contents of the tube were subjected to final sedimentation for 5 min at 10,000 rpm (10,000 g), 4℃, and the resulting clear supernatant, containing purified total bacterial DNA (genomic DNA, or gDNA, and plasmid DNA, or pDNA), was transferred to a clean tube. Note that thorough vortexing and use of wide-opening pipet tips is critical during the phenol/chloroform extraction steps to obtain a condensed interface, and to avoid contaminating the upper viscous aqueous phase with denatured proteins and polysaccharide from the interface. If the upper aqueous phase is still too viscous for pipetting, repeat the vortexing and spinning steps in the phenol/chloroform extraction.

Agarose Gel Electrophoresis and Scanning of Negative Film

Serial twofold dilutions of bacterial DNA sample (2,4,8,16,32,64,128, and 256-fold) were made with TE buffer. Ten microliters of undiluted and each diluted sample was mixed with 2 μl of 6× gel loading dye and loaded into the wells of 0.8% 0.5× TBE agarose gel. Two hundred nanograms of λ DNA/HindIH digest was used as a molecular weight marker. The gel was run in 0.5× TBE buffer using Bio-Rad WIDE MINI-SUB CELL at 100 V until the bromophenol blue reached approximately 2 cm from the bottom of the gel (about 1 h). The gel was stained with ethidium bromide (0.5 μg/ml) for 20 min and destained in water for 10 min. The photographs of the ethidium bromide-stained gels were taken under UV light with Polaroid positive/neg ative film No. 55 (exposure time is 1 min at F 4.5 with a yellow filter). The negatives were developed by immersing in 18% sodium sulfite solution. Four bands could be detected on the gel (see Figure 2): genomic DNA (gDNA, ~23 kb

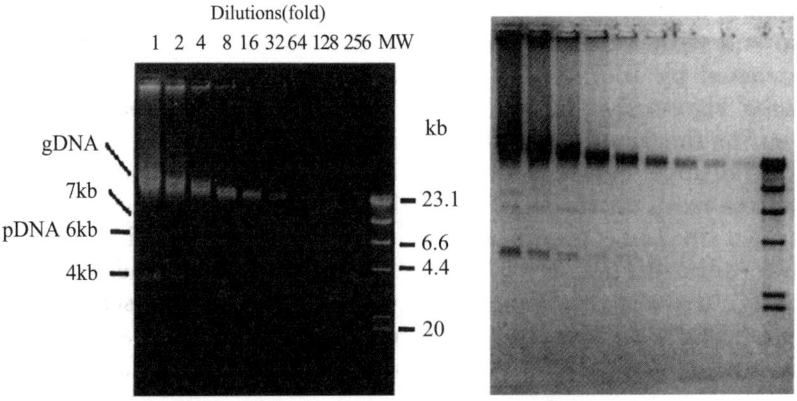

(a) (b)

Figure 2 Serial twofold dilutions of total bacterial DNA sample run in 0.8% agarose gel

(a) Ethidium bromide-stained gel: MW is molecular weight marker (λ DNA/HindIII digest); (b) Gel negative

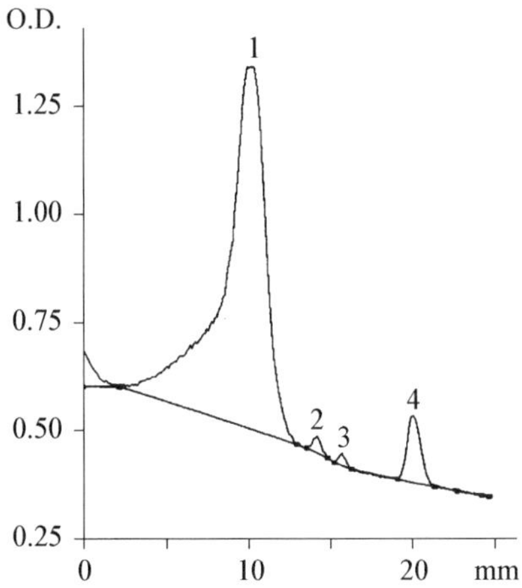

Area Quantitation Report

Peak#	Area
1	25.253
2	0.238
3	0.142
4	1.476

Figure 3 Typical negative scan result representing one agarose gel lane: peak 1 is gDNA; peak 2 is pDNA1; peak 3 is pDNA2; peak 4 is pDNA3

mobility); plasmid form with ~7 kb mobility (pDNA1, linear form); plasmid form with ~ 6 kb mobility (pDNA2, nicked circular form); and plasmid form with ~ 4 kb mobility (pDNA3, supercoiled form). The negatives were scanned using the Personal Densitometer SI (Molecular Dynamics). The baseline was selected using the All Valleys Method according to the Gel Scanner User's Guide, i. e. , by drawing a straight line between the lowest points of the left and right peak shoulders (see Figure 3). It is important to remember that using an incorrect baseline may significantly affect the resulting peak area and, consequently, the estimated plasmid copy number.

Scanner and image analysis software that can be used in this plasmid copy number assay is not limited to the Personal Densitometer SI (Molecular Dynamics). One may use, for example, those available from Bio-Rad or Alpha Innotech. In our experiments, scanning the same negative with three different scanners yielded the same plasmid copy number. With software other than Molecular Dynamics, however, some adjustments in the Data Used values range might be necessary, i. e. , the selected range might be different from 1. 00 to 6. 00 for genomic DNA, and 0. 10 to 1. 00 for plasmid DNA (see data reduction rules below).

Data Reduction

After scanning, the plasmid and genomic DNA peak area values were entered into the plasmid copy number calculation form (we find a Microsoft Excel table format to be the most convenient). See Table 1 for a representative example. The genomic DNA peak area for undiluted sample should be >20. If it is <20, the amount of DNA is not enough for quantitation, and the assay should be repeated with a double amount of cells. Adjusted genomic and plasmid DNA peak areas (@dil 1 line) were calculated by multiplying the peak area values by the corresponding dilution factors for all dilutions. To determine the linear range for the plasmid and genomic DNA peak areas at twofold sample dilutions, the percentage difference (% dif) between the adjusted value for each dilution and the adjusted value for each previous dilution (e. g. , 16→8, 32→16, 64→32) was calculated using the formula:

$$\% \text{ dif} = \frac{\text{Adjusted peak area-Adjusted previous peak area}}{\text{Adjusted previous peak area}} \times 100\%.$$

The linear range was considered to be the range of values that differed by less than 20%.

Table 1 Representative example of plasmid copy number calculation form with integrated scan data

Sample ID: F									
Dilutions:	1	2	4	8	16	32	64	128	256
gDNAarea	31.62	27.71	21.35	14.85	8.55	5.69	2.91	1.46	0.79
g@dill	32	55	85	119	137	182	186	187	202
% dif		75.3	54.1	39.1	15.2	33.1	2.3	0.3	8.2
Data used						182	186	187	
Average	185.07								
pDNA area 1	1.16	0.60	0.38	0.18	0.09				
p1@dill	1.16	1.20	1.52	1.44	1.44				
% dif		3.4	26.7	−5.3	0.0				
Data used		1.52	1.44						
Average	1.48								
pDNA area 2	1.17	0.86	0.60	0.31	0.17	0.09			
p2@dill	1.17	1.72	2.40	2.48	2.72	2.88			
% dif		47.0	39.5	3.3	9.7	5.9			
Data used				2.40	2.48	2.72			
Average	2.53								
pDNA area 3	4.61	2.89	1.83	0.95	0.51	0.24	0.14		
p3@dill	4.61	5.78	7.32	7.60	8.16	7.68	8.96		
% dif		25.4	26.6	3.8	7.4	−5.9	16.7		
Data used				7.60	8.16	7.68	8.96		
Average	8.10								
Total gDNA	185.07								
Total pDNA	12.11								
PCN	65								

The adjusted values in the linear range to be used for copy number calculation were entered into the Data Used line. For genomic DNA, the selected Data Used values in the linear range were limited only to the values originating from unadjusted peak areas in the range between 1.00 to 6.00. For plasmid DNA, the selected Data Used values in the linear range were limited only to the values originating from unadjusted peak areas in the range between 0.10 and 1.00. If only one peak value for plasmid DNA was detected, the number was used as it was. If no linear range could be determined, or more than one linear range was found, the solution in each particular case would be as follows: ① If no linear range for pDNA1 and pDNA2(~7 and ~6 kb pDNA) bands was determined, all the adjusted values for unadjusted peak areas in the range between 0.10 and 1.00 were taken into calculation. ② The absence of linear range for genomic gDNA or ~4 kb plasmid pDNA3 bands would indicate that an operational error must have occurred either during sample dilution or loading into the gel. Since the gDNA and pDNA3 peak area values are the most critical for the final result, in the latter case the sample must have been rediluted, and the assay repeated starting from the agarose gel electrophoresis step. ③ If the linear range of gDNA could not be found only because the sample DNA was too high, then further dilution (>256 - fold) was the solution. ④ If more than one linear range was found for pDNA bands, then the values in all the linear ranges, as well as the values between them, were entered into Data Used line. ⑤ If there was only one Data Used point in the linear range for genomic DNA, then %dif with respect to the previous value must have been between −10% and +20%. If it was not between-10% and +20%, the sample DNA was rediluted, and the assay was repeated starting from the agarose gel electrophoresis step.

The values from the Data Used lines were averaged. The average value for genomic DNA was entered into the Total gDNA line. The sum of average values for plasmid DNA bands was entered into the Total pDNA line. Finally, the plasmid copy number was calculated using the following formula:

Plasmid copy number (PCN) per genome

$$=\frac{\text{Size of chromosomal DNA (bp)} \times \text{Total pDNA}}{\text{Size of plasmid DNA (bp)} \times \text{Total gDNA}}$$

For example,

IL-2 plasmid copy number (PCN) per genome

$$=\frac{4.75\times10^6\times\text{Total pDNA}}{4,560\times\text{Total gDNA}}\cong\frac{1,000\times\text{Total pDNA}}{\text{Total gDNA}}$$

Results

Assay Precision

To determine the intraassay precision, bacterial transformant lots A and B were run through the whole procedure four times in one assay. The resulting plasmid copy numbers, as well as the standard deviation (STDEV) and the percentage coefficient of variation (% CV) of the replicates, as shown in Table 2. For determination of the interassay precision, the same samples were run through the whole procedure in four separate assays, and the resulting plasmid copy numbers were compared (see Table 3). The results show that the % CV in the intraassay precision test corresponded to 8.2%—9.9%, while the %CV in the interassay precision experiment ranged from 7.1%—9.8%.

Table 2 Intraassay precision of the plasmid copy number assay

Sample ID	PCN values	Mean PCN	STDEV	%CV
A	10 11 10 9	10	0.816	8.2
B	26 31 29 25	28	2.754	9.9

Table 3 Interassay precision of the plasmid copy number assay

Sample ID	PCN values	Mean PCN	STDEV	%CV
A	10 11 9 9	10	0.957	9.8
B	27 31 27 27	28	2,000	7.1

Accuracy/Recovery

Following the procedure, either 10 or 50 ng of purified pIL-2 plasmid DNA was spiked per 10 μl into cell suspensions of lots A and B either prior to DNA isolation (front spike), or after DNA isolation, prior to agarose gel electrophoresis (back spike). The unspiked bacterial DNA sample as well as 50 or 10 ng of pIL-2 plasmid DNA was run in the agarose gel along with the spiked sample DNA. The negative of the gel was scanned. The spike recovery for each sample dilution was calculated using the formula:

% spike recovery

$$=\frac{\text{Total pDNA area in spiked sample} - \text{Total pDNA area in unspiked sample}}{\text{Total pIL-2 pDNA area}}\times 100\%.$$

The results are summarized in Table 4. Values for each sample were obtained from the means of spike recovery values at dilutions 1, 2, 4, and 8 (see Materials and Methods). Percentage recovery values for sample lot A before and after DNA isolation were 88%±1% and 98%±4%, respectively. Percentage recovery values for sample lot B before and after DNA isolation were 90.5%±7.5% and 86.5%±6.5%, respectively.

Table 4 Accuracy/Spike recovery in plasmid copy number assay

Sample ID	pIL-2 Spiked	Spike amount/ (ng/10 μl)	Spike recovery/%
A	before DNA isolation	10	89
	after DNA isolation	10	102
	before DNA isolation	50	87
	after DNA isolation	50	94
B	before DNA isolation	10	98
	after DNA isolation	10	80
	before DNA isolation	50	83
	after DNA isolation	50	93

Assay Sensitivity

The sensitivity of the assay was determined as the minimum quantifiable plasmid copy number (PCN)①. The minimum quantifiable PCN was calculated as the sample PCN divided by the sample dilution factor at which the minimum plasmid DNA area value at the linear range (mimimum quantifiable pDNA peak area value) was detected for this sample. The quantitation limit of peak area for plasmid DNA is 0.1 to 1.0 (see Materials and Methods). The results shown in the Table 5 are based on five replicate experiments with sample lot A, and four replicate experiments for sample lot B. The standard deviation and the percentage coefficient of variation for the minimum quantifiable peak area values are included. The assay sensitivity for both samples tested was estimated as one pIL-2 plasmid copy per genome.

Table 5 Sensitivity of plasmid copy number assay

Sample	A	B	F
N replicates	5	4	1
Sample dilution factor at which the minimal quantifiable pDNA peak area value was			
detected	8	32	64
Sample PCN	10	28	65
Assay sensitivity (minimum quantifiable PCN) = Sample PCN/Sample dilution factor	1	1	1

① Abbreviation used: PCN, plasmid copy number.

Assay Ruggedness

For determination of assay ruggedness, bacterial transformant lots A and B were run through the whole procedure by four different people in two different labs at different times. The resulting plasmid copy numbers, as well as the standard deviation (STDEV) and the percentage coefficient of variation (%CV), as shown in Table 6. The resulting %CV was 9.3% to 17.5%.

Table 6 Ruggedness of plasmid copy number assay

Sample ID	PCN values	Mean PCN	STDEV	%CV
A	9 8 10 12	10	1.708	17.5
B	28 27 28 33	29	2.708	9.3

Discussion

The plasmid copy number assay provides important information on gene dosage, gene regulation, and protein expression. Quantitation of the plasmid copy number in recombinant protein production strains is also required by drug regulatory agencies. Current plasmid copy number assays described in the literature are mainly divided into nucleic acid hybridization-based assays and agarose gel electrophoresis-based assays.

Nucleic acid hybridization-based assays include Southern blot hybridization[3,12,13], and the Threshold sequence-specific assay[14]. In Southern blot hybridization the use of radioactively labeled probes[12] represents a safety problem. Some authors have employed nonradioactive labels, such as digoxigenin[3]. Generally, the Southern blot hybridization procedure includes digestion of plasmid DNA with a restriction enzyme, separation of plasmid fragment and genomic DNA by agarose gel electrophoresis, and blot DNA transfer to the membrane, followed by hybridization with labeled specific probes. The procedure is evidently laborious and time-consuming. Furthermore, it is a semiquantitative assay that has poor reproducibility. Some investigators have recently reported a plasmid copy number determination method using the Threshold sequence-specific assay[14]. It was reported as a sensitive, specific, and accurate assay, but the assay is complicated and also very costly.

The agarose gel electrophoresis-based assays include purification of either total bacterial (genomic and plasmid) DNA[4,15,16] or plasmid DNA only[7,17,18], followed by agarose gel electrophoresis and quantitation of DNA by densitometry. If the genomic DNA was not included in the assay as an internal DNA recovery control, the authors had to mix the standard transformant cells into the experimental cell culture to monitor the recovery of plasmid DNA[17,18]. This makes the assay more complicated. In cases where both genomic and plasmid DNAs were taken into analysis, the authors have run the agarose gel electrophoresis of undiluted sample only, followed by scanning either of negatives of the gel photographs[4] or of ethidium bromide-stained gels directly[15]. Sometimes an additional 1 : 10 dilution of the sample was analyzed to account for the difference in chromosomal and plasmid DNA concentrations[16]. However, our experience shows that the linear ranges for genomic and plasmid DNA peak areas lie within different dilution factors of the samples. Moreover, the linear range dilution factor for plasmid DNA varies depending on the particular plasmid form (linear, nicked circular, or supercoiled) and the plasmid copy number of the given sample. As a rule, in a particular sample dilution the genomic DNA band is more saturated than are the plasmid DNA bands. Therefore, the plasmid copy number is well overestimated if the genomic and plasmid DNAs are quantitated at the same dilution number.

To overcome the described problems we have developed an easy and simple plasmid copy number assay. This assay combines isolation of total bacterial DNA, separation ofgenomic and plasmid DNA by agarose gel electrophoresis, and calculation of the plasmid copy number in a linear range after scanning the negative of an ethidium bromide-stained gel. The validation results have demonstrated high precision, accuracy, and sensitivity of the assay. The unique feature of our assay, compared to similar published approaches, is the quantitation of plasmid and genomic DNAs in linear range of their respective peak areas. This is achieved by running a series of twofold dilutions of total DNA samples, and applying strict data reduction rules to select for the reliable peak area values in the linear range, which are subsequently used for calculation of the final plasmid copy number.

Another advantage of this assay is simultaneous isolation, separation, and quantitation of genomic and plasmid DNA. The plasmid copy number is calculated based on the ratio and sizes of genomic DNA and plasmid DNA. Here, genomic DNA is used as an internal control. This avoids an error caused by efficiency in cell lysis and increases the assay accuracy. Since the assay is based on the ratio of genomic DNA and plasmid DNA, it also greatly increases the assay sensitivity. The high sensitivity of a single copy of plasmid per genome was achieved not because of the sensitivity of detection by ethidium bromide staining, but rather because more cells could be used in the assay.

In conclusion, we believe that our easy and accurate assay for quantitation of bacterial plasmid copy numbers can be successfully applied to any other organism, with respective modifications in total DNA isolation procedure and data reduction rules.

Acknowledgments

We thank Carmen Arnold for cell culture preparation, Becky Elliott and Biqi Gao for their assistance in the ruggedness study, and Titania Buchholdt for help in preparation of the manuscript.

References

[1] Shin, C. S., Hong, M. S., Bae, C. S., and Lee, J. (1997) Enhanced production of human mini-proinsulin in fed-batch cultures at high cell density of *Escherichia coli* BL21 (DE3)[pET-3aT2M2]. Biotechnol. Prog. 13,249 - 257.

[2] Morino, T., Morita, M., Seya, K., Sukenaga, Y., Kato, K., and Nakamura, T. (1988) Construction of a runaway vector and its use for a high-level expression of a cloned human superoxide dismutase gene. Appl. Microbiol. Biotechnol. 28,170 - 175.

[3] Jones, K. L., and Keasling, J. D. (1998) Construction and characterization of F plasmid-based expression vectors. Biotechnol. Bioeng. 59,659 - 665.

[4] Birnbaum, S., and Bailey, J. E. (1991) Plasmid presence changes the relative levels of many host cell proteins and ribosome components in recombinant *Escherichia coli*. Biotechnol. Bioeng. 37,736 - 745.

[5] Carrier, T., Jones, K. L., and Keasling, J. D. (1998) mRNA stability and plasmid copy number effects on gene expression from an inducible promoter system. Biotechnol. Bioeng. 59,666 - 672.

[6] Bailey, J. E., Da Silva, N. A., Peretti, S. W., Seo, J. H., and Srienc, F. (1986) Studies of host-plasmid interactions in recombinant microorganisms. Ann. NY Acad. Sci. 469,194 - 211.

[7] Seo, J. H., and Bailey, J. E. (1985) Effects of recombinant plasmid content on growth properties and cloned gene product for mation in *Escherichia coli*. Biotechnol. Bioeng. 27,1668 - 1674.

[8] Peretti, S. W., Bailey, J. E., and Lee, J. J. (1989) Transcription from plasmid genes, macromolecular stability, and cell-specific productivity in *Escherichia coli* carrying copy number mutant plasmids. Biotechnol. Bioeng. 34,902 - 908.

[9] Nordstrom, K., and Uhlin, B. E. (1992) Runaway-replication plasmids as tools to produce large quantities of proteins from cloned genes in bacteria. Biotechnology 10,661 - 666.

[10] Makrides, S. C. (1996) Strategies for achieving high-level expression of genes in Escherichia coli. Microbiol. Rev. 60,512 - 538.

[11] Sambrook, J., Fritsch, E. F., and Maniatis, T. (1989) Molecular Cloning: A Laboratory Manual, 2nd ed., pp. B. 20-B. 24, Cold Spring Harbor Laboratory, Cold Spring Harbor, NY.

[12] Adams, C. W., and Hatfield, G. W. (1984) Effects of promoter strengths and growth conditions on copy number of transcription-fusion vectors. J. Biol. Chem. 259,7399 - 7403.

[13] Schendel, F. J., Baude, E. J., and Flickinger, M. C. (1989) Determination of protein expression and plasmid copy number from cloned genes in *Escherichia coli* by flow injection analysis using an enzyme indicator vector. Biotechnol. Bioeng. 34,1023 - 1036.

[14] Mermelstein, L. (1998) Plasmid copy number determination in *Escherichia coli* cultures using the Threshold sequence-specific assay. In Fifth Threshold Users' Group Meeting Summary, October 15th, Florence, Italy.

[15] Projan, S. J., Carleton, S., and Novick, R. P. (1983) Determination of plasmid copy number by fluorescence densitometry. Plasmid 9,182 - 190.

[16] Nugent, M. E., Primrose, S. B., and Tacon, W. C. A. (1983) The stability of recombinant DNA. In

Developments in Industrial Microbiology (Proceedings of the Thirty-Ninth General Meet ing of the Society of Industrial Microbiology), Vol. 24, pp. 271 - 285.

[17] Moser, D. R., and Campbell, J. L. (1983) Characterization and complementation of pMBl copy number mutant: Effect of RNA I gene dosage on plasmid copy number and incompatability. J. Bacteriol. 154, 809 - 818.

[18] Uga, H., Matsunaga, F., and Wada, C. (1999) Regulation of DNA replication by iterons: An interaction between the ori2 and incC regions mediated by RepE-bound iterons inhibits DNA replication of mini-F plasmid in *Escherichia coli*. EMBO J. 18, 3856 - 3867.

第七篇

科 普 文 章

植物的电世界*

为什么向日葵的花朵会随着阳光转？为什么含羞草的叶子经不得轻轻一碰，便会“低眉垂首”地害起羞来？

植物电的形形色色

早在18世纪，科学家第一次在电鳗中发现了生物电。以后发现，生物电在各种植物中也普遍存在，尤其是一些所谓敏感植物如含羞草、捕蝇草等。

以含羞草为例，只要轻轻碰触叶片，它便马上折合拢来。折合动作一路传递下去，最后连叶柄也垂了下来。含羞草的运动器官是生在主叶柄和茎杆连接处的叶褥。它离开叶片有相当长一段距离，当叶片被触碰的时候，叶褥是怎样“发觉”的呢？

科学家做了这样的实验：用两个电极接通叶片和叶褥，当叶片受到刺激时，电流计上马上显示有电流经过，叶柄也接着垂了下来。原来是植物电流当了通讯员。向日葵的向阳运动，有人认为也是生物电的作用。在阳光下，向日葵茎端的向阳面和背阳面带有不同电荷，向阳面带负电，背阳面带正电；而茎生长点所产生的生长素在溶液中是带负电的，它移向带正电的背阳面，加速了那部分细胞的生长，结果茎便弯向阳光了。有一种捕蝇草，从昆虫停留在叶面上时所产生的电流中获得信号，竟然还能捕蝇除害哩。

植物电的发生和传导

据研究，细胞内原生质对某些离子的结合能力很强，因此在细胞膜内外造成离子不平衡的分布。有一种轮藻（淡水藻类），节间细胞极大，长数厘米以上，宽也有几毫米，是研究植物电的好材料。在轮藻细胞中央的液泡中，钾离子的浓度很大（是细胞周围溶液中的1 065倍），它们拼命朝细胞外面挤；可是，由于细胞内蛋白质对钾离子有吸引力，钾离子只能挤到细胞的表面，便停止不前。这样，细胞表面便积聚了相当多带正电的钾离子，因而带正电，细胞内则相对地带负电。若把一根电极放在细胞表面，另一根插到细胞里面，电流计的指针就会偏转（见图1）。这就是所谓休止电位，是细胞内外离子浓度的不同所造成，在轮藻上可以达到100～200毫伏。

研究证明，植物在受到震击、刺伤、切割，各种辐射、光、电流、磁场、离心力、重力、热、盐溶液、毒物等等的刺激后，便会发生兴奋。这种兴奋有时以电波的形式传导，并且对外界产生相应的反应。

---

* 朱雨生

此文发表于1961年“科学画报”7期208－210页。

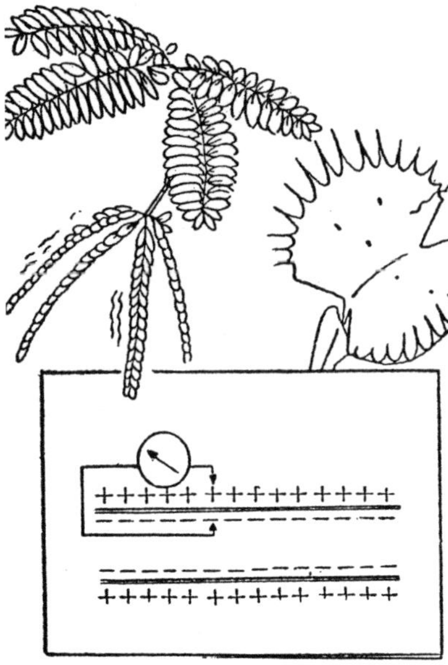

图 1　植物的休止电位

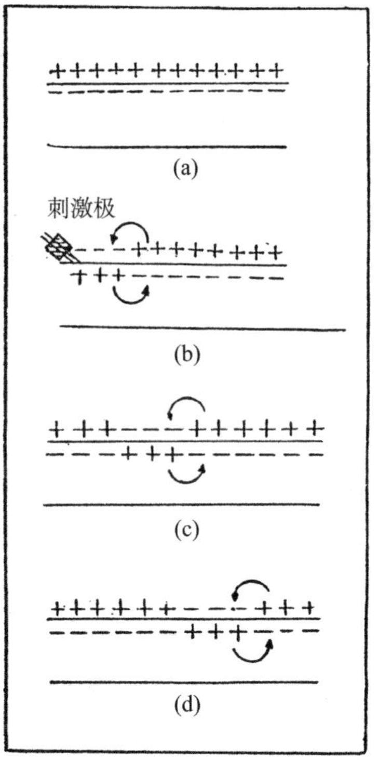

图 2　动作电流的产生

(a) 没有受刺激的细胞膜　(b)、(c)、(d) 受刺激后电流传布的过程

图注：箭头所示为局部电流产生之处。

动作电流的产生，是由于受到刺激处的细胞膜发生了化学变化，使对离子的透性起了变化。据测定，轮藻细胞在受到电极刺激时，细胞膜的电阻降低至平时的 1/200，因而造成细胞膜两边电荷相反的分布：膜外带负电，膜内带正电。这样，它同邻近的部位间便产生了局部电流，使那里的电荷分布也发生同样的变化，而原来刺激点的电荷随即又恢复了原状。电流便如此相继传导下去(见图 2)。这种电流能从细胞的一头传到另一头，但也可能中途便消失。这就是在含羞草中当"通讯员"的电流。葫芦科、葡萄科等攀缘植物的茎、卷须以及其他植物中，都能找到这种电流。但在不同植物中，它的传导速度是不同的：在葡萄和轮藻中，大约是每秒钟 1 厘米；而在含羞草中则每秒钟可达 30 厘米。

植物电和植物的生命活动

产生和维持植物电的必要条件是能量供应；到今天为止，研究者还没能从一株死的植物或一块死的组织中找到植物电。

如将一株植物放在抽真空的密闭罩内，把一个电极放在叶片上，另一个电极放在根的基部，接好线路，这时所测得的电位大约是 30 毫伏。然后抽气，当气压下降到 10～15 毫米水银柱高时，电位便消失了。这是因为，在近于真空的情况下，植物停止呼吸，能量的供应也就中断。就是在动作电流传导的路上，也必须有生理上的完整组织。假使跟它开个玩笑，中途用冰水或麻醉剂麻醉，或用火熨伤，便会“断电”。而那时只要用一张在导电溶液中浸过的滤纸，将“断路”的两端连接起来，动作电流便又能在纸桥上通过。

由此可见，植物电是与植物的生命活动紧密地联系在一起的。科学家正开始利用这种关系来解释植物的某些生命活动。

以前，人们是以细胞渗透压的差异、水分子的内聚力以及叶片蒸腾时向上的一股吸吮力等，来解释植物体内水份和矿物质的吸收和输送的。可是，这个理论解释不了为什么根部水分反而能从细胞汁浓度大的根细胞压向浓度小的导管；也不能解释为什么导管中的水是从下向上送，而筛管中的却能从上向下送。后来发现，根的表面和内部间有一电位差，根内部带负电，根外部带正电。根部原生质内有时存在着很大的电场，甚至达 500 伏特/厘米，而水分子在溶液中实际是以带正电的形式（H_3O^+）存在的，所以水以及溶解在水中的矿物质能被吸收到根的内部，压向中央的导管。又发现茎下部的电性要比上部的“负一些”，因此水分子在筛管中能从上向下运输；至于水在导管中的向上运输是由于受到了叶片蒸腾和根部压力所造成的更强的向上的力（见图 3）。

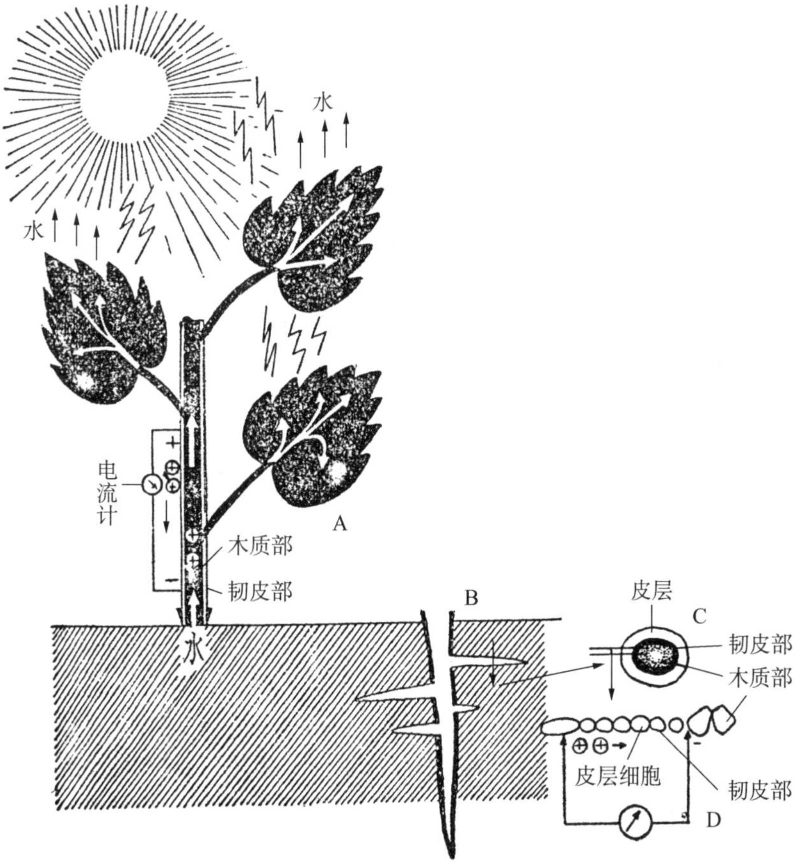

图 3　植物体内水分和矿物质的运输

A—在木质部向上输送，在韧皮部则向下输送；B—植物根；C—根的横切面；D—根部运输路线的放大图；⊕—代表带正电的水分子，沿着电场从根表面进入木质部绿色植物的光合作用过程，是在所谓叶绿素分子的参与下进行的。叶绿素分子好比生产的机器，整齐地“安装”在叶绿粒的厂房里，大约2,500个分子构成一组。每个叶绿素分子又可被看成一架巧妙的无线电接收器和发射器，能够接收太阳光的电磁波，并把它们变成波长长一些的电磁波，发射给邻近的分子。这样传递下去，直到最后一个分子。它利用了这种电磁波将原料水分解，加上碳酸气，经过十几套“把戏”便产生了葡萄糖。这里又可以看到电在起作用，电同电磁波连系在一起充作了生产葡萄糖的“燃料”。

植物体内还有许多生理过程如再生、愈伤以及生长发育等过程，也无不与植物电联系在一起。

植物电和植物的自动控制

不仅人和高等动物具有完善的自动控制系统，就是植物体，哪怕是海洋里的那种原始生物——藻类，也存在着类似的自动控制装置，只不过构造简单些罢了。

从自动技术的角度看来，植物体内至少存在着自动控制装置所必须具备的六个元件：①感受外界刺激并把它们转化成为兴奋冲动的接受器；②信号输入系统；③某种中央控制系统，它把外来的刺激信号加以改造，变成“命令”发布出去，筹划个体发育的过程；④信号输出系统；⑤效应器；⑥反馈装置。

目前对这些元件已有部分了解。如芽鞘、茎的顶端、根尖、表皮外壁的原生质突起、叶绿素分子的“头部”、核酸、蛋白质等大分子的某些基团等等，都可以派接受器用场；薄壁组织、韧皮部、胞间连丝、蛋白质周围的一层水分子等，可算是传导装置；含羞草和许多豆科植物的叶褥，向阳性运动中茎端的弯曲部分、细胞膜等，可以作为感应器；植物激素和其他一些生理上活跃的高分子以及植物电等，在整个自动控制系统中起了信号作用。

植物体是处在周围环境的包围中的。一天之中以及在它的整个生长阶段中，光照、温湿度、土壤中矿物质的种类和数量，甚至风霜雨雪等，无时无刻不在对植物体发生作用。即在白天和夜晚分别测得的植物体的电位，也各自不同。这些信号使植物体能随时适应外界环境的变化。植物体接受了这些信号并加以改造，作为自己生长发育的信号，也就是开花结果、生产粮食的信号。若我们一旦掌握了植物自动控制的结构和方法，听懂了这些“语言”，那么就可能加以控制和调节，生产更多更好的粮食、棉花……

电场中的植物

据研究，地球的表面带正电，而大气低层的云层带负电，因而造成了大气中一个方向的电场，加速了水份和矿物质从土壤向植物体的输送。不久前发现，在公路高压电线下面生长的植物，比一般的植物高大，产量也多些。原来，在高压电流周围形成了一个“人工”电场，加速了植物对带电的水份和矿物质的吸收。而若用一个接地的铜网使植物同外界电场隔离，则它们对于某些矿物质的吸收，就比不过那些生长于天然大气电荷影响下的同类伙伴们。

苏联科学家还做了这样的实验：用两个电极插在棉花田畦的两边，通电一定时间，以后棉花生长迅速，成熟早，产量增加了50%左右，而且纤维的质量也很好。如将棉花种子放在玻璃管中，两端通以电流，结果种子的发芽率提高，幼苗的生长速度也加快了。

把糖用甜菜和马铃薯栽培在埋有锌极和炭极的土壤中。通电一定时间后，在甜菜的试验部分，可使种子发芽率从11%提高到19%～20%，同时还增加了叶面积；块根产量增加了5%～40%，块根中含糖率也有提高的趋势，而且这些性状可以遗传给后代。在马铃薯的试验田中，发现它们的开花期提早了三到三天半，块根产量增加了16%～46%。

由于许多藻类，单细胞的低等植物、性细胞、游走子等，表面都带有电荷，因此在电场中能发生定向的移动，也就是趋电性。利用这一性质，我们可以像利用电流捕鱼一样，采集特殊的藻类或其他低等生物。如要得到纯粹的样品，可以利用它们表面所带电荷的性质和大小的不同，以及在不同强度电场下游动的趋向和速度的不同，在一个容器里制造一定强度的电场来将它们分开。

生物学中研究电磁场对植物的作用已成为一个新的方向。苏联生物学家亚历山大·克雷洛夫认为，电磁场在植物生命中的作用，比所有已知的物理和化学刺激剂的作用大得多。植物与磁场也有着异常奇妙的联系。

如果将两颗玉蜀黍种子放在黑暗中发芽，则不论胚根朝着哪一方向，新芽总伸向南方。这正是植物机体具有磁场和极性的明证。研究者还发现，植物机体的极性是不对称的，负极比正极强。这一点非常重要，它是植物发育的一个决定性条件；也只有在这样的条件下，生物化学变化才会引起生物电流。所以种子在磁场里的位置如果放得好，磁场就能加强它的负极，发芽就迅速而良好；反之，就会增强它的正极，种子发育便落后，甚至患病死亡。研究者进一步证明，某些生长激素所以能造成无子的番茄或早熟的果实，就是因为它们改变了果实细胞的极性，从而影响了进行正常发育所需的那些生物化学变化。

所有这些，植物的电现象和极性，将成为揭示光合作用、遗传性等秘密的钥匙，成为促进或控制植物的生长发育和提高作物产量的有力工具。

生物半导体*

用火烧一下金属棒的一头，另一头就热了。在一个电池和电灯构成的封闭电路中，把开关打开，灯就亮了。这些都是能量在导体中传导的例子。在生物体中，光合作用、神经传导、肌肉收缩、眼睛感光、蛋白质合成，甚至癌症的发生、遗传等都与能量传递有关。传导方式很多，有的靠光（主要是萤光和磷光），有的靠电子或电磁波。这里只来谈谈蛋白质、叶绿体是怎样传递能量的。

蛋白质是半导体吗

生物体的原生质、肌肉、酶、激素、抗体甚至我们的头发，都是由蛋白质组成的。可以这么说：没有蛋白质就没有生命。

蛋白质中能量是能够传导的。用 X 光从不同方向轰击从血液中提取的血清蛋白的不同部位，最后总是在一个一定的地方断成两半（见图 1(b)）。可见血清蛋白分子，能够把 X 光的能量，通过某种方式，沿着分子内部传到联系最薄弱的、隐蔽的地方，使那里断裂开来。

蛋白质是由许多小的单位——氨基酸组成的。机器的零件是靠螺丝联系起来的，蛋白质中的这些氨基酸手拉手地用所谓肽链连起来，构成一条长的骨架。这种骨架不是笔直的，由于骨架上每隔 3.7 个氨基酸可以靠氢原子和氧原子间的氢键联系，形成弹簧那样的螺旋结构（见图 1(a)）。这一点，我们不妨做一个实验：将头发的一端挂一重物吊起来，然后用热的水蒸汽

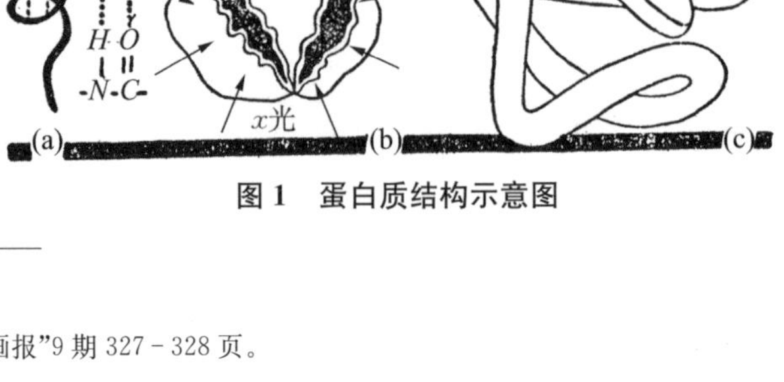

图 1　蛋白质结构示意图

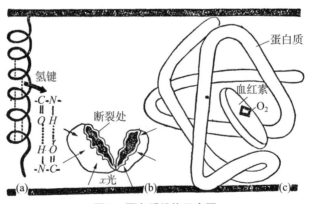

* 朱雨生
此文发表于 1962 年“科学画报”9 期 327 – 328 页。

去喷它，一会儿，头发蛋白中的氢键便断裂，“弹簧”变松，因而头发变得较原来长得多了。蛋白质的这种弹簧本身又是盘来扭去的，好似炸麻花。

蛋白质的结构有很高的规则性和重复性，正如某些晶体物质。用X光衍射方法已证实蛋白质具有晶体的性质。因此它也可能具有半导体的能带结构即满带、禁带和空带。如禁带较宽，满带中电子跳不过，便是绝缘体。如禁带较狭，满带中电子受光或热的激发便易跳到空带中而能导电。加入杂质可改变导电性质，成为空穴(P)型或电子(N)型半导体。

近年来科学家对几十种干蛋白质作了测定，发现它们的禁带宽度是在2.5～3.2电子伏特之间，须用很高的能量——相当于紫光和紫外光——才能激发，仅用简单的加热不能引起导电。所以说在一般情况下，纯碎的蛋白质是绝缘体，至多只能说是光导体。

但是蛋白质分子还可和其他成份结合起来，那么它可能是杂质半导体吗?

就血液中输送氧气的血红蛋白来说，它身上背着一只背包(见图1(c))，就是红色的血红素，装了许多很活跃的电子，很容易受光或热的激发而导电。测定的结果，血红蛋白的禁带变得窄一些了。

此外，还发现血红蛋白、白明胶和角蛋白，稍微吸附了一些水后，导电率迅速增加。这时水大概起了电子半导体的作用。又如将四氯苯醌加在血清蛋白中，导电率竟然增加了三万倍，大概它起了空穴半导体的作用。由此看来，蛋白质是杂质半导体，使它在能量传递中起重要的作用。

对蛋白质半导体性质的研究还是刚开始，实验结果也还有很多差异甚至是矛盾的。不过这项研究对生物化学、生物物理学和生物催化作用的探讨将有很大的帮助，这是一个崭新的科学方向。

绿色工厂内的半导体

植物绿色的秘密就是叶绿素，它们是安排在所谓叶绿体的绿色颗粒中。叶绿体是一座微妙的“工厂”，里面分成许多车间，包括从光能贮藏、燃料供应、原料送入直到葡萄糖制成和产品输出，相互密切配合。它能源源不断地从自然界取之不尽的廉价的水和二氧化碳中生产出昂贵的葡萄糖、淀粉来，把太阳的光能变成化学能贮藏起来。

有一位科学家做了这样一个实验：从新鲜的烟草、菠菜、甜菜或萝卜的叶子提取叶绿体，晾干后铺成一个薄层。他先用光照一下，然后放到暗处稍微加一下热，奇怪的事发生了，干的叶绿体和叶绿素制剂中发生了微弱的光线；假如不照光，在叶绿体制剂中放两个电极，然后对它们加热，导电率迅速地增加了。这正像半导体的现象。还有一位科学家，他把叶片放在专门用来检查电子讯号的顺磁共振仪旁边，用光照一下，马上在示波器的萤光屏上出现一个电信号，这是自由电子产生的标志。

现在，人们已经开始看清叶绿体工厂的秘密了。剖开这座工厂，我们可以看到里面还有一粒粒椭圆形的东西，称为“基粒”，周围是“间质”(见图2A)。间质主要是由蛋白质组成，里面有许多催化剂——酶，这里可算是生产葡萄糖的车间。基粒是供应燃料和还原剂的车间，是由一层脂肪、一层蛋白质相间排列而成层状的结构(见图2B)。图2的照片是玉蜀黍细胞中的一个叶绿体(电子显微镜放大40 000倍)，可看出其中很多片层结构的基粒。

基粒中有些脂肪层的二头连接起来，而形成一只“橡皮袋子”形状的构造，袋子的“橡皮”就

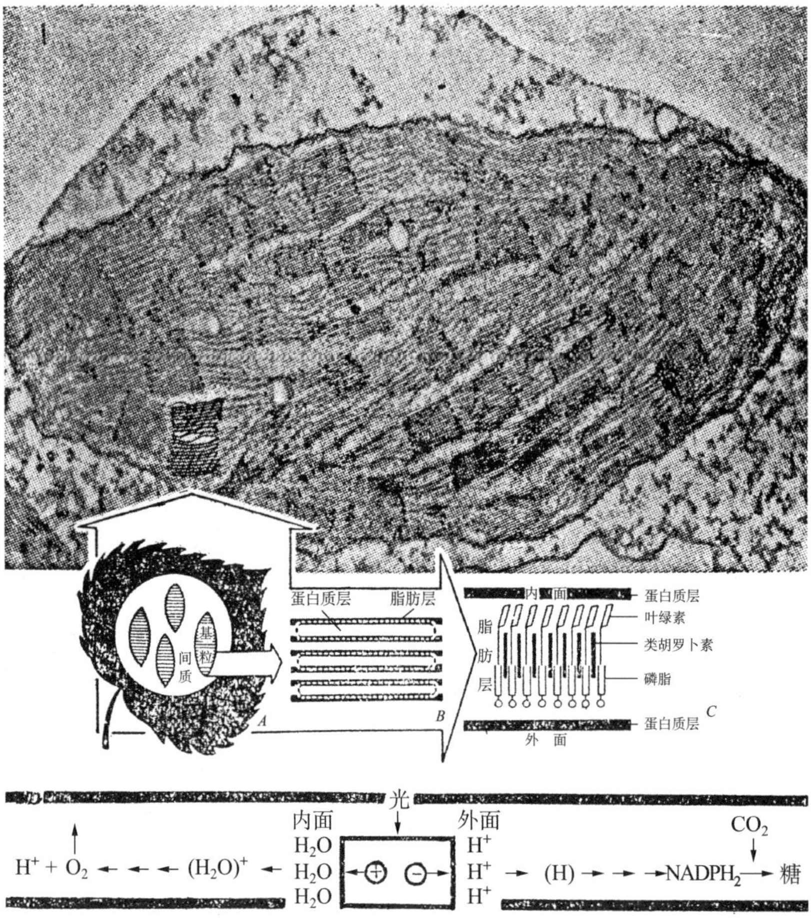

图 2 叶绿体半导体结构

是脂肪层，其中除了不能导电的脂肪外，还有能导电的导线——类胡萝卜素。袋里面和外面就是蛋白质。叶绿素是脂溶性的，溶解在袋子的“橡皮”中。这块橡皮是 p－n 型半导体(见图 2C)。

叶绿素头部也就是血红素的亲兄弟——卟啉环，这是一台最巧妙的电磁波接收器。太阳的电磁波照在它上面，里面的电子便恍了几下，然后丢出一个电子，留下一个空穴(带正电)，电子和空穴分别沿着类胡萝卜素的导线向袋子内外传导，袋外带正电的氢离子接受了电子变成中性的氢原子，再经过几步变化，依附于一个复杂的有机分子，形成一种还原剂($NADPH_2$)。而空穴呢？向袋子里面跑，使袋内的水变成带正电的水，再经过几套手续，就形成了带正电的氢离子，并放出氧气来(见图 2D)。

假如空穴和电子还没有传到两边，中途又结合时，就放出能量(以化学能形式)藏在生物体内的腺三磷(ATP)中。

基粒中这样产生的 ATP 和还原剂 $NADPH_2$，最后都运到生产葡萄糖的车间——间质中，加上 CO_2 作为原料，酶作催化剂，葡萄糖便合成了。

我们可以相信，随着对叶绿体半导体性质的研究，人工模仿光合工厂的任务，也将提到日程上来了。

生物的时辰节律——生物钟*

自然界的许多现象都是反复地、有节奏地进行着，如四季的更递，如昼夜的交替。一切有生之物的生命活动也各有一套与生俱来的自然节律。金鸡司晨，老鼠夜出。在百花盛开的夏日，你可以看到牵牛花破晓，太阳花当午，紫茉莉送夕阳，晚香玉迎素月。18 世纪著名的植物分类学家林奈会按开花的钟点顺序种了一些花，它们次第开来，宛如借自然之手拨弄的一只花钟（见图 1）。另外蜜蜂、鸟类等有跟随太阳飞行的本领。太阳的相对位置在一天之中是变化着的，它们跟随太阳飞行时必须斟酌太阳的相对位置，这说明它们对时间消逝的警觉。

生物体内的自然钟

是什么东西使生物灵敏地感到时光在流动呢？有两种假设：一是生物的时间感只不过是 24 小时中环境变化的反映；二是有某种内在的“生物钟”，它好象闹钟，在应该开始行动时发出指示，也可以随时查看时间。

为了找寻答案，200 多年前就有一位科学家将豆类植物放在与外界隔绝的黑暗山洞中，可是叶片的昼启夜合运动仍照常进行。近年来研究者又用蟑螂做实验，以看不见的红外线光束通过笼子悄悄地照射在光电管上，蟑螂行动时阻断了光束，从而自动做了长时间的行动记录。蝉螂在恒稳条件下仍能保持有节律的昼伏夜出的活动，周期是 23 小时 53 分。实验结果发现，生物各有不同，行为却很一致，它们虽处在温、湿度和气压不变的连续黑暗中，似仍能“看到”外界的日落日出（见图 2）。也就是说，通过内在的时钟而保持有节律的生命活动至数日、数星期甚或更久。一般时钟指针的走动，是表示钟内的机械在运转；生物有节律的行动，也正是内在时钟在运转的可见信号。通过实验，不仅证明生物钟的确存在，并发现它的周期相当稳，一般在 22～28 小时的范围内，同 24 小时的昼夜期虽稍有偏差，却相当“准”。

生物钟还可以对时，拨动和调整。通过一定时期的光、暗控制，我们可以将某一生物的时钟“指针”拨到任何一个位置上表示某一时刻。例如，白天含羞草复叶张开的时候，偏将它放在暗中；夜里复叶合拢时，偏给以光照。连续训练许多次以后，它便会日夜颠倒。但在正常的光暗交替恢复后，生命的节律同样也可以恢复。有些动物在经过连续的黑暗锻炼后，可能失去 24 小时节律的特性，如果蝇的孵化周期。可是，只要给以 1/2000 秒的闪光，它又可以恢复原来的节律。这种自动调整的本领是生物钟的特点之一。

一般说来，生物钟对温度变化不敏感。可是当温度下降到 0℃时，它也会骤然停摆；温度

---

* 雨生

此文发表于 1963 年“科学画报”5 期 190 - 195 页。

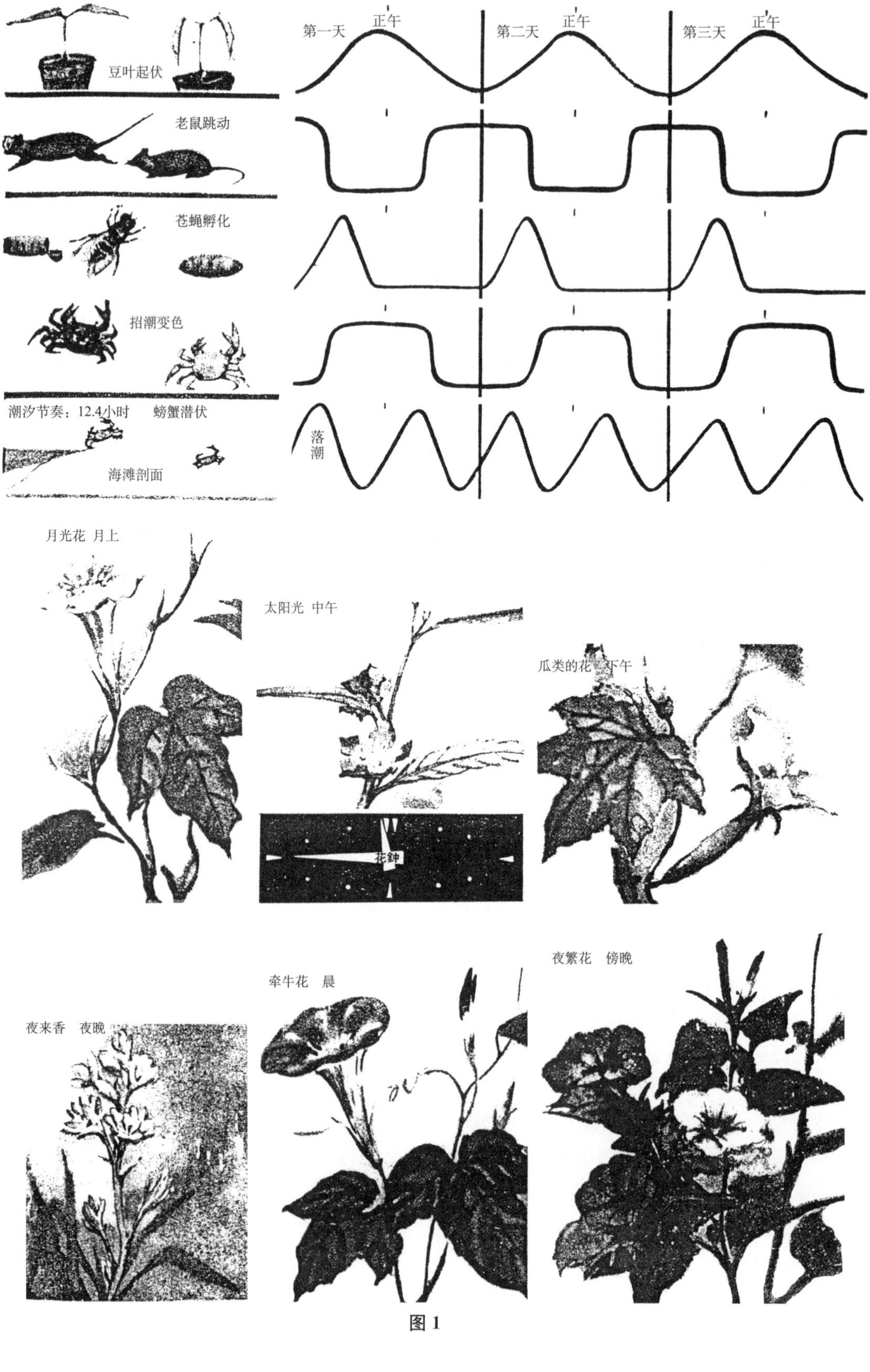

豆叶起伏
老鼠跳动
苍蝇孵化
招潮变色
潮汐节奏：12.4小时
螃蟹潜伏
海滩剖面
第一天
正午
第二天
正午
第三天
正午
落潮
月光花 月上
太阳光 中午
花钟
瓜类的花 下午
夜来香 夜晚
牵牛花 晨
夜繁花 傍晚

图1

升高时才又重新走动。当机械钟停走几小时后，如果在重新上发条时不加以调整拨快，它便会慢几小时；生物钟也如此，受冻几小时，它便也比正常的慢几小时。

生物钟的另一特点是可以遗传。据试验，果蝇在连续弱光下培育了15代，它从蛹中飞出来的昼夜节律不变。可见，外界因素虽可以拨动或中止钟的指针，但不能改变它的遗传特性。

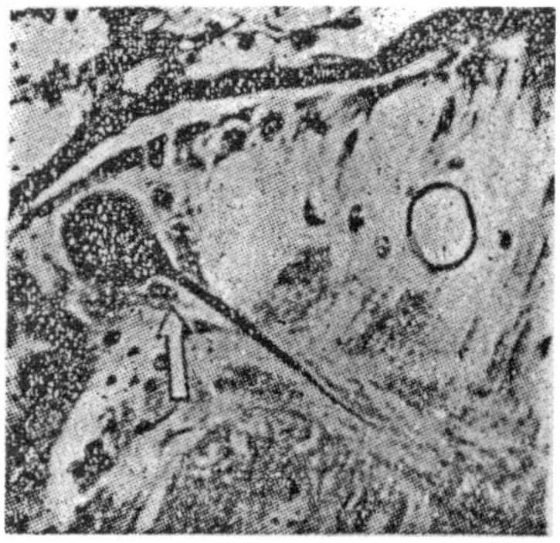

图2 蟑螂的大神经分泌细胞之一(已染色，箭头所指处)

生物钟的奥秘

钟到底在哪里？又是怎样调节的呢？动物的脑部是不是它认识时间机制的中心呢？经切除动物脑不同部位试验的结果，某些节律停止了，另一些则继续在进行。这样看来，生物体内可能不止存在一个钟，而是同时有几个钟在协调地活动着，赋予生命以总的节律。

研究者用一对蟑螂做试验，他们把其中一只放在全天光照的条件下，使不显示活动节律；另一只放在光暗各12小时的交替下，它有明显的24小时活动节律。然后使两只蟑螂血液相通，背对背连结一起，无节律的在下，有节律的驼在上面，把它们放在连续的光照下。这样，下面那只蟑螂也出现了24小时的活动节律。研究者因而认为，钟与血液内的激素有关。经过多次努力，终于在蟑螂食管下面一个大神经节上找到了四个专门分泌激素的神经细胞(见图2和3)。这种细胞可以移植到另一只动物的血流中，而仍保持其本身原有的分泌节律。

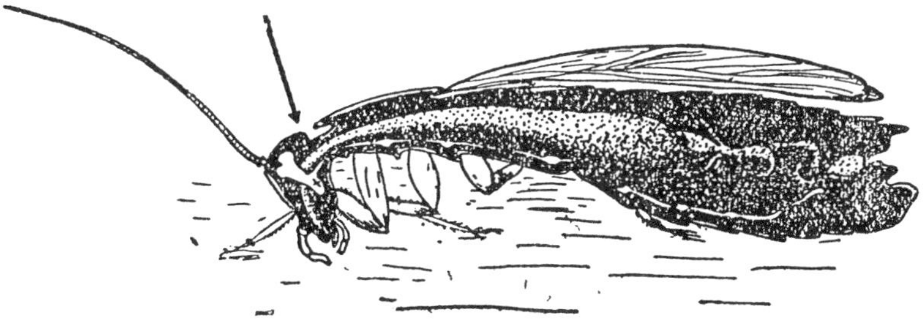

图3 蟑螂体内神经分泌钟所在的神经节(箭头所指×记号处)

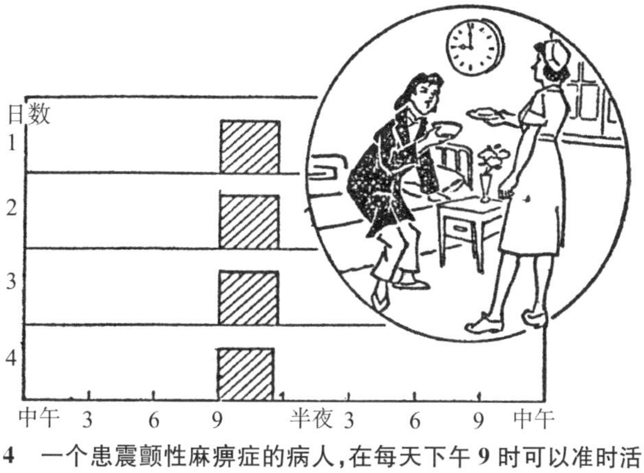

图4 一个患震颤性麻痹症的病人，在每天下午9时可以准时活动

别的处所还有没有钟呢？研究者把认为是钟的所在的这一神经组织摘去，而隔不了多久，钟又“开动”了。后来，发现这里的只不过是“子钟”，实际上还有一个调节它的“母钟”存在，这就是神经突轴分泌激素的节律性活动，它可以控制神经分泌细胞中激素的产生，因而可以调节“子钟”。外界环境对生物钟的影响可能先通过“母钟”，再调整“子钟”。生物钟可以说是神经-体液综合调节的结果，而且是定形的；但到底是什么样子，怎样调节，还不能做出圆满的解答。

至于植物计测昼夜长短的时钟，或许可以用光敏色素间的相互转化来解释。近年来植物学家发现，植物中有两种光敏色素，一种能吸收660毫微米的红光，简称为P_{660}；另一种能吸收735毫微米的红外光，简称P_{735}。它们在光暗交替下可以相互转化：在黑暗中P_{735}会逐渐转化为P_{660}；而经短期光照后，P_{660}又会转化为P_{735}。植物可以从这个转化所经历的时间来检查昼夜的长短。

研究生物钟的意义

在漫长的生物进化道路上，为了适应外界环境，生物机体内许多生理学的、生物化学和生物物理学的活动以及新陈代谢过程必须进行节律调整。像植物既要利用光照，又不能让水分大量散失，于是它们的生活也形成了一定的节律。蜜蜂一早就飞出来，因而蜂媒花开得也早；靠蛾子传粉的花，则要到晚上才开放。

生物钟的研究具有多方面的意义。首先，生物测时和钟的调整是研究个体自动控制和自动调节的大好材料。生物化学家、生物物理学家和生理学家在这里都将大有可为。

研究植物的时钟，对于增加作物的授粉率，进行人工定向培育和提高蜂蜜产量，都非常重要。有位科学家在研究中发现，植物根吸收离子也有昼夜节律。经用菜豆和向日葵实验，结果发现植物体内磷的数量白天增加，晚上减少。如在植物吸收离子（如钾离子）最少的时候，给以钾离子，则植物非但不吸收，甚至有时还会“消化不良”。这就暗示应该在植物“饥饿”时施肥。不难设想，如何适当地控制外界因素使符合植物内在的时钟，将是增产措施的新方面。

人体的许多生理过程也是有昼夜节律的，如体温的升降、血压的高低、激素分泌的多少等等（见图4）。生活于夏季无黑夜的北极的人们，刚开始时会引起排泄规律的失调；日夜轮班过于频繁的人、远出旅行者以及高空飞行人员，也可能引起生活节律的暂时失调。有人计算过，从美国飞行到日本，在有些人要过9天才能像激素分泌的节律适应于两地时差。

很多疾病的发生，是由于生理节律的紊乱、神经-体液调节的异常所致。研究者对生物钟与疾病之间的关系作了进一步实验，他们将一组蟑螂放在正常光暗条件下，并使另一组的日夜程序颠倒。这样，后一组蟑螂的时钟正常节奏被破坏了，直到它的时相比正常的落后12小时，然后将这只走慢了的钟移植到那组正常的蟑螂身上，结果所有正常蟑螂的肠上都长了癌。而把两个个体运行一致的生物钟移植在一起时，并不生癌。这一研究暗示，癌可能与失调的生物钟所分泌的激素浓度异常有关。

如前所述，一般是在正常的生理节律被扰乱后才出现疾病，可是也有相反的情况，有些疾病会加强或引发平时所不能测得的24小时节律。这在一些精神症状中得到了例证。据报道，有一患震颤性麻痹症的病人卧床9年，平时昏昏沉沉，连话都讲不清；可是在每日下午9时，她却能下床走路，吃饭讲话也清楚。可能是人体中有很多在正常情况下不协调的钟，而在某些疾病的不正常情况下却协调了，并产生了明确的体征和精神症状周期。不可否认，生物钟的研究对预防医学和临床医学也都非常重要。

试管中生成的植物*

一株高大的树木或一朵美丽的花,都是由功能不同的亿万个细胞所组成;但是,一花一木的生长发育,却是由一个微小的受精卵开始。受精卵是生命的桥梁。可是,现在有些研究者用实验方法打破了千百年来开花结果、传种接代的框框,他们从一个普通的、停止了增殖的植物细胞开始,在玻璃瓶里培养出新的一代来了。

生物学上的一个大问题

还是从受精卵说起。

受精卵中携带着从双亲遗传来的分子"样板",它能正确地复制出同双亲相似的无数细胞而成为一个新的生命。可是,当种子萌发成幼苗并长大以后,随着器官的分化,有些细胞停止了分裂,而只有根、茎和腋芽生长点的细胞仍保持着继续分生的能力,还有形成层的细胞也永无休止地增殖着。像有些已经生长了 4 000 年的水杉,它们的形成层还能每年增殖,从而加粗了树干。但同一株植物体内的其他亿万个细胞,则因为所在部位不同,虽然具有从亲本传下来的同样遗传信息,也同样有充足的养料与水分,却不再分裂了。

是什么因素控制着、决定着细胞的分生能力呢? 这是生物学上的一个大问题,科学家长期以来一直在努力探索其中的秘密。

胡萝卜块根的实验

让我们先来观察一下植物的种子。它可以分为胚和胚乳两个部分。在植物的受精过程中可以看到,由花粉管送进胚囊中的两个精核,其中之一跟卵细胞结合,以后发育成胚(即幼小的植物);另一个跟胚囊中的次生细胞(位于胚囊中央,与营养有关)相结合,以后发育成胚乳(见图 1)。胚乳的生长一般在胚之前,它供给胚以生长所需的养分和促进生长的物质。这说明受精卵的增殖能力受胚乳的直接影响。

在一棵长大的植物体内,如前所述,细胞因所在部分不同而会表现出不同的分生能力。看来,一个细胞能不能继续进行分裂,是受邻近细胞中的化学物质影响的。也就是说,细胞的分生能力主要受外在条件所控制。现在研究者已经成功地用实验证明,在适宜的营养条件下,一个成熟的植物细胞(如取自胡萝卜块根上的)可以像受精卵一样增殖发育,长成一株新的、同母体完全一样的植物(见图 2 和图 3)。

---

* 同生(赵同芳,朱雨生)。

此文发表于 1964 年"科学画报"7 期 89 - 90 页。

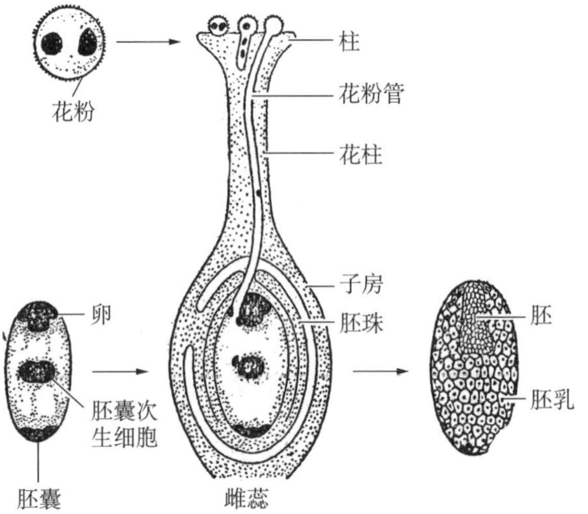

图 1　植物的受精过程示意图：由花粉管送进胚囊的两个精核，各自跟胚囊中的卵和次生细胞结合，前者发育成胚，后者成胚乳

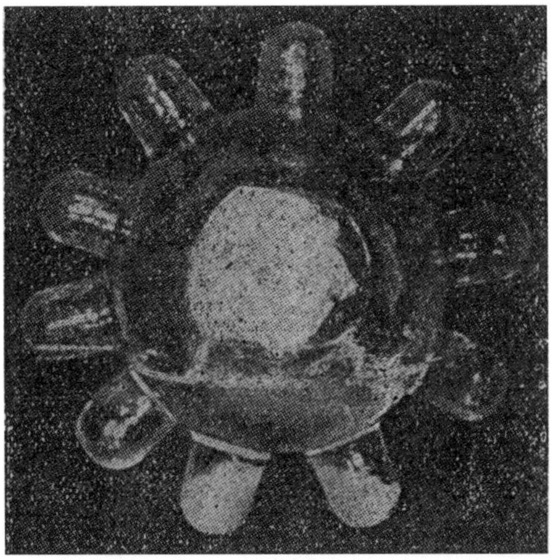

图 2　培养瓶转动时，细胞便从胡萝卜组织块上脱落

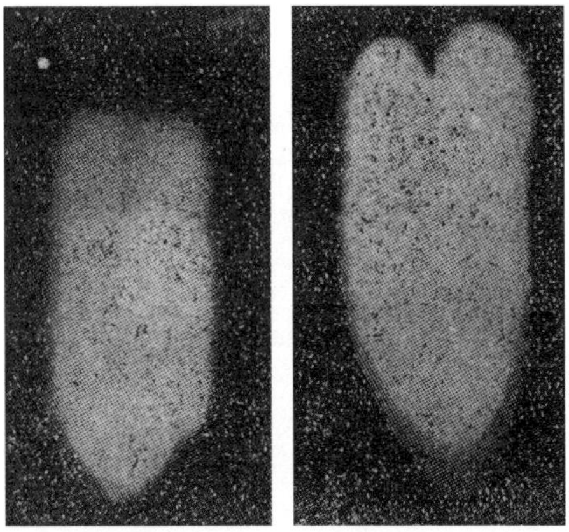

图 3　天然长成的(从种子上剥离下来)胡萝卜胚(图左)同实验培育成的(图右)有明显相似之处

实验者把胡萝卜块根的组织切块，放在一个装有培养液的特制玻璃瓶中(见图 4(a)～(c))，以研究在加或不加各种天然物质(如胚乳)，以及人工合成的物质(如激素)的条件下，它是否能够恢复分生的能力。从试用过的一些物质中，发现椰子汁(液体胚乳)对促进细胞分裂最有效。放在含有椰子汁培养剂中的胡萝卜块根组织小块，由于生长而在 20 天内增重 80 倍。

在不同时期内，研究者用铬酸和盐酸混合液使组织中的细胞离解，并用数血球的方法计算细胞的数目和体积。结果证明，在培养的初期细胞只增大体积而不分裂；当开始分裂后，细胞的平均体积又变小了(见图 5)；最后分裂速度变慢，细胞体积重趋增大。这种使块根中不再分裂的细胞又恢复增殖能力，是一件重大的科学成就，由此也不难推想，椰子汁中一定含有某种

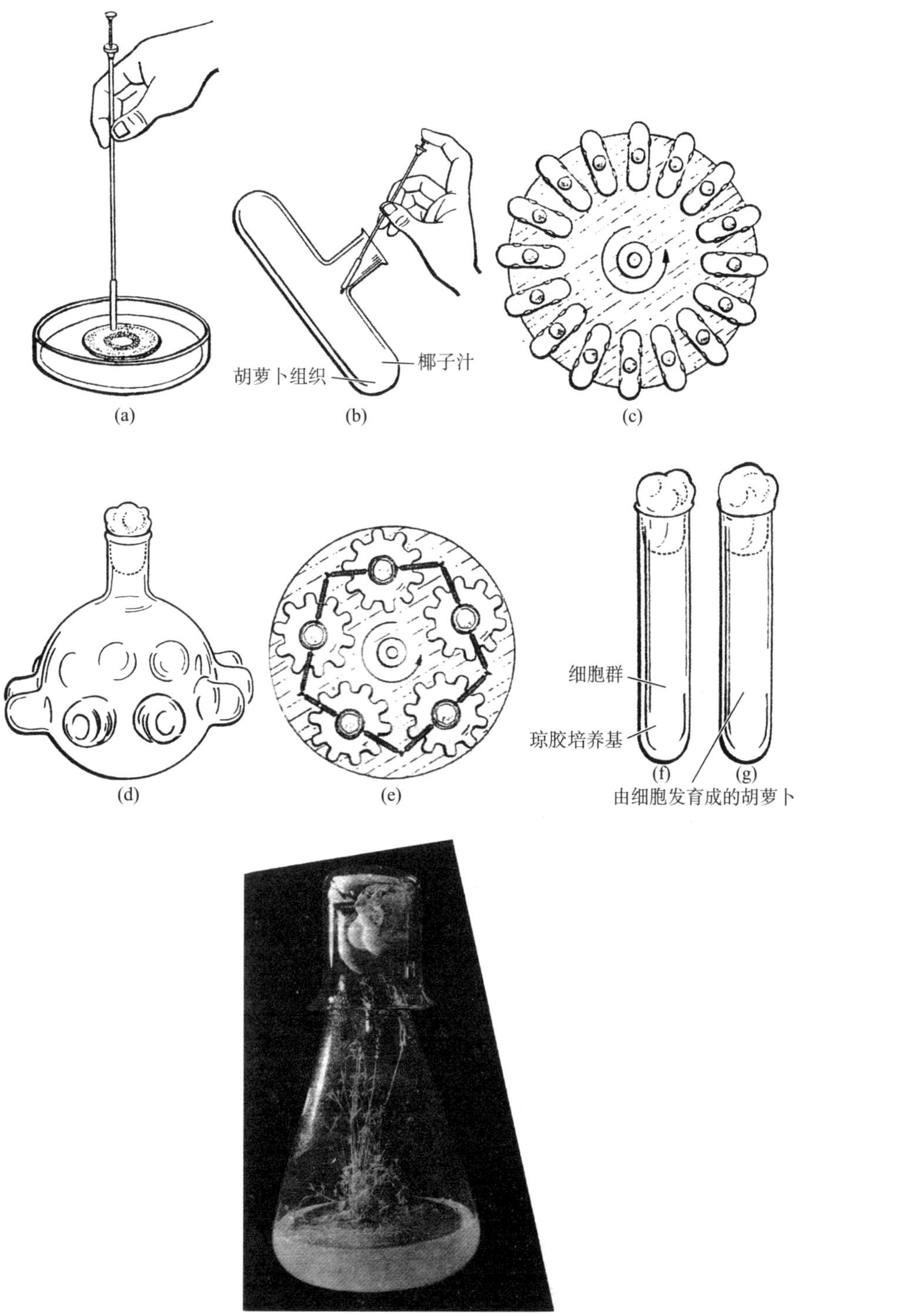

图 4　试管中生长的植物

(a) 在无菌操作的条件下，用细钻孔器从胡萝卜块根的横切片(离形成层 1～2 毫米的不生长区)上取出重约 2～3 毫克的圆柱形组织小块；(b)将一些组织小块转移到一种特制的圆柱形玻璃培养管中，内盛少量椰子汁培养液，管侧有一个颈口，以通空气；(c)把培养管固定在水平装置的旋转轴上作垂直于地面的转动(每分钟一转)，使附着在管壁上的胡萝卜组织可以轮替浸在培养液中和露在空气中，有利于生长；(d)将人工培养下长大的胡萝卜组织块移到一种特制的培养瓶中；(e)当把好几个培养瓶一起放在一个旋转着的平面上时，在离心力的作用下，胡萝卜细胞会从组织小块的边缘脱离下来；(f)将一些离体细胞移植到含新鲜椰子汁琼胶培养基中，它们会迅速以不同方式生长；(g)有些离体的细胞群会长成一株胡萝卜幼苗；(h)在试管内长成的完整的胡萝卜植株

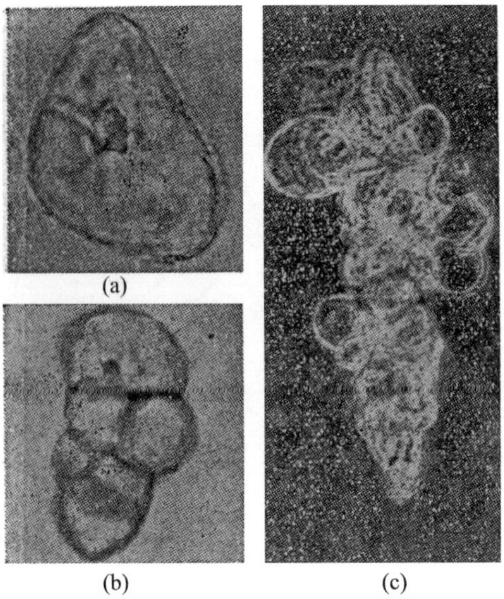

图 5　实验室中胡萝卜细胞的发育成长

(a) 一个离体细胞；(b) 离体细胞分裂长大；(c) 细胞群开始具有一般胡萝卜植株组织的极性

可以促使细胞恢复生长本能的物质。这种物质是什么呢?

研究结果还指明，对于有些植物的组织细胞，像马铃薯块茎，椰子汁并不能促使它恢复分生能力，而需要再加用其他物质做增效剂。如以百万分之六的 2,4-二氯苯氧基乙酸加入含有椰子汁的培养剂中，马铃薯块茎的细胞就开始增殖了。更复杂的是对一些落叶树木(如糖槭)的休眠芽，椰子汁也不能促进它们生长，但经过冬季低温解除休眠以后，椰子汁就能发挥促进作用了。据证明在休眠芽中含有抑制生长的物质，抵消了椰子汁的作用，而在低温解除休眠期间，这类物质转化消失了。这些事实说明，细胞的生长除了适宜的养料、水分与温度等条件之外，还受到刺激物质和抑制物质的含量以及它们间相互作用的控制。

试管中生成植物

培养在特制玻璃瓶中的胡萝卜组织小块，在玻璃瓶不断转动期间(见图 4(d)(e))，边缘的细胞可以完全分离下来。把这种单个细胞再进行分离培养，其中有些细胞便可分裂成一个像胚一样的细胞群，有些还会生出根来(见图 4(f)和(g))；把有根的细胞群移植在含有椰子汁的琼胶培养基上之后，就会再生出茎叶而长成一株幼苗(见图 4(g)和(h))。这样经过不断移栽，最后种在土中就会开花结籽，而实现了由一个成熟的体细胞再生成一株完整植物的理想。

更有趣的是，当取下胡萝卜的花、从子房中剥出胚，用同样方法培养时，胚的每个细胞可以被分离开来，且可以各自生长成一个独立的个体。一个胚具有十万个以上的细胞，也就可能长成为十万株幼苗，但一般用胚做种子时却只能生出一株幼苗。这是多么值得惊奇的一种繁殖方法，不有些像神话中的分身术么?

控制细胞生长的化学物质

现在我们再来考察一下，是什么物质控制着细胞的生长呢？首先应当明确，椰子汁能够促进多种植物的细胞生长，加用一些增效剂后有效范围更广；其他植物种子如乳熟期玉米籽粒的提取液，也具有椰子汁相似的作用。用化学方法分析检定后，察知促进生长是三类物质联合作用的结果，它们是：①对诱导细胞分裂有高度活性的物质，其中包括1,3-二苯尿(椰子汁)、吲哚乙酸(玉米)等；②对于活性物质有增加效应作用的成分(增效剂)，像山梨醇、甘露醇和环己六醇等，如果缺乏本类物质，上述①类物质便不起作用或效果很小；③还原性的有机氮化物，本类物质可用氨基酸代替。此外，用人工合成的激素可以代替植物产品以控制细胞的生长。

至于上述这些物质在细胞中究竟起了什么作用，迄今为止尚未完全揭穿。只知道当细胞吸取了这些物质之后，它的新陈代谢方式就会改变旧观；其中最突出的是合成蛋白质的能力大为加强。

从这些研究成果中我们看到了一个远景，生命将不再是神秘莫测，而是可以人为控制的。可以预计，一旦这些理论成就经过实践而应用于生产之后，经济价值将是十分宏大的。

附　录

UMBC receives Chinese grad. student*

Yu-sheng Zhu

UMBC's first scientist from the peoples Republic of China has arrived. Yu-sheng Zhu, a 40 year old plant biologist, will be working towards his Ph. D. under the direction of Dr. Shain Dow Kung, associate professor of biological sciences. He hopes to have his doctorate completed in about three years, after which he will return to China to demonstrate his newly acquired rechniques and knowledge of plant genetics and recombinant DNA to his fellow scientists.

Some of Mr. Zhu's work may be with Dr. Richard Wolf, assistant professor of biological sciences.

Mr. Zhu's visit here is the result of last year's trip to UMBC by a delegation of Chinese bontonists. Zhu, who is living in the dorms, finds students here friendly and helpful. He said, "I have been here only for 5 days, but I enjoy the life on campus very much, so many students are all very kind to me... Although I am already 40 years old... I feel younger than before (he arrived)."

Zhu does have one thing in common with many dorm students. "The first day the food in the cafeteria didn't make me feel very well." he said.

Zhu has met with many of the students studying here from Taiwan, China. "We are all Chinese: we should all be friends," he said.

Mr. Zhu chose to study in the U. S. because of the advanced science and technology here. He hopes to get a lot of knowledge in this country. " He turned down a fellowship from the the Carnegie Institution of Washington, Which is affiliated with Stanford. "I chose UMBC for Dr. Kung's reputation, his work is very interesting. His work is very important to solve many economic problems." he said.

Zhu, who left a wife and a seven year old son in China, says his family and he realize the separation is a sacrifice that must be made in order to advance his education.

Zhu studied English for a year in a university in China, but to really make progress he

---

* By Damian Jones and Mike Rifkin
此报道刊登在 1980 年 2 月 12 日 UMBC 校刊"The Retriever"上。

had to study and read English publications on his own. He finds the American pronunciation different from the more British English he was laught. However, he speaks English very well.

At UMBC, Zhu will take an English course as well as audit Advanced Topics in Plant Physiology taught by Dr. Kung.

"I never imagined I could come here, several years ago," he said. "Now our countries are very friendly." he added.

Zhu feels strongly that both the U. S. and the People's Republic of China can benefit from cultural and technical exchange. "Many people in China now favor the U. S.." he said.

East meets West at UMBC lab

Chinese scientist does research

The small, Oriental man never expected so much attention. But ever since he set foot in America this past January, he has been the object of much curiosity.

First of all he's a 40-year-old graduate student at the University of Maryland Baltimore County.

And then, of course. Yu-sheng Chu, whose name translated means "the first" and "to come out," is the first person from the People's Republic of China to attend UMBC. According to his sponsor, Dr. Shain-Dow Kung, associate professor of biological sciences at the university, Chu is also "one of the top scientists in China."

"Mr. Chu is a very intelligent man," says Kung, "I am very proud to have him here."

Chu, whose modesty allows him to greet such compliments with only "no, no," is pursuing his Ph. D. while working with Dr. Kung on experiments involving the tobacco plant.

Dr. Kung and Chu are experimenting with the tobacco plant because so much more is known about it than other plants. But they hope their findings will be applicable to green plants in general.

"We are trying to increase plant productivity," explains Chu, "Most of the energy produced, directly or indirectly, comes from green plants."

One of the implications of this research applies to world food production.

According to the World Almanac of 1980, between 1970 and 1978, there was a 2.5 percent increase in world food production, while there was an 18 percent increase in world population. These statistics paint a dim picture of future conditions.

Large amounts of energy, such as oil, are put, in ever-increasing proportions, into food production, processing, and distribution, while the amount of fertile land available remains the same. Because fossil fuels are limited, their use to expand food production is also limited. One method of combatting the problem is to increase the productivity of the plant itself.

Kung and Chu initially realized their common goal when Kung visited Chu's hometown of Shanghai in 1978. Kung was a guest lecturer at Shanghai's Institute of Plant Physiology, where Chu had been working since 1962.

"I met with Dr. Kung after his lecture," says Chu, "We discovered that our ideas and

Chinese scientist does research

East meets West at UMBC lab

Yu-Sheng Chu and Dr. Shain-dow Kung examine Chu's experiment on tobacco plants

opinions were very similar."

It was this sort of scientific 'love at first sight' which brought Chu to the U. S. to study and work under Dr. Kung.

But Chu doesn't mind the work. That's why he's here. He can be found in the lab seven days a week, where he conducts experiments using modern American equipment.

One thing that he still is not too familiar with, however, is the English language. But, with the help of his colleague, Dr. Kung, and an English writing course, Chu is learning.

But, when he is not in the lab, or practicing English, Chu may be found in front of a television.

"I don't like it too much though," admits Chu, "There is too much advertisement."

While that is a common complaint heard even from Americans, television programming in China includes no commercials. But the avid American TV watcher would hardly be satisfied by the limited viewing schedule in Shanghai.

Television programs operate for two hours in the morning, two hours in the afternoon, and hours in the evening. Kids can't turn on Captain Kangaroo or Sesame Street in the morning. But they can get instruction in English, Math, and Physics.

Television is not the only difference Chu's noticed.

"It is warmer inside houses here," says Chu. In China the houses have no central heating and depend upon fireplaces or summertime for heat.

On his first trip to the super market, Chu was amazed.

"There is so much food everywhere." exclaims Chu, "In China, it is much different."

Chu praises the U. S. for its part in helping foreign students to come to study in America which, in turn, helps develop technology in the countries to which they return.

"I know very many famous acientiats who studied in America," says Chu, "The United States has made the most contribution to scientific development in the world.

"Foreign students have their own culture, custom, and tradition, but they always have that common feeling and common concern about their future."

Chu's immediate concern is his own future. While the three years ahead of him seem long when he considers the wife and seven-year-old son he left in Shanghai, he admits that he doesn't feel as if there is time to do everything he wants.

And that includes further investigation of two of his moat recent and fascinating discoveries — Security square Mall and UMBC basketball games.

In some ways, however, America has given him a greater appreciation of home.

"When I come to America." says a grinning Chu, "I realize how good Chinese cooking is."①

① Judy Lee Morrison. 此报道刊登在1980年3月26日"巴尔的摩太阳报"上。

热爱科学　缅怀光明[*]

——光明中学毕业50周年之际的回忆

2004年2月29月在美洲版《世界日报》周刊上刊登了一篇著名专栏作家信怀南标题为“三条平行线”的专栏文章。它讲的是1940年生的三个同龄人的不同命运。其中第一位是政治家，第二位是政论家，另一位是科学工作者。文中分别介绍了他们在不同时代大变迁中的遭遇和命运。三个人的名字均未点明。其实，第一位，它讲的政治人物就是前民进党主席施明德。此人在国民党时期坐过二十年牢，现是中国台湾发动全民“反扁”的领袖之一，他一生饱受苦难。第二位，它讲的政论家就是专栏作家信怀南自己。他是1960年代来美的留学生，经历了餐馆打工、读工商管理等艰辛，最后变成了著名专栏作家、政论家，出了不少畅销书，并当上了旧金山电台脱口秀主持人。最后一位呢，就是我——一个在国内经历了朝气勃勃的新中国初创阶段，1957年的“反右”、1958年的“大跃进”和始于1960年代中叶的“文革”、始于1970年代后期的改革开放，直至1980年代初赴美留学深造的学者。文中关于我的说法是：“以他在学术界的成就来看，做学问，颇有成就”。其实，本人只能说是在科学上做过点研究，有些心得，想与大家分享(为说明这点，我曾专门给作者信怀南写过信)。

回顾一生，自小热爱科学。如果说对科学还有点贡献的话，与母校光明中学六年的培养教育是分不开的。

好学上进　争当“三好”学生

我是1951年进光明中学初一的。刚进学校，什么都觉得新鲜。那时对邢文钧老师的《植物学》特别有兴趣，第一次考试得了100分，从此我成了植物爱好者。邢先生邀我去他家花园采集木槿枝条做发育试验，也从此与邢先生及全家建立了深厚的友情。我大学毕业参加工作后一直住在集体宿舍，1972年要结婚时，他就将自家的房子腾出一套来让我们搬进去住。邢师母还亲自帮我们把厕所打扫干净。我们设在他家中既节俭又丰盛的三桌婚宴，还是由能做一手好菜的邢先生亲自掌的厨。我们的友谊一直维系到他1999年4月22日因脑瘤而离世的18年前。我与他的两个子女至今仍保持着密切的朋友关系。

因为爱好科学，特别是植物学，当时我把每月不多的零用钱省下来定《科学画报》杂志和买参考书(如《米丘林全集》《光与植物》等)。我还买了如嫁接刀、吸量管、温度计、花盆、种子等不少小工具、小仪器，在家中做试验。参加了光明中学“少年米丘林”小组(见图1)，任少先队辅

---

* □57届丁班　朱雨生

此文刊登在2007年庆祝光明中学成立120周年的“光明中学57校刊”2期19－25页上。

图 1　1956 年朱雨生(右)在母校“少年米丘林”小组

图 2　1956 年，少先队辅导员朱雨生(左)在母校对“少年米丘林”小组成员进行辅导

导员后，我就和大家开始在校园做起了试验(见图 2)。因地方太小，阳光也不好，我们就到对面的小学去借地。当时做的是研究微量元素如溴、碘对植物生长发育有什么影响的试验。后来“少年米丘林”小组活动扩大了，我就与附近的“外国坟山”(即现在的淮海中路淮海公园)联系，让他们允许我们利用“坟山”的空地来做土豆番茄嫁接等试验。记得那普安路大门口有位块头大得令人生畏的看门人，还记得碰到下雨天，嗖嗖冷风发出的尖声，像洋人阴魂在鬼嚎，呆在寂静荒凉的坟地里，心里实在有点害怕。但是为了研究成果，每次我们还是硬着头皮一直坚持到把试验做完。记得那时我们的“少年米丘林”小组是每周活动一次，试验多由我负责辅导，用的教材有的是大学的实验本。这个阶段对我学识及能力的增长影响很大。1982 年我被美国马里兰大学派到中国科技大学开展“暑期分子生物工程训练班”，在指导五六十位全国各地来的学员做实验时，我常常会不经意地回想起二十多年前在母校“少年米丘林”小组活动的情景。而且，觉得自己能将眼下的工作做得如此得心应手，与在母校“少年米丘林”小组得到的锻炼是紧密相关的。

母校“少年米丘林”小组活动培养了我们接触自然、热爱科学、动手做实验的兴趣，为我今后成为一名科学工作者打下了基础。记得当时对科学感兴趣的还有本班同学周焕椿和黄融。他们两人数学根底都很好，在上海的数学竞赛中得过名次。周焕椿家庭经济条件较好，在家里搞了一个化学实验室，曾经试着做过各种颜色的荧光粉、炸药等。黄融的兴趣是物理，而且他知识渊博，文史地理什么都通。玩笑时大家都尊称他为黄“博士”。因兴趣相投，我们三个人课前课后经常在一起讨论生物、化学和物理的问题，简直成了一个“三人帮”。巧的是，几年后还如愿以偿地以第一志愿分别考上了复旦大学的生物系、化学系和物理系。可惜黄融同学体弱多病，进大学不久患肺结核，被安排在复旦附近的叶家花园(即现在的肺结核病防治院)疗养，几进几出，不久就离世了。

基于在“少年米丘林”小组的收获，我为《辅导员》杂志写过一篇有关无籽西瓜的文章，出乎意料地居然还收到了稿费 12 元！这在当时一个中学生的眼里可是一笔不小的数字，因为它差不多相当于那时一个大学生每月的伙食费了。这是我人生中第一次拿到报酬，心里特别高兴。重要的是，它给了我很大的鼓励。进大学后我继续为《科学画报》等杂志写科普文章(内容涉及植物与光、植物半导体、植物电世界、生物钟等方面)，学习和写作的热情大概也源自于这第一次“成功”的激励吧。

回想起来，中学学习期间，不仅是我们学好文、史、地、数、理、化，也是培养我们人生观、价值观的重要时期。因为要求进步，努力学习，好好工作，立志将来在科学上为国家为人类多做贡献，初三时我参加了共青团，并兼做了少先队的干部工作。1956 年又荣幸地被评为了区的“三好”学生(当时邑庙区仅有的三名“三好”学生，光明中学占了两名)。如今，总结自己所走过的人生道路，应该说自己无论在学生生活中以及走向社会后，无论是在国内还是国外，还是坚持了诚实热忱、艰苦奋斗、助人为乐、忧国为民等这些做人的基本原则。这些原则就是中学时代在我头脑中扎下的根。由于参与共青团和少先队社会工作，与同年级甲班的易楚芬、王尔慧以及乙班的费兆馥等也有接触，她们工作中的认真、努力、活跃，在同学中有很好的口碑；与我一起被评为区“三好”生的同班同学吴宗善，他学习成绩很好(后来考上北京航空航天大学)，酷爱飞机模型，我们彼此学习、相互鼓励不断……这些，对我积极向上，做到“三好”中的“学习好”“工作好”都是有推动作用的。多年之后，我又与乙班曹炽康相识成了知交好友，他是光明中学一位懂科学又擅长写作的才子。

“三好”中最后一好是“身体好”。那时我除了和大家一起每天早晨在操场排队做广播操外，还进行长跑、练单双杠、攀竹竿和各种球类运动，加上有徐道群和王子文两位优秀的体育教师的悉心指导，身体也练得相当健壮。它使我能够在后来国内的工作中游刃有余，也能够从容应对在国外所面临的各种挑战；它也使我到了现今年过花甲之后，仍能精力充沛，对游泳、爬山及旅游等活动的兴趣依然不减。

开花结果 不忘栽树人

离开母校五十年，每当我们为自己和当年的学友们踏上社会后在各自的岗位上做出了一定的贡献感到欣慰时，总会怀着一颗感恩的心，首先思念起那些在我们成长关键时期，精心培育过我们、为我们打下了坚实基础的母校的一批好老师。

讲课生动条理清晰的邢文钧老师，思路清晰、逻辑性极强的乐嘉鑫物理老师，慈祥和蔼、声音洪亮的朱铨几何老师，过去的女国大代表、口才极好的崇启历史老师，热情洋溢的高士模和书法家邹梦禅中文老师，曾代过我们俄文课、慈祥严谨的董寿山外语老师，温文尔雅的陈宏济化学老师，讲话带有昆山地方乡音、和蔼可亲的曾肇勋历史老师……一个个都让我们永远难忘。

在此，我特别要提到的是我们的俄语老师马秀凤，她的教学提高了我学习俄语以及其他外语的兴趣并成绩不斐起了至关重要的作用。事情还得由我在光明中学支持下参加市少年宫活动认识了一些在上海就读的苏联孩子说起。他们的父母都是来援助中国建设的专家。我与其中的瓦洛佳等成了好朋友，常被邀去苏联领事馆看电影，去文化俱乐部参加宴会，去他们的学校参加舞会……他们也在周末骑车到光明中学来玩过。因为要交流，俄文变得特别重要，特别是口语。因为我发音较准，马老师就常让我在班上领读，我有不懂之处她总会给予耐心、热情的指点，使我的俄语水平有了长足的进步。我甚至都可以在杂志上发表一点自己翻译的科学小品文。与苏联友人的友谊也一直保持至中苏分裂后的 1960 年代初。1957 年当我以在 180 多新生中第一名成绩考上复旦大学生物系，在系里对一千多名新生进行的外语测验中，我又以突出的成绩被选进了快班。而且还是二年级的俄语快班。一个新生，外语可以直接跳至二年级去学的情况，这在当时的复旦是绝无仅有的。后来，我在大学里花了一年时间又学完了英语的相关课程。进入中国科学院，在该院对全院新科研人员进行的两小时或一门俄语或一门英语的外语考试中，我用一小时做完了英语试题后，又用一小时做完了俄语试题，而且，结果两门都得了第一。没多久，因为研究工作需要，我又用一年时间完成两年的日文课程，又前后学完了德文和法文的两年课程，成绩均名列前茅。回想这一切，我怎能不感激引领我在学外语上入门的马老师啊！

另外，还值得一提的是光明中学尊师爱生校风。老师对我们尽心尽力，我们也热爱和尊敬自己的老师。几乎每个春节我们都会自动组织团拜老师，还有去赵书文校长家拜年，关系亲密融洽。给我留下深刻印象的是老实巴交、对我格外亲切的曾肇勋老师，他甚至开玩笑地要我叫他为哥哥。

大学五年　打下了做科研的底子

进大学不久，就遇上了大炼钢铁、种试验田等运动和三年自然灾害时期。但是中学时期确立的人生观使我在最困难的时期仍对国家的未来抱有信心。我当时选攻的是植物生理专业，四年后我的毕业论文是研究棉铃和其他落花落果落叶的生理机制。由于我创立了一套新的研究技术，在良师薛应龙教授指导下，又有了一些重要新发现，因此曾在《复旦学报》和《植物生理学报》上发表过二篇论文。这在学生中是少有的，受到了学校领导的接见和表扬。这些都大大增强了我未来从事科研的自信心。然而，即使在我最兴奋的时刻，我都很清楚，我走上的这条光明之路，源头就在光明中学。光明中学的“少年米丘林”小组是点亮我事业前程的一盏永不熄灭的明灯。

另外，我还想借此再说一说复旦大学的教授谈家祯（见图 3）。谈教授是全国著名的遗传学家，是美国创立基因学说的摩尔根的门生。他的学术曾被苏联批判为“资产阶级伪科学”，但他毫不动摇。毛主席曾三次亲自接见了他，并提出了“百花齐放，百家争鸣”的双百方针。他后来担任了复旦大学的遗传所所长、生命科学院院长以及复旦大学副校长等职，又兼任过上海中

图 3　我的良师益友谈家祯教授及其夫人

国民盟的主席。他一生献身科学，是我非常敬仰、对我的人生有巨大影响的一位学者。

忘不了的是在 1957 年，我到复旦生物系报到的第一天，在迎接新生的大会上，当时任系主任的谈家祯教授向大家致的欢迎词。他说，今年考复旦很难，你们能脱颖而出很不容易。大学是高等学府，复旦又作为全国重点的高等学府之一，是培养人才出专家的重地，你们要不辱使命，好好学习，将来多为建设国家出力。他语重心长的这席话，令我刻骨铭心，可以说几十年来，无论走到哪里，它都让我想着培养自己成材的祖国，不忘尽自己之所能为祖国的发展和强大多多出力。同样忘不了的是，谈教授教的遗传学，他生动有趣的讲课，使我对遗传学也产生了浓厚的兴趣。毫无疑问，若干年后在面对人生新的十字路口时，我有勇气选择了分子遗传学的研究工作，就是因为有谈家祯教授当年在学术上给垫的底和培养的感情。

难得的是，谈家祯教授既是我的良师，又是我的益友。

他与我的师生情从我的大学时代一直延续了几十年。1980 年代他一到美国出差就会到旧金山我的居所来见我。1995 年还与夫人一起在我家住了一段时间。我清楚地记得，那次他带着的任务是想招引在美留学的人才回复旦发展。有一次竟召集了近三十名研究生、访问学者在我家聚会。现在复旦生物系的主任和生命科学院院长、当时在斯坦福大学留学的金力，就是谈教授当时看中并动员回国的。那年谈教授已 87 岁，耄耋之年还在为学校和祖国的兴旺奔波，崇高的爱国主义情怀真让我感动。

进中科院　艰苦奋斗出成果

我在中学时所交的苏联学生朋友，家住在建国西路的一幢洋楼里，旁边就是岳阳路中国科学院植物生理所等单位的豪华大院。透过大院铁门可以看到里面穿白大褂的科学家在活动，我很是羡慕他们，期盼将也来能到中科院植物生理所工作，穿上白大褂，做一名科学家。想不到，我进中国科学院的心愿在大学毕业时实现了。那年，该研究所在上海只招收一名研究生，我能被选中真是有幸。进所后，我被分派到脂肪代谢组，从事油料种子脂肪合成时代谢变化的研究。令我激动万分的是，当我把大学里做论文时的思路应用到这一研究课题，提出了自己的观点和设想时，立刻就得到了当时导师的赞同。为此我的干劲更足了，经常一干就是半夜。每遇集体宿舍关了大门，就爬墙头跳进院子，睡不了几小时，又匆匆投入到新一天的工作中去了。千辛万苦，终于赢得了累累硕果，我们发现了一种新的很奇妙的代谢调节机制，而且发现了是一种类黄酮在起调节剂的作用。我们又将这些结果写成了六篇论文公布在了学报上。其中有一篇还被学部委员（院士）推荐到《中国科学》杂志上用英文发表了。更令我们兴奋的是，1979年我国著名植物生理学家、北京大学教授和植物所所长汤佩松先生到美国接受"美国科学院通讯院士"称号，在全球植物生理年会上作题为"中国植物生理研究五十年"的演讲时，还特别提到了我们组在上海的工作。他说我们组的工作是建国以来植物呼吸代谢研究方面最为出色的。据报道，他的报告结束之后，全场起立鼓掌达数分钟之久。他回国不久，就将在美的那次演讲手稿送给了我，并鼓励我继续努力，我俩也因此成了忘年交（见图4）。

图4　汤佩松教授（右）成了我的忘年交

虽不在同一单位，还经常互相磋商，曾经还由我执笔合作发表了一篇具有前瞻性的文章，这篇文章得到我所殷宏章所长的高度重视，在全所高级研究员中进行了传阅。之后，殷老还让我做了他一段时间的秘书。

在中科院所取得的一些成绩，与那时具有的那股子特别爱钻研、特别能吃苦的劲头分不开，而这种热爱科学、艰苦奋斗的精神是早在母校时就初步形成了。

"文革"期间　科学研究初露端倪

"文革"期间，我们研究所关了门，理论工作不能做了，科研人员下乡下厂接受贫下中农"再教育"。我下乡当农民的几年间，养过猪、做过搬运工。后来提出"抓革命促生产"了，我们才回到了城里。一进实验室，我们马上想出了一个联系生产实际为工农兵服务的实验项目，即对纤维素酶进行科学研究。目的是将纤维素废料用微生物的纤维素酶分解成葡萄糖。葡萄糖可发酵成酒精成为能源。纤维素废料是取之不尽、用之不竭的植物光合作用的产物，这个课题可解决今后能源问题、粮食问题，具有重大的战略意义，此设想由我执笔向中科院打了报告。院部同意后，它成了科学院、农业部、轻工部等单位共同操办的全国性的协作项目。此间，我曾到南方许多深山老林中去寻找这种微生物。经过物理化学诱变，我的同事们培育出来的高活力菌株，超过了美国、日本的水平，而且在工农医及科研上所作的各种应用试验，都取得了可喜的成绩。我们研究组也因此获得了国家科技奖。1981 年我又应邀赴美到 Rutgus 大学围绕此项研究作了演讲，并在美国一研究所权威的建议和推荐下，将我们在国内的工作整理成文，用英文发表在了国外的杂志上。

“文革”后期，全国制订十年科研规划。那时学术权威重新被使用，参加制订规划的全是全国各地来的资深学者和老年科学家。而我们研究所还选了两位年轻的科研人员参加。我是其中之一，另一位即是现在的北大校长许智宏，他思路开阔，社会组织活动能力极强，让我深感钦佩。他后来去英国留学了，是回来后由室主任、所长一步步提升到中科院副院长再担当北大校长的。我们是很好的朋友，去年十月初他来旧金山开会，百忙中还抽空来看我，住我家，家事国事天下事一直畅谈到深夜。次日上午又召集两名年青精英在我家聚会，情景令我十分难忘(见图 5)。

图 5　与北大校长许智宏(左 3)召集两名年青精英在我家聚会

改革开放　首批出国圆留学梦

改革开放后，中国第一次派出 40 名学生去美国留学是在 1979 年。我也是较早被派出国并获美方的奖学金的一个，我出去的时间是 1980 年，一月初先到的北京，集训后经法国才到达美国纽约，终于圆了我中学时想出国留学的梦。我是从接受马里兰大学邀请作为访问学者开始，尔后才改读研究生的。后来斯坦福大学也向我发出过邀请，可惜因时间晚了没有去成。但我研究生三年毕业后又攻进了哈佛大学，对此遗憾也算做了弥补。

刚到美国一切都感新奇，美国巴尔的摩《太阳报》曾登了一篇专访我的文章。我在大学的三年内顺利地通过全部考试，门门得了 A，还完成了六篇论文并得以发表。当时我研究的课题是叶绿体的分子遗传。我首次将烟草的基因在细菌中表达，作了植物叶绿体基因光控制的开创性工作。这对于四十多岁年纪的我来说可不是一件容易的事。我之所以能经受得住研究过程中的各种磨难，大概是因为我的身体和意志有在母校时期打下好底子的缘故吧。当然，也得益于于我的导师对我的鼓励。我的导师当时是副教授,后升为正教授、系主任、教务长,香港科技大学副校长。他因为社交工作太忙,我就离开了他去了哈佛大学,在那里遇到了一流的科学家,受益匪浅。

以后我先后在依利诺大学、加州伯克利大学和劳伦斯研究所继续做光合作用分子遗传的研究。特别值得回忆的事情有以下几件:一次是,一个研究结果曾与一位诺贝尔奖金获得者Stanley Cohn的结果相争,我还发现了他们因实验条件不对而导致错误结论。一次是我走进加州伯克利分校卡尔文实验室的经历,那是在一次学术会上,著名诺贝尔奖金获得者卡尔文做了个专题报告(见图6),报告介绍了他发现并以他名字命名的卡尔文碳循环的科学原理(也就是植物怎么把二氧化碳和水利用光能变成葡萄糖的生化过程)。会后我向他谈了在国内做的可解决能源问题的纤维素酶的工作。他听了很有兴趣,把我引到他房间长谈,还把我介绍给了他妻子。离开时他还将当天演讲的手稿送给了我留作纪念。几年后我进了卡尔文实验室,有好几篇论文都是由卡尔文推荐到美国科学院院报上发表的。对此我一直心存感激。十年前我得到了卡尔文教授病逝的消息,去参加了加州大学为他组织的追悼会。在乐队演奏的哀乐声中,追忆着我与这位一代科学大师的相识以及他对我的帮助,联想到远在祖国的大植物学家汤老、中科院植物生理所的殷老都已相继离开了我们,不禁黯然泪下,无限惆怅!

图6 诺贝尔奖金获得者卡尔文

继续探索 寻找治疗艾滋病的药物

1995年,科学杂志上登了一个Chiron大公司招聘高研的广告,我去报名了。后知报名者有五百人左右,最后我中标,取得了这位置。由此,我的科研工作从基础理论研究,从农业转到了医学生物工程。我希望能用我的知识更多地造福人类。十多年间,我曾经研发过用非常灵敏栓测艾滋病的方法。还与艾滋病专家何大一合作过。我也参与了研发一种治疗艾滋病的药物,研究结果都在几次艾滋病国际会议上介绍和发表过。加起来,我一生中发表的论文约有50篇左右,也翻译过书,并撰书总结过我多年工作的成果。最近几年我转到了Roche(罗氏)医药大公司任高级研究员(见图7),直到去年秋决定辞职退休。从此过起春秋飞上海,夏冬飞美国的候鸟退休生活。

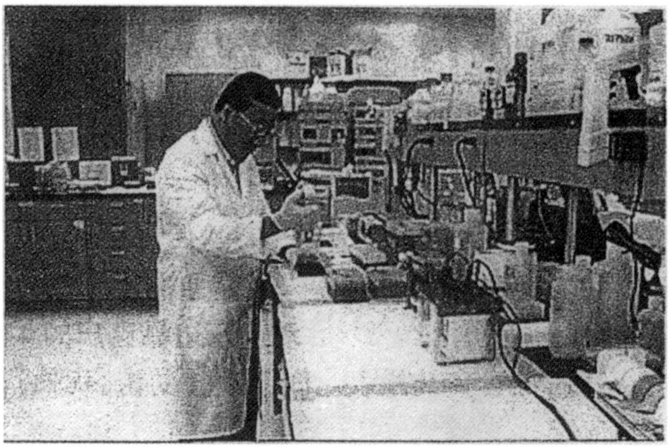

图 7 在罗氏医药大公司实验室

酷爱艺术 但成不了艺术家

母校光明中学贯彻德智体美全面发展的教育方针，很注意不让学生做书呆子。文体活动搞得很活跃，经常组织同学们在操场上跳集体舞，在教室礼堂里跳交谊舞。还鼓励我们放了学去看电影。记得看了印度电影《流浪者》后我们就常在一起唱《拉兹之歌》……对音乐兴趣也随之产生了。进大学后我学了小提琴，参加过校管弦乐团的演出；到了美国后我又学了三年钢琴，不为当音乐家，只为提高自己的艺术修养。那时儿子刚出生，大概是我的琴声激活了他幼小的音乐细胞，他对音乐很有感觉，五岁开始学钢琴，九岁就能与奥克莱交响乐团合作演奏海顿的协奏曲了。听到他的美妙琴声我常会热泪盈眶，为我有如此聪明可爱的儿子感到自豪。他多次在音乐比赛中获奖。现在进了加州伯克利分校主修商科，音乐列为副科，爱好仍不放弃。我既爱科学，也爱艺术，但成不了艺术家，却培养了孩子。但我对艺术的热情也不减当年，这几年又参加了旧金山有三十六年历史的龙吟合唱团。同时还热衷摄影艺术，在世界各地旅游中经常会留下一些精彩的瞬间，我的摄影作品曾被陈列在公司的大厅里。

我对艺术、对生活的爱好也受到过一些朋友的影响，说起来与光明中学也有点关系。在我到上海市少年宫参加科技活动时，有一位活泼可爱的小女孩给我留下了深刻的印象。后来打听到她是“小伙伴艺术团”朗诵队的，就读市二女中，再进一步了解，知道了她是有名的电影小童星，主演过《小梅的梦》(赵丹导演)和《小伙伴》《天罗地网》等电影，以后在《青春之歌》里与秦怡、谢芳等名演员合作，担任过重要角色。为结识她，我就设法通过邢文钧先生以光明中学名义，把她请到光明中学来在一个联欢活动上表演朗诵。从此我成了她家的常客，彼此还成了好朋友。以后她去了北京中央电视台，参与了中国早期电视剧的开创工作。后来她又在北京电影制片厂成了影视编导。她的一些影片和电视连续剧(《玻璃屋里的人》《爱之上》和《苏雅的故事》等)都得到了好评。有的还得了“飞天奖”。她现在在影视评论方面努力，也取得了可喜的业绩。我们见了面，经常会一起探讨人生和艺术方面的一些问题，友情一直保持到了今天。

光明校庆 离校五十年后的祝福

多少年来，光明中学始终让我念念不忘。每次从国外回来探亲、访问、开会，路过光明中学时总会进去看看。当我看到的尽是生面孔，想到与过去的老同学几乎都失去了联系，心中就会生起了一种莫名的失落感。三年前我找到了马秀风和高士模老师的家，重叙旧情，非常高兴。去年初我又接到滕宗兴和胡绳玉的邮信，更是喜出望外。得知10月28日是光明中学120周年校庆日，立即决定回去参加庆典，渴望与离别了近半个世纪的同学们好好欢聚一下。当我见到了吴宗善、李连生、李仕洪、李洁、梁福珍、蒋红娣、陈祖卿、林钦荣、周国元、陈其本、方存正等二十多位同学时，心情真是万分激动。我为不少同学成了教授、名医、工程师、处长、主任……成就非凡，许多还有了第三代而感到欣喜，同时也为大家都是白发苍苍、年近七旬的老人，有好多位还离开了人世而感慨人生苦短……好在回顾以往走过的路，我们可以告慰母校：您的学子多还没有辜负当年学校老师对我们的教导和期望！

（2007年1月）

后　记

我是1940年代出生，1950年代成长的一代。我在中学时代正是祖国蒸蒸日上的时代。在"向科学进军"的号召下，我们努力学习，立志做科学家，将来为国家做贡献。但是"向科学进军"的道路从来就不是平坦的，要实现自己的理想，总要克服各种障碍，排除各种干扰，其中有政治环境和学术环境的影响，也有研究工作本身战略上思考；大到选题，战术上的布局；细到每个实验的计划，技术方法的制订；还要仔细观察，及时总结，才能有所发现，有所创造。有时一个小小的现象观察可能孕育着一项重要的发现。当然，在研究工作中，虽然艰辛是常态的，毅力是必需的，坚持就是胜利会笑到最后的，但随时适应环境变化的应变能力也是需要的。大学毕业进入中国科学院的最初研究进行得十分顺利，但是不久遇到"文革"，下乡种田、养猪三年。当"抓革命，促生产"又回到实验室时，基础理论研究不能搞了，就想出转为纤维素酶研究，搞发酵在工农医领域的应用仍能有所作为。到了美国后因各种原因，从分子生物学、叶绿体DNA克隆分析，转到光合作用基因的调控、植物转基因研究，最后从学术界转到工业界进行HIV病毒检测、艾滋病药物开发和分子诊断等多项医药工程研究，跨度很大。其实我们做基础研究应随时抓到苗子，转化到工农医业生产上，将理论知识变为产品。另外，植物学、微生物学、人体医药都是息息相关的，尤其从细胞分子水平上互通有无，生命的本质有其同一性。

我有幸选择的生物学，从植物生理学开始，到植物生物化学、分子生物学、分子医学，都与人生最大的"食品和医药"两件事，即"吃得营养"和"活得健康"有关系。现在生命科学的发展日新月异，人类全基因分析已完成，DNA顺序分析越来越普及，基因分离、编辑技术越来越完善，分子育种技术越来越发展，这类研究无疑将为人类带来无限福祉。作为科学工作者，脑海中还总萦绕着随着世界发展，人口增加，如何解决伴随而来的粮食短缺、能源来源、环境污染以及人类健康问题，这也是我退休后一直关注的问题。我自小喜爱植物，省下有限的零用钱去买书，买实验用品，在上海弄堂狭小的天井里种植花花草草进行研究。以后进入工作岗位后又从植物、农业，转到微生物发酵，最后进入分子医学研究。但是心底里还是对植物、园艺情有独钟，怀有浓厚的兴趣。退休后正好有时间有条件在自己家园种花养草，培育果蔬，重温我幼时对园艺兴趣的梦。至今在旧金山湾区我的"观海山庄"中，经过多年努力种植了15种不同的果树：苹果、生梨、柑橘、蜜桔、柿子、樱桃、桃子、杏子、石榴、葡萄、无花果、弥猴桃……也种植了番茄、黄瓜、四季豆、辣椒、茄子、南瓜、生菜等十多种蔬菜。仅番茄就有红、黄、紫，樱桃番茄等多种品种。我的果园菜园也为我提供了重新做科学试验的良好机会，每天观察种子发芽，幼苗生长，开花结实。我施用有机肥加化肥，试用细菌肥料、植物生长调节素，促进发芽、生根、结实，每天记录观察到的试验结果。我用科学方法种植出来的蔬菜，新鲜、安全。我种的水果结实多，味道好。每年苹果的收成好几百斤，除了分送亲朋邻里，我研制了一个制作苹果干可以保存的工艺，用维生素C预处理，在低温下用脱水机烘焙的方法，制作味道好、营养价值高、无添加剂的天然果干，得到大家的好评。为了冬天也能吃到新鲜的蔬菜，我设计制作了人工培植箱，安装了暖气、人工光和自动浇水系统。也在家屋顶安装了太阳能板，除了足够满足家庭日常供电需要，多余的电能供应我温室加温照光的需要，这是我极力推崇提倡的一种简单、朴实、健康、快乐的生活方式，也可说实现了我幼时当科学家、过田园生活的梦想。

我们做科学研究的，增加兴趣，广开思路，是非常重要的。尤其科学与艺术有时也密不可分的。艺术的美感可体现在科学的成果中。我兴趣较广泛，在大学时拉小提琴，参加复旦大学管弦乐团。到美国读研时，学习钢琴并影响了我儿子日后成为金融专业人士，又是优秀钢琴演奏员，退休前后我又参加了旧金山湾区合唱团，对艺术爱好（音乐、美术、电影、摄影）的激情有时又会变成科研的动力。退休后我喜爱旅游和摄影，近期学习航拍，和家人过着和谐欢愉的生活。这里选择一些风景照和田园农家生活照放在本书与读者分享，也算是本书的结束语。

编 者 的 话

本著作是朱雨生研究员跨度60年的个人生平研究成果的总结，由于时间跨度大，不同期刊的格式要求有所不同，为尊重当初期刊格式，本书中参考文献的格式基本未做修改。

工 作 摄 影

大学时代(1958)

米丘林小组活动(1956)

复旦大学棉花试验田与谢志强同学讨论棉铃脱落问题(1961)

马里兰大学(UMBC)与孔宪铎教授(1980)

中国植物生理学会理事讨论全国规划扩大会(1977,上海)

第一排右 1 薛应龙,右 4 娄成后,右 5 汤佩松,右 6 罗宗洛,左 1 罗士韦,左 3 殷宏章,左 4 沈善炯;后排合计:右 2 黄维南,右 3 唐锡华,右 4 汤玉玮,右 7 余叔文,右 8 许智宏,右 9 朱雨生,右 10 王天铎,右 13 夏镇澳,左 5 施教耐,左 7 沈允钢,左 9 阎龙飞

马里兰大学生物系系主任 Schwards 欢迎酒会(1980)

马里兰大学做实验(1980)

与 Lovett 教授(1993)

马里兰大学谈家祯来访(1981)

左起 Kevin 陈、沈桂芳、谈家祯、Madeline 吴、朱雨生、吴仲荣、李立人

访问 Rutgus 大学 C. Price 教授(1980)

与美国陆军部纤维素酶研究室主任 Mandel 博士于 Brokheaven 研究室相聚(1980)

与诺贝尔将金得主卡尔文,迈阿密(1982)

汤佩松教授夫妇同游华盛顿白宫(1981)

中国科技大学遗传工程暑期训练班师生合影(1982)

左起 2 唐惕,4 朱雨生,5、6 Lovett 教授夫妇

旧金山机场迎接殷宏章(1987)

访植物生理所付所长沈善炯院士(2014)

访植物生理所光合室王天铎研究员(2014)

谈家祯、殷宏章、沈善炯、孔宪铎教授等欢聚在上海科学会堂(1982)

前排左起朱雨生夫妇，孔宪铎母亲，谈家祯夫妇，孔宪铎哥哥，后排左起孔宪铎，沈善炯，华东分院领导，殷宏章

马里兰大学获博士(1983)

哈佛大学校园

哈佛大学 Bogorad 实验室团队(1993)

后排左 2 朱雨生，左 6 Bogorad 教授，后排左 3 弈升

在伊里诺依大学实验室(1984)

在伊里诺依大学实验室培养光合细菌(1985)

Kaplan 教授来访(1984)

与华罗庚,伊利诺依大学(1984)

左 2 朱雨生,左 3 华罗庚

与外国学生,伊利诺依大学(1984)

加州大学卡尔文实验室(2015)

Hearst 教授夫妇来访(2016)

加州伯克利大学 Hearst 教授及团队(1999)

前排左 6 中间 J. Hearst 教授;左 7 他母亲 Lily,我的钢琴老师;左 8 David Cook;后排左 1 朱雨生;最后一排中间穿绿衣 T 袖者 Tom Chech 是师兄,诺贝尔化学奖得主

1986年参加国际光合作用大会(Brown大学)

前排左5李有则,左6匡廷云;后排左1魏家绵,左3邓兴旺,右1朱雨生,右2吴敏贤,右3沈允钢,右5李淑俊

在Roche(RMS)实验室(2004)

左1朱雨生,左2 Mike Geier博士,室主任

在 Roche(RMS)实验室(2005)

罗士公司(RMS)我的办公室(2005)

访问养老院休养的谈家祯教授和他夫人(2000)

许智宏,邓兴旺,周群来访(1988)

同学何康与，俞志洲(2012)

中学同学曹炽康(2017)

北大校长许智宏(中右)和我(中左)及二友相聚在我家(2006)

右1弈升，师弟，加州伯克利大学教授；左1刘诚，加州优瑞科生物技术公司董事长

北京参加 SCBA 国际讨论会与北大校长许智宏(2004)

青年画家丁文星访问植生所沈允钢院士(2017)

光明中学同学会(2007)

前排左起方存正(教授),林钦荣(电机工程师),吴宗善(航空工程师),后排左 1 朱雨生,右 1 李连生(班长,工程师)

复旦大学同学会(2007)

第二排左3朱雨生

复旦大学同学会(2017)

前排左 1 周国兴(古人类学家),左 2 莫鑫泉(遗传学家),左 3 朱雨生,右 1 段吉光(植物学家),右 6 李春芳(原年级支部书记),二排左 5 曾溢涛(生物化学院士),左 10 洪德元(植物学院士),三排左 4 宋昆衡(微生物学家),左 10 沈寅初(生物化学工程院士),右 3 潘重光(原年级长,遗传学教授)

植生所纪念殷宏章院士寿辰 110 周年(2018)

一排,右 1 许大全,右 2 朱雨生,右 3 殷蔚芷,右 4 王天铎,右 5 沈允钢,右 6 郭金华,右 7 韩斌,右 8 朱广新,右 9 魏加绵,右 11 徐春和,右 13 陈根云。二排,右 1 华雪珍,右 5 王维光,右 6 蔡可,右 7 陈一飞,左 1 程剑锋,三排右 2 李止正。最后一排右 1 韩琪

家庭田园

父亲

全家福照，前排左起2妻子 3母亲，4奶妈，6儿子晓峰，其余为哥，姐，妹夫妇及他们子女(1979)

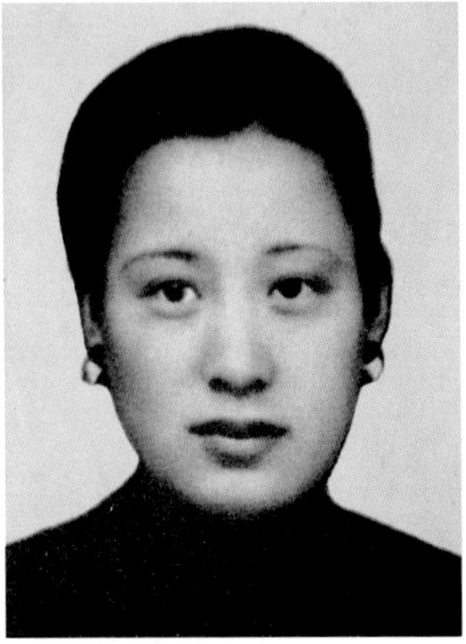

母亲

母亲

奶妈

全家照 88 年

哥嫂姐夫妇来访旧金山(1992)

左 1 哥渭生,左 2 姐夫殷鸣歧,左 3 姐兰珍,左 4 嫂周雅珍,左 5 妻调笙,左 6 儿子晓林

庆祝生日,妻子和二儿子

妹菊珍全家(2000)

前排左 1 妹菊珍,左 3 外甥孙冬琳,中间孙子孙震浩,后排右 1 妹夫孙庆尧

40 余年后与演员导演赵玉嵘相聚在北京(2004)

加州优山美地骑马(1982)

泰国开汽艇(2015)

泰国玩降落伞(2013)

阿拉斯加游轮上(父子)(2009)

花莲(2001)

日月潭(2001)

与妻子同遊阿拉斯加(2011)

在家阳台上(2014)

桂林(2010)

夏威夷(2012)

云南石林(2014)

苏州甪直古镇(2017)

张家界(2016)

实验记录(2017)

培育

培育箱

培育箱 LED 人工光照

李子

柑桔

蜜桔

葡萄

生梨

柿子

甜桃

家园丰收的果蔬

南瓜

黄瓜

番茄

制果干脱水机

家园集锦

家院

家院

家院

我的住屋

家院景色

屋顶安装太阳能板

庭园鲜花盛开

风 景 摄 影

阿拉斯加冰河

尼亚加拉大瀑布

加州纳帕酒乡

家门云雾观景

加州家邻野猫谷

旧金山金门公园

瑞士苏梨世公园

瑞士琉森八角塔和卡贝尔桥

瑞士琉森湖

迪拜

法国巴黎全景(埃菲尔铁塔顶摄)

南非的公路

泰国布吉岛

泰国布吉岛

南非企鹅

南非犀牛

南非野鹿

庭院早春

少女戏狗

漓江水上表演

早春的观海山庄

福建武夷山九曲溪

贵州稻田

贵州黄果树景

贵州荔波小七孔

贵州西江千户苗寨少女

贵州镇远古城

杭州花港观鱼

贵州红枫湖

湖南凤凰古镇

千岛湖夕

黄山美景

黄山梦笔生花

黄山云雾

江南庭院

江西婺源

江西婺源

九寨沟

四川黄龙

四川黄龙

小三峡人家

小三峡人家

小三峡人家

山峡人家

温州雁荡山

张家界

张家界

张家界少数民族

漓江美景

漓江山水

小山峡

周庄

周庄双桥

上海外滩的早晨

湖北恩施大峡谷

湖北神农架

三峡大瀑布